Nematology in South Africa: A View from the 21st Century

Hendrika Fourie • Vaughan W. Spaull
Robin K. Jones • Mieke S. Daneel • Dirk De Waele
Editors

Nematology in South Africa: A View from the 21st Century

Springer

vliruos
SHARING MINDS, CHANGING LIVES

Editors
Hendrika Fourie
North-West University Unit for
Environmental Sciences and Management
Potchefstroom, South Africa

Vaughan W. Spaull
Durban, South Africa

Robin K. Jones
Robin Jones Nematological Services
Murrayfield, Pretoria, South Africa

Mieke S. Daneel
Agricultural Research Council-Institute for
Tropical and Subtropical Crops
Mbombela, South Africa

Dirk De Waele
Laboratory of Tropical Crop Improvement
Department of Biosystems
University of Leuven
Heverlee, Belgium

ISBN 978-3-319-44208-2 ISBN 978-3-319-44210-5 (eBook)
DOI 10.1007/978-3-319-44210-5

Library of Congress Control Number: 2016959607

Printed on acid-free paper

This Springer imprint is published by Springer Nature
The registered company is Springer International Publishing AG
The registered company address is: Gewerbestrasse 11, 6330 Cham, Switzerland

Dedicated to the late Alex Mc Donald

Foreword I

What generation of parents and children has not been exposed at least rhetorically to the pernicious presence of roundworms in humans and the animals we associate with? Perhaps only a few of us still get to actually observe such worms, but we know very well that this does not mean the worms are no longer there. Those growing up in rural and agricultural environments will be sure to have witnessed them on many occasions. Nematodes have been a fact (and factor) of life since the dawn of human consciousness and will continue to be so, certainly from a biological and perhaps also from a cultural point of view.

In 2016, South Africa, along with every country in the world, is faced with an accelerating race to sustain and increase food production. This is driven by the need to feed an ever-increasing population with both enhanced quantity and quality of food, advance human and animal health by controlling a myriad of pests and diseases and meet these challenges in a sustainable and secure manner. Within each of these endeavours, the science of Nematology plays a significant role, and for many years, South Africans have made major contributions to this field.

The challenges outlined above, especially those related to food security and safety and human health, are among South Africa's national priorities, and many different elements of the public sector are seized with addressing them. The Plant Protection Research Institute of the Department of Agriculture, Forestry and Fisheries Agricultural Research Council (ARC) undertakes extensive research on the impact of plant-parasitic nematodes on various agricultural crops. With the support of the Department of Science and Technology (DST), the ARC is also the custodian of the national collection of nematodes. Through the National Research Foundation, the DST has established a national Centre of Excellence in Food Security, co-hosted by the universities of the Western Cape and Pretoria. This centre undertakes transdisciplinary research and capacity building to promote a sustainable food system that brings about food security for poor, vulnerable and marginal populations. In the area of human health, the Medical Research Council under the National Department of Health has an extensive set of programmes looking at the role of nematodes.

Nematology in South Africa: A View from the 21st Century brings into a single publication the latest reference data on nematode-induced crop losses, nematode-caused parasitic diseases of humans, the use of nematodes as biocontrol agents and many other important areas of application. The publication is focused strongly on the science of Nematology and is of significance to local and international researchers, crop commodity organisations, agricultural advisers and a broad range of readers involved in crop production and environmental and human health. The dissemination and utilisation of this publication in these communities will provide an important basis for the application of scientific knowledge to the very real socio-economic challenges of poverty and food security and thereby contribute to the improved quality of life of many South Africans.

The book reviews progress in a wide range of highly relevant scientific research challenges in the field of Nematology, but it also addresses critical ancillary issues such as succession planning and postgraduate training. This is of great importance as we strive to induct a growing number of young people into science in general and into this field in particular.

South Africa maintains a very productive national research system and excels globally in a number of fields. For instance, according to a 2014 Thomson Reuters analysis, the impact of the most highly cited South African research papers in the field of environmental sciences and ecology was on average 120 % higher than the global average over the period 2002–2011.

Our globally competitive research performance in these areas is a consequence of a number of factors, including long-term and sustained financial investment by government in the public research sector, coupled with certain geographical advantages we enjoy – and which my department is systematically seeking to exploit.

Although the Thomson Reuters study, referred to above, indicates that South Africa's average research impact in the biological and agricultural sciences is about 20 % below the world average, its most highly cited papers in microbiology, immunology and plant and animal sciences have an impact above the world average. Clearly there is a strong base of nematological research in South Africa. This base will be strengthened by this book, which collects seminal reviews of progress in the field.

The very important nematological contributions made by South African scientists, in collaboration with their international partners, in this publication will be a welcome source of current information and intelligence to those whose role is to combat the harmful, and maximise the useful, impacts of nematodes in agriculture, human health and the environment at large. I encourage the relevant scientific community and practitioners to read this book and make its knowledge known to a wider audience. I thank and congratulate the editors and all contributing authors for producing this book and encourage them in the continued research that will emanate from this record of current research.

April 2016

Naledi Pandor
Minister of Science and Technology
Pretoria, South Africa

Foreword II

This publication has been produced to complement and expand on the book compiled by the late Prof J. Heyns and myself in 1982 entitled *Nematology in southern Africa*. A great deal of research has been done in South Africa since 1982, and it is most opportune that it should be recorded, not only to document the information that has been generated over the past 34 years but also to present both a challenge and encouragement to future nematologists. Since most of the authors of this new book are based in South Africa, the focus has naturally been placed on this part of the continent. However, as the crops covered are mostly grown under similar climatic conditions in southern Africa, the information should be helpful to nematologists and other interested persons in the entire region. In addition, the 6th International Congress of Nematology held in Cape Town in 2014 was attended by delegates from about 38 different countries and 100 students. This indicates that the developments in South African nematology are attracting greater attention from the international community – a situation which also makes this book of particular interest to nematologists in other parts of the world.

This new publication contains chapters that update our previous knowledge of nematode morphology and classification, techniques and procedures, nematode control as well as the nematode pests of sugarcane, citrus, grapes, deciduous fruit, bananas, pineapple and potato. However, a number of new chapters have been added including a history of Nematology in southern Africa, alternative nematode management strategies, nematode pests of cereals, leguminous and oil seed crops, industrial crops and minor tropical and subtropical crops. There are also chapters on nematodes associated with grasses and weeds; non-parasitic, terrestrial and aquatic nematodes; entomopathogenic nematodes; nematodes associated with terrestrial slugs; marine and estuarine nematodes; and nematode parasites of humans.

I think Prof Heyns would have been very pleased with the appearance of this publication. For my part, I have great pleasure in writing a Foreword to this comprehensive book edited by Hendrika Fourie, Vaughan W. Spaull, Robin K. Jones, Mieke S. Daneel and Dirk De Waele, and I would like to thank them and all those who have contributed chapters for their enthusiasm and hard work. I hope that this book will not only be of use to those who are seeking more information on the biology of nematodes in southern Africa but it will also encourage others to further expand our understanding of these fascinating organisms.

April 2016

David P. Keetch
Pretoria, South Africa

Preface

The need for this book was first identified in 2009 when the then president of the Nematological Society of Southern Africa (NSSA), Hendrika (Driekie) Fourie, raised the idea at a meeting of members of the NSSA. When many potential authors voiced support for the idea, Driekie Fourie then approached Alex Mc Donald to be the chief editor. By mid-2012, most of the content of the book had been received and significant editorial input completed. At this point, Alex Mc Donald's health began to deteriorate to the point where progress was halted. Alex, after a long battle with his health, finally passed away in October 2014. Soon after, Driekie Fourie decided to reinvigorate the project and assumed the role of chief editor and called upon colleagues to form a new editorial grouping.

The objectives of this book are:

(a) To provide a record of the advances in Nematology that occurred since 1982 when Keetch and Heyns published the book, *Nematology in southern Africa*. Over the years, Keetch and Heyns' book provided the only comprehensive reference for scientific and commercial researchers on the subject of nematode problems of crops in the region. With the passage of time, its value became limited. Many developments called for this new publication, not least of which are the significant changes that have occurred in taxonomic techniques, management strategies and the discovery of new pest problems.

(b) To broaden the subject matter away from the emphasis on plant-parasitic nematode problems to include new fields of knowledge and research. In this regard, the disciplines of entomopathogenic nematodes and the nematode parasites of slugs have matured to play key roles in pest control, and new chapters have been added on alternative nematode management strategies; nematodes associated with grasses and weeds; the non-parasitic, terrestrial and aquatic nematodes; marine and estuarine nematodes; and nematode parasites of humans.

(c) To broaden the access to the local knowledge base. Nematologists in the region have made significant contributions to the science of Nematology and this contribution is poorly known internationally. In 1982, the NSSA had few members who all were from southern Africa. Student participation was uncommon. In

2014, the NSSA hosted the 6th International Congress of Nematology which was attended by 450 delegates from 38 countries and of which 100 delegates were students. Thus, the interaction between local and international nematologists is now frequent, and the need for a comprehensive publication on local research data is acute.

Finally, the editors would like to express their sincere thanks to all those who have assisted in this project, particularly the authors who are the leading experts in their field. Due to the prolonged genesis of the book, these authors have been asked and repeatedly accepted the task to revise the content of their chapters to keep the content current. In addition, the editors would like to thank Roland Perry of the School of Life and Medical Sciences (University of Hertfordshire, Hatfield, United Kingdom) for kindly agreeing to read all the chapters and use his extensive knowledge of Nematology to enhance the content of the book, Kirk West for being the official photographer and visiting many of the authors to record new photographic images and bring old photographs into the digital age, Hannes Visagie (Graphikos, North-West University) for skilfully capturing key subject content in clear and creative illustrations, Liesl de Swartdt (School for Geo- and Spatial Sciences, North-West University) for the basic map template of South Africa, Wiltrud du Randt (Agricultural Research Council–Grain Crops Institute) for constructing the maps included in Chap. 1 and Ebrahim Shokoohi (postdoctoral fellow, Unit for Environmental Sciences and Management, North-West University) for his inputs in creating and enhancing various illustrations. In addition, the editors acknowledge and thank our financial backer, the Flemish Interuniversitary Council (VLIR), Belgium, for funding the publication as part of a grant for conducting research in gardens and fields of smallholding farmers in South Africa. Ultimately we acknowledge our publisher Springer (Germany) and in particular Lars Koerner for agreeing to publish the book and Martina Himberger for acting as the project manager to enhance and produce a publication that meets international standards.

The editors are (*from left to right in the back row*) **Vaughan Spaull, Robin Jones, Dirk De Waele and Driekie Fourie and Mieke Daneel** (*front row*).

About the Editors

Hendrika (Driekie) Fourie started her career in nematology as a research technician at the Grain Crops Institute of the then Department of Agriculture in 1991 after she had been awarded an honours qualification by the then Potchefstroom University for Christian Higher Education. Thereafter she was awarded an MSc (North-West University) and a PhD (University of Leuven, Belgium) in Nematology, focusing her research on identifying and introgressing root-knot nematode resistance in local soybean material. Research by Driekie and her teams at the Grain Crops Institute, and later at North-West University, integrated breeding, botany, molecular biology and plant and animal biodiversity. At present, Driekie is involved with both training of postgraduate students and research, now mainly directed at the development of nematode management strategies on various food crops (potato, maize, soybean and vegetables). She enjoys running as an extramural activity and travelling across the globe.

Vaughan Spaull was awarded his PhD in Nematology from the University of Reading in England in 1973. Prior to that, he was employed as nematologist in the South Orkney Islands with the British Antarctic Survey. Later he worked as a biologist on the Aldabra atoll with the Royal Society. Thereafter he joined the South African Sugarcane Research Institute at Mount Edgecombe, where, for a period of 32 years, his primary interest was the control of plant-parasitic nematodes. He is now retired but keeps a vial of preserved specimens of massive, 10 mm long *Paralongidorus* sp. collected from around the roots of sugarcane, with which to impress and educate his grandchildren (and, at every opportunity, anyone else).

Robin Jones started his Nematology career at the Imperial College London from where he was awarded his PhD in 1976. His first post was at the then Citrus and Subtropical Research Institute in Nelspruit (now Mbombela) where, under the guidance of the late Lindsey Milne, he worked on the nematode problems of bananas. In 1980 he returned to the United Kingdom for about 1 year where he worked as an abstractor for CABI. South Africa then called him back and he assumed a series of posts in the Crop Protection Industry dealing with the technical development and marketing of nematicides. He left the industry in 2006 when he started his consulting and analytical services. He is an active Rotarian and keeps fit by cycling.

Mieke Daneel was awarded her PhD at the Rand Afrikaans University in 1989, after which she started working at the Research Institute for Subtropical and Tropical Fruit of the Department of Agriculture, Forestry and Fisheries (now Agricultural Research Council–Institute for Tropical and Subtropical Crops). She is still at the same institute based in Mbombela. The research of Mieke and her team is focused on alternative control strategies for nematodes in bananas, subtropical crops and vegetables.

Dirk De Waele received MSc degrees in Zoology (1976) and international co-operation development (1984) and a PhD degree in Nematology (1983) from the University of Gent, Belgium. His MSc research was on the taxonomy and ecology of freshwater nematodes of high-mountain lakes of Mount Kenya. His PhD research was on the taxonomy and ecology of virus-vector nematodes (Trichodoridae and Longidoridae) in Belgium. From 1984 to 1989, he worked at the Grain Crops Institute in Potchefstroom, South Africa, mainly on nematode problems of maize (*Pratylenchus* spp.) and peanut (*Ditylenchus africanus*). Upon his return to Belgium in 1989, he worked as a senior scientist at Plant Genetic Systems, a private company and one of the pioneers of GMO research. After a diversion working as science advisor for the minister-president of Flanders, in 1994 he joined the Laboratory of Tropical Crop Improvement at the University of Leuven, Belgium, to start his own research group. His research has focused mainly on the development of low-input management strategies for tropical nematodes (especially *Radopholus similis* and *Pratylenchus* spp. on banana and *Meloidogyne graminicola* on rice) and on the effect of arbuscular mycorrhizal fungi on plant-parasitic nematodes. Since 1996, he has been a professor in Plant Pathology at the University of Leuven. He is currently also an extraordinary professor at North-West University in South Africa.

Contents

1 **Introduction** 1
Hendrika Fourie, Robin K. Jones, Vaughan W. Spaull, Mieke S. Daneel, and Dirk De Waele

2 **History of Nematology in Southern Africa** 13
Albertus J. Meyer†

3 **Nematode Morphology and Classification** 33
Esther van den Berg, Mariette Marais, and Antoinette Swart

4 **Techniques and Procedures** 73
Mariette Marais, Antoinette Swart, Hendrika Fourie, Shaun D. Berry, Rinus Knoetze, and Antoinette P. Malan

5 **Quarantine Nematodes** 119
Rinus Knoetze, Antoinette Swart, and Moses Lesufi†

6 **Nematode Control and Nematicides: Developments Since 1982 and Future Trends** 129
Robin K. Jones

7 **Alternative Nematode Management Strategies** 151
Phatu W. Mashela, Dirk De Waele, Zakheleni Dube, Mbokota C. Khosa, Kgabo M. Pofu, Grace Tefu, Mieke S. Daneel, and Hendrika Fourie

8 **Nematode Pests of Maize and Other Cereal Crops** 183
Alexander H. Mc Donald†, Dirk De Waele, and Hendrika Fourie

9 **Nematode Pests of Leguminous and Oilseed Crops** 201
Hendrika Fourie, Alexander H. Mc Donald†, Sonia Steenkamp, and Dirk De Waele

10 **Nematode Pests of Potato and Other Vegetable Crops** 231
Robin K. Jones, Sheila G. Storey, Rinus Knoetze, and Hendrika Fourie

11 **Nematode Pests of Sugarcane** 261
Shaun D. Berry, Patrice Cadet, and Vaughan W. Spaull

12 **Nematode Pests of Tobacco and Fibre Crops** 285
Elizabeth R. van Biljon

13 **Nematode Pests of Citrus** 311
Mathys C. Pretorius and Hendrik F. Le Roux†

14 **Nematode Pests of Grapevine** 325
Sheila G. Storey, Antoinette P. Malan, and Hans J. Hugo

15 **Nematode Pests of Deciduous Fruit** 345
Hans J. Hugo and Sheila G. Storey

16 **Nematode Pests of Banana** 359
Mieke S. Daneel and Dirk De Waele

17 **Nematode Pests of Minor Tropical and Subtropical Crops** 373
Mieke S. Daneel

18 **Nematode Pests of Pineapple** 395
Elmarie C. Rabie

19 **Nematodes of Grasses and Weeds** 409
Keikantsemang N. Ntidi, Suria Bekker, and Hendrika Fourie

20 **Non-parasitic, Terrestrial and Aquatic Nematodes** 419
Antoinette Swart, Mariette Marais, Caroline Mouton, and Gerhard C. du Preez

21 **Overview of the South African Plant-Parasitic Nematode Survey (SAPPNS)** 451
Mariette Marais, Antoinette Swart, and Naomi Buckley

22 **Entomopathogenic Nematodes** 459
Antoinette P. Malan and Tiarin Ferreira

23 **Nematodes Associated with Terrestrial Slugs** 481
Jenna L. Ross and Antoinette P. Malan

24 **Marine and Estuarine Nematodes in South Africa** 495
Mathys C. Vosloo and Martin G.J. Hendricks

25 **Nematode Parasites of Humans in South Africa** 521
Christopher C. Appleton

26 **Achievements and Challenges** 559
Hendrika Fourie, Robin K. Jones, Vaughan W. Spaull, Mieke S. Daneel, and Dirk De Waele

Index 565

Contributors

Christopher C. Appleton School of Life Sciences, University of KwaZulu-Natal, Durban, South Africa

Suria Bekker North-West University, Unit for Environmental Sciences and Management, Potchefstroom, South Africa

Shaun D. Berry BASF Corporation, Research Triangle Park, NC, USA

Naomi Buckley Nematology Unit, Biosystematics Division, Agricultural Research Council–Plant Protection Research, Queenswood, South Africa

Patrice Cadet St Forgeux Lespinasse, France

Mieke S. Daneel Agricultural Research Council–Institute for Tropical and Subtropical Crops, Mbombela, South Africa

Dirk De Waele Laboratory of Tropical Crop Improvement, Department of Biosystems, University of Leuven, Heverlee, Belgium

Zakheleni Dube University of Limpopo, Sovenga, South Africa

Gerhard C. du Preez Unit for Environmental Sciences and Management, Potchefstroom, South Africa

Tiarin Ferreira Department of Conservation Ecology and Entomology, Faculty of AgriSciences, Stellenbosch University, Matieland, Stellenbosch, South Africa

Hendrika Fourie North-West University, Unit for Environmental Sciences and Management, Potchefstroom, South Africa

Martin G.J. Hendricks Department of Biodiversity and Conservation Biology, University of the Western Cape, Bellville, South Africa

Hans J. Hugo Agricultural Research Council–Infruitec/Nietvoorbij, Stellenbosch, South Africa

Robin K. Jones Robin Jones Nematological Services, Murrayfield, South Africa

Mbokota C. Khosa Agricultural Research Council–Institute for Tropical and Subtropical Crops, Mbombela, South Africa

Rinus Knoetze Department of Agriculture, Forestry and Fisheries, Plant Quarantine Station, Stellenbosch, South Africa

Hendrik F. Le Roux Citrus Research International, Mbombela, South Africa

Moses Lesufi Plant Health, Department of Agriculture, Forestry and Fisheries, Pretoria, South Africa

Antoinette P. Malan Department of Conservation Ecology and Entomology, Faculty of AgriSciences, Stellenbosch University, Stellenbosch, South Africa

Mariette Marais Nematology Unit, Biosystematics Division, Agricultural Research Council–Plant Protection Research, Queenswood, South Africa

Phatu W. Mashela University of Limpopo, Sovenga, South Africa

Alexander H. Mc Donald North-West University, School for Environmental Sciences and Development, Potchefstroom, South Africa

Albertus J. Meyer Department of Conservation Ecology and Entomology, Faculty of AgriSciences, Stellenbosch University, Matieland, South Africa

Caroline Mouton Nemconsult, Upington, South Africa

Keikantsemang N. Ntidi Agricultural Research Council–Grain Crops Institute, Potchefstroom, South Africa

Kgabo M. Pofu Agricultural Research Council–Vegetable and Ornamental Institute, Queenswood, South Africa

Mathys C. Pretorius Citrus Research International, Mbombela, South Africa

Elmarie C. Rabie Agricultural Research Council–Institute for Tropical and Subtropical Crops, Mbombela, South Africa

Jenna L. Ross Department of Conservation Ecology and Entomology, Faculty of AgriSciences, Stellenbosch University, Matieland, Stellenbosch, South Africa

Vaughan W. Spaull Glenwood, Durban, South Africa

Sonia Steenkamp Agricultural Research Council–Grain Crops Institute, Potchefstroom, South Africa

Sheila G. Storey Nemlab, Nematode Advisory Service, Durbanville, South Africa

Antoinette Swart Nematology Unit, Biosystematics Division, Agricultural Research Council–Plant Protection Research, Queenswood, South Africa

Grace Tefu Agricultural Research Council–Institute for Tropical and Subtropical Crops, Mbombela, South Africa

Elizabeth R. van Biljon Agricultural Research Council–Institute for Industrial Crops, Rustenburg, South Africa

Esther van den Berg Nematology Unit, Biosystematics Division, Agricultural Research Council–Plant Protection Research, Queenswood, South Africa

Mathys C. Vosloo Zitholele Consulting, Midrand, South Africa

Chapter 1
Introduction

Hendrika Fourie, Robin K. Jones, Vaughan W. Spaull, Mieke S. Daneel, and Dirk De Waele

This book documents what we know about the nematodes that parasitise the crops of South Africa (SA). It also provides information on the free-living nematodes that occur in the soil and inland waterbodies, those found on the sea shore and in estuaries around the coast, the nematodes that parasitise insects and terrestrial slugs and those that parasitise animals and humans.

South Africa is a diverse country with regard to its climate, soil types, biomes and fauna and flora. Situated between the 22 and 35 °S latitudes in the southern hemisphere, SA has a wider variety of climatic conditions than most sub-Saharan African countries. These range from Mediterranean (south-western parts) to temperate (interior plateau) and subtropical (north-eastern parts). Except for the Western

H. Fourie (✉)
North-West University, Unit for Environmental Sciences and Management,
Private Bag X6001, Potchefstroom 2520, South Africa
e-mail: driekie.fourie@nwu.ac.za

R.K. Jones
Robin Jones Nematological Services, 24 Victor Street, Murrayfield, South Africa

V.W. Spaull
170 Edmonds Road, Glenwood, Durban 4001, South Africa

M.S. Daneel
Agricultural Research Council–Institute for Tropical and Subtropical Crops,
Private Bag X11208, Nelspruit 1200, South Africa

D. De Waele
North-West University, Unit for Environmental Sciences and Management,
Private Bag X6001, Potchefstroom 2520, South Africa

Laboratory for Tropical Crop Improvement, Department of Biosystems,
University of Leuven, Willem de Croylaan 42, 3001 Heverlee, Belgium

H. Fourie et al. (eds.), *Nematology in South Africa: A View from the 21st Century*, DOI 10.1007/978-3-319-44210-5_1

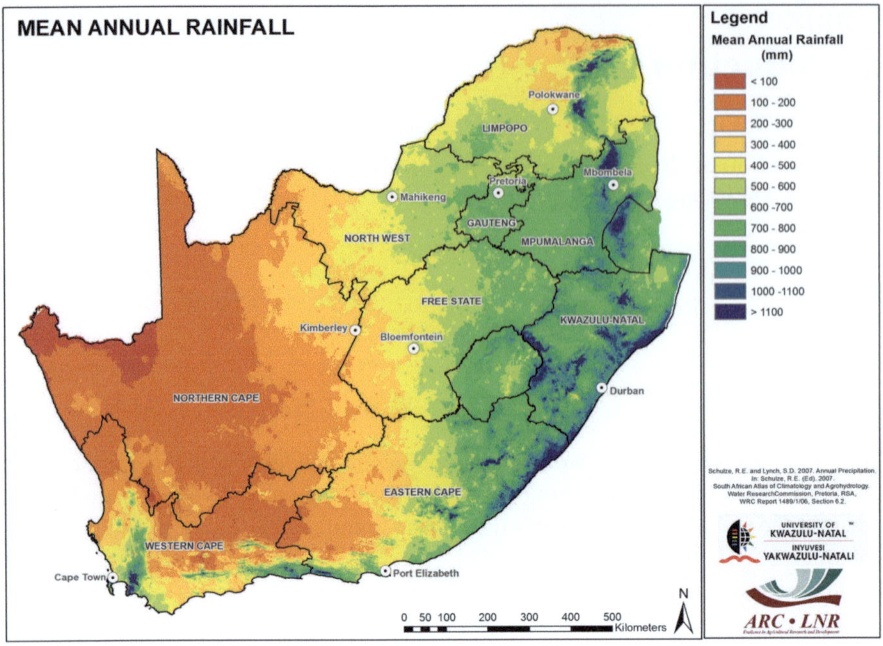

Fig. 1.1 Distribution of the mean annual rainfall in South Africa (Wiltrud Durand, Agricultural Research Council–Grain Crops Institute, South Africa)

Cape Province, which is a winter (June–September) rainfall area, the greatest part of SA receives rainfall in summer (December–March). Rainfall varies considerably from the western to eastern parts in particular, as indicated in Fig. 1.1.

South Africa has lower average temperatures than other countries situated within the same latitude range (e.g. Australia and some South American countries) since the greatest part of the interior of the country is at a higher elevation, forming a plateau. However, little variation exists in average temperatures for the interior parts of the southern and northern parts of the country that stretch across 10 ° of latitude (Fig. 1.2).

The vegetation types occurring in SA are represented by nine terrestrial biomes (Fig. 1.3), which are classified according to the vegetation and climate that dominate in each geographic region. These determine the plant and animal life each biome hosts. The nine biomes occurring in SA are referred to below with regard to the soil type as well as the flora and fauna:

(i) Desert (soil mostly consists of sand, gravel and rocks; little vegetation present; fauna represented mainly by insects and reptiles)
(ii) Indian Ocean Coastal Belt (nutrient-stressed soils; dominantly grass, trees and shrubs; grass and leaf-eating herbivores, birds)
(iii) Forest (drained soils representing virtually all soil types; trees dominated with grass, herbaceous and bulbous plants also occurring; small mammals, birds)

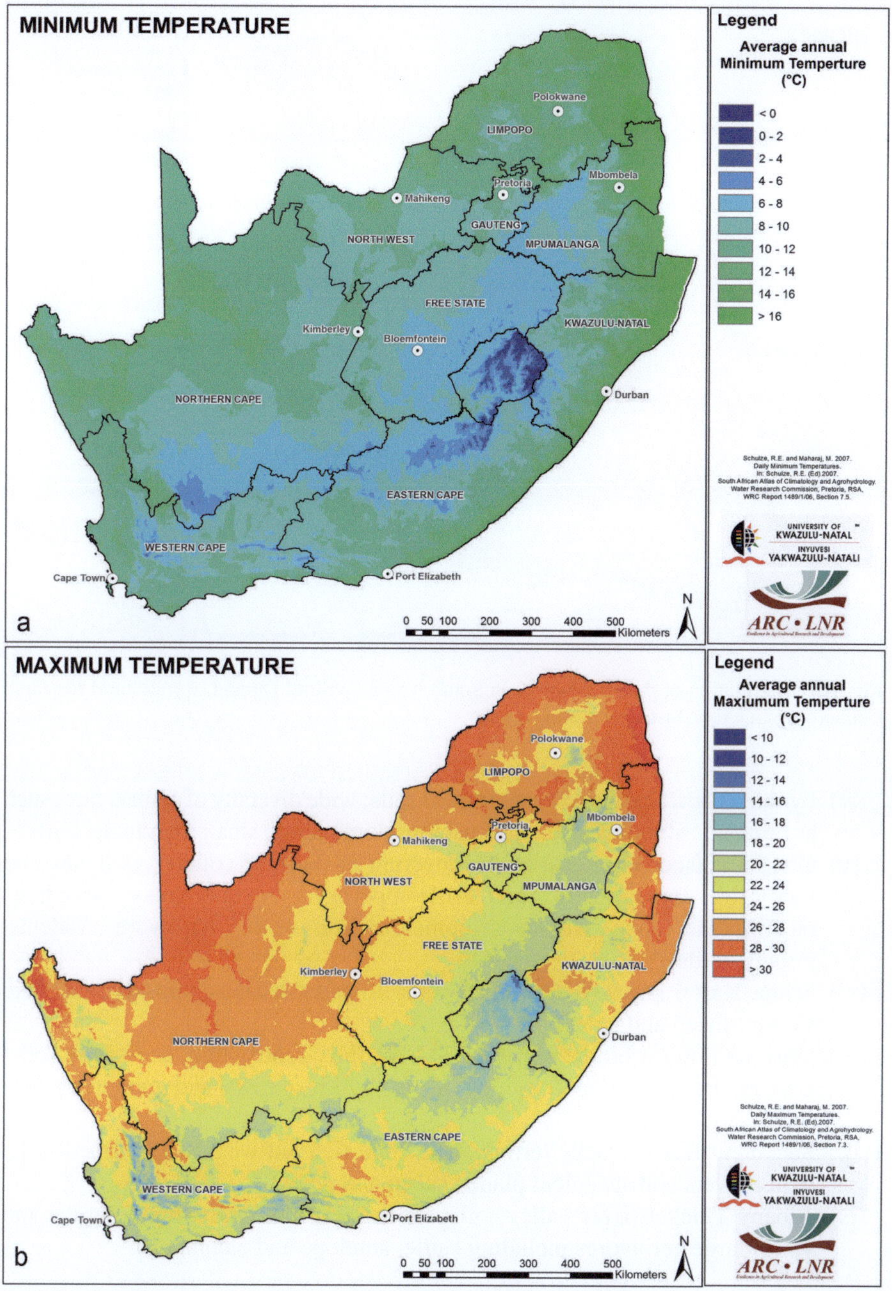

Fig. 1.2 (**a**, **b**) Mean minimum (**a**) and maximum (**b**) annual temperatures recorded in South Africa (Wiltrud Durand, Agricultural Research Council–Grain Crops Institute, South Africa)

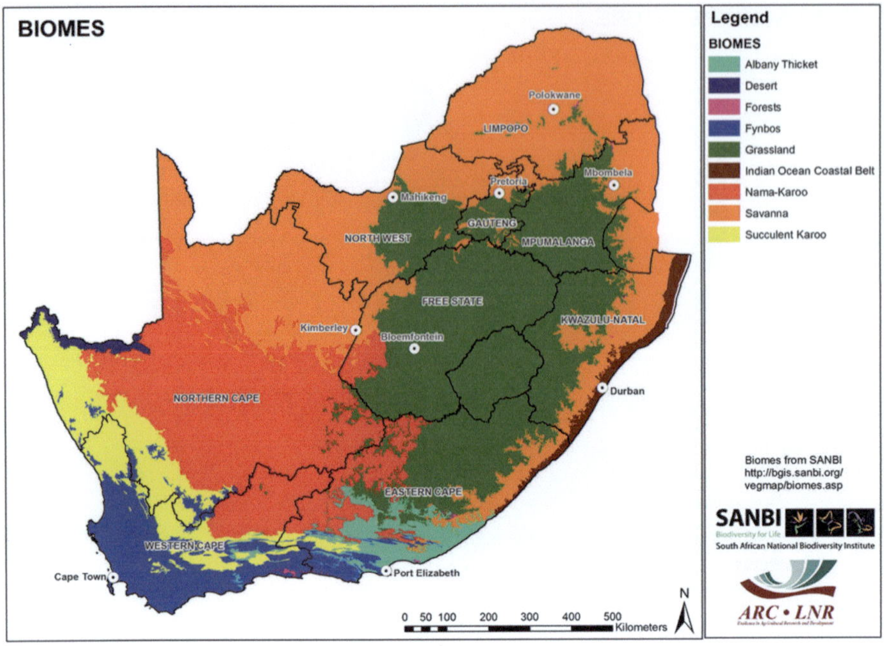

Fig. 1.3 Distribution of the nine biomes in South Africa (Wiltrud Durand, Agricultural Research Council–Grain Crops Institute, South Africa)

(iv) Fynbos (poor, acid and coarse-grained soils; wide diversity of fynbos flora such as proteas, 'silver trees' and pincushions dominating; small mammals, birds)
(v) Grassland (second-largest biome covering 28 % of the country with clay and sandy soils containing rich and fertile upper layers; mainly grasses with trees on hills and along river beds; numerous grass-eating herbivores, rodents, insects, birds)
(vi) Nama Karoo (weakly developed soils in rocky terrains; grassy dwarf-like shrubs; sheep and goats)
(vii) Savanna (largest biome covering 34 % of the country; clay and sandy soils; grasses, herbaceous and woody plants; wild animals and numerous bird species)
(viii) Succulent Karoo (rocky terrains with lime-rich but weakly developed soils; wild flowers and succulent plants; sheep and goats)
(ix) Albany Thicket (river valley soils; short trees, low intertwining shrubs and vines; large herbivores including cattle, antelope and elephant)

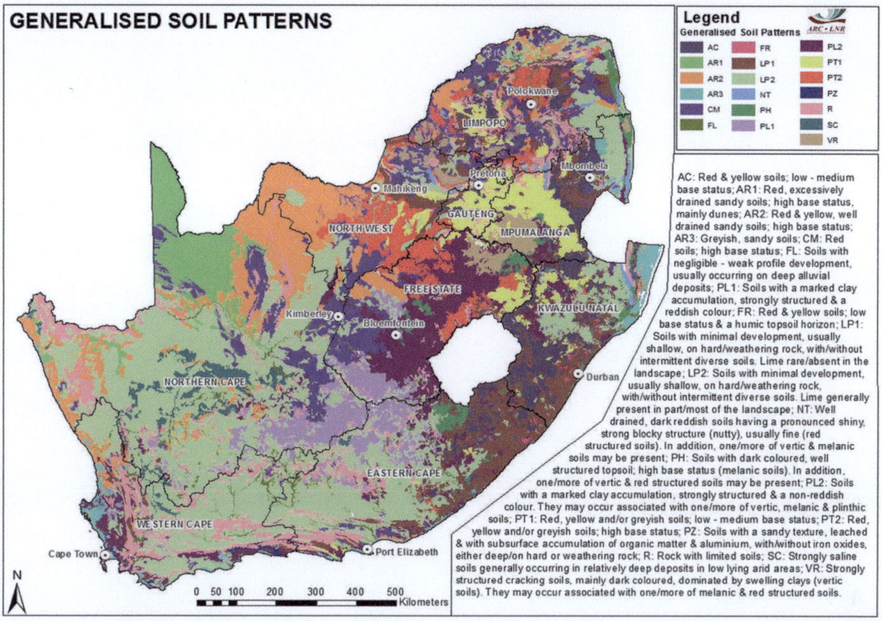

Fig. 1.4 Distribution of the main soil types in South Africa (Wiltrud Durand, Agricultural Research Council–Grain Crops Institute, South Africa)

A very wide range of soil types occurs in SA (Fig. 1.4) and they are strongly related to the geology of each biome. The geology of the country, which is interlinked with topography and climate, is complex.

The diverse geography, geology and climates in SA enable the cultivation of a variety of crops (Figs. 1.5a, b, 1.6, 1.7a, b, and 1.8). The maps are only an indication of the production area of the major crops and do not represent the extent of such areas or the mass produced. To put things into perspective, hectares (ha) harvested and metric tonnes (MT) produced of the important agricultural and horticultural crops in SA, averaged over the past five years, are listed in Table 1.1.

Associated with the various crops, below ground and out of sight, is an abundance of plant-parasitic nematodes, sometimes referred to as the 'hidden enemy'. The greater part of this book is devoted to these plant-parasitic nematodes, with Table 1.2 listing those that are the most common and abundant in SA.

Nematodes that feed from the outside of the roots or other plant parts are termed ectoparasites and include migratory and sessile forms. Migratory ectoparasites (e.g. dagger and stubby-root nematodes) remain outside the roots or other plant parts and

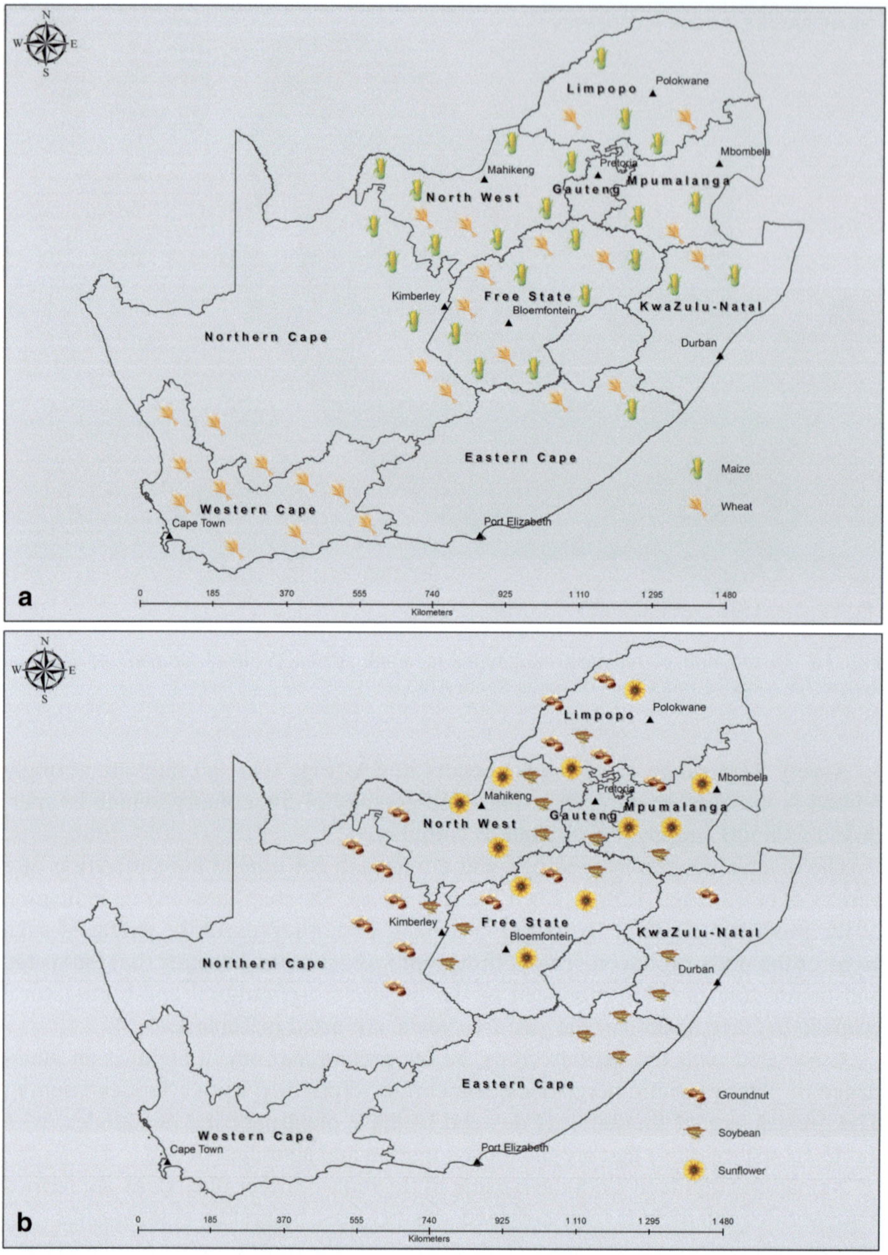

Fig. 1.5 Production areas of maize and wheat (**a**), and groundnut, soybean and sunflower (**b**) in South Africa (Liesl de Swardt and Hannes Visagie, North-West University, Potchefstroom, South Africa)

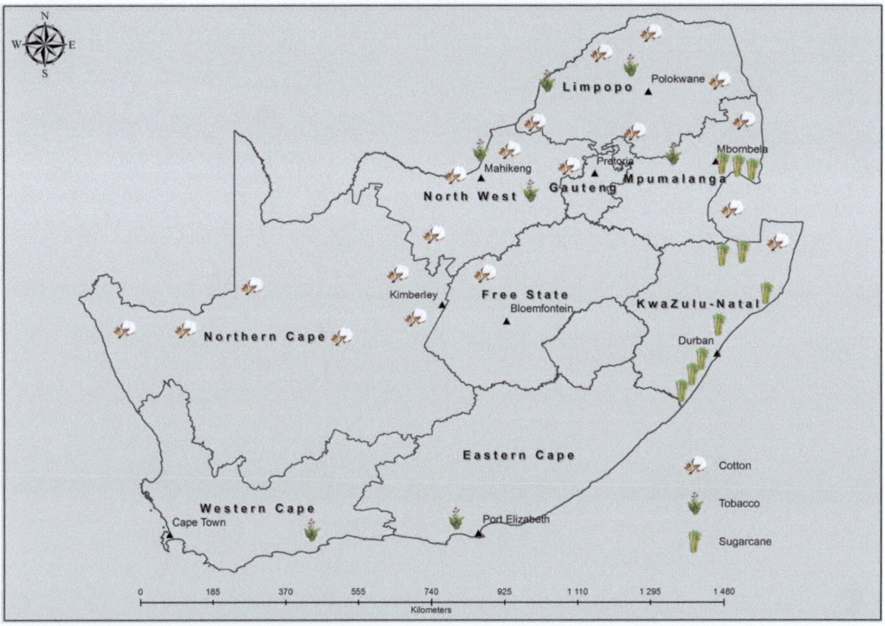

Fig. 1.6 Production areas of cotton, tobacco and sugarcane crops in South Africa (Liesl de Swardt and Hannes Visagie, North-West University, Potchefstroom, South Africa)

feed on epidermal cells or cells deeper in the plant tissues (Fig. 1.9). They retain the ability to move and find new feeding sites. Sessile ectoparasites (e.g. ring nematodes) remain outside the roots and feed on one cell or group of cells for an extended period. Nematodes that invade and feed within the roots or other plant parts are referred to as endoparasites. Migratory endoparasites move about and feed on numerous cells within the root (e.g. lesion and burrowing nematodes). Sedentary endoparasites feed at one site on enlarged and modified cells (e.g. root-knot and cyst nematodes). Some nematodes are semi-endoparasites as only the anterior part of their bodies are located within the root (e.g. citrus, reniform and some spiral nematodes). These categories are, however, not mutually exclusive.

An overview of the nematodes that are of concern and of interest to SA is given in the following chapters. The book concludes with an outline of the achievements made during the past three decades and the challenges that face nematologists, plant breeders and farmers in the country (see Chap. 26).

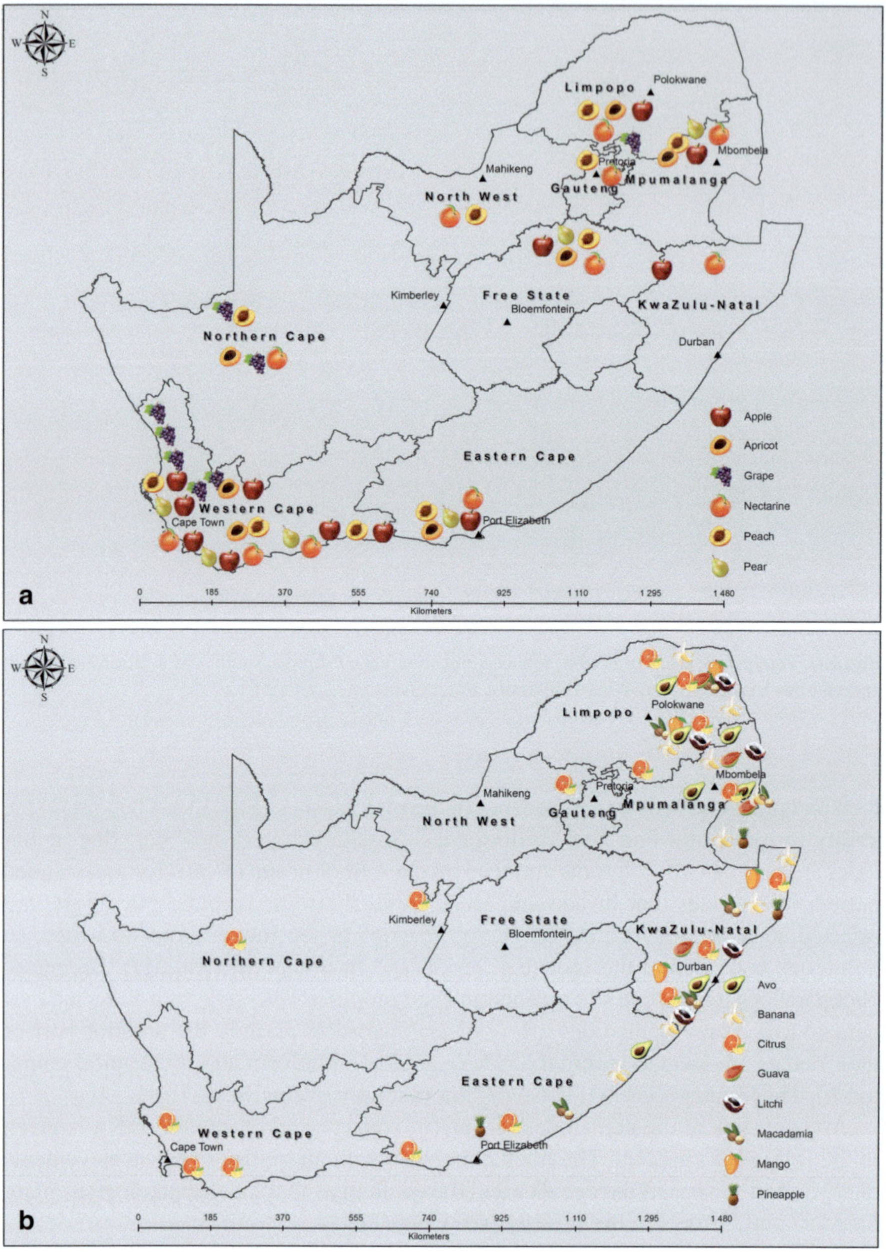

Fig. 1.7 Production areas of the major deciduous (**a**) and tropical and subtropical crops as well as citrus (**b**) in South Africa (Liesl de Swardt and Hannes Visagie, North-West University, Potchefstroom, South Africa)

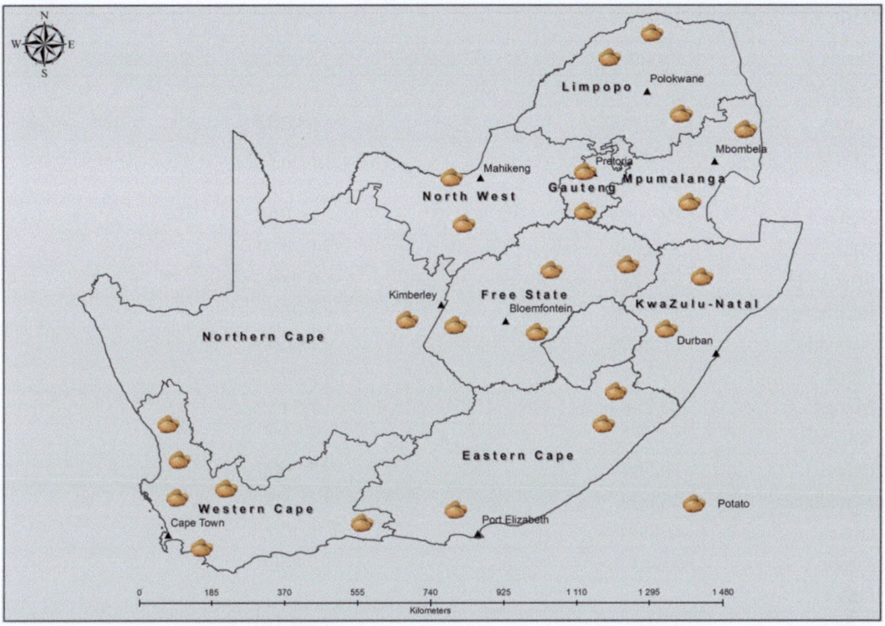

Fig. 1.8 Production areas of potato in South Africa (Liesl de Swardt and Hannes Visagie, North-West University, Potchefstroom, South Africa)

Table 1.1 Average production figures (mass and area) for the major crops grown in South Africa over a 5-year period (2009/10–2013/14)

Crops	Production mass (metric tonnes)	Area harvested (hectares)[a]
Grain, legumes and oilseed		
Maize	12,271,140	2,656,700
Wheat	1,826,600	564,400
Sunflower	652,200	519,480
Soybean	732,220	444,170
Groundnut	65,610	51,415
Dry bean	56,852	45,024
Vegetable		
Potato	1,730,973	62,440
Onion	573,665	25,182
Tomato	545,944	7,605
Cucurbit	175,125	12,194
Carrot	167,447	5,832
Brassica	140,290	2,393
Green bean	24,672	3,931
Green pea	11,490	5,401

(continued)

Table 1.1 (continued)

Crops	Production mass (metric tonnes)	Area harvested (hectares)[a]
Fruit		
Citrus	2,077,600	64,202
Grapevine	1,827,019	132,538
Apple	807,054	22,214
Banana	385,031	10,000
Pear	380,068	11,769
Pineapple	286,781	8,285
Peach, nectarine	173,998	3,869
Avocado	112,000	18,000[a]
Plum	69,199	4,777
Mango	57,788	7,517[a]
Apricot	56,713	3,237
Pecan	10,500	22,000[a]
Macadamia	46,000	21,500[a]
Guava	45,000	1,200[a]
Litchi	6,340	1,800[a]
Other		
Sugarcane	2,050,601	265,010
Cotton	2,682	6,259
Tobacco	14,177	4,638

[a]For fruit crops, the area refers to the ha planted but not necessarily harvested

Table 1.2 The most common and abundant plant-parasitic nematode genera and species that occur in South Africa, their common names, trophic groups they belong to and major host crops they parasitise

Nematode genera/ species	Common name of nematode pest	Trophic group	Host crop
Aphelenchoides arachidis	Testa nematode	Migratory endoparasite	Groundnut
Anguina and *Subanguina*	Seed, gall and leaf nematodes	Migratory and sedentary[b] endoparasite	Grass, wheat
Criconemoides xenoplax	Ring nematode	Sessile[c] ectoparasite	Apricot, grapevine, nectarine, peach, plum
Ditylenchus africanus	Groundnut pod nematode	Migratory endoparasite	Groundnut
Globodera rostochiensis[a]	Golden cyst nematode	Sedentary endoparasite	Potato

Table 1.2 (continued)

Nematode genera/ species	Common name of nematode pest	Trophic group	Host crop
Helicotylenchus	Spiral nematodes	Ectoparasite, semi-endoparasite	Banana, guava, maize, pear, soybean, sugarcane
Hemicriconemoides strictathecatus	Ring nematode	Ectoparasite	Litchi, mango
Hemicycliophora	Sheath nematodes	Sessile ectoparasite	Citrus, litchi, spinach
Heterodera	Cyst nematodes	Sedentary endoparasite	Brassica crops, wheat, beetroot, carrot, spinach
Longidorus	Needle nematodes	Ectoparasite	Grapevine, maize
Meloidogyne	Root-knot nematodes	Sedentary endoparasite	Bambara, banana, carrot, cotton, cowpea, dry bean, ginger, granadilla, grapevine, green pepper, groundnut, guava, kenaf, lupin, maize, peach, pecan, pineapple, potato, soybean, sugarcane, sunflower, tobacco, tomato
Nanidorus minor	Stubby-root nematode	Ectoparasite	Carrot, grapevine, maize, onion, pear, potato, tomato, soybean
Paratrichodorus	Stubby-root nematodes	Ectoparasite	Apple, cotton, pear, sugarcane, sweet potato, tobacco
Pratylenchus	Lesion nematodes	Migratory endoparasite	Apple, canola, grapevine, grain sorghum, groundnut, lupin, maize, pineapple, sugarcane, tobacco, wheat
Radopholus similis	Burrowing nematode	Migratory endoparasite	Banana
Rotylenchulus	Reniform nematodes	Sedentary semi-endoparasite	Banana, maize, soybean, sunflower, sugarcane, wheat
Scutellonema	Spiral nematodes	Ectoparasite	Maize, soybean, sugarcane, wheat
Tylenchorhynchus	Stunt nematodes	Ectoparasite	Bean, carrot, groundnut, maize, vegetable crops
Tylenchulus semipenetrans	Citrus nematode	Sedentary semi-endoparasite	Citrus, grapevine, persimmon
Xiphinema	Dagger nematodes	Ectoparasite	Apple, apricot, grapevine, litchi, peach, pear, plum, sugarcane

[a]Not widely distributed, but listed as a quarantine pest
[b]The terms 'sedentary' is used for endoparasitic nematodes
[c]'sessile' for ectoparasitic nematodes

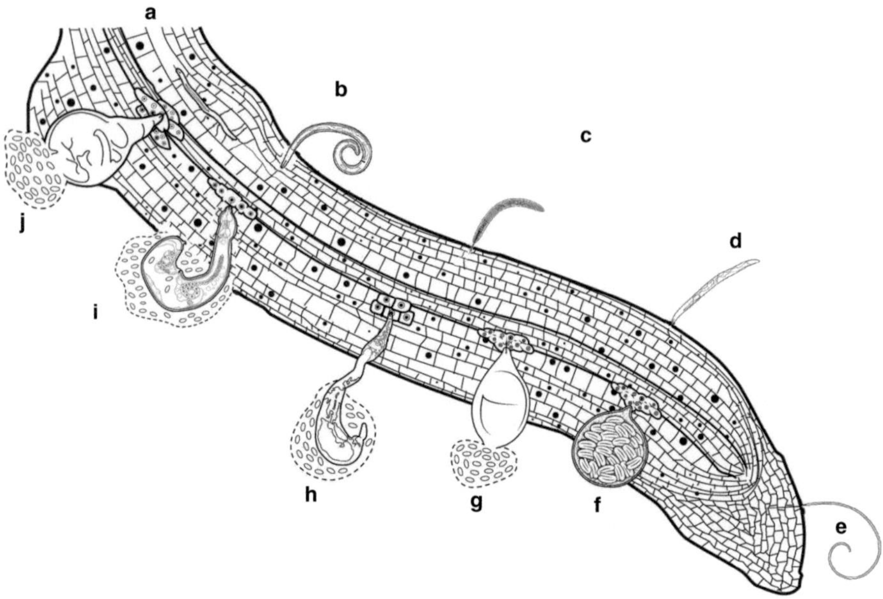

Fig. 1.9 A longitudinal section through the root of a plant indicating the feeding sites of different commonly occurring nematode genera: a migratory endoparasite (**a**: e.g. *Pratylenchus*), a semi-endo-/ectoparasite (**b**: e.g. *Helicotylenchus*), sessile ectoparasites (**c**, **d**: e.g. *Criconema* and *Nanidorus*, respectively), ectoparasite (**e**: e.g. *Longidorus*) and sedentary endoparasites, *Globodera* cyst filled with eggs (**f**), *Heterodera* (**g**), *Tylenchulus* (**h**), *Rotylenchulus* (**i**) and *Meloidogyne* (**j**) female (Hannes Visagie and Ebrahim Shokoohi, North-West University, Potchefstroom, South Africa)

Chapter 2
History of Nematology in Southern Africa

Albertus J. Meyer†

2.1 Introduction

When the idea for this publication was first mooted, Albertus J. Meyer, or Bertus as he was affectionately known, proposed to draft a historical record of the science of nematology in southern Africa. Unfortunately, he was only able to complete the first draft. This draft remains the core of the chapter, but in order to bring the content up to date and broaden its scope, several authors of chapters of this book have contributed to the final draft. All have sought to honour Bertus' original intent. The chapter is loosely structured around a timeline and consolidates the events on a location, activity basis and is focused mainly on research and training aspects related to plant-parasitic nematodes.

2.2 Research

2.2.1 The Origins: Cape Town (Western Cape Province)

The first report of plant-parasitic nematodes in South Africa (SA), indeed in the whole of Africa south of the Sahara, was published more than 100 years ago, in 1904, by CP Lounsbury, the then government entomologist based in Cape Town. This and subsequent publications by Lounsbury over the next 20 years reported root-knot nematodes (*Meloidogyne* spp.) on potato and other vegetables as well as

†Author was deceased at the time of publication

A.J. Meyer
Department of Conservation Ecology and Entomology, Faculty of AgriSciences,
Stellenbosch University, Private Bag X1, 7602, Matieland, South Africa
e-mail: driekie.fourie@nwu.ac.za

H. Fourie et al. (eds.), *Nematology in South Africa: A View from the 21st Century*, DOI 10.1007/978-3-319-44210-5_2

on fruit trees, grapevine and ornamental plants. He also reported the stem and bulb nematode (*Ditylenchus dipsaci* (Kühn, 1857) Filipjev, 1936) on lucerne and the citrus nematode (*Tylenchulus semipenetrans* Cobb, 1913) on citrus.

2.2.2 Department of Agriculture, Forestry and Fisheries (DAFF), Pretoria (Gauteng Province)

Willem van der Linde played a founding role in establishing the science of nematology within DAFF in southern Africa. He obtained his PhD from Cornell University in the United States of America (USA) in 1935, describing new species of *Diphtherophora*, *Funaria*, *Tylencholaimus* and other genera collected from Ithaca in New York State. On his return he worked in Pretoria for the next 30 years, mainly on root-knot nematodes. One of van der Linde's co-workers, Victoria Coetzee, also worked on root-knot nematodes, and in 1956 she described a new species, *Meloidogyne acronea* Coetzee, 1956, from sorghum. This species was later found to be an important pest of cotton in Malawi. She also described several new species of Mononchida.

2.2.3 Fruit and Fruit Technology Research Institute (FFTRI), Stellenbosch (Western Cape Province)

Llewellyn J. Koorts was another co-worker of Willem van der Linde. He was the first researcher who, besides his entomology tasks, investigated root-knot nematodes on peach trees at FFTRI in Stellenbosch. He left the institute to join Union Carbide in the early 1960s. Nematode research of deciduous fruit was revived by Bertus Meyer at the University of Stellenbosch, who was awarded a PhD for his thesis on nematodes of peach trees in 1973. This led to the appointment of Hans Hugo in 1978, one of Bertus' students, as the first full-time nematologist at FFTRI.

2.2.4 Agricultural Research Council–Infruitec-Nietvoorbij Institute (Previously, Viticulture and Oenology Research Institute (VORI), Nietvoorbij that Amalgamated with the Fruit and Fruit Technology Research Institute), Stellenbosch

On his return from nematology studies at the University of Wageningen in the Netherlands, Peter Smith started as a nematologist at VORI in Stellenbosch. He later moved to the Plant Quarantine Station. In the early 1980s, Jan Loubser was appointed to research root-knot nematodes of grapevine and was succeeded by Gerd

Hoepner in the 1990s. When VORI and FFTRI amalgamated to become the ARC–Infruitec-Nietvoorbij Institute, Hans Hugo became responsible for nematology research on both deciduous fruit crops and grapevine. In 2011, he was joined by Nomakholwa Stokwe, a student of Antoinette Malan at Stellenbosch University, with experience of entomopathogenic nematodes (EPN). She was tasked to investigate the use of these nematodes to control insect pests of deciduous fruit.

2.2.5 Department of Agriculture, Forestry and Fisheries (DAFF), Directorate of Plant and Quality Control, Stellenbosch

In 1979, having moved from VORI, Peter Smith, together with Ross Urban and Antoinette Malan, worked at what was then the Plant Quarantine Station in Stellenbosch. Their research focused on nematodes of quarantine importance. The team was expanded with the appointment of Welma Pieterse, Erina van Reenen, Rinus Knoetze and Sharon Roos. In 1983, the group was joined by Mary Barbercheck from the USA who worked on root-knot and dagger nematodes (*Xiphinema* spp.) on grapevine. Erina van Reenen left in 1984 to join Juan Heyns at the University of Johannesburg to do her MSc. Rinus Knoetze, who was awarded a PhD degree for his thesis on cyst nematodes (*Globodera* spp.) in SA, now heads the Nematology Section at the Directorate of Plant and Quality Control, Stellenbosch. Caroline Mouton joined the team after completing her MSc at the University of Gent (UGent, Belgium). In 2007, she left to set up a nematode diagnostic laboratory, Nemconsult, in the Western Cape Province (see Sect. 2.3) and was replaced by Lené van der Walt. Lené, in turn, left the Quarantine Station, in 2012, to work at Nemlab (see Sect. 2.3). Later, Ria Wentzel and Anthony Lategan joined the team.

2.2.6 Agricultural Research Council: Institute for Industrial Crops (ARC–IIC), Rustenburg (Previously, Tobacco and Cotton Research Institute), Rustenburg (North West Province)

Both Juan Heyns and Lindsey Milne began their careers as entomology students working on Coleoptera at the Tobacco and Cotton Research Institute in Rustenburg. Later, impressed by the damage done to tobacco by root-knot nematodes, they focused their studies on nematodes. Juan Heyns departed for the USA in 1960 to study nematodes under Gerald Thorne at the University of Wisconsin. Lindsey Milne stayed on in Rustenburg to work on root-knot nematodes instead of his first love, the cigarette beetle (*Lasioderma serricorne*). He made notable contributions on root-knot nematode control on tobacco before leaving to join the Research

Institute for Citrus and Subtropical Fruit in Mbombela (previously, Nelspruit). Here he continued his research on nematodes, focusing on litchi, citrus, banana and other subtropical fruit crops. Lindsey Milne left the Nelspruit Research group in the mid-1970s to become general manager of Westfalia Estates in the Limpopo Province.

Martie Botha-Greeff succeeded Lindsey Milne at the ARC–IIC and was later joined by Jeannie van Biljon. They developed integrated nematode management programmes for both cotton and tobacco and also conducted research on the nematode problems of a number of other crops, including hemp, flax, kenaf and cassava. Similar research, but aimed at smallholding farmers, was initiated in 2000.

Martie left the service of the ARC in 2003. Jeannie van Biljon carried on with the research work and mentored Martha Pofu, who completed her MSc at the University of Limpopo. Jeannie retired in 2011 and has been the editor of the newsletter of the Nematological Society of Southern Africa (NSSA) for most of the past 15 years.

2.2.7 Agricultural Research Council–Plant Protection Research (ARC–PPR), (Previously, Division of Plant Protection, Plant Protection Research Institute), Pretoria

Juan Heyns returned to SA with a PhD from Wisconsin in 1963 and headed the Nematology Section of the Division of Plant Protection in Pretoria. Here he assembled younger co-workers including Esther Van den Berg, Kent Kleynhans and Johan (Fursti) Furstenberg. They were later joined by Antoinette Swart and Mariette Marais and, assisted by Naomi Buckley, they formed what became a centre of taxonomic excellence at the ARC–PPR. Among their accomplishments is the ongoing South African Plant-Parasitic Nematode Survey (SAPPNS), coordinated by Mariette Marais (see Chap. 21). This database comprises a vast collection of nematode specimens from natural and cultivated soils throughout SA. It was initiated in 1987 to produce a comprehensive assessment of the nematode biodiversity in SA. The institute is also the custodian of the National Collection of Nematodes (NCN), the largest nematode reference collection in Africa.

In 1991, Kent Kleynhans compiled a monograph on the root-knot nematodes of SA. In 1996, together with his colleagues Esther Van den Berg, Antoinette Swart, Mariette Marais and Naomi Buckley, he published the handbook *Plant Nematodes in South Africa*, which 20 years on is still used by nematologists in the region.

In the early days at the ARC–PPR, besides taxonomic studies, there was also work done on the nematode problems of potato. This research was conducted by Ida Botha, Hoffie Koen and Kent Kleynhans and focused on root-knot nematodes, the lesion nematode *Pratylenchus brachyurus* (Godfrey, 1929) Filipjev & Schuurmans Stekhoven, 1941, and the golden cyst nematode *Globodera rostochiensis* (Wollenweber, 1923) Behrens, 1975. Hoffie Koen, a co-worker of Heyns, had graduated with a PhD from UGent and made several publications on root-knot and lesion nematodes before he moved from nematology to a teaching post and then to administration at the Technical College in Pretoria.

2.2.8 *Nelson Mandela Metropolitan University (NMMU) (Previously, University of Port Elizabeth), Port Elizabeth (Eastern Cape Province)*

Fursti Furstenberg moved from PPR to the University of Port Elizabeth in 1968 to become a lecturer in zoology. There, in collaboration with Anton McLachlan and AH Dye, he worked on marine and estuarine nematodes around the coast of SA. In the 1980s and 1990s, in collaboration with Magda Vincx from UGent, he described a number of new species. Further work on marine nematodes was carried out by Martin Hendricks at the University of the Western Cape and by Mathys Vosloo at NMMU. Mathys Vosloo is now an environmental scientist with a private company.

2.2.9 *University of Johannesburg (UJ) (Previously in part, the Rand Afrikaans University, RAU), Johannesburg (Gauteng Province)*

In 1971, Juan Heyns moved to RAU in Johannesburg, first as senior lecturer and then as professor. He trained 25 MSc and 13 PhD students in various nematode groups and by the end of 2000 he had published a total of 235 articles (as sole author and with his students and colleagues). Juan Heyns described at least 520 new and known nematode species, 27 new genera and two new families from SA and other countries. His book *A Guide to Plant and Soil Nematodes of South Africa*, published in 1971, is still used by nematologists and in the training of students at various institutions. He was a founder member of the NSSA and attended the first 15 biennial symposia of the society. In 1975, he organised the first short course in nematology at RAU. Through his endeavours two specialist nematology courses were presented at the university in 1984 by August Coomans (Belgium) and Jon Eisenback (USA). Juan Heyns collaborated with many nematologists overseas and was instrumental in arranging visits of many prominent nematologists to SA, including August Coomans and Michel Luc (French West Africa). The scientific community honoured him for his excellent work when he received the prestigious Havenga Prize from the South African Academy for Science and Arts in 1984. He was also rated as a 'Class A' researcher by the Foundation for Research Development (now the National Research Foundation). Juan Heyns started specialising in the genus *Xiphinema* Cobb, 1913 and became internationally recognised as an expert of this group. The comprehensive book *Character Analysis, Phylogeny and Biogeography of the Genus Xiphinema (Nematode: Longidoridae)* was published by August Coomans, Rony Huys, Juan Heyns and Michel Luc in 2001. Lastly, posthumously, he contributed to the *Guides to the Freshwater Invertebrates of Southern Africa* by JA Day and IJ de Moor (2002) by writing the chapter on nematodes. Juan Heyns can truly be regarded as one of the fathers of nematology in SA, not least due to the large number of students he trained, many of whom went on to valuable careers in nematology. A doctoral degree was awarded to Esther Van den Berg (whose thesis in 1974 on the taxonomy of some genera within the superfamily Tylenchoidea Örley, 1880, was the first PhD in nematology in SA). Other recipients of PhDs included Kent Kleynhans (taxonomy of several dorylaimid and tylenchid nematodes), Martie Botha-Greeff (detection of a nematode infection using a spectroradiometer), Antoinette Swart (morphology and taxonomy of species from five nematode orders), Mieke Daneel (morphology and taxonomy of some *Xiphinema* spp.), Sandra de Bruyn (taxonomic studies of nematodes from Botswana), Annelize Botha (freshwater nematodes from the Kruger National Park), Amand Dassonville (taxonomic and ecological study of freshwater nematodes), PJF (Billy) Jacobs (morphology and taxonomy of Longidoridae Thorne, 1935) and JC de W Kruger (morphology and taxonomy of *Xiphinema* spp.). Other recipients of Master degree qualifications included Gideon (Don) Loots, Ida Botha, WJJ (Willie) Vermeulen, Elsie D'Hollander, Antoinette Swart, Mariette Marais, Gerhard Stocker and Nadia Nell. These students worked mainly on the taxonomy of various soil and freshwater nematodes. Heyns retired at the end of 1994 but continued his taxonomic studies and published a number of works after his retirement. For his huge contribution to the science of nematology, he was awarded a DSc from the University of Stellenbosch in 1995. He died in 2001.

2.2.10 University of Stellenbosch (US), Stellenbosch

Nematology at the University of Stellenbosch was initiated by Bertus Meyer in 1973. He had received his training in nematology at Silwood Park, Imperial College (University of London, UK), in Antibes (France) and in Wageningen (the Netherlands). Peter C. Smith was his first MSc student in Stellenbosch, writing a thesis on *Xiphinema* and *Longidorus* Micoletsky, 1922, in vineyards in the Western Cape. Other students to be awarded their MSc degree under the guidance of Bertus Meyer included Selmarie Basson (dynamics and survival of *Ditylenchus destructor* Thorne, 1955, later reclassified as *Ditylenchus africanus* Wendt, Swart, Vrain & Webster, 1995), Hans Hugo (nematodes associated with apple trees in the Grabouw area), Rinus Knoetze (the use of the polymerase chain reaction to identify nematodes), Antoinette Malan (the potential of *in vitro* culturing of *Meloidogyne* spp.), Nick Neethling (biocontrol of plant-parasitic nematodes with *Purpureocillium lilacinum* on potato and citrus), Welma Pieterse (biology of *Criconemella xenoplax* (Raski, 1952) Luc & Raski, 1981), Sheila Storey (fluctuations in numbers of two *Pratylenchus* spp. in maize fields) and Piet Willers (distribution, pathogenicity and control of the citrus nematode in the Sundays River Valley). Caroline Du Preez did her MSc in Nematology in 2003 in collaboration with UGent on the effect of lectins on migratory and sedentary endoparasitic nematodes.

Students awarded PhD qualifications by the University of Stellenbosch under the supervision of Bertus Meyer were Jan Loubser (root-knot nematodes on grapevine), Antoinette Malan (*Xiphinema index* Thorne & Allen, 1950 on grapevine), Mariette Marais (taxonomy of the genus *Helicotylenchus* Steiner, 1945), Jacques van Zyl (*Heterodera schachtii* Schmidt, 1871 on vegetable crops) and Cheryl Venter (*Ditylenchus destructor* Thorne, 1945, later reclassified as *Ditylenchus africanus*

Wendt, Swart, Vrain & Webster, 1995, on groundnut). Bertus Meyer also assisted Juan Heyns with the first nematology short course at RAU in 1975 and Don Loots with the nematology short courses at North-West University (previously, Potchefstroom University for Christian Higher Education; PU for CHE).

Research at Stellenbosch also centred on nematode problems of deciduous fruit and grapevine, with ring (*Criconemoides* Taylor, 1936), dagger and lesion nematodes receiving the most attention. Annual ryegrass toxicity and *Ditylenchus dipsaci* (Kühn, 1857) Filipjev, 1936 on lucerne were the main focus of Bertus' research during 1979–1985. Shortly before his retirement in 2002, Bertus initiated projects on nematodes in smallholding farming and on EPN. The latter project was subsequently extended by his successor, Antoinette Malan and her PhD students, Jeanne de Waal (EPN to control the codling moth) and Tiarin Ferreira (EPN and their symbiotic bacteria). The contribution from MSc students included work on EPN for the control of codling moth (Jeanne de Waal), storage of EPN and control of the obscure mealy bug (Nomakholwa Stokwe), rearing of banded fruit weevil and its control using EPN (Tiarin Ferreira), *in vivo* production of EPN (Carolien van Zyl), EPN to control the citrus mealy bug (Sonnica van Niekerk), the distribution and diversity of EPN (Isaiah Nthenga), EPN to control the vine mealy bug (Patrique Le Vieux) and field trials with EPN to control the codling moth (Deidré Odendaal). Other MSc research on nematodes at US included the use of soil nematodes as indicators of soil health by Caro Kapp and biofumigation as a means of controlling ring and root-knot nematodes by Daniel Kruger.

Jenna Ross, from the Institute of Biological and Environmental Sciences, University of Aberdeen, UK, joined the research group as a postdoctoral fellow in 2012, bringing with her expertise in the diversity and mass production of slug-parasitic nematodes.

2.2.11 University of the Free State (UFS), Mangaung (Previously, Bloemfontein) (Free State Province)

The focus of nematology at UFS is primarily the taxonomy of aquatic nematodes. Originally a parasitologist, Candice Jansen van Rensburg received her nematology training under the supervision of Wilfrida Decraemer and her team at UGent. There she worked on nematodes originally collected from the Bakwena Cave near Pretoria. Under her supervision MSc degrees have been awarded to Henda Pretorius (a taxonomic revision of the genus *Histotylenchus* Siddiqi, 1971) and Ayesha Mobara (nematodes of the Seekoeivlei Nature Reserve, Free State Province).

2.2.12 University of Pretoria (UP), Pretoria

Nematology at UP was part of the plant pathology course in the Department of Plant Pathology and Microbiology. Frans van der Vegte lectured the practical nematology course. Under the guidance of Nico Labuschagne, Baone Kwerepe was awarded a PhD for her thesis on the integrated management of *Meloidogyne incognita* (Kofoid & White, 1919) Chitwood 1949 on Bambara groundnut. Whilst at the university she also worked on biofumigation with *Brassica* spp. to control *M. incognita.* Other students to be awarded PhD degrees under Nico Labuschagne were Shahasi Athman, for his work on the interaction between the endophytic fungus *Fusarium oxysporum* and *Radopholus similis* (Cobb, 1893), Thorne, 1949 on banana, and Hugues Baimey for his work on *Scutellonema bradys* (Steiner & LeHew, 1933) Andrássy 1958 as a pest of yam in Benin. Francois Viljoen completed an MSc study under Nico's supervision focusing on the control of *Meloidogyne incognita* on carrot using plant growth promoting rhizobacteria.

2.2.13 University of Limpopo (UL) (Previously, University of the North; Polokwane; Previously, Pietersburg) (Limpopo Province)

Nematology at UL was initiated by William Mashela. William obtained his BSc Agriculture Honours (Nematology & Horticulture) under the guidance of Mike Scott (Fort Hare) and Don Loots (Potchefstroom), MSc (Nematology) under Robert McSorley (Florida, USA) and PhD (Nematology & Horticulture) under Larry Duncan (Florida, USA). He initiated the use of botanicals in nematode management in SA using fruits from indigenous *Cucumis* spp., which has since resulted in an internationally recognised training niche in nematology. Under the supervision of William, a large number of students obtained postgraduate degrees. Those that completed MSc studies included Kinsley L. Mabuka (host status and host sensitivity of sweet sorghum to root-knot nematodes), Pontsho E. Tseke (responses of tomato plant growth and root-knot nematodes to phytonematicides from fermented fruits of *Cucumis* spp.), Kagisho D. Maile (responses of *T. semipenetrans* to crude extracts of indigenous fruits with and without effective microorganisms in citrus production), Raisibe V. Matabane (aggressiveness and identification of *T. semipenetrans* biotypes in SA), Anastatia Aluvilu (influence of salinity, *M. incognita* and time on the productivity of two pepper cultivars), Kgabo M. Pofu (host status of hemp cultivars to *Meloidogyne javanica* (Treub, 1885) Chitwood, 1949), Simon B. Radebe (effective microbe farming on tomato production and suppression of *M. incognita*), Hangwani Muedi (bioactivity of *Cucumis myriocarpus* fruit extracts on nematodes and bacteria), Alinah M. Mofokeng (host status of *C. myriocarpus* to *M. incognita*), Mbokota C. Khosa (efficacy of ground *C. myriocarpus* fruits, aldicarb and fenamiphos on suppression of *M. incognita* on tomato), David M. Mabitsela (combined effect of *C. myriocarpus*, *Lippia javanica* and *Ricinus communis* on *M. incognita*, tomato productivity and soil properties), Joseph M. Mphosi (influence of *Bacillus* spp. and *C. myriocarpus* on *M. incognita* suppression and tomato productivity), Anna

M. Mangena (interaction responses of *M. incognita* and effective microbes with *Brassica* organic amendment on growth of tomato) and Lucas G. Ngobeni (*L. javanica, M. incognita* and *Bacillus* interaction on tomato and soil properties). Those that completed PhD studies included Given K. Shadung 'Quality protocols for nemarioc-A and nemafric-B phytonematicides', Osvaldo M. Pelinganga 'Developing phytonematicide from *Cucumis africanus* and *C. myriocarpus* fruits for use in tomato production systems', Paulus T. Mafeo 'Responses of economically important crops to crude extracts of *Cucumis myriocarpus* fruit when used as a pre-emergent bio-nematicide', Kgabo M. Pofu 'Potential of *C. africanus* and *C. myriocarpus* as root-knot nematode-resistant rootstocks in watermelon (C*itrullus lanatus*) husbandry', Raisibe V. Mathabatha 'Developing nemarioc-AL and nemafric-BL phytonematicide dosage models for managing *T. semipenetrans* in citriculture' and Zakheleni P. Dube 'Nemarioc-AL and nemafric-BL phytonematicides: Bioactivities in nematodes, crops and soil'.

2.2.14 North-West University (NWU) (Previously, Potchefstroom University for Christian Higher Education PU for CHE), Potchefstroom (North-West Province)

Nematology at NWU was initiated by Don Loots. Don began his scientific career as an acarologist at NWU, but after obtaining his PhD in Acarology, he enrolled for an MSc degree in Nematology with a study of *Xiphinema americanum* sensu stricto Cobb, 1913, under the supervision of Juan Heyns. He completed the study in 1982, and in 1984 he undertook a study visit to the University of Guelph in Canada and expanded his knowledge on nematology. Under the supervision of Don Loots, a number of students obtained postgraduate degrees. Those that completed MSc studies included Elizabeth Jordaan (association of *Pratylenchus* spp. and *Fusarium moniliforme* on maize), Alexander (Alex) Mc Donald (evaluation of nematicides for the control of maize nematodes), Johannes Luïs (evaluation of nematicides for their efficacy in controlling nematodes of maize), Cheryl Bolton (a survey of nematodes associated with sunflower), John Swanepoel (a study on nematode anhydrobiosis), Elmarie Rabie (influence of *M. incognita* on the occurrence of False Panama Disease on banana), Sonia Steenkamp (biology and survival strategies of *Pratylenchus zeae* Graham, 1951) and Hendrika (Driekie) Fourie (effect of plant-parasitic nematodes on soybean production in SA with special reference to *Meloidogyne* spp.). Alex Mc Donald completed his PhD thesis (management of *D. africanus* on groundnut), whilst Sonia Steenkamp later obtained hers (host-plant resistance as a tool to manage *D. africanus* on groundnut) under Don's supervision.

Don Loots retired in 2009, after which Alex Mc Donald and Driekie Fourie (whilst still employed by the ARC–GCI; see Sect. 2.2.16) were appointed by

NWU, first on an extraordinary and later (2010) on a permanent basis. Alex supervised Mbokota C. Khosa during his PhD study on the use of plant-derived extracts to control *M. incognita*. Under Driekie's guidance, the following students completed their MSc studies: Caroline Leswifi (host status of cowpea genotypes to root-knot nematodes), Tshiamo Mothata (host suitability of vegetable crops to root-knot nematodes and verification of resistance), Nancy Ntidi (plant-parasitic nematodes associated with weeds in cropping systems of smallholding farmers), Suria Bekker (prevalence and identification of seed, leaf and gall nematodes infecting grasses), Ernst de Beer (root-knot nematode control on cotton using abamectin), Emil Engelbrecht (occurrence, identification and control of root-knot nematodes in potato cropping sequences), Gerhard du Preez (biodiversity of nematodes in a karst ecosystem and effect of pollution), Chanté Venter (traditional and new-generation techniques to identify root-knot nematode resistance in soybean cultivars), Chantelle Jansen (concomitant occurrence of nematodes and microbes in genetically modified and conventional soybean crops) and Melissa Agenbag (molecular and morphological identification of root-knot nematodes). Driekie also supervised Suria Bekker who obtained a PhD degree for a study titled 'Nematode communities: Bio-indicators of soil quality in conventional and conservation agricultural cropping systems'.

2.2.15 Pineapple Research Station (East London) and the Agricultural Research Council–Institute for Tropical and Subtropical Crops (ARC–ITSC), Hluhluwe (KwaZulu-Natal Province)

David (Dave) Keetch qualified as an entomologist from Rhodes University in Grahamstown (SA) and obtained in 1973 an MSc in nematology from Silwood Park (University of London), where he worked on the mode of action of nematicides. He served nematology and the NSSA from 1972 until the early 1990s, first at the Pineapple Research Station in East London and from the early 1980s as chief of DAFFs Plant Quarantine Services in Pretoria. In 1982, he and Juan Heyns co-edited 'Nematology in Southern Africa', a landmark publication for the science in this region. In 1984, Dave Keetch and co-author Naomi Buckley published a checklist of plant-parasitic nematodes associated with numerous crops, grasses and weeds titled 'A check list of the plant-parasitic nematodes in Southern Africa'. He retired early and went into business.

Further work on the control of nematodes on pineapple, primarily *Meloidogyne javanica*, was conducted by Graham Petty at Bathurst in the Eastern Cape and by Flip du Toit and Piet Willers in Hluhluwe in northern KwaZulu-Natal. Since 1989, after the establishment of a research station by the ARC–ITSC at Hluhluwe, Elmarie Rabie continued the work on nematode control (particularly *M. javanica* and *P. brachyurus*) on pineapple. Previously she had worked at the ARC–ITSC in Nelspruit (now Mbombela) where she did research on nematode control on banana and obtained her MSc under the supervision of Don Loots.

2.2.16 *Agricultural Research Council–Grain Crops Institute (ARC–GCI) (Previously, Grain Crops Institute of the Department of Agriculture, Forestry and Fisheries, DAFF), Potchefstroom*

Nematology research at the Grain Crops Institute in Potchefstroom was initiated by Mike Walters and Bertus Meyer. They laid out the first demonstration trials in the 1970s to enlighten producers on the importance of plant-parasitic nematodes (root-knot and lesion nematodes in particular) on maize and other grain crops. This led to the appointment of Sheila Storey as the first full-time nematologist working on lesion nematodes on maize. In 1983, she was transferred to the Small Grain Centre of DAFF in Elsenburg to concentrate on a survey of plant-parasitic nematodes of wheat in the Western Cape. Two years later she left to set up her own nematode diagnostic laboratory (see Sect. 2.3).

Sheila Storey was succeeded at GCI by Elizabeth Jordaan, Alex Mc Donald, Cheryl Venter and Selmarie Basson. Hendrik Riekert joined the team in the mid-1980s. A noteworthy impetus to nematology research came with the appointment of Dirk De Waele from Belgium. With fellow nematologists at GCI, he spent 1985–1989 conducting surveys of plant-parasitic nematodes on a number of field crops, including maize, sorghum, groundnut, sunflower and wheat. During this time he and his group described a new disease on groundnut caused by the groundnut-pod nematode (*D. africanus)*. This new species is endemic to SA and has become a serious pest of groundnut in the country. Alex Mc Donald and colleagues investigated a number of management options to control *D. africanus*. Alex passed away in 2014 whilst working as the chief editor on an earlier draft of this book.

During the early 1990s, GCI gained the services of Driekie Fourie and Sonia Steenkamp. Both of them completed their MSc degrees under the supervision of Don Loots at NWU (see Sect. 2.2.14). Driekie obtained her PhD from the University of Leuven (Belgium) under the supervision of Dirk De Waele. Her study focused on genetic host-plant resistance of soybean to *M. incognita* race 2. Whilst at the ARC–GCI, Alex and Driekie collaborated with Caroline Zijlstra (Plant Research International, Wageningen), an expert on the molecular identification of root-knot nematode species. Also, collaboration between Dirk De Waele, Alex Mc Donald, Driekie Fourie, Mieke Daneel (ARC–ITSC) and William Mashela (UL) enabled a project entitled 'Mobilising IPM for sustainable nematode management in household and community gardens of resource-poor farmers in South Africa'. This project was funded by the Flemish Interuniversity Council (VLIR). The outcomes from this initiative included the identification and implementation of environmentally friendly management strategies, including the use of permaculture, incorporation of organic manures and identification and use of root-knot nematode-resistant vegetable cultivars in rotation and intercropping systems. At present, Sonia Steenkamp and Nancy Ntidi are the nematologists that conduct nematology research at the ARC–GCI.

2.2.17 Agricultural Research Council–Institute for Tropical and Subtropical Crops (ARC–ITSC) (Previously, Citrus and Subtropical Fruit Research Institute), Mbombela

In Mbombela, nematology research was initiated by Lindsey Milne and Etienne de Villiers. Their work concentrated on the citrus nematode but also included studies on the ring nematode, *Hemicriconemoides mangiferae* Siddiqi, 1961, on litchi and various nematodes on granadilla, mango, papaya and other tropical and subtropical crops. Eli Cohn, a specialist in citrus nematology from the Volcani Centre in Israel, was a visiting scientist at the institute from 1975 to 1976. Robin Jones, who joined the institute in 1976, obtained his PhD at Silwood Park (University of London). He worked until 1980 largely on the problems of nematodes on banana, before joining the private sector in 1981. Peter (Piet) Willers later joined the team and extended the research activities to include work on *Meloidogyne enterolobii* Yang & Eisenback, 1983 (previously, known as *Meloidogyne mayaguensis* Rammah & Hirschmann, 1988) and the spiral nematode *Helicotylenchus dihystera* (Cobb, 1893) Sher, 1961 on guava. Mieke Daneel came from Belgium to SA as an exchange student to do her PhD under the supervision of Juan Heyns at RAU during 1986–1988. Afterwards she settled in Mbombela to become an active member of the ITSC's team conducting research on alternative management practices on banana, guava, granadilla and vegetables. Mieke contributed a great deal to the initiatives of the NSSA, locally and internationally.

2.2.18 Citrus Research International (CRI) (Previously, Outspan Centre), Mbombela

Hennie le Roux, who passed away in 2016, MC Pretorius and Laura Huisman, at the Outspan Centre (later CRI), concentrated their efforts almost exclusively on the citrus nematode. They pioneered work on the control of this pest in nurseries (largely eradicating the problem), in commercial orchards, and also highlighted the problems of accelerated microbial degradation of nematicides. They also successfully exploited the concept of repeated applications of a nematicide to break the life cycle of the citrus nematode in relation to the timing of new root flushes and thus reduce the pest burden. MC Pretorius obtained his MSc on the epidemiology and control of *Pseudocercospora angolensis*, the angular leaf spot fungus of citrus. His current research is focused on citrus decline and its spatial and temporal distribution patterns. The Diagnostic Centre at CRI also offers a commercial nematode diagnostic service, mainly for citrus growers. Elaine Basson is the current diagnostician.

2.2.19 Agricultural Research Council–Vegetable and Ornamental Plants Institute (ARC–VOPI), Roodeplaat Near Pretoria, Gauteng Province

Studies on the nematodes of vegetables were initially conducted by Karl Daiber and later by Flip Steyn, Anelia Steyn and Rosanne Jansen van Vuuren at the ARC–VOPI. Flip and Anelia subsequently moved to the private sector. More recently, Martha Pofu joined VOPI after she graduated from UL. Her research is focused on root-knot nematode-resistant medicinal plants, wild cucumber and African pumpkin and their inclusion in low-input agricultural systems.

2.2.20 South African Sugarcane Research Institute (SASRI), Mount Edgecombe, Durban (KwaZulu-Natal Province)

Studies on the nematodes of sugarcane in SA began with John Dick at Mount Edgecombe. By fumigating sandy soils with ethylene dibromide, he showed that nematodes were the most probable cause of poor yields of sugarcane on these soils. After his retirement, his work on chemical control of nematodes was continued, in the early 1970s by Dick Harris and then by several agronomists, in particular Rob Donaldson. When Dick Harris left the institute in 1976, he was replaced by Vaughan Spaull, who, like Robin Jones, Dave Keetch and Bertus Meyer, received his nematology training at Silwood Park (University of London). Collaboration in the 1980s with Patrice Cadet, who was working on nematodes of sugarcane in west Africa with the French Institute of Research for Development (IRD), led to a better understanding of the relationships between nematodes and sugarcane. This was especially the case for *M. javanica*, *H. dihystera* and *Xiphinema elongatum* Schuurmans Stekhoven & Teunissen, 1938. Patrice Cadet spent 5 years (1999–2005) on secondment from IRD at SASRI, bringing with him skills in multivariate analysis that resulted in a productive period of research. This was enhanced by the addition to the team of Shaun Berry, a molecular biologist by training. Shaun was awarded a PhD for his study of the effect of various management practices on the nematodes associated with sugarcane and their molecular identification. A colleague, Jeh-han Omarjee, obtained her PhD with a study of the interaction between the species of rhizosphere bacteria belonging to the genus *Burkholderia* and nematodes associated with sugarcane. Pelisa Dana received her PhD for her study of the relationship between soil factors and nematode communities in sugarcane fields. Both Vaughan Spaull and Patrice Cadet have now retired, and Shaun Berry has moved to the USA. Research on the nematodes of sugarcane is now continued by Prabashnie Ramouthar.

2.2.21 Zimbabwe

In 1912, not long after Lounsbury's first report of nematodes in SA, the government entomologist in Zimbabwe, RW Jack, reported on the dangers associated with importing root-knot nematode-infected seed potato tubers from SA. He later turned his attention to the control of root-knot nematodes on tobacco.

Work on the control of nematodes attacking tobacco in Zimbabwe was continued over a 28-year period (1947–1975) by RAC Daulton and George C. Martin. Each was the author of numerous reports and papers on the nematode problem, predominantly caused by *M. javanica*. The attention given to controlling this nematode was perhaps not surprising given that, according to Daulton, *M. javanica* was responsible for an annual loss of tobacco worth £3 million which at today's value amounts to £62 million or more than ZAR1,3 billion.

George Martin was born and educated in the UK, but settled in Zimbabwe to work for the government and the tobacco industry shortly after World War II. On arrival in the country, he laid out various experiments on several commercial farms comparing the efficacy of chloropicrin, ethylene dibromide and dichloropropene/dichloropropane as nematicides. By 1953, he undertook an extensive study tour of Europe (Wageningen and Rothamsted) and the USA. On his return he continued his work on the damage caused by root-knot nematodes on a wide range of host plants and also did extensive host-parasite studies. Upon retirement at the age of 60, he moved to Stellenbosch and for a few years he was employed by DAFF.

Howard Ferris came from the UK to work at the Tobacco Research Board in Harare on the nematodes of tobacco from 1964 to1968. He moved to the USA to what is now the Department of Entomology and Nematology at the University of California, Davis.

John Shepherd also came from the UK to Zimbabwe, around 1967, to work for the Tobacco Research Board. He, like George Martin, was a founding member of the NSSA. He retired in the mid-1990s and returned to his native England in 2001. A long-time colleague of both George Martin and John Shepherd was Jennifer Way, who was educated at the University of KwaZulu-Natal (UKZN) in SA but moved to Zimbabwe to work on tobacco nematodes.

Zibusiso Sibanda (from Zimbabwe) and Herbert Talwana (from Uganda) were among several nematologists from eastern and southern Africa that were involved in the formation of the Nematology Initiative for eastern and southern Africa (NIESA), under the leadership of the late Brian Kerry (Rothamsted Research, UK). The project was funded by the Gatsby Charitable Foundation, with technical support from CABI Bioscience, Rothamsted Research and the University of Reading in the UK. It was designed to build capacity in Plant Nematology and to develop a network of expertise in sub-Saharan Africa. Since the inception of the project, a number of MSc and PhD students have completed their studies, a plant nematode training manual was compiled, 60 scientists were trained through workshops, and six nematology laboratories have been equipped.

2.2.22 *Malawi*

Studies of the nematodes in Malawi, in the 1950s and 1960s, were performed by Don Corbett who was at the then Rothamsted Experimental Station (now Rothamsted Research) in the UK. His endeavours lead to the description of a number of new nematode species. In the 1970s, Vincent Saka, at the Bvumbe Research Station in Limbe, worked on the nematode pests of various crops. More recently, at the same institute, Andrew Daudi worked on the control of *Radopholus similis* (Cobb, 1893) Thorne, 1949 on banana. John Bridge and Samantha Page from CABI Bioscience investigated the biology and pathogenicity of *M. acronea* on cotton in the Shire Valley.

2.2.23 *Mozambique*

Work on nematodes in Mozambique has been mostly limited to surveys on various crops. Such nematode research was done for banana, citrus and sugarcane by Louis Reis, Rinie van den Oever, HAM van den Oever, Serafina Mangana and Jorge Chirruco (Plant Protection Department, Maputo) in the 1990s.

2.3 Private Industry and Commercial

From the onset of regular meetings organised by the NSSA, in 1973, a close association developed with individuals from companies such as Bayer SA, Dow Chemicals, Fisons, DuPont, Illovo and many more. They included Keith Brown, Vivian de Villiers, Frik de Beer, Frans van der Vegte, Llewellyn J Koorts, Malcolm Mason, Johan Bothma and Greg Burger. Through their association with the NSSA, support came from the companies in sponsoring the NSSA symposia, held biennially. Notable contributions came from Frans van der Vegte, a student of Willem van der Linde, who later joined the private sector. He was a founding member of the NSSA and almost always presented a paper at the symposium on his novel approaches to extraction techniques. Robin Jones also joined the private sector and was a regular contributor to the activities of the NSSA. He retired from Bayer in 2006 to initiate his own nematology consultancy business.

Almost 30 years ago, in 1987, Sheila Storey established a privately owned nematode diagnostic laboratory, Nemlab, in Durbanville (Western Cape Province). She had seen the need for an independent laboratory that focused on nematode analyses for commercial farmers. Although most of the samples sent in for analysis are from fruit and vine growers in the Western Cape, the laboratory also receives samples from a wide range of crops from across SA. A sister company of Nemlab, NemaBio, was recently set up to conduct research on the commercial use of EPN as biocontrol agents for use in the deciduous fruit and wine industries. Another diagnostic laboratory, Nemconsult, was established by Caroline Mouton. It was first located in Caledon, Western Cape Province, but recently moved to Upington in the Northern Cape Province. More recently, Suria Bekker opened her diagnostic laboratory, EcoNemaria, in Potchefstroom.

2.4 Training

Except for formal training of postgraduate students at various universities, short courses and graduate training in Nematology have been and still are conducted in SA. Following a proposal at a meeting of the NSSA in the mid-1970s, Juan Heyns presented the first short course on Plant Nematology at RAU. Subsequently the short courses were moved to the PU for CHE where they were first presented annually and then biennially by Don Loots. Through these courses, more than 240 newcomers to nematology were trained in the basic principles of Nematology. Over the years, Don was ably assisted by Bertus Meyer, Dirk De Waele, Alex Mc Donald, Driekie Fourie, as well as taxonomists Juan Heyns, Esther Van den Berg, Antoinette Swart and Mariette Marais. After his retirement in 2009, the courses were organised by Alex Mc Donald and Driekie Fourie assisted by Esther Van den Berg, Mariette Marais, Antoinette Swart, Jeannie van Biljon, Mieke Daneel, Sonia Steenkamp and Nancy Ntidi. Recently, a postdoctoral fellow at the NWU, Ebrahim Shokoohi, also assisted as well as some students (Suria Bekker, Gerhard du Preez and Milad Rashidifard).

A number of short courses in nematology were also presented in the 1990s by Dutch nematologists AF (Ton) van der Wal and Jan van Bezooijen at the Pathology Department of the UKZN, Pietermaritzburg Campus. The training was initiated by plant pathologist Mark Laing, and the students were mostly postgraduate plant breeders from various countries in Africa.

Graduate courses in nematology at UP are presented by Nico Labuschagne and more recently by Mariette Marais and Antoinette Swart. Ongoing graduate courses are also presented by Antoinette Malan at US, by William Mashela at UL and since 2008, by Jo Van As and Candice Jansen van Rensburg at UFS.

2.5 Nematological Society of Southern Africa (NSSA)

The NSSA was born in a pub in Pretoria in 1971, through the inspirational leadership of Lindsey Milne. He assembled a few colleagues one evening and over a glass of beer, Peter Smith suggested ‘Let’s form a society of nematologists in South Africa’. The idea was taken seriously by Lindsey Milne and together with Etienne de Villiers, they set out to work on the idea. Resulting from this informal gathering was the first symposium of the NSSA, which took place in Mbombela (Nelspruit) in

Fig. 2.1 Delegates that attended the first symposium of the Nematological Society of Southern Africa (NSSA) in 1973 in Nelspruit (from left to right). *First row*: Fresno A, Chamberlain J, Koorts LJ, Kinsman J, Martin GC, Milne DL, Pienaar, AP, Heyns J, Prinsloo MA, Meyer AJ and Argo A. *Second row*: Furstenburg JP, De Villiers EA, Smith PC, Naude G, Harris RHF, Gonggrÿp F, Shepherd JA, Cairns RO, Way JI, Watt EJ, Van der Vegte FA, Kleynhans KPN, Brown KF and Van den Berg E (Anonymous)

April 1973. In addition to about 25 members from SA, it was attended by George Martin, John Shepherd and Jennifer Way from Zimbabwe (Fig. 2.1). In the early days, when funding was hard to come by, the meetings were held at the institutes where research on nematodes was conducted. As more and more people attended the symposia and more sponsors were found, the meetings moved to larger venues at more enticing locations (see Table 2.1). They also took on a more international character with frequent contributions at symposia by nematologists from around the world including Dirk De Waele (University of Leuven), Danny Coyne (IITA, Kenya), Patrice Cadet (IRD, France), Patrick Quénéhervé (IRD, Martinique), Herbert Talwana (Makerere University, Uganda), Roger Fogain (Regional Centre for Research on Bananas and Plantains, Cameroon), the late Paul Speijer (IITA, Uganda), Howard Ferris (University of California, USA), Roland Perry (University of Hertfordshire, UK), John Bridge (CABI Bioscience), the late Brian Kerry (Rothamsted, UK), Larry Duncan (Citrus Research, USA), Maurice Moens (UGent) and Richard Sikora (University of Bonn, Germany).

Local nematologists, led by Mieke Daneel and her organising committee, were honoured to be selected to organise and host the 6th International Congress of Nematology (6ICN) held in Cape Town in 2014. The congress was attended by more than 440 nematologists from around the world (Fig. 2.2). It turned out to be a triumph for the organisers and enhanced the status of the NSSA and south African Nematology.

In the early 1980s, two biennial awards of the NSSA were instituted, namely, the Willem van der Linde medal for the best paper presented at a NSSA symposium and the George Martin Memorial Scholarship. Recipients of the latter scholarship must use the funds to further their studies to make nematology his/her career. Also, the Bayer CropScience Trophy (previously the Rhône-Poulenc Trophy) is awarded

Table 2.1 Venues where symposia of the Nematological Society of Southern Africa (NSSA) were hosted since the inception of the society in 1971

Year	Venue (institution, town and province)
1973	Institute for Tropical and Subtropical Crops, Mbombela (Nelspruit) Mpumalanga
1975	University of Stellenbosch, Stellenbosch, Western Cape
1977	Agricultural Research Station, East London, Eastern Cape
1979	University of Port Elizabeth, Port Elizabeth, Eastern Cape
1981	South African Sugarcane Research Institute, Mount Edgecombe, KwaZulu-Natal
1983	Agrivaal Building, Pretoria, Gauteng
1985	University of Stellenbosch, Stellenbosch, Western Cape
1987	North-West University, Potchefstroom, North-West Province
1989	Golden Gate Highlands National Park, near Clarens, Free State
1991	Wilderness National Park, Knysna, Western Cape
1993	Wigwam Resort, Rustenburg, North-West Province
1995	Kruger Gate, Kruger National Park, Mpumalanga
1997	San Lameer, Southbroom, KwaZulu-Natal
1999	Dikhololo, Rustenburg, North-West Province
2001	Skukuza, Kruger National Park, Mpumalanga
2003	Strand Beach Hotel, Strand, Western Cape
2005	Hans Merensky Estate, Phalaborwa, Limpopo
2007	Boardwalk Conference Centre, Port Elizabeth, Eastern Cape
2009	Casa Do Sol Hotel, Hazyview, Mpumalanga
2011	Spier Estate, Stellenbosch, Western Cape
2014	6th International Congress of Nematology, Cape Sun, Cape Town, Western Cape

Fig. 2.2 Delegates that attended the 6th International Congress of Nematology (6th ICN) in Cape Town from 4 to 9 May 2014 (Anonymous)

during symposia to an individual or team for the advancement of nematology in southern Africa.

Honorary membership of the NSSA is awarded to members for exceptional contributions to Nematology or to the society over an extended period. Recipients of the award are Ethné Cameron, Dirk De Waele, Juan Heyns, Robin Jones, Kent Kleynhans, Don Loots, Bertus Meyer, John Shepherd, Vaughan Spaull, Jeannie van Biljon and Esther Van den Berg.

Acknowledgements The editors acknowledge the original input of Bertus Meyer and the later contributions from numerous SA nematologists. The editors endeavoured to include all pertinent information about individual nematologists, groups and activities that contributed to nematology in SA and some southern African countries. No person or activity has been intentionally excluded.

Chapter 3
Nematode Morphology and Classification

Esther van den Berg, Mariette Marais, and Antoinette Swart

3.1 Introduction

This chapter is adapted from that by Heyns (1982) and is concerned with those morphological characters that are most useful for nematode identification and classification, which are being used since the first description of nematodes (Box 3.1).

Box 3.1 First Records of Morphological Descriptions of Plant-Parasitic Nematodes

The first report of a nematode being described dates back to 235 BC when an ancient Chinese symbol, resembling the shape of a typical adult-female cyst nematode, was made from an organism found on soybean roots. According to text books, Needham in 1743 reported the existence of the first tylenchid, apparently in an anhydrobiotic state in infected wheat seeds. The disease caused by this tylenchid in wheat was reported 30 years later by Roffrredi and described by Scopoli in 1799 as *Vibrio tritici* (now known as *Anguina tritici*). Nearly 60 years later, in 1855, root-knot nematode (*Meloidogyne* spp.) galls were identified on cucumber roots by Berkeley. This root-knot nematode species was described by Cornu in 1879 as *Anguillula marioni* (later reported as *Meloidogyne marioni* (Cornu,1879) Chitwood & Oteifa, 1952, and is now considered a species *inquirendae*). Shortly after this discovery, Schacht, in 1859, identified cysts (*Heterodera* spp.) causing "beet-tired" disease on sugar beet (Wouts and Baldwin 1998).

E. van den Berg • M. Marais (✉) • A. Swart
Nematology Unit, Biosystematics Division, Agricultural Research Council–Plant Protection Research Institute, Private Bag X134, Queenswood 0121, South Africa
e-mail: VanDenBergE@arc.agric.za; maraism@arc.agric.za; swarta@arc.agric.za

H. Fourie et al. (eds.), *Nematology in South Africa: A View from the 21st Century*, DOI 10.1007/978-3-319-44210-5_3

A brief outline of the classification of plant-parasitic nematodes of South Africa (SA) is also given in this chapter. The emphasis is placed on the more common terrestrial nematode forms, particularly the plant-parasitic groups. The marine nematodes as well as the vast diversity of animal-parasitic forms are not taken into consideration here. For these nematodes see Chaps. 20 and 24. Nematode genera and species identified this way by taxonomists of the Agricultural Research Council–Plant Protection Research (ARC–PPR) since the early 1960s are listed in the National Collection of Nematodes (NCN) (Box 3.2).

Box 3.2 National Collection of Nematodes
The National Collection of Nematodes (NCN), housed at Biosystematics of the Agricultural Research Council–Plant Protection Research (ARC-PPR) in Pretoria, is the largest nematode reference collection in Africa and compares well with other important international collections. The collection consists of three parts, namely the core, *Meloidogyne* and Juan Heyns Nematode Collections. The core collection was established in 1961 and consists of 50,000 microscope slides that contain approximately 180,000 preserved plant-parasitic, free-living, aquatic and entomopathogenic nematodes. The *Meloidogyne* Collection was established in 1981 and consists of 14,750 slide-mounted specimens from various countries. An important acquisition was the Juan Heyns Nematode Collection that was donated in 1999 by the late Juan Heyns of the University of Johannesburg (UJ: previously Rand Afrikaans University or RAU). This comprehensive collection contains 7,300 slides, with about 21,900 specimens of plant-parasitic as well as free-living and aquatic nematodes. The NCN currently consists of more than 200,000 specimens (7,209 type specimens) and is fully digitised. The oldest holotype was collected in 1959 by Heyns at Rustenburg in the North-West Province. Specimens from 27 other African countries and regions outside Africa (such as Antarctica, Micronesia, Europe, Asia and North, South and Central America) are deposited in the NCN. Material deposited in the NCN also contributes to the fulfillment of South Africa's obligations as part of international agreements such as the Convention on Biodiversity, the Nagoya Protocol and National Environmental Management: Biodiversity Act (Act 10 of 2004). The NCN was relocated in 2009 from ARC-PPRI (Rietondale) to new custom-built facilities at ARC-PPRI (Roodeplaat) that also house the National Collections of Fungi, Arachnida and Insects.

3.2 Morphology

Nematodes are unsegmented animals, typically with a cylindrical thread-like body, which tapers towards both ends and do not have any locomotory or other appendages. In the females of plant-parasitic forms the cylindrical body shape may be modified and the body may vary from sausage- to pear-shaped in the different species

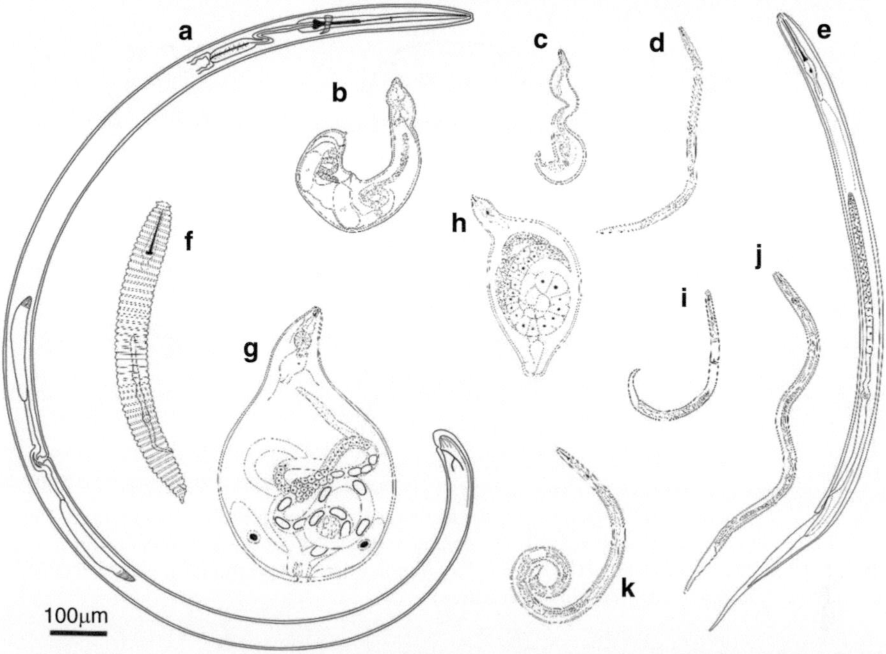

Fig. 3.1 Body shapes and sizes of various plant-parasitic nematode genera: *Xiphinema* (**a**), *Rotylenchulus* (**b**), *Tylenchulus* (**c**), *Pratylenchus* (**d**), *Hemicycliophora* (**e**), *Criconema* (**f**), *Meloidogyne* (**g**), *Heterodera* (**h**), *Paratylenchus* (**i**), *Radopholus* (**j**), *Helicotylenchus* (**k**); adult sedentary females are represented by **b**, **c**, **g** and **h** (Ebrahim Shokoohi, North-West University, Potchefstroom, South Africa)

(Fig. 3.1). Terrestrial and aquatic nematodes vary from less than 0.5 to approximately 10 mm in length. Most of the plant-parasitic nematodes, with the exception of the Longidoridae, which are slightly larger, fall within the range 0.4–1.5 mm. This means that they are barely visible to the naked eye when suspended in water and hence their presence cannot be easily detected in the soil or in plant tissues.

3.2.1 Head

Poorly defined terms are used rather loosely for certain parts of the nematode body. Head, for instance, is a general term to denote the front part of the body but is sometimes used more specifically to denote that part of the body from the anterior end to the base of the stylet. Even more often, head is used as a synonym for the lip region. The latter term is used for that portion of the front part of the body that is often set off by a constriction (Fig. 3.2a) or a depression (Fig. 3.2b), or supplied with a sclerotised framework (Fig. 3.2c). The lip region may also be confluent (Fig. 3.2d) or set off, angular, rounded or cap-like in outline (Fig. 3.2e) and may on occasions

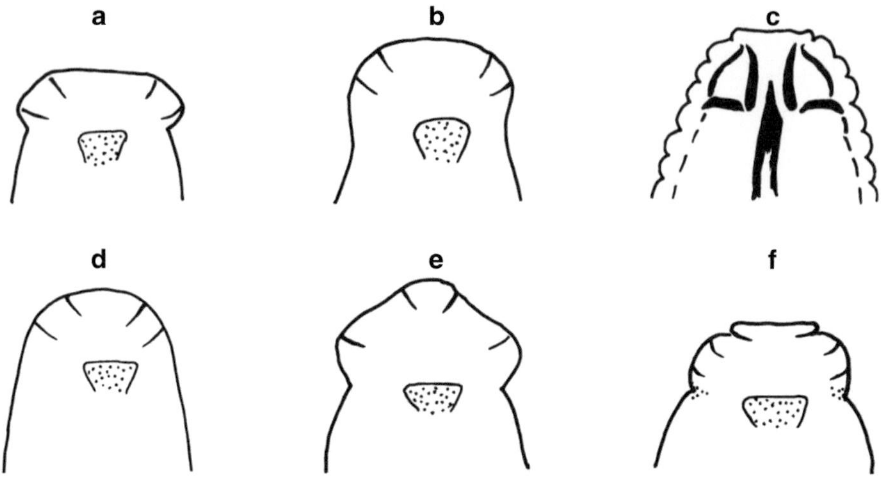

Fig. 3.2 Lip regions set off from the body by: a constriction (**a**), depression (**b**), or with a sclerotised framework (**c**). Lip regions may also be confluent (**d**), set off or cap-like (**e**) and with a labial disc (**f**) (Redrawn from Heyns (1982) by Esther van den Berg, Agricultural Research Council–Plant Protection Research, Pretoria, South Africa)

have a labial or oral disc (Fig. 3.2f). Another vague term that is sometimes used is neck, which is usually understood to mean the anterior part of the body as far as the base of the oesophagus.

3.2.2 Tail

The tail is the part of the body behind the anus. Tail shape varies greatly and is often a useful diagnostic character. Commonly occurring shapes are illustrated in Fig. 3.3.

3.2.3 Body Wall

Basically, the body wall of a nematode consists of three main layers: the cuticle; the subcuticle, hypodermis or epidermis and the somatic muscles.

3.2.3.1 Cuticle

The cuticle exhibits several features that are of diagnostic value. Most important of these are the transverse striae, which give rise to annulations that should not be confused with segments of insects or earthworms since striae are limited to the

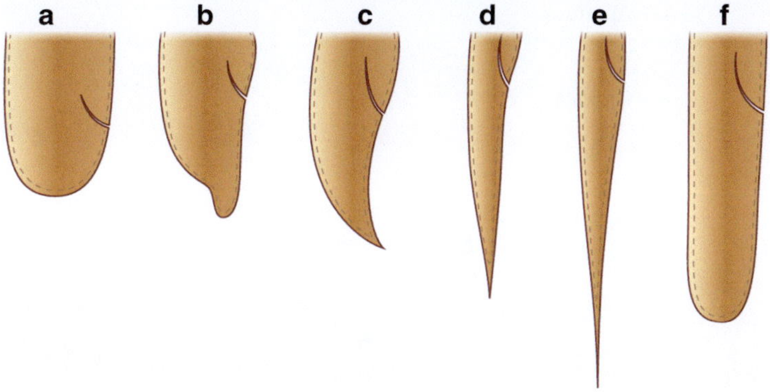

Fig. 3.3 Tail shapes: rounded or hemispherical (**a**), with a projection (**b**), convex-conoid (**c**), elongate-conoid (**d**), filiform (**e**) and cylindroid (**f**) (Redrawn from Heyns (1982) by Esther van den Berg, Agricultural Research Council–Plant Protection Research, Pretoria, South Africa)

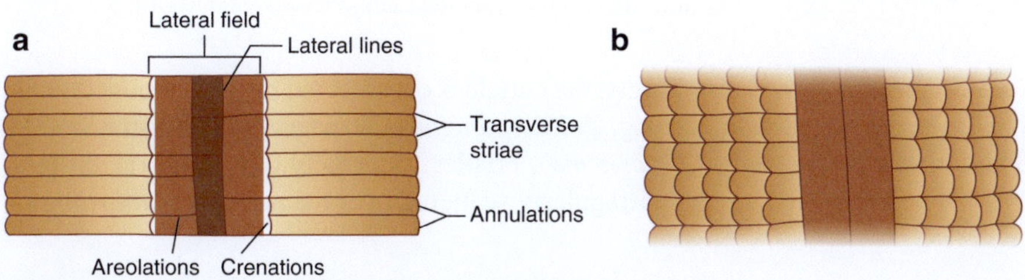

Fig. 3.4 Cuticular ornamentation with striae, lateral lines as well as areolations and crenations (**a**). Cuticular ornamentation of *Stegelleta* with the longitudinal lines along the whole length of the body either as involutions or ridges (**b**) (Redrawn from Heyns (1982) by Esther van den Berg, Agricultural Research Council–Plant Protection Research, Pretoria, South Africa)

cuticle. The striae may be minute and barely visible (as in many Dorylaimida), pronounced (many Tylenchida and Cephalobidae) or very pronounced (Criconematidae). Usually there are also longitudinal striations, the lateral lines, which occur laterally on the body and demarcate the lateral fields (e.g. those areas overlying the lateral cords) (Fig. 3.4a). These lateral lines are usually two or four in number, but may vary from 0 to 20. The lines are actually involutions of the cuticle, which allow for a certain amount of stretch in the circumference of the body (e.g. gravid females). When the outer lines of the lateral field are impinged upon by the transverse striae, the lines are said to be crenate (Fig. 3.4a). When transverse lines enter the lateral field, either regularly or irregularly, the spaces between the lines are known as areolae and the lateral field is said to be areolated (Fig. 3.4a). Occasionally, longitudinal lines may occur over the entire circumference of the body, either as involutions or as ridges as in some Cephalobidae (*Stegelleta* and some *Acrobeles* spp.), *Coslenchus costatus* (de Man, 1921) Siddiqi, 1978 and species of *Dorylaimus* and *Actinolaimus*. When these occur on species with prominent transverse striae, the entire cuticle with the exception of the lateral field is divided into rectangular blocks

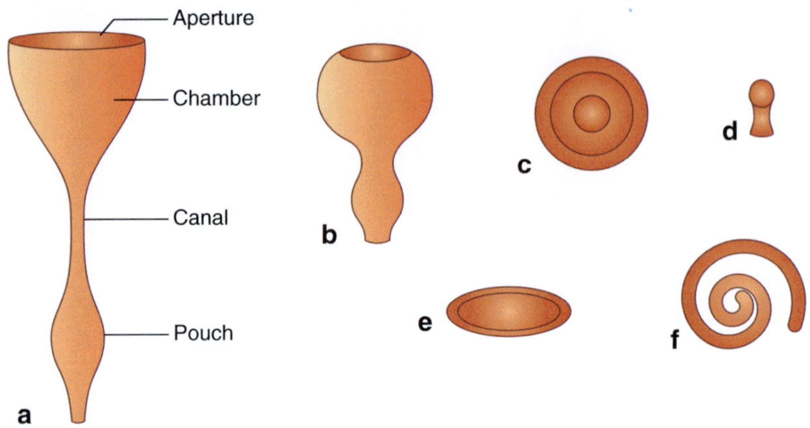

Fig. 3.5 Shape and structure of amphids: stirrup- or funnel-shaped (**a**), cyathiform (**b**), circular (**c**), pore-like (**d**), oval (**e**) and spiral-shaped (**f**) (Redrawn from Heyns (1982) by Esther van den Berg, Agricultural Research Council–Plant Protection Research, Pretoria, South Africa)

(e.g. *Stegelleta*) (Fig. 3.4b). Rarely, the cuticle is ornamented by dots or punctations, one or two rows per annule (e.g. *Zeldia punctata* (Thorne, 1925) Thorne, 1937). There may also be scales (*Criconema*), bristles, especially on the lip region (e.g. *Tripyla* and *Tobrilus*), or setae irregularly scattered over the body (e.g. *Tobrilus*).

Cuticular Organs

Organs visible on or in the cuticle include the amphids, phasmids, deirids, papillae, lateral, ventral and dorsal pores, hemizonid, hemizonion and excretory pore.

The amphids are a pair of sense organs (or sensilla) situated laterally on or near the lip region. Their shape varies among groups, and they are of great diagnostic value. The more common types are stirrup- or funnel-shaped (Fig. 3.5a), cyathiform (Fig. 3.5b), circular (Fig. 3.5c), pore-like (Fig. 3.5d), oval (Fig. 3.5e) and spiral (Fig. 3.5f). The amphid consists of the following parts: the aperture or opening, to which the above-mentioned types mainly refer; the amphidial chamber containing nerve fibres and the amphidial pouch containing *the* sensilla (a group of rod-like nerve structures). It is further connected to the lateral nerve cords. The phasmids are paired lateral caudal organs located between the lateral lines, situated from anterior to posterior to the anus (Fig. 3.6a). In some genera of the Hoplolaimidae the phasmids are extremely enlarged and are termed scutella (single: scutellum) (Fig. 3.6b). In other genera of the Hoplolaimidae the scutella are far forward towards the middle of the body (e.g. *Hoplolaimus*). Each phasmid consists of a small tube opening externally between the lateral lines and connected internally to a unicellular gland.

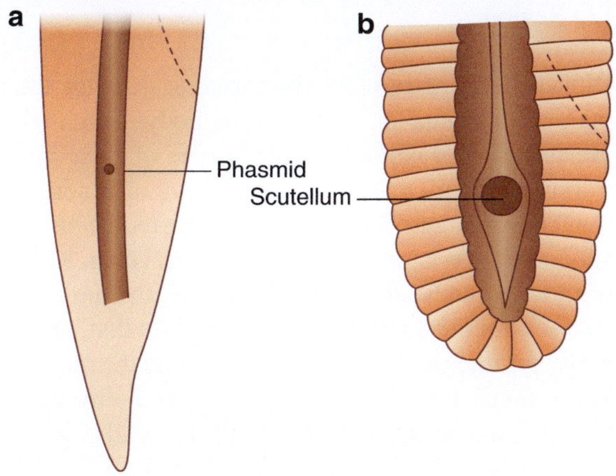

Fig. 3.6 Phasmid situated posterior to the anus (**a**) and the scutellum in *Scutellonema* on the tail (**b**) (Redrawn from Heyns (1982) by Esther van den Berg, Agricultural Research Council–Plant Protection Research, Pretoria, South Africa)

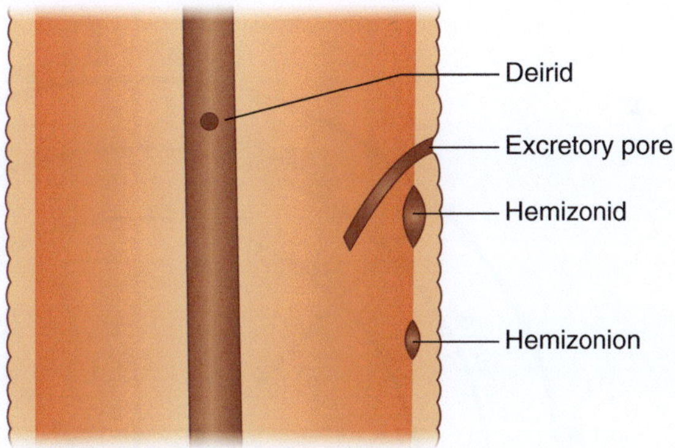

Fig. 3.7 Position of deirid, excretory pore, hemizonid and hemizonion (Redrawn from Heyns (1982) by Esther van den Berg, Agricultural Research Council–Plant Protection Research, Pretoria, South Africa)

The deirids are sensory organs resembling the phasmids. They are found in some genera of the Tylenchida and occur as paired papillae (singular: papilla) in the lateral field in the vicinity of the nerve ring. They are also known as cervical papillae (Fig. 3.7).

The hemizonid (Fig. 3.7) is a nerve commissure extending around the ventral half of the body in the vicinity of the nerve ring. It is usually seen as a small refractive structure under the cuticle near the excretory pore in the Tylenchida, or slightly posterior to the nerve ring in nematodes without an excretory pore. The excretory pore (Fig. 3.7) is the external opening of the excretory system. It is usually located ventrally in the region of the oesophageal isthmus, but it is sometimes far forward on the neck area (e.g. some *Acrobeles* spp.). It occurs occasionally far back on the body, as in *Tylenchulus*. The pore itself and a short section of the duct are usually all that is visible of the excretory system in free-living nematodes. It is commonly seen only in the Tylenchida and the Cephalobidae. The hemizonion (Fig. 3.7) is a small nerve commissure similar to the hemizonid but situated posteriorly to it.

Lip Region

Bilateral symmetry is the term used to describe the arrangement of the basic body plan of an organism, or part of its body, which when divided along a central axis yields two approximate mirror images of each other (Anonymous 2015). An illustration to identify the different sectors of a nematode body is shown in Fig. 3.8. Typically, the anterior end of the nematode consists of six sectors or lips, viz. two subdorsal, two lateral and two subventral (Fig. 3.9). The lips may be reduced to four or three or the head may be so reduced that no lips can be distinguished at all. The six (or three) lips reflect the phenomenon that, although the nematode body is bilaterally symmetrical, the structures around the mouth opening and the stoma are radially symmetrical. The stoma appears to be triradial, as is also seen in the triradial lumen of the oesophagus (Fig. 3.9). Typically, each subdorsal and subventral lip has

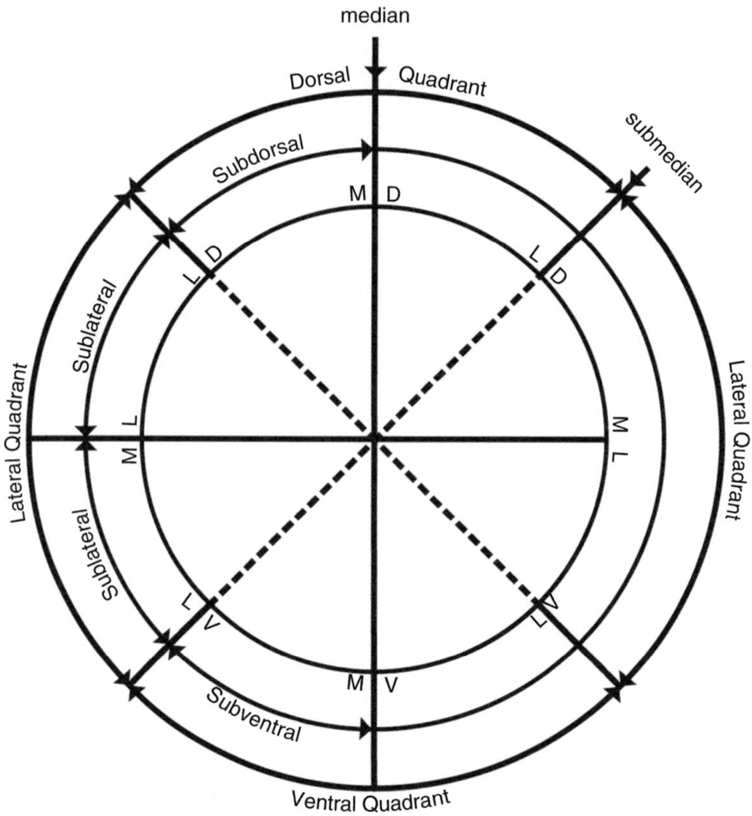

Fig. 3.8 Bilateral symmetry of a nematode body as illustrated by different planes and quadrants used to identify the different body sectors (Elsa van Niekerk, Agricultural Research Council–Plant Protection Research, Pretoria, South Africa)

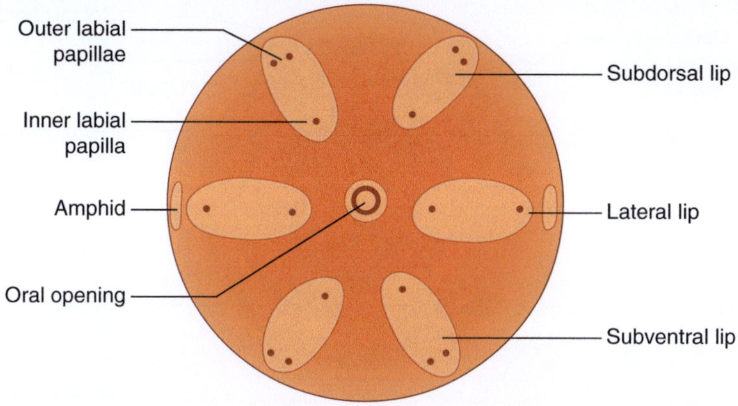

Fig. 3.9 *En face* view of a nematode lip region that usually consists of six sectors, or lips, of which two each are situated subdorsally, laterally and subventrally (Redrawn from Heyns (1982) by Esther van den Berg, Agricultural Research Council–Plant Protection Research, Pretoria, South Africa)

two outer and one inner papilla, while each lateral lip has only one outer and one inner papilla. In this way, two circles of papillae are formed: an outer circle of ten and an inner circle of six papillae.

3.2.3.2 Hypodermis

Beneath the cuticle lies the hypodermis, also called the subcuticle or epidermis. This is a thin acellular layer of which all the nuclei are found in four longitudinal thickenings (Fig. 3.10), the so-called lateral, ventral and dorsal cords, which may also contain reserves of fats and glycogen.

3.2.3.3 Somatic Muscle

In the four areas between the four cords lie the somatic muscles (Fig. 3.10), consisting of a single layer of longitudinal spindle-shaped muscle cells, each with a nerve connection from either the dorsal or ventral nerve cord. Nematodes are unusual (not unique) in sending a process from the non-contractile portion of the muscle cell to the dorsal or ventral nerve, rather than the nerve process to the muscle cell.

3.2.4 Alimentary Canal

From the point of view of the taxonomist, the digestive tract is the most important body part of the nematode. Variations found in the digestive tract, especially in its anterior region, provide taxonomists with some of the most useful diagnostic characters for the

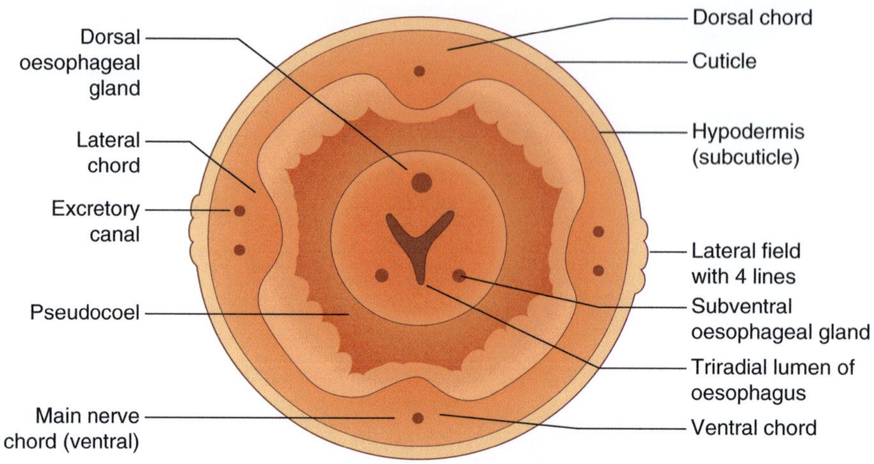

Fig. 3.10 Cross section through the oesophageal region of the body of a nematode (Redrawn from Heyns (1982) by Esther van den Berg, Agricultural Research Council–Plant Protection Research, Pretoria, South Africa)

separation of nematodes into major groups. The head and mouth structures, for example, are closely related to feeding habits and present ecologists with a tool to recognise the different nematode feeding or trophic groups (see Chap. 20).

Basically, the alimentary canal consists of the stomodaeum (mouth opening, stoma and oesophagus, mesenteron or intestine) and proctodaeum (rectum and anus or cloaca).

There is some confusion regarding the terminology of the various parts of the stomodaeum. The reader is referred to the glossary of Eisenback (1998) for the meaning of terms such as buccal capsule, buccal cavity, stoma, oesophagus and pharynx. The mouth opening leads to the stoma, which may assume a variety of shapes, depending on whether it is unarmed, armed with a tooth or teeth, with a stylet, or both.

3.2.5 *Stylet*

The stylet (spear) is an extrusible long and slender, heavily sclerotised organ used to pierce plant cell walls, fungal hyphae or small animals. Some stylets (stomato- and odontostylets) are hollow, whilst others (onchiostylets) are solid. The content of plant cells or body contents of the prey of nematodes are sucked up through hollow stylets. Stylets are found in four groups, viz. the Aphelenchida, Tylenchida, Dorylaimida and Triplonchida. The stylets of Aphelenchida and Tylenchida are called a stomatostylet since they are supposed to be derived through sclerotisation of the walls of the stoma. The stylet of the Dorylaimida is called an odontostylet (or mural tooth) because it is formed more posteriorly in the wall of the oesophagus. The basal part of the onchiostylet of the Trichodoridae is called the oesophagostomal region and the anterior part the cheilostomal region.

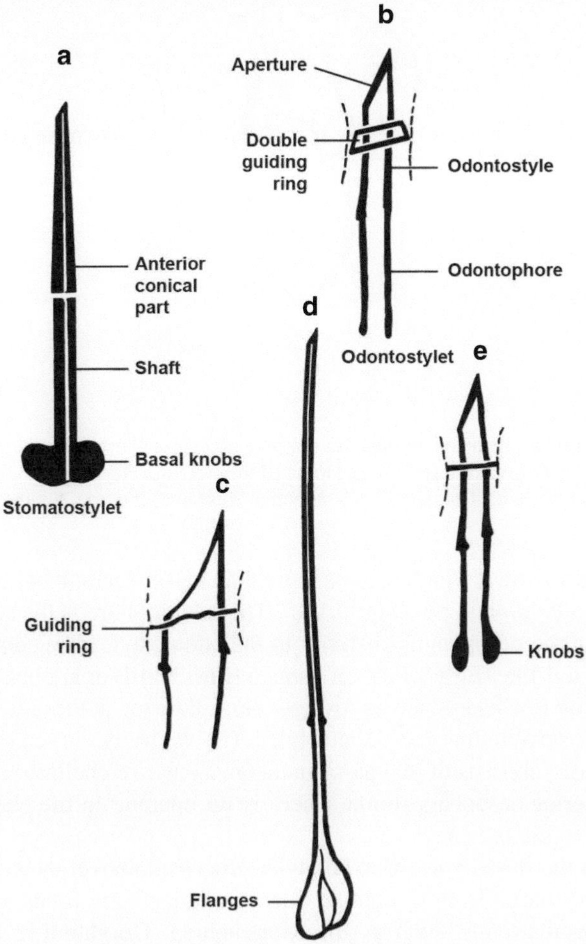

Fig. 3.11 Typical appearance of a stomatostylet in the Tylenchida (**a**) and an odontostylet in the Dorylamida as well as some free-living nematodes (**b**–**e**) (Redrawn from Heyns (1982) by Esther van den Berg, Agricultural Research Council–Plant Protection Research, Pretoria, South Africa)

The stomatostylet consists of three parts: the anterior conical part, the shaft and the three basal knobs, of which one lies dorsally and two situated subventrally (Fig. 3.11a). The three knobs, which correspond in number with the triradial nature of the lumen (Fig. 3.9), serve for attachment of protractor muscles.

Although the odontostylet (Figs 3.11b–e) differs in origin and structure from the stomatostylet, it is used in the same way and serves the same purpose in feeding. The odontostylet consists of two parts, namely the odontostyle (formed by a specialised cell in the anterior oesophagus) and the stylet extension or odontophore to which it attaches itself (Fig. 3.11d). The odontostyle itself varies from exceedingly long and slender as in the Longidoridae (Fig. 3.11d) to short and broad with a

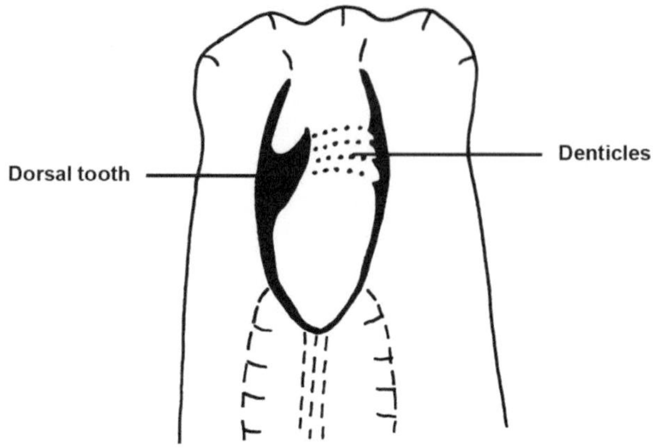

Fig. 3.12 Stoma of a predacious nematode (*Mylonchulus* sp.) illustrating the dorsally situated mural tooth with denticles in the wall of the stoma (Redrawn from Heyns (1982) by Esther van den Berg, Agricultural Research Council–Plant Protection Research, Pretoria, South Africa)

relatively large aperture, which may be more than half the length of the odontostyle, as in some *Aporcelaimus* spp. (Fig. 3.11c). The odontophore is the sclerotised part of the lumen of the oesophagus posterior to the odontostyle. The odontophore may be simple and rod-like (Figs 3.11b, c), flanged (Fig. 3.11d) or knobbed (Fig. 3.11e). The guiding ring is a sclerotised or strongly muscular ring at the anterior end of the guiding sheath surrounding the odontostylet. The ventrally curved onchiostylet of the Trichodoridae consists of two parts, namely an anterior cheilostoma with a solid tip and a posterior oesophagostoma. There is no opening in the cheilostoma that could connect to the lumen.

Apart from the three types of axial stylet described above, the stoma of nematodes may carry mural teeth or onchia. Such mural teeth are found sporadically in various unrelated groups, such as the Mononchida, Dorylaimida, Enoplida and Rhabditida. These are sclerotised outgrowths from the wall of the stoma. They may be single and ventral or dorsal, or they may be opposed by other smaller teeth, or supplemented by groups of denticles (Fig. 3.12).

3.2.6 Oesophagus

The portion of the alimentary canal between the stoma and the intestine is called the oesophagus. The word oesophagus literally means "food transporter", but because of its muscularity some nematologists prefer the term pharynx. The term oesophagus is considered more appropriate because of the structure of this organ, whilst pharynx is more appropriate functionally (Jairajpuri and Ahmad 1992). Because of its general acceptance through usage over a long period of time, the term oesophagus will be used in this chapter. The appearance and structure of the oesophagus is subject to great diversity and is of major diagnostic value in many nematode groups. In most nematodes, the oesophagus is divisible into three regions, namely the

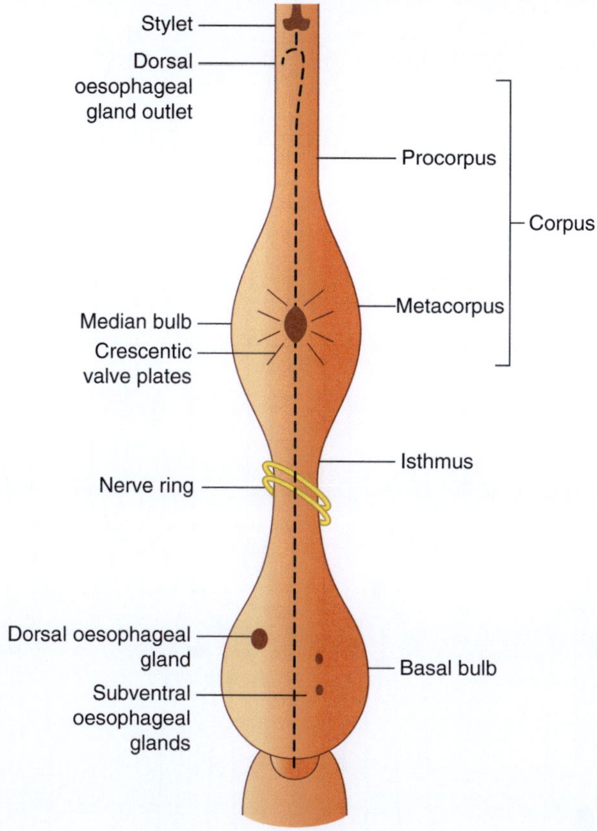

Fig. 3.13 Structure of an oesophagus of Tylenchida (Redrawn from Heyns (1982) by Esther van den Berg, Agricultural Research Council–Plant Protection Research, Pretoria, South Africa)

corpus (more or less cylindrical and the anterior part), the basal, posterior or terminal bulb (mostly pyriform in shape) and the slender isthmus connecting these two parts (Fig. 3.13). The posterior part of the corpus (metacorpus) may be swollen to form a median bulb containing the crescentic valve plates.

The oesophageal glands (salivary glands) are single-celled uninucleate glands situated within the oesophagus, secreting digestive fluids through small ducts into the lumen of the oesophagus. Most conspicuous is the dorsal oesophageal gland and the dorsal oesophageal gland duct. Usually there are two subventral oesophageal glands; sometimes there may be four (Figs. 3.13 and 3.14).

Recognition of the various oesophagus types, as depicted in Fig. 3.14, is the first step in nematode taxonomy. The simplest type of oesophagus is the cylindrical or cylindroid type, as found in the Mononchida (Fig. 3.14a). Two other easily recognisable types are the bulboid type, consisting of a slender, cylindroid anterior part and an ovoid basal bulb (Fig. 3.14b), and the dorylaimid type in which the anterior part is narrower than the posterior part but both are roughly cylindrical (Fig. 3.14c).

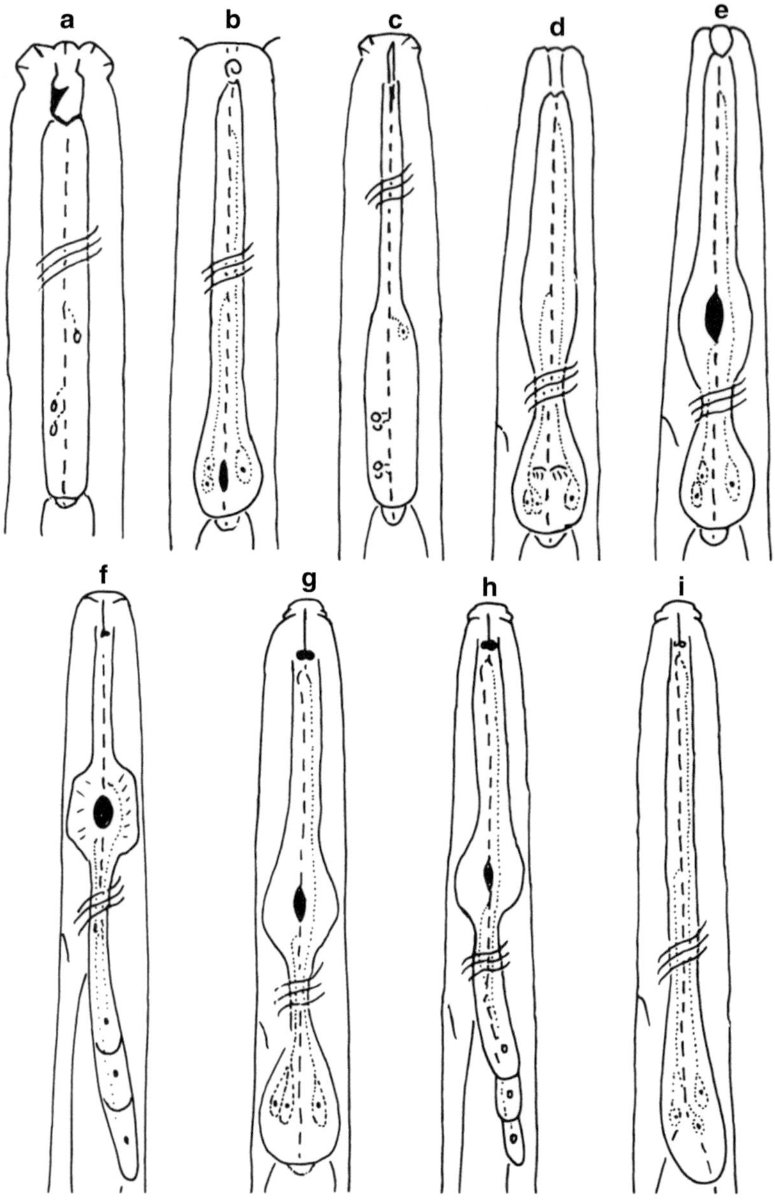

Fig. 3.14 Oesophagus types: (**a**) a cylindrical or cylindroid type (Mononchida), (**b**) a bulboid type with a slender, cylindroid anterior part and an ovoid basal bulb, (**c**) a dorylaimid type, (**d**) a rhabditoid type, (**e**) a diplogasteroid type, (**f**) an aphelenchid type: the dorsal gland opening (DGO) opens at the level of the median bulb, (**g–i**) tylenchid types: the DGO opens at the level of the base of the stylet. In Tylenchida, the oesophageal glands may either be contained within the basal bulb (**g**) or they may be lobe-like and overlap the intestine, dorsally, ventrally or laterally (**h**). In some nematode groups the oesophagus does not contain a valved median bulb (**i**) (Redrawn from Heyns (1982) by Esther van den Berg, Agricultural Research Council–Plant Protection Research, Pretoria, South Africa)

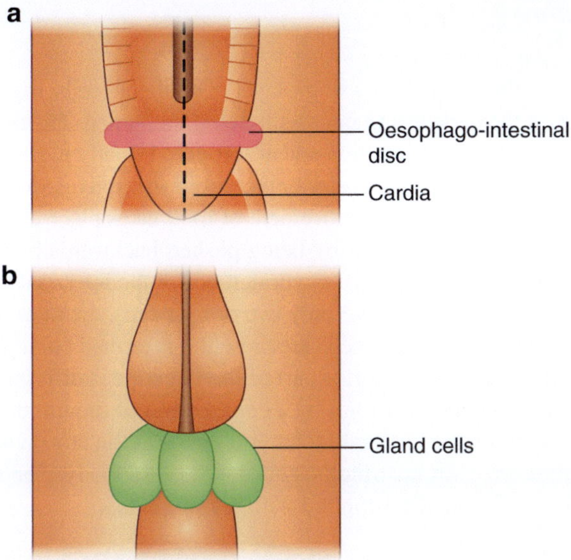

Fig. 3.15 Cardia situated as junction between the oesophagus and the intestine (**a**) and gland cells associated with the cardia (**b**) (Redrawn from Heyns (1982) by Esther van den Berg, Agricultural Research Council–Plant Protection Research, Pretoria, South Africa)

The rhabditoid type has a slightly swollen anterior part (the corpus), a distinct isthmus and then a pyriform basal bulb containing plate-like valves (Fig. 3.14d). The diplogasteroid oesophagus (Fig. 3.14e) has two bulbs, a median bulb (the metacorpus) with crescentic valve plates and a pyriform basal bulb. In the Tylenchida and Aphelenchida, the basic structure of the oesophagus is the same. However, in the Aphelenchida (Fig. 3.14f) the dorsal oesophageal gland opening (DGO) is at the level of the median bulb, whereas in the Tylenchida it is at the level of the base of the stylet (Figs 3.14g–i). Within the Tylenchida itself further variation can be seen in the oesophageal glands, which may either be contained within the basal bulb (Fig. 3.14g) or be lobe-like and overlap the intestine dorsally, ventrally or laterally (Fig. 3.14h). In some of the Tylenchidae (e.g. Neotylenchidae) the valved median bulb is absent (Fig. 3.14i).

A valve called the oesophago-intestinal valve or cardia controls the junction between the oesophagus and intestine. It is a muscular organ that projects from the base of the oesophagus into the lumen of the intestine and can be heart-shaped, conoid or elongated (Fig. 3.15a). Structures such as a disc, basal shield, epithelial sheath and glands may be associated with this valve (Fig. 3.15a). In some nematodes, e.g. *Nygolaimus*, there are three conspicuous gland cells at the base of the oesophagus (Fig. 3.15b).

3.2.7 Intestine

The wall of the intestine consists of a single layer of epithelial cells, usually with a basement membrane on the outside and microvilli lining the lumen. The microvilli enlarge the surface and serve both for secretion and absorption. Usually the anterior part of the intestine is more secretory and the posterior part more absorptive. Since the wall of the intestine has no muscular layer, the movement of food takes place either as a result of movements of the organism itself or by being pushed backwards by the ingestion of more food. The number of cells in the circumference of the intestinal wall varies from two to as many as 16. When there are only two or four cells, the intestine is called oligocytous and when more than four cells are present it is termed polycytous.

In some Dorylaimida, the posterior part of the intestine is differentiated from the intestine proper and separated from it by a constriction of the lumen. The rectum is separated from the intestine by a sphincter and opens at the anus, which appears as a mid-ventral transverse slit near the posterior end of the body. The rectum is lined with a cuticle and is frequently supplied with rectal glands.

3.2.8 Reproductive System

The reproductive system in both sexes consists of one or two tubular genital branches (the organ that produces either sperm or ova), which do not only vary greatly in length but may also be straight, reflexed or coiled. The nomenclature of female genital branches refers to the position, number and direction of the uteri. A female with one uterus is called monovarial, while those that possess two uteri are diovarial (Fig. 3.16). Amphidelphic refers to opposed uteri, prodelphic to uteri parallel and anteriorly directed and opistodelphic to the uteri parallel and posteriorly directed. The term antepudendum refers to the genital branch that proceeds anterior to the vulva, while the term postpudendum refers to the genital tube that proceeds posteriorly from the vulva. They are applied to ancestrally amphidelphic species that were proposed by Maggenti (1981). The terms digonic and monogonic refer to the presence of two or one ovary, respectively.

Males may have one testis (monorchic) or two opposed testes (diorchic). Only rarely are both testes anteriorly directed as in the genus *Anticoma* (Enoplida) (Fig. 3.17). The genital tubes of both sexes are lined with a layer of epithelial cells. The gonads of nematodes are usually telogonic since germ cells are produced only in the distal end of either ovary or testis. Hologonic gonads, where germ cells are formed along the entire length of gonads, are rare in nematodes.

3.2.8.1 Female

The external opening of the female reproductive system, called the vulva, is situated ventrally on the body. This opening is usually near the middle of the body, but can be further posteriad or anteriad. Mostly it takes the form of a transverse slit, but it

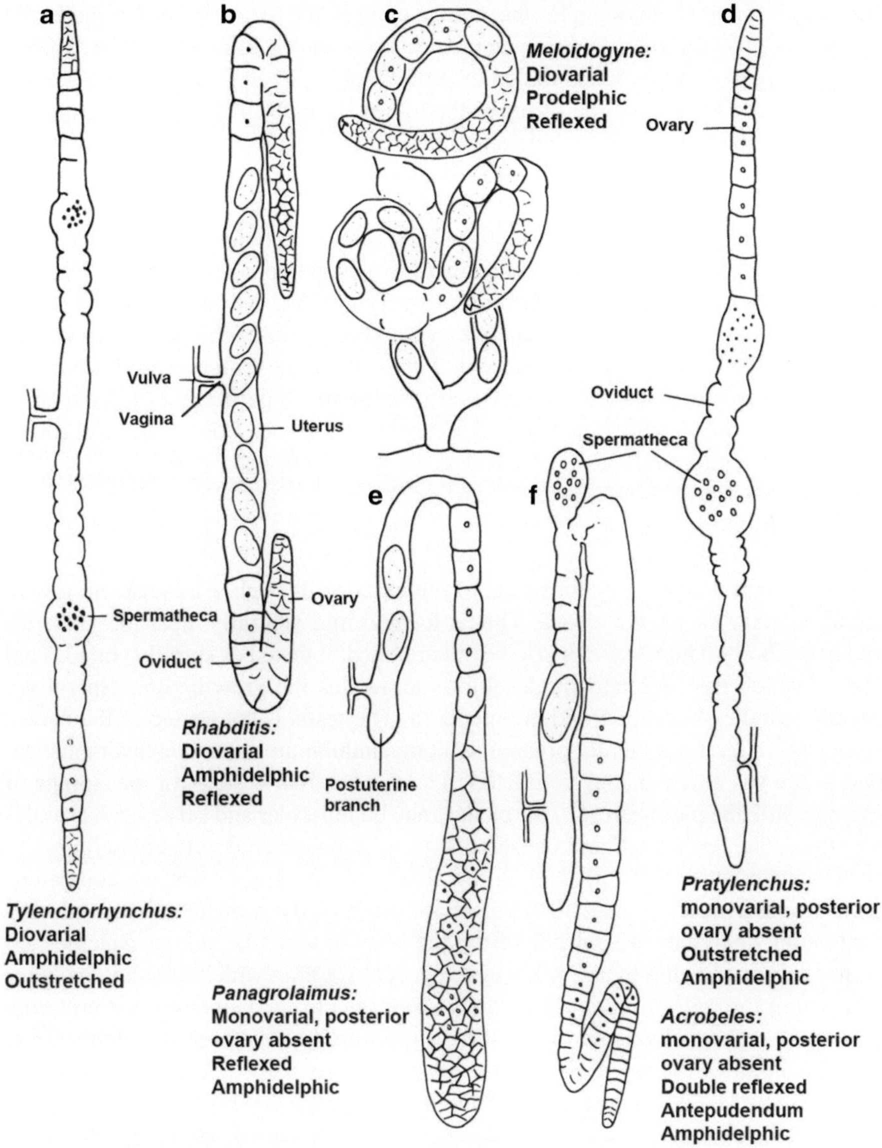

Fig. 3.16 (**a–f**) Female reproductive systems (Redrawn from Heyns (1982) by Esther van den Berg, Agricultural Research Council–Plant Protection Research, Pretoria, South Africa)

may also be a longitudinal slit as in *Dorylaimellus* or a sunken pore as in *Aporcelaimellus*. The vulva opens into the vagina, a strong muscular tube that is lined on the inside with a cuticle. The vagina leads to the uterus (plural: uteri), which is a dilated tube consisting of cuboid epithelial cells surrounded by a layer of longitudinal and oblique muscle fibres. The distal portion of the uterus, furthest from the vagina, may be dilated to serve as a spermatheca (Fig. 3.16a). In other cases, the entire uterus serves as a spermatheca (e.g. *Rhabitis*) (Fig. 3.16b). The uterus also serves as an organ for the storage of eggs, the formation of eggshells and sometimes even for embryonic development. The uterus is followed by the oviduct, representing a narrower thick-walled tube of cuboid epithelial cells that is used for the transfer of eggs from the ovary to the uterus. Usually it is separated from the uterus by a sphincter muscle, but sometimes it is distinguished functionally rather than morphologically. The ovary itself consists of elongated, flat, spindle-shaped epithelial cells, with a terminal cap cell from which both epithelial and germ cells originate. The germinal zone of the ovary sometimes contains a central rachis to which the oogonia are attached. When there is a single gonad, a postvulval sac (or postvulval branch) is usually present in many Tylenchida and represents the uterus of a rudimentary posterior reproductive organ (Fig. 3.16d). The genus *Xiphinema* (Longidoridae) exhibits the most structural diversity in the female reproductive system of all dorylaim genera. For terms such as unipartite, bipartite and tripartite uteri, z-differentiation, pars dilatata uteri and bipartite oviduct, see Coomans et al. (2001).

3.2.8.2 Male

The male reproductive system opens to the exterior through a joint opening with the alimentary canal, the cloaca. This is located mid-ventrally, near the posterior end of the body. The testes can also be outstretched, reflexed or spirally coiled. That part in which the spermatozoa develop is known as the growth zone. Sperm are usually spindle-shaped, conoid or amoeboid. The testis is connected to the cloaca by the vas deferens, a tubular organ in which glandular areas may be distinguished. Part of the vas deferens may be dilated to form a seminal vesicle for the storage of sperm, while the part nearest to the cloaca may be muscular and serve as an ejaculatory duct. The testis itself and the seminal vesicle are covered with epithelium.

Associated with and surrounding the cloaca are various sclerotised copulatory organs such as spicules. Such structures are used for the transfer of sperm during copulation when they are extruded from the body to enter the vagina. Spicules are paired organs secreted by the walls of the pouches within which they lie, the latter originating through invagination of the wall of the cloaca. Generally the proximal portion of each spicule is known as the manubrium. In the Aphelenchoididae (e.g. *Aphelenchoides*), the spicules are typically rose thorn-shaped. The spicules of this group comprise a dorsal and ventral limb, a dorsal apex and a ventral rostrum (see Hunt 1993). The gubernaculum is a short, movable, unpaired, grooved, sclerotised structure formed in the dorsal wall of the spicular pouch. It acts as a guide for the spicules, which slide along its groove when it is extended. In some groups there is a telamon associated with the gubernaculum. This is an immovable structure formed from the lateral and ventral walls of the cloaca. Lateral guiding pieces (Fig. 3.17a) are usually found in members of the Dorylaimida which lack the gubernaculum, though they may sometimes be present in addition to the gubernaculum. They are elongate sclerotised structures situated on either side of the spicules.

The males of many nematodes (especially those belonging to the Rhabditidae, Tylenchida and some Aphelenchida) have a bursa (copulatory bursa), which con-

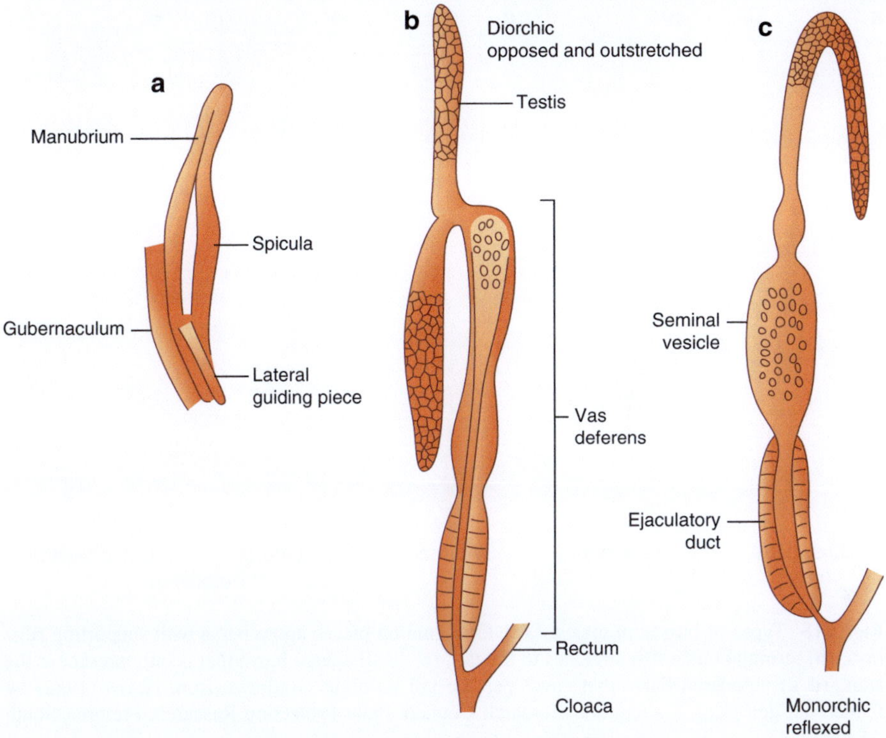

Fig. 3.17 Male reproductive systems: spicula and gubernaculum (**a**), and a diorchic (**b**) and monochic (**c**) system (Redrawn from Heyns (1982) by Esther van den Berg, Agricultural Research Council–Plant Protection Research, Pretoria, South Africa)

sists of paired lateral extensions of the cuticle at the posterior end of the body. The bursa is termed adanal when it is small and present only at the anus (Fig. 3.18c), terminal when it reaches to the tail terminus (Fig. 3.18b) and peloderan when it envelopes the tail terminus (Fig. 3.18d, e). A bursa, which does not reach the tail terminus is called leptoderan. In Rhabditidae and some Aphelenchida, the bursa has supporting ribs or rays, also known as genital papillae (Fig. 3.18a).

In most dorylaims and other groups that lack a bursa, there are so-called supplementary organs, copulatory papillae or genital papillae. These are located ventromedially anterior to the anus. Supplements may be regularly or irregularly spaced, some distance apart (Fig. 3.19a) or they may be contiguous (Fig. 3.19b). They may also be in fascicles, i.e. arranged in several distinct contiguous groups (Fig. 3.19c). Often there are two adjacent supplements close to the anus and somewhat separate from the ventromedian series, called the adanal pair (Figs 3.19a–c). The ventromedian series varies in number from two to three, to more than 30 supplements. Supplements are called either papilloid or mammiform when they have protruding tubercles (Fig. 3.19d) or tuboid when they have distinctly sclerotised ducts (Fig. 3.19e).

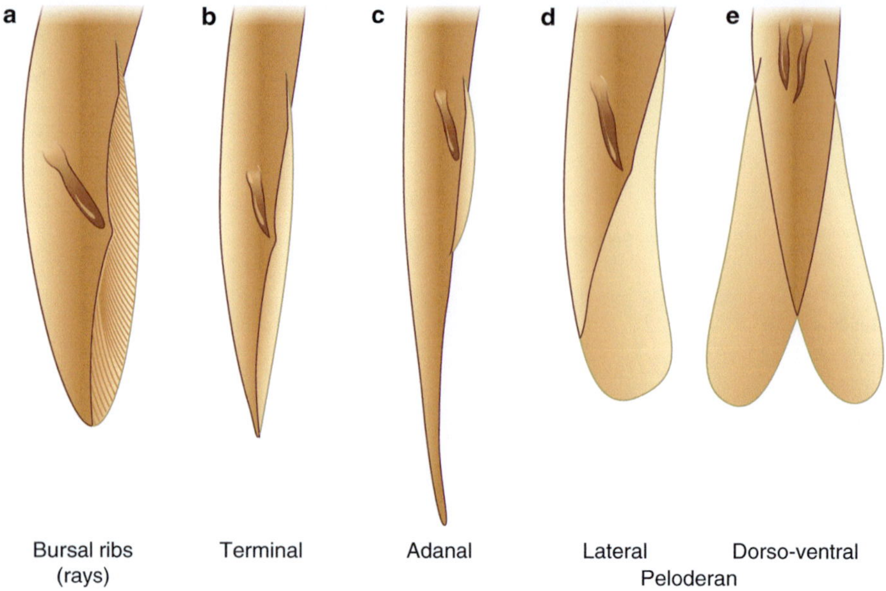

Fig. 3.18 Types of bursae in males: (**a**) lateral guiding pieces and a bursa with supporting ribs/rays, (**b**) terminal bursa that stretches to the tail, (**c**) small adanal bursa that is only present at the anus, (**d–e**) peloderan bursa that envelopes the tail terminus (Redrawn from Heyns (1982) by Esther van den Berg, Agricultural Research Council–Plant Protection Research, Pretoria, South Africa)

3.3 Classification

Decraemer and Hunt (2013) stated that nematode classification was in a state of flux, as more molecular phylogenies were being published. They used the classification of De Ley and Blaxter (2002) for the higher systematic categories, i.e. family level and above. We have decided to use the following classification as we consider it more stable and it is also based on the morphology of nematodes. In this chapter we will deal with the plant-parasitic nematodes as defined by Yeates et al. (1993). The free-living nematodes are dealt with in Chap. 20. The classification of South African plant-parasitic nematodes followed here (Table 3.1) is a synthesis of the classifications of Maggenti et al. (1988) and Geraert (2011) for Tylenchina, Hunt (1993) for Aphelenchida and Decraemer (1995) and Duarte et al. (2010) for Trichodoridae. The synonymisation of the genera *Longidoroides* and *Siddiqia* with *Paralongidorus* as proposed by Escuar and Arias (1997) will be followed in this chapter. Although cladistics and molecular phylogenies were and are being used in the understanding of relationships between members of the Heteroderidae, no clear results have been reached (Ferris 1998). Consequently the classification of the Heteroderidae as proposed in Kleynhans et al. (1996) will be used. Authorities for genera here regarded as valid are Marais (2001) for *Helicotylenchus*, Maggenti et al. (1988) for *Subanguina*, Brzeski (1998) for *Paratylenchus*, Loof and De Grisse (1989) for *Xenocriconemella*, Siddiqi (2000) for *Neolobocriconema* and Hunt et al. (2005) for *Criconemoides*. For other useful

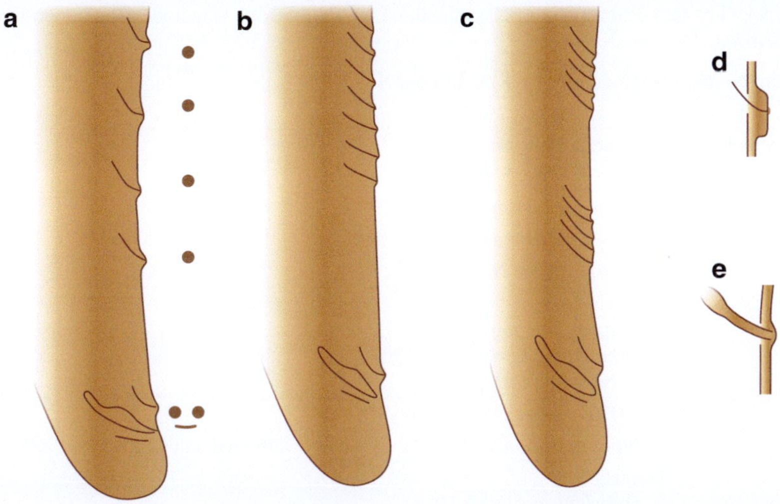

Fig. 3.19 Supplementary organs: supplements spaced apart (**a**) or contiguous (**b**), arranged in several distinct contiguous groups (**c**), adanal pair (two adjacent supplements close to the anus and somewhat separate from the ventromedian series (**a**–**c**), papilloid supplements with protruding tubercles (**d**) and tuboid supplements with distinctly sclerotised ducts (**e**) (Redrawn from Heyns (1982) by Esther van den Berg, Agricultural Research Council–Plant Protection Research, Pretoria, South Africa)

literature relating to the classification of cyst nematodes see Subbotin et al. (2010a, b). Notes as well as the taxonomic ranks used are elaborated on in Box 3.3.

Box 3.3 The Phylum Nematoda and Taxonomic Ranks Used in Identification

The phylum name remains a debate among nematode taxonomists (Anonymous 2006; Maggenti 1981), with both names Nemata Cobb, 1919, and Nematoda (Rudolphi 1808) Lankester, 1877, being recognised by different nematologists. Decraemer (2000) urged nematologists to decide on the use of one name for the phylum and suggested that as Nematoda is the name most used, nematologists should accept it.

Various taxonomic ranks with their different suffixes:

Class	-ea
Subclass	-ia
Order	-ida
Suborder	-ina
Superfamily	-oidea
Family	-idae
Subfamily	-inae

Table 3.1 The classical format for classification of plant-parasitic nematode genera reported from South Africa

Phylum Nematoda (Rudolphi, 1808) Lankester, 1877
Class Secernentea von Linstow, 1905
Subclass Diplogasteria Maggenti, 1982
Order Tylenchida Thorne, 1949
Suborder Tylenchina Chitwood, 1950
Superfamily Tylenchoidea Örley, 1880
Family Anguinidae Nicoll, 1935
Genus *Anguina* Scopoli, 1777 *Subanguina* Paramonov, 1967 *Ditylenchus* Filipjev, 1936
Family Dolichodoridae Chitwood, 1950
Subfamily Dolichodorinae Chitwood in Chitwood and Chitwood, 1950
Genus *Dolichodorus* Cobb,1914 *Neodolichodorus* Andrássy, 1976
Subfamily Brachydorinae Siddiqi, 2000
Genus *Brachydorus* de Guiran and Germani, 1968
Subfamily Merlininae Siddiqi, 1971
Genus *Geocenamus* Thorne and Malek, 1968 *Amplimerlinius* Siddiqi, 1976
Subfamily Telotylenchinae Siddiqi, 1960
Genus *Histotylenchus* Siddiqi, 1971 *Neodolichorhynchus* Jairajpuri and Hunt, 1984 *Paratrophurus* Arias, 1970 *Quinisulcius* Siddiqi, 1971 *Telotylenchus* Siddiqi, 1960 *Trichotylenchus* Whitehead, 1960 *Trophurus* Loof, 1956 *Tylenchorhynchus* Cobb, 1913
Family Pratylenchidae Thorne, 1949
Subfamily Pratylenchinae Thorne, 1949
Genus *Apratylenchoides* Sher, 1973 *Hirschmanniella* Luc and Goodey, 1964 *Pratylenchoides* Winslow, 1958 *Pratylenchus* Filipjev, 1936 *Radopholus* Thorne, 1949 *Zygotylenchus* Siddiqi, 1963
Family Hoplolaimidae Filipjev, 1934
Subfamily Hoplolaiminae Filipjev, 1934

Table 3.1 (continued)

Genus *Hoplolaimus* von Daday, 1905 *Rotylenchus* Filipjev, 1936 *Helicotylenchus* Steiner, 1945 *Scutellonema* Andrássy 1958
Subfamily Rotylenchulinae Husain and Khan, 1967
Genus *Rotylenchulus* Linford and Oliveira, 1940
Family Heteroderidae Filipjev and Schuurmans Stekhoven, 1941
Subfamily Heteroderinae Filipjev and Schuurmans Stekhoven, 1941
Genus *Heterodera* A Schmidt, 1871 *Afenestrata* Baldwin and Bell, 1985 *Globodera* Skarbilovich, 1959 *Cactodera* Krall and Krall, 1978 *Punctodera* Mulvey and Stone, 1976
Subfamily Meloidogyninae Skarbilovich, 1959
Genus *Meloidogyne* Goeldi, 1892
Superfamily Criconematoidea, Taylor 1936
Family Criconematidae Taylor, 1936
Subfamily Criconematinae Taylor, 1936
Genus *Criconema* Hofmänner and Menzel, 1914 *Criconemoides* Taylor, 1936 *Discocriconemella* De Grisse and Loof, 1965 *Hemicriconemoides* Chitwood and Birchfield, 1957 *Ogma* Southern, 1914 *Neolobocriconema* Mehta and Raski, 1971 *Xenocriconemella* De Grisse and Loof, 1965
Subfamily Hemicycliophorinae Skarbilovich, 1959
Genus *Caloosia* Siddiqi and Goodey, 1964 *Hemicycliophora* de Man, 1921
Family Tylenchulidae Skarbilovich, 1947
Subfamily Paratylenchinae Thorne, 1949
Genus *Paratylenchus* Micoletzky, 1922
Family Tylenchulidae Skarbilovich, 1947
Genus *Meloidoderita* Pogosyan, 1966 *Sphaeronema* Raski and Sher, 1952 *Trophotylenchulus* Raski, 1957 *Tylenchulus* Cobb, 1913
Order Aphelenchida Siddiqi, 1980
Suborder Aphelenchina Geraert, 1966

(continued)

Table 3.1 (continued)

Superfamily Aphelenchoidoidea, Skarbilovich,1947
Family Aphelenchoididae Skarbilovich, 1947
Subfamily Aphelenchoidinae Skarbilovich, 1947
Genus *Aphelenchoides* Fischer, 1894
Family Parasitaphelenchidae Rühm, 1956
Subfamily Bursaphelenchinae Paramonov, 1964
Genus *Bursaphelenchus* Fuchs, 1937
Class Adenophorea von Linstow, 1905
Subclass Enoplia Pearse, 1942
Order Triplonchida Cobb, 1920
Suborder Diphtherophorina Coomans and Loof, 1970
Superfamily Diphtherophoroidea Micoletzky, 1922
Genus *Trichodorus* Cobb, 1913 *Nanidorus* Siddiqi, 1974 *Paratrichodorus* Siddiqi, 1974
Order Dorylaimida Pearse, 1942
Suborder Dorylaimina Pearse, 1936
Superfamily Dorylaimoidea de Man, 1876
Family Longidoridae Thorne, 1935
Subfamily Longidorinae Thorne, 1935
Genus *Longidorus* Micoletzky, 1922 *Paralongidorus* Siddiqi Hooper and Khan, 1963
Subfamily Xiphinematinae Dalmasso, 1969
Genus *Xiphinema* Cobb, 1913
Superfamily Tylencholaimidae Filipjev, 1934
Family Tylencholaimidae Filipjev, 1934
Genus *Xiphinemella* Loos, 1950

3.3.1 Plant-Parasitic Nematodes Reported from South Africa

The following section adds to the list of plant-parasitic nematodes previously reported in SA, with new information being included according to the general lay-out used in Kleynhans et al. (1996).

3.3.1.1 Family Anguinidae

Anguina woodi Subbotin, Tiedt and Riley, 2004

Described from galls on dune grass, *Ehrharta villosa* var. *villosa*, Western Cape Province. The mature galls varied in colour from purplish to brown and formed elongated to round elevations on the stems, leaf sheaths and occasionally also the leaf blades. The galls occurred singly or in clusters and part of the stem or leaves might be covered with galls. In such extreme cases the leaf was distorted (Swart et al. 2004). Some of the galls were found to contain bacteria, similar in appearance to *Rathaibacter* spp. (Riley and Swart 2004).

Ditylenchus triformis Hirschmann and Sasser, 1955
Described from *Gladiolus* sp. in the USA. Recorded from tomato, North West Province.

3.3.1.2 Family Dolichodoridae

Neodolichorhynchus estherae (Kleynhans, 1992) Siddiqi, 2000
Previously reported in SA as *Tylenchorhynchus estherae* Kleynhans, 1992.

Quinisulcius capitatus (Allen, 1955) Siddiqi, 1971
Previously reported in SA as *Tylenchorhynchus capitatus* Allen, 1955.

Telotylenchus avaricus Kleynhans, 1975
Previously reported in SA as *Tylenchorhynchus avaricus* (Kleynhans 1975) Fortuner and Luc, 1998.

Telotylenchus dewaelei (Kleynhans, 1992) Siddiqi, 2000
Described from pearl millet, groundnut and maize in Namibia. Recorded from around grass, dune and indigenous vegetation in the Northern Cape and Western Cape provinces.

Telotylenchus namibiensis (Rashid and Heyns, 1990) Siddiqi, 2000
Described from a riverbed in the Namib Desert in Namibia. Recorded from around fynbos and indigenous vegetation in the Northern Cape and Western Cape provinces.

Telotylenchus ventralis Loof, 1963
Previously reported in SA as *Tylenchorhynchus ventralis* (Loof, 1963) Fortuner and Luc, 1987.

Telotylenchus verutus Kleynhans, 1975
Previously reported in SA as *Tylenchorhynchus verutus* (Kleynhans, 1975) Fortuner and Luc, 1987.

3.3.1.3 Family Pratylenchidae

Hirschmanniella kwazuna Van den Berg, Subbotin, Handoo and Tiedt, 2009

Described from grasses in a wetland and also recorded from sugarcane in the KwaZulu-Natal Province.

Hirschmanniella spinicaudata (Schuurmans Stekhoven, 1944) Luc and Goodey, 1964

Described from Lake Edouard in the Democratic Republic of Congo. Recorded from a wetland in the KwaZulu-Natal Province. *Hirschmanniella spinicaudata* has been recorded parasitising rice in Africa (Luc and Fortuner 1975).

Pratylenchus fallax Seinhorst, 1968

Originally described from apple orchard with grass undergrowth on loam soil in The Netherlands. Recorded from around maize and indigenous vegetation in the KwaZulu-Natal and Northern Cape provinces. Damage attributed to *P. fallax* has been reported on cereals in Europe (Rivoal and Cook 1993).

Pratylenchus teres Khan and Singh, 1975

Originally described from mustard (*Brassica juncea*) in India. Recorded from around carrot, cucumber, cotton, hemp, tobacco, sorghum, soybean, citrus, grapevine and *Grubbia tomentosa* in the Eastern Cape, Gauteng, KwaZulu-Natal, Limpopo, Mpumalanga, Northern Cape, North-West and Western Cape provinces.

Pratylenchus teres vandenbergae Carta, Handoo, Skantar, Van Biljon and Botha, 2002

Described from cotton and mustard in the Northern Cape and Kwazulu-Natal provinces.

Radopholus antoni Van den Berg, Heyns and Tiedt, 2000

Described from around grasses in the Eastern Cape Province.

Zygotylenchus natalensis Van den Berg and Tiedt, 2003

Described from around potato in the KwaZulu-Natal Province.

3.3.1.4 Family Hoplolaimidae

Rotylenchus aqualamus Van den Berg, Marais and Tiedt, 2007

Described from around indigenous vegetation in the Goegap Nature Reserve in the Northern Cape Province.

Rotylenchus abnormecaudatus Van den Berg and Heyns, 1974

Originally described from around potato in SA. Recorded from cotton, potato and ornamental shrubs in the KwaZulu-Natal and Mpumalanga provinces.

Rotylenchus cypriensis Antoniou, 1980
Originally described from grapevine in Cyprus. Recorded from *Melianthus comosus* in the Western Cape Province.

Helicotylenchus anhelicus Sher, 1966
Originally described from around willow (*Salix* sp.) in the USA. Recorded from grass, dune vegetation, indigenous forest and vegetation in the Eastern Cape and KwaZulu-Natal provinces.

Helicotylenchus curatus Marais, Van den Berg, Swart and Tiedt, 2004
Described from around *Cyclopia plicata* in the Western Cape Province.

Helicotylenchus delanus Marais, 1998
Described from around a pine tree (*Pinus patula*) in the Mpumalanga Province. Previously reported in SA as *Helicotylenchus intermedius* (Luc, 1960) Siddiqi and Husain, 1964.

Helicotylenchus egyptiensis Tarjan, 1964
Originally described from around sugarcane in Egypt. Recorded from around maize, okra, *Eucalyptus* trees, *Bambusa* sp. and indigenous vegetation in the Limpopo Province.

Helicotylenchus hydrophilus Sher, 1966
Originally described from swamp soil in the USA. Recorded from around fynbos in the Eastern Cape Province.

Helicotylenchus marethae Marais, Quénéhervé, Tiedt and Meyer, 2003
Described from around grasses in the uKhahlamba-Drakensberg Park in the KwaZulu-Natal Province. Previously reported in SA as *Helicotylenchus labiosdiscinus* Sher, 1966.

Rotylenchulus brevitubulus Van den Berg, 1990
Originally described from indigenous vegetation in Namibia. Recorded from around grasses and indigenous vegetation in the Northern Cape Province.

Rotylenchulus macrodoratus Dasgupta, Raski and Sher, 1968
Originally described from grapevine in Italy. Recorded from around fynbos and dune vegetation in the Western Cape Province. Females of *R. macrodoratus* induce giant cells in both cortical and stellar root tissue (Vovlas and Inserra 1976).

Rotylenchulus macrosomoides Van den Berg, Palomares-Rius, Vovlas, Tiedt, Castillo and Subbotin, 2016.
Described from around sugarcane in the Mpumalanga Province.

3.3.1.5 Family Heteroderidae

Specific diagnostic features are used to identify individuals belonging to the Family Heteroderidae, especially the cysts (Box 3.4).

Box 3.4 Definitions of Diagnostic Features of the Vulval Cones of Cyst Nematodes

Cysts are swollen into a spherical, sub-spherical or lemon shape. The vulva lies at the opposite pole of the body and in lemon shape species, e.g. *Heterodera* and *Cactodera*, is raised on a projection, the vulval cone. The morphology of the vulval cone is of the utmost importance for the identification of cyst nematodes, as is the correct description of its features and terminology, which were defined in the pioneering work of Cooper (1955).

Definitions of the features of lemon-shaped cysts:

(i) Fenestrae: The vulval slit runs transversely to the anal vulval axis. A thin-walled area surrounds the vulva, and the cuticle from this region is lost in older cysts, forming an opening or fenestra
(ii) Vulval bridge: The vulva lies in a strip of thicker-walled cuticle, the vulval bridge that crosses the fenestral region and divides it into two semi-fenestrae
(iii) Underbridge: The vaginal wall may remain in the vulval cone and stays connected to the sides of the cone forming a transverse structure, the underbridge
(iv) Bullae: Thickening of the inner surface of the vulval cone may be present, forming variously shaped knobs called bullae

In spherical cysts, e.g. *Globodera* and *Punctodera*, bullae may also occur around the vulval basin. In these cysts the vulva does not occur on a raised cone, but lies in a shallow, circular depression, the vulval basin.

Genus *Cactodera* Diagnosis: Mature female and cyst with a posterior protuberance. Vulva terminal, vulval region circumfenestrate in cyst. Anus without fenestration. Bullae and underbridge absent, vulval denticles usually present. Cuticle with D-layer. Second-stage juveniles (J2) with four lines in the lateral field, phasmids punctiform. Labial disc and six lips present according to an emended diagnosis by Sturhan (2002). The genus *Cactodera* currently contains nine species, almost all parasitising Cactaceae, Chenopodiaceae and Amaranthaceae. Members of this genus are believed to have evolved in Mexico.

Remark: Unidentified species have been recorded from around potato and *Leucadendron* sp. in the Western Cape Province.

Genus *Punctodera* Diagnosis: Cyst ovoid to spheroid, without vulval protuberance. Circumfenestrate fenestra surrounding a very small vulva with another circumfenestrate fenestra of similar size surrounding the anus. Bullae present or absent, underbridge absent. Eggs retained within female body, no egg mass is formed. Three or four lines in the lateral field of J2. Stylet less than 30 µm long. This genus currently contains only three species, of which *Punctodera chalcoensis* Stone, Sosa Moss and Mulvey, 1976, is the most important as it causes yield losses of maize in Mexico. *Punctodera punctata* Thorne, 1928, is sometimes found in temperate areas where solanaceous crops are grown and can be confused with the much more important *Globodera* spp. (Evans and Rowe 1998).

Remark: Unidentified species have been recorded from around potato in the Eastern Cape and Western Cape provinces.

Globodera capensis Knoetze, Swart and Tiedt, 2013

Described from a potato field in the Western Cape Province and recorded from pig's foot (*Conicosia pugioniformis*) in the same province.

Heterodera trifolii Goffart, 1932

Originally described from around red clover (*Trifolium pratense*ae) in Germany. Recorded from broccoli, buchu (*Agathosma* sp.), cabbage, carrot, cauliflower, lettuce and potato in the Eastern Cape, Gauteng and Western Cape provinces.

Meloidogyne enterolobii Yang and Eisenback, 1983

Originally described from roots of the pacara earpod tree (*Enterolobium contortisiliquum*) in China. *Meloidogyne enterolobii* was previously reported in SA as *Meloidogyne mayaguensis* Rammah and Hirschmann, 1988, and recorded from guava, green pepper, potato and common blackjack (*Bidens pilosa*) in the KwaZulu-Natal, Limpopo and Mpumalanga provinces. *Meloidogyne enterolobii* has been declared a quarantine pest in Europe and has been reported from Africa, South, Central and North America, the Caribbean, Asia and Europe (Castagnone-Sereno 2012).

Meloidogyne fallax Karssen, 1996

Originally described from black-salsify (*Scorzonera hispanica*) in The Netherlands. Recorded from peanut and tomato in the Northern Cape and Mpumalanga provinces. *Meloidogyne fallax* has been declared a quarantine pest in Europe and has also been reported outside Europe in Australia, New Zealand and SA (Fourie et al. 2001; Marshall et al. 2001; Nobbs et al. 2001).

3.3.1.6 Family Criconematidae

Criconema ericius Van den Berg and Tiedt, 2001

Described from around a *Vechellia* tree in the Limpopo Province.

Criconema ihlathum Van den Berg and Tiedt, 2001

Described from around indigenous vegetation in the Limpopo Province.

Criconema javonnense Van den Berg and Tiedt, 2001

Described from around *Cynodon dactylon* in the Limpopo Province.

Criconema mundum Van den Berg and Tiedt, 2001
Described from around maize in the Limpopo Province.

Criconema princeps (Andrássy, 1962) Raski and Luc, 1985
Originally described from onion in Hungary and reported from various European countries and the USA. Recorded from around moss and grapevine in the Limpopo Province.

Criconema simplex Marais and Van den Berg, 1996
Described from fynbos in the Western Cape Province. Recorded from around fynbos, other indigenous vegetation including *Zantedeschia aethiopica* in the Eastern Cape, Mpumalanga and Western Cape provinces.

Criconema zantene Van den Berg and Tiedt, 2001
Described from around a *Podocarpus* tree in the Limpopo Province. Recorded from around a *Quercus* sp. in a *Podocarpus falcatus* plantation and indigenous forest in the Limpopo Province.

Criconemoides azania (Van den Berg, 1979) Hunt, Luc and Manzanilla-López, 2005
Previously reported in SA as *Mesocriconema azania* (Van den Berg, 1979) Loof and De Grisse, 1989.

Criconemoides britsiensis (Heyns, 1970) Hunt, Luc and Manzanilla-López, 2005
Previoulsy reported from SA as *Mesocriconema britsiense* (Heyns, 1970) Loof and De Grisse, 1989.

Criconemoides curvatus Raski, 1952
Previously reported from SA as *Mesocriconema curvatum* (Raski, 1952) Loof and De Grisse, 1989.

Criconemoides dherdei (De Grisse, 1967) Luc, 1970
Originally described from *Persica* sp. in Belgium. Recorded from around pear and a *Ficus* sp. in the Limpopo and Western Cape provinces.

Criconemoides ferniae Luc, 1959
Reported in SA as *Mesocriconema ferniae* (Luc, 1959) Loof and De Grisse, 1989.

Criconemoides incisus Raski and Golden, 1966
Reported in SA as *Mesocriconema incisum* (Raski and Golden, 1966) Loof and De Grisse, 1989.

Criconemoides ixhaphozi Van den Berg, Scroeder and Tiedt, 2007
Described from wetlands in the KwaZulu-Natal Province.

Criconemoides jessiensis (Van den Berg, 1992) Hunt, Luc and Manzanilla-López, 2005

Previously reported in SA as *Mesocriconema jessiense* (Van den Berg, 1992) Van den Berg, 1994.

Criconemoides kirjanovae Andrássy, 1962
Previously reported in SA as *Mesocriconema annulatiforme* (De Grisse and Loof, 1967) Brzeski, 1998.

Criconemoides maskaka (Heyns, 1970) Hunt, Luc and Manzanilla-López, 2005
Previously reported in SA as *Mesocriconema maskaka* (Heyns, 1970) Loof and De Grisse, 1989.

Criconemoides neli (Van den Berg, 1994) Hunt, Luc and Manzanilla-López, 2005
Previously reported in SA as *Mesocriconema neli* Van den Berg, 1994.

Criconemoides obtusicaudatus Heyns, 1962
Previously reported in SA as *Mesocriconema obtusicaudatum* (Heyns, 1962) Loof and De Grisse, 1989.

Criconemoides silvicola Van den Berg, 1996
Described from an indigenous forest in the Tsitsikamma National Park in the Eastern Cape Province.

Criconemoides sphaerocephaloides (De Grisse, 1967) Hunt, Luc and Manzanilla-López, 2005
Originally described from around sugarcane in Kenya and subsequently also reported from Sudan (Zeidan and Geraert 1989). Recorded from indigenous vegetation in the Free State, Limpopo and North West provinces.

Criconemoides sphaerocephalus Taylor, 1936
Previously reported in SA as *Mesocriconema sphaerocephalum* (Taylor, 1936) Loof and De Grisse, 1989.

Criconemoides thabaus (Van den Berg, 1996) Hunt, Luc and Manzanilla-López, 2005
Previoulsy reported in SA as *Mesocriconema thabaum* Van den Berg, 1996.

Criconemoides xenoplax Raski, 1952
Previously reported in SA as *Mesocriconema xenoplax* (Raski, 1952) Loof and De Grisse, 1989.

Criconemoides zulu Van den Berg, Schroeder and Tiedt, 2007
Described from a wetland in the KwaZulu-Natal Province.

Discocriconemella degrissei Loof and Sharma, 1980
Originally described from around coffee trees in Brazil. Recorded from grass in the Eastern Cape Province.

Hemicriconemoides strictathecatus Esser, 1960
Reported in SA as *Hemicriconemoides mangiferae* Siddiqi, 1961.

Neolobocriconema gariepense (Van den Berg, 1996) Siddiqi, 2000
Previously reported in SA as *Criconema gariepense* Van den Berg, 1996.

Ogma civellae civellae (Steiner, 1949) Reay and Davies, 1998
Previously reported in SA as *Criconema civellae* Steiner, 1949.

Ogma formosum Van den Berg, Heyns and Tiedt, 2000
Described from around dune vegetation in the Eastern Cape Province.

Ogma palmatum (Siddiqi and Southey, 1962) Siddiqi, 1986
Originally described from around strawberry in England. Recorded in a *Pinus* plantation and from indigenous vegetation in the Limpopo Province. *Ogma palmatum* has a worldwide distribution including Europe, India, New Zealand and the USA (Chitambar 1992; Orton Williams 1974).

Ogma tripartitum Van den Berg and Tiedt, 2001
Described from a citrus orchard in the North-West Province.

Ogma tuberculatum Van den Berg, 1996
Described from an indigenous forest in the Tsitsikamma National Park in the Eastern Cape Province.

Caloosia exigua Van den Berg, Marais and Tiedt, 2003
Described from around *Protea punctata* in the Swartberg Nature Reserve in the Western Cape Province.

Hemicycliophora demani Edward and Rai, 1971
Previously reported in SA as *Hemicycliophora nullinca* Van den Berg, 1987.

Hemicycliophora dulli Van den Berg and Tiedt, 2001
Described from around indigenous vegetation in the Mpumalanga Province.

Hemicycliophora epicharoides Loof, 1968
Described from sand dunes in The Netherlands. Recorded from around potato, fynbos and indigenous vegetation in the Mpumalanga and Western Cape provinces.

Hemicycliophora koreana Choi and Geraert, 1971
Described from an apple orchard in Korea. Recorded from around indigenous vegetation in the North West Province.

Hemicycliophora stiaani Van den Berg and Tiedt, 2000
Described from around fynbos in the Eastern Cape Province.

3.3.1.7 Family Tylenchulidae

Paratylenchus aquaticus Merny, 1966

Described from around rice, Ivory Coast. Recorded from *Paspalum* sp. in the KwaZulu-Natal Province.

Paratylenchus arculatus Luc and de Guiran, 1962

Previously reported in SA as *Paratylenchus nainianus* Edward and Misra, 1963.

Paratylenchus baldaccii Raski, 1975

Described from around grapevine in Italy. Recorded from a *Pinus radiata* plantation in the Eastern Cape Province.

Paratylenchus crenatus Corbett, 1966

Previously reported in SA as *Gracilpaurus crenatus* (Corbett, 1966) Ganguly and Khan, 1990.

Paratylenchus goodeyi Oostenbrink, 1953

Previously reported in SA as *Gracilacus goodeyi* (Oostenbrink, 1953) Raski, 1962.

Paratylenchus microdorus Andrássy, 1959

Described from around grass roots in an alpine meadow in Montenegro. Recorded from around banana, *Paspalum* sp., *Psilocaulon coriarum* in the KwaZulu-Natal and Western Cape provinces.

Paratylenchus nanus Cobb, 1923

Described from around grass roots in the USA. Recorded from around creeping bent grass (*Agrostis stolonifera*) in the Western Cape Province.

Paratylenchus steineri Golden, 1961

Previously reported in SA as *Gracilacus steineri* (Golden, 1961) Raski, 1962.

Paratylenchus straeleni (De Coninck, 1931) Oostenbrink, 1960

Described from moss in Belgium. Recorded from around *Ficus* sp., *Hydrangea* sp., *Khaya anthotheca*, *Quercus* sp., moss, ferns, a *Pinus* plantation and an indigenous forest in the Eastern Cape and Limpopo provinces.

Paratylenchus teres (Raski, 1976) Siddiqi, 1986

Previously reported in SA as *Gracilacus teres* Raski, 1976.

3.3.1.8 Family Aphelenchoididae

Aphelenchoides arachidis Bos, 1977

Originally described from the testa, pod shells, roots and hypocotyls of groundnut in Nigeria. It also feeds ectoparasitically on the roots of groundnut (Bridge et al. 1977). *Aphelenchoides arachidis* was found in the Northern Cape Province in the testas and shells of groundnuts, together with high infections of *Ditylenchus africanus* Wendt, Swart and Webster, 1995. The infected seeds were shrivelled and discoloured (see Chap. 9, Fig. 9.8a), and the infected pod shells showed blackish discolouration on the inside. These symptoms are identical to those caused by *D. africanus* in groundnut (Lesufi 2007). In March 2012, the testa nematode (*A. arachidis*) was again identified from infected groundnut pods and seeds on farms near Hartswater (Northern Cape Province) (Lesufi et al. 2015). This time, *D. africanus* was not found in these samples, but the seeds were also shrivelled, small, discoloured and malformed. Some seeds within infected pods germinated early, while some, coupled with reddish brown discoloration of the seed stalk, showed arrested formation and did not germinate. Maize, sorghum, pearl millet, sugarcane and rice are considered hosts of *A. arachidis* (Anonymous 2001). Severe infections of this nematode can have an adverse effect on the appearance and size of the seeds, decreasing the value of the groundnut (Bridge et al. 1977).

Aphelenchoides ritzemabosi (Schwartz, 1911) Steiner, 1932

Originally described from *Chrysanthemum* sp. in the USA (neotype of Allen, 1952). In SA, Wager (1972) reported that *A. ritzemabosi* was identified by typical symptoms that appeared reddish angular leaf spots on unknown plants in KwaZulu-Natal in 1941, 1956 and 1957 as well as in the Western Cape Province in 1957. Since 1957, however, it has never been reported from SA, until 2004 when *Nerine* bulbs showing faint to well-defined brown rings in crosscuts were found to be infected with *A. ritzemabosi*. This represented the first report on *A. ritzemabosi* infected bulbs of a plant and also the first identification of specimens of the nematode by morphological and morphometrical means in SA (Swart et al. 2008). So far, *A. ritzemabosi* has been reported from *Zinnia* sp. and *Nerine* sp. in the KwaZulu-Natal, Mpumalanga and Western Cape provinces.

Bursaphelenchus africanus Braasch, Gu, Burgermeister, Brandstetter and Metge, 2006

Bursaphelenchus africanus was isolated from packaging wood made of *Pinus radiata* that was exported from SA to China and is the second *Bursaphelenchus* sp. known to occur in SA (Braasch et al. 2006).

Bursaphelenchus leoni Baujard, 1980

Originally described from *Pinus pinaster* in France. Recorded from *Pinus radiata* in the Western Cape Province. The trees from which the nematode was extracted had a greyish appearance and short, very sparse needles before dying. About 3 % of the trees in the Silvermine Nature Reserve (Table Mountain National Park) showed these symptoms (Braasch et al. 1998).

3.3.1.9 Family Trichodoridae

Trichodorus iuventus Decraemer and Marais, 2000
Described from around fynbos in the Western Cape Province.

Nanidorus minor (Colbran, 1965) Siddiqi, 1974
Previously reported in SA as *Paratrichodorus minor* (Colbran, 1965) Siddiqi, 1974.

Nanidorus renifer Siddiqi, 1974
Previously reported in SA as *Paratrichodorus renifer* Siddiqi, 1974.

3.3.1.10 Family Longidoridae

Longidorus jagerae Heyns and Swart, 1998
Described from around *Galenia africana* in the Swartberg Nature Reserve in the Western Cape Province. Recorded from peach, *Eragostis* sp. and *Melianthus comosus* in the Limpopo and Western Cape provinces.

Longidorus pini Andres and Arias, 1988
Described from around *Pinus sylvestris*, *Quercus pyrenaica*, *Juncus* sp. and pastures in Spain. Recorded from plum in the Western Cape Province.

Paralongidorus hooperi (Heyns, 1966) Escuer and Arias, 1997
Previously reported in SA as *Longidorus hooperi* (Heyns, 1966) Jacobs and Heyns, 1982.

Paralongidorus jacobsi (Heyns, 1998) New combination
Described from around sugarcane in the KwaZulu-Natal Province. Heyns (1998) described two *Longidoroides* species from SA but the authors of this chapter accept the synonymisation of the genus *Longidoroides* with *Paralongidorus* (Escuar and Arias 1997) for the genus *Paralongidorus* and transfer *Longidoroides jacobsi* to the genus *Paralongidorus*.

Paralongidorus sandellus (Heyns, 1966) Escuer and Arias, 1997
Described from around beans in the Limpopo Province. Also recorded from beans and grass in the Limpopo and Mpumalanga provinces. Previously known in SA as *Xiphinema sandellum* Heyns, 1966, *Longidorus sandellus* (Heyns, 1966) Khan, Chawla and Saha, 1987, *Brevinema sandellum* (Heyns, 1966) Chaves and Coomans, 1984, and *Paralongidorus (P.) sandellus* (Heyns, 1966) Hunt, 1993.

Paralongidorus seinhorsti (Heyns, 1998) New combination
Originally described from reeds and grasses in the North-West Province. Heyns (1998) described *Longidoroides seinhorsti* from SA, but the authors of this chapter

accept the synonymisation of the genus *Longidoroides* with *Paralongidorus* (Escuar and Arias, 1997) for the genus *Paralongidorus* and transfer *Longidoroides seinhorsti* to the genus *Paralongidorus.*

Xiphinema aaba Heyns, 2000

Described from around grasses and *Vachellia xantophloea* in the KwaZulu-Natal Province.

Xiphinema capriviense Hutsebaut, Heyns and Coomans, 1989

Described from indigenous vegetation in Namibia. Recorded from orange trees and wheat in the Limpopo and North-West provinces.

Xiphinema paritaliae Loof and Sharma, 1979

Described from around granadilla in Brazil. Recorded from guava, maize and indigenous veld in the Eastern Cape, Free State and Limpopo provinces. Reported from bean, maize, sunflower and tomato in Mozambique (Van den Oever et al. 1998).

Xiphinema sharonae Malan, Swart, Meyer and Heyns, 1997

Described from around *Brunia albiflora* in Western Cape Province. Also recorded from grapevine in the Western Cape Province.

Xiphinema simile Lamberti, Choleva and Agostinelli, 1983

Described from around *Populus alba* in Bulgaria. Recorded from grass and indigenous vegetation in the Gauteng and Mpumalanga provinces.

Xiphinema swartae Stocker and Kruger, 1988

Previously reported in SA as *Xiphinema swarti* Stocker and Kruger, 1988.

Xiphinema vuittenezi Luc, Lima, Weischer and Flegg, 1964

Previously reported in SA as *Xiphinema petersmithi* Malan, Swart, Meyer and Heyns, 1997. Recorded from grapevine in the Western Cape Province.

Xiphinema zyzy Heyns and Swart, 2002

Described from around grasses (indigenous vegetation) in Limpopo Province.

3.3.1.11 Family Tylencholaimidae

Xiphinemella christiae De Bruin and Heyns, 1991

Described from around hops in the Western Cape Province. Recorded from around *Acacia mearnsii*, apple trees, *Cynodon dactylon*, pineapple, *Chloris gayana* in the Eastern Cape and Western Cape provinces.

Xiphinemella eversa (Heyns, 1963) Siddiqi, 1966

Described from around cowpea in the Free State Province. Previously reported in SA as *Botalium eversum* Heyns, 1963. Also recorded from around cereals, cherry, citrus, *Chloris gayana*, *Cynodon dactylon*, *Eucalyptus* sp., peach, pineapple, *Populus* sp., sweet potato and tomato in the Eastern Cape, Free State, KwaZulu-Natal, North-West and Mpumalanga provinces.

Xiphinemella marindae De Bruin and Heyns, 1991

Described from around grasses in the uKhahlamba-Drakensberg Park in KwaZulu-Natal Province. Also recorded from around conifers, *Eucalyptus* sp., *Hypoxis rigidula* and indigenous vegetation in the KwaZulu-Natal Province.

References

Allen MW (1952) Taxonomic status of the bud and leaf nematodes related to *Aphelenchoides fragariae* Ritzema Bos, 1891. P Helm Soc Wash 19:108–120

Anonymous (2001) *Aphelenchoides arachidis* Bos. Distribution maps of plant diseases, Map No. 833. CAB International, Wallingford

Anonymous (2006) Introduction to nematodes: Nemaplex. University of California, Davis. http://plpnemweb.ucdavis.edu/Nemaplex/Taxadata/Nemata.htm. Accessed 3 Sept 2009

Anonymous (2015) Thesaurus. Bilateral symmetry. http:\\thesaurus.com. Accessed 1 Dec 2015

Berkeley MJ (1855) Vibrio forming cysts on the roots of cucumbers. Gard Chron 14:220

Braasch H, Swart A, Tribe G et al (1998) First record of *Bursaphelenchus leoni* in South Africa and comparison with some other *Bursaphelenchus* spp. EPPO Bulletin 28:211–216

Braasch H, Gu J, Burgermeister W et al (2006) *Bursaphelenchus africanus* sp. n. (Nematoda: Parasitaphelenchidae) – found in packaging wood exported from South Africa to Ningbo/China. J Nematode Morphol System 9:71–81

Bridge J, Bos W, Page L et al (1977) The biology and possible importance of *Aphelenchoides arachidis*, a seed-borne endoparasitic nematode of groundnuts from northern Nigeria. Nematologica 23:253–259

Brzeski M (1998) Nematodes of Tylenchina in Poland and temperate Europe. Museum Instytut Zoologii Polska Akademia Nauk, Warzawa

Castagnone-Sereno P (2012) *Meloidogyne enterolobii* (= *M. mayaguensis*): profile of an emerging, highly pathogenic, root-knot nematode species. Nematol 14:133–138

Chitambar JJ (1992) SEM observations of species of *Ogma* Southern, 1914 and *Criconemella* De Grisse & Loof, 1965 (Nematoda: Criconematidae). Fundam Appl Nematol 15:297–303

Coomans A, Huys R, Heyns J et al (2001) Character analysis, phylogeny and biogeography of the genus *Xiphinema* Cobb, 1913 (Nematoda: Longidoridae). Royal Musem for Central Africa, Tervuren, Belgium. Annalen Zoölogische Wetenschappen 287:239

Cooper BA (1955) A preliminary key to British species of *Heterodera* for use in soil examination. In: Kevan DKM (ed) Soil Zoology: Proceedings of the University of Nottingham Second Easter School in Agricultural Science. Butterworths, London, pp 269–280

Decraemer De Ley P, Blaxter ML (2002) Systematic position and phylogeny. In: Lee DL (ed) The biology of nematodes. Taylor and Francis, London, pp 1–30

Decraemer W (1995) The family Trichodoridae: Stubby root and virus vector nematodes. Kluwer Academic Publishers, Dordrecht

Decraemer W (2000) Letters to the editor. Nematology Newsletter, June 2000. Phylum Nematoda or Nemata. Nemaplex. University of California, Davis. http://plpnemweb.ucdavis.edu/Nemaplex/Taxadata/Nemata.htm. Accessed 2 Sept 2009

Decraemer W, Hunt D (2013) Structure and classification. In: Perry RN, Moens M (eds) Plant nematology, 2nd edn. CAB International, Wallingford, pp 3–39

Duarte IM, De Almeida M, Brown MT et al (2010) Phylogenetic relationships, based on SSU rDNA sequences, among the didelphic genera of the family Trichodoridae from Portugal. Nematol 15:171–180

Eisenback JD (1998) Glossary of plant nematology and related terms, CD-ROM. CAB International, Wallingford

Escuar M, Arias M (1997) *Paralongidorus iberis* sp. n. and *P. monegrensis* sp. n. from Spain with a polytomous key to the species of the genus *Paralongidorus* Siddiqi, Hooper & Khan, 1963 (Nematoda: Longidoridae). Fundam Appl Nematol 20:135–148

Evans K, Rowe JA (1998) Distribution and economic importance. In: Sharma SB (ed) The cyst nematodes. Kluwer Academic Publishers, Dordrecht, pp 1–30

Ferris VR (1998) Evolution, phylogeny and systematics. In: Sharma SB (ed) The cyst nematodes. Kluwer Academic Publishers, Dordrecht, pp 57–82

Fourie H, Zijlstra C, Mc Donald AH (2001) Identification of root-knot nematode species occurring in South Africa using SCAR-PCR technique. Nematol 3:675–689

Geraert E (2011) The Dolichodoridae of the world. Identification of the family Dolichodoridae. Academia Press, Gent

Heyns J (1982) Nematode morphology and classification. In: Keetch DP, Heyns J (eds) Nematology in southern Africa, Science Bulletin No. 400. Department of Agriculture and Fisheries, Pretoria, pp 1–11

Heyns J (1998) Description of two new *Longidoroides* species (Nematoda: Dorylaimida) from South Africa, with a note on *L. strelitziae* (Heyns, 1966). Fundam Appl Nematol 21:491–499

Hunt DJ (1993) Aphelenchida, Longidoridae and Trichodoridae: Their systematics and bionomics. CAB International, Wallingford

Hunt DJ, Luc M, Manzanilla-López RH (2005) Identification, morphology and biology of plant parasitic nematodes. In: Luc M, Sikora RA, Bridge J (eds) Plant parasitic nematodes in subtropical and tropical agriculture, 2nd edn. CAB International, Wallingford, pp 11–52

Jairajpuri MS, Ahmad W (1992) Dorylaimida. Free-living, predacious and plant-parasitic nematodes. Brill, Leiden

Kleynhans KPN, Van den Berg E, Swart A et al (1996) Plant nematodes in South Africa. Plant Protection Research Institute Handbook No 8. ARC – Plant Protection Research Institute, Pretoria

Lesufi MM (2007) A study of the systematics and implications of the presence of the testa nematode *Aphelenchoides arachidis* Bos, 1977 in South Africa. MSc Dissertation, North-West University, Potchefstroom

Lesufi MM, Swart A, Mc Donald AH et al (2015) Morphological and molecular studies on *Aphelenchoides arachidis* Bos, 1977 (Tylenchina: Aphelenchoididae) from groundnuts in South Africa. Nematol 17:433–445

Loof P, De Grisse A (1989) Taxonomic and nomenclatorial observations on the genus *Criconemella* De Grisse & Loof, 1965 sensu Luc & Raski, 1981 (Criconematidae). Meded Fac Landbouwwet Rijksuniv Gent 54:53–74

Luc M, Fortuner R (1975) *Hirschmaniella spinicaudata.* CIH. Descriptions of plant-parasitic nematodes. Set 5, No. 68

Maggenti AR (1981) General Nematology. Springer, New York

Maggenti AR, Luc M, Raski DJ et al (1988) A reappraisal of Tylenchina (Nemata). 11. List of generic and supra-generic taxa, with their junior synonyms. Rev Nématol 11:177–188

Marais M (2001) A monograph of the genus *Helicotylenchus* Steiner, 1945 (Nemata: Hoplolaimidae). PhD. Agric. Dissertation, University of Stellenbosch, Stellenbosch

Marshall JW, Zijlstra C, Knight KWL (2001) First record of *Meloidogyne fallax* in New Zealand. Australas Plant Pathol 30:283–284

Needham T (1743) A letter concerning certain chalky tubulous concretions called malm; with some microscopical observations on the farina of the red lily, and of worms discovered in smutty corn. Philos Trans R Soc Lond 42:173, 174, 634–641

Nobbs JM, Liu Q, Hartley D et al (2001) First record of *Meloidogyne fallax* in Australia. Austral Plant Pathol 30:373

Orton Williams KJ (1974) *Criconema palmatum*. C.I.H. Descriptions of plant-parasitic nematodes, Set 4, No 57

Riley I, Swart A (2004) Aspects of reproduction and survival of *Anguina woodi*. Australas Plant Pathol 33:595–596

Rivoal R, Cook R (1993) Nematode pests of cereals. In: Evans K, Trudgill DL, Webster JM (eds) Plant parasitic nematodes in temperate agriculture. CAB International, Wallingford, pp 259–303

Siddiqi MR (2000) Tylenchida parasites of plants and insects, 2nd edn. CAB International, Wallingford

Sturhan D (2002) Notes on the genus *Cactodera* Krall & Krall, 1978 and proposal of *Betulodera betulae* gen. nov., comb. nov. (Nematoda: Heteroderidae). Nematol 4:875–882

Subbotin SA, Mundo-Ocampo M, Baldwin JG (2010a) Systematics of the cyst nematodes (Nematoda: Heteroderinae). Nematology monographs and perspectives 8A. Hunt DJ, Perry RN (Series eds) Systematics of cyst nematodes (Nematoda: Heteroderinae). Brill, Leiden

Subbotin SA, Mundo-Ocampo M, Baldwin JG (2010b) Systematics of the cyst nematodes (Nematoda: Heteroderinae). Nematology monographs and perspectives 8B. Hunt DJ, Perry RN (Series eds) Systematics of cyst nematodes (Nematoda: Heteroderinae). Brill, Leiden

Swart A, Subbotin SA, Tiedt LR et al (2004) *Anguina woodi* sp.n. (Tylenchida: Anguinidae) from dune grass, *Ehrharta villosa*, in South Africa. Nematol 6:129–144

Swart A, Lesufi MM, Tiedt LR (2008) Report on the chrysanthemum leaf nematode, *Aphelenchoides ritzemabosi* from South Africa. S Afr J Plant Soil 24:253

Van den Oever R, Van den Berg E, Chirruco JA (1998) Plant parasitic nematodes associated with crops grown by smallholdings in Mozambique. Fundam Appl Nematol 21:645–654

Vovlas N, Inserra RN (1976) Istopatologia di radici di olivo infestate da *Rotylenchulus macrodoratus* Dasgupta, Raski et Sher. Nematol Mediterr 4:223–230

Wager VA (1972) Records of diseases of ornamental plants not previously reported in South Africa. Technical Communication No. 100, Department of Agricultural Technical Services, Pretoria

Wouts WM, Baldwin JG (1998) Taxonomy and identification. In: Sharma BS (ed) The cyst nematodes. Kluwer Academic Publishers, Dordrecht, pp 83–122

Yeates GW, Bongers T, De Goede RGM et al (1993) Feeding habits in soil nematode families and genera – an outline for soil ecologists. J Nematol 25:315–331

Zeidan A, Geraert E (1989) Plant parasitic nematodes from Sudan: Criconematidae, Tylenchulidae, Hoplolaimidae (Tylenchida). Meded Fac Landbouwwet Rijksuniv Gent

Chapter 4
Techniques and Procedures

Mariette Marais, Antoinette Swart, Hendrika Fourie, Shaun D. Berry, Rinus Knoetze, and Antoinette P. Malan

4.1 Introduction

A prerequisite for sound and trustworthy nematology research and diagnostics is the accurate identification of nematode genera and species. A wide range of specialised techniques and procedures are available and have been published for use in Nematology. However, in this chapter only those techniques and procedures that are used on a routine basis in nematology laboratories in South Africa (SA) are described. An important aspect to keep in mind is safety prerequisites that have to be applied in a laboratory set-up (Box 4.1).

M. Marais (✉) • A. Swart
Nematology Unit, Biosystematics, Agricultural Research Council–Plant Protection Research Insitute, Private Bag X134, Queenswood 0121, South Africa
e-mail: maraism@arc.agric.za

H. Fourie
North-West University, Unit for Environmental Sciences and Management, Private Bag X6001, Potchefstroom 2520, South Africa

S.D. Berry
BASF Corporation, 26 Davis Drive, Research Triangle Park, Durham, NC 27709, USA

R. Knoetze
Department of Agriculture, Forestry and Fisheries, Plant Quarantine Station, Polkadraai Road, Stellenbosch 7600, South Africa

A.P. Malan
Department of Conservation Ecology and Entomology, Faculty of AgriSciences, Stellenbosch University, Private Bag X1, Matieland 7602, Stellenbosch, South Africa

H. Fourie et al. (eds.), *Nematology in South Africa: A View from the 21st Century*, DOI 10.1007/978-3-319-44210-5_4

Box 4.1 Health and Safety in Research Environments
Health and safety in the laboratory, glasshouse, microplot and field is usually regulated by law. In South Africa, health and safety are regulated by the Occupational Health and Safety Act No. 85 of 1993 as amended by the Occupational Health and Safety Act No. 181 of 1993. Many of the chemicals, apparatus and devices used in nematology laboratories, glasshouses, microplots and fields are potentially dangerous or hazardous. Therefore, all devices and chemicals must be handled with due diligence, while standard procedures must be in place (e.g. the use of appropriate personnel protective equipment). These recommendations and regulations must be strictly adhered to by all research and support staff.

Detailed descriptions of specific methodologies used in nematology research and diagnostics are available in numerous chapters of books including Hooper et al. (2005), Been and Schomaker (2013), Bridge and Starr (2007), Khan (2008), Nguyen and Hunt (2007), Hunt and Handoo (2009), Manzilla-López (2012), Powers and Ramírez-Suárez (2012) and Subbotin et al. (2013). Summaries of basic methodologies used in laboratories in SA have been reported previously by Koen (1969), Keetch (1982), Kleynhans et al. (1996) and Van den Berg and Furstenberg (1982).

4.2 Nematode Sampling

One of the aims of nematode sampling is to determine the diversity and population densities of both plant-parasitic and non-parasitic nematodes for research, advisory or regulatory purposes. Since nematodes typically occur in patches (horizontally and vertically) in the soil (Fig. 4.1), sampling should be planned well in advance to enable the highest level of accuracy. This will minimise the inherent variability in the data obtained. To diagnose nematode problems accurately, the following information is crucial: (i) sampling techniques and apparatus used, (ii) sample size and area sampled, (iii) sampling depth, (iv) time of sampling and (v) sampling pattern. Also important is the handling, labelling, transport and storage of the samples.

4.2.1 Sampling Techniques

Several techniques are used to collect nematode samples from fields (Kleynhans et al. 1996; Kleynhans 1997; Hooper et al. 2005; Bridge and Starr 2007; Coyne et al. 2007; Khan 2008; Been and Schomaker 2013). Popular and widely used methods for sampling annual row crops or fallow fields (i.e. for pre-planting sampling) are the 'zig-zag', 'W' or 'grid-pattern' methods. Large fields should be divided in

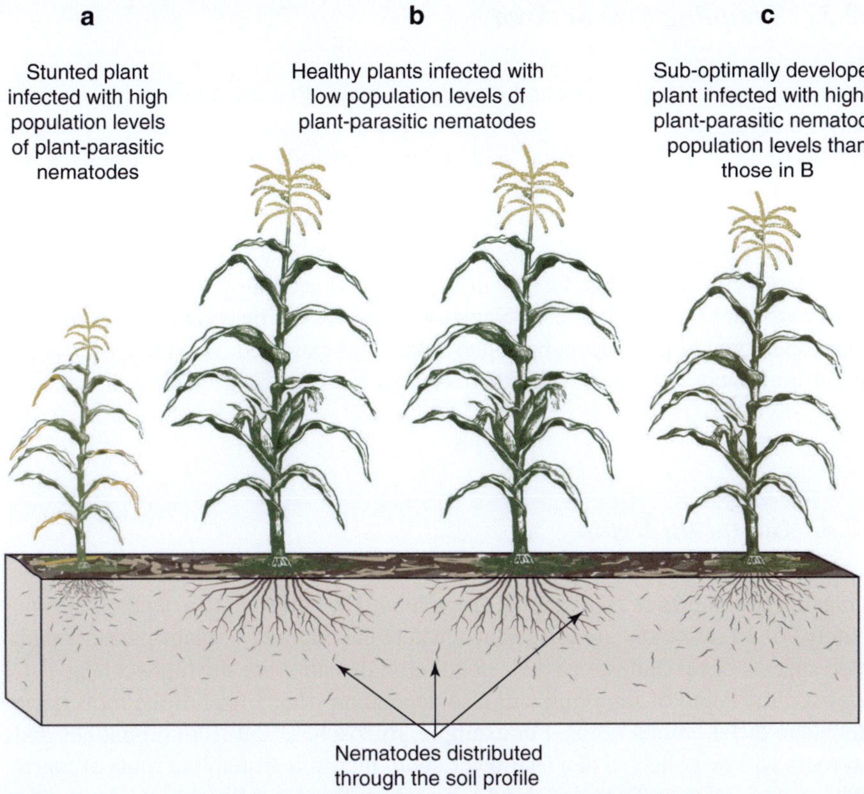

Fig. 4.1 Horizontal and vertical distribution of plant-parasitic nematodes in the soil in a maize field (Hannes Visagie, North-West University, Potchefstroom, South Africa)

smaller units of at least 1 hectare (ha) for sampling purposes. Soil samples should be collected systematically, not randomly, at equally-spaced points that cover the whole sampling unit, e.g. a 2×2 m or a 10×10 m pattern grid should be used. The smaller the distance between such points, the higher the sampling precision will be. Another sampling technique includes sampling a field at equally-spaced points along a diagonal line that runs across the field. Also, in experimental plots, annual and perennial row crops can be sampled within as well as across the crop rows. Root and soil samples from shrubs, trees and grapevine are usually taken underneath the canopy areas and on or near to the drip lines. Sampling is done on both sides of the stems and grapevine rootstocks to ensure that representative samples are obtained.

4.2.2 Sampling Equipment

Various apparatus are used for nematode sampling, including a garden spade, a trowel, a soil corer or an Edelman soil auger (Kleynhans et al. 1996; Kleynhans 1997).

4.2.3 Sampling Size or Area

The more samples (soil, roots and other plant parts) that are taken the greater the chance that they will be representative of a field or orchard. The number of samples depends mainly on the size of the field and the nematode extraction capacity available. Nematodes occur in uneven and aggregated patches in fields and hence sampling errors must be minimised by taking as many samples as practical possible. To increase the probability of detecting nematode species that may occur in low numbers, a large number of samples should be obtained according to a systematic pattern across a field (see Sect. 4.2.1). Nematode samples can be combined, thoroughly mixed and a single or a few subsamples taken and extracted. A typical subsample should preferably contain no less than 100 g of soil or 50 g of plant tissue (e.g. roots).

4.2.4 Sampling Depth

Soil and root samples of most annual crops should be collected to a depth of approximately 20–30 cm below the soil surface, where the majority of plant-parasitic nematode species occur and where their population densities are the highest (Fig. 4.1). However, the depth of sampling will also depend on where most of the roots occur, which can differ among crops. For example, rhizosphere soil from groundnut pods and roots will be collected at a shallower depth than soil around the roots of maize, soybean or sunflower. Samples should, however, also be collected at a depth of at least 30–45 cm in regions where hot, dry summers are experienced to ensure that also nematodes that migrated downwards in the soil profile are collected. Samples collected from perennial crops, trees and grapevine should be sampled in the soil layer were feeder roots are formed. For deep-rooted perennial crops, samples should be collected to a depth of approximately 20–30 cm as well as at lower depths up to 50–60 and even 100 cm. Feeder roots as well as roots that are present deeper in the soil profile are obtained this way.

4.2.5 Sampling Time

The timing of sampling depends on the purpose of the sampling. Taking samples before planting or during the early vegetative plant growth phase can provide evidence of a relationship between the nematode population densities in the soil or in the roots and the yield of annual crops. This information can be used to decide if treatment is necessary or not. Taking samples at regular intervals for the duration of a crop cycle can provide information on the population dynamics of the nematodes over time. For example, Riekert (1996) found that population densities of plant-parasitic nematodes peaked during the flowering of maize plants.

4.2.6 *Handling, Labelling, Transport and Storage of Samples*

Kleynhans et al. (1996) and Kleynhans (1997) published useful protocols on the handling, labelling, transport and storage of nematode samples. Soil and plant tissue samples should be placed in sturdy plastic bags, which can be sealed. The bags should be labelled with a permanent marker on the outside, preferably on masking tape or a sticky label, as ink of permanent markers can rub off. A bagged sample may be placed inside a second bag and a label inserted between the bags. Care should be taken that no labels are placed directly in contact with the soil or plant tissues. It is important to clearly label each sample. Information about the precise location, date of sampling, crop sampled, plant growth stage, name of farmer, Global Positioning System (GPS) coordinates where the samples were taken and crop history should be recorded in a field book (hard copy or electronic) or analysis form. This is an important aspect of the nematode sampling process.

It is preferable to place samples of rhizosphere soil and roots together in a bag. The plant tissues will be preserved for a longer time when stored this way. Other plant tissue samples, particularly aerial parts such as stems, leaves and seeds, should be wrapped in moist paper towels after sampling and placed in a separate bag. Samples should neither be dropped nor placed in direct sunlight or in a warm place. It is crucial that samples collected from fields containing dry soils not be moistened until they are processed in the laboratory. Processing of nematode samples should be done as quickly as possible after arrival at the laboratory. However, samples can be stored in a cold room or refrigerator at 5–10 °C, for as short a time as possible.

4.3 Visual Examination of Plant Material

It is advisable to examine both above- and below-ground plant parts for visual symptoms as a first measure to recognise nematode infections. Although symptoms are not always visible to the naked eye, small pieces of plant material can be inspected for nematode infection using a dissecting microscope.

The most typical symptom of nematode infection of plant tissues and an indication of potential damage are root galls induced by root-knot nematodes (*Meloidogyne* spp.) (Fig. 4.2a). Egg masses of root-knot nematode females may also be visible as small, white or brown mass protruding from the roots or visible just below the surface of infected roots/other plant parts. The egg mass can be stained pinkish-red with Phloxine B to facilitate observation and counting of egg-laying females (Fig. 4.2b).

On other plant parts, e.g. potato tubers, protuberances on the surface can also indicate the presence of root-knot nematodes (Fig. 4.3a). As in the roots of other plants, root-knot nematode females are generally visible as small, white, roundish organisms that produce white and brown egg masses below the skin of potato tubers (Fig. 4.3b).

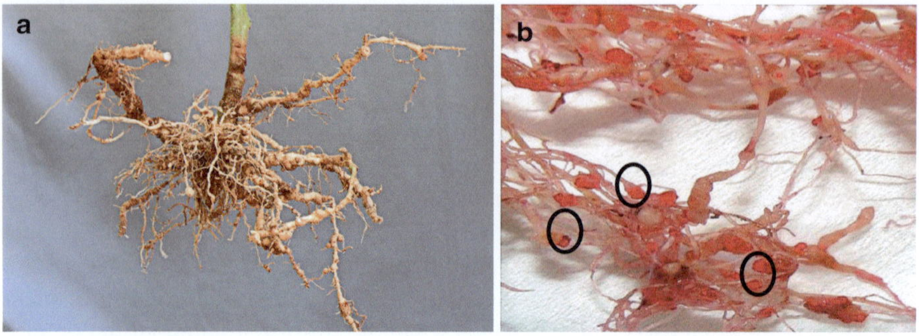

Fig. 4.2 Severe galling of tomato roots infected by root-knot nematodes (**a**) and pinkish-red stained egg masses (*black circles*) produced by females on tomato roots (**b**) (Kirk West, Port Elizabeth, South Africa)

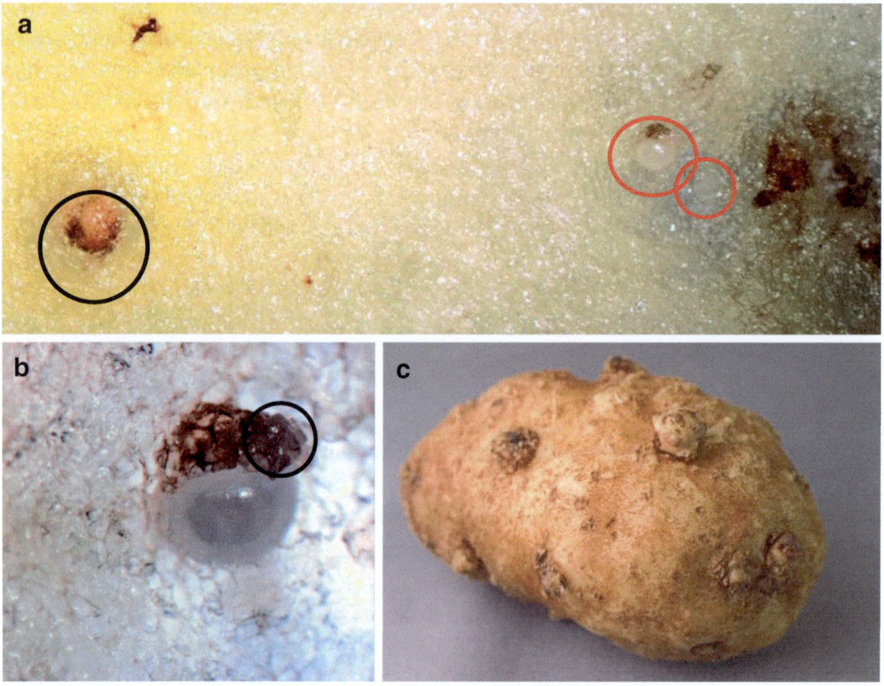

Fig. 4.3 White, roundish root-knot nematode females (*red circles*) and a brown egg mass (*black circle*) (**a**: 13× magnification), with a close-up view of a female and her brown egg mass (*black circle*) (**b**; 100× magnification) visible approximately 1 cm below the skin of an infected, galled tuber (**c**) (**a**: Kirk West, Port Elizabeth, South Africa; **b**, **c**: Gerhard du Preez, North-West University, Potchefstroom, South Africa)

Fig. 4.4 Yellowish (*yellow circle*) and brown (*red circle*) cysts of golden cyst nematodes on potato roots (Caroline Mouton, Department of Agriculture, Forestry and Fisheries, Stellenbosch, South Africa)

Cyst nematodes can also be observed on roots and tubers of infected crop plants. Towards mid-season, small, white to yellow, roundish young females (which may contain egg masses in some genera) may be visible on infected below-ground plant parts and at the end of the crop cycle as white to yellow and brown, roundish to oval-shaped cysts (Fig. 4.4).

Symptoms that result from infection by *Ditylenchus* spp. are also usually visible to the naked eye. Although the majority of species within this genus are mycetophagous, a few species are of great importance as parasites of higher plants. These include three species that are present in SA, namely the groundnut pod nematode *Ditylenchus africanus* Wendt, Swart, Vrain and Webster 1995 (see also Sect. 9.3.1, Chap. 9), *Ditylenchus dipsaci*, Kuhn, 1857 and *Ditylenchus destructor* Thorne, 1945 (Kleynhans et al. 1996). Groundnut pods and seeds infected with *D. africanus* show characteristic symptoms that are visible as corky, darkened brown-blackish tissue on the outside of the pods (Fig. 4.5), with brown and necrotic sections on the inside. Feeding of groundnut pod nematodes near or in the vascular bundles of the seed testa furthermore results in brown discolouration of the testa, with distinctly darkened veins being visible. Early germination of seeds is another distinctive symptom of *D. africanus* infection (Jones and De Waele 1990).

Symptoms caused by *D. dipsaci* infection generally are distorted and discoloured stems and leaves. Leaves may also develop small, yellowish galls or the lower parts of infected leaves may turn white ('white flagging') as is common in infected lucerne crops (Griffin 1998; Kleynhans et al. 1996). An *Aphelenchoides* sp. that can also cause damage to groundnut is the groundnut testa nematode *Aphelenchoides arachidis* Bos, 1977 (Lesufi et al. 2015). The symptoms caused by this nematode generally resemble those caused by the groundnut pod nematode and are illustrated in Chapter 9 (see Sect. 9.3.1.1, Fig. 9.8a, b).

Fig. 4.5 (**a**) Below-ground symptoms caused by *Ditylenchus africanus* on infected groundnut pods, visible as brown/grey/black discolouration on pegs and hulls (*white circles*), and (**b**) darkened veins of the seed testa (*yellow circles*) and early germination of seeds (*green circles*) (Johan Els, Agricultural Research Council–Grain Crops Institute, Potchefstroom, South Africa)

Below-ground symptoms caused by root lesion nematodes (*Pratylenchus* spp.) are usually not easy to spot with the naked eye, but in some crops, such as banana, these symptoms can be observed as short to long, dark brown to black, lesions sometimes bordered by reddish cells as is the case for *Radopholus similis* Cobb, 1913 and some *Pratylenchus* species that parasitise banana. Dark, dry spots or pimples on the surface of potato tubers may also represent symptoms caused by lesion nematodes.

The examples mentioned above represent only a few of the symptoms caused by nematodes that can be detected through visual inspection of plant parts. Photographs of galled seeds as a result of infection by seed-gall nematodes (*Anguina* and *Subanguina* spp.) are included in Chap. 19. Photographs of the damage caused by some other nematode are included in some of the other chapters (see Chaps. 7–18).

4.4 Nematode Extraction

Extraction of nematodes from soil and plant tissue is a crucial step in determining the population densities of plant-parasitic nematode species or genera during the various growth stages of a crop. Although a wide variety of nematode extraction methods have been described, the most appropriate method should be used. The quantity of the (sub) sample, be it 100 g soil, 100 ml soil or 50 g roots, should

Table 4.1 A summary of the most common techniques and procedures used in laboratories in South Africa to extract nematodes from soil and plant tissue samples and the principles upon which they are based

Method	Principle	Soil	Plant tissue
Baermann funnel and tray	Active movement of nematodes	x	x
Decanting	Specific gravity of nematodes	x	
Sieving	Size of nematodes	x	
Flegg's method	Active movement of nematodes	x	
Insect-baiting technique (specifically for entomopathogenic nematodes)	Active movement of nematodes	x	
Maceration	Size of nematodes		x
Mistifier	Active movement of nematodes	x	x
Modified NaOCl method	Dissolving of egg masses by NaOCl		x
Oostenbrink's elutriation and sieving	Specific gravity and size of nematodes	x	
Seinhorst's cyst elutriator	Specific gravity and size of nematodes	x	
Sugar centrifugal flotation	Specific gravity of nematodes	x	x
Soaking	Active movement of nematodes		x

always be prepared in exactly the same way. Also, the processing of samples from the same experiment or study should be carried out by the same operator because even small differences in handling by different operators may result in differences in extraction efficiencies. The extraction method should also not be changed during an experiment or study. An extensive, useful and detailed document has been published by the European and Mediterranean Plant Protection Organization (EPPO) in which a wide range of methods used for extracting nematodes from soil and plant tissues are described (EPPO 2013). A few of these methods are also used in laboratories in SA on a daily basis. These methods can be divided into those used for the extraction of nematodes from soil (Sect. 4.4.1) and from plant material (Sect. 4.4.2). The most common methods used in SA as well as the principles upon which they are based are listed in Table 4.1.

4.4.1 Extraction of Nematodes from Soil

A number of methods used to extract nematodes from soil are based upon the specific gravity and the size of nematodes. During a specific period of flotation or elutriation, nematodes and soil particles with a similar specific gravity can be separated from heavier soil particles in water (or another substance) and then collected on one or more stacked sieves. Other methods are based on the active movement of nematodes. Often, different methods are combined.

4.4.1.1 Decanting and Sieving Method

This method is based on the specific gravity of nematodes (Jenkins 1964). The specific gravity of terrestrial nematodes is approximately 1.08 g cm^{-3} and that of marine nematodes about 1.13 g cm^{-3}. In a water layer, terrestrial nematodes will slowly sink to the bottom but not as fast as soil particles with a higher specific gravity such as sand particles. This difference allows the separation of terrestrial nematodes from a large part of the particles present in a soil. The method described below is a modified method based on descriptions by inter alia Jenkins (1964), Kleynhans (1997) and Hooper et al. (2005). It allows the extraction of both active and inactive nematodes.

A soil sample is soaked in tap water and washed through a coarse-meshed 2-mm-aperture sieve into a plastic bucket. The residue on the sieve (that may include plant tissues) is discarded and the bucket filled up to 5 l with tap water. The soil within the bucket is then thoroughly mixed with the tap water and the mixture allowed to stand for 30 s. The sediment that settles at the bottom of the bucket is left behind when only the supernatant is decanted through stacked 45- and 25-μm-aperture sieves. The nematodes that are suspended in the supernatant are retained on the sieves and washed into a beaker with distilled water. The entire procedure is repeated twice but each time with a shorter settling time (20 and 10 s). Each time, the residue on the sieves is washed into the beaker. The content of the beaker is then mixed thoroughly and the suspension divided over 100-ml centrifuge tubes to be centrifuged.

4.4.1.2 Sugar Centrifugal Flotation Method

Different flotation methods are commonly used for the extraction of nematodes. These methods are also based on the specific gravity of nematodes. They involve the use of solutions with different densities to separate unwanted materials from nematodes by either flotation or precipitation. While water typically represents the low specific gravity (1 g cm^{-3}) solution, sugar or NaCl can be used to create the high specific gravity (1.15–1.3 g cm^{-3}) solution. After the centrifugation process in water, only organic material with a specific gravity <1 g cm^{-3} will remain in suspension and can be discarded. When centrifuged in a sugar solution with a specific gravity of 1.15 g cm^{-3}, the nematodes will remain in suspension and can be separated from soil particles with a higher specific gravity. This method also allows the extraction of both active and inactive nematodes. It is a rapid technique which enables the processing of a large number of samples within a relative short time. A limitation of this method is that the required equipment is expensive. Exposure of the nematodes to the sugar solution also represents a risk since this may cause plasmolysis if the exposure time is too long. The use of Ludox offers an alternative in this respect (see Sect. 4.4.1.9).

First, 5 ml kaolin (see Sect. 4.4.2.3) is added to each centrifuge tube, mixed with the nematodes suspended in distilled water, and the suspension centrifuged for 4 min at 1,800 *g*. The supernatant is decanted and discarded. From this step onwards, the nematodes are subjected to sugar centrifugal flotation. A sucrose solution with a specific gravity of 1.15 g cm^{-3} is added to each of the centrifuge tubes. This solution is prepared by adding 624 g of sugar to 1,000 ml of tap water. The sugar solution and the sediment, containing the nematodes, are thoroughly stirred with a spatula and re-centrifuged for 3 min at 1,800 g. The spatula is rinsed every time after it has been used to prevent cross contamination among the centrifuge tubes. After the centrifugal flotation has been completed, the supernatant that contains the nematodes is decanted through a 25-µm-aperture sieve and the nematodes washed into a beaker with distilled water for counting and identification. The time that the nematodes are exposed to the sugar solution should be as limited as possible to prevent plasmolysis of the nematodes and minimise osmotic stress.

4.4.1.3 Flegg's Method

The method described by Flegg (1967) is based on the active migration of nematodes and especially suited to extract larger nematodes (e.g. *Xiphinema, Longidorus* and *Paralongidorus* spp.) from soil samples. This method is, however, not suited to extract cysts, sluggish or inactive and small nematodes.

A known soil volume, e.g. 500 ml, is soaked in water for 1 h (for high clay content soils) while it is stirred intermittently. The resulting suspension is washed through a 4-mm-aperture sieve into a 5 l plastic bucket to separate coarse soil particles from the nematodes. The water in the bucket is stirred vigorously by hand to suspend all soil particles and nematodes, and then allowed to stand for 30 s. The supernatant is decanted through three stacked 150-µm-aperture sieves. The residue on the sieves is thoroughly rinsed with a gentle stream of tap water before it is washed with distilled water into a beaker. Water is again added to the 5 l bucket and the process repeated once more. The residue collected in the beaker is gently stirred before decanting through a 90-µm-aperture polyethylene-supported nylon sieve. This sieve is placed in a Petri dish and distilled water added in such a way that it just touches the residue. After 24 h the nematodes will have moved through the sieve in to the Petri dish and are ready for counting and identification.

4.4.1.4 Baermann Funnel Method

In contrast to Flegg's method, the Baermann funnel and tray methods are not suited to extract large nematodes from soil. A known volume of soil, e.g. 50 ml, is evenly spread on paper tissue which is placed on a 710-µm-aperture sieve. The sieve is rested onto the wide top of a funnel which is supported in a ring stand or rack. A pinch clamp is attached to a rubber or plastic tube clipping the funnel stem to seal it off during the extraction process. The funnel is filled with distilled water until the soil on the filter is slightly submerged. Nematodes will migrate from the soil through the paper tissue and the meshes of the sieve into the water in the funnel and

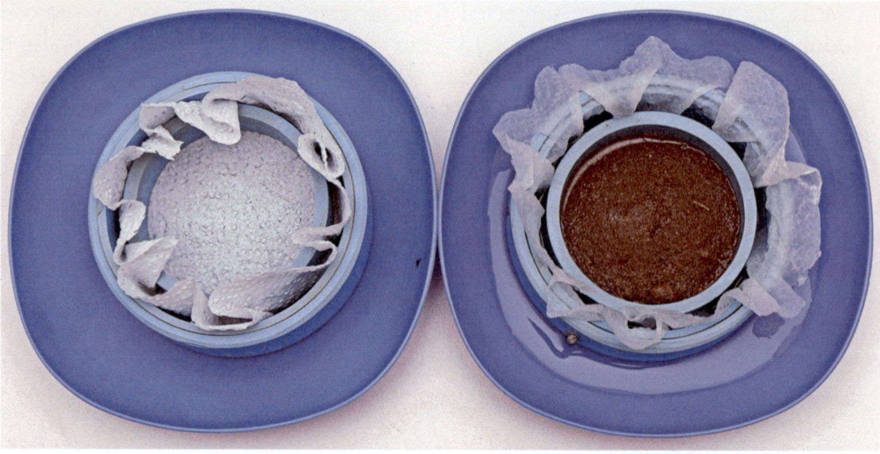

Fig. 4.6 Modified Baermann trays used for the extraction of nematodes from soil, consisting of 1-mm-aperture sieves lined with paper tissue (Kirk West, Port Elizabeth, South Africa)

sink to the funnel stem. After 24 h at room temperature, the nematodes can be collected by opening the pinch clamp. The water in the funnel can then be filled up to the original mark to replace the water that had evaporated and after another 48 h the nematodes can be collected again in a beaker. This method allows extraction of nematodes without using any substance that can be harmful to them. This method is, however, also only suited to allow extraction of only actively moving nematodes.

4.4.1.5 Baermann Tray Method

The Baermann tray method is a modification of the Baermann funnel method (Hooper and Evans 1993). The funnel is replaced by a shallow tray in which a 1-mm-aperture sieve or a kitchen sieve is placed and lined with paper tissue (Fig. 4.6). A known volume of soil is evenly spread on the paper tissue. Nematodes that migrated from the soil into the water can be collected after 24–72 h (Kleynhans 1997).

4.4.1.6 Oostenbrink's Elutriation and Sieving Method

Vermiform nematodes can be extracted from both wet and dry soil samples by keeping them in suspension in water using a controlled upward current of water in a so-called elutriator. Oostenbrink's elutriation and sieving method combines this principle with sieving (Oostenbrink 1960) (Fig. 4.7). This method is only suited for the extraction of active nematodes and is not suited to extract cysts, sluggish or inactive and large nematodes.

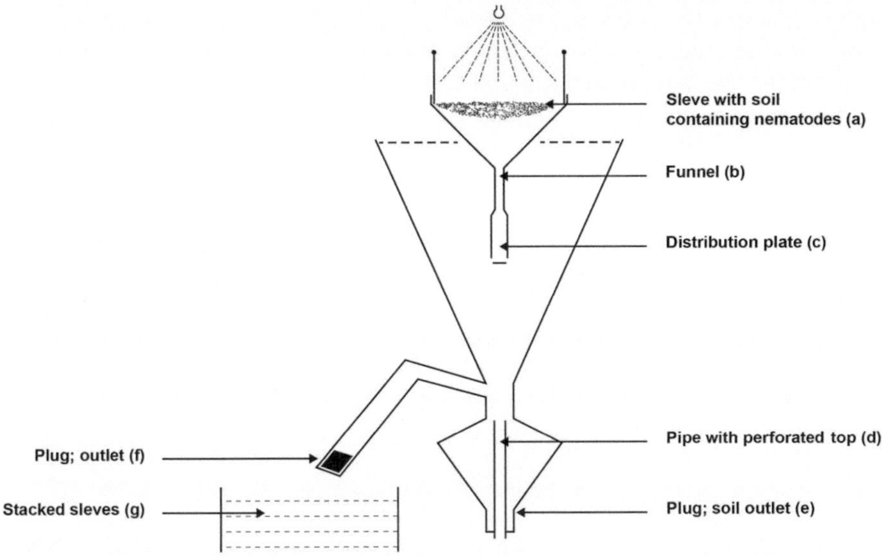

Fig. 4.7 Oostenbrink's elutriation apparatus used for the extraction of nematodes from soil; sieve with soil containing nematodes (**a**), funnel (**b**), distribution plate (**c**), pipe with perforated top (**d**), plug to control soil outlet (**e**), plug to control water outflow in which nematodes are suspended (**f**), stacked sieves on which nematodes are collected (**g**) (Redrawn by Hannes Visagie, North-West University, Potchefstroom, South Africa)

Water enters the bottom of the apparatus at about 1,000 ml min^{-1} through a pipe with a perforated top (d). When the water reaches the small plate, a 250 ml soil sample is placed on a 4-mm-aperture sieve (a) and washed into the flotation apparatus via the funnel (b). The upward water current prevents nematodes and fine soil particles from entering the neck of the apparatus. Heavy soil particles settle at the bottom of the apparatus. When the apparatus is almost full, the outlet plug (f) is removed and the water containing the nematodes is drawn off and transferred to four stacked 20-cm-diameter sieves (one 90-µm-, two 53-µm- and one 45-µm-aperture sieve) (g). The outlet pipe (f) is regulated with a clamp to prevent excessive water flow clogging the sieves. The nematodes and the fine soil particles are immediately washed with distilled water from the sieves into beakers and then transferred to paper tissue on a sieve placed in a Petri dish and filled with distilled water. After 24–48 h, the nematodes in the Petri dish can be counted and identified. Alternatively, the suspension of nematodes and small soil particles may be subjected to the centrifugal flotation method.

4.4.1.7 Seinhorst's Cyst Elutriator Method

Cysts of the Heteroderidae can also be extracted from both wet and dry soil samples using an elutriator (Seinhorst 1964). This apparatus consists of a cylinder with a bowl at the top (Fig. 4.8). Below the rim of the bowl is a sloping collar with an upright rim (a). The collar tapers towards an outlet (b), and the cylinder is closed with a rubber plug at the bottom (e). A water inlet (d), which is supplied from a constant-flowing current from a header tank, is present at the bottom of the cylinder. A lower outlet is situated in the lower half of the cylinder (c). After the cylinder and lower outlet are closed with the rubber plug, the stopcock of the header tank is opened to the desired setting (3.5 l min^{-1}). As water rises in the cylinder, a 500 ml soil sample is washed through a 2-mm-aperture sieve into the bowl. Water that spills over the collar is collected in two stacked sieves (840- and 250-μm-aperture). After approximately 30 s, the lower outlet is opened and the residue collected on the bucket sieves. The bowl and cylinder are then rinsed until the water in the cylinder is clear. The content of the 840-μm-aperture sieve is washed gently, but thoroughly, onto the 250-μm-aperture sieve. The residue on the latter sieve is then transferred to filter paper that lines a funnel and is left to dry. The dried debris containing the cysts is ready to be examined and the cysts counted using a dissecting microscope.

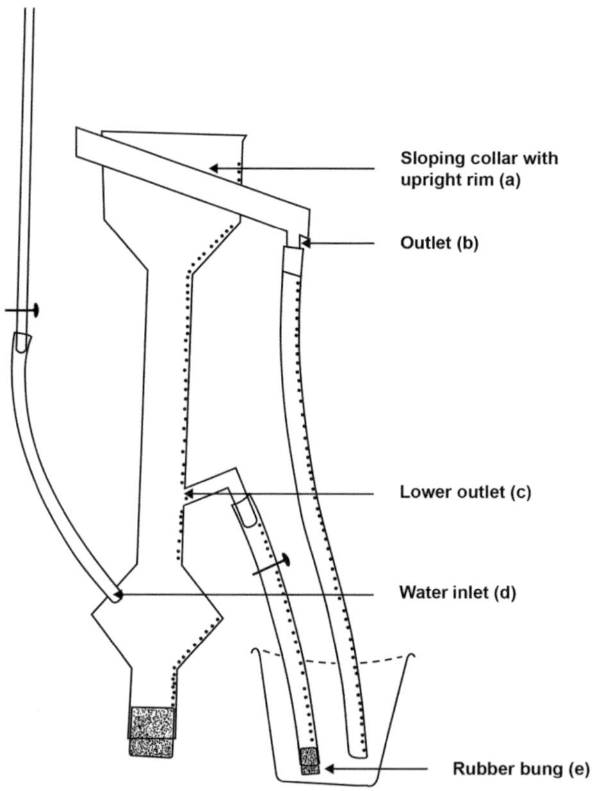

Fig. 4.8 Seinhorst's elutriator apparatus used for the extraction of cysts from soil; sloping collar with upright rim (**a**), outlet (**b**), lower outlet (**c**), water inlet (**d**) and rubber bung (**e**) (Redrawn by Hannes Visagie, North-West University, Potchefstroom, South Africa)

4.4.1.8 Alternative to Seinhorst's Cyst Elutriator Method

With this alternative method cysts of the Heteroderidae can be extracted from both wet and dry soil, but better results are obtained with dry soil (Seinhorst 1964).

A large glass trough, about 19-cm-diameter, is lined with a 7×60 cm strip of blotting paper. Tap water is added to the trough until the lower half of the paper is submerged. A known volume or mass of air-dried soil, e.g. 50 ml or 50 g, is crumbled and submerged in the water. The heavier soil particles will settle at the bottom of the trough whereas some light debris and the cysts will float. After stirring the suspension, the cysts adhere to the blotting paper. To separate the cysts from the suspended particles, the level of the suspension has to be raised momentarily. To achieve this, a large conical flask, half-filled with water, is slowly pushed down into the container and after about 15 s pulled out of the water. Adhering particles are washed into the container with a wash bottle. The water in the trough is siphoned off, and the paper strip with adhering cysts and debris is removed and laid flat on a strip of Perspex. The cysts are then separated from the debris with a needle using a dissecting microscope. The cysts are collected in a dish with a wet camel-hair brush. This method is illustrated in Kleynhans (1997).

4.4.1.9 LUDOX® Centrifugal Flotation Method for Extracting Nematodes from Sediment

As mentioned above (see Sect. 4.4.1.2), a sugar solution is very viscous and a NaCl solution causes osmotic stress which can severely damage the nematodes. As an alternative LUDOX®, a colloidal silica solution with low viscosity and osmolarity, can be used to create the required specific gravity solution. For a single extraction an efficacy rate of 70 % in terms of nematode recovery can be achieved. However, if the procedure is repeated three or more times, between 95 and 100 % of the nematodes can be extracted from a sample (Hodda and Eyualem-Abebe 2006).

A commonly used flotation method for the extraction of freshwater nematodes from sediment, which can also be used to extract nematodes from soil, is as follows: The sample is washed with distilled water into centrifuge tubes up to one-third of the total volume. Each tube is then filled with an equal amount of water to balance the centrifuge. If necessary, kaolin (see Sect. 4.4.2.3) is added to separate and settle the sediment from the organic material fraction. The tube is then shaken manually or using a mechanical device and centrifuged at 400 *g* for 5 min. After discarding the supernatant, the tube is filled with a LUDOX® solution (with a specific gravity of 1.15 g cm^{-3}) and the content resuspended. After centrifugation at 300 *g* for 5 min, the supernatant (now containing the nematodes) is decanted through a 38-μm-aperture sieve and rinsed well with distilled water. The content of the sieve is finally washed with distilled water into a beaker.

4.4.1.10 Isolation of Entomopathogenic Nematodes from Soil Samples: Insect-Baiting Technique

The same principles for the sampling and handling of plant-parasitic nematodes are used for the sampling of entomopathogenic nematodes (EPN) belonging to the Steinernematidae and Heterorhabditidae families. Individuals are recovered from soil samples by using an insect-baiting technique (Bedding and Akhurst 1975). The soil sample is placed in 250 ml plastic containers. Any suitable insect can be used as a trapping host, but in general 5 larvae of either the greater wax moth (*Galleria mellonella*) or the mealworm (*Tenebrio molitor*) are placed on the soil surface of each container, which is then closed (Fig. 4.9a). The two trapping hosts are added together or the one followed by the other, depending on their availability. During a 7-day period, for wax moth, and a 14-day period, for mealworm, the samples are periodically checked for the presence of dead insects. Dead, non-putrefied insects are washed with water to remove all surface scavenger nematodes and especially mites.

Insect cadavers that may be infected with EPN are washed and placed on a modified White trap (Figs 4.9b, c) (Kaya and Stock 1997). Nematodes are harvested within the first week of emergence. Nematode isolates are maintained in 150 ml of filtered tap water in vented culture flasks, which are kept horizontal at 14 °C and shaken weekly. Infective juveniles (IJ) are maintained by recycling through wax moth larvae within 3 months for heterorhabditids and within 6 months for steinernematids (Nguyen and Hunt 2007).

4.4.2 *Extraction of Nematodes from Roots*

4.4.2.1 Modified Baermann Funnel and Baermann Tray Methods

These methods are the same as used for the extraction of nematodes from soil samples (see Sect. 4.4.1.5) (Hooper and Evans 1993). Instead of soil, a small sample of plant tissues, e.g. 5 g roots, is placed on a paper tissue that is placed on a 2-mm-aperture sieve. The same procedures as described in Sects. 4.4.1.4 and 4.4.1.5 are followed from this step onwards. Plant material can also be first macerated in 100 ml

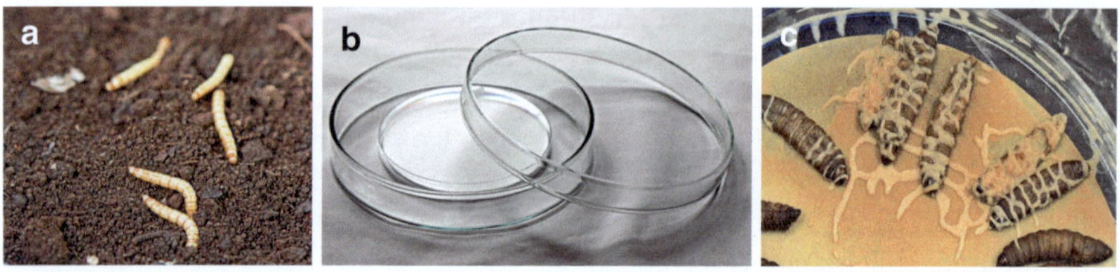

Fig. 4.9 Insect larvae of *Tenebrio molitor* used as bait to recover entomopathogenic nematode infective-stage juveniles (IJ) from soil (**a**), a modified White trap consisting of a small Petri dish covered with a filter paper and placed in a larger Petri dish partly filled with water (**b**) and IJ emerging from a wax moth larva on the moistened filter paper (**c**) (Kirk West, Port Elizabeth, South Africa)

of tap water in a kitchen blender for 30 s. The suspension containing the nematodes is then gently poured onto the paper tissue.

4.4.2.2 Modified Mistifier Method

This is a popular method to extract active nematodes from plant tissues. It was originally described by Seinhorst (1950). In principle it is another version of the Baermann funnel method, but in this method the funnels are placed in a chamber wherein a fine mist of water is sprayed over the plant tissue. The fine mist is produced by nozzles similar to those used in the irrigation systems of glasshouses. Debris (soil and plant material) is washed off the plant tissue and allows the extraction of nematodes in clean water. The water also remains well oxygenated which is a major advantage over the Baermann funnel method if used for 48 h or longer. The incubation time can be chosen, e.g. should 24-h-old second-stage juveniles (J2) of root-knot nematodes be needed for a specific experiment, they can be collected at the appropriate time. The advantages of this method outweigh the disadvantages. For example, extraction of nematodes can be done over a prolonged period which allows *Meloidogyne* J2 to hatch from the eggs, a high extraction efficiency is obtained and nematodes are usually in a better condition compared to those extracted using other methods. However, this method is time consuming and uses high volumes of water, although this can be reduced by regulating spray-time intervals. The use of a water softener is advised in areas where the water has a high mineral content. A version of the mistifier apparatus that is used at the Nematology Laboratory of North-West University is shown in Fig. 4.10.

4.4.2.3 Centrifugation

Nematodes can also be extracted from plant tissues using an adapted sugar flotation method (Coolen and D'Herde 1972; Hooper et al. 2005).

The Importance of Kaolin

Kaolin is a clay mineral with a specific gravity of 2.6 g cm^{-3} and consists of particles that range from 2–3 μm in size (Coolen and D'Herde 1972). Although the specific gravity of kaolin is higher than that of nematodes, kaolin particles are small and flat and therefore sink to the bottom of a centrifuge tube more slowly than nematodes. This way kaolin spreads out to form a layer over the loose sediment on the bottom of the centrifuge tubes and seals it off when the supernatant is decanted. When the sucrose solution is added to the sediment the mixture must be stirred again thoroughly to break the kaolin layer. Another advantage of kaolin is that it also precipitates during the sugar centrifugation, thus preventing the remixing of the sedimented debris when the sugar solution with the nematodes in suspension is decanted. Thus, a more clear suspension of nematodes can be obtained.

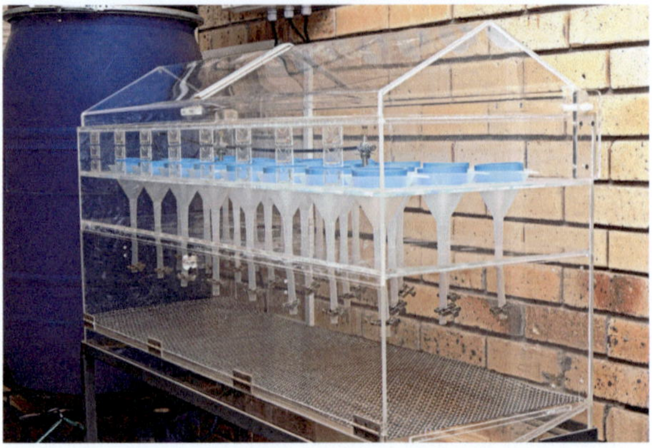

Fig. 4.10 Mistifier apparatus used to extract nematodes from soil and plant tissues (Kirk West, Port Elizabeth, South Africa)

An Adapted Sugar Centrifugal Flotation Method

This method has been adapted from those described by De Waele et al. (1987) and Kleynhans (1997). A plant tissue sample, not less than 5 g, is cut into 2–5-mm pieces. It is then macerated in 250 ml tap water at medium speed in a kitchen blender for 30–45 s to release the nematodes from the tissue. The water suspension, containing the nematodes and root fragments, is decanted through a 1-mm-aperture sieve that is stacked on 150-, 45-, 38- and 25-μm-aperture sieves. The use of several sieves prevents clogging of the 25-μm-aperture sieve. The root pieces on the 1-mm-aperture sieve are washed thoroughly with tap water and discarded. The residue on the stacked sieves is collected and decanted into centrifuge tubes. Kaolin is added to each of the tubes, the content of the tubes stirred well and the tubes centrifuged at 3,484 g for 7 min. The supernatant that contains mainly debris is decanted and a sucrose solution with a specific gravity of 1.15 g cm^{-3} added to each tube. The content of the tubes, containing the nematodes, is stirred well and again centrifuged at 3,484 g for 3 min. The supernatant, containing the nematodes, is decanted through a 25-μm-aperture sieve and rinsed well with distilled water to remove the sucrose solution. The residue on the sieve, containing the nematodes, is washed with distilled water in a beaker for counting and identification.

4.4.2.4 Extraction of Eggs and Second-Stage Juveniles of Sedentary Endoparasitic Nematodes from Plant Tissue

Eggs and J2 from nematodes that produce gelatinous egg masses (e.g. *Meloidogyne* spp., *Rotylenchulus* spp., *Tylenchulus semipenetrans* Cobb, 1913) can be extracted from roots of infected plants using the NaOCl method. This method was originally described by McClure et al. (1973), is used routinely in nematology laboratories across the globe and has been adapted by Riekert (1995). It is important to note that there is no known method, except molecular analyses (Bekker et al. 2016), to distinguish between the eggs of different plant-parasitic nematode genera or species.

Riekert's Adapted NaOCl Method

A known mass of roots, e.g. 50 g, is taken from the root system of an infected plant, cut into 10-mm pieces and mixed thoroughly. The root sample is shaken thoroughly for 4 min in 400 ml of a 1 % NaOCl solution. The bleach solution breaks down the gelatinous matrix surrounding the eggs. The solution, containing the eggs and J2, is decanted through a range of stacked sieves, from top to bottom: a 720-, 75-, 25- and 20-μm-aperture sieve. This procedure ensures less clogging on the bottom of the 20-μm-aperture sieve. A vacuum pump is connected to the 20-μm-aperture sieve and suction is applied. This enhances the passing of the suspension containing the eggs and J2 through the sieves. Root fragments are washed thoroughly through the stacked sieves for about 4 min with tap water before the eggs and J2 are collected on the 20-μm-aperture sieve. Eggs and J2 are finally washed with distilled water in a beaker for identification and counting. The set-up of the sieves and vacuum pump are illustrated in Fig. 4.11.

4.4.2.5 Extraction of Nematodes from Groundnut Pods

Bolton et al. (1990) published the following protocol for the extraction of *D. africanus* of all developmental stages from groundnut hulls and seeds. A 5 g subsample of the hulls and the seeds are broken by hand or cut with scissors into smaller pieces of about 3-mm-in-diameter. The hull and seeds tissues are then soaked for 24 h at room temperature in a Petri dish containing 20 ml of distilled water. This allows ample time for the nematodes to move out of the hull and seed tissues into the water. The content of the Petri dish is then washed through a 750-μm-aperture sieve, which is stacked on a 25-μm-aperture sieve. The nematodes retained on the 25-μm-aperture sieve are washed with distilled water in a beaker for counting. The same protocol is also followed for the extraction of nematodes from grass seeds and other aerial plant parts (Bekker 2009).

4.4.2.6 Extracting Nematodes from Onion Seed

An adapted method to extract nematodes from onion seeds was described by Southey (1965). Onion seeds are mixed, weighed and a subsample of 14 g soaked in distilled water for 48 h at room temperature in a Baermann tray (see Sect. 4.4.2.1). The water in the tray is then poured off in a beaker and allowed to stand for at least

Fig. 4.11 Stacked sieves, vacuum pump and shaker set-up for the extraction of eggs and second-stage juveniles (J2) of sedentary endoparasitic nematodes that deposit their eggs in gelatinous egg mass (Kirk West, Port Elizabeth, South Africa)

1 h, thereby allowing the nematodes to sink to the bottom of the beaker. The bottom 10 ml suspension, containing the nematodes, of each beaker is removed using a pipette and examined for the presence of nematodes.

4.4.2.7 Extraction of Nematodes from Carrot Tissue

This method was developed by the Nematology Unit, ARC-PPR (Marais and Shubane 2012). Carrot roots are washed clean of soil and peeled to a depth of about 2 mm using a vegetable peeler. The peeled strips, and if present the secondary roots, are cut into 2-mm pieces, thoroughly mixed and a 100 g subsample then macerated in water for 40 s in a kitchen blender. The macerated carrot tissues and the nematodes are washed into a 250 ml bottle and 10 ml of a 1 % NaOCl solution is added before the bottle is closed with a lid through which an aquarium tube is threaded. The suspension is aerated for 72 h and then washed through a stack of 1,000-, 150-, 45- and 38-µm-aperture sieves. The residue on the top sieve is discarded, and the residues on the other sieves are washed with distilled water into a centrifuge tube and processed by the sugar centrifugal flotation method (see Sect. 4.4.1.2). This technique is also used to extract nematodes from beetroot, groundnut and potato tissues.

4.4.2.8 Extraction of Nematodes from Wood

Swart (1997) developed a method whereby standing trees may be sampled by a forestry pressler borer or, in the case of dead or dying trees, by felling and taking stem discs at various levels. Imported coniferous wood should be randomly sampled using a low-speed drill, borer or saw. However, preference must be given to select wood pieces with grub holes or fungal growth, especially blue stained wood, which is a typical symptom of *Bursaphelenchus xylophilus* (Steiner and Buhrer 1934) Nickle, 1970 infection. The presence of flat-headed larval stages of the *Monochamus* beetles or pupae in galleries with an oval diameter and galleries sometimes blocked with wood particles may also indicate the presence of *B. xylophilus*. The emergence holes of the *Monochamus* beetles are round. Live nematodes can be extracted from infected wood using a Baermann tray (see Sect. 4.4.2.1).

4.4.2.9 Assessing Numbers of Citrus Nematode Females in Roots

This method was adapted from the method used by the Diagnostic Centre of Citrus Research International in Mbombela, Mpumalanga Province (Laura Huisman 2010, Mbombela, personal communication). Citrus roots shaken free from soil are cut into approximately 1-cm pieces, suspended in 150–250 ml of water and macerated at a low speed for 20 s using a kitchen blender. The macerated suspension is then decanted through stacked sieves with apertures of 1,000, 150 and 38 μm. The residue on the top sieve is washed using a gentle stream of water. *Tylenchulus semipenetrans* Cobb, 1913 females are collected on the 38-μm-aperture sieve and washed with distilled water into a beaker for identification and counting.

Alternatively, a staining solution containing 10 ml of a 1 % acid fuchsin solution, 10 ml of a 1 % orcein solution, 200 ml distilled water, 188 ml phenol, 165 ml lactic acid and 318 ml glycerol is prepared. The nematodes collected on the 38-μm-aperture sieve (as described above) are washed through another 38-μm-aperture sieve, using a 2 % acetic acid solution. The sieve is then submerged into the staining solution for 60 min. Thereafter, the root tissues are rinsed with tap water and washed into a beaker to which a few drops of 5 % acetic acid are added. The stained citrus nematode females are visible and can be counted using a dissecting microscope.

4.5 Counting of Nematodes

After nematodes have been extracted from soil or plant tissues, they can either be counted and identified to genus level using a counting dish or counting slide or fixed prior to mounting on microscope slides for species identification. Counting and identification of live nematodes is usually more effective in terms of time saved by not fixing and mounting the specimens. Also, the characteristic movement of certain nematode species may aid in their identification. Substantial practice and experience are required to identify nematodes to genus level with a dissecting microscope. The task of counting a large number of nematodes present in one sample can be eased by counting several subsamples (Kleynhans 1997).

De Grisse (1963) designed a counting dish that is used when counting the number of nematodes extracted from soil or plant material. The process generally followed to count nematodes is as follows: a 10 ml subsample taken from a 100 ml

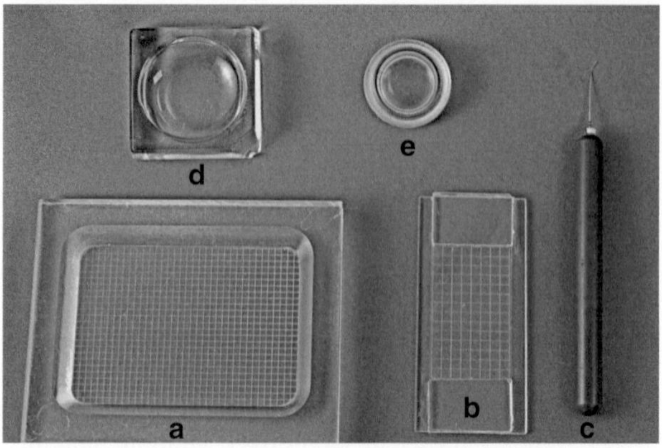

Fig. 4.12 A modified version of a De Grisse counting dish (**a**) and a Peters counting slide (**b**), fishing needle (**c**), block glass dish (**d**) and Syracuse dish (**e**) used for handling, counting, fixing or staining nematodes for identification purposes (Kirk West, Port Elizabeth, South Africa)

nematode suspension is transferred to a De Grisse counting dish (Fig. 4.12a) using a pipette. A dissecting microscope, fitted with a transmitted light source, is used to identify and count the nematodes to genus level. The subsample is returned to the nematode suspension, the suspension stirred well and another subsample taken to repeat the counting procedure The same process described above are also used to count nematodes with a Peter's slide (Fig. 4.12b), although only a 1ml subsample is used for such purposes.

4.6 Handling of Nematodes

Because of their small size, nematodes are handled in a fluid medium using a dissecting microscope. For individual specimens a handling tool is used such as a sharpened bamboo splinter, a mounted eyebrow hair, a No. 0 insect mounting pin with a re-curved tip, the fine tip of a 10 μl glass pipette or a tailor-made fishing needle (Fig. 4.12c), etc. Once picked up with the handling tool, the individual specimen is transferred to a drop of distilled water on a glass slide or in a small Syracuse dish (Fig. 4.12d) or watch glass (Fig. 4.12e) containing distilled water.

4.7 Killing, Fixing and Mounting of Nematodes

Fixation kills and hardens animal tissues and preserves the cellular structure of organisms. Nematode specimens should be fixed immediately after killing, or killed and fixed simultaneously (Kleynhans 1997). Proper methods of killing, fixing and mounting nematodes are essential to ensure that specimens remain in good condition for many years.

4.7.1 *Preparation of Vermiform Nematodes for Temporary Mounting*

During routine identification of vermiform nematodes, it is often necessary to make temporary slides. After the nematodes have been picked up from the liquid in which they were suspended for inspection and counting, they are transferred to a glass microscope slide with a drop of liquid and mounted by applying a coverslip onto the drop. The specimens can then be examined for their finer morphological features using a compound microscope.

4.7.2 *Preparation of Vermiform Nematodes for Permanent Mounting*

4.7.2.1 Killing and Fixing

One method to kill nematodes is to heat them in water in a small Syracuse dish over a spirit flame. Care must be taken not to boil the water. Specimens should be checked frequently and heating stopped as soon as the specimens stop moving. Most of the water in the dish is then pipetted off cautiously and the specimens are mounted on glass slides for identification. Alternatively, nematodes can be killed and fixed simultaneously by adding an equal volume of a boiling (at approximately 90 °C) 8 % formaldehyde solution. Because of the dilution, the final concentration of the formaldehyde will be 4 %. Another fixative, containing 10 ml of a 40 % formaldehyde solution, 1 ml glacial acetic acid and 2 ml glycerol, made up to 100 ml with distilled water, can be heated at 90–100 °C and used to kill and fix nematodes simultaneously.

To study their finer morphological structures (Box 4.2) individual nematodes can be mounted in a number of media. Glycerol is a medium with a refractive index that is nearly the same as that of glass. However, if live or fixed nematodes are suddenly placed in pure glycerol, they will plasmolyse. Therefore, the transfer of nematodes to glycerol must be done gradually. The Syracuse dish with the nematodes suspended in water should be partly covered and kept for 2–3 weeks at room temperature to allow the fixative (that contains a small amount of glycerol) to slowly evaporate, leaving the nematodes in pure glycerol. The Syracuse dish with the nematodes is then stored in a desiccator, with Calcium chloride as the desiccant, until the specimens can be mounted (Kleynhans 1997).

Another method, an adaptation of the method originally described by De Grisse (1965), can also be used to kill and fix vermiform nematodes. The nematodes are placed in distilled water in a small glass dish with a volume of 5 ml. Most of the

Box 4.2 Morphometric and Morphological Characters Commonly Used in Nematode Systematics

Symbols and Abbreviations:	Habitus	Width of annules at excretory pore and at mid-body
L = Total body length	Width of amphid aperture	Width of lateral field at mid-body and number of lines in lateral field
a = Body length divided (/) by body width	Lip height and lip width	Lateral field width/body width × 100
b = Body length/ oesophageal length	Basal ring width	Position of phasmid/scutellum
b' = Body length/ anterior end to posterior end of oesophageal gland	Stylet/ondontostyle/onchiostyle length	Diameter of phasmids/ scutellum
c = Body length/tail length	Length of odontophore	Tail length
c' = Tail length/body width at anus or cloaca	Telenchium and metenchium length in Tylenchida	Length of peg on tail terminus
o = Distance of dorsal oesophageal gland opening from stylet knobs/stylet length × 100	Stylet knob height and width	Number of ventral annules on tail
m = Length of conus/ stylet length × 100	Position of dorsal gland opening behind stylet knobs	Position of vulva
V = Distance of vulva from anterior end/ length of the body × 100	Position of guiding ring	Length of vagina
OV_1 = Length of anterior gonad/body length × 100	Position of median bulb	Length of anterior and posterior genital branches in females
OV_2 = Length of posterior gonad/body length × 100	Length and width of median bulb	Position and form of spermatheca
T = Length of testis/ body length × 100	Length and width of median bulb valve	Diameter of sperm in spermatheca
h = Hyaline length on tail	Length of oesophagus	Length of testis
	Position of excretory pore	Spicule length
	Position of hemizonid and hemizonion	Gubernaculum length
	Position of dorsal gland nuclei	
	Body width at excretory pore, mid-body, vulva and anus or cloaca	

water in which the nematodes are suspended is carefully removed with a pipette. A solution containing 100 ml of 40 % formaldehyde, 10 ml propionic acid, 890 ml distilled water and a pinch of picric acid (to stain the solution citrus yellow) (FPG) is prepared and heated in a water bath to 60–70 °C. The hot FPG fixative is added to the glass dish containing the nematodes and placed in a Petri dish with a closed lid, which is then placed in a desiccator with an atmosphere saturated with the FPG solution. The latter is obtained by adding FPG to the bottom of the desiccator. The desiccator is placed in an oven at 38–40 °C for at least 3 days, but this period could be extended to 3 months. After this time, the lid of the Petri dish is removed and half of the FPG is carefully withdrawn with a pipette from the glass dish in which the nematodes are suspended. A small amount of a Solution 1 that contains 200 ml of a 95 % ethanol solution, 10 ml glycerol and 790 ml distilled water is then added to the glass dish. Without replacing the cover lid, the Petri dish with the nematodes is returned to the desiccator for 12 h. A 95 % ethanol solution is then added to the bottom of the desiccator. The desiccator is returned to the oven during this step for 12 h after which half of Solution 1 is carefully removed from the glass dish and replaced with Solution 2. The latter solution contains 950 ml of a 95 % ethanol solution and 50 ml glycerol. After this step, the Petri dish is partially covered with the lid to ensure that evaporation occurs slowly. The desiccator is returned to an incubator at 38–40 °C for 2–3 days or until all the ethanol had evaporated. The Petri dish containing the glass dish is then removed from the incubator and the glass dish with the nematodes placed in a desiccator with silica on the bottom for another 2 days. After that time, the nematodes are ready to be mounted on microscope slides.

A method to prepare nematodes for permanent mounting involves the transfer of individual specimens from a fixative solution to a small glass dish containing 0.5 ml of Seinhorst's first solution (20 parts of 96 % ethanol, one part glycerol and 79 parts of distilled water) (Seinhorst 1959). The glass dish with the nematodes and the fixative is then placed in a closed glass vessel containing an excess (e.g. 1/10 volume of the vessel) of a 96 % ethanol solution and left in this saturated atmosphere for at least 12 h at 35–40 °C. The evaporation of the water results in the nematodes becoming suspended in a mixture of glycerol and ethanol. The glycerol-ethanol solution is then carefully removed using a pipette. The container is filled with Seinhorst's second solution, which contains five parts glycerol and 95 parts of 96 % ethanol, and placed in a partially covered Petri dish in an oven at 40 °C until the ethanol has evaporated. The duration for the latter step is at least 3 h. After this time the nematodes are suspended in pure glycerol and are ready to be mounted on slides. Using this method, nematodes can be preserved and mounted permanently. If shrinkage of the nematode specimens occurs after the first incubation, more water should be added and allowed to evaporate slowly until a 98 % ethanol solution remains. If shrinkage of nematodes occurs after the second incubation, the process should be repeated from the very beginning.

4.7.2.2 Mounting of Vermiform Nematodes

Melted paraffin wax is mixed with liquid paraffin to prevent the solid wax from becoming brittle. The end of a 7-cm-long copper tube with an internal diameter of 10 mm and wall thickness of 1 mm is heated for a few seconds in the flame of a spirit lamp and then pressed into the solid wax to a depth of approximately 2 mm. The wax will melt and cling to the tip of the copper pipe which is then pressed onto a glass slide to produce a wax ring which will cool down and solidify rapidly. A small drop of distilled water for temporary slides, 5 % formaldehyde for semi-permanent slides or glycerol for permanent slides is added in the middle of the ring and the fixed nematodes placed therein using a fine handling tool such as a fishing needle. A circular or rectangular cover slip is placed on the wax ring and the glass slide is then heated from below on a warm plate at 50 °C or by using a Bunsen burner flame to melt the wax. Alternatively a small drop of glycerol can be added to the centre of a microscope slide and several small fibreglass rods arranged around the edges of the glycerol drop. The rods must match the nematodes in thickness. Several fixed nematode specimens are then placed in the glycerol. A cover slip is heated over a spirit flame and is gently lowered onto the glycerol drop and fixed at three points with glyceel or clear nail varnish. The coverslip is finally sealed with glyceel or clear nail varnish and the specimens are ready for identification (Kleynhans 1997).

4.7.3 Preparation of Swollen Nematodes

4.7.3.1 Temporary Mounting of Nematode Cysts

The method described by Turner (1998) can be used for the temporary mounting of nematode cysts. The cysts are soaked in tap water for 24 h at room temperature. Each cyst is then transferred to a drop of water mixed with pure glycerol. The posterior end of the cyst is cut off with a scalpel and the body contents removed with a dissecting needle without damaging the structures in the vicinity of the vulva. The cuticle around the vulval area is also trimmed away using a scalpel. The cuticle piece, containing the vulval cone is then transferred to a drop of pure glycerol that is placed on a glass slide with the outer surface facing upwards. The cuticle piece is gently pressed down onto the surface of the slide and a cover slip lowered onto the drop of glycerol, which is heated gently over a spirit flame. Glass rods are used as cover slip supports to prevent the vulval cone being flattened and damaged.

4.7.3.2 Permanent Mounting of Perineal Patterns and Anterior Ends of Root-Knot Nematode Females

Methods for the preparation and mounting of perineal patterns and anterior ends of mature root-knot nematode females for identification have been described by various authors including Kleynhans (1991, 1997). Plant material infected with root-knot nematodes is washed gently with tap water to remove adhering soil and debris. The plant material is then placed for approximately 3 min in a fuming (not boiling) lactophenol solution that contains 20 ml liquid phenol, 20 ml lactic acid, 40 ml glycerol and 20 ml distilled water. Inhalation of phenol fumes is dangerous and should be avoided by conducting this activity in a fume hood. Otherwise, the phenol can be replaced with glycerol. After removing the plant material from the fuming lactophenol solution, it is transferred to a cold, clear solution of lactophenol (or glycerol). Mature root-knot nematode females are removed from the plant material using a scalpel (Fig. 4.13a) and stored in clear lactophenol or glycerol in labelled vials until mounting.

To mount the females they are first incubated with lactophenol in an oven at 40 °C for 4 days. One by one the root-knot nematode females are then transferred to a small drop of 100 % lactic acid in a plastic Petri dish. The body is punctured using a sharp dissecting or insect needle (Fig. 4.13b) and cut in half just above mid-body using a ocular scalpel or insect needle. The anterior half of the body is transferred to a small drop of glycerol on a glass microscope slide. The body contents spilling out from the posterior half of the female body are removed with an insect or dissecting needle. The anterior half of the body is transferred to a small drop of glycerol on a glass microscope slide, and the cuticle around the perineal pattern is trimmed to a neat square using an ocular scalpel or insect needle. This cuticle piece, containing the perineal pattern, is transferred to the drop of glycerol in which the anterior part

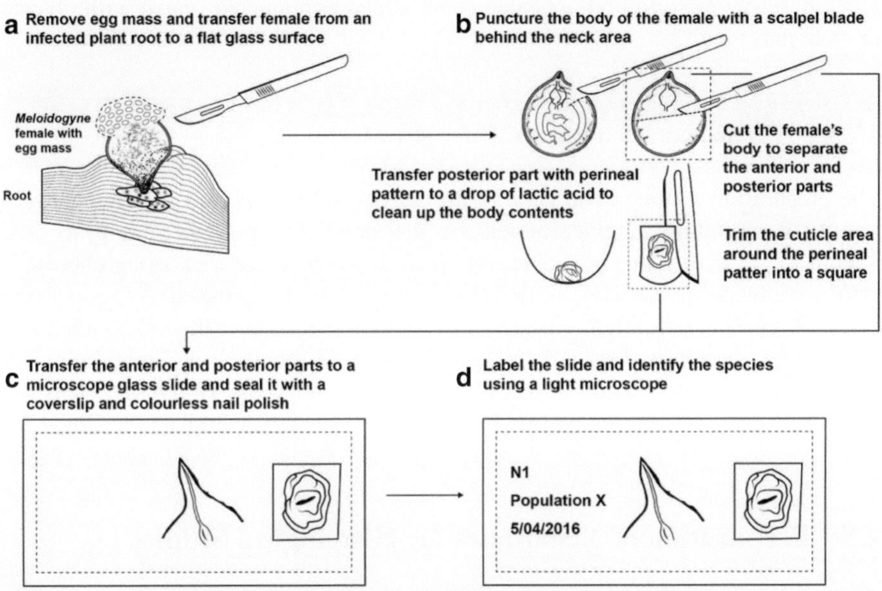

Fig. 4.13 Procedure used for mounting the perineal and anterior ends of a mature root-knot nematode females: the female is removed from a root fragment (**a**), the body is punctured, cut in half and the body content gently removed (**b**), the cuticle around the perineal pattern is trimmed (**c**) and the oesophageal and perineal pattern of the females are mounted in glycerol on glass microscope slides (**d**) (Hannes Visagie, North-West University, Potchefstroom, South Africa)

of the body has been placed (Fig. 4.13c). The cuticle piece is carefully pressed down to ensure that the outer surface of the cuticle is facing upwards. The anterior and posterior parts of several individuals are mounted on the same slide, covered with a cover slip and sealed with either glyceel or clear nail varnish (Fig. 4.13c-d). Care must be taken when using lactic acid as it is corrosive.

4.7.3.3 Permanent Mounting of *Globodera* Cysts

The procedure described below has been adapted from Kleynhans (1997). Dried cysts are soaked in distilled water for 24 h, transferred to a fuming (not boiling) lactophenol solution and then to 100 % lactic acid in a cavity glass slide. While suspended in the lactic acid, the posterior end of the cyst is cut off with a scalpel and the body contents removed. These handlings should be carried out without damaging the structures that are situated in the vicinity of the vulva and anus (the so-called terminal pattern) that are crucial for species identification. The cuticle around the terminal pattern is trimmed away and the piece of cuticle that contains the terminal pattern is first transferred to a drop of distilled water, then to a drop of a 96 % ethanol and finally to a drop of xylene for clearing. From the xylene, the cuticle piece is finally transferred to a drop of gently heated glycerol jelly that is placed in the middle of a glass slide. The cuticle piece is positioned while the jelly is still liquefied in such a way that the outer surface faces upwards. As soon as the glycerol jelly has solidified, a heated cover slip is applied and sealed with glyceel or clear nail varnish.

4.7.3.4 Permanent Mounting of *Heterodera* Cysts

The preparation of the vulval cone (with the vulva, fenestra and associated structures) for mounting in glycerol jelly is the same as for *Globodera* cysts (see Sect. 4.7.3.3), except that the cover slip needs to be supported (e.g. using glass slivers from a broken cover slip) so that the vulval cone is not crushed. As soon as the glycerol jelly has solidified, a heated cover slip is placed over the vulval cone and sealed with glyceel or clear nail varnish. Various morphometric and morphological characters are then measured and observed to identify the *Heterodera* species. Characters that are commonly used in nematode systematics are listed in Box 4.2.

4.8 Preparation of Nematodes for Histological Studies

4.8.1 Transmission Electron Microscopy

The method described below for preparing nematodes for transmission electron microscopy (TEM) studies has been described by Fourie et al. (2013) and has been modified for the preparation of root-knot nematode-infected soybean roots. Infected plant tissues are first inspected using a dissecting microscope to identify swollen areas on the roots where various developing stages of root-knot nematodes (all juvenile stages and females) are situated. Infected areas are then cut into approximately 5-mm-sections and fixed in Todd's fixative (pH 7.5) in 20 ml glass flasks at room temperature for 12 h (Todd 1986). The flasks are sealed with perforated plastic plugs and placed in a vacuum for approximately 2 min to remove all air trapped inside the root tissues. From this point onwards the fixation and infiltration methods of Murphy et al. (1974) and Spurr (1969) are used. The nematode-infected sections are washed three times for 15 min in a 0.05 M sodium cacodylate buffer (pH 7.4), post-fixed in 1 % osmium tetroxide for 1 h and washed three times in distilled water. The root pieces are subsequently dehydrated in an acetone series of 50, 70, 90, 100 and again 100 % acetone, each for a 15 min period. The sections are then infiltrated with a 1:1 mixture of acetone and resin and left for 3 h during which the root sections sink to the bottom of the flasks. The 1:1 acetone-resin mixture is replaced with a freshly mixed resin solution (100 %) and the root sections transferred to it for 5–12 h, again allowing the sections to sink to the bottom of the flasks. The root sections are subsequently placed in the acetone:resin mixture for 2 h, followed by embedding of each root section in resin (100 %) in a flat mould. Moulds containing the root sections are then cured in an oven at 65 °C for 8–12 h to allow the resin to harden. Embedded resin blocks containing the root-knot nematode-infected root sections are inspected using a dissecting microscope and excessive resin removed with sandpaper to position the sections containing the root-knot nematodes for optimal sectioning. The root sections embedded in resin are sectioned transversely using an Ultracut Microtome (±100 nm) and fixed on glass slides. Fixed sections are stained with toluidine blue (Reynolds 1963) for light microscope observations, and with 2 % uranyl acetate and lead citrate (pH 12) for TEM observations. Slides containing the fixed root sections are examined and photographed using a light microscope at 80, 100, 200, 400 and 1000× magnifications. Root sections are examined with TEM using a Philips CM 10 at 80 kv at a range of 2,200–21,000× magnification. The same procedure as described above can be used for preparing individual nematodes, not contained within plant tissues, for examining changes at the ultrastructural level. A detailed account of the use of TEM to study nematodes is given by Carta (1991).

4.8.2 Scanning Electron Microscopy

Live individual nematodes to be prepared for scanning electron microscopy (SEM) are transferred into tap water in a small Syracuse dish. The dish containing the nematodes is heated over a spirit flame until all specimens stop moving. The dish is left to cool down completely after which most of the water is drawn off with a syringe. TAF fixative is poured over the specimens and they are left at room temperature for a week. The specimens are then transferred to a range of ethanol solutions viz. 70, 80, 90 and 96 % at 3 h intervals. Rinsing in 96 % ethanol is repeated three times. The specimens are then critical-point dried using liquid carbon dioxide and stuck up against a short hair stuck on copper foil conductive tape on a SEM viewing stub. The stub containing the nematodes is then sputter-coated with gold (66 %) palladium (33 %) pieces ranging 21–25 nm in size (Minagawa 1986). Mounted nematode specimens can also be prepared for SEM investigations. This is done by removing the specimens from the slides, rehydration of the specimens in a graded series of glycerol-thinning media and finally suspension in distilled water. The thinning medium consists of 30 % absolute ethanol mixed with distilled water. The percentage glycerol in the thinning medium in the graded series is 85, 65, 45, 25, 5 and 0 %. The various thinning medium concentrations are replaced at 15–30 min intervals. Rinsing of the nematode specimens during the last step of this series with a 70 % ethanol solution is repeated three times. The nematodes are then hydrated by placing the specimens for 15–30 min in a 15 % ethanol solution, followed by a 5 % ethanol solution and then distilled water. Rinsing of the nematodes in distilled water is repeated three times. The specimens are then transferred to a capsule containing a 30 % ethanol solution and transferred to a 50, 75, 95 and 100 % ethanol solution for 20 min each. The last step is repeated twice. The specimens are then critical-point dried and coated with gold palladium as described above.

The external (e.g. cuticular pattern of body and vulval regions) and internal (e.g. bullae and underbridge) structures of *Globodera* and *Heterodera* cysts can also be studied with SEM (Lax and Doucet 2002). Other procedures for the preparation of nematodes for SEM studies have been described by Eisenback (1985) and Hooper (1998).

4.8.3 Interference Microscopy

Roots, pegs, hulls and seeds of groundnut plants infected with *D. africanus* have been studied by Jones and De Waele (1990) to establish the time and mode of entry of these nematodes. Infected plant tissue sections are fixed in FAA (20 ml 96 % ethanol, 6 ml formaldehyde, 1 ml glacial acetic acid and 40 ml distilled water) and dehydrated in a series of alcohol solutions. The sections are transferred to propanol and butanol and embedded in paraplast. Transverse and longitudinal sections (12-μm-thick) are cut with a microtome, stained with Mallory triple stain (Basson et al. 1991) and mounted in DPX (BDH Ltd, Broom Rd, Poole, Dorset BH12 4 NN, England). These sections can then be examined using interference microscopy using the protocol described by Lille and Fullmer (1976). Other accounts of the use of interference microscopy to study nematodes are given by Shaham (2006) and Seacor et al. (2015).

4.9 Nematode Identification using Molecular Techniques

The use of molecular techniques to complement the morphological and morphometric identification of nematode species has increased considerably during the last two decades. Protein electrophoresis was the first molecular-based technique applied in

Nematology (Esbenshade and Triantaphyllou 1985a, b, 1987; Karssen et al. 1995). Since, new techniques were developed and further optimised for the efficient and accurate identification of nematodes including the polymerase chain reaction (PCR), PCR-restriction fragment length polymorphism (PCR-RFLP), multiplex PCR, random amplified polymorphic DNA (RAPD), amplified fragment length polymorphism (AFLP), reverse dot blot hybridisation, sequencing of DNA, DNA bar-coding and real-time PCR (Carneiro et al. 2000; Zijlstra 2000; Zijlstra et al. 2000; Hernandez et al. 2004; Adams et al. 2009; Blok and Powers 2009; Subbotin et al. 2013).

Currently, DNA-based methods are used on a regular basis as an efficient, diagnostic tool to confirm and supplement the identification of a wide range of nematode species, both parasitic and non-parasitic. In SA, the first molecular characterisation of a new plant-parasitic nematode was the description of *D. africanus* by Wendt et al. (1995). During the beginning of the 21st century, Fourie et al. (2001) differentiated six root-knot nematode species using the SCAR-technique. Numerous papers using molecular techniques to characterise nematodes from SA have since been published (Table 4.2).

For DNA extraction from nematode individuals, various protocols are used by different researchers (see, for instance, Zijlstra et al. 2000; Hooper et al. 2005; Subbotin et al. 2013). For entomopathogenic nematode species, DNA is generally extracted using the method described by Nguyen (2007). For example, one first generation female for steinernematids or one hermaphrodite for heterorhabditids is placed in 30 ml lysis buffer (16 mM $[NH_4]_2SO_4$, 67 mM Tris-HCl pH 8.8, 0.1 %

Table 4.2 Nematode genera occurring in South Africa of which morphological and molecular characterisation have been done

Nematode genera	Reference
Anguina spp.	Swart et al. (2004)
Aphelenchoides spp.	Lesufi (2007); Lesufi et al. (2015)
Criconemoides spp.	Van den Berg et al. (2012)
Ditylenchus spp.	Wendt et al. (1995)
Globodera spp.	Knoetze et al. (2006, 2013)
Helicotylenchus spp.	Subbotin et al. (2011)
Hemicycliophora spp.	Van den Berg et al. (2010); Subbotin et al. (2014)
Heterorhabditis spp.	Malan et al. (2008, 2011, 2014)
Hirschmanniella spp.	Van den Berg et al. (2009)
Meloidogyne spp.	Fourie (1998); Fourie et al. (2001); Berry et al. (2008); Ntidi et al. (2012); Onkendi and Moleleki (2013a, b)
Paratylenchus spp.	Van den Berg et al. (2014)
Pratylenchus spp.	Berry et al. (2008)
Rotylenchulus spp.	Van den Berg et al. (2016); Bekker et al. (2016)
Scutellonema spp.	Van den Berg et al. (2013)
Steinernema spp.	Nguyen et al. (2006); Çimen et al. (2014a, b); Malan et al. (2011, 2015); Nthenga et al. (2014); Stokwe et al. (2011); Malan et al. (2016)
Subanguina spp.	Bekker (2009)
Xiphinema spp.	Knoetze et al. (2000); Berry et al. (2008)

Tween 20) containing 60 μg ml^{-1} proteinase K on the side of a 0.5 ml Eppendorf tube. The nematode is cut into pieces with the sharp side of a syringe needle and immediately put on ice and frozen overnight at –80 °C. The Eppendorf tubes are transferred to a thermocycler at 65 °C for 1 h followed by 95 °C for 10 min and then centrifuged for 2 min at 11,600 *g*. The top 20 μl of the solution is transferred to a clean Eppendorf tube and kept at 20 °C. The 18S and 26S primers suggested by Vrain et al. (1992) are used for amplification of the ITS region. If a good sequence is not obtained, the primers TW81 and AB28 (Hominick et al. 1997) are used. The technique of Nguyen (2007) for PCR amplification is then also followed, with purified DNA being sequenced at the Analytical Centre of the Department of Genetics at Stellenbosch University using the BigDye 3.1 chemistry (PE Applied Biosystems). The base-pair calls of the sequences are verified and edited, using the software CLC DNA Workbench, Version 6. To indicate the phylogenetic position of the nematode isolates, sequences of *Heterorhabditis* and *Steinernema* isolates are compared with sequences obtained from Genbank. Phylogenetic and molecular analyses are conducted and based on maximum parsimony of the ITS region using the software ClustalX ver. 1.83 (Thompson et al. 1997) and PAUP Version 4.08b (Swofford 2002) or maximum likelihood using Mega5 (Tamura et al. 2013).

4.10 Preparation of Nematode Inoculum and Inoculation of Plants

The basic procedure used for preparing inoculum of the groundnut pod nematode (*D. africanus*) for experimental purposes has been described in detail by Mc Donald (1998) but can be used for other nematode species too. Mc Donald used *D. africanus* individuals of all developmental stages obtained from in vitro cultures (see Sect. 4.11.2). The nematode eggs, juveniles and mature adult stages were placed in a calibrated glass flask and kept in suspension by means of a magnetic stirrer throughout the inoculation process. Air could be bubbled through the nematode suspension with an aquarium pump should a magnetic stirrer not be available.

At least 10 aliquots of 10 ml each of the nematode suspension are collected with a 10 ml pipette, poured into a counting dish and counted to check for at least 95 % accuracy.

Roots or other below-ground plant parts to be inoculated are exposed by removing the soil. The prepared nematode inoculum is pipetted directly onto the plant part(s). The inoculation flask is rinsed and the process repeated. The soil that was removed before inoculation is then replaced. Excessive watering of plants should be avoided for at least 48–96 h after nematode inoculation was done to ensure optimal penetration of nematodes into the plant tissue. By preference, watering should be added to the saucers in which the pots are placed. This inoculation procedure is suitable for potted plants, microplots and small field trials.

4.11 Mass Rearing of Nematodes

4.11.1 In Vivo *Mass Rearing of Root-Knot Nematodes*

The protocol described here is based on Fourie et al. (2012). Root-knot nematode populations are obtained from various localities. The GPS coordinates as well as relevant information, including crop sampled and field history, should be listed in a field book. This information is necessary to ensure that the origin of the root-knot nematode population is precisely known. Populations of single species of *Meloidogyne* are established by collecting individual egg masses from infected plant material. Each egg mass is placed near the roots of a susceptible tomato seedling (for instance, the cultivars (cvs) Moneymaker, Rodade, Floradade or UC82B) or susceptible cvs of other crops that are easy to maintain in a steam-sterilised or fumigated soil (>90 % sand) in 25 l pots in a glasshouse. By rotating the host plant on a routine basis with susceptible cvs of different crops (e.g. soybean cv. LS6248R, maize cv. DKC77-77BR, *Amaranthus cruentus* accession Arusha), the virulence of the different single species populations can be maintained.

Nutrients are added according to soil nutrient analyses. The root-knot nematode population should be well established within 56–90 days after inoculation with a single egg mass. At this stage, root-knot nematode-infected roots can be removed, and the eggs and J2 extracted. As a precaution, different *Meloidogyne* spp. populations should be kept well separated, preferably in different glasshouses. An ambient temperature range of 18–26 °C and a 14:10 (light:dark) photoperiod should be maintained in the glasshouse throughout the culturing period of termophylic species. Single species populations or a mixture of more than one species can be used, depending on the objective of the research. According to Hussey and Jansen (2002), a combined inoculum is a precautionary measure to ensure that germplasm is screened against a spectrum of nematode populations. This reduces the risk of finding resistance that is not durable and can be overcome by the concommittant occurrence of more than one nematode species in the field.

4.11.2 In Vitro Mass Rearing of the Groundnut Pod Nematode

Ditylenchus africanus can be cultured on groundnut callus tissue, derived from groundnut leaves that have been placed on an agar medium containing growth promotors (usually 2,4-D) (Van der Walt and De Waele 1989). The procedure is conducted in a sterile environment (laminar flow cabinet). Callus tissue is initiated from 4-week-old groundnut (cv. Sellie) leaves. The leaves are sterilised in 70 % ethanol for 30 s and then suspended in a 0.5 % NaOCl solution for 15 min. The leaves are rinsed four times in sterile, distilled water and cut into 1-cm-pieces. They are then transferred to 9-cm-diameter Petri dishes containing 25 ml of a nutrient medium. The pH of the medium is adjusted to 5.7 and the nutrient autoclaved for

15 min at 121 °C. The callus tissue cultures are incubated in the dark in a growth chamber at 25 °C for 4 weeks. Callus tissue that increased in volume after this period is transferred to Petri dishes containing freshly-prepared medium and inoculated with surface-sterilised *D. africanus* individuals. The Petri dishes are sealed with parafilm and incubated in a growth chamber at 25 °C for 1–5 weeks. Nematodes are extracted by transferring each infected callus tissue to a Petri dish containing 25 ml water and cutting the tissue into smaller pieces. The Petri dishes are left at room temperature to allow *D. africanus* individuals to emerge from the infected callus tissue and migrate into the water. The nematode suspension from each Petri dish is decanted through a 25-μm-aperture sieve and the nematodes collected in a beaker.

4.11.3 In Vitro Mass Rearing of Lesion and Burrowing Nematodes

The monoculture of the lesion nematode species *Pratylenchus brachyurus* (Godfrey 1929) Filipjev & Schuurmans Stekhoven, 1941 on carrot discs reported by Fourie et al. (2003) is described below. This method has been adapted from the method used to rear the burrowing nematode *R. similis* on carrot discs (Stoffelen et al. 1999). Specimens of *P. brachyurus* are extracted from infected material, surface-sterilised in 2 % streptomycin sulphate and rinsed twice in sterile water prior to inoculation. Carrots are sterilised with 99.9 % ethanol, peeled, flamed, cut in discs (approximately 20-mm-diameter and 5-mm-thick) and placed in Petri dishes. An inoculation density of 50 *P. brachyurus* juveniles and adults carrot disc^{-1} is used. Discs that were kept in sealed Petri dishes and incubated at 30 °C yielded a large number of this species, viz. 3,580 individuals disc^{-1} in 60 days and 48,928 individuals disc^{-1} in 80 days. At the end of the incubation period, *P. brachyurus* individuals were extracted by macerating individual carrot discs for 15 s in a kitchen blender and passing the suspension through a range of stacked sieves, viz. 710-, 250-, 75-, 25- and 20-μm-aperture sieves. The nematodes were collected on the 25- and 20-μm-aperture sieves.

4.11.4 Mass Rearing of Entomopathogenic Nematodes

Large numbers of EPN can be cultured using either in vivo or in vitro culture techniques. In vivo production is a low-technology and labour-intensive process. However, it is easily implemented in research laboratories and can be used for small-scale nematode production for field trials. Compared to in vivo production, the in vitro production of EPN demands a high level of technology input and capital investment, thus making it more suitable for commercial mass production. Recent reviews of the methods used for the culturing of EPN have been reported by Ferreira and Malan (2014), Shapiro-Ilan et al. (2014) and Van Zyl and Malan (2014a).

4.11.4.1 In Vivo Production of Entomopathogenic Nematodes

In vivo production of IJ of EPN is based on the White trap method (White 1927) and on modifications of the method (see Sect. 4.4.1.10). This procedure begins with the inoculation of a susceptible insect host and ends with the harvesting and concentration of the IJ. This method makes use of the natural migration of the IJ away from the infected cadaver at the end of the life cycle. Nematode-infected insects are placed on moist filter paper in a 9-mm-diameter Petri dish, which, in turn, is placed in a 15-cm-diameter Petri dish, the bottom of which is filled with just enough water so that the smaller dish still rests on the bottom, without causing it to float. The lids of both dishes are closed to maintain a moist atmosphere during the development of the nematodes. The length of the cycle, which depends on the size of the host, tends to last from 7 to 14 days. During this period, small drops of water should periodically be added to the filter paper to keep it moist.

When the food source in the insect cadaver is depleted, the IJ move over the side of the Petri dish in search of a new host, whereupon they are trapped in the surrounding water contained in the bigger Petri dish. Just before emergence, the lid of the smaller Petri dish should be removed and the nematodes that will have by this time aggregated on the inside of the lid should be washed off into the larger Petri dish. High-quality IJ can be obtained using this technique. The most widely used insect hosts are wax moth larvae and mealworms. Wax moth larvae, highly susceptible to EPN, can easily be cultured in large numbers, but unfortunately are not easily stored. Mealworms are more easily cultured and stored, but they tend to be less susceptible to infection and fewer IJ are produced per host (Van Zyl and Malan 2014a, b).

4.11.4.2 In Vitro Production of Entomopathogenic Nematodes

The in vitro production of nematodes is essential for the commercial use of EPN. In vitro cultures can be reared on a solid medium or in a liquid. For in vitro production of EPN, a monoxenic culture is a prerequisite. This requires that the symbiotic bacteria are isolated and that the nematodes are produced free of bacteria. The procedure for culturing symbiotic bacteria is described by Kaya and Stock (1997) and Ferreira et al. (2014, 2016). The establishment of nematodes without their symbiotic bacteria can be achieved by harvesting eggs from gravid females (Han and Ehlers 1998; Ferreira et al. 2014, 2016).

In vitro techniques for culturing EPN on a solid medium involve the preparation of the medium, inoculation of the bacteria and then the nematodes and, finally, harvesting of the progeny. A culture on a solid medium can be accomplished on various agar media in the same Petri dish or in a so-called three-dimensional rearing system consisting of a nematode culture that is reared on crumbled polyether polyurethane foam coated with a diet (Bedding 1981).

In a liquid culture, the bacteria are generally introduced first followed by the nematodes. Nematodes can be produced in Erlenmeyer flasks on orbital shakers or in bioreactors with a capacity of up to 80,000 l under sterile conditions. As the process time (meaning the completion of the life cycle to the IJ stage) can last up to 3 weeks, the maintenance of sterile conditions throughout the culturing poses a challenge (Ehlers and Shapiro-Ilan 2005).

4.11.5 Surface Sterilisation of Nematodes

Nematodes are routinely sterilised before they are used to establish in vitro populations (Van der Walt and De Waele 1989). Nematode individuals used for this purpose should preferably be extracted using a method such as the Baermann funnel or Baermann tray methods to ensure that they are not damaged in any way. The nematodes are then transferred to a Syracuse dish containing sterile water and then into sterilised centrifuge tubes containing 2 ml of a 0.1 % $HgCl_2$ solution and a piece of milk filter at the bottom. The tubes containing the nematodes are closed with aluminium covers and centrifuged for 2 min at 1,750 *g*. The $HgCl_2$ solution is decanted and the nematodes are trapped on the milk filter. The milk filter is rinsed three times in 2 ml of sterile water. The nematodes are now surface-sterilised and ready for use.

4.12 Reproductive Potential of Root-Knot Nematodes

Root-knot nematode egg masses are stained to facilitate counting and thus establish the reproductive potential of a population on a specific crop cv. The method described here is adapted from Hussey and Boerma (1981).

Plants of the crop to be examined are inoculated with a given number of eggs and J2 or only J2 (e.g. 5,000 per seedling). Fifty-six days after inoculation, the root systems are lifted and excised, rinsed free of adhering soil and debris with running tap water. The root systems are blotted dry with paper towel and weighed, and the number of egg masses per root system counted. Egg masses are stained by immersion of the roots for 20 min in a 0.1 % Phloxine B solution (Hussey and Boerma 1981). Each root system is cut into approximately 10-mm pieces and transferred to a container filled with approximately 200 ml water. The root pieces are individually inspected and the red-stained egg masses counted. Egg masses coming loose from the root pieces during the staining and counting process are collected using a Pasteur pipette and are also counted. The egg-laying-female index for each root system is determined according to the method of Hussey and Boerma (1981). This index is based on a 0–5 scale where 0 = no egg masses; 1 = 1–2 egg masses; 2 = 3–10 egg masses; 3 = 11–30 egg masses; 4 = 31–100 egg masses and 5 => 100 egg masses system^{-1}. Subsequently, the eggs are extracted from the stained egg masses using the modified NaOCl method (Riekert 1995) and counted using a dissecting microscope. The reproduction potential of the nematodes is determined using Oostenbrink's Rf value as published in Windham and Williams (1987): Rf = Pi (initial number of nematodes inoculated per root system/Pf (final number of nematodes per root system). The number of eggs per g of fresh roots is determined for each plant.

4.12.1 *Estimation of Juvenile Numbers and Viability of Cyst Nematodes*

Depending on the age and state of heteroderid cysts, their egg and J2 content can vary considerably. The viability of cysts can be determined by examining the biological status of the eggs and J2 they contain. A technique to do this is to place cysts, that had been soaked for 12 h in water, into a 0,05 % aqueous solution of Meldola's blue (Shepherd 1962) for 1 week. The cysts are then crushed in a microcentrifuge tube with a homogeniser, and the eggs and J2 are released by vortexing. The eggs are separated from the cyst fragments by washing them through a 90-µm-aperture sieve and collecting them on a 25-µm-aperture sieve. Eggs and J2 are left in a water suspension for up to 12 h to remove excess stain. Two aliquots are then pipetted onto a counting chamber and observed using a light microscope. The viability of the eggs are estimated by counting stained (non-viable) and non-stained (viable) eggs. Spontaneous hatching of J2 are estimated by counting empty eggs. Turner (1998) and Le Roux (2000) also described useful methods to estimate the viability of cysts.

4.13 Staining of Roots to Establish Nematode Penetration and Development

Staining of roots or other plant parts is a useful tool to study, e.g. the penetration and development of a target plant-parasitic nematode species over time. To do such research, root systems infected with plant-parasitic nematodes (e.g. root-knot nematodes) are obtained, washed free of soil and stained with an acid fuchsin-lactoglycerol solution (875 ml of lactic acid, 63 ml of glycerol, 62 ml of water and 0.1 g acid fuchsin) to facilitate nematode counting (Byrd et al. 1983; Hooper et al. 2005). The solution is heated on a warm plate that is placed inside a fume hood to avoid inhaling the lactic acid vapour. Root-knot nematode-infected root pieces are wrapped inside a muslin cloth and submerged into the solution as soon as it starts to boil. After 1–2 min, the muslin cloths with the stained roots are removed and rinsed with running tap water to get rid of the excess stain solution. The next step entails the transfer of the stained root systems to a solution that contains equal volumes of glycerol and distilled water to allow destaining of the roots, but not of the nematodes inside the roots. A few drops of lactic acid can be added to the glycerol:distilled water solution to enhance destaining of the roots. Nematodes are usually visible as red-coloured organisms within the roots and can be counted, using a dissection microscope, after the roots were left for 2–5 days in the destaining solution. Chemicals other than acid fuchsin, e.g. methyl blue, can also be used to stain plant tissue to enable identification and counting of nematodes within plant tissues (Bridge et al. 1982)

This chapter concludes with an inventory (Box 4.3) of the basic and important infrastructure and apparatus needed to set up and run a small nematology laboratory. Box 4.4 demonstrates the basic order and flow of activities from the moment a nematode sample is received by personnel of a nematology laboratory until the identification and nematode counts have been completed.

Box 4.3 An Inventory of the Basic Infrastructure and Equipment Needed to Set Up and Run a Small Nematology Laboratory for Diagnostic and Research Purposes

Glasshouse, field and microplot research	Preparation of nematode samples for extraction, extraction and counting of nematodes	Staining of plant tissues, killing, fixation and mounting of nematodes	Inoculation of plants for experiments
Infrastructure: Glasshouse or tunnel; microplots; room for the storage of chemicals including nematicides; vehicle to access trial sites; automated irrigation system for glasshouses and microplots *Tools/small apparatus*: Knapsack sprayer; plastic measuring cylinders; field-operatable balance to record crop yield and weight of aerial or below-ground plant parts; water cans; garden spades; hoes; oil augers; arden rakes; plastic pots (1, 5, 10 or 20 l); plot markers; pegs to mark field plots; insulated cool boxes, field book; protective clothing for when using nematicides *Consumables*: Plastic bags for soil and root samples; cable ties; measuring tape, string; permanent marker pen; fertiliser; disposable gloves	*Infrastructure*: Room to use as laboratory; working area; cool room or refrigerator to store samples at 5–10 °C; two to four basins with taps fitted with latex tubing (basins must be deep enough to place stacked sieves); mistifier; running water; sand-trap outside the laboratory *Equipment/big apparatus*: Dissecting- and light microscopes; inverted microscope; camera system that fits on microscopes; centrifuge; orbital shaker; kitchen blender; balance (4 decimals); two sets of sieves (1000-, 750-, 500-, 250-, 75-, 63-, 38-, 25- and 20-μm-apertures); elutriator; Baermann funnels or trays; water distiller *Small equipment/apparatus*: Stop watch; multiple or single tally counters (e.g. blood-cell counter); De Grisse counting dishes; Peter's counting slides; laboratory clock; measuring cylinders; flasks; pipettes; 5 l plastic buckets; plastic wash bottles; plastic sample bottles; plastic pot trays; plastic funnels; marking tape; scissors; Syracuse dishes; needles to fish nematodes; tissue paper *Chemicals*: Sugar; kaolin, NaOCl; other chemicals needed for specific extraction methods	*Infrastructure*: Room to use as laboratory *Equipment/big apparatus*: Fume cupboard; oven *Smaller equipment/ apparatus*: Hot plate; dissection set (needles, scalpel, etc.); Petri dishes; Syracuse dishes; glassware (e.g. beakers, measuring cylinders) *Consumables*: Microscope slides; cover slips; transparent nail polish; labels; holders for slides *Chemicals*: Acid fuchsin/cotton blue; glycerol, lactic acid; ethyl alcohol, formaldehyde; lactophenol; others	*Smaller equipment/ apparatus*: Micropipettes and tips; Socorex 411 Stepper used for accurate preparation of inoculum; plastic beakers; magnetic stirrer; magnetic stir bars (different sizes)

Box 4.4 The General Process Followed by Personnel of a Nematology Laboratory from the Receiving of Nematode Samples Until Recording of Data and Report Writing

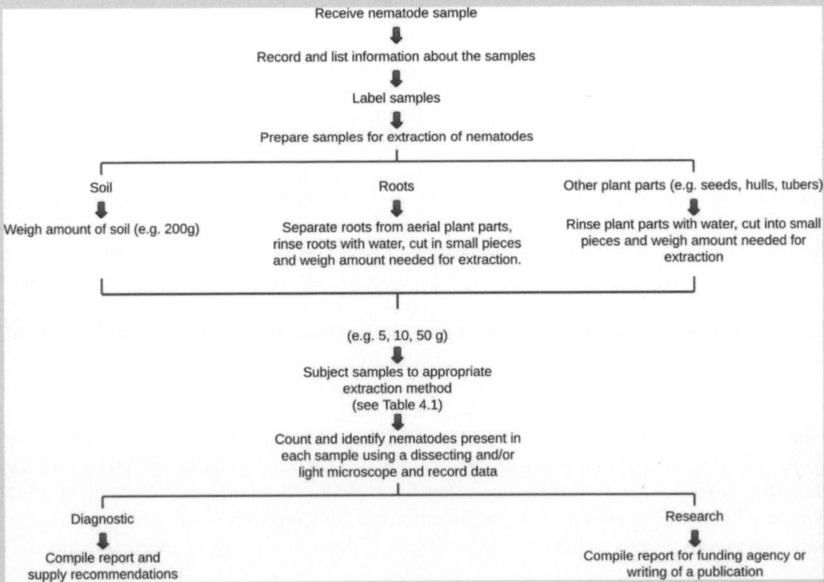

References

Adams BJ, Dillman AR, Finlinson C (2009) Molecular taxonomy and phylogeny. In: Perry RN, Moens M, Starr JL (eds) Root-knot nematodes. CAB International, Wallingford, pp 119–138

Basson S, De Waele DGMA, Meyer AJ (1991) Population dynamics of *Ditylenchus destructor* on peanut. J Nematol 23:485–490

Bedding RA (1981) Low cost *in vitro* production of *Neoaplectana* and *Heterorhabditis* species (Nematoda) for field control of insect pests. Nematologica 27:109–114

Bedding RA, Akhurst RJ (1975) A simple technique for the detection of insect parasitic rhabditid nematodes in soil. Nematologica 21:109–110

Been TH, Schomaker CH (2013) Plant growth and population dynamics. In: Perry R, Moens M (eds) Plant nematology. CAB International, Wallingford, pp 301–330

Bekker S (2009) Assessment of the identity, distribution and control options for seed-and leaf-gall nematodes on grass seeds in South Africa. MSc Dissertation. North-West University, Potchefstroom

Bekker S, Fourie H., Rashidi M et al (2016) Discriminating between the eggs of two egg-mass-producing nematode genera using morphometric and molecular techniques. Nematol 18:1119–1123. doi:10.1163/15685411-00003022

Berry SD, Fargette M, Spaull VW et al (2008) Detection and quantification of root-knot nematode (*Meloidogyne javanica*), lesion nematode (*Pratylenchus zeae*) and dagger nematode (*Xiphinema elongatum*) parasites of sugarcane using real-time PCR. Mol Cell Probes 22:168–176

Blok VC, Powers TO (2009) Biochemical and molecular identification. In: Perry RN, Moens M, Starr JL (eds) Root-knot nematodes. CAB International, Wallingford, pp 98–118

Bolton C, De Waele D, Basson S (1990) Comparison of two methods for extracting *Ditylenchus destructor* from hulls and seeds of groundnut. Rev Nématol 13:33–235

Bridge J, Starr JL (2007) Plant nematodes of agricultural importance. A color handbook. Academic Press, Boston

Bridge J, Page S, Jordan S (1982) An improved method for staining nematodes in roots. Report of the Rothamsted Experimental Station for 1981. Part 1. Rothamsted, Harpenden

Byrd DW, Kirkpatrick T Jr, Barker KR (1983) An improved technique for clearing and staining plant tissues for detection of nematodes. J Nematol 15:142–143

Carneiro RMDG, Almeida MRA, Quénérhervé P (2000) Enzyme phenotypes of *Meloidogyne* spp. populations. Nematol 2:645–654

Carta LK (1991) Technical methods for collection and preparation of nematodes. Part 3. Preparation of nematodes for transmission electron microscopy. In: Nickle WR (ed) Manual of agricultural nematology. Marcel Dekker Inc, New York, pp 97–106

Çimen HC, Lee M-M, Hatting et al (2014a) *Steinernema innovationi* n. sp. (Panagrolaimomorpha: Steinernematidae), a new entomopathogenic nematode species from South Africa. J Helminthol 89:415–427

Çimen HC, Lee M-M, Hatting J et al (2014b). *Steinernema tophus* sp. n. (Nematoda: Steinernematidae), a new entomopathogenic nematode from South Africa. Zootaxa 3821:337–353

Coolen WA, D'Herde CJ (1972) A method for the quantitive extraction of nematodes from plant tissue. State Agricultural Research Centre, Merelbeke

Coyne DL, Nicol JM, Claudius-Cole B (2007) Practical plant nematology: a field and laboratory guide. SP-IPM Secretariat, International Institute of Tropical Agriculture (IITA), Cotonou

De Grisse A (1963) A counting dish for nematodes excluding border effect. Nematologica 9:162

De Grisse AT (1965) Redescription ou modifications de quelques techniques ultilisees dans l'etude des nematodes phytoparasitaires. Medelingen Rijksfaculteit Landbouwetenschappen Gent 34:351–359

De Waele E, De Waele D, Wilken R (1987) Effect of root mass on the efficacy of methods for extracting root-lesion nematodes from maize roots. Phytophylactica 19:473–474

Ehlers R-U, Shapiro-Ilan DI (2005) Mass production. In: Grewal PS, Ehlers R-U, Shapiro-Ilan DI (eds) Nematodes as biocontrol agents. CAB International, Wallingford, pp 65–78

Eisenback JD (1985) Techniques for preparing nematodes for electronmicroscopy. In: Barker KR, Carter CC, Sasser JN (eds) An advance treatise on *Meloidogyne*, Vol III, Methodology. A cooperative publication of the Department of Plant Pathology and the United States Agency for International Development. North Carolina State University Graphics, Raleigh, pp 79–105

EPPO (2013) Diagnostics PM 7/119 (1) Nematode extraction. European and Mediterranean Plant Protection Organization, European and Mediterranean Plant Protection Organization. Bull OEPP/EPPO Bull 43:471–495. doi:10.1111/epp.12077

Esbenshade PR, Triantaphyllou AC (1985a) Use of enzyme phenotypes for identification of *Meloidogyne* species (Nematoda: Tylenchida). J Nematol 17:6–20

Esbenshade PR, Triantaphyllou AC (1985b) Identification of major *Meloidogyne* species employing enzyme phenotypes as differentiating characters. In: Sasser JN, Carter CC (eds) An advanced treatise on *Meloidogyne*. Volume I: Biology and control. A cooperative publication of the Department of Plant Pathology and the United States Agency for International Development. North Carolina State University Graphics, Raleigh, pp 135–140

Esbenshade PR, Triantaphyllou AC (1987) Enzymatic relationships and evolution in the genus *Meloidogyne* (Nematoda: Tylenchida). J Nematol 19:8–18

Ferreira T, Malan AP (2014) *Xenorhabdus* and *Photorhabdus*, bacterial symbionts of the EPNs *Steinernema* and *Heterorhabditis* and their *in vitro* liquid mass culture: a review. Afr Entomol 22:1–14

Ferreira T, Addison MF, Malan AP (2014) *In vitro* liquid culture of a South African isolate of *Heterorhabditis zealandica* for the control of insect pests. Afr Entomol 22:80–92

Ferreira T, Addison MF, Malan AP (2016) Development and population dynamics of *Steinernema yirgalemense* and growth characteristics of its associated *Xenorhabdus* symbiont in liquid culture. J Helminthol 90:364–371

Flegg JJM (1967) Extraction of *Xiphinema* and *Longidorus* species from soil by a modification of Cobb's decanting and sieving technique. Ann Appl Biol 60:429–437

Fourie H (1998) The influence of plant-parasitic nematodes on soybean production in South Africa with special reference to root-knot nematodes (*Meloidogyne* species). MSc Dissertation, Potchefstroom University for Christian Higher Education, Potchefstroom

Fourie H, Zijlstra C, Mc Donald AH (2001) Identification of root-knot nematode species occurring in South Africa using the SCAR – PCR technique. Nematol 3:675–680

Fourie H, Venter GA, Mc Donald AH (2003) Optimum temperature and inoculation requirements for *in vitro* production of *Pratylenchus brachyurus* on carrot discs. S Afr J Plant Soil 20: 44–45

Fourie H, Mc Donald AH, Mothata TS et al (2012) Indications of variation in host suitability to root-knot nematode populations in commercial tomato varieties. Afr J Agric Res 7:2344–2352

Fourie H, Mc Donald AH, De Waele D et al (2013) Comparative cellular responses in resistant and susceptible soybean cultivars infected with *Meloidogyne incognita*. Nematol 15:695–708

Griffin GD, Barker KR, Pederson GA et al (1998) Plant and nematode interactions. Number 36 in the Agronomy series. American Society of Agronomy, Inc., Crop Science Society of America, Inc., Soil Science Society of America, Inc., Madison, USA, p 381–397

Han R, Ehlers R-U (1998) Cultivation of axenic *Heterorhabditis* spp. dauer juveniles and their response to non-specific *Photorhabdus luminescens* food signals. Nematologica 44:425–435

Hernandez A, Fargette M, Sarah J-L (2004) Characterization of *Meloidogyne* spp. (Tylenchida: Meloidogynidae) from coffee plantations in Central America and Brazil. Nematol 6:193–204

Hodda M, Eyualem-Abebe (2006) Techniques for processing freshwater nematodes. In: Eyualem-Abebe, Traunspurger W, Andrássy I (eds) Freshwater nematodes: ecology and taxonomy. CAB International, Wallingford, pp 31–45

Hominick WM, Briscoe BR, del Pinto EG et al (1997) Biosystematics of entomopathogenic nematodes: current status, protocols and definitions. J Helminthol 71:271–298

Hooper DJ (1998) Use of the scanning electronmicroscope to identify nematodes. In: Fortuner R (ed) Nematode Identification and Expert System Technology. Plenum Press, New York, pp 301–311

Hooper DJ, Evans K (1993) Extraction, identification and control of plant-parasitic nematodes. In: Evans K, Trudgill DL, Webster JM (eds) Plant parasitic nematodes in temperate agriculture. CAB International, Wallingford, pp 1–59

Hooper DJ, Hallman J, Subbotin S (2005) Methods for extraction, processing and detection of plant and soil nematodes. In: Luc M, Sikora RA, Bridge J (eds) Plant parasitic nematodes in subtropical and tropical agriculture. CAB International, Wallingford, pp 53–86

Hunt DJ, Handoo ZA (2009) Taxonomy, identification and principal species. In: Perry RN, Moens M, Starr JL (eds) Root-knot nematodes. CAB International, Wallingford, pp 55–97

Hussey RS, Boerma HR (1981) A greenhouse screening procedure for root-knot nematode resistance in soybeans. Crop Sci 21:794–796

Hussey RS, Jansen GJW (2002) Root-knot nematodes: *Meloidogyne* species. In: Starr JL, Cook R, Bridge J (eds) Plant resistance to parasitic nematodes. CAB International, Wallingford, pp 43–70

Jenkins WR (1964) A rapid centrifugal-flotation technique for separating nematodes from soil. Plant Dis Rep 48:692

Jones B, De Waele D (1990) Histopathology of *Ditylenchus destructor* on peanut. J Nematol 22:268–272

Karssen G, Van Hoenselaar T, Verkerk-Bakker B et al (1995) Species identification of cyst and root-knot nematodes potato by electrophoresis of individual females. Electrophoresis 16:105–109

Kaya HK, Stock SP (1997) Techniques in insect nematology. In: Lacey SA (ed) Manual of techniques in insect pathology. Academic, London, pp 281–324

Keetch DP (1982) Notes on the field evaluation of nematicidal materials. In: Keetch DP, Heyns J (eds) Nematology in Southern Africa, Science Bulletin No. 400. Department of Agriculture and Fisheries, Pretoria, pp 151–157

Khan MR (2008) Plant nematodes: methodology, morphology, systematics, biology and ecology. Science Publishers Inc, Enfield

Kleynhans KPN (1991) The root-knot nematodes of South Africa. Technical Communication 231. Department of Agricultural Development, Pretoria

Kleynhans KPN (1997) Collecting and preserving nematodes. Manual for a SAFRINET course in practical Nematology, 1st print. SAFRINET, the Southern African (SADC) LOOP of BioNet-International, Pretoria

Kleynhans KPN, Van den Berg E, Swart A et al (1996) Plant nematodes in South Africa. Plant Protection Research Institute Handbook No 8. ARC-Plant Protection Research Institute, Pretoria

Knoetze R, Burger JT, Meyer AJ (2000) Discrimination of some *Xiphinema* species from South Africa by rDNA-RFLP analysis. Afr Plant Prot 6:25–30

Knoetze R, Malan AP, Mouton C (2006) Differentiation of South African potato cyst nematodes (PCN) by analysis of the rDNA internal transcribed spacer region. Afr Plant Prot 12:103–110

Knoetze R, Swart A, Tiedt LR (2013) Description of *Globodera capensis* n. sp. (Nematoda: Heteroderidae) from South Africa. Nematol 15:233–250

Koen H (1969) Die voorkoms en beheer van plantparasitiese aalwurms in Suid-Afrika, Bladskrif No 35. Departement van Tegniese Dienste, Pretoria

Lax P, Doucet ME (2002) New process for preparing heteroderid cysts for scanning electron microscopy. Nematol 4:445–447

Le Roux S (2000) Life cycle, epidemiology and control of the golden cyst nematode *Globodera rostochiensis* (race: Ro1) under South African conditions at Ceres, Cape. ARC-Report 1999–2000. Project No. 103402. ARC-Vegetable and Ornamental Institute, Pretoria

Lesufi MM (2007) A study of the systematics and implications of the presence of the testa nematode, *Aphelenchoides arachidis* Bos, 1977 in South Africa. MSc Dissertation. North-West University, Potchefstroom

Lesufi M, Swart A, Mc Donald AH et al (2015) Morphological and molecular studies on *Aphelenchoides arachidis* Bos, 1977 (Tylenchina: Aphelenchoididae) from groundnuts in South Africa. Nematol 17:433–445

Lille RD, Fullmer HM (1976) Histopathologic techniques and practical histochemistry, 4th edn. McGraw-Hill Inc., New York

Malan AP, Nguyen KB, De Waal JY et al (2008) *Heterorhabditis safricana* n. sp. (Rhabditida: Heterorhabditidae), a new entomopathogenic nematode from South Africa. Nematol 10:381–396

Malan AP, Knoetze R, Moore SD (2011) Isolation and identification of entomopathogenic nematodes from citrus orchards in South Africa and their biocontrol potential against false codling moth. J Invertebr Pathol 108:115–125

Malan AP, Knoetze R, Tiedt L (2014) *Heterorhabditis noenieputensis* n. sp. (Rhabditida: Heterorhabditidae), a new entomopathogenic nematode from South Africa. J Helminthol 88:138–151

Malan AP, Knoetze R, Tiedt LR (2016) A new entomopathogenic nematode from South Africa, *Steinernema jeffreyense* n. sp. J Helminthol 90:262–278

Manzilla-López RH (2012) Methodology and symptomatology. In: Manzilla-López RH, Marbán-Mendoza N (eds) Practical plant nematology. BiBliotheca Básica de Agricultura, Monticillo, pp 89–129

Marais M, Shubane A (2012) 'n Tegniek om nematodes uit geelwortels te onttrek. In: Abstracts of the Annual Symposium of the Biological Sciences Division of the South African Academy of Arts and Science, North-West University, Potchefstroom, 25–26 October 2012

McClure MA, Kruk TH, Misaghi I (1973) A method for obtaining quantities of clean *Meloidogyne* eggs. J Nematol 5:230

Mc Donald AH (1998) Management of *Ditylenchus africanus* (Nematoda: Tylenchida) ongroundnut (*Arachis hypogaea*). PhD Thesis, Potchefstroom University for Christian Higher Education, Potchefstroom

Minagawa N (1986) Taxonomic studies of Criconematidae (Nematoda: Tylenchida) of Japan. Genera *Neolobocriconema*, *Paralabocriconema* n.gen. and *Macrocriconema* n.gen. Bulletin of the National Institute of Agro-Environmental Sciences, Japan No 1, pp 95–126

Murphy JA, Pappelis AJ, Thompson MR (1974) Morphological aspects of *Diplodia maydis* and its role in the stalk rot of corn. Scanning electron microscopy. Part II. In: Abstracts of of the workshop of scanning electron microscope and the plant sciences. IIT Research Institute, Chicago

Nguyen KB (2007) Methodology, morphology and identification. In: Nguyen KB, Hunt DJ (eds) Entomopathogenic nematodes: systematics, phylogeny and bacterial symbionts. Nematology monographs and perspectives, vol 5. Brill, Leiden

Nguyen KB, Hunt DJ (2007) Entomopathogenic nematodes: systematics, phylogeny and bacterial symbionts. Nematology monographs and perspectives, vol 5. Brill, Leiden

Nguyen KB, Malan AP, Gozel U (2006) *Steinernema khoisanae* n. sp. (Rhabditida: Steinernematidae), a new entomopathogenic nematode from South Africa. Nematol 8:157–175

Nthenga I, Knoetze R, Berry S et al (2014) *Steinernema sacchari* n. sp. (Rhabditida: Steinernematidae), a new entomopathogenic nematode from South Africa. Nematol 16:475–494

Ntidi N, Fourie H, Mc Donald AH et al (2012) Plant-parasitic nematodes associated with weeds in developing agriculture in South Africa. Nematol 14:875–887

Onkendi EM, Moleleki LN (2013a) Distribution and genetic diversity of root-knot nematodes (*Meloidogyne* spp.) in potatoes from South Africa. Plant Pathol. 62:1184–1192

Onkendi EM, Moleleki LN (2013b) Detection of *Meloidogyne enterolobii* in potatoes in South Africa and phylogenetic analysis based on intergenic region and the mitochondrial DNA sequences. Eur J Plant Pathol 136:1–5

Oostenbrink M (1960) Estimating nematode populations by some selected methods. In: Sasser JN, Jenkins WR (eds) Nematology. University of North Carolina Press, Chapel Hill, pp 85–102

Powers TO, Ramírez-Suárez Á (2012) Molecular diagnostics of nematodes: some practical considerations. In: Manzilla-López RH, Marbán-Mendoza N (eds) Practical plant nematology. BiBliotheca Básica de Agricultura, Monticillo, pp 131–145

Reynolds ES (1963) The use of lead citrate at high pH as an electron opaque stain in electronmicroscopy. J Cell Biol 17:208–213

Riekert HF (1995) An adapted method for extraction of root-knot nematode eggs from maize root samples. Afr Plant Prot 1:41–43

Riekert HF (1996) Economic feasibility of nematode control in dryland maize in South Africa. Afr Crop Sci J 4:477–481

Seacor T, Striluk M, Havek E (2015) Classification of wild isolate nematode strains using classical systematics and molecular sequencing. J Student Res 4:45–52

Seinhorst JW (1950) De betekenis van de toestand van de grond voor het optreden van aantasting door het stengelaaltjie (*Ditylenchus dipsaci* (Kühn) Filipjev). Tijdschrift over Plantenziekten 56:289–348

Seinhorst JW (1959) A rapid method for the transfer of nematodes from fixative to anhydrous glycerine. Nematologica 15:67–69

Seinhorst JW (1964) Methods for the extraction of *Heterodera* cysts from not previously dried soil samples. Nematologica 10:87–94

Shaham S (2006) Methods in cell biology. In: Shaham S (ed) Wormbook: the online review of *C. elegans* biology. http://wormbook.org.doi:101895/wormbook.1.49.1

Shapiro-Ilan DI, Han R, Qiu X (2014) Production of entomopathogenic nematodes. In: Morales-Ramos JA, Rojas MG, Shapiro-Ilan DI (eds) Mass production of beneficial organisms. Academic Press, Amsterdam, pp 321–356

Shepherd AM (1962) New blue R, a stain that differentiates between living and dead nematodes. Nematologica 8:201–208

Southey JF (1965) The incidence of stem eelworms on onion seed. Plant Pathol 14:55–59

Spurr AR (1969) A low-viscosity epoxy resin embedding medium for electron microscopy. J Ultrastruct Res 26:31–43

Stoffelen R, Jimenez MI, Dierckxsens C et al (1999) Effect of time and inoculum density on the reproductive fitness of *Pratylenchus coffeae* and *Radopholus similis* populations on carrot disks. Nematol 1:243–250

Stokwe N, Malan AP, Nguyen KB et al (2011) *Steinernema citrae* n. sp. (Rhabditida: Steinernematidae), a new entomopathogenic nematode from South Africa. Nematol 13:569–587

Subbotin SA, Inserra RN, Marais M et al (2011) Diversity and phylogenetic relationships within the spiral nematodes of *Helicotylenchus* Steiner, 1945 (Tylenchida: Hoplolaimidae) as inferred from analysis of the D2-D3 expansion segments of 28S rRNA gene sequences. Nematol 13:333–345

Subbotin SA, Wayenberge L, Moens M (2013) Molecular taxonomy and phylogeny. In: Perry R, Moens (eds) Plant nematology. CAB International, Wallingford, pp 40–72

Subbotin SA, Chitambar JJ, Chizkov VN et al (2014) Molecular phylogeny, diagnostics, and diversity of plant-parasitic nematodes of the genus *Hemicycliophora* (Nematoda: Hemicycliophoridae). Zool J Linn Soc 171:475–506

Swart A (1997) Description of *Laimaphelenchus patulus* sp. n. (Nematoda: Aphelenchoididae) from *Pinus pinaster* Ail in South Africa. Afr Plant Prot 3:23–28

Swart A, Subbotin SA, Tiedt L et al (2004) *Anguina woodi* sp. n. (Tylenchida: Anguinidae) from dune grass, *Ehrharta villosa*, in South Africa. Nematol 6:129–144

Swofford D (2002) PAUP*. Phylogenetic analysis using parsimony (*and other methods). Version 4. Sinauer Associates, Sunderland

Tamura K, Stecher G, Peterson D et al (2013) MEGA6: molecular evolutionary genetics analysis version 6.0. Mol Biol Evol 30:2725–2729

Thompson JD, Gibson TJ, Plewniak F et al (1997) The CLUSTAL_X windows interface: flexible strategies for multiple sequence alignment aided by quality analysis tools. Nucleic Acids Res 25:4876–4882

Todd WJ (1986) Effects of specimen preparation on the apparent ultrastructure of microorganisms. In: Aldrich HC, Todd WJ (eds) Ultrastructure techniques for microorganisms. Plenum Press, New York, pp 87–100

Turner SJ (1998) Sample preparation, soil extraction and laboratory facilities for the detection of potato cyst nematodes. In: Marks RJ, Brodie BB (eds) Potato cyst nematodes: biology, distribution and control. CAB International, Wallingford, pp 75–90

Van den Berg E, Furstenberg JP (1982) Extraction and handling of plant and soil nematodes. In: Keetch DP, Heyns J (eds) Nematology in Southern Africa, Science Bulletin No. 400. Department of Agriculture and Fisheries, Pretoria, pp 158–161

Van den Berg E, Subbotin SA, Handoo ZA et al (2009) *Hirschmanniella kwazuna* sp. n. from South Africa with notes on a new record of *H. spinicaudata* (Schuurmans Stekhoven, 1944) Luc & Goodey, 1964 (Nematoda: Pratylenchidae) and on the molecular phylogeny of the genus *Hirschmanniella* Luc & Goodey, 1964. Nematol 11:523–540

Van den Berg E, Subbotin SA, Tiedt LR (2010) Morphological and molecular characterisation of *Hemicycliophora lutosa* Loof & Heyns, 1969 and *H. typica* de Man, 1921 from South Africa (Nematoda: Hemicycliophoridae). Nematol 12:303–308

Van den Berg E, Tiedt LR, Subbotin SA (2012) Morphological and molecular characterisation of *Criconemoides brevistylus* Singh & Khera, 1976 and *C. obtusicaudatus* Heyns, 1962 from South Africa (Nematoda: Criconematidae) with first description of a male *C. obtusicaudatus* and proposal of new synonyms. Nematol 14:961–976

Van den Berg E, Tiedt LR, Coyne DL et al (2013) Morphological and molecular characterisation and diagnostics of some species of *Scutellonema* Andrássy, 1958 (Tylenchida: Hoplolaimidae) with a molecular phylogeny of the genus. Nematol 15:1–27

Van den Berg E, Tiedt LR, Subbotin SA (2014) Morphological and molecular characterisation of several *Paratylenchus* Micoletzky, 1922 (Tylenchida: Paratylenchidae) species from South Africa and USA, together with some taxonomic notes. Nematol 16:323–358

Van den Berg E, Palomares-Rius JE, Vovlas N et al (2016) Morphological and molecular characterisation of one new and several known species of the reniform nematode, *Rotylenchulus* Linford & Oliveira, 1940 (Hoplolaimidae: Rotylenchulinae), and a phylogeny of the genus. Nematol 18:67–107

Van der Walt P, De Waele D (1989) Mass culture of the potato rot nematode *Ditylenchus destructor* on groundnut callus tissue. Phytophylactica 21:79–80

Van Zyl C, Malan AP (2014a) Optimisation of inoculation techniques for *in vivo* mass culture of entomopathogenic nematodes through nematode and insect host manipulation. Afr Entomol 24:405–416

Van Zyl C, Malan AP (2014b) The role of nematodes as biological control agents, the history of nematode mass culture and *in vivo* nematode production. Afr Entomol 22:236–249

Vrain TC, Wakarchuck DA, Lèvesque AC et al (1992) Intraspecific rDNA restriction fragment length polymorphism in the *Xiphinema americanum* group. Fundam Appl Nematol 15:563–573

Wendt KR, Swart A, Vrain TC et al (1995) *Ditylenchus africanus* sp. n. from South Africa; a morphological and molecular characterization. Fundamen Appl Nematol 18:241–250

White GF (1927) A method for obtaining infective nematode larvae from cultures. Science 66:302–303

Windham GL, Williams WP (1987) Host suitability of commercial corn hybrids to *Meloidogyne arenaria* and *Meloidogyne incognita*. J Nematol 19(S):13–16

Zijlstra C (2000) Identification of *Meloidogyne chitwoodi*, *M. fallax* and *M. hapla* based on SCAR-PCR: a powerful way of enabling reliable identification of populations or individuals that share common traits. Eur J Plant Pathol 106:283–290

Zijlstra C, Donkers-Venne DTHM, Fargette M (2000) Identification of *Meloidogyne incognita*, *M. javanica* and *M. arenaria* using sequence characterised amplified region(SCAR) based PCR. Nematol 2:847–853

Chapter 5
Quarantine Nematodes

Rinus Knoetze, Antoinette Swart, and Moses Lesufi†

5.1 Introduction

In South Africa (SA), plant quarantine is a crop protection strategy enforced by legislation. Quarantine legislation is promulgated by governments worldwide under the auspices of the International Plant Protection Convention (IPPC), founded in Rome in 1951 at a special meeting of the Food and Agricultural Organization (FAO) of the United Nations. The IPPC is a multilateral treaty for cooperation in plant protection. Plant protection strategies are most effective when countries or regions coordinate their activities. South Africa became a contracting member of the IPPC on 21 September 1956. In ratifying this convention, the SA Government in effect agreed to draft and approve legislation by which effective phytosanitary procedures can be established in order to prevent the introduction and spread of harmful organisms (including nematode pests) across regional and national borders. In SA, the Directorates of Plant Health and Inspection Services of the Department of Agriculture, Forestry and Fisheries (DAFF) are responsible for the administration of the Agricultural Pests Act 1983 (Act 36 of 1983). The purpose of this act is to prevent

†Author was deceased at the time of publication.

R. Knoetze (✉)
Department of Agriculture, Forestry and Fisheries, Plant Quarantine Station,
Polkadraai Road, Stellenbosch 7600, South Africa
e-mail: RinusK@daff.gov.za

A. Swart
Biosystematics Division, Agricultural Research Council–Plant Protection Research Institute,
Private Bag X134, Queenswood 0121, South Africa

M. Lesufi
Department of Agriculture, Forestry and Fisheries,
Private Bag X258, 0001, Pretoria, South Africa

H. Fourie et al. (eds.), *Nematology in South Africa: A View from the 21st Century*, DOI 10.1007/978-3-319-44210-5_5

the introduction of agricultural pests which are not already present or are still of restricted presence in the country. This legislation provides for measures whereby agricultural pests already in the country can be controlled and also provides facilities where imported and exported goods are inspected, subject to certain conditions. Hockland et al. (2013) referred to the different terms that are used by regulatory instances that control the transport and spread of harmful organisms. These are 'plant health', 'plant quarantine', 'plant protection' and 'plant biosecurity'. For the purpose of this chapter, the terms 'quarantine nematodes' and 'plant quarantine' will be used.

Plant quarantine (Box 5.1) restricts the entry of plants, plant products, soil and other substrates, cultures of living organisms, packing materials and commodities as well as their containers and other means of transportation. The aim is to protect agriculture and the environment from avoidable damage that can be caused by harmful organisms inadvertently introduced by man. Measures employed to control plant pests through regulatory action can be grouped into (i) import quarantine controls, (ii) domestic quarantine controls and (iii) certification schemes.

Box 5.1 Plant Quarantine

Quarantine refers to a restriction imposed on the production and movement of plants and plant products in order to prevent or limit the introduction or spread of an alien pest or to limit the spread of an established pest (Cotten and Van Riel 1993). Based on definitions given by various authorities on this topic, a quarantine pest is an organism that is based on scientific evidence to be of potential economic importance to the area endangered thereby and not yet present there or present but not widely distributed and officially controlled (FAO 2010; Hockland et al. 2013).

5.2 Import Quarantine Controls

Import quarantine controls are intended to reduce the risk of introducing pests that are thought to present the greatest threat to the agriculture, horticulture and forestry industries. These are either pests that are thought to be absent from that country (see Box 5.2 for SA) or pests that, though present, are of limited distribution and are under a domestic quarantine control programme designed to eradicate them. The effective implementation of the articles of the IPPC is based on the plant importation permit and the phytosanitary certificate (IPPC 2006).

5.2.1 Plant Importation Permit

A plant importation permit is an official document authorising importation of a commodity in accordance with specific phytosanitary requirements. They are the passports to facilitate the entry of plant materials. The risk of introducing a new pest can be reduced by specifying in the permit, as a condition of entry, that the importer must meet certain entry requirements.

Box 5.2 Some Economically Important Plant-Parasitic Nematodes that Are Stated on South African Plant Importation Permits

Aphelenchoides fragariae (Ritzema Bos, 1890) Christie, 1932: a pest of strawberry that can cause 'summer dwarf' disease and is widely spread throughout Europe and the United States of America (USA). Crop losses of up to 60 % have been reported as a result of parasitism by this nematode pest. The nematode is also associated with *Corynebacterium fascians* and can cause 'cauliflower disease' of strawberry (Evans et al. 1993; Duncan and Moens 2013).

Bursaphelenchus xylophilus (Steiner & Buhrer, 1934) Nickle, 1970: a pest of pine trees occurring in Japan, China and the USA that can cause 'pine wilt' disease. Two longhorn beetles (Coleoptera: Cerambycidae) are the major vectors of this nematode (Hunt 1993; Mota and Vieira 2008).

Nacobbus aberrans (Thorne, 1935) Thorne & Allen, 1944 (also known as the false root-knot nematode): a pest of potato in South America. This nematode occurs in temperate agro-ecological regions (Brodie et al. 1993; Scurrah et al. 2005; Bridge and Starr 2007).

5.2.2 Phytosanitary Certificate

Under national and international plant protection and quarantine regulations, many kinds of plants and plant products must be accompanied by a valid phytosanitary certificate. Such a certificate will state that the commodity imported meets the requirements of the importing country. Furthermore, additional information on the pest(s) in question or procedures to be followed to comply with phytosanitary regulations may be stated on a phytosanitary certificate (Hockland et al. 2013). To comply with the specific requirements of the IPPC, inspections should be carried out and certificates issued only by or under authority of technically qualified and duly authorised officers. This will be done under such circumstances and with such knowledge and information available to the designated officers that the authorities of importing countries may accept such certificates. These officers must therefore be properly trained and have extensive knowledge of plant pests and the phytosanitary requirements of importing countries. Confirmation is required that internationally accepted standard procedures have been used in establishing the absence of the specified pest(s) on the growing crops or in samples of the plant materials. Test results must usually accompany the phytosanitary certificate.

5.3 Domestic Quarantine Controls

Domestic quarantine controls may be imposed to reduce the risk of the spread of a pest from infested areas into areas in which the pest is not known to occur. These controls may include actions such as the destruction of infected plants, prohibition on the cultivation of certain plants in certain areas and control over the movement of growth medium and plant material from certain areas. Box 5.3 lists the quarantine nematodes that occur in SA.

Box 5.3 Plant-Parasitic Nematodes Occurring in South Africa that Are Listed as Quarantine Pests and Thus Regulated by the Agricultural Pests Act (Act 36 of 1983)

Meloidogyne partityla: a serious pest of pecan nut.

Tylenchulus semipenetrans Kleynhans, 1986: the citrus nematode, also infects grapevine and persimmon.

Ditylenchus dipsaci (Kühn, 1857) Filipjev, 1936: a stem nematode pest of lucerne in the Western and Eastern Cape provinces.

Globodera rostochiensis (Wollenweber, 1923) Behrens, 1975: the golden potato cyst nematode; infested fields are placed under quarantine.

Radopholus similis (Cobb, 1893) Thorne, 1949: a serious pest of banana; the movement of planting material is regulated by quarantine measures.

5.3.1 Identification of Quarantine Nematodes

The identity of quarantine nematode pests needs to be known to enable the correct phytosanitary action to be taken. Concerns, however, exist about the identification of quarantine nematode pests caused by the dwindling number of well-trained classical taxonomists and a lack of resources to curate and conserve nematode collections. The use of both morphological and molecular techniques to identify nematode pests is a powerful tool to detect quarantine nematodes. However, the development and application of molecular tools for nematode quarantine purposes are still not optimal and should be approached cautiously (Hockland et al. 2013).

In SA, interrelated activities and cooperation exist among the different authorities and nematology laboratories for the detection, identification and certification of quarantine nematodes during import and export of infected plant material and infested substrates (see Box 5.4).

Box 5.4 Interrelated Activities Among the Various Authorities and Nematology Laboratories in South Africa to Detect, Identify and Certify Quarantine Nematode Pests

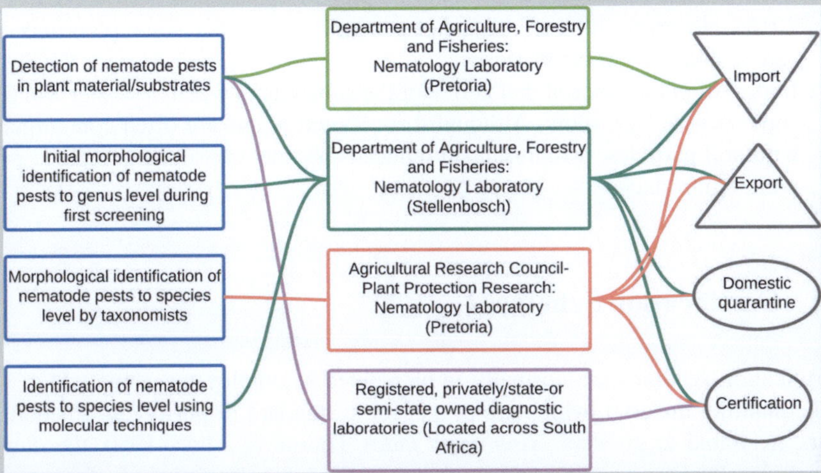

5.3.2 *Measures to Reduce the Risks of Importing Quarantine Pests*

To reduce the risk of importing quarantine pests, countries (including SA) will take into account the following aspects before an importation permit is issued (Cotten and Van Riel 1993; Hockland et al. 2013):

(i) The cropping history of the field or area where the crop, of which plant material is being imported or exported, has been cultivated. This information gives a reliable indication of the risks of infection by a nematode pest. Postharvest soil sampling of the field(s) where the crop was cultivated may also be required to detect specific quarantine nematode pests

(ii) Inspections of crops during the growing season may be useful to detect specific symptoms indicating nematode infection. An example is *Aphelenchoides fragariae* (Ritzema Bos, 1890) Christie, 1932, which produces leaf necrosis and distortions of leaf margins of infected host plants such as strawberry and various ornamentals

(iii) Postharvest or pre-export inspection of plant material to detect symptoms of nematode infection and to reduce adhering infested substrate. For example, this measure is effective in limiting the spread of potato cyst nematodes with seed potato tubers

(iv) Specific treatment(s) to disinfect imported bulbs, rootstocks, seeds, etc. Import regulations may require that consignments of plants be treated with an approved chemical to control a specific quarantine nematode pest. For example, hot water treatment has been used to control nematodes in a range of bulbs of ornamentals, rootstocks and tubers of crops (e.g. cassava) (Hockland et al. 2013). However, this method is not generally used since some plant material can be damaged in the process

(v) Import of plant material that is free of the substrate in which it was grown (e.g. soil) such as bare roots. Although bare-rooted plants are often contaminated with soil particles, it represents a reduced risk that unwanted nematode pests will be introduced

5.4 Certification Schemes

Certification schemes are normally implemented to employ measures which will make planting material, which are of a specific standard of purity and freedom of pests, available to growers. *Xiphinema index* Thorne & Allen, 1950, the dagger nematode attacks grapevine and is also a vector of grapevine fanleaf nepovirus (GFLV). This nematode is regulated by regulations in the grapevine certification scheme. *Globodera rostochiensis* (Wollenweber, 1923) Behrens, 1975, is also regulated under the potato certification scheme, whereby potato seed production is not allowed on fields infested with this nematode.

5.5 Pest Risk Analysis

A Pest Risk Analysis (PRA) is the process (Box 5.5) of evaluating biological or other scientific and economic evidences to determine whether an organism could become or is a pest and should be regulated and the strength of the phytosanitary measures to be taken (IPPC 2006; FAO 2010). This analysis may be triggered when (i) a request is made to consider a pathway (means by which the pest can be introduced into the country) that may require phytosanitary measures, (ii) a pest is identified that may justify phytosanitary measures, (iii) a decision is made to review or revise phytosanitary measures or policies or (iv) a request is made to determine whether an organism could become or is a pest (IPPC 2006).

Since only quarantine pests can be regulated in international trade, it is of utmost importance that the nematode pest in question meets the definition of a 'quarantine pest'. Only a quarantine nematode pest needs to be subjected to a complete risk assessment and management (Hockland et al. 2013).

5.6 Clean Handling Procedures in Research Laboratories

If economically important nematodes that are not yet present in an area are reared (in vitro or in vivo) for research purposes, maximum effort should be made to contain such cultures and avoid accidental release of the nematodes into the surrounding environment. Therefore, the following facilities should be in place in such a laboratory (De Ley and Mundo-Ocampo 2004):

(i) Hot water (preferably with a temperature >65 °C) should be available at all sinks in the laboratory. Rinsing all used equipment with hot water will kill nematodes that will otherwise be flushed down the drain alive
(ii) The tank that collects the water used for nematode extraction should be treated, for instance, chemically, to kill all nematodes that could have been washed in it
(iii) An autoclave for sterilising discarded nematode-infested soil. Infected plant material should be collected in plastic bags
(iv) Restricted access to quarantine nematode cultures, i.e. by keeping containers, etc., locked and clearly labelled as under quarantine

Box 5.5 The Basic Procedure Followed During a Pest Risk Analysis (PRA)

A Pest Risk Analysis (PRA) is a complex, extensive and tedious activity. The emerging threat of *Meloidogyne enterolobii* Yang and Eisenback, 1983 (at present associated with green pepper, potato and tomato in South Africa), is a good example of a nematode pest that poses a risk as published by the European and Mediterranean Plant Protection Organization (EPPO) (NAPPO 2015). *Meloidogyne enterolobii* attained an A2 phytosanitary categorisation (List no. 361) in 2010 (NAPPO 2015) and is represented by EPPO code MELGMY (NAPPO 2011). The three important stages to be completed during a PRA are illustrated below together with the actions required for each stage

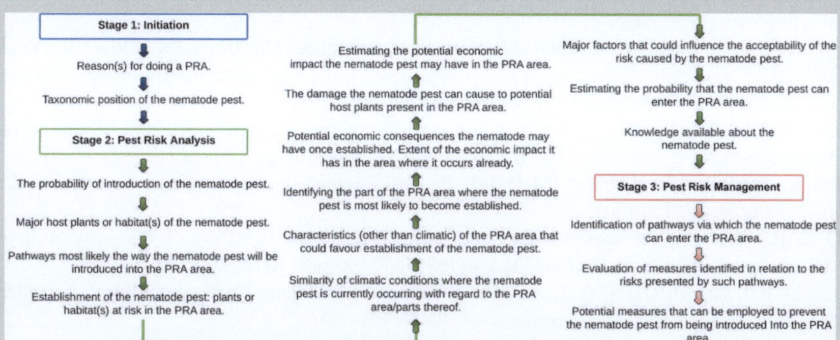

5.7 Conclusions

The principle role of quarantine legislation is to prevent the introduction of serious nematode pests through importation from abroad and to prevent dispersal of local nematode pests into non-contaminated areas within a country. Once a specific nematode pest has been identified as a phytosanitary risk, regulatory actions can be applied to inhibit its spread and multiplication and, in this way, limit the damage it can cause. Legislation aimed at preventing the introduction and spread of such nematodes is, however, highly dependent on the practicality of enforcing such legislation and on the attitude of the farmers and general public towards such measures. The availability of sufficient, qualified human resources to attend to border posts, airports and harbours is dependent on government policies. Moreover, a high level of responsibility also rests on the shoulders of the public sector since it would be almost impossible to stop harmful organisms from entering the country if there was no commitment to follow the correct protocols with regard to the import and export of plant material. This, in turn, may only be accomplished by proper education and by disseminating information regarding the dangers of breaching these protocols. Despite the difficulties involved in the maintenance of effective quarantine services, some regulations have been highly successful in many countries (Ferraz and Brown 2002). The implementation of regulatory actions should be regarded as essential and beneficial.

References

Bridge J, Starr JL (2007) Plant nematodes of agricultural importance. Academic Press, Boston

Brodie BB, Evans K, Franco J (1993) Nematode parasites of potatoes. In: Evans K, Trudgill DL, Webster JM (eds) Plant parasitic nematodes in temperate agriculture. CAB International, Wallingford, pp 87–132

Cotten J, Van Riel H (1993) Quarantine: problems and solutions. In: Evans K, Trudgill DL, Webster JM (eds) Plant parasitic nematodes in temperate agriculture. CAB International, Wallingford, pp 593–607

De Ley P, Mundo-Ocampo M (2004) Cultivation of nematodes. In: Chen ZX, Chen SY, Dickson DW (eds) Nematology. Advances and perspectives Vol I. Nematode morphology, physiology and ecology. CAB International, Wallingford, pp 541–619

Duncan LW, Moens M (2013) Migratory ectoparasitic nematodes. In: Perry RN, Moens M (eds) Plant nematology, 2nd edn. CAB International, Wallingford, pp 179–214

Evans K, Trudgill DL, Webster JM (1993) Parasitic nematodes in temperate agriculture. CAB International, Wallingford

FAO (2010) International standards for phytosanitary measures. ISMP No. 5. Glossary of phytosanitary terms. http://eppo.int/QUARANTINE/ISPM05.pdf. Accessed 6 Apr 2016

Ferraz LCCB, Brown DJF (2002) An introduction to nematodes. Plant nematology. A student's textbook. Pensoft Publishers, Sofia

Hockland S, Inserra RN, Kohl LM (2013) International plant health – putting legislation into practice. In: Perry RN, Moens M (eds) Plant nematology, 2nd edn. CAB International, Wallingford, pp 359–382

Hunt DJ (1993) Aphelenchida, Longidoridae and Trichodoridae: their systematics and bionomics. CAB International, Wallingford

IPPC (2006) What we do. https://www.ippc.int/en/what-we-do/. Accessed 6 Apr 2016

Mota MM, Vieira P (2008) Pine wilt disease: a worldwide threat to forest ecosystems. Springer, Berlin

NAPPO (2011) *Meloidogyne enterolobii*. Eur Mediterranean Plant Protect Organizat Bull 41:329–339

NAPPO (2015) EPPO and pest risk analysis. European and Mediterranean Plant Protection Organization. http://www.eppo.int/QUARANTINE/Pest_Risk_Analysis/PRA_intro.htm. Accessed 28 Dec 2015

Scurrah ML, Niere B, Bridge J (2005) Nematode parasites of solanum and sweet potato. In: Luc M, Sikora RA, Bridge J (eds) Plant parasitic nematodes in subtropical agriculture. 2nd edn. CAB International, Wallingford, pp193–220

Chapter 6
Nematode Control and Nematicides: Developments Since 1982 and Future Trends

Robin K. Jones

6.1 Introduction

Nematologists generally agree that the need to control nematodes originated directly after man's first attempts to cultivate crops. The resident plant-parasitic nematodes that were originally present in low numbers were selected and increased to populations that, year by year, increasingly limited crop production. Clearly the causal organisms were unknown, but growers would have tested many avenues to prevent or even reverse this trend. Thus, without realising the details of the soil-borne interactions, farmers in the pursuit of enhancing crop production would have tested and begun the 'selection' process to implement the three major foundation pillars of nematode control that are still in use today, viz. (i) crop rotation, (ii) selection of crop varieties that show genetic resistance against pests such as nematodes and (iii) application of soil amendments, e.g. green and animal manures, which possessed both fertility and nematicidal properties and ultimately evolved as the precursors of modern chemical control.

Over the centuries, particularly in the last 50 years, the need to enhance crop production has resulted in farming becoming increasingly specialised. It is now a global industry with specialisation occurring at producer, regional and country levels. South Africa (SA) is just such a specialised producer of fresh citrus, subtropical and deciduous fruits, maize and sugarcane, to name but a few crops. Such specialisation has only been possible due to the continual advances in refining yield potential, fertilisation, irrigation, variety breeding and selection, pest control and others. Set alongside these developments within the discipline of nematode control, crop rotations have been refined, progress has been made in resistance breeding and

R.K. Jones
Nematological Services, 24 Victor Street, Murrayfield, South Africa
e-mail: robin.rsa@gmail.com

H. Fourie et al. (eds.), *Nematology in South Africa: A View from the 21st Century*, DOI 10.1007/978-3-319-44210-5_6

chemical control options have flourished. These have all become established as essential inputs.

The reliance on nematicides contributed to the expansion of monoculture (specialised) production. This led to the increased frequency of planting high-value crops (placing further dependency on nematicides) and the retreat of crop rotation as the primary nematode control tool. In perennial crops, nematicides were able to limit the nematode-induced component of decline problems, improving yield and quality, as well as extending the 'healthy life' of the crop and reducing the frequency of replanting. The reliance on nematicides at the expense of crop rotation and, to a lesser degree, resistance breeding is currently being challenged by the demands of consumers to promote a 'safe' food supply, i.e. devoid of pesticide residues, and with limited adverse environmental impact. This need has re-established a research emphasis based on soil amendments, natural remedies, crop tolerance and crop rotations to create a balance mitigating the development and impact of pest populations. In this scenario, the use of chemicals will be reduced. They will become part of an integrated approach and not the major input for nematode control as is still the case.

Aside from these consumer-driven pressures, biotechnology breakthroughs are also opening up new avenues of research. This is represented by the identification of species and races of nematode pests, identification of resistant genes and the ability to select rapidly new resistance varieties. These trends could well lead to a return to the traditional control options of resistance breeding and soil amendments coming to the fore in nematode control. Despite the trends referred to above, the production of many crops is still dependent on chemical nematode control. It needs to be repeatedly reaffirmed that nematicides are currently an essential component of secure food production in SA and that their loss would undermine our agricultural competitiveness. Keetch (1989) estimated that the average loss caused by plant-parasitic nematodes in SA for the 42 crops listed then was 14 %. Losses on staple food crops were approximately 10 %, a figure very similar to losses caused by plant-parasitic nematodes in world agriculture (Khan 2008). In specific situations, crop production has become dependent on nematicidal inputs and without their use, the cropping patterns at such sites would have to change. A concise overview on the use of nematicides is included in Box 6.1.

Aside from crop rotation, genetic host-plant resistance and, chemical control, several other approaches have been used to effectively control plant-parasitic nematodes. Flooding, steam treatment of soil, heat treatment of plant material, soil solarisation, biological control and plant quarantine are all in use in specific areas. These practices have been utilised in specific niche markets, notably in the intensive nursery market. In this sector, steam sterilisation of soil, heat treatment of plant material and plant quarantine procedures provide valuable control options. In addition, considerable progress is being made in the harnessing of many biological control options such as the application of specific control agents and the manipulation of the soil environment with soil amendments.

Box. 6.1 Nematicides: The Predominant Nematode Management Strategy

The application of nematicides typically occurs in an integrated nematode management (IPM) approach where crop rotation and host-plant resistance also play a part. The interactions of the pest, the host and the environment (the so-called disease triangle) will affect the choice and application method of the product or technique used. Nonetheless, in commercial agriculture, the predominant control strategy employed is dependent on the use of nematicides. For most nematode-induced crop loss situations, several products with nematicidal characteristics are registered. Each situation, however, has its specific recommendations but the physical, chemical and biological properties of each product are fixed; the broad-stroke guidelines outlined in this chapter apply in most situations, but for crop-specific recommendations, reference should be made to Chaps. 7–18.

6.2 Classical Nematicides

6.2.1 History

Rich et al. (2004) tabulated a chronological history of the development of nematicides, citing carbon disulphide in 1881 as the first product identified with nematicidal properties. The use of chemicals to control nematode pests gained widespread acceptance only after the discovery of the halogenated aliphatic hydrocarbon fumigant products in the early to mid-1900s. This was followed by the identification of the carbamates (aldicarb was identified in 1966) and organophosphates (fenamiphos was identified in 1967) from the mid-1960s through to the 1990s, when fosthiazate was identified. The introduction of these products led to a realisation of the damage caused by nematodes, which resulted in a steady increase in their use on more and more crops, with a constant introduction of refinements in the management practices. In SA, from about 1970, much of this developmental work was driven by a core of research workers who were employed in specific commodity-driven research centres. After this initial flurry, the use of nematicides had grown to be an extensive and relatively stable industry. Review articles of nematode control practices abound (Bromilow 1988; Bunt 1987; Hague and Gowen 1987; Hough et al. 1975; Keetch 1974; Keetch and Milne 1982; Rich et al. 2004; Whitehead 1998; Wright 1981; Sikora et al. 2005; Nyczepir and Thomas 2009; Haydock et al. 2013). The reviews 'The control of plant-parasitic nematodes' by Keetch and Milne (1982) and 'The use of nematicides' by (Keetch 1982) remain very thorough sources and important introductions to the applied aspects of plant nematology in southern Africa.

6.2.2 *Currently Registered Products*

Nematicides are generally classified according to three groups as presented in Box 6.2. Table 6.1 (Van Zyl 2013) lists the crops and the active substances (a.s.) of registered nematicides for use on those crops, whilst fumigant- and non-fumigant products available are displayed in Tables 6.2 and 6.3, respectively. These lists comprise fumigant, organophosphate (OP), carbamate, biological actives as well as furfural. There are over 50 'crop descriptions' currently listed in terms of the Fertilizers, Farm Feeds, Agricultural Remedies and Stock Remedies Act 36 of 1947 (Van Zyl 2013).

Box. 6.2 The Three Major Nematicide Groups (Fumigants, Non-fumigants and Biological Products)

Fumigant products (Table 6.2) represent two main subgroups, viz.:

(i) The halogenated, aliphatic hydrocarbons, methyl bromide, 1,3-dichloropropene, chloropicrin (never registered in South Africa as a stand-alone product) and ethylene dibromide (EDB)
(ii) The methyl isothiocyanate (MITC) generators dazomet, metam sodium and metam potassium

Non-fumigant products (Table 6.3) represent three main subgroups, viz.:

(i) The organophosphates
(ii) The carbamates, both of which act as inhibitors of the enzyme acetyl cholinesterase
(iii) Other chemical classes that include the aldehyde furfural, the dicarboximide iprodione and others

Biological products of which various are registered (Van Zyl 2013). Chapters 7–18 outline more crop-specific details of the use of these products.

Situations where these products can be used are largely defined by their physical, chemical and biological properties. Thus, the fumigants all need a delivery system that reduces the influence temperature can have on the upward movement and loss of the gas phase either by sealing the soil with a light irrigation and/or rolling or by covering the soil with plastic. Moreover, all these products have to dissolve into water to act on the target nematode pest; thus temperature and water solubility constraints act on where and how a specific product is used. These restrictions are related to soil type constraints and thus apply for all crops as referred to in Chaps. 7–18.

Table 6.1 Crops on which nematicides are registered for use in South Africa (Van Zyl 2013)

Crop	Active substances registered
Banana	Cadusafos, fenamiphos, fosthiazate, oxamyl and *Purpureocillium lilacinum* strain 251
Bean (dry)	Terbufos
Bean (green)	Ethoprophos
Carrot	*Beauveria bassiana*, furfural and *Trichoderma harizium*
Chicory	Cadusafos and furfural
Citrus	Cadusafos, ethoprophos, fenamiphos, fosthiazate, *P. lilacinum* strain 251 and terbufos
Cotton	Fenamiphos
Cruciferae	Ethoprophos
Cucurbit	Ethoprophos
Deciduous fruit	Fenamiphos
Flowers only	Furfural
Flowers and ornamentals	Fenamiphos and furfural
Ginger	Fenamiphos
Grape	Cadusafos and fenamiphos
Green chilli	Furfural
Groundnut	Fenamiphos, furfural, oxamyl and terbufos
Guava	Cadusafos and fenamiphos
Hop	Aldicarb withdrawn. No product currently registered
Lawns and turf	Fenamiphos
Lettuce	Ethoprophos and furfural
Litchi	Cadusafos and fenamiphos
Maize	Carbofuran, carbosulfan, furfural and terbufos
Nectarine	Fenamiphos and oxamyl
Onion	Fenamiphos and furfural
Papaya	Fenamiphos and *P. lilacinum* strain 251
Paprika	Furfural
Peach	Cadusafos, fenamiphos and oxamyl
Pea	Ethoprophos and fenamiphos
Pecan	Fenamiphos
Pepper	Furfural
Pineapple	1,3-D, cadusafos, fenamiphos and oxamyl
Plum	Cadusafos
Potato	1,3-D, cadusafos, ethoprophos, fenamiphos, fosthiazate, furfural and oxamyl
Seed beds	1,3-D, 1,3-D+chloropicrin and metam sodium
Soil	Dazomet, EDB, furfural, metam potassium and sodium and methyl bromide/chloropicrin
Sorghum	Carbofuran
Spinach	Ethoprophos
Sports fields	Fenamiphos
Stone fruit	Cadusafos

(continued)

Table 6.1 (continued)

Crop	Active substances registered
Sugarcane	Carbofuran, furfural and oxamyl
Sunflower	Carbofuran and terbufos
Sweet corn	Carbofuran
Sweet potato	Aldicarb withdrawn. No product currently registered
Tobacco	1,3-D, cadusafos, ethoprophos, fenamiphos, fosthiazate, methyl bromide/chloropicrin, oxamyl and *P. lilacinum* strain 251
Tomato	1,3-D, fenamiphos, furfural, oxamyl and *P. lilacinum* strain 251
Turf	Fenamiphos and furfural
Various crops	1,3-D, EDB and metam sodium
Various, fruit orchard, vineyard establishment	Metam sodium and methyl bromide/chloropicrin
Various, soil fumigant	EDB and methyl bromide/chloropicrin
Various crops, vegetables, fruit trees and strawberries	1,3-D
Wheat	Carbofuran

Table 6.2 Fumigant products registered in South Africa (Van Zyl 2013)

Active substance	Trade name	Concentration	Application rate of active substance
1,3-dichloropropene	Telone® II	1110 g l^{-1}	60–225 l ha^{-1}
1,3-dichloropropene	DD 92®	920 g l^{-1}	NA
Ethylene dibromide	EDB 1800VP®	1800 g l^{-1}	20–60 l ha^{-1}
1,3-D plus chloropicrin	Telopic®	850/465 g kg^{-1}	40.6–52.8 g m^{-2}
Metam potassium	Tamifume K 690®	690 g l^{-1}	550–1000 l ha^{-1}
Metam sodium	Ag-Fume®	510 g l^{-1}	650–900 l ha^{-1}
Metam sodium	Metafume®	510 g l^{-1}	650–900 l ha^{-1}
Metam sodium	Nemasol®	510 g l^{-1}	650–900 l ha^{-1}
Metam sodium	HERBIFUME®	510 g l^{-1}	650–900 l ha^{-1}
Methyl bromide/chloropicrin	Metabrom®	980/20 g kg^{-1}	0.5–1.0 kg 10 m^{-2}

When Keetch and Heyns published a book in 1982 entitled *Nematology in southern Africa*, the halogenated fumigant products were well established. The methyl isothiocyanate (MITC) generator dazomet was registered in SA but was not commercially released. There were three carbamates and one organophosphate registered. Aldicarb and fenamiphos were each registered for use on seven crops (Keetch 1982). Currently all the aliphatic hydrocarbons remain registered. Dazomet, though registered, is still not commercially available, but two MITC generators, metam sodium and metam potassium, are now commercially available. There are still two carbamates registered (aldicarb was withdrawn from the market by Bayer in 2011), but there are now five organophosphates registered. In addition, there has been a marked increase in the extent of their use. Aldicarb was registered on 13 crops and fenamiphos is currently registered for use on 18 crops (Van Zyl 2013).

Table 6.3 Non-fumigant products registered in South Africa (Van Zyl 2013)

Active substance	Trade name	Concentration applied as g kg^{-1}	Application rate (kg) of active substance ha^{-1}
Carbofuran	Curaterr®, Furadan®, Carbosan 100GR®, Carbofuran 100GR®, Carbodan®	100	1–3
Cadusafos	Rugby 10ME®	100	3–10
Cadusafos	Rugby 110GR®	100	4–10
Ethoprophos	Mocap 100GR®	100	5.25–7
Fenamiphos	Nemacur 100GR®	100	1.5–12
Fenamiphos	Nemacur 40EC®, Spitfire, Fenamiphos®	400	1.2–12
Fenamiphos	Nemacur 240CS®	240	10
Fosthiazate	Nemathorin 100GR®	100	3
Fosthiazate	Nemathorin 900EC®	900	3
Furfural	Crop Guard®	900	22.5–67.5
Furfural	MultiGuard Protect®	990	396–594
Oxamyl	Vydate GR®, Oxagran®	100	1.24–4
Oxamyl	Vydate SL®, Platoon®, Blockade®, Stetson®	310	1.24–4
Oxamyl	OxiDate®, Oxatak®	250	1.24–4
Terbufos	Counter®, Ortofos®, Terbufos, Terburops®, Terrafos®	100	1–5
Terbufos	Counter®, Ortofos®, Terbufos, Terfos®	150	1–5

Several reasons account for the widespread adoption of nematicides, but essentially commercially viable products were registered that provided returns very attractive to the producer industries. Aside from the commercial developments, the science of plant nematology in SA blossomed and built up an increasing body of information, highlighting the extent of the plant-parasitic nematode problems that were previously unrealised. Some of the key reasons for this explosion of knowledge and application of nematicides were as follows:

(i) The regional agricultural research institutes developed into centres of expertise and researched the potential nematode-induced crop losses that the nematicides had controlled. New, more extensive surveys were undertaken and, for example, resulted in the discovery of new nematode pests, refinement of the distribution of known pests, refinement of pest thresholds and identification of circumstances where treatments were needed. The citrus, cotton, deciduous fruit, maize, groundnut, sugar and tobacco industries benefited significantly. The potato industry was one of the largest users of nematicides, but until recently, it was poorly served during this period.

(ii) The SA regulatory system and the size of the potential markets were such that the suppliers of nematicides were attracted to develop the products. In this regard, the suppliers acted as another centre of knowledge, refining further an understanding of losses, particularly on potato, which became the largest market for nematicides for a long time.

(iii) Once these markets had been opened, the SA regulatory systems also permitted ready access for suppliers of generic products. Market forces were hence able significantly to reduce the input costs of treatment and further expand the market (e.g. currently there are eight registration holders of carbofuran).

(iv) The diverse and export-orientated agricultural industry in SA furthermore favoured the occurrence of nematode pest problems, notably in the specialised production of vegetables and cut flowers.

(v) The preponderance of poorly structured and frequently coarse soils, especially favourable to the establishment of nematode pest populations.

(vi) The widespread adoption of irrigation practices typically in the normally drier, warmer and more sandy soil conditions.

(vii) The introduction and spread of exotic nematode pests, e.g. *Radopholus similis* (Cobb 1983) Thorne, 1949, and *Tylenchulus semipenetrans* Cobb, 1913, as well as the presence of the emerging root-knot nematode *Meloidogyne enterolobii* Yang and Eisenback, 1983, which is recognised as an emerging pest worldwide (Jones et al. 2013).

All of these factors contributed to the expansion of the use of nematicides in terms of crops and hectares applied and the versatility of treatment options. More recently, the expansion and use of nematicides have been questioned, and the period of growth has ended. The challenges faced by the global chemical industry are summarised in Box 6.3.

Box. 6.3 Early Challenges Experienced During the Introduction of Nematicides

Research aimed at studying the interactions between the nematode pest, the crop, the environment and the product poses a great challenge. Application timing, placement and rates, all had to be tested for each crop-specific situation (see Chaps. 7–18 for the crop-specific details of product use). This culminated in a registration and then a growth phase of market acceptance dependent on the benefits accrued. The challenge is still existent due to a recent surge in the discovery of new chemistry. But because most of the established products are classified either as toxic or extremely toxic, the defence of these products by their manufacturers, especially in the export markets, became an additional challenge. The established products increasingly face restrictions due to:

(i) Consumer organisation pressure restricting the producers' options of chemical control measures

(ii) Regulatory pressures, both locally and internationally, acting to restrict the use of products

(iii) Misuse of several of the products and the resultant adverse media coverage

Fumigant nematicides are applied to the soil either as a solid, liquid or gas before planting. The product then vaporises and disperses by fumigant action in the soil air spaces after which it dissolves in the soil moisture where it will kill nematodes. However, not all individuals of the nematode population present are killed, and not all fumigants are active on nematode eggs. Fumigants generally provide a quick kill and knockdown of the nematode population prior to planting.

Soil preparation, soil moisture levels, soil temperatures and product placement all affect dispersion and water solubility of nematicides. Thus, a poorly-cultivated, cold, saturated soil with large clogs will limit dispersion and reduce the final product concentration in the soil and reduce its efficacy. Conversely, a well-cultivated, hot, dry soil with no posttreatment surface sealing, especially of the shear fissures, will lead to rapid loss of the vapour from the soil surface and result in ineffective nematode control. As all the registered fumigants are phytotoxic, an interval between application and planting is required to allow the fumigant to dissipate. This interval is dependent on fumigant type, temperature, soil moisture and soil type.

The non-fumigant, neurotoxic organophosphates and carbamate nematicides that are applied at or just prior to planting and/or after planting are available either as granules or liquids. Their activity depends on correct placement and distribution in the soil. Thereafter, the active substance (a.s.) dissolves in the soil moisture, and toxic concentrations act on the nematodes over a prolonged period of time. Typically the products are not crop sensitive and thus can be used at and/or near or after planting of the crop. Their action is dependent on their solubility (e.g. to kill nematodes the product has to dissolve in the soil moisture), chemical stability (e.g. affecting the persistency of action), systemic movement (i.e. affecting the products redistribution by the plant) and their chemical toxicity to nematodes. Thus, application rates, placement and application timings will vary depending on the chemical. As with fumigants, close liaison with a knowledgeable representative is important when first utilising these products. The mode of action of all these products is essentially similar. Directly after planting high concentrations of the products will exist in portions of the soil that will kill mobile nematode stages and eggs directly, but typically concentrations will be below this level. At these sublethal concentrations, the nematicides will act to disrupt nematode movement and the ability to invade plant tissue. Should nematode invasion occur despite the application of nematicides, higher concentrations are needed to kill the nematodes in the roots and/or other plant parts; thus corrective control is not achieved.

The less persistent nematicides, e.g. oxamyl and furfural, can be applied during the growth of the plant, extending the application window for control. This approach must always follow effective control at or pre-plant because post-plant treatment only has a limited efficacy on nematodes that have already invaded the roots and/or other plant parts. The successful application of post-plant nematicides depends on moving the nematicide to the root system and tubers. If the nematicide is translocated basipetally, a foliar application can be used, e.g. oxamyl, but if this is not the case, the product has to be incorporated either mechanically (not easily feasible on potato) or by physical movement in water. The latter is an approach that delivers variable results due to the difficulty in determining the correct water volume to achieve the desired result. The timing of application before harvest is very important with these products, and application closer to harvest must not occur as is clearly indicated on the label of such registered products.

The lack of consistent efficacy of these nematicides due to a variety of cultural and environmental factors is such that crop losses still occur. This particularly applies for the warmer production areas on sandy soils where regular application of irrigation both dilutes and activates the breakdown of the products, resulting in short persistency and the subsequent invasion by plant-parasitic nematodes. Recently, due to environmental and consumer pressures, concerns have been raised about the use and overuse of these remedies, and additional options are being investigated and becoming available. Furfural, a by-product of sugarcane, is one such product. It is a natural product present in many food crops and has a preharvest interval (PHI) of 14 days allowing application practically up to lifting. It is a short-acting contact product and needs to be used in conjunction with other products.

The fungus *Purpureocillium lilacinum* that parasitises nematode eggs represents an interesting biological-based product that acts directly against root-knot nematode eggs that are contained in egg masses. Other products listed in this chapter have recently been registered, e.g. those containing fluopyram and iprodione as a.s. More are likely to be registered in the near future. All such products bring new options for the grower as well as challenges of application, but adherence to label recommendations is essential if reliable results are to be obtained.

6.3 The Impact of the Vienna Convention and the Montréal Protocol on the Protection of the Ozone Layer

South Africa is a signatory of both the Vienna Convention and the Montréal Protocol on the regulation of ozone-depleting products. Within this protocol, the use of methyl bromide (a general soil biocide) was due for local phase-out in 2015, opening up the markets for alternative products. Prior to the phase-out process, the principle nematicide markets for methyl bromide, where broad-spectrum control was required, were:

(i) Apples for control of the specific replant problem
(ii) Flowers, strawberry, tobacco seedbeds and vegetables for the control of soil-borne fungal diseases, nematode pests and weeds

6.3.1 New Chemistry to Replace Methyl Bromide

The enforced reduction in the use of methyl bromide has meant that these markets have had to develop new control practices. In the United States of America (USA), the multipurpose fumigant carbon disulphide (originally discovered as a soil biocide in

1881), sodium tetrathiocarbonate (a generator of carbon disulphide in the soil), dimethyl disulphide and methyl iodide have been registered for use in some of the old methyl bromide nematicide markets. It is likely that some of these products are under development by several suppliers to the SA market and will ultimately be registered.

Locally, the general-purpose biocides metam sodium and metam potassium, with soil herbicidal, fungicidal and nematicidal activity, have gained significant use. 1,3-D plus chloropicrin, with nematicidal concentrations of 1,3-D and fungicidal concentrations of chloropicrin, was registered several years ago and is now entering commercial usage.

6.3.1.1 Methyl isothiocyanate (MITC) Generators

Aside from the use of these products to replace methyl bromide, several changes have occurred in the MITC generator market since 1982. These include:

(i) The registration and commercial use of metam sodium and metam potassium and the withdrawal of the granular product
(ii) On seed potato and vegetables, especially in the Western Cape Province, metam sodium and metam potassium control root-knot nematodes (*Meloidogyne*), seed-borne, pathogenic fungi, *Rhizoctonia* spp. The combination product of 1,3-D plus chloropicrin has a similar spectrum and may be increasingly used in this market

6.3.1.2 Change in the Use of the Specific Fumigants 1,3-D and Ethylene Dibromide (EDB)

The largest markets for these products in 1982 were on potato, pineapple and tobacco where EDB dominated. Since 1982, there has been (i) a marked reduction in the plantings of tobacco from 10,000 to 5,139 ha in 2013, (ii) a significant tightening of supply and increased input costs of EDB and (iii) a broadening of the registration holders of 1,3-D (Van Zyl 2013).

The increased costs of EDB has prompted users to switch to non-fumigants and 1,3-D. Initially, when this switch of products occurred, 1,3-D was positioned in the market as a direct replacement for EDB. However, the physical properties of the two products demand different use practices. For example, 1,3-D is more volatile and less water soluble than EDB. This is not a problem in cool areas, but it is a problem in the warmer areas of SA, particularly in summer. However, the loss of 1,3-D from the soil surface prior to its dissolving in the water phase requires either the use of plastic sheeting, application of 1,3-D in cooler seasons as a long-term rotational control strategy for that site or the use of alternative products. Thus, the application timing of 1,3-D in crop rotation sequences becomes an important part of the overall nematode control strategy.

6.3.1.3 Organophosphate (OP) and Carbamate Nematicides

The principal non-fumigant nematicides belong to the carbamate and OP groups of pesticides. These products have a similar mode of action (MOA). They entered the SA market from the early 1970s, with the most recent entry, fosthiazate, occurring in the late 1990s. They are essentially old products in the field of chemical pest control.

Various excellent reviews of their MOA, which is based on the inhibition of the enzyme acetyl cholinesterase, have been published (Keetch 1974; Wright 1981; Hague and Gowen 1987; Rich et al. 2004). All the OP and carbamate nematicides are highly active neurotoxins, with the MOA in nematodes being similar to vertebrates and arthropods. Their acute oral LD_{50} ranges from 0.6 (for aldicarb) to 57 mg kg^{-1} (for fosthiazate) (Table 6.4).

Table 6.4 Toxicity classification and regulatory status in the European Union (EU) and United States of America (USA) of nematicides registered in South Africa (SA) (Anonymous 2011)

Active substance	LD_{50} active mg kg^{-1}	Formulation	SA toxicity group	EU 91/414/ EEC; Status 4/1/9	USA, Reregistration Eligibility Decision (RED), Interim RED (IRED), Tolerance RED (TRED)
Dichloropropene (1:3)	130	Emulsifiable concentrate, undiluted liquid	1b	Not approved	Approved
Aldicarb	0.6	Granule	1b	Not approved	Approved
Cadusafos	30	Granule, emulsion	2, 1b	Not approved	TRED approved
Ethoprophos	33	Granule	2	Approved	Approved
Ethylene dibromide (EDB)	Not listed	Emulsifiable concentrate, undiluted liquid	2	Not approved	Cancelled
Fenamiphos	6	Granule, emulsifiable concentrate, capsule suspension	2, 1a	Approved	Cancelled
Fosthiazate	57	Granule, emulsifiable concentrate	2, 1b	Approved	Conditional registration
Metam Na/K	896	Soluble liquid	3	Pending	Approved
Oxamyl	2.5	Granule, soluble liquid	2, 1a	Approved	IRED

Unlike in the European Union (EU) and the USA, groundwater penetration of these products has not been identified as a problem in SA. Shallow groundwater is uncommon in SA, and where it occurs, e.g. the Hex River Valley (Western Cape Province), Viljoenskroon and Bothaville areas (North West Province) and the Vaalharts Irrigation Scheme (Northern Cape Province), no nematicides have been reported to occur in the groundwater. The high ambient temperatures of these locations and the resultant rapid degradation of the products probably limit the movement of the products to groundwater. Table 6.4 lists the regulatory status in the EU and the USA of the products registered for use in SA.

6.4 Accelerated Microbial Degradation

Soil-applied nematicides degrade typically by chemical degradation, but a wide range of soil microbes (bacteria and fungi) also actively degrade the products. With repeated use, soil microbe populations that degrade the products are selected leading to an increasingly rapid breakdown of the product and a resultant control failure. The latter scenario is referred to as accelerated microbial degradation (AMD) of which the main causes as well as measures to prevent or alleviate the process are illustrated in Box 6.4. High soil temperatures and pH levels as well as high organic matter content favour such microbial activity. Although the organic content of the majority of agricultural soils in SA is very low (<0.5 %) (Du Preez et al. 2011), AMD has been reported from certain vineyard and fruit-orchard soils (Le Roux et al. 2002; Hugo et al. 2014). Populations of these microbe complexes increase when multiple applications of nematicides at high rates are applied over a short period. Selection of microbe populations that degrade the pesticide occurs, and the residual activity of nematode control is reduced. When the population density of such a microbe complex is allowed to increase beyond a certain level, the microbial degradation takes place so fast that the residual control decreases to days or even a few hours. Thus, AMD leads to nematode control failures (Smelt et al. 1987; Smelt and Leistra 1992).

This occurrence of AMD is, however, not limited to non-fumigant nematicides but also takes place with soil-applied fungicides, insecticides, herbicides and soil fumigants (Ou 1998). Accelerated microbial degradation is most problematic in fields on which perennial crops, such as citrus and banana, are cultivated with two or more applications of nematicides each year and in fields and glasshouses used for intensive vegetable, flower and nursery production. In SA, Le Roux et al. (2002) reported that in citrus fields, fenamiphos developed AMD in both a clay and sandy soil where it was tested. Ethoprophos developed AMD in a clay soil more quickly under drip irrigation than microjet. Cadusafos only developed AMD in sandy soils under drip irrigation. For this reason, it is recommended that nematicides are not applied through drip irrigation systems.

Box. 6.4 An Overview of Accelerated Microbial Degradation Occurring for Some Synthetically Derived Nematicides

Hugo et al. (2014) characterised accelerated microbial degradation (AMD) of organophosphate and carbamate nematicides as a phenomenon whereby biodegradation in the soil is increased, leading to a dramatically shortened persistence of nematicides and resultant control failures. In South Africa, AMD has been identified as a problem on vines and orchards and on citrus.

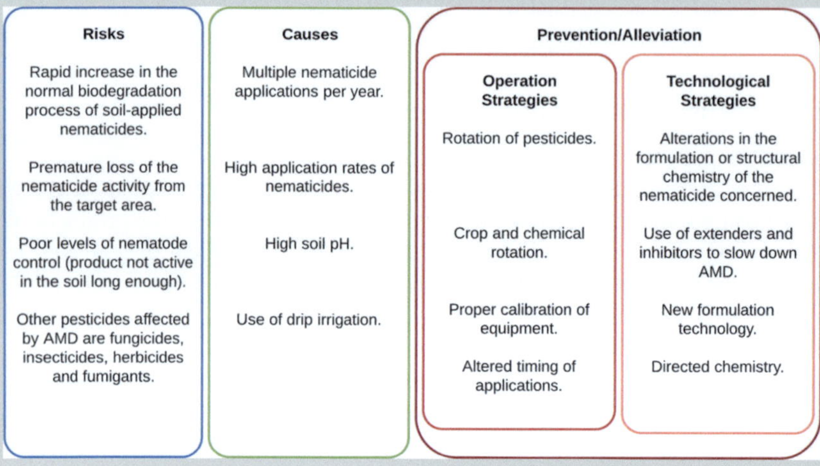

6.5 Furfural

Furfural is classified as an aldehyde product and is manufactured from the waste fibre of sugarcane. It is a contact nematicide that was identified in the mid-1990s as a possible replacement for methyl bromide. Since its initial launch, it has become a widely registered product in SA with 17 end usages. It is also registered in the USA and Europe. Though the product is water soluble and has a low octanol-water partition coefficient, the very short half-life precludes leaching and systemic activity. These properties, especially the <1-day half-life, allows the product great versatility in application timing and crop use situations. Thus it can be used following fumigation or OP and carbamate products, very close to harvest in order to protect late infections of nematode pests. The MOA is not fully explained, but it does react directly with the nematode cuticle, adversely affecting mobility and root penetration, thus immobilising and/or killing the infective second-stage juveniles (J2) of root-knot nematodes (Fourie et al. 2014).

6.6 Regulatory Issues

The established products discussed above are essentially all old chemistry. They belong to a period of chemical control that has largely been eclipsed in the fungicide, insecticide and herbicide fields. Typically chemical pest control has entered an era during which the old products are being re-evaluated, and risk criteria are far more stringent. Some trends are as follows:

(i) High quantities of a.s. per ha are no longer acceptable. Products are now applied at g a.s. per ha and not in kg as is the case for most of the nematicides
(ii) The environmental impact of newly introduced products in terms of 'chemical load' and the measurable biotic effects must be known and be within strict parameters
(iii) The toxicity of the product is no longer exclusively assessed based on the toxicity of the a.s. to rats but must be acceptable in relation to worker and environmental exposure and risks for the consumer

Most of the existing nematicides fail these criteria. The fumigants are typically applied at 100 kg ha^{-1} and the OPs and carbamates at up to 12 kg ha^{-1}. In many instances they pose high levels of risk to nontarget organisms. For example, in the USA carbofuran has been withdrawn due to bird fatalities. Also in the USA and the EU, groundwater has been contaminated with aldicarb and 1,3-D, respectively, in some areas, and its use was strictly controlled in these areas. Furthermore, the acute oral activity of their a.s. placed them in the most hazardous category of products. In response to these issues, the USA and the EU authorities have, over the last 10 years, largely reassessed the use of these products, and whilst the process is not complete, it is evident that many products have been or will be lost.

The regulatory environment within SA, unlike the USA and the EU, currently remains relatively unchanged since 1982. All products to control nematodes must still be registered with the National Department of Agriculture, Forestry and Fisheries (DAFF), in terms of Act 36 of 1947 (Van Zyl 2013). Aldicarb was withdrawn by the manufacturer in 2011 and methyl bromide in 2015. No regulatory controls have been introduced to mitigate environmental, worker, consumer or other concerns, except in the case of aldicarb, where an escalating level of misuse led to it being classified and labelled as a restricted-use pesticide (RUP) in 2002. This placed stringent controls on the supply chain to legal users of aldicarb, but did little to stem the levels of illegal misuse.

South Africa does not operate in isolation of global environmental pressures. Aside from the Vienna Convention, it is a signatory of:

(i) The Stockholm Convention on persistent organic pollutants (POPs), which recognises the need to take global action on all chemicals with POP-like characteristics, such as their:
    - Persistency in the environment
    - Ability to be distributed long distances via air and water
    - High levels of toxicity
    - Bioaccumulation in living tissues of animals and humans in particular
(ii) The Rotterdam Convention on listings of chemicals that fall within the prior informed consent (PIC) category, which limits trade of products

Currently none of the nematicides are listed as a POP or PIC, although aldicarb was proposed to be included on the PIC list of products. In 2006, the DAFF published a position paper (Anonymous 2006) in which it is stated that measures will be implemented to remove Group 1 products in line with international trends.

In March 2009, the DAFF called for companies to state their support for retaining registration of Group 1 products and a data call-in was enforced, which required the applicant to supply reasons for retention of the products. Similarly in December 2010, DAFF published Notice 1120 of 2010 titled 'Adoption of Pesticide Management Policy for SA' (Anonymous 2010) that further reinforced the principles outlined in the position paper. As yet no measures have been implemented locally. Nonetheless, export market forces are compelling reduced usages of all the registered nematicides used in SA. This way, they are effectively dictating the availability of our current registered nematicides, both in the export and increasingly in the local markets.

The toxicological profile and the US Environmental Protection Agency (EPA) and EU classification of the major nematicides in SA are listed in Table 6.4. According to this, it is obvious that many of the more widely used products will be lost if the measures as stated above are implemented. Clearly such a development would be to the detriment of nematode control and crop production in SA.

6.7 Misuse

Due to the inherent toxicity of many of the nematicides, a major local threat to the continued use of several products is that they are extensively misused. This resulted in aldicarb being classified as a RUP, but the stringent measures introduced have had no real impact on the problem. The position paper No. 28711 (Anonymous 2006) calls for the removal of products where misuse cannot be controlled, and several of the most widely used nematicides, notably carbofuran, are at additional risk of withdrawal.

6.8 Consumer-Driven Standards

Aside from the regulatory standards, consumer groups have introduced many 'secondary regulations', which have led to further constraints on the use of nematicides. The greatest impact in SA has come from Europe where consumer-market-driven organisations sets and harmonises standards and procedures that follow the requirements of good agricultural practices (GAP). Initially EurepGAP (see Box 6.5) and now GlobalG.A.P., set voluntary standards for the certification of agricultural and horticultural products around the world. The GlobalG.A.P. standard is primarily designed to reassure consumers about how food is produced on the farm by minimising detrimental environmental impacts of farming operations. This way the use of chemical inputs is reduced, and a responsible approach to worker health and safety, as well as animal welfare, is ensured (Globalgap 2009). The net effect of these consumer-driven standards is firstly to ensure strict compliance with local and export regulations and secondly to prohibit the use of specific practices that fall outside the standards set by that organisation. These standards are both tangible, where pesticide residue levels are stated, and intangible, when criteria deal with sustainability. Ill-defined organic or natural production standards are furthermore prescribed. Similar pressure groups are slowly developing within SA, and it is likely that many nematicides will be excluded for use on crops destined for the premium consumer outlets.

Box. 6.5 GlobalG.A.P.(GG)

GlobalG.A.P.(GG) is an international body that represents producer, retailer and consumer organisations that sets and harmonises standards and procedures that follow the requirements of good agricultural practices (GAP). It evolved from EurepGAP in 2007 and now functions in more than 100 countries. Specific roles of GlobalG.A.P. include:

(i) Producers: certification that demonstrates to retailers and consumers a commitment to advancing GAP
(ii) Retailers: an assurance that the producers belong to a network of reliable producers where produce is certified and traceable
(iii) Consumers: a means to verify by using a 13 digit GLOBALG.A.P. Number (GGN) that the product meets the standards for safe and sustainable food production

6.9 The Future for Conventional Nematicides

The future availability of nematicides in SA is largely dependent on the policies laid down by the local regulatory authorities and to an increasing extent on the market forces led by consumer groups. The environmental fate, the exposure during application and the residue load in the harvested commodities have become critical issues, even if the environmental and toxicological standards are in place. In the EU, a policy has been adopted whereby more stringent safety standards have led to the withdrawal of products (see Table 6.4) and very few options remain. The only general biocides that will remain are the metam sodium and potassium-based products. The specific nematicidal fumigants will be lost. Only three of the OPs and one of the carbamates will remain. This has resulted in a situation in which nematode problems exist for which there is no registered product available. This has already caused growers to switch to untested and undoubtedly less effective options. Alternatively, land-use practices will need to change with the introduction of longer rotations between growing susceptible crops. In the USA, the adoption of new standards has been aimed at the removal of non-essential registrations and specific risk factors of essential registrations. Thus, products that pose a risk to groundwater cannot be used in sensitive localities. Furthermore, products that have applicator safety concerns were required to improve the safety of the application process. This approach has resulted in many more products having been approved for continued use within the USA, albeit with additional use restrictions. The suppliers of nematicides have continued to screen for new products and re-evaluate existing products. Several new candidates, in varying stages of development, were reported on at the 6th International Congress of Nematology (6ICN) that was held in Cape Town, SA, during 2014 (see Box 6.6).

Companies have also developed 'safer' formulations of their existing products, e.g. encapsulated and other formulations are replacing the old emulsion products. Additives have been incorporated into formulations, e.g. aversion and emetic agents were incorporated in aldicarb. Old but less toxic chemistry has also been rescreened, as was the case of abamectin and iprodione. The latter has been commercialised as a nematicide in Turkey and the USA and is registered for use on potato in SA. In addition, the use of seeds as a carrier of a nematicide is being widely researched. Abamectin applied as a packaged seed treatment in combination with a fungicide and insecticides is registered for use on maize in SA, the USA and Brazil. Yield increases averaged 5 % over 12 trials when compared to standard seed treatments (Slaats et al. 2014). The bacterium *Bacillus firmus*, applied as a seed treatment with other pesticides, is also reported to reduce damage by plant-parasitic nematodes infecting maize (Riggs et al. 2014). Alongside these developments, there has been an extensive commercialisation of organic products that claim to control nematodes in a natural way. These products are based on animal waste, plant extracts and exudates, seaweed-based products or nematode pathogens, which typically either amend the soil environment, break down to active products or directly target nematodes. Many of these products are unregistered and untested, but they can be readily identified and obtained over the Internet and from owner companies. Such products are unlikely to be able to duplicate the activity of the traditional nematicides but will find a very receptive market.

Box. 6.6 New Products with Nematicidal Activity

At the 6th International Congress of Nematology (6ICN) that was held in Cape Town, South Africa, in 2014, several papers were presented on new products with nematicidal properties. Most of these new chemicals exhibit novel modes of action, have low mammalian and environmental toxicity and are applied at much lower dose rates than the established products. One of these products contains fluopyram as the a.s. and is currently registered in SA on potato and tobacco. It is new chemistry and has a novel mode of action, acting as a mitochondrial respiratory chain inhibitor (Broeksma 2014). Fluensulfone, also new chemistry, is reported to have irreversible nematicidal activity against *Meloidogyne* spp. (Karmon 2014; Kearn et al. 2014). This product appears not to act as a broad-range nematicide with limited activity on migratory nematodes, notably *Aphelenchoides* and *Ditylenchus* spp. (Oka 2014). Hafez and Pudasaini (2014) reported on the use of spirotetramat and fluensulfone in controlling *Meloidogyne chitwoodi* Golden, O'Banon, Santo & Finley, 1980 on potato, sugar beet and onion using multiple applications and combinations of the products. Spirotetramat was also reported to reduce *Rotylenchulus reniformis* Linford and Oliveira, 1940, population levels in pineapple (Sipes 2014) and that of *Xiphinema* spp. in grapevine by 70 % 18 days after treatment (McKenry et al. 2009). This product acts as a lipid metabolism inhibitor and has been registered in the USA.

It is evident that the future for the continued use of many of the conventional products is in jeopardy and that a gap will develop whereby nematode problems will not adequately be controllable with the remaining products. Already the chemical industry, researchers and the grower commodity organisations have recognised this problem and have adopted research projects that test new options. The old pillars of nematode control, crop rotation and host-plant resistance are likely to receive far greater emphasis than has occurred in the last 20–30 years. Additional alternative control options (Box 6.7), such as using soil amendments, will be refined with the objective of creating a far more diverse fauna and flora in the soil. This will act to reduce the potential for plant-parasitic nematodes to exploit the vacuum of interactions that currently exist in our heavily utilised and degraded soils. Fortunately, scientists are still involved in applied nematology, both in SA and other countries, and are busy trying to overcome the common problems that exist in nematode control. In this regard this new set of challenges will hopefully lead to a resurgence in research funding.

This chapter is concluded with a list of terms (Box 6.7) related to nematode management and will be referred to in the chapters that follow.

Box. 6.7 Nematode Management Practices

This glossary defines the key components of nematode management strategies, which have the objective of maximising the efficiency of chemical, agronomic and biological inputs to control nematode pests over an extended time period.

Biological control, natural enemies: The use of nematode pathogens (protozoa, bacteria, fungi, etc.) to infect and control nematodes. A number of these products are now commercially available in South Africa (SA).

Biological control, organic amendments: Such treatments involve the addition of waste materials and green manures into the soil. These have been used for a long time as effective nematode control strategies, e.g. animal manures (from cows, chickens, etc.), plant parts (aerial parts and roots of Brassicaceae, etc.), oil cakes (from soybean, etc.) and compost (plant waste such as peels, leaves, fruits, etc.).

Biological control, allelopathic plants: The use of a rotation crop that leaches or releases nematicidal compounds (e.g. *Tagetes* spp.) during decomposition and in this way aids in reducing nematode pest population densities.

Chemical control: The use of synthetic products with known nematicidal properties to control nematode pests. This practice is regulated in SA under Act 36 of 1949.

Fallow: Leaving a field after cultivation unsown resulting in a reduction in nematode pest population densities. This practice is not widely used during the summer growing seasons in SA due to the high value of commercial farmland. However, due to limited water resources, many fields are left fallow in winter. This does not usually represent 'clean fallow', and the presence of weeds in such fields is a problem since a wide range of these plants acts as hosts for plant-parasitic nematodes and in this way supports their population levels.

Flooding: Despite the fact that nematodes are aquatic organisms, where water is abundant, the flooding of soil for an extended period (several months) can lead to the control of nematode pests. As with fallow, this practice is not suited for conditions because of the shortage of water and the value of farmland.

Genetic host-plant resistance: The planting of a crop variety that is a poor or non-host of the target nematode pest. To be an acceptable variety, the growth and yield properties have to be comparable to those varieties typically grown.

Heat, hot water treatment: The immersion of plant material in hot water for a specific time-temperature period, e.g. 55 °C for 20 min. The time-temperature window needs to be refined for each crop situation to prevent temperature damage to the crop.

Heat sterilisation and solarisation: Heating soil by infusing steam to raise temperatures to 80 °C for 1 h is a very effective method of sterilising soil. It is particularly used in the nursery industry where it is important to control a broad spectrum of fungal, nematode and weed pests. For solarisation a transparent plastic sheet is placed over the soil to trap heat from the sun and raise the soil temperature to a level that is lethal for the nematodes in the soil. Typically, the soil needs to be covered with a transparent plastic cover for several days during times of the year when the sunshine and temperatures reach their highest levels.

Quarantine/sanitation: The exclusion of pest and disease organisms and preventing any infection from establishing (see Chap. 5).

Rotation: The practice whereby non-hosts or poor host crops are grown in sequence with a susceptible (good host) crop for a long enough period to allow the nematode pest population densities to decline below the damage threshold level before planting the susceptible crop again.

Tillage: The cultivation of soil to physically kill nematode pests when combined with the drying out of soil is particularly effective in reducing nematode population densities. This practice is declining in SA due to the dangers of soil erosion and the growing popularity of conservation agriculture.

Trap cropping: Planting a good host crop to attract a pest nematode followed by destroying this crop before the nematodes have had the chance to reach the egg-laying stage. Also not a widely used practice is SA due to the high value of farmland and lack of water.

References

Anonymous (2006) Draft pesticide management policy for South Africa. Government Gazette No. 28711, Notice 499 of 2006. Department of Agriculture, Pretoria

Anonymous (2010) Adoption of pesticide management policy for South Africa. Government Gazette No 33899, notice 1120. Department of Agriculture, Pretoria

Anonymous (2011) Functional tools for pesticide risk assessment and management in Europe. University of Hertfordshire Agriculture and Environment Research Unit. Footprint portal. http://sitem.herts.ac.uk/aeru/projects/ppdb/index.htm. Accessed 21 Nov 2011

Broeksma A (2014) Velum® – an evolutionary nematicide for efficient crop protection. J Nematol 46:140–141

Bromilow RH (1988) Physico-chemical properties and pesticide placement. In: Taylor T (ed) Application to seeds and soil, British Crop Protection Monograph No. 39, Martin T (ed), held at Univ. of Surrey, pp 295–308

Bunt JA (1987) Mode of action of nematicides. In: Veech JA, Dickson DW (eds) Vistas on nematology: a commemoration of the twenty-fifth anniversary of the Society of Nematologists. Society of Nematologists, Hyattsville, pp 461–468

Du Preez CC, Van Huyssteen CW, Mnkeni PNS (2011) Land use and soil organic matter in South Africa 1: a review on spatial variability and the influence of rangeland stock production. S Afr J Sci 107. doi:10.4102/sajs.v107i5/6.354

Fourie H, van Aardt WJ, Venter C et al (2014) Investigating the effects of Cropguard® on the motility, ultrastructure and respiration of two *Meloidogyne* species. Nematropica 44:86–92

Globalgap (2009) http://www.globalgap.org/cms/front_content. Accessed 12 Dec 2009

Haydock PF, Woods SR, Grove IG et al (2013) Chemical control of nematodes. In: Perry RN, Moens M (eds) Plant nematology, 2nd edn. CAB International, Wallingford, pp 459–480

Hafez SL, Pudasaini MP (2014) New chemistries, modes of action, and different formulations for nematode management in Idaho. J Nematol 46:171

Hague NGM, Gowen SR (1987) Chemical control of nematodes. In: Brown RH, Kerry BR (eds) Principles and practice of nematode control in crops. Academic Press, Sydney, pp 131–178

Hough A, Thomason IJ, Farmer WJ (1975) Behaviour of aldicarb in soil relative to control of *Heterodera schachtii*. J Nematol 7:214–221

Hugo HJ, Mouton C, Malan AP (2014) Accelerated microbial degradation of nematicides in vineyard and orchard soils. S Afr J Enol Vitic 35:157–167

Jones JT, Haegeman A, Danchin EGJ et al (2013) Top 10 plant-parasitic nematodes in molecular plant pathology. Mol Plant Pathol 14:946–961

Karmon D (2014) NIMITZ™: a paradigm change in nematode control. J Nematol 46:186–187

Kearn J, Ludlow E, O'Connor V et al (2014) Mode of action studies on fluensulfone, a new nematicide. J Nematol 46:186

Keetch DP (1974) The effect of nematicides on feeding, posture and dispersal of *Aphelenchus avenae*. Nematologica 20:107–118

Keetch DP (1982) The use of nematicides. In: Keetch DP, Heyns J (eds) Nematology in Southern Africa, Science Bulletin No. 400. Department of Agriculture and Fisheries, Pretoria, pp 130–144

Keetch DP (1989) A perspective of plant nematology in South Africa. S Afr J Sci 85:506–509

Keetch DP, Milne DL (1982) The control of plant-parasitic nematodes. In: Keetch DP, Heyns J (eds) Nematology in Southern Africa, Science Bulletin No. 400. Department of Agriculture and Fisheries, Pretoria, pp 113–129

Khan MR (2008) Plant nematodes: methodology, systematics, biology and ecology. Science Publishers, Enfield, New Hampshire

Le Roux HF, Pretorius MC, Huisman L (2002) A comparison of accelerated degradation rates of nematicides applied through drip and micro irrigation. Afr Plant Prot 8:6–7

McKenry M, Kaku S, Buzo T (2009) Spirotetramat foliar, a new nematicide to replace drip-applied fenamiphos. http://www.uckac.edu/files/90841.pdf. Accessed 12 Dec 2009

Nyczepir AP, Thomas SH (2009) Current and future management strategies in intensive crop production systems. In: Perry RN, Moens M, Starr JL (eds) Root-knot nematodes. CAB International, Wallingford, pp 412–443

Oka Y (2014) The new nematicide Nimitz™ (fluensulfone): its advantages and limitations. J Nematol 46:214

Ou LT (1998) Enhanced degradation of the volatile fumigant-nematicides 1,3-D and methyl bromide. J Nematol 30:56–64

Rich JR, Dunn RA, Noling JW (2004) Nematicides: past and present uses. In: Chen ZX, Chen SY, Dickson DW (eds) Nematology – advances and perspectives, Volume II, nematode management and utilization. CAB International, Wallingford, pp 1179–1200

Riggs J, Andersch W, Plischkel U, Bugg K (2014) VOTiVO®: innovation driven, nematode protection and plant health benefits. J Nematol 46:227

Sipes B (2014) The effect of spirotetramat on *Rotylenchulus reniformis* infecting pineapples. J Nematol 46:236

Slaats B, Dreyer J, Watrin C (2014) Avicta® Complete Corn – a preventative solution against the hidden pest. J Nematol 46:230

Smelt JH, Leistra M (1992) Availability, movement and accelerated transformation of soil-applied nematicides. In: Gommers FJ, Maas PW (eds) Nematology from molecules to ecosystems. Dekker and Huisman, Wildervank, pp 281–293

Smelt JH, Crum SJH, Teunisson W et al (1987) Accelerated transformation of aldicarb, oxamyl and ethoprophos after repeated soil treatments. Crop Prot 6:295–303

Sikora RA, Bridge J, Starr JL (2005) Management practices: an overview of integrated nematode management technologies. In: Luc M, Sikora RA, Bridge J (eds) Plant parasitic nematodes in subtropical and tropical agriculture, 2nd edn. CAB International, Wallingford, pp 793–825

Van Zyl K (2013) A guide to crop pest management in South Africa. A compendium of acaracides, insecticides, nematicides, molluscicides, avicides and rodenticides. A CropLife South African Compendium, 1st edn. VR Print, Pinetown

Whitehead AG (1998) Plant nematode control. CAB International, Wallingford

Wright DJ (1981) Nematicides: mode of action and new approaches to chemical control. In: Zuckerman BM, Rhode RA (eds) Plant parasitic nematodes, 3rd edn. Academic Press, New York, pp 421–449

Chapter 7
Alternative Nematode Management Strategies

Phatu W. Mashela, Dirk De Waele, Zakheleni Dube, Mbokota C. Khosa, Kgabo M. Pofu, Grace Tefu, Mieke S. Daneel, and Hendrika Fourie

7.1 Introduction

South Africa (SA) has approximately nearly 56 million inhabitants of which 48 % live in rural areas (Anonymous 2016). A large portion (35 %) of this rural population lives below the poverty line (Labadarios et al. 2012). For their food, most of these communities depend on vegetables as well as grain and leguminous crops produced mainly in household or communal gardens (Fig. 7.1) (Fourie and Schoeman 1999; Van der Berg 2006; Aliber 2009; Coyne et al. 2009; Fourie et al. 2012; Ntidi et al. 2012). Available land is often limited and, therefore, frequently reused, which aggravates soil degradation as well as soil disease and pest problems (Van der Berg 2006; Aliber 2009). All these factors have a direct and negative effect on food security and cash income (Aliber 2009; Ntidi et al. 2012).

P.W. Mashela (✉)
University of Limpopo, Private Bag X1106, Sovenga 0720, South Africa
e-mail: phatu.mashela@ul.ac.za

D. De Waele
Laboratory of Tropical Crop Improvement, Department of Biosystems, University of Leuven, Willem de Croylaan 42, 3001 Heverlee, Belgium

Z. Dube
University of Limpopo, Private Bag X1106, Sovenga 0720, South Africa

M.C. Khosa • G. Tefu • M.S. Daneel
Agricultural Research Council–Institute for Tropical and Subtropical Crops, Private Bag X11208, Nelspruit 1200, South Africa

K.M. Pofu
Agricultural Research Council–Vegetable and Ornamental Plants Institute, Private Bag X134, Queenswood 0121, South Africa

H. Fourie
North-West University, Unit for Environmental Sciences and Management, Private Bag X6001, Potchefstroom 2520, South Africa

H. Fourie et al. (eds.), *Nematology in South Africa: A View from the 21st Century*, DOI 10.1007/978-3-319-44210-5_7

Fig. 7.1 A typical setup of intercropping by smallholding farmers in South Africa, with most of the crops being good hosts to plant-parasitic nematodes (Johnnie van den Berg, North-West University, Potchefstroom, South Africa)

In a commercial agricultural production, more than 10 % of crop yields can be lost due to diseases and pests, including plant-parasitic nematodes (Keetch 1989; Kleynhans et al. 1996; Sikora et al. 2005). However, in rural areas these percentages are much higher, and, in some circumstances, diseases and pests can cause total crop failures (Kleynhans et al. 1996; Sikora et al. 2005; Coyne et al. 2009).

During the past decade, substantial crop losses as a result of nematode infection were frequently reported from the local smallholding farming sector. A preliminary survey was conducted in 1999–2001 in collaboration with provincial governments, selected communities and non-governmental organisations (Mtshali et al. 2002a). The survey revealed that in 49 out of 51 rural and peri-urban home, community and school gardens in the Eastern Cape, North-West, Limpopo, Mpumalanga and KwaZulu-Natal provinces, root-knot nematodes (*Meloidogyne* spp.) were the major biotic constraint for food production. This was especially the case in gardens where vegetables were the primary food crop grown.

Numerous surveys have shown that many crops, as well as weeds and non-food crops occurring in and around crop fields, grown by rural and peri-urban communities are highly susceptible to nematode pests and can maintain exceptionally high population densities of particularly *Meloidogyne* spp. (Fourie and Schoeman 1999; Mtshali et al. 2002a; Tefu et al. 2005; Marais and Swart 2007; Fourie et al. 2012; Ntidi et al. 2012). In many cases, especially in locations where root-knot nematodes occurred in high population densities, the production of vegetables had to be abandoned despite acceptable cultivation practices by the farmers (Mtshali et al. 2002a; Ntidi et al. 2012). In rural areas, communities and households depend on these crops for their food security, and the adverse impact caused by plant-parasitic nematodes on the livelihood of the rural population is, therefore, considerable (Fourie and Schoeman 1999; Mtshali et al. 2002a; Coyne et al. 2009; Ntidi et al. 2012).

Fig. 7.2 In South Africa, as in most African countries, the awareness of smallholding farmers of the presence and damage caused by plant-parasitic nematodes is either totally lacking or very limited (Originally drawn by Koos van Rensburg, Agricultural Research Council–Grain Crops Institute, Potchefstroom, South Africa; Redrawn by Ebrahim Shokoohi, North-West University, Potchefstroom, South Africa)

The nematode problem is further aggravated by the lack of knowledge among smallholding farmers of diseases and pests, including plant-parasitic nematodes (Fig. 7.2). Typically, farmers believe that the damage caused by plant-parasitic nematodes are symptoms of nutrient deficiency, drought or other abiotic harmful factors (Chitwood 2002; Sikora et al. 2005; McSorley 2011; Ntidi et al. 2012). Also, extension officers advising smallholding farmers often have a limited knowledge of nematode pests. Therefore, effective nematode management is impossible without raising the awareness of smallholding farmers and increasing the knowledge of extension officers of nematode problems (Fourie and Schoeman 1999; Mtshali et al. 2002a; Fourie et al. 2012; Ntidi et al. 2012). To this end, already in 2000, a leaflet was produced by the Agricultural Research Council–Grain Crops Institute (ARC–GCI) and disseminated in rural areas (Fourie and Mc Donald 2000).

The use of nematicides to limit the build-up of nematode pest population densities below a damage threshold is, in general, effective but unsuitable for the smallholding farming sector as these chemicals are expensive, usually toxic and hence hazardous to humans, livestock and the environment. In addition, farmers need expensive equipment and protective clothing to apply such chemicals, and a prerequisite is that a secure location is available to store these toxic substances. Therefore, to alleviate the nematode problem, alternative low-input, cost-effective and environmentally friendly management strategies need to be developed and made available. Only this approach will enable smallholding farmers to regain and maintain acceptable levels of food production and to generate some cash income (Chitwood 2002; Sikora et al. 2005; McSorley 2011).

In SA, research on these alternative nematode management strategies for the smallholding farming sector has focused on:

(i) The discovery of local botanical nematicides (phytonematicides) and their use as soil amendments to manage root-knot nematodes
(ii) The use of natural sources of resistance (the ability of a plant to limit nematode reproduction) or tolerance (the ability of a plant to limit nematode damage)
(iii) The application of crop rotation and intercropping
(iv) The use of organic amendments
(v) The use of cover crops as biofumigants

7.2 Agricultural Research in South Africa and Government Policies

In SA, agricultural research is strongly influenced by government policies. Agriculture in SA is dualistic in nature since it includes, at the same time, a strong established commercial sector and a large developing sector. This situation is still prominent despite initiatives to uplift the smallholding, developing sector since the implementation of the new democratic dispensation in 1994. Unfortunately, in the beginning, a top-down approach was applied resulting in non-sustainable projects that typically failed. Recently, however, a bottom-up approach is being followed. Today, the policies and strategies of the ARC are guided by the national priorities referred to in Box 7.1.

Box. 7.1 National Priorities of the Agricultural Research Council (ARC)
The ARC is the convenor of a task force on rural development, which is one of the seven Presidential Imperative Programmes. The other six imperatives are job creation, regional integration, urban renewal, human resource development, human immunodeficiency virus - acquired immunodeficiency syndrome (HIV/AIDS) and crime prevention.

The national minister of agriculture together with nine members of the executive committee (MECs), representing the nine provincial departments of agriculture, developed seven priorities for public entities in the agricultural sector to deliver on, viz.:

(i) Farmer settlement
(ii) Promotion of sustainable agricultural production of food gardens and opening up opportunities for industrial crop production
(iii) Supporting agribusiness development aimed at job creation, black economic empowerment and expanding income opportunities by focusing on value-adding and responsible exploitation of indigenous resources

(iv) Improving the availability of services to support diverse types of farming systems
(v) Restructuring irrigation schemes in ex-homelands
(vi) Nurturing of human and natural resources through capacity building and training and implementation of sustainable farming practices
(vii) Rural development, alleviation of poverty and job creation

The delivery of research priorities of the Department of Agriculture, Forestry and Fisheries (DAFF) must also be in accordance with the policy directives in the White Paper on Agriculture (1995) which are summarised as follows:

- Promoting competitiveness and employment creation
- Enhancing the quality of life
- Developing human resources
- Working towards environmental sustainability
- Promoting an informed society

According to the White Paper on Agriculture (1995), a major constraint in the SA agricultural structure was the unequal distribution of income between commercial and smallholding farmers. Therefore, not only reallocation of land to smallholding farmers needed to be considered and implemented but also appropriate and relevant research methodologies needed to be developed to ensure optimal food production and food security on both national and household/community levels. The White Paper on Agriculture (1995) further listed that one of the main technical problems experienced by the smallholding agricultural sector was a lack of awareness of diseases and pests and, consequently, an inability to manage these production constraints adequately. According to the document, this problem was related to a lack of knowledge and understanding of crop management in general.

Also, according to the White Paper on Agriculture (1995), imperatives applicable to technology development in the smallholding agricultural sector should ensure research to be need driven. Furthermore, a so-called farming systems approach (in which all components of farming, land, labour, capital, etc. are integrated) should be used in collaboration with the extension services and farmers, and research capacity should be expanded to include applied on-farm research.

7.3 Plant-Parasitic Nematodes Associated with Crops Grown by Smallholding Farmers in South Africa

The same nematode pests that are of economic importance in commercial agriculture (see Chaps. 8–19) are also present in the fields of smallholding farmers, resulting in yield and quality losses. The most abundant and damaging genus is *Meloidogyne* (Mtshali et al. 2002a; Daneel et al. 2003; Ntidi et al. 2012).

During a survey in the early 2000s (see Sect. 7.1) in the Mpumalanga, KwaZulu-Natal and North-West provinces, the highest root-knot nematode population density (about 1 million 50 g^{-1} roots) was recorded from potato (*Solanum tuberosum*) at Tshetshe near Ventersdorp (Mtshali et al. 2002a). The lowest root-knot nematode population density recorded (from turnip – *Brassica rapa* subsp. *rapa* roots at Bethal near Coligny) was 9,350 g^{-1} roots, which is still high. Identification of the root-knot nematode species found during this survey revealed that *Meloidogyne javanica* (Treub, 1885) Chitwood, 1949, was predominant, followed by *Meloidogyne incognita* (Kofoid and White, 1919) Chitwood, 1949. Other root-knot nematode species identified from smallholding farming areas were *Meloidogyne arenaria* (Neal, 1889) Cobb, 1890; *Meloidogyne chitwoodi* Golden, O'Bannon, Santo and Finley, 1980; *Meloidogyne fallax* Karssen, 1996; and *Meloidogyne hapla* Chitwood, 1949 (Coyne et al. 2009; Ntidi et al. 2012).

Together with *Meloidogyne*, many other plant-parasitic nematode genera identified from smallholding fields and gardens, as well as adjacent areas, were also reported (Tefu et al. 2005; Marais and Swart 2007; Ntidi et al. 2012). Surveys in the Bizana, Lusikisiki and Port St Johns area (Eastern Cape Province), for example, recorded the presence of 11 plant-parasitic nematode families, represented by 27 genera and 105 species (Marais and Swart 2007) on food and non-food crops, and natural vegetation. During these surveys the crops most sampled were maize (*Zea mays*), bean (*Phaseolus* spp.) and soybean (*Glycine max*) (38, 36 and 30 fields, respectively). Various vegetable and fruit crops were also sampled. The genera *Helicotylenchus*, *Meloidogyne*, *Scutellonema* and *Xiphinema* were present in more than 30 % of the localities sampled, whereas *Criconema*, *Criconemoides*, *Discocriconemella*, *Dolichodorus*, *Hemicycliophora*, *Longidorus*, *Nanidorus*, *Ogma*, *Paralongidorus*, *Paratrichodorus*, *Paratylenchus*, *Pratylenchus*, *Rotylenchulus*, *Rotylenchus*, *Trophotylenchulus* and *Tylenchorhynchus* were present to a lesser extent. The genera *Aphelenchoides*, *Ditylenchus*, *Hemicriconemoides*, *Hoplolaimus*, *Radopholus*, *Trichodorus* and *Tylenchulus* were each identified from a single locality only. *Discocriconemella degrissei* Loof and Sharma, 1980, was reported as a first record for SA, while various nematode-plant associations were also reported for the first time. For example, for food crops only, associations of *Helicotylenchus dihystera* (Cobb, 1893) Sher, 1961, with butternut squash (*Cucurbita moschata*), chickpea (*Cicer arietinum*), lablab (*Dolichos* sp.), mung bean (*Vigna radiata*), pigeon pea (*Cajanus cajan*), sweet potato (*Ipomoea batatas*) and taro (*Colocasia esculenta*) represented first records for SA. The same applied to *Helicotylenchus martini* Sher, 1966, on maize; *Helicotylenchus multicinctus* (Cobb, 1893) Golden, 1956, on bean; *Helicotylenchus paraplatyurus* Siddiqi, 1972, on soybean and sweet potato; and *Helicotylenchus serenus* Siddiqi, 1963, on butternut squash.

7.3.1 Root-Knot Nematodes

Vegetables, including beetroot (*Beta vulgaris*), brinjal (*Solanum melongena*), peppers (*Capsicum* spp.), spinach (*Spinacia oleracea*) and tomato (*Solanum lycopersicon*), host a wide variety of plant-parasitic nematode species, but predominantly

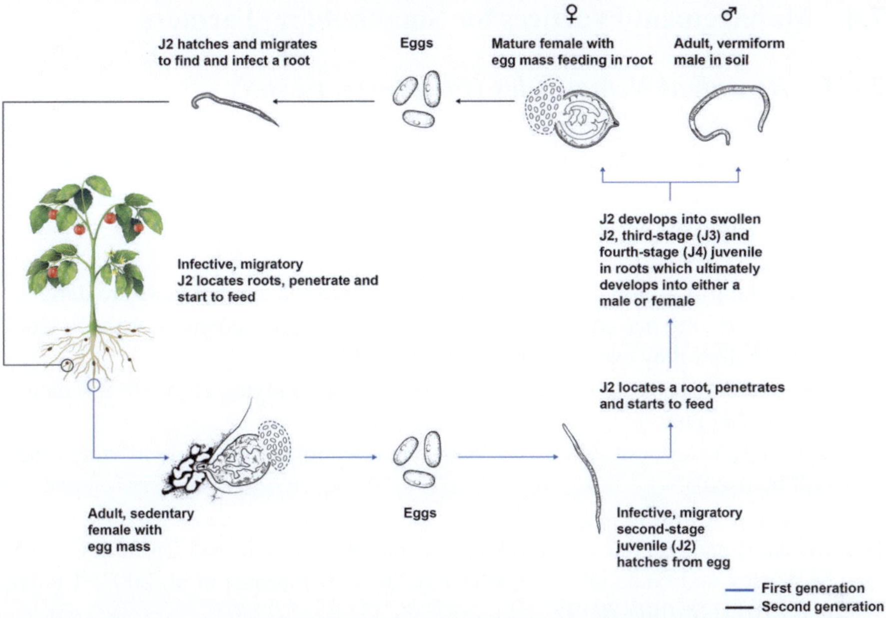

Fig. 7.3 Life cycle of *Meloidogyne* (Hannes Visagie, North-West University, South Africa)

Meloidogyne spp. (see Chap. 10) (Mtshali et al. 2002a; Sikora and Fernandez 2005; Bridge and Starr 2007; Ntidi et al. 2012). Severe infection of good host crops by root-knot nematodes usually results in significant yield and/or quality loss and in various cases even in total crop failure (Mtshali et al. 2002a; Tefu et al. 2005). Worldwide in tropical and subtropical regions, *M. incognita* and *M. javanica* are the predominant root-knot nematode species on tomato (Nono-Womdim et al. 2002). Both species infect vegetable crops wherever they are grown which may result in substantial yield losses when proper nematode management strategies are not applied (Sikora and Fernandez 2005; Bridge and Starr 2007; McSorley 2011). Yield losses of 20–40 % (Bridge and Starr 2007) and in excess of 50 % (Nono-Womdim et al. 2002) have been reported on tomato because of infection by *Meloidogyne* spp. The ability of root-knot nematodes to produce high numbers of offspring, particularly in warmer areas where they go through several life cycles (Fig. 7.3) in one season, is a major contributing factor to the high crop losses caused by them.

Many smallholding farmers either buy commercial seed or use second- and even third-generation seed for planting vegetables in their home or communal gardens in which nematode pests may thrive. Most of the commercial crop varieties used by these farmers are susceptible to root-knot nematodes (Fourie et al. 2012). Roots of seedlings often become already infected in so-called home nurseries before they are transplanted in the field. The impact of root-knot nematode infection on tomato (and other crops) depends upon many factors, including abiotic and biotic factors, agronomic practices, etc., and therefore can vary considerably among localities (Nono-Womdim et al. 2002; Bridge and Starr 2007; Coyne et al. 2009).

7.4 Management Practices for Smallholding Farmers

7.4.1 Botanical Nematicides (Phytonematicides)

Research on phytonematicides was initiated inter alia to mitigate the drawbacks of conventional organic amendments in suppressing plant-parasitic nematode population densities (Mashela 2002). These drawbacks include:

(i) Unavailability of the basic materials from which the amendments are derived
(ii) The rather long period needed for the microbial decomposition of the amendments before they can be applied on the field
(iii) The need for large quantities of the amendments (ranging from 10-500 metric tonnes (MT) ha^{-1})
(iv) The high transportation costs to haul the amendments to the field where they will be used
(v) Excessive lowering of soil pH
(vi) Inconsistency of the results (Jafee et al. 1994; Bélair and Tremblay 1995; McSorley and Gallaher 1995; Mashela 2002; Kimpinski et al. 2003; Thoden et al. 2011; Stirling 2014)

The basic materials for the development of phytonematicides were usually collected from locally available plants (Muller and Gooch 1982; Akhtar and Malik 2000; Oka 2010; Mashela et al. 2011; Ahmad et al. 2013). The phytochemical compounds of these phytonematicides can be broadly classified as alkaloids, alkamides, carbohydrates, cyanogenic glycosides, fatty acids, glucosinolates, nonprotein amino acids, phenolic compounds (coumarins, flavonoids, phenylpropanoids, tannins), polyacetylenes, polyketides, terpenoids, thiophenes and waxes (Zasada and Ferris 2003; Wuyts et al. 2006; Oka 2010; Okwute 2012; Wink and Van Wyk 2014). Van Wyk et al. (2002) listed 390 plant species endemic to SA that are potentially toxic to humans and animals. For the purpose of this chapter, we can classify these potential toxins in seven groups based on their toxic effects (Table 7.1).

Table 7.1 Plants with potential toxicities for uses in various industries in South Africa

Toxic effect	Number of plant species	percentage of total number of plant species
Deadly to humans and animals	25	6
Very poisonous to humans and animals	68	17
Poisonous to humans alone	84	22
Poisonous to animals alone	163	42
Poisonous to humans and animals	18	5
Skin allergies or contact dermatitis	18	5
Not poisonous to human and animals	14	4
Total	390	

Adapted from Van Wyk et al. (2002)

Wild cucumber (*Cucumis myriocarpus*) and castor bean (*Ricinus communis*), which produce phytonematicides that can suppress nematode population densities using the ground leaching technology (GLT) system (Mashela 2002; Mashela and Nthangeni 2002), are classified as being poisonous and very poisonous, respectively (Van Wyk et al. 2002). By contrast, crude extracts from oleander (*Nerium oleander*) leaves and tamboti (*Spirostachys africana*) bark did not have nematicidal properties against *Meloidogyne* spp. or the citrus nematode (*Tylenchulus semipenetrans* Cobb, 1913) (Mashela et al. 2011). The bark, twigs and leaves of oleander and tamboti are deadly to both humans and animals (Van Wyk et al. 2002). On the other hand, certain plants listed as not poisonous by Van Wyk et al. (2002), for example, fever tea (*Lippia javanica*) and *Brassica* spp., produce potent nematicides against *Meloidogyne* spp. (Mashela et al. 2007, 2012). The degree of toxicity to humans and animals, therefore, is not an indication that a plant or a plant organ also contains nematicidal substances.

7.4.1.1 Chemical Properties of Cucurbitacin A and B

Cucurbitacins are oxygenated tetracyclic triterpenoids and are the bitterest chemical compounds known to humankind (Chen et al. 2005). Cucurbitacin A ($C_{32}H_{46}O_8$) is unstable and disintegrates rapidly to cucumin ($C_{27}H_{40}O_9$) and leptodermin ($C_{27}H_{38}O_8$) which are both stable compounds (Jeffrey 1978). Cucurbitacin A is concentrated in the roots and fruits of *Cucumis myriocarpus* (Jeffrey 1978), with leaves being used as a vegetable food source (Mashela et al. 2011; Mashela and Dube 2014). By contrast, cucurbitacin B ($C_{32}H_{48}O_8$) is concentrated in all organs of *Cucumis africanus* and is stable (Jeffrey 1978). The chemical is not equally concentrated in all organs, and significantly higher concentrations occur in the fruit (Shadung and Mashela 2016). Cucurbitacins are widely used in African traditional medicine (Mphahlele et al. 2012). Pure cucurbitacins can be highly toxic to animals (Lee et al. 2010).

The efficacy of phytonematicides derived from *Cucumis* spp. in suppressing nematode pest population densities was similar to the efficacy of aldicarb and fenamiphos, which are systemic chemical nematicides (Mashela et al. 2008). *Meloidogyne incognita* eggs exposed to 0–2.5 $\mu g\ ml^{-1}$ cucurbitacin A and B dilutions in distilled water exhibited negative curvilinear quadratic relations (Dube et al. 2016). In other words, at low concentrations the cucurbitacins inhibited the hatching of second-stage juveniles (J2), whereas at high concentrations J2 hatching was stimulated. These results confirmed observations by Yu (2015) who reported that *M. incognita* eggs exposed to increasing crude extracts of mustard (*Sinapis arvensis*) levels stimulated the hatching of J2. Hatching of J2 of root-knot nematodes at various concentrations of phytochemicals can be explained on the basis of the extent to which J2 pick up chemical cues and their 'interpretation' of these cues. At low concentrations, the chemoreceptors of first-stage juveniles (J1) in eggs may interpret these phytochemicals as being analogous to chemical signals sent by plants entering senescence, resulting in survival stage setting of the nematode juveniles (McSorley 2003). In contrast, at high concentrations the J1 may be tricked to interpret the phytochemicals as being analogous to chemical signals sent by actively growing plants, thereby resulting in further development of such juveniles and the hatching of the infective J2 in high numbers.

7.4.1.2 Application Through Ground Leaching Technology (GLT)

The ground leaching technology (GLT) system was developed to ameliorate the above-mentioned drawbacks of the use of conventional organic amendments (Mashela 2002). Following an extensive search for botanical materials that would not reduce soil pH when used in small quantities (2–4 g plant^{-1}) in powdered or fine particulate form, fruits of *C. myriocarpus* and *C. africanus* were observed to fit this requirement. This discovery was followed by in vitro (Muedi 2005) and in vivo (Mashela 2002; Mashela et al. 2008) trials to examine the effects of these materials on *Meloidogyne* spp. and *T. semipenetrans*. Preparations for the development of the GLT system involved collecting plants from fields under cultivation, cutting fruits into small pieces followed by drying of these pieces at 52 °C for 72 h. Thereafter the dried material was ground to pass through a 1-mm-aperture sieve (Mashela 2002). Drying of the plant material at a lower temperature will result in decay of the fruit pieces, whereas drying at temperatures higher than 52 °C will result in substantially lower cucurbitacin levels (Shadung et al. 2015).

In granular (G) formulation, the two cucurbitacins are referred to as nemarioc-AG and nemafric-BG, with the first part referring to the cucurbitacin species name and the second part to the formulation type. In tomato production, the granular material is applied at 2 g plant^{-1} (Fig. 7.4). The material is applied around the crop

Fig. 7.4 Application of crude material of a phytonematicide (nemarioc-AG) after transplanting of a tomato seedling according to the ground leaching technology (GLT) system to suppress population densities of plant-parasitic nematodes (Zakheleni Dube, University of Limpopo, Polokwane, South Africa)

seedling at transplanting or postemergence after which it is covered by soil. Not all plant species are suitable for use in a GLT system. *Brassica* spp., which are widely used in biofumigation for nematode management, require microbial degradation first to release the active substance (a.s.) with nematicidal properties (Bello 1998). Along with chilli (*Capsicum* spp.), oleander and tamboti, materials derived from *Brassica* spp. were not effective in suppressing root-knot nematode population densities when used in a GLT system (Thovhakale et al. 2006; Mashela et al. 2012). Generally, in a GLT system, the efficacy of the plant materials to suppress nematode population densities depends upon the solubility of the a.s. to enable its leaching into the soil rhizosphere (Mashela and Nthangeni 2002).

7.4.1.3 Application Through Botinemagation

Although the GLT system is labour intensive and requires at least two applications per tomato growing season, the system has become popular with the smallholding farmers. In large commercial farming systems, phytonematicides can be applied through drip irrigation, a technology which is being referred to as botinemagation (Mashela 2014). By definition, botinemagation is using botanicals (phytonematicides) to manage nematode population densities through irrigation. The a.s. is extracted from plant organs through fermenting with anaerobic microorganisms. Technically, fermentation specifically refers to the anaerobic breakdown of glucose through pyruvic acids to lactic acids. To prepare phytonematicides in liquid formulation, the basic inputs are simple (Table 7.2). In Nemalan, a biologically-derived product developed locally and which exhibits antinematodal properties, the plant materials are used in fresh form (see Sect. 7.4.1.7) (Daneel et al. 2014a), whereas in *Cucumis* spp. fruits are used in ground or granular form (Mashela 2002). Care should be taken to ensure that the entire fermentation process is taking place in an airtight container since a large quantity of gases are released during fermentation. An escape route for the gas should be provided by inserting a small tube into a bottle half filled with water in order to avoid explosions. In an airtight container, the

Table 7.2 Ingredients of selected phytonematicides, before fermentation, to be used for botinemagation

Ingredient	Amount (g) or volume (ml or l) of ingredient added	Nemalan[a]	Nemafric-BL	Nemarioc-AL
Plant material	g	120	40	80
Brown sugar	g	100	100	100
Molasses	ml	300	300	300
EM[b] stock	ml	300	300	300
Chlorine-free H_2O	l	16	16	16

[a]A biologically-derived product with antinematodal properties developed by the South African leading fresh produce company ZZ2, Polokwane, South Africa; [b]A product based on micro-organisms produced by Microzone, Polokwane, South Africa

process is usually completed within 14 days, with the pH of the mixture having declined to at least 3.7 (Mashela et al. 2011; Mashela 2014).

7.4.1.4 Non-phytotoxic Concentrations

Under in vitro conditions, the use of phytonematicides usually provides consistent results (Mashela et al. 2011; Mashela 2014), but under field conditions most phytonematicides (being allelochemical compounds; Rice 1984) are highly phytotoxic (Mafeo and Mashela 2010; Mafeo et al. 2011a, b). In many countries, as exemplified by the European and Mediterranean Plant Protection Organisation regulations (EPPO 2010), there is a zero tolerance for agricultural inputs that have a phytotoxic effect on crops. Consequently, non-phytotoxic concentrations for each phytonematicide and its application time interval(s) should be empirically established. Generally, at low concentrations cucurbitacins stimulate cell division of the affected plant part (Lee et al. 2010). In a number of studies, it was consistently demonstrated that various plant variables, when subjected to increasing concentrations of phytonematicides, invariably showed quadratic relations which characterise the existence of density-dependent growth (DDG) patterns (Zasada and Ferris 2003; Wuyts et al. 2006; Mafeo and Mashela 2010; Mafeo et al. 2011a, b; 2012; Pelinganga and Mashela 2012; Pelinganga et al. 2012, 2013a, b, c).

Curve-Fitting Allelochemical Response Dosage Model

Using data collected to assess the responses of organisms to increasing allelochemical concentrations, Liu et al. (2003) developed a three-phase curve-fitting allelochemical response dosage (CARD) model. The CARD model quantifies three phases (i.e. stimulated, neutral and inhibited phase) and three zones (i.e. nematicidal, neutral and herbicidal zone) using biological indices (Fig. 7.5). Mashela (2014) conceptualised the three phases in terms of plant growth responses to increasing concentrations of phytonematicides using the mean values of the biological indices (Table 7.3). The latter was feasible since indices are numbers without units and could therefore be added and averaged. Mashela et al. (2015) demonstrated that plant responses can either be stimulated, neutral or inhibited, with the degree of the response being dependent on the concentration of the phytonematicides. In the neutral zone, growth of untreated and treated plants can statistically not be differentiated resulting in the conclusion that the phytonematicide has no effect.

Mean Concentration Stimulation Point

Using the 50 % inhibition (D_{50}) concept, Mashela (2014) investigated the possibility of the existence of its counterpart (50 % stimulation) in the stimulated growth phase, which was possibly situated midway between D_m and R_h (Fig. 7.5). The values

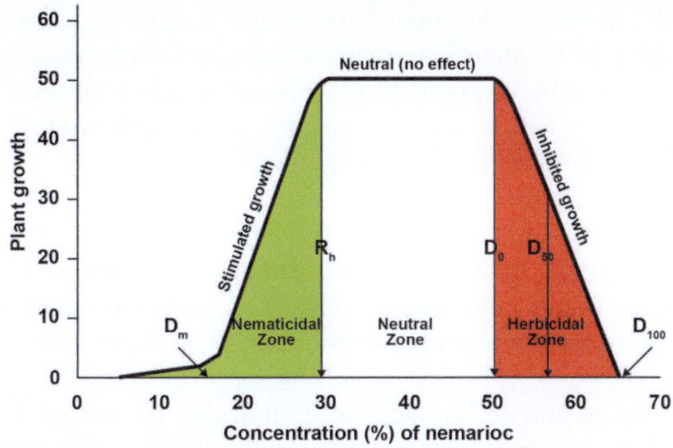

Fig. 7.5 The curve-fitting allelochemical response dosage (CARD) model that quantifies the effect of phytonematicides on plant growth (Mashela 2014)

Table 7.3 Mean biological indices generated from dry shoot mass (DSM), dry root mass (DRM), plant height (PHT) and stem diameter (SDM) of tomato seedlings over six nemarioc-AL phytonematicide concentrations (Pelinganga et al. 2013a)

Biological index[a]	DSM	DRM	PHT	SDM	Mean
Threshold stimulation (D_m)	2.53	2.20	2.73	1.53	2.22
Saturation point (R_h)	0.71	0.32	2.00	0.08	0.77
0 % inhibition (D_0)	11.48	9.21	19.94	5.42	10.96
50 % inhibition (D_{50})	164.90	59.00	2899.74	1603.20	957.17
100 % inhibition (D_{100})	703.50	170.30	2902.47	1604.74	1110.84
k (sensitivity ranking)	4	1	2	4	–
Overall sensitivity ranking: $\sum k = 11$					
$P \leq$	0.01	0.01	0.01	0.05	–

[a]All biological indices are without units and can be added to calculate the integrated mean

between the actual D_m and R_h define the concentration stimulation range (CSR) which is representative of the concentration range at which plant growth is stimulated (Fig. 7.5). The midpoint of CSR is arbitrarily referred to as the mean concentration stimulation point (MCSP) and can be considered as the concentration that will stimulate plant growth while consistently suppressing nematode population densities (Mashela 2014). Using the relation MCSP = $D_m + (R_h/2)$, MCSP values for nemarioc-AL and nemafric-BL were established at 2.63 (Table 7.3) and 2.89 %, respectively (Pelinganga and Mashela 2012; Pelinganga et al. 2013a). A point worth mentioning is that with an MCSP of 2.63 %, application of nemarioc-AL on tomato resulted in a D_0, D_{50} and D_{100} of 13.96 ($D_m + R_h + D_0$), 971.12 ($D_m + R_h + D_0 + D_{50}$) and 2081.97 % ($D_m + R_h + D_0 + D_{50} + D_{100}$), respectively (Table 7.3). Incidentally, when the MCSP concept is correctly used, it is unlikely that the concentrations of phytonematicides that will suppress nematode population densities will be phytotoxic at the same time. In principle, for every 263 ml nemarioc-AL and 289 ml nemafric-BL phytonematicide applied, respectively, 10 l of chlorine-free water should be used for the botinemagation of a tomato field. However, these values differ for each crop. For African geranium (*Pelargonium sidoides*), MCSP values were 6.2 and 2.9 % for nemarioc-AL and nemafric-BL, respectively (Sithole 2016). Similarly, on *Citrus volkameriana* seedling rootstocks, the values were 8.6 and 6.3 %, respectively (Mathabatha et al. 2016). These values, although they would stimulate plant growth, were rather high when compared to those calculated for tomato (Pelinganga et al. 2012, 2013a, b). In both studies, lower MCSP values than the actual stimulating values consistently suppressed nematode population densities and therefore were adopted instead of the relatively higher empirically derived MCSP values.

The MCSP values should be interpreted alongside the overall *k*-values of the phytonematicide on the crop (Table 7.3). Generally, the closer the overall sensitivity index ($\sum k$) of the plant is to zero, the higher the sensitivity of the plant to the phytonematicide and vice versa (Liu et al. 2003). Basically, $\sum k$ is specific for several factors including the phytonematicide concentration, application rate, plant species and growth phase and nematode life stage. For example, experimental data show that seedlings are less sensitive than mature plants to phytonematicides derived from fruits of *Cucumis* spp. (Mafeo et al. 2011a, b; Pelinganga et al. 2013a, b, c).

7.4.1.5 Application Interval of Phytonematicides

Maile et al. (2013) demonstrated that the response of plants to phytonematicides, in addition to being concentration specific, was application time interval specific as well. Trial results suggested that treatment of tomato plants infected with root-knot nematodes with nemarioc-AL and nemafric-BL should be made at 17- and 19-day intervals, respectively, over a 56-day period (Pelinganga et al. 2013a, c).

7.4.1.6 Dosage in Phytonematicides

Dosage (D) is the product of concentration (equivalent to MCSP value) and the application frequency (T_f), summarised as D (%) = C (%) × T_f. The T_f is the proportion of the crop cycle (days) to the application interval (days) and is constant for a particular crop. For example, when the crop cycles of two tomato varieties were 56 and 112 days, T_f values for 2.63 % nemarioc-AL would be 3.3 and 6.6, respectively. Therefore, for nemarioc-AL in the 56- and 112-day tomato crop cycles, the dosage would be 8.68 and 17.36 %, respectively. Bearing in mind that MCSP values are empirically derived to avoid phytotoxicity, increasing the concentration, e.g. to 4 %, would not have any added effect on the suppression of the nematode population densities, but would increase the residues of the phytonematicide in the soil, which is undesirable in terms of the soil allelochemical residue (SAR) concept (Mashela and Dube 2014). The SAR concept assesses the effects of soil allelochemical residues of phytonematicides on the following crop(s). It has been shown that allelochemical residues in the soil had phytotoxic effects on cowpea (*Vigna unguiculata*) as well as inhibitory effects on the development of *Bradyrhizobium japonicum* nodules and residual nematode population densities (Mashela and Dube 2014; Mashela et al. 2012).

7.4.1.7 Phytonematicides Used on a Commercial Scale

Effective application of phytonematicides in large commercial and smallholding farming systems in SA involves the use of crude extracts in granular and liquid formulations (Nzanza and Mashela 2012; Mashela 2014). Plant materials commonly used in the phytonematicide Nemalan include lantana (*Lantana camara*) shoots and wild garlic (*Tulbaghia violacea*) (Daneel et al. 2014a). Fruits from *C. africanus* and *C. myriocarpus* (Mashela 2014) are used in the production of nemafric-BL and nemarioc-AL, respectively (Mashela. 2014). The centre of biodiversity of the two *Cucumis* spp. is the Limpopo Province (Kristkova et al. 2003), whereas lantana is an invader plant (Daneel et al. 2014a). Active substances in lantana leaves are saponins and cucurbitacins in fruits of the two *Cucumis* spp. (Van Wyk and Wink 2004). Combining plant organs with different a.s., for example, *C. myriocarpus, L. javanica* and *R. communis*, resulted in synergistic effects on the suppression of nematode population densities (Mashela et al. 2007).

7.4.2 Medicinal Plants ('Muti') Used as Phytonematicides

Plant parts (powdered leaf meals) of non-crop plant species used in traditional medicine in SA were selected and examined for their nematicidal activity as soil amendments on *M. incognita* race 2 (Khosa 2013). Parts of these non-crop species are locally known as 'muti' as they are considered to have certain medicinal properties. Traditional healers in SA frequently use these mutis to treat human and domestic animals for various ailments. Living specimens of these plant species, as well as supplies of dried and finely ground material made from them, can be found in abundance in SA in the rural areas and communities of the lowveld in the Mpumalanga, Limpopo and KwaZulu-Natal provinces. These plant species contain chemicals such as alkaloids, diterpenes, diterpenoids, esters, fatty acids, ingenol, oxalic acid and terpenoids. The observed general effects of these traditional medicines on humans at prescribed dosage rates suggested that they might be toxic to small multicellular organisms such as plant-parasitic nematodes.

Nine plant species were identified, collected and examined for their nematicidal activity and plant growth enhancement in glasshouse trials. Five of these plant species, namely, cactus vine (*Cissus cactiformis*), Candelabra tree (*Euphorbia ingens*), Bushveld bead-bean tree (*Maerua angolensis*), Dead-man's tree (*Synadenium cupulare*) and Toad tree (*Tabernaemontana elegans*) were further tested under field conditions (Table 7.4) (Khosa 2013). Soil amendments of powdered leaf meals of

Table 7.4 Effect of soil amendments of powdered leaf meals of five plant species on the number of root-knot nematode eggs and second-stage juveniles (J2) in roots of tomato grown under glasshouse and field conditions (Khosa 2013)

Source of amendment	Rate (g)	Glasshouse No. of eggs and J2 root system^{-1}	Field No. of eggs and J2 root system^{-1}
Control (no amendment)	0	4.42[a] (30,063)[b] a[c]	4.4 (26,800) a
Cissus cactiformis	5	3.20 (1,688) bcd	3.5 (4,267) d
	10	3.20 (1,950) bcd	3.6 (5,867) cd
	15	2.82 (2,975) cd	3.7 (6,933) cd
Euphorbia ingens	5	3.10 (1,838) cd	3.6 (5,067) cd
	10	3.04 (1,350) cd	4.0 (13,600) bc
	15	2.63 (1,113) d	3.4 (2,733) d
Maerua angolensis	5	3.77 (6,688) b	3.7 (8,200) cd
	10	3.79 (6,625) ab	3.6 (5,150) cd
	15	3.37 (3,763) bc	3.7 (7,333) cd
Synadenium cupulare	5	3.25 (2,588) bcd	3.8 (8,200) cd
	10	3.20 (1,838) bcd	3.7 (9,633) cd
	15	3.01 (1,300) cd	3.6 (6,933) cd
Tabernaemontana elegans	5	3.44 (3,588) bc	3.7 (7,067) cd
	10	3.32 (2,275) bc	3.5 (6,200) cd
	15	3.09 (1,375) cd	3.7 (7,300) cd
Cucumis myriocarpus	5	2.86 (2,550) cd	3.4 (4,267) d
Fenamiphos	5	1.17 (138) e	3.3 (4,267) d
LSD	–	0.64	0.45
P-value	–	0.001	0.002
F-ratio	–	7.79	2.59

[a]Log transformed (x + 1)
[b]Non-transformed means in parenthesis
[c]Means followed by different letters in the same column differ significantly at LSD ($P \leq 0.05$)

C. cactiformis, *M. angolensis* and *T. elegans* were shown to reduce the number of eggs and J2 of *M. incognita* race 2 in tomato (Khosa 2013). The effect of the leaf meals was similar to that of fenamiphos and the *C. myriocarpus* treatment (Mashela 2002) which was also included in the study.

A previous study showed that ground fruit of *C. myriocarpus* suppressed root-knot nematode population densities under glasshouse, microplot and field conditions (Mashela 2002; Mashela and Mphosi 2002). Tomato plants treated with soil amendments of powdered leaf meals of *C. cactiformis*, *M. angolensis* and *T. elegans* had larger root systems, greater shoot mass and were taller when compared with the untreated control plants. This effect in turn resulted in a higher number of fruit and greater fruit mass. Similar results with botanical soil amendments were obtained by Akhtar (2000), Akhtar and Malik (2000), Mashela and Mpati (2002), Mashela and Nthangeni (2002) and Sikora and Fernandez (2005). Miami and Rodríguez-Kábana (1982) demonstrated that improved plant growth responses could be due to the absorption of carbon compounds that leached from plant residues into the soil. The soil C:N ratio of the amendments varied, which could explain why plant growth responses were different. Tomato growth response observed in the glasshouse, microplots and field indicated some phytotoxic effects with the applications of *E. ingens* and *S. cupulare*. This was, however, not consistent in all the trials and with all variables measured. These negative effects casted doubt on the possible large-scale use of such phytonematicides as soil amendments in crop production. After screening and further selection, the extracts of *M. angolensis* and *T. elegans* that showed significant nematoxic and/or growth-stimulating potential are being tested at a more advanced level of chemical separation (MC Khosa, 2016, personal communication).

Possible human, animal and environmental risks of the use of phytonematicides in crop production should be determined, preferably in consultation with the traditional healers that use these materials for medicinal purposes. However, these materials have a potential to be used as soil amendments in integrated nematode management strategies, in domestic garden, smallholding and large commercial farming in SA. Although some of these plants are quite common, e.g. *C. cactiformis*, *M. angolensis* and *T. elegans*, and grow abundantly in the Mopani and Vhembe districts of the Limpopo Province as well as in the Lowveld regions of the KwaZulu-Natal and Mpumalanga provinces, overexploitation of these natural resources should be avoided.

7.5 Organic Amendments

As stated above, organic amendments have several drawbacks limiting their use in smallholding farming. Nonetheless, when planting crops in small areas such as home and communal gardens and small fields, the use of organic amendments is an important alternative practice to assist in the management of nematode pests and to increase crop yields. Nematologists of the ARC-Institute for Tropical and Subtropical Crops (ITSC) conducted a series of trials to determine the effects of several pre-planting treatments on natural occurring plant-parasitic nematode complexes (dominated by mixed *M. incognita* and *M. javanica* populations). The treatments represented chicken (45 MT ha^{-1}) and cattle manure (45 MT ha^{-1}), compost (30 l $(m^2)^{-1}$, marigold (*Tagetes minuta*), oilcakes of sorghum (*Sorghum bicolor*) and soybean (2–4 kg $plot^{-1}$), plastic cover (solarisation) and permaculture which consists of a pit (about 60 cm deep) filled with waste material from branches, leaves, fruits and weeds (Fig. 7.6). In SA, the main sources of fertiliser used by smallholding farmers include kraal manure, chicken litter, organic waste and compost (Masarirambi et al. 2002). The trials were conducted during 2004–2005 with vegetable/maize/vegetable crop rotation sequences. Maize was planted in all plots, while the same vegetable crops were planted in the original plots in the 2005 season. Vegetables included tomato, spinach and beetroot/green pepper. The application of animal manures resulted in yield increases in both seasons, while nematode

Fig. 7.6 Permaculture is practised by filling trenches (approximately 2.5–3 m long × 1 m wide × 1 m deep) with different types of organic material (Mieke Daneel, Agricultural Research Council–Institute for Tropical and Subtropical Crops, Mbombela)

control was inconsistent. However, permaculture provided substantial better yields especially in the second season, viz. between 63 and 81 % compared to the untreated plants. Nematode control varied among the different crops being unsatisfactory in especially the tomato crop (Daneel 2007; Tefu et al. 2009). Following these initial results, permacultures with different waste materials were compared mainly to attempt improving nematode control. Permaculture that contained mainly fruit waste materials resulted in the highest vegetable yields and nematode control ranging between 50 and 80 % (Tefu et al. 2009). Further investigation showed that permaculture with citrus fruit gave the best yield and nematode control, and this permaculture is now under further investigation (Tefu et al. 2014). It is important to mention that permaculture is renewed (i.e. the pit filled with fresh waste material) once every 3–4 seasons and not every time a new crop is planted. Also, citrus and other fruits are widely available in the subtropical areas of SA throughout the year. Most smallholding farmers living in the subtropical areas of SA possess one or a few citrus, avocado, guava, mango or other fruit trees in their home gardens (Fig. 7.7). Waste fruits can hence be used for permaculture and contribute towards both higher crop yields and improved nematode management.

Nematologists of the ARC-Grain Crops Institute (ARC-GCI) conducted a series of on-farm (smallholdings) trials during several summer-growing seasons at Coligny, Potchefstroom and Morokweng (North-West Province) and Dingleydale (Mpumalanga Province) to investigate the effect of animal manures, compost and

Fig. 7.7 Smallholding farms with fruit trees growing in small gardens in the Mpumalanga and Limpopo provinces of South Africa (Mieke Daneel, Agricultural Research Council–Institute for Tropical and Subtropical Crops, Mbombela)

marigold amendments as well as solarisation on root-knot nematode population densities (i.e. single species as well as mixed populations of either *M. incognita* or *M. javanica*) (Fourie and Schoeman 1999; Mtshali et al. 2002a; Ntidi et al. 2012). A 52 % reduction of the root-knot nematode population densities in tomato roots (susceptible cv. Rodade) (Fourie et al. 2012), 60 days after planting, in plots amended with chicken manure (equivalent to 5 MT ha^{-1}) was recorded at Potchefstroom. Root-knot nematode population densities after cattle manure amendment (20 MT ha^{-1}) were reduced by 28 and 92 % at Potchefstroom and Morokweng, respectively, and 41 and 74 % (45 MT ha^{-1}) at Dingleydale and Coligny, respectively. When cattle manure amendment was combined with solarisation, a 99 % reduction in the root-knot nematode population densities was obtained at Morokweng, which was 7 % higher than when cattle manure amendment was applied alone. At Morokweng, the amendment of only compost (equivalent to 20 MT ha^{-1}), consisting of residues of various plants grown by the community, as well as a combination of compost and solarisation reduced the root-knot nematode population densities by 90 %. Reductions of 49 and 99 % in the root-knot nematode population densities in roots of maize (cv. SR52), 60 days after planting, in plots amended with cattle manure (equivalent to 45 MT ha^{-1}) were observed at Dingleydale and Coligny, respectively.

Unfortunately, yield data could only be obtained at Morokweng where a 370 and 500 % increase in tomato yield was recorded for plots amended with compost and cattle manure, respectively. For the combined treatment of either cattle manure or compost and solarisation, a 17 and 146 % increase, respectively, in tomato yield was obtained. Another amendment tested consisted of shoot cuttings of mature marigold plants (equivalent to 12 MT ha^{-1}) growing alongside agricultural fields on the ARC-GCI campus. This treatment resulted in a 71 % reduction in the root-knot nematode population densities 60 days after planting of the tomato plants (cv. Rodade).

7.6 Host Plant Resistance

Smallholding farmers buy seeds or seedlings of crops that are readily available. Commercially available varieties of grain, leguminous and vegetable crops were hence obtained from seed companies and evaluated for their host response to *M. incognita* and *M. javanica*. Crops included in these evaluations were *Amaranthus* spp., beetroot, cabbage (*Brassica oleracea* var. capitata), dry bean (*Phaseolus vulgaris*) (Mtshali et al. 2002b), chilli, carrot (*Daucus carota*), cowpea (Riekert 1999), green bean (*Phaseolus* spp.), pumpkin (*Cucurbita* spp.), maize (Ngobeni et al. 2011), spinach and tomato (Riekert 1999; Mothata 2006; Steyn et al. 2013, 2014). A limited number of genotypes were identified with resistance to *M. incognita* race 2, viz. *Amaranthus* sp. accession Local 33 (Steyn et al. 2013), green pepper cv. Tabasco (Steyn et al. 2013) and tomato cvs. Rhapsody, MFH 9324, FA 1454 and FA 593 (Fourie et al. 2012). The superior *M. incognita* resistance in the tomato cv. Rhapsody was further verified in a microplot trial using a range of initial nematode population densities (Pi). The results of this trial showed that final nematode population densities (Pf) and reproduction factor (Rf) values were consistently significantly lower compared to those recorded for the susceptible tomato cv. Moneymaker.

Resistance to *M. incognita* and *M. javanica* was observed in some maize (Ngobeni et al. 2011; see also Sect. 8.2.4.2) and cabbage (Mothata 2006) genotypes and to *M. javanica* in two cowpea lines (T182b-889 and R6A; Riekert 1999). Nonetheless, no root-knot nematode resistance was identified in the beetroot, carrot, green bean, pumpkin and spinach germplasm screened (Mothata 2006; Steyn et al. 2014).

The results of all this research underline the importance of the continuous screening of new crop varieties that enter the local market so that farmers can be updated about the host status of these crops to abundant root-knot nematode species. This up-to-date information about genotypes, together with the accurate identification and monitoring of the root-knot nematode species populations in fields, will enable farmers and extension officers to make informed and thus the best possible decisions regarding nematode management.

7.7 Crop Rotation and Intercropping

Limited research has been conducted in the smallholding farming sector in particular with regard to the use of crop rotation sequences and intercropping as practices to manage important nematode pests.

The effect of crop rotation on grain, legume and vegetable crops was examined during a long-term on-farm microplot trial carried out on the premises of the ARC-GCI (Venter et al. 2004) over four (consecutive) growing seasons. At the start of the trial, microplots with either a low or high infestation level of *M.*

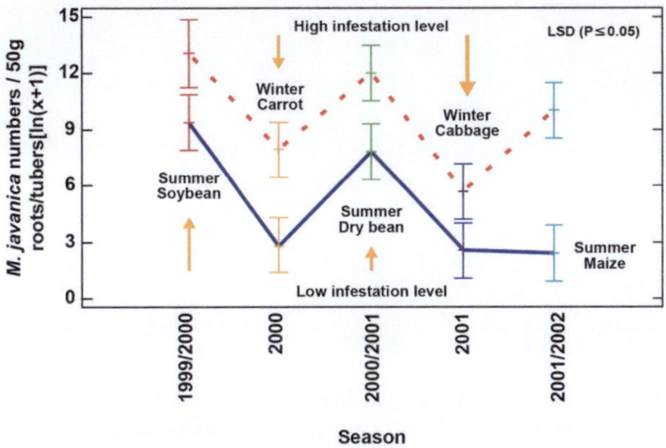

Fig. 7.8 The effect of crop rotation over four consecutive growing seasons on *Meloidogyne javanica* population densities (Venter et al. 2004)

javanica were created by growing resistant (cv. A7119 - low infestation) and susceptible (cv. Prima2000 - high infestation) soybean cvs in microplots (Fourie et al. 1999, 2006). The following crop sequence was examined: carrot (cv. Chantenay Karoo), dry bean (cv. Mkuzi), cabbage (cv. 3306) and maize (cv. SC701). At the start of the trial, the plants were inoculated with a range of Pi levels (viz. 0, 100, 500, 1000, 5000, 10,000 and 20,000 *M. javanica* eggs and J2 plant^{-1}) to study also the effect of pest pressure on nematode reproduction and yield. Final nematode population densities in the rhizosphere and roots were determined for each crop 60 days after nematode inoculation during each of the growing seasons. Data were ultimately pooled over Pi levels to demonstrate the collective value of the crop sequence. The results (Fig. 7.8) showed that the first winter planting of carrot resulted in a significant 99 % decrease (from a Pf of 12,625 eggs and J2 50 g roots^{-1} in soybean cv. A7119 to a Pf of 75 in carrot) in *M. javanica* population densities in the low infestation microplots. A similar trend was evident for the high infestation microplots with a 91 % reduction (from a Pf of 268,786 eggs and J2 in soybean cv. Prima2000 to a Pf of 24,033 in carrot). Growing of cabbage during the second winter growing season resulted in a reduction of 54 and 50 % in *M. javanica* population densities in cabbage compared with dry bean grown during the preceding summer season in the low and high infestation microplots, respectively. The high residual Pf in both carrot and cabbage roots in microplots with a high infestation showed that, even in winter, inclusion of these crops is not adequate to reduce the population densities of *M. javanica*. However, the value that the inclusion of winter plantings of carrot and cabbage may have in the reduction of *M. javanica* population densities in fields infested with a lower population density of *M. javanica* was demonstrated. The results also showed that inclusion of dry bean during the second summer season

Fig. 7.9 (**a**, **b**) Healthy dry bean plants (cv. Mkuzi) (**a**) in microplots where the *Meloidogyne javanica*-resistant soybean cultivar A7119 was grown the previous season compared to stunted plants (**b**) with chlorotic leaves in plots where the susceptible soybean cultivar Prima was grown (Driekie Fourie, Agricultural Research Council–Grain Crops Institute, Potchefstroom, South Africa)

increased the *M. javanica* population densities to levels comparable to those in the resistant and susceptible soybean cvs at the start of the trial. Interestingly, inclusion of maize during the third summer season kept *M. javanica* population densities in the low infestation microplots similar to those recorded on the preceding cabbage crop. By contrast, in the high infestation microplots inclusion of maize during the third summer season increased the *M. javanica* population densities significantly by 37 %. At the termination of the trial, the four season crop sequence had reduced the *M. javanica* population root densities by 70 and 20 % in low and high infestation microplots, respectively. The yield of dry bean grown in the high infestation microplots was significantly lower compared with the low infestation microplots, but for carrot, cabbage and maize, no yield differences were observed (Fig. 7.9a, b).

The effect of intercropping with *Tagetes erecta* (cv. Lemon Drops) on root-knot nematode population densities was studied in two on-farm experiments at Coligny and Dingleydale (Fourie and Schoeman 1999). The results showed a substantial reduction in nematode numbers in the roots of tomato cv. Rodade (by 57 and 30 % at Coligny and Dingleydale, respectively) and maize cv. SR52 (by 91 % at Dingleydale) when the crops where intercropped with *T. erecta*.

The effect of intercropping on soil and plant fertility, nematode population composition and yield of sugarcane grown in a sandy soil on a smallholding farm was also studied (Berry et al. 2009). Groundnut (*Arachis hypogaea*) and sugar bean (*Phaseolus limensis*) were intercropped between the sugarcane rows in an irrigated experiment, while the same was done with velvet bean (*Mucuna deeringiana*) and sweet potato (*Ipomoea batatas*) in a rain-fed experiment. These practices were compared with a standard nematicide (aldicarb) treatment and an untreated control. Intercropping with velvet bean, groundnut and sweet potato increased the population densities of *M. javanica* and *Pratylenchus zeae* Graham, 1951, in sugarcane sett roots. By contrast, intercropping with sugar bean reduced the nematode numbers. Intercropping with velvet bean, sugar bean and sweet potato had no effect on sugarcane yield, whereas intercropping with peanut reduced sugarcane yield by 22 % and sucrose yield by 29 %. Also, intercropping with velvet bean increased the levels of some nutrients in the soil and leaves of sugarcane. These results showed that intercropping can be used by smallholding farmers to manage nematode pests, in this case on sugarcane. Furthermore, intercropping provided nutrients to the sugarcane crop when velvet bean in particular was used. Ultimately, the use of intercropping provided an alternative food source and/or income, viz. sweet potato, and improved the overall productivity of the land without being detrimental to sugarcane cultivation.

7.8 Cover Crops Used for Biofumigation

Brassicaceae crops have been and are continuously evaluated for their cover and biofumigation properties as an alternative strategy to control plant-parasitic nematode pests and more specifically root-knot nematodes worldwide (Fourie et al. 2016). Biofumigation is defined as a process during which volatile compounds (e.g. isothiocyanates and thiocyanates) with pesticidal effects are released during the decomposition and biodegradation of plant parts or animal products incorporated in the soil. Brassicaceae crops contain such compounds that are collectively referred to as glucosinolates (Youssef 2015).

Climate, soil conditions, plant density and other factors play an important role in the efficacy of biofumigation using Brassicaceae crops. In Box 7.2, the modes of action of biofumigation are summarised. Cultivars of *Brassica* spp. that are commercially available showed variable results in terms of reducing nematode population densities (Fourie et al. 2015). A field experiment with mustard (*Eruca sativa* cvs. Rocket Trio and Nemat), Indian mustard (*Brassica juncea* cvs. Calienté and Fumigreen) and radish (*Raphanus sativus* cvs. Doublet and Terranova) showed an increase (115 %) in *M. incognita* population densities in the roots of potato (cv. Mondial) at tuber initiation when Indian mustard cv. Calienté was first grown, and its aerial parts subsequently incorporated in the soil for biofumigation. At this specific experimental site, the high root-knot nematode population (±25,000 eggs and J2 50 g^{-1} potato tubers in the season preceding the experiment) proved to be too high for biofumigation to be successful (Engelbrecht 2012). Nonetheless, in another experiment (at Mbombela), when tomato (cv. Monica) plants were grown after biofumigation with the same Brassicaceae cvs. as mentioned above, substantial reductions in the number of eggs and J2 50 g $root^{-1}$ (a mixture of *M. javanica* and *M. incognita*) and significantly higher yields were recorded compared with the uninfected control plants. The increases in tomato yield were, for example, between 103 and 163 % higher in plots cover cropped and subsequently biofumigated with aerial parts of the radish cvs. Terranova and Doublet, respectively, compared with the untreated control. By contrast, for plots cover cropped with cv. Calienté, a yield decrease of 26 % was observed (Daneel et al. 2014b).

Box. 7.2 Modes of Action of Biofumigation with Brassicaceae to Protect Crops Against Plant-Parasitic Nematodes

Modes of action:

(i) Production of nematotoxic glucosinolate (GSL) degradation products, viz. isothiocyanates (ITCs), thiocyanates, nitriles or oxazolidinethiones
(ii) Stimulation of antagonistic microbial communities
(iii) Production of nitrogenous compounds that are toxic to plant-parasitic nematodes

Degradation products are formed as a result of the hydrolysation of sulphur-containing secondary metabolites by the enzyme myrosinase (stored separately in plant cells), yielding nitriles, epithionitriles and thiocyanates. To optimise ITC release from aerial parts of Brassicaceae crops, the plant cells of these aerial parts must be damaged by slashing and/or rupturing and immediately incorporated into the soil. Such actions are most effective when ITC levels are highest which is usually during flowering in aerial plant parts of Brassicaceae crops. Popular Brassicaceae crops are *Brassica oleracea* (broccoli, Brussels sprouts, cabbage, cauliflower), *Brassica oleracea acephala* (kale), *Brassica napus* (canola and rape seed), *Brassica rapa* (turnip), *Raphanus sativus* (radish) and a variety of mustards such as *Brassica juncea* (Indian mustard) and *Sinapis alba* (white mustard). More than 200 GSLs, contained in the vegetative parts and seeds of cultivated and wild plant species, have been identified mainly from plants belonging to the Brassicaceae family. The majority of these GSLs are unique in their chemical characteristics, and the types and quantities of GSL vary among individual plant species, plant organs, developmental stages and environmental factors (e.g. drought, availability of sulphur in the soil, sulphate and nitrogen nutrients, seasonal and diurnal cycles). Canola cv. Hyola 401, for example, contains low levels of GSLs, while rapeseed cv. Dwarf Essex, turnip cv. Purple Top and yellow mustard cv. Ida Gold contain moderate levels. The growth stage of Brassicaceae crops and the amount of biomass slashed and incorporated into the soil are the two main factors that contribute towards the success of biofumigation. Also, the Brassicaeae crop planted and its adaptability to the environment depict its biofumigation efficacy. Except for green manures, seed meals of Brassicaceae crops also may have biofumigation effects. Seed meals can be easily spread and incorporated into soil with no risk of frost damage as is the case with green manure crops. Also, seed meals do not serve as hosts to plant-parasitic nematodes as the roots and tubers of cover crops may do.

Literature consulted: Zasada and Ferris (2003), Bellostas et al. (2004), Kirkegaard and Matthiessen (2004), De Pascale et al. (2007), Larkin and Griffin (2007), Van Dam et al. (2009), Winde and Wittstock (2011), Borgen et al. (2012), Lelario et al. (2012), Kruger et al. (2013) and Fourie et al. (2016).

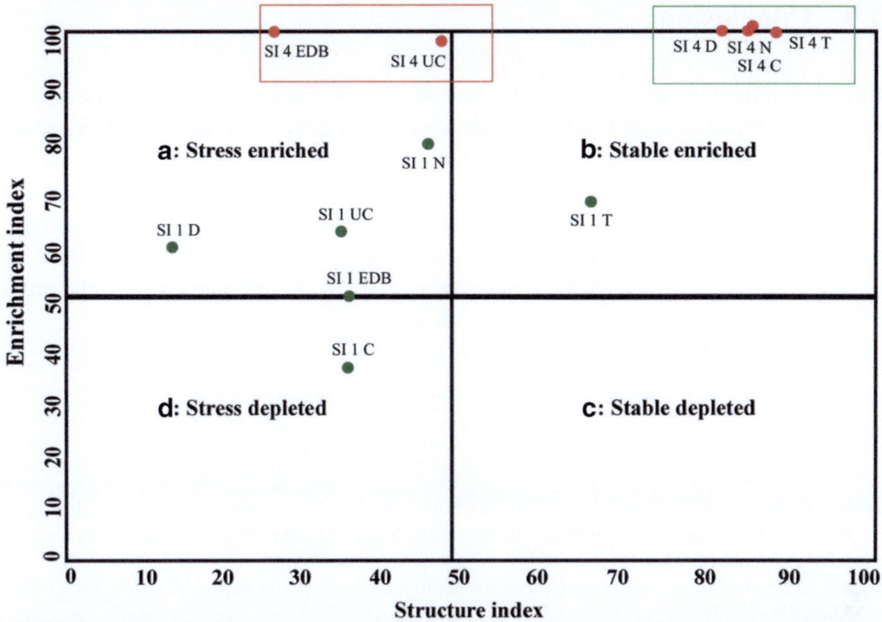

Fig. 7.10 Enrichment index (*EI*) and structure index (*SI*) of free-living nematode assemblages (according to soil food web analyses – Ferris et al. 2001) demonstrating the effect of cover cropping and biofumigation with four Brassicaceae crops after four consecutive sampling intervals during the 2010/2011 growing season. Soils of treated plots (*green box*) plotted in the 'stable and enriched' quadrant opposed to soils of untreated and EDB-fumigated plots (*red boxes*). Only data for sampling interval (SI) 1 (SI 1, *green dots* in *red block*) and 4 (SI 4, *red dots* in *green block*) are shown: *EDB* ethylene dibromide treated, *UC* untreated control, *C* cultivar Calienté, *D* cultivar Doublet, *N* cultivar Nemat, *T* cultivar Terranova (Adapted from Engelbrecht 2012)

Results obtained from local studies demonstrated that Brassicaceae crops could be used for their cover and biofumigation effects to reduce root-knot nematode population densities, increase crop yields and contribute towards soil health (Engelbrecht 2012; Daneel et al. 2015). In Fig. 7.10, improvements of the quality of soils (assessed in terms of free-living nematode assemblages) in which Brassicaceae crops were grown and biofumigated are shown. For example, at the termination of the experiment, the soil of plots biofumigated with Brassicaeae crops attained high so-called enrichment (EI) and structure indices (SI). The number of free-living nematodes increased in the soils cover cropped and biofumigated with the Brassicaceae compared with the untreated and EDB-treated soils. The correct choice of the Brassicaceae crop species and cvs is crucial since different species and cvs have different levels of GLS, which are the a.s. responsible for the nematicidal effects obtained as a result of biofumigation (Daneel et al. 2015; Fourie et al. 2016).

7.9 Conclusions

Research aimed at managing the adverse impact of plant-parasitic nematode pests in fields/gardens of smallholding farmers are being done on a relatively large scale and contributed already to combatting these pests in this agricultural farming sector. Furthermore, knowledge generated to date on the incidence and abundance of nematode pests in the smallholding agricultural sector yielded value data also on nematode-plant interactions. Practically applied research should be continued to ensure that these farmers reap the benefits and be able to produce crops in the presence of reduced nematode pest population levels.

References

Ahmad F, Siddiqui MA, Babalola OO (2013) Characterization of nematicidal activity of plant residues and their application with moisture approach against *Meloidogyne incognita* in tomato. Afr J Agric Res 8:93–101

Akhtar M (2000) Nematicidal potential of the neem tree *Azadirachta indica* (A. Juss). Integr Pest Manag Rev 5:57–66

Akhtar M, Malik A (2000) Role of organic soil amendments and soil organisms in the biological control of plant-parasitic nematodes: a review. Biores Tech 74:35–47

Aliber M (2009) Exploring statistics of South Africa's national household surveys as sources of information about food security and subsistence agriculture. Unpublished report. Centre for Poverty Employment and Growth, Human Sciences Research Council, Pretoria

Anonymous (2016) Statistics South Africa: Mid-year population estimates. http://www.statssa.gov.za. Accessed 17 Nov 2016

Bélair M, Tremblay N (1995) The influence of chitin-urea amendments applied to an organic soil on a *Meloidogyne hapla* population and on the growth of greenhouse tomato. Phytoprotect 76:75–80

Bello OA (1998) Biofumigation and integrated crop management. In: Bello OA, Gonzalez JA, Arias M, Rodriguez-Kabana R (eds) Alternatives to methyl bromide for the Southern European countries. DG XI, EU, CSIC, Madrid

Bellostas N, Sørensen JC, Sørensen H (2004) Qualitative and quantitative evaluation of glucosinolates in cruciferous plants during their life cycles. Agroindustria 4:267–272

Berry SD, Dana P, Spaull VW et al (2009) Effect of intercropping on nematodes in two small-scale sugarcane farming systems in South Africa. Nematropica 39:11–33

Borgen BH, Ahuja I, Thangstad OP (2012) 'Myrosin cells' are not a prerequisite for aphid feeding on oilseed rape (*Brassica napus*) but affect host plant preferences. Plant Biol 14:894–904. doi:10.1111/j.1438-8677.2012.00578.x

Bridge J, Starr JL (2007) Plant nematodes of agricultural importance. A color handbook. Academic, Boston

Chen JC, Chiu MH, Nie RL et al (2005) The cucurbitacins and cucurbitane glycosides: structures and biological activities. Nat Prod Rep 22:386–399

Chitwood DJ (2002) Phytochemical-based strategies for nematode control. Ann Rev Phytopathol 40:221–249

Coyne D, Moens M, Fourie H (2009) Current and future management strategies in resource-poor farming. In: Perry RN, Moens M, Starr JL (eds) Root-knot nematodes. CAB International, Wallingford, pp 444–475

Daneel M, De Jager K, Mdluli D (2003) IPM strategy for nematode control in small-scale farming systems in South Africa. Afr Plant Prot 10:127

Daneel MS (2007) Nematode control in rural farming. Progress Report 2007/571020. Agricultural Research Council–Institute for Tropical and Subtropical Crops, Mbombela

Daneel MS, Fourie H, McLeod A et al (2014a) Nemalan products as alternative nematicidal agents for the control of *Meloidogyne* species in tomato. J Nematol 46:150–151

Daneel MS, Lammers J, Fourie H et al (2014b) Brassicaceae-based management strategies to reduce *Meloidogyne* species damage in crops. J Nematol 46:152

Daneel MS, Fourie H, Engelbrecht E et al (2015) Influence of *Brassica* cover crops on root knot nematodes, soil abiotic factors and plant data in potato and tomato production under South African conditions. J Nematol 47:231–232

De Pascale S, Maggio A, Pernice R et al (2007) Sulphur fertilization may improve the nutritional value of *Brassica rapa* L. subsp. *sylvestris*. Eur J Agron 26:418–424

Dube ZP, Mashela PW, De Waele D (submitted) Density-dependent growth patterns of nematode egg hatch exposed to active ingredients of Nemarioc-AL and Nemafric-BL phytonematicides. Natl Prod Res

Engelbrecht E (2012) Nematode (Phylum Nematoda) community assemblages: a tool to implement environmentally-sound management strategies for root-knot nematodes (*Meloidogyne* spp.) in potato-based cropping systems. MSc Dissertation, North-West University, Potchefstroom

EPPO (2010) Efficacy of evaluation of plant protection products: phytotoxicity assessment. European and Mediterranean Plant Protection Organisation EPPO PP1/135(3), Wiley Blackwell Publishing Inc., New York

Ferris H, Bongers T, De Goede RGM (2001) A framework for soil food web diagnostics: extension of nematode faunal analysis concept. Appl Soil Ecol 18:13–29

Fourie H, Mc Donald AH (2000) Nematodes. Crop protection series no. 18. ARC-Grain Crops Institute, Potchefstroom

Fourie H, Schoeman MJ (1999) Nematode management in home gardens and small-scale farming systems: a preliminary study. Afr Plant Prot 5:12–13

Fourie H, Mc Donald AH, Loots GC (1999) Host suitability of South African commercial soybean cultivars to two root-knot nematode species. Afr Plant Prot 2:119–124

Fourie H, Mc Donald AH, De Waele D (2006) Host suitability of South African and foreign soybean cultivars to *Meloidogyne incognita* race 2. SA J Plant Soil 23:132–137

Fourie H, Mc Donald AH, Mothata TS et al (2012) Indications of variation in host suitability to root-knot nematode populations in commercial tomato varieties. Afr J Agric Res 7:2344–2352

Fourie H, De Waele D, Mc Donald AH et al (2015) Nematode pests threatening soybean production in South Africa, with reference to *Meloidogyne*. S Afr J Sci 111. doi:10.17159/sajs.2015/20140212

Fourie H, Ahuja P, Lammers J et al (2016) Brassicaceae-based management strategies as an alternative to combat nematode pests: a synopsis. Crop Prot 80:21–41

Jafee BA, Ferris H, Stapleton JJ et al (1994) Parasitism of nematodes by the fungus *Hirsutella rhossiliensis* as affected by certain organic amendments. J Nematol 26:152–161

Jeffrey C (1978) Cucurbitaceae. In: Launert E (ed). Flora Zambezi, London

Keetch DP (1989) A perspective of plant nematology in South Africa. SA J Sci 85:506–508

Khosa MC (2013) An investigation into the potential of crude and partially separated material of selected non-crop plant species as control agents of root-knot nematodes (*Meloidogyne incognita*) in tomato. PhD Thesis, North-West University, Potchefstroom

Kimpinski J, Gallant CF, Henry R et al (2003) Effect of compost and mature soil amendments on nematodes and yields of potato and barley: a 7-year study. J Nematol 35:289–293

Kirkegaard JA, Matthiessen J (2004) Developing and refining the biofumigation concept. Agroindustria 3:233–239

Kleynhans KPN, van den Berg E, Swart A et al. (1996) Plant nematodes in South Africa. Plant Protection Research Institute handbook no. 8. Agricultural Research Council–Plant Protection Research Institute, Pretoria

Kristkova E, Lebeda A, Vinter V et al (2003) Genetic resources of genus *Cucumis* and their morphological description. Horticul Sci 30:14–42
Kruger DHM, Fourie JC, Malan AP (2013) Cover crops with biofumigation properties for the suppression of plant-parasitic nematodes-a review. S Afr J Enol Vitic 34:287–295
Labadarios D, Michiza ZJ, Steyn NP et al (2012) Food security in South Africa: a review of national reviews. Bull World Health Organ 89:853–928
Larkin RP, Griffin TS (2007) Control of soilborne potato diseases using brassica green manures. Crop Prot 26:1067–1077
Lee DH, Lwanski GB, Thoennissen NH (2010) Cucurbitacin: ancient compound shedding new light on cancer treatment. Sci World J 10:413–441
Lelario F, Bianco G, Bufo SA et al (2012) Establishing the occurrence of major and minor glucosinolates in Brassicaceae by LC-ESI hybrid linear ion-trap and Fourier-transform ion cyclotron resonance mass spectrometry. Phytochemistry 73:74–83
Liu DL, An M, Johnson IR et al (2003) Mathematical modelling of allelopathy. III. A model for curve-fitting allelochemical dose responses. Nonlinearity Biol Toxicol Med 1:37–50
Mafeo TP, Mashela PW (2010) Allelopathic inhibition of seedling emergence in dicotyledonous crops by *Cucumis* bio-nematicide. Afr J Biotech 9:8349–8354
Mafeo TP, Mashela PW, Mphosi MS (2011a) Sensitivity of selected Alliaceae seedlings to crude extracts of *Cucumis myriocarpus* fruits. Afr J Agric Res 6:158–164
Mafeo TP, Mashela PW, Mphosi MS et al (2011b) Modeling responses of maize, millet and sorghum seedlings to crude extracts of *Cucumis myriocarpus* fruit as pre-emergent bio-nematicide. Afr J Agric Res 6:3678–3684
Mafeo TP, Mashela PW, Mphosi MS (2012) Allelopathic responses of various seeds to crude extracts of *Cucumis myriocarpus* fruits when used as a pre-emergent bio-nematicide. Acta Horticult 938:409–414
Maile KD, Mashela PW, Tseke PE (2013) Responses of the citrus nematode to a phytonematicide Nemarioc-AG with and without micro-organisms in citrus production. Afr Crop Sci Confer Proc 11:333–337
Marais M, Swart A (2007) Plant nematodes in South Africa. 8. Bizana, Lusikisiki and Port St Johns area, Eastern Cape Province. Afr Plant Prot 13:16–27
Masarirambi MT, Hlawe MM, Oseni OT et al (2002) Effects of kraal manure, chicken manure and inorganic fertilizers on growth and yield of lettuce (*Lactuca sativa* L. var. Commander) in a semi-arid environment. Asian J Agric Sci 4:58–64
Mashela PW (2002) Ground wild cucumber fruits suppress numbers of *Meloidogyne incognita* on tomato in microplots. Nematropica 32:13–19
Mashela PW (2014) Soil allelochemical residue effects in a tomato cowpea rotation – nodulation and productivity of cowpea and nematode suppression. Acta Agric Scand Sect B Soil Plant Sci 64:372–375
Mashela PW, De Waele D, Pofu KM (2011) Use of indigenous *Cucumis* technologies as alternative to synthetic nematicides in management of root-knot nematodes in low-input agricultural farming systems: a review. Sci Res Essays 33:6762–6768
Mashela PW, Dube ZP (2014) Soil allelochemical residue effects of nemafric-BL and nemarioc-AL phytonematicides on soil health, growth of sweet sorghum and *Meloidogyne* species. Acta Agric Scand Sect B Soil Plant Sci 64:79–84
Mashela PW, Mpati KW (2002) Effects of castor bean fruit residues on population densities of *Meloidogyne incognita*, soil properties and growth of tomato plants. Afr Plant Prot 10:125–142
Mashela PW, Mphosi MS (2002) Wild cucumber fruit residues reduce population densities of *Meloidogyne incognita* on tomato plants. Afr Plant Prot 10:125–142
Mashela PW, Mphosi MS, Shimelis H et al (2007) Interactions of *Cucumis myriocarpus*, *Lippia javanica* and *Ricinus communis* organic amendments on suppression of *Meloidogyne incognita*. J Phytopathol 55:690–693
Mashela PW, Nthangeni ME (2002) Efficacy of *Ricinus communis* fruit meal with and without *Bacillus* species on suppression of *Meloidogyne incognita* and growth of tomato. J Phytopathol 150:399–402

Mashela PW, Pofu KM, Dube ZP (2015) Managing phytotoxicities in phytonematicides: the dosage model. J Nematol 46:201–201

Mashela PW, Pofu KM, Nzanza B (2012) Suitability of *Brassica oleracea* leaves in managing *Meloidogyne incognita* through the ground leaching technology system under microplot conditions. Acta Agric Scand Sect B Soil Plant Sci 63:19–24

Mashela PW, Shimelis HA, Mudau FN (2008) Comparison of the efficacy of ground wild cucumber fruits, aldicarb and fenamiphos on suppression of the root-knot nematode in tomato. J Phytopathol 156:264–267

Mathabatha RV, Mashela PW, Mokgalong NM (2016) Sensitivity of Nemarioc-AL and Nemafric-BL phytonematicides to volkameriana citrus rootstock seedlings. Transylvanian Rev 7:969-972

McSorley R (2003) Adaptations of nematodes to environmental extremes. Florida Entomol 86:138–142

McSorley R (2011) Overview of organic amendments for management of plant-parasitic nematodes, with case studies from Florida. J Nematol 43:69–81

McSorley R, Gallaher RN (1995) Cultural practices improve crop tolerance to nematodes. Nematropica 25:53–60

Miami IH, Rodríguez-Kábana R (1982) Organic amendments with high tannin and phenolic contents for control of *Meloidogyne arenaria* in infested soil. Nematropica 12:221–234

Mothata TS (2006) Evaluation and verification of resistance in selected vegetable crops for sustainable root-knot nematode management in developing agriculture. MSc Dissertation, North-West University, Potchefstroom

Mphahlele RR, Mashela PW, Pofu KM (2012) Post-harvest fruit decay-inducing pathogen in medicinally important *Cucumis* species indigenous to South Africa. Afr J Agric Res 6:3786–3791

Mtshali M, Fourie H, Mc Donald AH (2002a) Incidence of root-knot nematodes in resource-poor agriculture in South Africa. Afr Plant Prot 8:14–15

Mtshali M, Fourie H, Mc Donald AH (2002b) Host suitability of commercial South African dry bean cultivars to *Meloidogyne javanica* and *M. incognita* race 2. Afr Plant Prot 8:14–15

Muedi HTH (2005) Bioactivity of *Cucumis myriocarpus* fruit extracts on nematodes and bacteria. MSc Dissertation, University of Limpopo, Sovenga

Muller R, Gooch PS (1982) Organic amendments in nematode control: an examination of the literature. Nematropca 12:319–326

Ngobeni GL, Fourie H, Mc Donald AH et al (2011) Host suitability of selected South African maize genotypes to the root-knot nematode species *Meloidogyne incognita* race 2 and *Meloidogyne javanica*. SA J Plant Soil 28:49–54

Nono-Womdim R, Swai IS, Mrosso LK et al (2002) Identification of root-knot nematode species occurring on tomatoes in Tanzania and resistant lines for their control. Plant Dis 86:127–130

Ntidi NK, Fourie H, Mc Donald AH et al (2012) Plant-parasitic nematodes associated with weeds in developing agriculture in South Africa. Nematol 14:875–887

Nzanza B, Mashela PW (2012) Control of whiteflies and aphids in tomato (*Solanum lycopersicum* L.) by fermented plant extracts of neem leaf and wild garlic. Afr J Biotechnol 11:16077–16082

Oka Y (2010) Mechanisms of nematode suppression by organic soil amendments: a review. Appl Soil Ecol 44:101–115

Okwute SK (2012) Plants as potential sources of pesticidal agents: a review. In: Soundararajan RP (ed) Pesticides: advances in chemical and botanical pesticides. Academic, Detroit

Pelinganga OM, Mashela PW (2012) Mean dosage stimulation range of allelochemicals from crude extracts of *Cucumis africanus* fruit for improving growth of tomato plant and suppressing *Meloidogyne incognita* numbers. J Agric Sci 4:8–12

Pelinganga OM, Mashela PW, Mphosi MS et al (2013a) Using density-dependent growth patterns of tomato plants to establish application intervals for 3% Nemarioc-A phytonematicide. Afr Crop Sci Confer Proc 11:343–347

Pelinganga OM, Mashela PW, Mphosi MS et al (2013b) Using computer-based model to determine phytotoxicity concentration of Nemarioc-A phytonematicide in tomato production. Afr Crop Sci Confer Proc 11:349–353

Pelinganga OM, Mashela PW, Mphosi MS et al (2013c) Optimizing application frequency of diluted (3%) fermented *Cucumis africanus* fruit in tomato production and nematode management. Acta Agric Scand Sect B Soil Plant Sci 63:278–282

Pelinganga OM, Mashela PW, Nzanza B et al (2012) Baseline information on using fermented crude extracts from *Cucumis africanus* fruit for suppression of *Meloidogyne incognita* and improving growth of tomato plant. Afr J Biotech 11:11407–11413

Rice EL (1984) Allelopathy. Academic, New York

Riekert HF (1999) Screening of local cowpea cultivars and lines for resistance to *Meloidogyne javanica*. Afr Plant Prot 5:11

Shadung KG, Mashela PW (2016) Suitable organ(s) in wild *Cucumis africanus* for the production of nemafric-BL phytonematicide. Acta Agric Scand Sect B Soil Plant Sci 66:381–383

Shadung K, Mashela P, Mulaudzi V et al (2015) Optimum harvest time of *Cucumis africanus* fruit using concentration of cucurbitacin B as a maturity standard. J Agric Sci 7:181–186

Sikora RA, Bridge J, Starr JL (2005) Management practices: an overview of integrated nematode management technologies. In: Luc M, Sikora RA, Bridge J (eds) Plant parasitic nematodes in subtropical and tropical agriculture. CAB International, Wallingford, pp 259–318

Sikora RA, Fernandez E (2005) Nematode parasites of vegetables. In: Luc M, Sikora RA, Bridge J (eds) Plant parasitic nematodes in subtropical and tropical agriculture. CAB International, Wallingford, pp 319–392

Sithole NT (2016) Mean concentration stimulation point of nemarioc-AL and nemafric-BL phytonematicides on *Pelargonium sidoides*: an indigenous future cultigen. MSc Dissertation, University of Limpopo, Sovenga

Steyn WP, Daneel MS, Slabbert MM (2013) Evaluation of *Amaranthus* species for their host suitability to the root-knot nematodes, *Meloidogyne incognita* race 2 and *Meloidogyne javanica* in South Africa. Acta Hort 1007:403–407

Steyn WP, Daneel MS, Slabbert MM (2014) Host suitability and response of different vegetable genotypes to *Meloidogyne incognita* race 2 and *Meloidogyne javanica* in South Africa. Inter J Pest Manag 60:59–66

Stirling GR (2014) Biological control of plant parasitic nematodes: soil ecosystem management in sustainable agriculture. CAB International, Wallingford doi:10.1079/9781780644158.0000

Tefu G, Daneel MS, Steyn WP et al (2005) The effect of organic amendments in integrated crop production for vegetables in rural farming. Afr Plant Prot 12:11

Tefu G, Daneel MS, Steyn WP et al (2009) The effect of organic amendments in integrated crop production for vegetables in rural farming. S Afr J Plant Soil 27:256–290

Tefu G, Daneel MS, Steyn WP et al (2014) Different juices and oils as soil amendments used in control of *Meloidogyne incognita* on tomato. J Nematol 46:246

Thoden TC, Korthals GW, Termorshuizen AJ (2011) Organic amendments and their influences on plant-parasitic and free-living nematodes: a promising method for nematode management? Nematol 13:133–153

Thovhakale LP, Van der Westhuizen C, Mashela PW (2006) Responses of root-knot nematode densities to aqueous extracts of chili and tamboti. J Elsenburg 2:10–12

Van der Berg S (2006) Public spending and the poor since the transition in democracy. In: Bhorat J (ed) Poverty and policy in post-apartheid South Africa. Human Research Council, Pretoria, pp 201–231

Van Dam NM, Tytgat TOG, Kirkegaard JA (2009) Root and shoot glucosinolates: a comparison of their diversity, function and interactions in natural and managed ecosystems. Phytochem Rev 8:171–186

Van Wyk BE, Wink M (2004) Medicinal plants of the world. Briza Publications, Pretoria

Van Wyk BE, Van Heerden F, Van Oudtshoorn B (2002) Poisonous plants of South Africa. Briza Publications, Pretoria

Venter GA, Fourie H, Mc Donald AH (2004) The effect of initial population density and growing season on root-knot nematodes of popular small-scale farmer crops. Afr Plant Prot 10:141–142

White Paper on Agriculture (1995) Department of Agriculture. Directorate Agricultural Information, Pretoria

Winde I, Wittstock U (2011) Insect herbivore counter adaptations to the plant glucosinolate-myrosinase system. Phytochemistry 72:1566–1575

Wink M, Van Wyk BE (2014) Mind-altering and poisonous plants of the world. Briza Publications, Pretoria

Wuyts N, Swennen R, De Waele D (2006) Effects of plant phenylpropanoid pathway products and selected terpenoids and alkaloids on the behaviour of the plant-parasitic nematodes *Radopholus similis*, *Pratylenchus penetrans* and *Meloidogyne incognita*. Nematol 8:89–10

Youssef MMA (2015) Biofumigation as a promising tool for managing plant parasitic nematodes. A review. Sci Agri 10:115–118. doi:10.15192/PSCP.SA.2015.10.3.115118

Yu Q (2015) Biofumigation using mustard for nematode disease control: Canadian contribution. Abstracts of the XVIII international Plant Protection Congress, 24–27 August, Berlin

Zasada IA, Ferris H (2003) Sensitivity of *Meloidogyne javanica* and *Tylenchulus semipenetrans* to isothiocyanates in laboratory assays. Phytopathol 93:747–750

Chapter 8
Nematode Pests of Maize and Other Cereal Crops

Alexander H. Mc Donald†, Dirk De Waele, and Hendrika Fourie

8.1 Introduction

The three major cereals in South Africa (SA), in terms of production volume or area planted, are maize (*Zea mays*), wheat (*Triticum aestivum*) and grain sorghum (*Sorghum bicolor*). Of these crops, maize dominates with approximately 9.95 million metric tonnes (MT) being produced from 2.6 million hectares (ha) planted during 2015 (Grain 2016). Half of the produce is used as a primary food source and the remainder as animal feed. Maize production in terms of area harvested shows a steady decline from 1980 (4.6 million ha) to 2015 (2.6 million ha) and fluctuated between 2.0 and 3.6 million ha during this period (FAO 2016; Grain 2016). Despite such fluctuations, mainly due to the periodic droughts, the gross production of maize increased from the 1980s since the productivity of the crop per ha increased markedly. A similar scenario is true for wheat production since the gap between the area under cultivation and total yield also widened substantially since the 1980s (FAO 2016; Grain 2016).

The upward trends in maize and wheat production could most probably be attributed to continuous and significant improvements in crop production

†Author was deceased at the time of publication.

A.H. Mc Donald • H. Fourie (✉)
North-West University, Unit for Environmental Sciences and Management,
Private Bag X6001, Potchefstroom 2520, South Africa
e-mail: driekie.fourie@nwu.ac.za

D. De Waele
North-West University, Unit for Environmental Sciences and Management,
Private Bag X6001, Potchefstroom 2520, South Africa

Laboratory for Tropical Crop Improvement, Department of Biosystems, University of Leuven,
Willem de Croylaan 42, 3001 Heverlee, Belgium

H. Fourie et al. (eds.), *Nematology in South Africa: A View from the 21st Century*, DOI 10.1007/978-3-319-44210-5_8

technology, including the adoption of superior cultivars (cvs). By contrast, the overall trend for grain sorghum production was downwards. In real terms sorghum production was 15 % of the gross annual production of wheat and only 3–4 % of annual maize production at the end of the first decade of the 21st century. Two factors can explain this phenomenon: (i) a decrease in demand for grain sorghum and (ii) reduced research inputs that sorghum received relative to those for the other two crops. Production of other cereals, e.g. oat (*Avena sativa*), barley (*Hordeum vulgare*), pearl millet (*Pennisetum glaucum*), rice (*Oryza sativa*) and rye (*Secale cereale*), is also low in SA. These crops largely serve niche markets, such as bird feed, brewing and traditional foods. Due to limited funding resources and only small and diverse industries to support these commodities, almost all technology and genetic sources for these crops are acquired from abroad.

Since most nematology research related to cereal crops in SA has focused on maize, the major part of this chapter is devoted to this crop. The limited information on nematode research available for barley, grain sorghum, millet, wheat and rice is briefly summarised.

8.2 Maize

8.2.1 Plant-Parasitic Nematodes Associated with Maize

In the 1970s, Walters (1979a, b) reported Hoplolaimidae; *Pratylenchus zeae* Graham, 1951; *Paratrichodorus*; and *Trichodorus* spp. as the most commonly occurring and abundant plant-parasitic nematodes in local maize fields. Before 1995, *Pratylenchus* was generally perceived as the economically most important nematode pest genus that infected maize (Walters 1979a, b; Louw 1982; Zondagh and Van Rensburg 1983; De Waele and Jordaan 1988a; Jordaan et al. 1989). Other plant-parasitic nematodes identified in association with maize crops included Criconematidae; *Ditylenchus*; *Helicotylenchus*; *Hemicycliophora*; *Longidorus*; *Meloidogyne*; *Rotylenchus*; *Scutellonema*; *Telotylenchus*; *Tylenchorhynchus; Quinisulcius; Xiphinema* spp.; *Hoplolaimus pararobustus* (Schuurmans Stekhoven and Teunissen, 1938) Sher, 1963; *Paratrichodorus lobatus* Colbran, 1965; and *Rotylenchulus parvus* (Williams, 1960) Sher, 1961 (Keetch and Buckley 1984; Kleynhans et al. 1996; Riekert 1996a; Riekert and Henshaw 1998; SAPPNS[1]). A concise summary of the most important nematodes of maize is given below.

[1] Dr Mariette Marais of the Nematology Unit, Biosystematics Division, Agricultural Research Council–Plant Protection Research Institute is thanked for the use of data from the South African Plant-Parasitic Nematode Survey (SAPPNS) database; E-mail: maraism@arc.agric.za

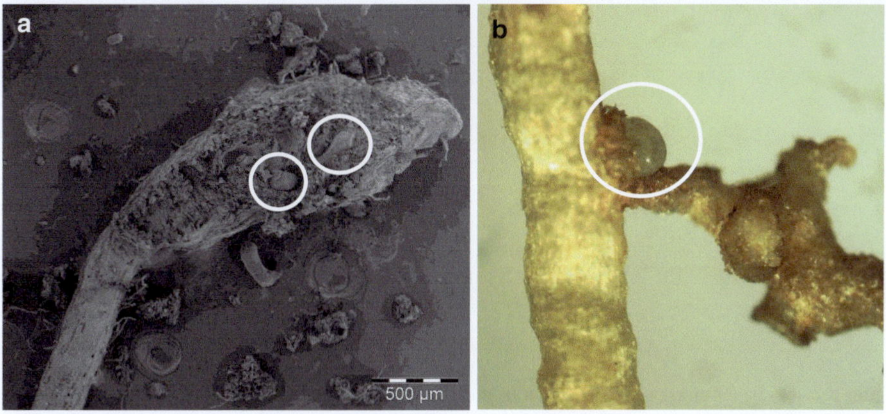

Fig. 8.1 (**a**, **b**) Root-knot nematode females (indicated by *white circles*) visible in the root tissue of a swollen, infected maize root tip (**a**) and at the junction of the tap and secondary roots (**b**) of a maize plant (**a** Louwrens Tiedt and **b** Driekie Fourie, North-West University, Potchefstroom, South Africa)

8.2.1.1 Root-Knot Nematodes

The predominant root-knot nematode species that parasitise local maize crops are *Meloidogyne javanica* (Treub, 1885) Chitwood, 1949, and *Meloidogyne incognita* (Kofoid and White, 1919) Chitwood, 1949 (Riekert 1996a; Riekert and Henshaw 1998) (Fig. 8.1a, b).

Meloidogyne arenaria (Neal, 1889) Chitwood, 1949, has also been recorded from maize fields (Kleynhans et al. 1996; Agenbag 2016; SAPPNS). The introduction of a more specialised extraction method (Riekert 1995) resulted in more accurate assessments of root-knot nematode infections in plant roots. The modified sodium hypochlorite (NaOCl) technique hence brought new perspectives to maize nematology. During the earlier work on nematode surveys in maize fields, this technique was not used, and the relative importance of *Meloidogyne* spp. on maize was not appreciated.

8.2.1.2 Lesion Nematodes

Pratylenchus zeae is generally listed as the major lesion nematode species that dominates in terms of abundance in local maize fields (Walters 1979a, b; De Waele and Jordaan 1988a), followed by *Pratylenchus brachyurus* (Godfrey, 1929) Filipjev and Schuurmans Stekhoven, 1941 (Mc Donald and De Waele 1987a, b; De Waele and Jordaan 1988a). Other lesion nematode species identified locally from maize crops are *Pratylenchus crenatus* Loof, 1960; *Pratylenchus delattrei* Luc, 1958; *Pratylenchus neglectus* (Rensch, 1924) Filipjev and Schuurmans Stekhoven, 1941; *Pratylenchus penetrans* (Cobb, 1917) Filipjev and Schuurmans Stekhoven, 1941;

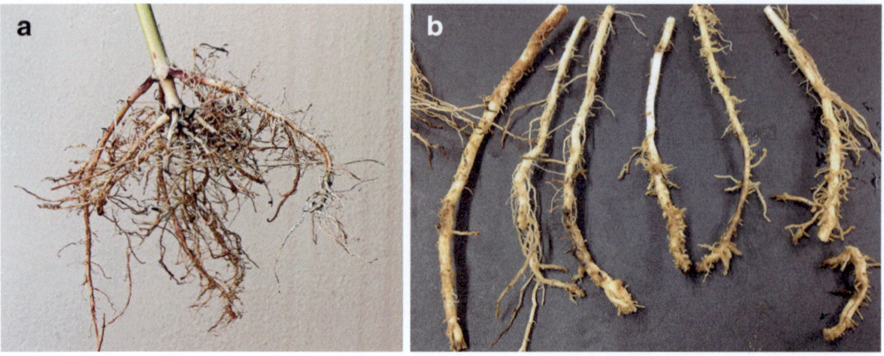

Fig. 8.2 (**a**, **b**) Root-knot nematode damage on roots of a maize plant (**a**) and a close-up of galled and stunted maize roots due to high population densities of this nematode genus (**b**) (**a** Kirk West, Port Elizabeth, South Africa and **b** Suria Bekker, North-West University, Potchefstroom, South Africa)

Pratylenchus pratensis (De Man, 1880) Filipjev, 1936; and *Pratylenchus vulnus* Allen and Jensen, 1951 (Kleynhans et al. 1996; SAPPNS).

8.2.1.3 Other Nematodes

The plant-parasitic nematode genus *Rotylenchulus* is worth mentioning here. Although *R. parvus* has been associated with maize plantings in earlier years (Louw 1982; Zondagh and Van Rensburg 1983; Keetch and Buckley 1984; Kleynhans et al. 1996), the impact and pathogenicity of this genus on the crop remain unknown (De Waele and Jordaan 1988a; Marais et al. 2009). Interestingly, exceptionally high egg and second-stage juvenile (J2) population levels (>10,000 50 g roots^{-1}) of this genus have been recorded during the past few seasons from maize under both conservation and conventional agricultural practices. However, since the identity of plant-parasitic nematode genera/species cannot be determined using morphological/morphometrical techniques, molecular analyses of eggs present in maize root samples (which represented both that of *Meloidogyne* and *Rotylenchulus*) was applied to confirm the identity of *Rotylenchulus* (Bekker et al. 2016). Routine use of the modified NaOCl method revealed that this phenomenon warrants further investigations (e.g. distribution of species involved and their pathogenicity), which are currently underway.

8.2.2 Symptoms

Symptoms of damage caused by plant-parasitic nematodes are usually not visible on below- or above-ground parts of infected maize plants (Mc Donald and Nicol 2005). However, root-knot nematode galling has been increasingly observed during the last decade on roots of maize (Fig. 8.2a, b). This is especially the case where exceptionally high infection levels of this nematode pest occur, e.g. 101,500 eggs and J2 50 g roots^{-1} from a field near Orkney (North-West Province).

Fig. 8.3 (a, b) Areas of 'poorly growing' maize plants in a root-knot nematode-infested field (a) near Viljoenskroon (Free State Province), showing stunting, and (b) phosphate deficiency visible as purple discolouration of the leaf edges of infected plants (Kirk West, Port Elizabeth, South Africa)

Anhydrobiotic *P. zeae* individuals were recorded from destroyed parenchymal cells of maize plants (cv. Pioneer 473). Their activity had resulted in small canals being formed in the plant tissue (Swanepoel et al. 1987). Also, infection by lesion nematodes can result in the formation of brown/black lesions on roots, which is difficult to identify when other pests and/or diseases are present (Mc Donald and Nicol 2005).

Generally no typical above-ground symptoms are visible in plant-parasitic nematode-infested maize fields. The occurrence of stunted and poorly developed plants (often chlorotic) (Fig. 8.3a) is, however, often attributed to high infection levels of root-knot and/or other nematode pests. Nevertheless, damage by pests and diseases other than nematodes, as well as nutrient deficiencies (Fig. 8.3b), drought conditions, excessive rainfall (water logging) and/or even plant-genetic disorders, may make it difficult to distinguish nematode-induced symptoms (Mc Donald and Nicol 2005).

8.2.3 Damage Potential

The damage potential of nematode pests is dependent on the length of their life cycle, but it is also affected by various abiotic and biotic factors. The life cycles of the two predominant nematode pests of maize are illustrated, that of *Meloidogyne* spp. in Chap. 7 (Sect. 7.3.1, Fig. 7.3) and that of *Pratylenchus* spp. below (Fig. 8.4).

The first publications on local maize nematode research (Walters 1979a, b) created an awareness of the incidence and damage potential of plant-parasitic nematodes on maize. The high sand and low organic matter contents of soils in most of the maize production areas, as well as the practice of monoculturing maize, were the main factors argued to predispose maize crops to nematode pests.

Keetch (1989) estimated a 12 % reduction in maize yields as a result of nematode damage. However, this figure referred to plant-parasitic nematodes collectively and not to a specific genus or species. Riekert (1996a, b) and Riekert and Henshaw (1998) subsequently reported maize yield losses of up to 60 % as a result of root-knot nematode parasitism, present as either single or mixed populations of *M. incognita* and *M. javanica*, in sandy soils in the North-West and Free State provinces.

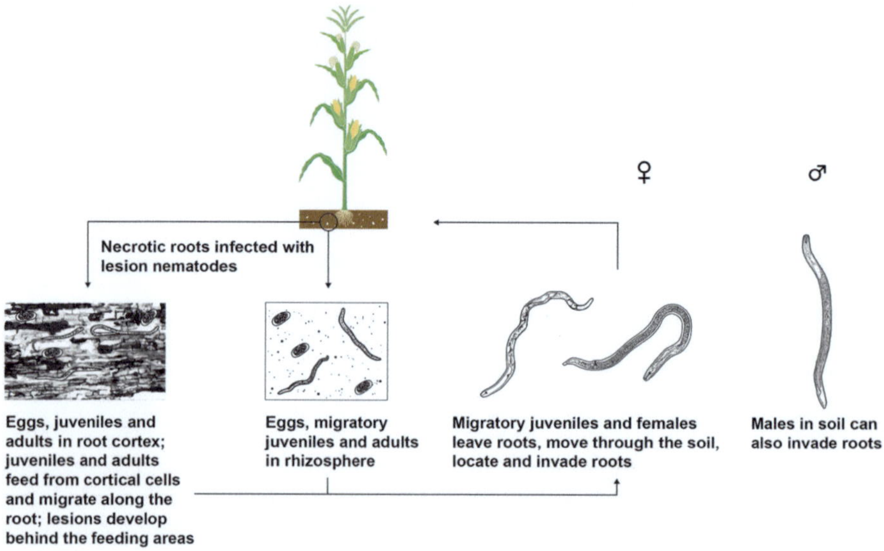

Fig. 8.4 The life cycle of lesion nematodes (Hannes Visagie, North-West University, Potchefstroom, South Africa)

Quantification of the adverse effects of nematode pests on maize is, however, difficult since the yield effect is confounded by the ability of maize plants to compensate for root damage by growing new roots to replace the damaged ones (Walters 1979b; Riekert 1996a, b). Also, due to the microscopic nature of plant-parasitic nematodes, farmers are sceptical of the extent of damage that these nematodes can cause to maize in particular. This is because maize is a so-called 'low cash' crop, with the income per MT grain being relatively small (Grain 2016) in relation to other crops such as potato or table grape (Anonymous 2016a). Any production inputs on maize that could not be related to an increase in yield would hence be considered a risk. Therefore, nematode control and particularly the application of nematicides fall into this category (see Chap. 6). This scenario is especially applicable to rain-fed maize production.

8.2.4 Management Strategies

8.2.4.1 Chemical Control

Field Studies

The application of synthetically derived nematicides was shown to substantially alleviate plant-parasitic nematode problems in maize production on sandy soils. Walters (1979b) reported yield increases (ranging from 28 to 42 %) as a result of carbofuran application in the Free State Province where *P. zeae* dominated. In the same study, 14–60 % increases in maize yields were recorded in plots that were fumigated with DD®. However, no mention was made regarding the economic implications of such nematicide treatments on the crop. Research by Zondagh and Van Rensburg (1983) showed a considerable variation in maize yield increases (ranging between 0 and 129 %) as a result of various fumigant and non-fumigant nematicide applications in the same area where Walters (1979b) did his research.

A few years later, Mc Donald and De Waele (1987b) demonstrated in five field experiments, conducted in the Free State and North-West provinces, that yield increases after nematicide applications were substantially lower than those reported by Walters (1979b) and Zondagh and Van Rensburg (1983). Mc Donald and De Waele (1987b), however, recorded yield data from one site to be significantly higher (791 kg ha^{-1}) for EDB® treated compared to untreated control plots. The plant-parasitic nematode complex at this site constituted *Criconemoides sphaerocephalus* Taylor, 1936; *Nanidorus minor* (Colbran, 1956) Siddiqi, 1974 (then reported as *Paratrichodorus minor*); a mixed population of *P. zeae* and *P. brachyurus* (95:5 ratio); *R. parvus*; and *Scutellonema brachyurus* (Steiner 1938) Andrássy, 1958. Conversely, in another site, plots treated with EDB® yielded significantly less than that of the untreated control plots, which was ascribed to a phytotoxic effect of bromide residues. Application of aldicarb, however, did not result in significantly higher yields but suppressed plant-parasitic nematode population levels significantly at two of the sites (Mc Donald and De Waele 1987b).

In another study, application of products with active substances (a.s.) cloethocarb and carbofuran were shown to reduce nematode pest complexes significantly (constituting Dorylaimidae, *Meloidogyne* spp., *Paratrichodorus* spp., and *Pratylenchus* spp.) in two field experiments in the Mpumalanga Province (Van Rensburg 1988). Concurrently, maize yield increases as a result of cloethocarb applications ranged from 52 to 110 % compared to the untreated control, whilst that for carbofuran was 59 %.

Riekert (1996a) later conducted seven field experiments with granular nematicides over four seasons in the western maize production areas in SA where *M. incognita* and *M. javanica* dominated as either single or mixed populations. Yield increases, ranging from 50 to 500 kg ha^{-1}, were recorded in these experiments as a result of nematicide application. The major feature of Riekert's nematode control research was, however, inconsistency in the results. Significant yield increases were obtained in some experiments, but in the majority the cost of the nematicide treatment was greater than the monetary value of the increase in yield. Unfortunately, these field experiments covered a period (1991–1994) during which seasonal rainfall patterns fluctuated substantially. Erratic rainfall is a reality in local maize production and thus presents an inherent risk when using a nematicide. This risk has considerable financial implications since nematode damage in maize is more commonly disregarded until severe infestation build-up becomes a reality.

Glasshouse Studies

Glasshouse experiments during which nematicides were evaluated for their efficacy on maize were also conducted. However, no correlations were evident between population levels of a nematode pest complex comprising *Criconemoides*,

Helicotylenchus, *Meloidogyne*, *Pratylenchus*, *Rotylenchulus* and *Xiphinema* spp. and plant variables (height, shoot mass and root mass) after planting with carbofuran (Meintjies 1993).

Riekert (1996b), however, recorded significant maize yield increases (ranging from 8 to 23 %) due to nematicide applications in a glasshouse experiment although inconsistencies occurred.

The period of the late 1980s and early 1990s was dominated not only by major political changes in SA but also in agriculture, and, along with it, applied plant nematology research took on new dimensions. The era, during which the frequent use of nematicides in low cash crops such as cereals boomed, ended due to an array of external factors. Some of the most effective nematicides came under immense global pressure not only due to environmental concerns but also because the benefits of chemical control did not exceed the cost for rain-fed maize production in particular.

8.2.4.2 Genetic Host Plant Resistance

During the initial testing of nematicides on maize, Walters (1979b) included different genotypes in some of his experiments and found variable genetic responses. All genotypes (e.g. cvs., hybrids and open-pollinated varieties), viz. PNR 95, SA 4, SSM 48, SA 11, SR 52 and A 471 W, were evaluated for their host suitability to a nematode pest complex that mainly consisted of *P. zeae*. Most were identified as susceptible.

Resulting from another study, Zondagh and Van Rensburg (1983) were surprised by the lack of resistance to plant-parasitic nematodes in a local composite (referred to as composite PWA). This composite was developed by maize breeders over many seasons in field sites where soils contained high population levels of plant-parasitic nematodes, including root-knot nematodes.

In a glasshouse experiment, Van Biljon and Meyer (2000) reported that maize cv. SNK 2340 supported the highest population levels of *P. zeae* and *P. delattrei* compared to tobacco (*Nicotiana tabacum*), weeping love grass (*Eragrostis curvula*), Rhodes grass (*Chloris gayana*), oat, soybean (*Glycine max*), pearl millet and wheat. High reproduction factor values obtained for *P. zeae* (ranging from 3 to 25) and *P. delattrei* (ranging from 30 to 143) demonstrated the high susceptibility of this maize cv.

Jordaan and De Waele (1987) compiled a comprehensive review of nematode resistance in maize in SA. They noted that nematodes were not considered a priority input-related item on breeders' or marketers' agendas for various reasons. At the time of the review, it was generally accepted that, as in other countries, *Pratylenchus* spp. were predominant and the major causal nematode pests that damaged maize plants. It was suggested that the interaction between lesion nematodes and maize needs to be better understood before major inroads could be made in terms of maize-nematode resistance (Jordaan and De Waele 1987).

More recently, the host status of root-knot nematodes in local commercially available maize genotypes was investigated (Ngobeni et al. 2011). Numerous genotypes planted by both commercial and smallholding farmers were screened against local populations of *M. incognita* race 2 and *M. javanica,* along with an inbred line (MP712W) from the USA with proven resistance (Aung et al. 1990; Windham and Williams 1987). Various cvs, e.g. DKC80-10 and AFG4410, proved highly susceptible to both nematode species, whilst others such as DKC78-15B, PHB3203 and DKC61-25B were resistant to one but not to the other. This scenario poses a problem since *M. incognita* and *M. javanica* often occur in mixed populations in local maize-based production areas (Kleynhans 1991; Riekert 1996a; Riekert and Henshaw 1998). Also, cvs resistant to one of these species but susceptible to the other can stimulate one to dominate in a particular field and hence adversely affect successive crops. The genetic variability in local maize germplasm with regard to resistance to the two predominant *Meloidogyne* spp. was demonstrated in the study by Ngobeni et al. (2011). In addition, the potential use of the USA line as a resistant donor parent in local maize breeding was realised.

The main concern about developing nematode resistance in maize is the lack of incentives to breeders of seed companies or even public breeding institutions (Mc Donald and Nicol 2005). The occurrence and effect of nematodes on the crop are generally still not regarded as a priority. However, this perception might soon change as a result of several interrelated factors starting to dominate in modern-day agriculture. These include the decreasing availability of effective nematicides (see Sect. 6.3) for use on maize. Another major new development is the number of maize farmers that are forced by constant economic pressures to revert to conservation or precision agriculture to reduce input costs. These two types of production are of such a nature that maize nematology research will require a new approach. Nematode control will require much more intensive and regular management inputs and closer interaction between nematology advisor and grower. Hence nematode resistance in maize genotypes will become much more important, based on its usefulness in nematode management systems. An important consideration in this sense is that sufficient nematode-resistant material has to be available for introgression into popular, high-yielding and mostly genetically modified maize genotypes. It would be unwise to rely on too few resistance donors, even should they be genetically modifiable. In SA, ways also need to be found to provide nematode resistance in the informal seed market, where maize is commonly rotated or intercropped with crops that are highly susceptible to root-knot nematodes. Such crops are soybean (Fourie et al. 2015), sunflower (*Helianthus annuus*) (Bolton and De Waele 1989; Bolton et al. 1989), tomato (*Solanum lycopersicon*) (Fourie et al. 2012), cowpea (*Vigna unguiculata*) (Riekert and Henshaw 1998) and Bambara (*Vigna subterranea*) (Mc Donald and De Waele 1989).

8.2.4.3 Crop Rotation and Alternative Hosts

Since the large-scale expansion of the local maize market following the success of the green revolution (Borlaug et al. 1969), monoculturing of the crop became common practice in SA. However, Louw (1982) concluded that maize monoculture was not successful in suppressing most plant-parasitic nematode populations that infected the crop.

The negative effects of crop rotation were highlighted by Riekert and Henshaw (1998) when rotating maize (cv. PAN6043) with oilseed crops in a sandy soil in the Free State Province. Significant increases (up to 189-fold) in levels of a mixed *M. incognita* and *M. javanica* population (70:30 ratio) were demonstrated in this study when cowpea (cv. Glenda) and soybean (cv. Knap) were included once in a maize-based rotation sequence over four consecutive growing seasons. Where initial root-knot nematode populations were already relatively high (>20,000 50g roots^{-1}), a 3-fold increase in root-knot nematode numbers in maize-soybean rotations was recorded. Concomitant yield losses of 44 % after one groundnut rotation (cv. Sellie), 55 % after one soybean rotation and 60 % after one cowpea rotation were recorded for maize in this study.

The importance of the above data is to demonstrate that nematode pest population composition and levels need to be assessed and monitored regularly in maize-based cropping systems. The intention is not to discourage crop rotation. Several other crops that are commonly rotated with maize have been demonstrated either to host nematode species that could damage maize or that maize could be an intermediary host to species that could damage crops that are rotated with maize. Bolton and De Waele (1989), however, reported that maize could be rotated with sunflower to reduce *P. zeae* populations. The same authors cautioned that when *M. incognita* and *M. javanica* are present in such fields, rotation of maize and sunflower is not advisable. Ntidi et al. (2012, 2015) also demonstrated that weeds commonly found in maize fields are susceptible to root-knot nematodes and serve as reservoirs of these pests.

8.2.4.4 Alternative Control Options

Safer and less expensive alternatives were also investigated for their effects on nematode pests associated with local maize crops. These included evaluation of a seaweed concentrate (De Waele et al. 1988) as well as that of various popular herbicides (Jordaan and De Waele 1988).

In terms of the seaweed product, the reproduction of *P. zeae* in an in vitro experiment was significantly reduced (47–63 %) when compared to an untreated control. However, in a glasshouse experiment, the reduction range was substantially lower (22–31 %). Furthermore, a phytotoxic effect was evident on maize plants, which apparently made plants more susceptible to attack by *P. zeae*. The authors concluded that the correct time and method of application as well as the concentration of the seaweed product influenced the reproduction of this lesion nematode species. From the herbicide study, it was concluded that products that contained different a.s., viz. atrazine, alachlor and 2,4-D, did not reduce *P. zeae* population levels in roots of maize plants (Jordaan and De Waele 1988).

At present only two biologically based products are registered on maize in South Africa. Both are seed coat products, viz. the one being Avicta® 500FS (Van Zyl

2013) that contains secondary metabolites of the soil-inhabiting bacterium *Streptomyces avermitilis* as a.s. The other biological product with nematicidal properties registered is Poncho®VOTiVo® with *Bacillus firmus* as the a.s. against nematode pests (Anonymous 2016b).

8.2.5 Interaction of Plant-Parasitic Nematodes with Soil-Inhabiting Micro-organisms

Several authors emphasised the potential adverse impact of soilborne pathogens, other than nematodes, that occur in local maize fields and limit production. These primarily include a range of root rots caused by *Fusarium* spp. that occur concomitantly with nematode pests (Walters 1979a, b; Zondagh and Van Rensburg 1983; Mc Donald and De Waele 1987a, b). Already during the early years of maize nematode research, Walters (1979a, b) warned that the contribution of root pathogens in terms of losses in local maize plantings would increase.

Results from a glasshouse study showed that the combined effect of two lesion nematode species (*P. brachyurus* and *P. zeae*) and the root rot fungus *Fusarium moniliforme* was greater than that of the individual organisms (Jordaan et al. 1987). This suggested the existence of a synergistic effect between the nematode pests and fungal pathogen. Plant height and stalk length of plants inoculated with both the fungus and lesion nematodes were significantly lower 2 weeks after planting compared to those where the organisms were applied individually. Furthermore, a treatment that contained both lesion nematodes and fungus suppressed plant growth more during the seedling stage than did the separate treatments with the individual organisms. The latter study also showed that *P. brachyurus* or *P. zeae* did not enhance fungal infection when they were inoculated prior to the fungus. The inoculation of lesion nematodes after fungi inoculation, however, resulted in an overall lower plant growth index 12 weeks after planting, indicating that *F. moniliforme* infection possibly facilitated nematode attraction/penetration. Results from this study showed that inoculation of approximately 500 lesion nematodes seedling^{-1} induced severe root rot symptoms, ranging from less than 10 but up to 60 %.

8.3 Grain Sorghum

Grain sorghum production was estimated at 265,000 MT during the 2014/2015 growing season from 71,000 ha being harvested. This crop is mainly cultivated in drier areas of SA (Grain 2016), with the Free State and Mpumalanga provinces representing the major production areas (Du Plessis 2008).

A range of plant-parasitic nematodes are associated with sorghum in SA (De Waele and Mc Donald 2000; SAPPNS). According to an extensive nematode survey from eight sorghum production areas, *Pratylenchus* spp. were the most abundant in root samples followed by *R. parvus*, *Meloidogyne* spp. and individuals from the Hoplolaimidae (De Waele and Jordaan 1988b). The predominant lesion nematode species was *P. zeae*, followed by *P. penetrans*, *P. crenatus* and *P. brachyurus*. In terms of the Hoplolaimidae, *Rotylenchus devonensis* Van den Berg, 1976; *Rotylenchus mabelei* Van den Berg and De Waele, 1989; and *Scutellonema brachyurus* and *Scutellonema sorghi* Van den Berg and De Waele, 1989, were reported (De Waele and Jordaan 1998a; Van den Berg and De Waele 1989a; Kleynhans et al. 1996). The latter authors also listed the root-knot nematode species *Meloidogyne acronea* Coetzee, 1956; *M. arenaria; M. incognita*; and *M. javanica* as infecting grain sorghum.

Basson et al. (1990) identified grain sorghum as a host for the peanut pod nematode *Ditylenchus africanus* Wendt, Swart, Vrain and Webster, 1995 (then reported as *Ditylenchus destructor* Thorne, 1945). Furthermore, *Longidorus pisi* Edward, Misra and Singh, 1964; *Paralongidorus lutosus* (Heyns, 1965) Escuer and Arias, 1997; *Paratrophurus anomalus* Kleynhans and Heyns, 1983; *N. minor*; *Xiphinema bourkei* Stocker and Kruger, 1988; *Xiphinema limpopoensis* Heyns, 1977; and *Xiphinema mluci* Heyns, 1976, were identified from soil samples obtained from sorghum fields (De Waele and Jordaan 1998a; Kleynhans et al. 1996; SAPPNS).

Mc Donald and Van den Berg (1993) reported that no effect on plant growth variables was recorded when *P. zeae*-infected sorghum (cv. NK304) was exposed to water stress in a glasshouse experiment. However, *P. brachyurus*-infected plants were significantly longer and had significantly higher root masses compared to uninfected plants.

8.4 Wheat

Wheat was produced on 477,000 ha during the 2014/2015 growing season, with approximately 1.8 million MT being harvested (Grain 2016). The major wheat production areas are in descending order: the southwestern parts of the Western Cape (Swartland and Rûens), Northern Cape, Free State, North-West, Mpumalanga, Limpopo, KwaZulu-Natal, Gauteng and Eastern Cape provinces (DAFF 2010a).

Numerous plant-parasitic nematodes have been associated with wheat crops (Jordaan et al. 1992; Kleynhans et al. 1996; SAPPNS). *Pratylenchus* spp. dominated as reported from a nematode survey that was conducted in seven major wheat production areas of SA, with *P. neglectus* being the most abundant. Other lesion nematode species identified during this study included *P. brachyurus*, *P. crenatus* and *P. zeae*, whilst *P. penetrans* and *Pratylenchus thornei* Sher and Allen, 1953, were also listed to infect wheat (Kleynhans et al. 1996; SAPPNS). Jordaan et al. (1992) also recorded the presence of *D. africanus* (then reported as *D. destructor*); *Hoplolaimus pararobustus*; *Heterodera avenae* Wollenweber, 1924; *Geocenamus brevidens* (Allen, 1955) Brzeski, 1991; *N. minor*; *Rotylenchulus parvus*; *Rotylenchus unisexus* Sher, 1965; *Rotylenchus mabelei*; *Scutellonema brachyurus; Scutellonema dreyeri* Van den Berg and Heyns, 1973; *Paratylenchus minutus* Linford, Oliveira and Ishii, 1944; *Tylenchorhynchus* sp.; and *Xiphinema* sp. Kleynhans et al. (1996) listed the root-knot nematode species *M. arenaria*; *Meloidogyne chitwoodi* Golden, O'Bannon, Santo and Finley, 1980; *M. incognita*; and *M. javanica* in association with wheat, whilst *Criconema*, *Criconemoides*, *Dorylaimellus*, *Geocenamus*, *Helicotylenchus*, *Hemicycliophora*, *Longidorus*, *Paralongidorus*, *Paratylenchus*, *Pratylenchoides, Rotylenchulus, Quinisulcius, Scutellonema* and *Xiphinema* spp. are also associated with wheat (SAPPNS).

In a glasshouse host suitability experiment, Van Biljon and Meyer (2000) reported that wheat cv. SST 825 maintained low population levels of both *P. zeae* and *P. brachyurus* (reproduction factor values <1), indicating the poor host status of the cv.

8.5 Rice

Only 1,150 ha of rice were planted in SA during 2013 from which 3,000 MT were produced (FAO 2016).

The first records of plant-parasitic nematodes associated with local rice plants listed *Ditylenchus angustus* (Butler, 1913) Filipjev, 1936; *M. arenaria; M. incognita*; and *M. javanica* (Keetch and Buckley 1984). Added to this list were *Brachydorus tenuis* De Guiran and Germani, 1968; *Criconema corbetti* (De Grisse, 1967) Raski and Golden, 1966; *Criconemoides incisus* Raski and Golden, 1966; *Criconemoides obtusicaudatus* Heyns, 1962; *C. sphaerocephalus*; *Helicotylenchus digonicus* Perry in Perry, Darling and Thorne, 1959; *H. dihystera*; *Helicotylenchus erythrinae* Zimmermann, 1904; *Hemicriconemoides brachyurus* (Loos, 1949) Chitwood and Birchfield, 1957; *Hemicriconemoides cocophilus* (Loos, 1949) Chitwood and Birchfield, 1957; *Hemicycliophora oryzae* (De Waele and Van den Berg, 1988); *Hemicycliophora typica* de Man, 1921; *H. pararobustus*; *L. pisi*; *N. minor*; *P. lobatus*; *P. brachyurus*; *P. zeae*; *Rotylenchus gracilidens* (Sauer, 1958) Sauer, 1958; *R. unisexus; S. brachyurus*; and *Trichodorus petrusalberti* De Waele, 1988 (De Waele and Van den Berg 1988; Van den Berg and De Waele 1989b; SAPPNS).

8.6 Millet

Pearl millet represents a small grain crop in terms of its local production and is mainly grown by subsistence farmers who use it as a staple food source and a beverage (DAFF 2011). Only a few plant-parasitic nematodes have been associated with pearl millet in SA, namely, *Meloidogyne acronea*; Coetzee, 1956; *M. incognita*; *M. javanica*;

Pratylenchus scribneri Steiner, 1943; *R. parvus*; and *Tylenchorhynchus brevilineatus* Williams, 1960 (Kleynhans et al. 1996; SAPPNS).

During 2000, Van Biljon and Meyer reported that an undisclosed pearl millet cv. supported medium to high reproduction factor values (up to 1.9) for *P. zeae* and (up to 13) *P. delattrei* in a glasshouse experiment. This illustrated the potential of high population level build-ups of lesion nematode species in fields where such nematode pests occur.

8.7 Barley, Rye and Oat

In 2013, barley production in SA amounted to 268,000 MT from 80,000 ha. Equivalent figures for oat were 59,000 MT from 27,000 ha and for rye 1,950 MT from 3,600 ha (FAO 2016). Barley is produced in various areas in the Western Cape, Northern Cape, Free State, Eastern Cape, KwaZulu-Natal, North-West and Limpopo provinces. Rye is a winter cereal which prefers subtropical to temperate areas in terms of its cultivation (Anonymous 2016c). Oat is suitable for all regions of SA due to its adaptability to a wide range of environmental conditions and high biomass production (DAFF 2010b).

An unidentified *Meloidogyne* sp. (SAPPNS), *M. incognita* and *R. incultus* have been reported to infect local barley crops (Kleynhans et al. 1996; Keetch and Buckley 1984).

The following plant-parasitic nematodes have been associated with rye: *C. sphaerocephalus*; *Ditylenchus* spp.; *H. dihystera; Hoplolaimus capensis* Van den Berg and Heyns, 1970; *M. arenaria*; *P. zeae*; *Rotylenchus incultus* Sher, 1965; *R. unisexus* as well as *S. brachyurus* (Keetch and Buckley 1984; Kleynhans et al. 1996; SAPPNS).

For oat several plant-parasitic nematodes have been recorded, viz. *Ditylenchus equalis* Heyns, 1964; *G. brevidens*; *H. dihystera; Hemicycliophora* spp.; *Meloidogyne* sp.; *M. javanica*; *M. hapla; M. incognita*; *P. crenatus*; *P. brachyurus*; *P. zeae*; *R. parvus*; and *S. brachyurus* (Keetch and Buckley 1984; Kleynhans et al. 1996; SAPPNS).

In a glasshouse experiment, Van Biljon and Meyer (2000) showed the poor host susceptibility of oat cv. Maluti to *P. zeae* (reproduction factors <1). For *P. delattrei*, the reproduction factors ranged from approximately 2 to 5. This illustrated that cv. Maluti can support high population levels of *P. delattrei* in fields where this nematode pest occur.

8.8 Conclusions

The most important challenge to local nematologists remains to find suitable, effective and sustainable measures for producers to manage cereal crops in ways that would keep plant-parasitic nematode populations below damage threshold levels. An important nematological aspect relating to all cereal crops produced in this country is the uncertainty of how widely economically important plant-parasitic nematodes are distributed and what the effects of different species have on each crop. Concrete proof and sound economic bases of damage could change general perceptions and bring greater benefits to producers and related concerns in the respective crop industries. Differences in crop-genotype susceptibility to the main nematode pests should also receive high priority in terms of research.

Another prominent knowledge gap is the effect that different forms of soil tillage might have on total nematode community structures and compositions under various cropping systems and abiotic conditions. Closely related to this is the need for investigations on interactions between root pathogens (fungi in particular) and nematode pests. Environmental conditions are another important variable to investigate and involve all possible crops and rotation-system variations. Nematode population dynamics and several other aspects of plant nematology relating to conservation and precision agriculture would not only contribute to support the development and adoption of these approaches but also would provide invaluable basic information about plant nematology that would previously have been very difficult to justify investigating.

Other aspects that are also often raised, speculated about but rarely been exploited include the reciprocal effects and/or dynamics between soil nematodes and important soil elements such as nitrogen (N), phosphate (P), potassium (K), microelements or even gas exchanges.

References

Anonymous (2016a) Potatoes South Africa. http://www.potatoes.co.za/home.aspx? IsHome=true. Accessed 9 Apr 2016

Anonymous (2016b) Poncho®Votivo®. http://www.cropscience.bayer.co.za/en/SeedGrowth/Poncho-Votivo.aspx. Accessed 20 Mar 2016

Anonymous (2016c) Pulses 2016: International year of pulses – Rye (*Secale cereale*). http://www.advanceseed.com/site/page?view=rye. Accessed 6 Feb 2016

Agenbag (2016) Identification and reproduction potential of South African *Meloidogyne* species. MSc Dissertation, North-West University, Potchefstroom

Aung T, Windham GL, Williams WP (1990) Reproduction of *Meloidogyne incognita* on open-pollinated maize varieties. J Nematol 22:651–653

Basson S, De Waele D, Meyer AJ (1990) An evaluation of crop plants as hosts for *Ditylenchus destructor* isolated from peanut. Nematropica 20:23–29

Bekker S, Fourie H, Rashidi M et al (2016) Discriminating between the eggs of two egg-mass producing nematode genera using morphometrical and molecular techniques. Nematol 18: 1119–1123. doi:10.1163/15685411-00003022

Bolton C, De Waele D (1989) Host suitability of sunflower hybrids to *Pratylenchus zeae*. J Nematol 21:682–685

Bolton C, De Waele D, Loots GC (1989) Plant-parasitic nematodes of field crops. 3. Sunflower. Rev Nématol 12:69–76

Borlaug NE, Aresvik O, Narvaez I et al (1969) A green revolution yields a golden harvest. Colombia J World Bus 4:9–19

DAFF (2010a) Wheat Production Guideline. Department of Agriculture, Fisheries and Forestry, Pretoria

DAFF (2010b) Oats Production Guideline. Department of Agriculture, Fisheries and Forestry, Pretoria

DAFF (2011) Pearl millet Production Guideline. Department of Agriculture, Fisheries and Forestry, Pretoria

De Waele D, Jordaan EM (1988a) Plant-parasitic nematodes on field crops in South Africa. 1. Maize. Rev Nématol 11:65–74

De Waele D, Jordaan EM (1988b) Plant-parasitic nematodes on field crops in South Africa. 2. Sorghum. Rev Nématol 11:203–211

De Waele D, Mc Donald AH (2000) Diseases caused by nematodes. In: Frederiksen RA, Odvody GN (eds) Compendium of sorghum diseases, 2nd edn. APS Press, St. Paul, pp 50–53

De Waele D, Van den Berg E (1988) Nematodes associated with upland rice in South Africa, with a description of *Hemicycliophora oryzae* sp. n. (Nemata: Criconematoidea). Rev Nématol 11:45–51

De Waele D, Mc Donald AH, De Waele E (1988) Influence of seaweed concentrate on the reproduction of *Pratylenchus zeae* (Nematoda) on maize. Nematologica 34:71–77

Du Plessis J (2008) Sorghum production. Department of Agriculture, Pretoria

FAO (2016) Statistics at FAO. http://faostat3.fao.org/faostat-gateway/go/to/home/E. Accessed 13 Feb 2016

Fourie H, Mc Donald AH, Mothata TS et al (2012) Indications of variation in host suitability to root-knot nematode populations in commercial tomato varieties. Afr J Agric Res 7:2344–2352

Fourie H, De Waele D, Mc Donald AH et al (2015) Nematode pests threatening soybean production in South Africa, with reference to *Meloidogyne*. S Afr J Sci 111(9/10), art. #2014-0212:9. doi:http://dx.doi.org/10.17159/sajs.2015/20140212

Grain SA (2016) Industry reports: report documents. http://www.grainsa.co.za/report-documents?cat=14. Accessed 1 Feb 2016

Jordaan EM, De Waele D (1987) Resistance of maize to plant-parasitic nematodes: a literature review. Technical Communication No. 212. Department of Agriculture, Pretoria, pp 60–66

Jordaan EM, De Waele D (1988) Effects of four herbicides on the infestation of maize by the nematode *Pratylenchus zeae*. S Afr J Plant Soil 5:165

Jordaan EM, Loots GC, Jooste WJ et al (1987) Effects of root-lesion nematodes (*Pratylenchus brachyurus* Godfrey and *P zeae* Graham) and *Fusarium moniliforme* Sheldon alone or in combination, on maize. Nematologica 33:213–219

Jordaan EM, De Waele D, Van Rooyen PJ (1989) Endoparasitic nematodes in maize roots in the Western Transvaal as related to soil texture and rainfall. J Nematol 21:356–360

Jordaan EM, Van den Berg E, De Waele D (1992) Plant-parasitic nematodes of field crops in South Africa. 5. Wheat. Fundam Appl Nematol 15:531–537

Keetch DP (1989) A perspective of plant nematology in South Africa. S Afr J Sci 85:506–508

Keetch DP, Buckley NH (1984) A check-list of the plant-parasitic nematodes of southern Africa. Technical Communication. Department of Agriculture, Plant Protection Research Institute, Pretoria

Kleynhans KPN (1991) The root-knot nematodes of South Africa. Technical Communication No. 231. Department of Agricultural Development, Pretoria

Kleynhans KPN, Van den Berg E, Swart A et al (1996) Plant nematodes in South Africa. Plant Protection Research Institute Handbook No 8. Agricultural Research Council–Plant Protection Research Institute, Pretoria

Louw IW (1982) Nematode pests of maize. In: Keetch DP, Heyns J (eds) Nematology in Southern Africa. Science Bulletin No. 400. Department of Agriculture and Fisheries, Republic of South Africa, Government Printer, Pretoria, pp 69–72

Marais M, Swart A, Lamprecth SC et al (2009) The influence of tillage on nematodes. S Afr J Sci Technol 28:165–166

Mc Donald AH, De Waele D (1987a) Control strategies against nematode pests on maize: a literature review. Technical Communication No. 212. Department of Agriculture, Pretoria, pp 55–60

Mc Donald AH, De Waele D (1987b) Effect of two nematicides on nematode populations associated with maize. Phytophylactica 19:475–478

Mc Donald AH, De Waele D (1989) Effect of *Meloidogyne javanica* on Bambara groundnut in South Africa. Phytophylactica 21:429–431

Mc Donald AH, Nicol J (2005) Nematode parasites of cereals. In: Luc M, Sikora RA, Bridge J (eds) Plant parasitic nematodes in subtropical and tropical agriculture, 2nd edn. CAB International, Wallingford, pp 131–191

Mc Donald AH, Van den Berg EH (1993) Effect of watering regimen on injury to corn and grain sorghum by *Pratylenchus* species. J Nematol 25:654–658

Meintjies S (1993) The interaction between carbofuran, maize and nematodes. Phytophylactica 25:253–255

Ngobeni GL, Fourie H, Mc Donald AH et al (2011) Host suitability of selected South African maize genotypes to the root-knot nematode species *Meloidogyne incognita* race 2 and *Meloidogyne javanica*. S Afr J Plant Soil 28:49–54

Ntidi KN, Fourie H, Mc Donald AH et al (2012) Plant-parasitic nematodes associated with weeds in developing agriculture in South Africa. Nematol 14:875–887

Ntidi KN, Fourie H, Daneel M (2015) Greenhouse and field evaluations of commonly occurring weed species for their host suitability to *Meloidogyne* species. Int J Pest Manag. doi:10.1080/09670874.2015.1087602

Riekert HF (1995) A modified sodium hypochlorite technique for the extraction of root-knot nematode eggs and larvae from maize root samples. Afr Plant Prot 1:41–43

Riekert HF (1996a) Economic feasibility of nematode control in dryland maize in South Africa. Afr Crop Sci J 4:477–481

Riekert HF (1996b) Greenhouse assessment of maize growth and yield response to nematode control with aldicarb. Afr Crop Sci J 4:471–475

Riekert HF, Henshaw GE (1998) Effect of soybean, cowpea and groundnut rotations on root-knot nematode build-up and infestation of dryland maize. Afr Crop Sci J 6:377–383

Swanepoel JVS, Hamilton-Attwell VL, Loots GC (1987) The use of polyethylene glycol in studying anhydrobiosis of *Pratylenchus zeae* in infested maize roots. Phytophylactica 19:367–370

Van Biljon ER, Meyer AJ (2000) Reproduction of *Pratylenchus zeae* and *P. delattrei* on crops resistant to *Meloidogyne javanica* and *M. incognita* races 2 and 4. Afr Plant Prot 6:43–45

Van den Berg E, De Waele D (1989a) *Scutellonema sorghi* n. sp., *S. dreyeri* n. sp. and *Rotylenchus mabelei* n. sp. (Nemata: Tylenchina) from sorghum in South Africa. Rev Nématol 12:357–364

Van den Berg E, De Waele D (1989b) Further observations on nematodes associated with rice in South Africa. Phytophylactica 21:125–130

Van Rensburg JBJ (1988) Efficacy of cloethocarb for control of *Busseola fusca*, *Cicadulina* spp. and nematodes in maize. Appl Plant Sci 2:63–67

Walters MC (1979a) The possible status of parasitic nematodes as limiting factors in maize production in South Africa. Technical Communication No. 142. Department of Agricultural Technical Services, Pretoria, pp 62–68

Walters MC (1979b) Present status of knowledge of nematode damage and control in South Africa. Technical Communication No. 152. Department of Agricultural Technical Services, Pretoria, pp 62–68

Windham GL, Williams WP (1987) Host suitability of commercial corn hybrids to *Meloidogyne arenaria* and *M. incognita*. J Nematol 1:13–16

Zondagh SG, Van Rensburg JBJ (1983) Progress in nematode research in maize. Technical Communication No. 182. Department of Agriculture and Fisheries, Pretoria, pp 64–66

Chapter 9
Nematode Pests of Leguminous and Oilseed Crops

Hendrika Fourie, Alexander H. Mc Donald†, Sonia Steenkamp, and Dirk De Waele

9.1 Introduction

Although a wide variety of oilseed and leguminous crops are grown in South Africa (SA), extensive nematology research has been conducted during the past three decades particularly on soybean (*Glycine max*) and groundnut (*Arachis hypogaea*). This knowledge as well as limited information available on the association and impact of plant-parasitic nematodes on other leguminous and oilseed crops (e.g. Bambara groundnut (*Vigna subterranea*), cowpea (*Vigna unguiculata*), dry bean (*Phaseolus vulgaris*), lupin (*Lupinus albus*) and sunflower (*Helianthus annuus*)) are presented in this overview.

†Author was deceased at the time of publication.

H. Fourie (✉) • A.H. Mc Donald
Unit for Environmental Sciences and Management, North-West University,
Private Bag X6001, Potchefstroom 2520, South Africa
e-mail: driekie.fourie@nwu.ac.za

S. Steenkamp
Agricultural Research Council–Grain Crops Institute,
Private Bag X 1251, 2520 Potchefstroom, South Africa

D. De Waele
Unit for Environmental Sciences and Management, North-West University,
Private Bag X6001, Potchefstroom 2520, South Africa

Laboratory of Tropical Crop Improvement, Department of Biosystems,
Willem de Croylaan 42, 3001 Leuven, Belgium

H. Fourie et al. (eds.), *Nematology in South Africa: A View from the 21st Century*, DOI 10.1007/978-3-319-44210-5_9

9.2 Soybean

Soybean is a major source of protein and oil and is used for both animal and human consumption (Liebenberg 2012). The production of the crop in SA was first recorded in 1961, being a mere 2,631 metric tonnes (MT) (Shurtleff and Aoyagi 2009). It, however, increased to 75,000 MT in 1987/88 and 1,070,000 MT during the 2015/2016 growing season (Grain 2016). Concurrently, the crop progressively increased in hectares (ha) planted from 40,000 ha in 1987/1988 to 687,300 during the 2015/2016 growing season. This is a direct reflection of the increasing and urgent need for oil and protein sources to feed a growing nation and its cattle (Liebenberg 2012). Furthermore, the expansion of local soybean production was expedited due to the introduction of genetically modified (GM), Roundup® Ready (RR) cultivars (cvs) since the 2001/2002 growing season (Anonymous 2001).

The expansion of soybean production outside the traditional growing areas, namely, the Free State, KwaZulu-Natal, Mpumalanga and Gauteng provinces, results in the exposure of this crop to new diseases and pests (Liebenberg 2012; PRF 2016). To be competitive on world markets and optimise crop production, local cvs need to be adapted to typical environmental conditions that prevail locally and exhibit resistance to endemic diseases (e.g. bacteria, fungi and viruses) and pests (e.g. insects and plant-parasitic nematodes). The most important nematode pests of soybean crops in SA are *Meloidogyne* and *Pratylenchus*, with various other genera also listed as being associated with the crop, as referred to below.

9.2.1 Plant-Parasitic Nematodes Associated with Soybean

9.2.1.1 Root-Knot Nematodes

Meloidogyne spp. pose a serious constraint to soybean production worldwide (Sikora et al. 2005a, b; Bridge and Starr 2007), including SA (Fourie et al. 2015). The peanut root-knot nematode *Meloidogyne arenaria* (Neal, 1889) Chitwood, 1949, was the first nematode pest reported to parasitise the crop locally (Van der Linde et al. 1959; SAPPNS[1]). About a decade later, Coetzee (1968) listed also *Meloidogyne hapla* Chitwood, 1949; *Meloidogyne incognita* (Kofoid and White, 1919) Chitwood, 1949; and *Meloidogyne javanica* (Treub, 1885) Chitwood, 1949, as infecting soybean crops. Resulting from an extensive survey conducted at 17 localities in local soybean production areas, Fourie et al. (2001) identified *M. incognita* as the predominant root-knot nematode species, followed by *M. javanica*; to a much lesser extent, *Meloidogyne ethiopica* Whitehead, 1968, and *M. hapla* were also identified, but at only a few localities.

[1]Dr Mariette Marais of the Nematology Unit, Biosystematics Division, Agricultural Research Council–Plant Protection Research Institute is thanked for the use of data from the South African Plant-Parasitic Nematode Survey (SAPPNS) database; E-mail: maraism@arc.agric.za

9.2.1.2 Lesion Nematodes

Various *Pratylenchus* spp. have been identified in association with local soybean crops (Keetch and Buckley 1984; Kleynhans et al. 1996; SAPPNS). In the survey conducted by Fourie et al. (2001), the predominant lesion nematodes identified were *Pratylenchus thornei* Sher and Allen, 1954; *Pratylenchus crenatus* Loof, 1960; and *Pratylenchus teres* Khan and Singh, 1974. No research has been conducted on the impact of lesion nematodes on soybean in SA.

9.2.1.3 Other Plant-Parasitic Nematodes

Criconemoides, Geocenamus, Helicotylenchus, Longidorus, Nanidorus, Radopholus, Scutellonema, Tylenchorhynchus and *Xiphinema* are also associated with local soybean crops (Keetch and Buckley 1984; Kleynhans et al. 1996; Fourie et al. 2001; SAPPNS). Although the soybean-cyst nematode *Heterodera glycines* Ichinohe, 1952, is a major pest of soybean crops in various countries (Sikora et al. 2005a, b; Bridge and Starr 2007), it has not been identified from local soybean production areas (M Marais, Agricultural Research Council–Plant Protection Research, Pretoria, 2016, personal communication).

9.2.2 Symptoms

Galling as a result of root-knot nematode parasitism is characterised by elongated to roundish, knot-like protuberances that form on soybean roots (Bridge and Starr 2007) (Fig. 9.1). The size of the galls is generally determined by many factors such as the species, host status of the cv. infected and the number of infective second-stage juveniles (J2) (Karssen et al. 2013) that developed to third- (J3) and fourth-stage juveniles (J4) and finally mature females or males. Important to bear in mind is that root-knot nematode galls can be mistaken for rhizobium nodules that are also present on soybean roots and are crucial for nitrogen fixing by the plant (Liebenberg

Fig. 9.1 Galls on soybean roots caused by root-knot nematode infection (Phillip Holtzhausen, South Africa)

Fig. 9.2 (**a**, **b**) Patches of poorly growing soybean plants in a field where high population densities of root-knot nematodes occurred (**a**) and a stunted plant with a galled root system and yellow leaves (*white circle*) next to a non-infected, healthy plant (**b**) (**a** Danie Du Plessis, Viljoenskroon, South Africa; **b** Anonymous, South Africa)

Fig. 9.3 (**a**, **b**) Necrotic, brown soybean roots infected with lesion nematodes (**a**), and a stunted plant with a reduced and necrotic root system (*white circle*) and suboptimally filled pods (*white circle*) compared to a noninfected plant (**b**) (**a** Suria Bekker and **b** Driekie Fourie, North-West University, Potchefstroom, South Africa)

2012). Galls formed by root-knot nematodes are distinguished from rhizobium nodules since the latter can easily be rubbed off the roots, while galls form an integral part of the root and cannot be removed without damaging the root. Furthermore, the inside tissue of rhizobium nodules has a pinkish colour when they are in an active developmental stage, as opposed to a greenish colour when inactive.

Above-ground symptoms of root-knot nematode infection in soybean plants are usually seen as stunting, yellowing of leaves, and patches of poor growth of infected plants (Fig. 9.2a, b).

Damage by *Pratylenchus* spp. results in brown, necrotic areas on roots (Fig. 9.3a) and a smaller root system compared to that of an uninfected soybean plant (Fig. 9.3b). Above-ground symptoms include stunting of plants with leaf chlorosis and yield loss (Sikora et al. 2005b; Bridge and Starr 2007). Pods of heavily infected plants may also be smaller and only partially filled compared to those of uninfected plants (Fig. 9.3b).

9.2.3 Damage Potential

In terms of damage caused by various plant-parasitic nematodes, Keetch (1989) reported soybean monetary yield losses of approximately 9 %. However, severe infection of soybean plants by root-knot nematodes can lead to total destruction of crops as occurred during the 1998 growing season in Mpumalanga Province (Smit and De Beer 1998). Yield losses, ranging from 25 to 70 %, have also been reported for soybean as a result of root-knot nematode parasitism (Riekert and Henshaw 1998; Fourie and Mc Donald 2001, 2007; Fourie et al. 2010).

Various factors favour increased parasitism of soybean and other rotation crops by root-knot nematodes, viz: (i) the progressive increase in area planted to soybean, (ii) current expansion of the crop into areas where maize was traditionally planted (PRF 2016), (iii) conduciveness of traditional and current cropping systems to an increase in plant-parasitic nematode population densities (Riekert and Henshaw 1998; Riekert 1996) and (iv) the lack of nematicides registered for use on the crop (Van Zyl 2013). Diagnostic analyses and research activities confirm the occurrence of exceptionally high levels of nematodes in root and rhizosphere samples of soybean crops during the past few growing seasons (Bekker et al. 2007; Fourie et al. 2011, 2015). Aggravating and complicating this situation is the occurrence of two predominant root-knot nematode species, *M. incognita* and *M. javanica*, in maize and soybean production areas, either as single or mixed populations (Riekert 1996; Fourie et al. 2001; Agenbag 2016). In addition, only one available commercial cultivar (cv. Egret) exhibits genetic resistance to *M. incognita* (Fourie et al. 2006). The GM, RR cvs that currently dominate the local market (De Beer and Bronkhorst 2015) are, in general, susceptible to populations of *M. incognita* and *M. javanica* (Venter 2014).

9.2.4 Management Strategies

The progressive withdrawal of Class 1 nematicides (Anonymous 2012; Verdoorn 2012) and the likelihood that no/limited synthetic nematicides will be registered on soybean locally in the near future are causes for concern. The wide host range of root-knot nematodes to agricultural and horticultural crops further complicates the management of these pests using crop rotation.

9.2.4.1 Host Plant Resistance

Genetic host plant resistance to root-knot nematodes represents one of the few cost-effective and environmental-friendly management strategies for reducing yield losses caused by these pests in soybean crops (Bridge and Starr 2007). Different types and mechanisms of genetic host plant resistance are presented in Box 9.1.

Box. 9.1 Glossary

Genetic host plant resistance: the prevention and/or limitation/restriction of feeding, development, reproduction and fecundity of the target plant-parasitic nematode as a result of the expression of specific genes by the host plant. Resistance is measured along a continuum, with plants usually ranging from low to moderate to highly resistant.

Highly resistant genotypes generally maintain <10 % of nematode's reproduction compared to that in a susceptible genotype and have an Rf of <1.

Vertical resistance: specific resistance that is only effective against one species/race of the target nematode and is also referred to as qualitative or monogenic/oligogenic resistance: usually controlled by one to three genes.

Horizontal resistance: general resistance that is effective against multiple species/races of the target nematode and is also referred to as quantitative or polygenic resistance: controlled by many genes.

Antibiosis: all adverse effects exerted by the host on the biology (survival, development and reproduction) of the target nematode.

Antixenosis: non-preference, representing a host plant that lacks characteristics to serve as a proper host for the target nematode.

Pre-infectional resistance: resistance exhibited by a host plant that is effective before nematode infection occurs, usually as: (i) a result of toxic substances that are present in the roots/other plant parts infected by a nematode pest and/or (ii) a barrier (e.g. thick cell wall) hindering or preventing nematode penetration.

Post-infectional resistance: resistance exhibited by a host plant after nematode infection occurs, usually as a result of substances present in roots/other plant parts that: (i) result in the formation of necrotic lesions that localise nematode infection, (ii) suboptimal development of nematode individuals, (iii) emigration of nematode individuals from the infected plant tissue and (iv) the formation of a skew proportion in male and female individuals.

Induced host plant resistance: host plants infected with plant-parasitic nematodes that subsequently develop enhanced resistance as a result of the application of an abiotic and/or biotic inducer. These include activation of defence mechanisms by the host such as enzymes and can represent:

(i) Systemic acquired resistance (SAR): generally induced by nematode pests that cause necrosis during parasitism
(ii) Induced, systemic acquired resistance (ISR): triggered by nonpathogenic organisms such as rhizobacteria

Hypersensitivity: a host plant that has a high level of damage/yield loss even when lightly infected with nematode pests.

Tolerance: host plants that withstand nematode infection and experience little yield loss.

> Intolerance: opposite of tolerance; nematode-infected plants develop sub-optimally and may even die.
>
> Definitions given above are from Roberts (2002), Aryal et al. (2011) and Starr et al. (2013).

Van den Berg and Mc Donald (1991) reported low to moderate levels of genetic resistance in 19 of the most popular, commercially available cvs screened for their host status to *M. incognita* and *M. javanica.* Fourie et al. (1999) reported substantial variation among 38 soybean cvs in terms of their host status to *M. javanica* and *M. incognita* races 1, 2 and 4; local cv. Gazelle was identified with superior genetic resistance against *M. javanica*. Although these authors identified several cvs with resistance to *M. javanica* and *M. incognita* races 1 and 4 using reproduction factor (Rf) as the main criterion (Windham and Williams 1988), none proved to exhibit resistance to *M. incognita* race 2. Van Biljon (2004), however, reported the poor host status of various commercially available soybean cvs to several races of *M. incognita* and also *M. javanica.* During another study, genetic resistance to *M. incognita* race 2 was identified by Fourie et al. (2006) in six local cvs, with that in cv. LS5995 being superior. The authors emphasised that screening for resistance should be done with different populations and races of *Meloidogyne* spp. to allow development of breeding lines with broad resistance (Hussey and Janssen 2002).

The introgression of root-knot nematode resistance into local, popular commercial genotypes was done using molecular-marker technology (Mienie et al. 2002; Fourie et al. 2008). The application of marker-assisted selection (MAS) in local nematode resistance breeding programmes expedited such efforts (Box. 9.2). This was done using the *M. javanica*-resistant cv. Gazelle and the *M. incognita*-resistant cv. LS5995. For *M. javanica*, two AFLP markers E-AAC/M-CAT1 and E-ACC/M-CTC2 mapped close to and bracketed the *M. javanica* resistance trait in cv. Gazelle. These markers respectively accounted for 25 and 42 % of the variation in *M. javanica* gall rating indices. Both markers were successfully converted to the two sequence-characterised amplified regions (SCAR), namely, SOJA7 (E-AAC/M-CAT1) and SOJA6 (E-ACC/M-CTC2), and were employed in a breeding population using MAS. In terms of *M. incognita*, a major quantitative trait loci (QTL) was identified in cv. LS5995 on Linkage Group M between markers Satt201 and Satt590 and, respectively, accounted for 62 % of the variation in gall rating indices and 80 % of those for eggs and J2 per root system in the segregating population. A minor QTL, accounting for 37 % of the variation in gall rating indices, was furthermore identified between markers Satt487 and Satt358 on LG-O.

The resistance trait exhibited by cv. Gazelle for *M. javanica* was identified as monogenic, while that for *M. incognita* in cv. LS5995 has been determined as being polygenic. Using the latter root-knot nematode resistance QTL in the respective cvs for the two *Meloidogyne* species, MAS has been successfully applied in a breeding programme of the ARC-GCI.

Box. 9.2 A schematic representation of the process generally followed during the screening of soybean genotypes for identification of root-knot nematode resistance, introgression of the trait and development of cultivars that exhibit the trait

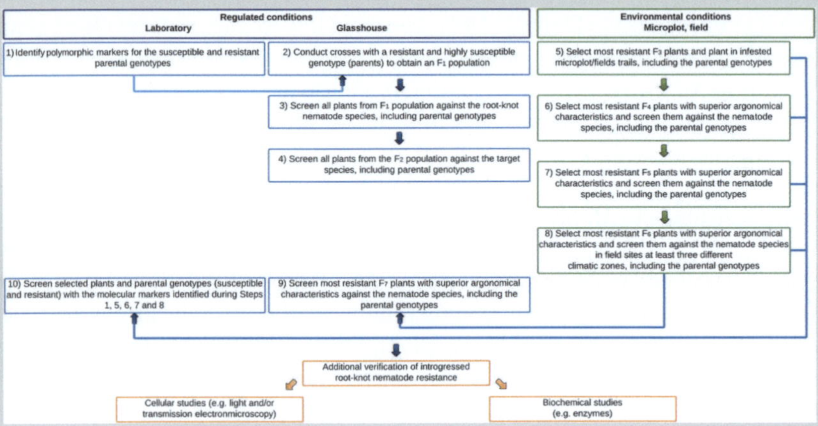

9.2.4.2 Mechanism of Resistance

The mechanism of *M. incognita* resistance exhibited by cv. LS5995 was determined by means of pre- and post-infectional as well as histopathological investigations (Fourie et al. 2013a). Data on the comparative penetration, development and reproduction rate of *M. incognita* in the roots of LS5995 and a susceptible counterpart showed that *M. incognita* J2 initially penetrated roots of both cvs in equal numbers. However, J2 numbers were significantly lower in roots of LS5995 10 days after inoculation (DAI). Moreover, roots of LS5995 maintained significantly higher numbers of sexually differentiated J2, J3 and J4 males but lower numbers of the latter two female-life stages, compared to those in roots of the susceptible cv. Concurrently, adult females in roots of LS5995 produced significantly fewer egg masses and also eggs per egg mass at both 30 and 45 DAI compared to those for the susceptible cv. The same trend was observed for the number of eggs per root system, with LS5995 maintaining 99 % fewer eggs per root system compared to that of the susceptible cv. at both 30 and 45 DAI. The resistance expressed by LS5995 was identified as typical post-infectional antibiosis.

Concurrent with pre-and post-penetration studies, cellular investigations by means of light and transmission electron-microscopy on giant cell formation in *M. incognita*-infected roots of LS5995 and its susceptible counterpart were conducted (Fourie et al. 2013a). This study confirmed that *M. incognita* J2 from 2 DAI migrated intercellularly to parenchyma cells (located in the rest meristem of the vascular cylinders) of both cvs. Giant cells in roots of LS5995 were smaller and fewer compared to those in roots of the susceptible cv., with three different types of irregular giant cells being observed in the roots of LS5995. The first type was empty and advanced degenerated giant cells. The second was characterised by the inclusion of strand-like structures, filling the cytoplasm of the cells. The thickness of cell walls of the third type of giant cells varied considerably. Information generated during histopathological studies helped explain the presence of antibiotic resistance in cv. LS5995.

Verification of Resistance

Meloidogyne incognita resistance exhibited by cv. LS5995 and six other soybean genotypes (collectively used for cvs and lines) was verified under natural environmental conditions in semi-field and field experiments at three different localities in SA (Fourie et al. 2013b). Plant host and yield responses of these genotypes were compared using non-inoculated vs nematode-inoculated treatments. Population levels of *M. incognita* in roots of LS5995 plants were significantly ($P \leq 0.05$) lower than those for a susceptible counterpart. Also in all experiments, yield of the resistant genotypes was not significantly different to the non-inoculated plants. Yield response and nematode resistance indices (Ri) were furthermore substantially lower in the resistant genotypes than in the susceptible cv., except in one field experiment. Since LS5995 consistently had Ri <1, indicating a high level of resistance, this cv. was suggested to have the most sustainable *M. incognita* resistance compared to that identified in the six other genotypes. However, yield response was generally dependent on environmental effects and thus limited further qualification of genotypes evaluated as tolerant, intolerant and hypersensitive. Due to the superiority of *M. incognita* resistance present in LS5995, it was used as the primary resistance source in local breeding programmes.

As a result of breeding efforts in which cv. LS5995 was used as the resistant parent, eight F_8 lines with superior levels of *M. incognita* resistance were developed. These lines were evaluated for their agronomic performance during two growing seasons at eight different localities within the soybean production areas. The yield data obtained for them were similar to that of cv. Egret (De Beer et al. 2014). However, these genotypes do not contain the RR gene. The use of such material and conversion thereof to genetically modified germplasm will hence add value to the local soybean industry and support sustainable production of the crop.

Effects of Increasing Population Density on Population Development and Yield

Effects of increasing initial population density levels (Pi) of *M. incognita* on nematode population development and yield of the resistant cv. LS5995 yielded valuable data. Strong non-linear relationships between all nematode variables measured as well as between Pi and percentage yield loss in cv. LS5995 and a susceptible counterpart were evident (Fourie et al. 2010). Significantly higher ($P \leq 0.05$) values for all nematode parameters measured were evident in the roots of the susceptible cv. compared to that in LS5995 from Pi = 100 and higher. Furthermore, yield losses in cv. LS5995 were at least six times higher than that of the susceptible cv., illustrating the substantial difference in monetary terms. Genetic host plant resistance to *M. incognita* may, however, not be sufficient as the only management tool in rotation systems in which root-knot susceptible crops are included.

9.2.4.3 Crop Rotation

Crop rotation is generally not an effective strategy to reduce root-knot or lesion nematode populations in local soybean-based cropping systems. The majority of crops that are or can be rotated with soybean are susceptible to the predominant root-knot nematodes, *M. incognita* and *M. javanica* (see Sects. 8.2.4.2, 8.2.4.3 and 10.2.1). The use of poor host or root-knot nematode-resistant cvs of rotation crops should, however, be the main strategy to combat *Meloidogyne* problems in soybean-based cropping systems (H Fourie, North-West University, Potchefstroom, 2016, personal communication).

9.2.4.4 Chemical Control

The use of synthetically-derived nematicides on soybean crops is seldom cost-effective as was demonstrated by Fourie and Mc Donald (2001, 2007). No/limited new synthetically-derived a.s. is foreseen to be registered on the crop in the near future.

Nematicides evaluated for their efficacy to reduce *M. javanica* population levels in two experiments showed that of the granular and fumigant nematicides evaluated; EDB® (AL; 1800 g^{-1}) consistently resulted in the lowest *M. javanica* population levels and highest yields (Fourie and Mc Donald 2001). Although aldicarb and terbufos treatments generally followed EDB®, they did not always differ significantly ($P \leq 0.05$) from the untreated control. Biological control agents (Pl Plus; a.s. *Purpureocillium lilacinum* and Biostart; a.s. *Bacillus* spp.) used in these experiments, however, failed to reduce *M. javanica* population levels significantly ($P \leq 0.05$) compared to the untreated control. Although Pl Plus is registered in SA, it is not currently available.

Four field experiments conducted where natural infestations of *M. incognita* occurred showed that significant differences ($P \leq 0.05$) existed among the 14 nematicide treatments evaluated with regard to population levels in roots (Fourie and Mc Donald 2007). Only oxamyl SL, terbufos GR and an abamectin seed treatment resulted in a significant ($P \leq 0.05$) reduction of root-knot nematode population levels. These nematicides also generally resulted in a higher income per hectare compared to the untreated control. Cost-efficacy analyses, however, did not allow registration of these products on soybean at that stage.

9.3 Groundnut

Groundnut is cultivated under both rain-fed and irrigation conditions in summer rainfall regions of SA. The main production areas are located in the western and northwestern Free State, North-West, Northern Cape, Limpopo and Mpumalanga provinces. It is grown by commercial and smallholding producers (northern and eastern parts of the country). Groundnut is primarily consumed as a vegetable by smallholding producers and their families (DAFF 2010a), but used as a cash crop by commercial producers. Protein and oil contents of the pods are in the order of 65 and 91 %, respectively (Swanevelder 1997).

Groundnut production slowly decreased from the start of the 1990s, with the exception of the 1992 growing season (Grain 2016). After a once-off increase in area planted during 2001, totalling 200,000 ha, a steep decline was consistently experienced until the 2014/2015 season when only 58,000 ha were planted. Production figures of the crop followed the same trend, declining from 676,000 MT in 1990 to 62,300 MT in the 2015/2016 growing season.

9.3.1 Plant-Parasitic Nematodes Associated with Groundnut

Numerous plant-parasitic nematodes are associated with groundnut crops. These include the economically most important groundnut-pod nematode *Ditylenchus africanus* Wendt, Swart, Vrain and Webster, 1995 (previously reported as *Ditylenchus destructor* Thorne, 1945) (Jones and De Waele 1988), the testa nematode *Aphelenchoides arachidis* Bos, 1977 (Lesufi et al. 2015) as well as various *Meloidogyne* and *Pratylenchus* spp. (Keetch and Buckley 1984; Venter et al. 1992; Kleynhans et al. 1996; SAPPNS). Other plant-parasitic nematodes identified and listed in the SAPPNS database in association with groundnut are *Criconemoides*, *Helicotylenchus*, *Longidorus*, *Rotylenchus*, *Nanidorus*, *Neodolichorhynchus*, *Rotylenchus*, *Rotylenchulus*, *Scutellonema, Tylenchorhynchus* and *Xiphinema* spp.

Groundnut-Pod Nematode

Ditylenchus africanus was discovered in severely damaged groundnut pods (hulls and seeds) collected from a rain-fed field in the Schweizer-Reneke district (North-West Province). A molecular study on the comparative taxonomy between some *Ditylenchus* spp. populations (Wendt 1992) and analysis of the ribosomal deoxyribonucleic acid (rDNA) of several geographic and host isolates of *Ditylenchus dipsaci* Filipjev, 1936; *Ditylenchus myceliophagus* Goodey, 1958; and *D. destructor* (Wendt et al. 1993) casted doubt on the original identification and classification of the *D. destructor* identified from local groundnut crops. Individuals of this species did not damage potato tubers (De Waele et al. 1991) or other crops

(De Waele et al. 1989) and were consequently considered to belong to a different ecotype that formed a distinct *D. destructor* race with a limited host range. Wendt et al. (1993) also indicated that the rDNA of originally described *D. destructor* specimens from local groundnut crops differed from that of *D. destructor* specimens from the United Kingdom and Wisconsin in the USA. Based on morphology and restriction length polymorphisms (RFLPs) of rDNA, the local groundnut-pod nematode was ultimately described as a new species of *Ditylenchus*, viz. *D. africanus* (Wendt et al. 1995). It parasitises various crops (Basson et al. 1990) and weeds (De Waele et al. 1990, 1997; Ntidi et al. 2012) but causes damage only to groundnut (De Waele et al. 1989) (Box 9.3). *Ditylenchus destructor* and *D. africanus* represent a classical example of cryptic species since they cannot easily be distinguished from one another using morphometrical and especially

Box. 9.3 The basic procedure followed to identify and assess the pathogenicity, distribution and economic impact of a nematode pest on a crop as illustrated for the groundnut pod and testa nematodes that parasitise groundnut in South Africa

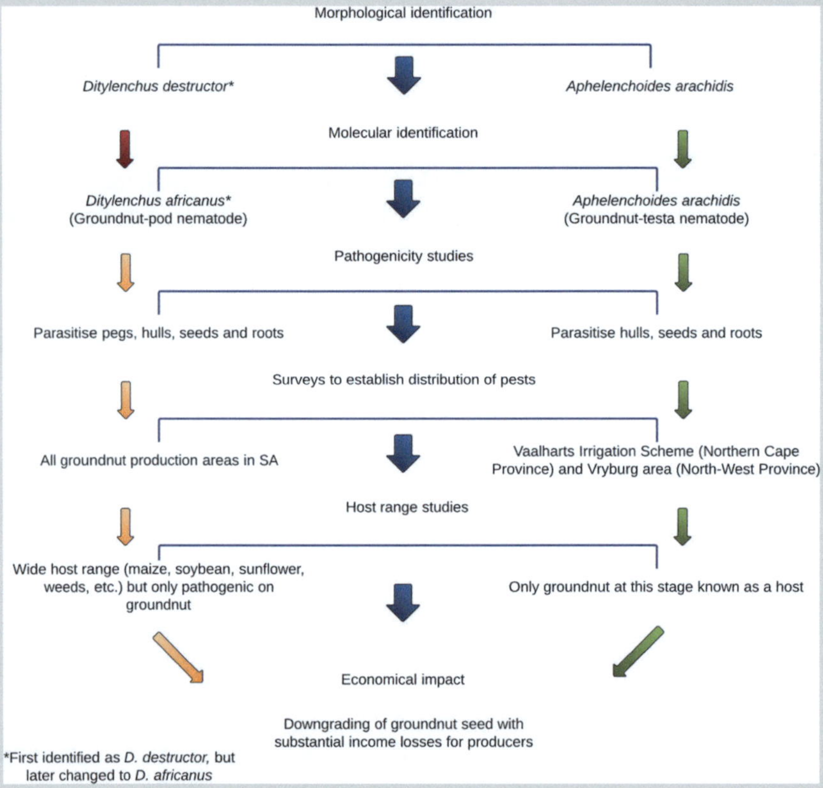

morphological identification. Fortunately they can be easily distinguished from each other by their different DNA composition.

Distribution, Symptoms, Damage Potential and Economic Importance

Data, obtained from the first official national nematode survey in groundnut fields and from later diagnostic analyses, showed that *D. africanus* was present in all production areas (De Waele et al. 1989; SAPPNS). Of the 877 damage-graded seed samples collected during this survey, 73 % were infected with *D. africanus*. The groundnut-pod nematode has not been reported from groundnut in other parts of the world and is considered endemic to SA (Dickson and De Waele 2005). Symptoms of damage inflicted by *D. africanus* on groundnut pods resemble black pod rot caused by the fungus *Chalara elegans* (Fig. 9.4) that is often found in irrigated groundnut fields (Labuschagne et al. 1980; Prinsloo 1980).

At cellular level the feeding behaviour of *D. africanus* results in initial symptoms that appear as brown and necrotic lesions on the inside of pods upon removal of the peg. The primary infection site appears to be located on the peg near the connection point at the base of the pod (De Waele et al. 1989; Jones and De Waele 1990). The outside tissue infected with *D. africanus* appears dark brown to blackish and corky on pods, with darkened veins (Fig. 9.5a). Individuals of the groundnut-pod nematode usually penetrate the hull endocarp through openings at the base of the exocarp or at the pod apex. Infected seeds are usually shrunken, with dark brown to black micropyles and flaccid testae with darker vascular strands (Fig. 9.5b). The testae of infected seed can be easily removed by gentle rubbing and reveals a distinct yellow discolouration on its inner layer.

At cellular level the feeding behaviour of *D. africanus* individuals causes collapse, malformations and cell-wall degradation of kernel and pod tissues (Venter

Fig. 9.4 Groundnut pods infected with groundnut-pod nematodes and the fungus *Chalara elegans*, showing blackish discoloured tissue (Johan Els, Agricultural Research Council–Grain Crops Institute, Potchefstroom, South Africa)

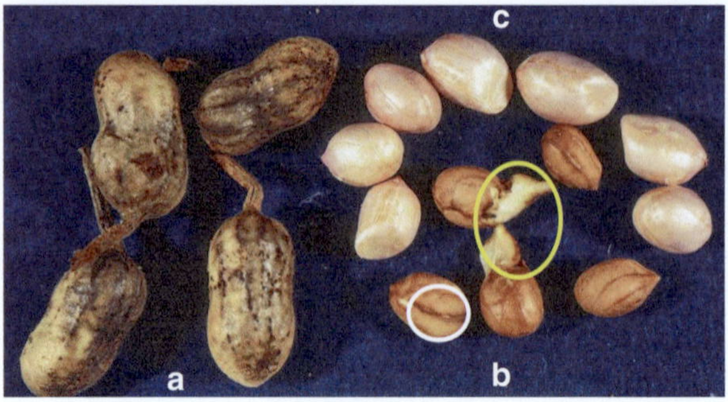

Fig. 9.5 Symptoms of infection by groundnut-pod nematodes, showing darkened (*blackish/greyish*) areas on pods with darkened veins (**a**) and dark testae of kernels (**b**) that have darkened veins (*white circle*) and germinate early (*yellow circle*) compared to noninfected kernels (**c**) (Johan Els, Agricultural Research Council–Grain Crops Institute, Potchefstroom, South Africa)

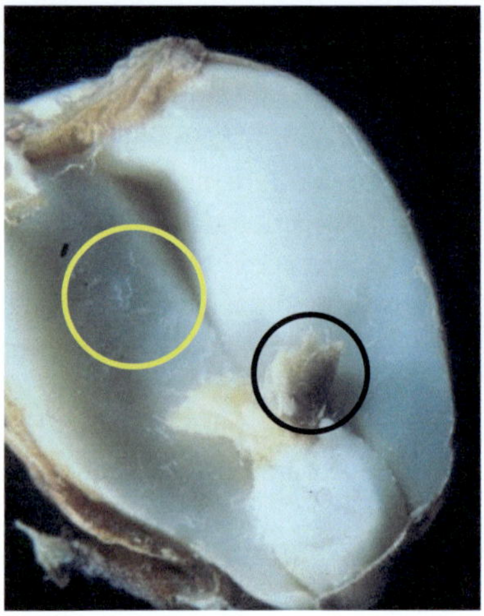

Fig. 9.6 Groundnut-pod nematodes feeding in the embryo of kernels (*black circle*) that turned olive green, while others are visible within the tissue of the kernel (*yellow circle*) (Johan Els, Agricultural Research Council–Grain Crops Institute, Potchefstroom, South Africa)

et al. 1995). These nematodes feed on parenchyma cells surrounding vascular bundles just below the surface of pods (Jones and De Waele 1990). At advanced stages of parasitism, *D. africanus*-infected pods appear dead and have dark brown to black veins. Feeding of *D. africanus* individuals near or in vascular bundles of the seed testa results in darkened veins. Groundnut-pod nematodes do not penetrate the cotyledons but feed on embryos (Fig. 9.6), causing them to turn olive green to brown.

Groundnut-pod nematodes cause severe yield and quality losses of groundnut (Mc Donald et al. 2005). At harvesting, 90 % of *D. africanus* populations are present in the pods (consisting of pegs, kernels and hulls) (Basson et al. 1991; Dickson and De Waele 2005). Penetration of *D. africanus* individuals at the basis of the pod causes weakening of the peg and pod connection (De Waele et al. 1989). This results in pods breaking off during lifting of the crop and being left behind in the soil, which may range from 40 to 60 % of the total number of pods (Jones and De Waele 1988).

Fig. 9.7 Kernels infected with groundnut-pod nematodes with dark testae and veins, with some kernels that germinated (**a**) compared to infected pods of which the kernel has also germinated (**b**) (Johan Els, Agricultural Research Council–Grain Crops Institute, Potchefstroom, South Africa)

The main effect of *D. africanus* on groundnut is qualitative (Mc Donald et al. 2005). The breakdown of the hull as a result of *D. africanus* damage increases water penetration into the pod (Venter et al. 1995). Weakened pods often split open during severe groundnut pod infections. The breakdown of hull tissue and the presence of split pods result in the occurrence of second-generation seedlings due to untimely seed germination (Fig. 9.7). Destruction of the seed testae caused by feeding of *D. africanus* leads to leaching of chemical compounds that function as inhibitors of seed germination (Svamv and Narasimhareddy 1977) and results in the initiation of hypocotyl growth (De Waele et al. 1997). Feeding of the nematodes near or in vascular bundles of the seed testae also results in an unattractive appearance of infected seed).

Symptoms caused by *D. africanus* infections increase the percentage of unsound, blemished and soiled (UBS) kernels and are highly correlated with the number of nematodes present in the testa of groundnut seeds (Venter et al. 1991; Mc Donald et al. 2005). For this reason, *D. africanus* infections can have substantial financial implications for producers. Grading of groundnut consignments in SA is specified by law, and kernels are classified into: (i) choice edible, (ii) standard edible, (iii) diverse or (iv) crushing grade (Anonymous 1997). Supply and demand dictate the prices of each grade and net gain increases with an increase in kernel grading. The economic importance of *D. africanus* is determined by the loss in income from infected groundnut consignments, which in turn depends on current prices for each grading class.

9.3.1.1 Testa Nematodes

Individuals of *A. arachidis* were first isolated from groundnut kernels in Nigeria (Bos 1977; Bridge et al. 1977; Bridge and Hunt 1985). To date, this nematode pest was identified from groundnut hulls and kernels in the Vaalharts Irrigation Scheme

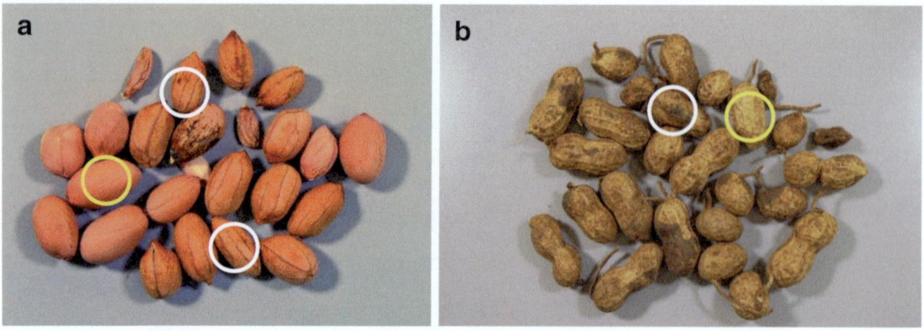

Fig. 9.8 (**a**, **b**) Kernels of groundnut infected with testa nematodes (*Aphelenchoides arachidis*) with darkened testae and veins (**a** *white circles*) compared to non-infected ones (*yellow circles*) and pods infected with the pest displaying black/grey areas (**b** *white circles*) compared to non-infected ones (*yellow circles*) (Antoinette Swart, Agricultural Research Council–Plant Protection Research, South Africa)

(Northern Cape Province) and Vryburg areas (North-West Province) of SA (Lesufi et al. 2015; SAPPNS). Interestingly, the latter study showed the presence of the two pathogenic fungi (*Thielaviopsis basicola* and *Neocosmospora vasinfecta*) in *A. arachidis*-infected hulls. No *D. africanus* specimens were found in this material. The spread of *A. arachidis* in local groundnut production areas other than the two localities mentioned above is not known at present. Symptoms of damage caused by *A. arachidis* (Fig. 9.8a, b) are similar to those caused by *D. africanus* (see Sect. 9.3.1). Studies aimed at determining the distribution of *A. arachidis* in local groundnut production areas and quantifying the damage potential of this pest should receive priority. Compared to *D. africanus*, information about the testa nematode is limited (Box 9.3).

9.3.1.2 Root-Knot Nematodes

Thermophilic *Meloidogyne* spp. listed to infect and damage local groundnut crops are *M. incognita* and *M. javanica,* while cryophilic nematodes are represented by *M. chitwoodi, Meloidogyne fallax* Karssen, 1996, and *M. hapla* (Keetch and Buckley 1984; Kleynhans et al. 1996; Fourie et al. 1998, 2001; SAPPNS).

Below-ground symptoms of root-knot nematode infection include gall formation on pegs, pods and roots of groundnut plants and are generally more pronounced at crop maturation (Bridge and Starr 2007). Root-knot nematode infections are routinely reported for groundnut (Bekker et al. 2007; Fourie et al. 2011; S Steenkamp, Agricultural Research Council–Grain Crops Institute, Potchefstroom, 2016, personal communication), and research should be initiated to establish the pathogenicity and damage potential of root-knot nematodes in local groundnut production areas.

9.3.1.3 Lesion Nematodes

Pratylenchus brachyurus (Godfrey, 1929) Filipjev and Schuurmans Stekhoven, 1941, and *Pratylenchus zeae* Graham, 1951, have been recorded to parasitise groundnut in SA (Keetch and Buckley 1984; Venter et al. 1992; SAPPNS). *Pratylenchus brachyurus*, reported to be most damaging on the crop globally (Bridge and Starr 2007), has also been identified as the predominant plant-parasitic nematode in groundnut pods and roots (Venter et al. 1992).

Typical below-ground symptoms of lesion nematode infection are distinct, necrotic lesions on roots and pods (Bridge and Starr 2007), but may be confused with symptoms caused by other soilborne pathogens. As with *A. arachidis* (see Sect. 9.3.1.2 and *Meloidogyne* spp. - see Sect. 9.3.1.3), no research has been done to establish the impact of *Pratylenchus* spp. in local groundnut plantings.

9.3.2 Management Strategies

Nematode pests of groundnut are difficult to control because of their wide host range. For example, the economically most important *D. africanus* as well as *Meloidogyne* and *Pratylenchus* spp. infect a wide range of crops and hence have the ability to survive in the absence of groundnut (Basson et al. 1990; De Waele and Jordaan 1988; De Waele et al. 1990, 1991). Furthermore, these nematode pests have high reproductive potentials, e.g. *D. africanus* in particular has a short life cycle of 7 days (De Waele and Wilken 1990). These characteristics enable nematode pest communities to build up quickly to large populations that cause severe damage to groundnut crops.

9.3.2.1 Chemical Control

Nematicides that are currently registered in SA for use on groundnut during planting include fenamiphos (EC; 400 g l^{-1} and GR; 100 g kg^{-1}), furfural (EC; 900 g l^{-1}) and terbufos (GR; 100 and 150 g kg^{-1}) (Van Zyl 2013). Oxamyl (SL; 310 g l^{-1}) is registered for application at planting and/or at the onset of peg formation (Van Zyl 2013). Nematicides are often used on groundnut to prevent damage to pods later in the season, but long-term suppression of nematode pest populations is impossible to achieve with chemical control (Sikora et al. 2005a). Nematicides are often ineffective in sufficiently reducing nematode population densities (Haydock et al. 2013), especially those of migratory, endoparasitic *D. africanus* that produces numerous generations during a single growing season (De Waele and Wilken 1990). Most of the nematicides currently registered on groundnut are, furthermore, generally effective for only 8 weeks after application. Effective control of nematode pests requires the use of a nematicide that remains active for at least 12 weeks after application (Basson et al. 1992). Mc Donald (1998) investigated the effect that fumigants, ethylene dibromide (EDB®) and methyl bromide had on reducing *D. africanus* damage in groundnut crops. These two fumigants were highly successful in limiting quality losses and concurrently resulted in yield increases. The conventional application rates tested with these fumigants were, however, not cost-effective.

Host Plant Resistance

Rotation with resistant plants is one of the most effective management tools to manage root-knot nematodes that parasitise groundnut (Bridge and Starr 2007). However, for lesion nematodes no resistant cultivars are available.

Rotation with resistant groundnut cvs should also be applicable for the control of *D. africanus*. No groundnut genotypes were identified with resistance to the groundnut-pod nematode during the 1900s (Basson et al. 1991; Venter et al. 1993). Subsequently, Steenkamp et al. (2010a) identified genotypes PC254K1 and GC7 with resistance to this pest that proved to be successful under field conditions. This was particularly the case for PC254K1 that showed resistance even where high *D. africanus* population densities occurred. The suggested mechanism of resistance in PC254K1 is antibiosis. Electronmicroscopy investigations showed that cell walls are broken down by enzymes secreted by *D. africanus* individuals (Steenkamp et al. 2010b). This resistance mechanism is similar to that of *D. dipsaci* reported by Seinhorst (1957). For both these *Ditylenchus* spp., successful parasitism is initiated by the destruction/dissolution of the middle lamella of plant cells by means of enzymatic secretions.

The resistant genotype PC254K1 consistently produced yields with a low UBS percentage of kernels at various Pi levels and could be used as a major source of resistance to *D. africanus* in the development of commercial cvs. Another breeding line, PC287K5, also maintained low groundnut-pod nematode population levels in some experiments, but its level of resistance does not seem to be as strong or as sustainable as that of genotypes PC254K1 or CG7.

9.3.2.2 Cultural Control

Cultural management of plant-parasitic nematodes that infect groundnut kernels includes the use of certified seed or nematode-free planting material (Bridge and Starr 2007). However, the local production of, for example, *D. africanus*-free certified seed is hampered by factors such as the omnipresence of this pest in all groundnut production areas (De Waele et al. 1989), unpredictable efficacy of nematicides under harsh field and environmental conditions (Mc Donald 1998) and the unavailability of commercial groundnut cvs that are resistant to the groundnut-pod nematode. The low damage threshold of groundnut-pod nematodes on the crop and its potential to spread by seed thus require stricter management strategies and improved quality control (Mc Donald 1998). Other cultural strategies, such as heat and mechanical methods, are not suitable for the treatment of groundnut kernels infected with *D. africanus*. Infected kernels are soft, moisture sensitive and easily damaged by such strategies that also invariably affect the germination of kernels.

9.3.2.3 Biological Control

Harsh field and environmental conditions that prevail in local production areas complicate the successful use of biological agents to control *D. africanus* and other nematode pests on groundnut. Extensive research on the use of biocontrol agents to manage groundnut pod infections under local conditions is needed before biological control agents can be registered for commercially use.

9.3.2.4 Crop Rotation

Although crop rotation is typically central in planning management strategies to combat plant-parasitic nematodes (Viaene et al. 2013), the effective management of *D. africanus* and other nematode pests using this strategy is limited. This is due to the ability of groundnut pod and other nematode pests to survive in roots and the rhizosphere of many crops that are planted in rotation with groundnut. Also, weeds such as cocklebur (*Xanthium strumarium*), feathertop chloris (*Chloris virgata*), goose grass (*Eleusine indica*), jimson weed (*Datura stramonium*), khaki weed (*Tagetes minuta*), purple nutsedge (*Cyperus rotundus*) and white goosefoot (*Chenopodium album*) are commonly found in groundnut fields and can also serve as temporary hosts of, for example, groundnut-pod nematodes (De Waele et al. 1990, 1997). Ntidi et al. (2012) recorded *D. africanus*, *Meloidogyne* and *Pratylenchus* spp. from a number of weeds during a nematode survey in the developing agricultural sector.

9.4 Sunflower

Sunflower is an important source of vegetable oil (DAFF 2010b). According to local crop estimates, 576,000 ha of sunflower were planted during the 2015/2016 growing season, with a crop yield of 742,750 MT being realised (Grain 2016). Fluctuations in sunflower production are mainly caused by uncertain price expectations, high input costs and high stock levels. In SA, sunflower is produced mainly in the summer rainfall areas such as the North-West, Free State, Limpopo and Mpumalanga provinces (DAFF 2010b). Among the various pests and diseases that parasitise the crop, plant-parasitic nematodes form an integral part of the complex (Bolton et al. 1989; Keetch and Buckley 1984; Kleynhans et al. 1996; Bekker et al. 2007; Fourie et al. 2011; SAPPNS).

9.4.1 Plant-Parasitic Nematodes Associated with Sunflower

Sunflower hosts a wide range of plant-parasitic nematodes. Early reports about the impact of nematode pests on local sunflower production indicated that root-knot nematodes, in particular *M. incognita*, *M. arenaria* and *M. javanica,* are the predominant parasites that damaged the crop (Van der Linde et al. 1959), while *M. hapla* has also been listed in association with sunflower (Keetch and Buckley 1984; Kleynhans et al. 1996). Except for root-knot nematodes, other plant-parasitic nematodes such as *Helicotylenchus*, *Longidorus*, *Nanidorus*, *Paratrophurus*, *Pratylenchus*, *Quinisulcius*, *Rotylenchulus Rotylenchus*, *Scutellonema*, *Trophurus*, and *Xiphinema* spp. have been reported to attack local sunflower crops (Keetch and Buckley 1984; Bolton et al. 1989; SAPPNS).

9.4.2 Symptoms

Below-ground symptoms as a result of root-knot nematode infections include galls on poorly-developed root systems of infected plants. Aerial symptoms usually include loss and chlorosis of leaves and stunting of seedlings in fields where severe infestations occur.

9.4.3 Management Strategies

Products that contain the a.s. carbofuran are registered on sunflower as an insecticide (Van Zyl 2013) and are hence also used by local producers to reduce plant-parasitic nematode populations.

Bolton and De Waele (1989) screened four sunflower hybrids for their host status to *P. zeae* in a glasshouse and in field experiments where natural infestations of this nematode species occurred. Results from both studies showed that all sunflower genotypes tested were identified as non- or poor hosts of *P. zeae*, while susceptible maize and grain sorghum genotypes maintained significantly higher ($P \leq 0.05$) population levels of this lesion nematode species. Pretorius et al. (2014) reported that 20 commercially available cvs that were screened for their host status to *M. incognita* were all susceptible to the pest.

9.5 Dry Bean

Dry bean is one of the most important field crops in SA due to its high protein content and dietary benefits (Liebenberg 2002). The crop was planted on 64,000 ha and yielded 73,390 MT during the 2015/2016 growing season (Grain 2016). The main production areas for dry bean are the eastern Free State and Mpumalanga provinces, while lesser quantities are also grown in the KwaZulu-Natal, North-West, Limpopo and Eastern Cape provinces. According to Liebenberg (2002) the incidence of diseases and pests contributed to unstable dry bean production in the past. Incidence and severity of pests and disease complexes vary between seasons because of environmental and management practices (Fig. 9.9).

9.5.1 Plant-Parasitic Nematodes Associated with Dry Bean

Plant-parasitic nematodes associated with dry bean crops include *Criconema, Criconemoides, Heterodera*, *Hemicycliophora*, *Longidorus*, *Meloidogyne*, *Nanidorus*, *Paralongidorus*, *Paratylenchus, Pratylenchus*, *Tylenchorhynchus* and *Xiphinema* spp. (Keetch and Buckley 1984; Kleynhans et al. 1996, SAPPNS).

Fig. 9.9 The heavily galled root system of a dry bean plant infected with root-knot nematodes (Johan Els, Agricultural Research Council–Grain Crops Institute, Potchefstroom, South Africa)

Root-knot nematodes are generally perceived as the most common and widespread nematode pests that parasitise dry bean locally (Fourie et al. 2002, 2011; Bekker et al. 2007). All four economically important root-knot nematode species, viz. *M. javanica*, *M. incognita*, *M. hapla* and *M. arenaria*, damage dry bean crops grown in SA. Under exceptionally high *M. javanica* infestations, crop losses that ranged between 30 and 87 % for cv. Mkuzi have been reported in a microplot experiment (Fourie et al. 2002).

9.5.2 Symptoms

Galls induced by root-knot nematode infection are clearly visible on roots of damaged plants (Fig. 9.9). Although above-ground symptoms are seldom visible in nematode-infested dry bean fields, patches where exceptionally high root-knot nematode infestations occur can generally easily be identified. Plants in such areas are stunted and yellowish and can die off.

9.5.3 Management Strategies

Two synthetic nematicides, viz. those with ethoprophos (GR; a.s. 150 g kg^{-1}) and terbufos (GR; a.s. 100 and 150 g kg^{-1}) as a.s., are registered on dry bean in SA (Van Zyl 2013).

The majority of dry bean breeding lines evaluated against root-knot nematode populations of *M. incognita* and *M. javanica* were classified as being susceptible. Although cv. Mkuzi had the lowest Rf for both root-knot nematode species, it remains susceptible and will lead to a build-up in population levels of these pests when planted in infested soil (Fourie et al. 2002).

9.6 Bambara Groundnut

Bambara production is limited in SA (DAFF 2011). This crop is similar to common groundnut since it also forms pods and seeds on or just below the soil surface. The two crops are, however, not botanically related. The protein content of ripe, dried seeds of Bambara groundnut is approximately 20 %, while kernels contain 6–12 % oil (Anonymous 1979). Bambara plants can withstand harsh environmental conditions that are generally too arid for cultivation of other legume or cereal crops. The crop produces satisfactory yields under both hot and dry conditions (Linneman 1987). Growing interest in the crop has been experienced since the 1980s, particularly by producers in the smallholding agricultural sector in the northern and eastern parts of the KwaZulu-Natal and Mpumalanga provinces.

9.6.1 Plant-Parasitic Nematodes Associated with Bambara Groundnut

Limited information about the association of nematode pests with Bambara groundnut exists for SA, with only a few species having been recorded. These include *Meloidogyne* and *Xiphinema* spp. (Mc Donald and De Waele 1989; Kleynhans et al. 1996; Kwerepe and Labuschagne 2004; SAPPNS). Nevertheless, plant-parasitic nematodes infect and cause significant yield losses to this crop as has been reported for semi-arid African countries (Ogbuji 1979; Mc Donald and De Waele 1989). The latter authors identified a severe *M. javanica* infection on roots of Bambara plants during the 1986/1987 growing season on an experimental site of the ARC-GCI in Potchefstroom (North-West Province). Aerial symptoms included poorly growing and stunted plants with rolled up leaves. Below-ground symptoms showed the typical formation of galls on roots of infected plants.

9.6.2 Management Strategies

No nematicides or biological agents are registered for use on Bambara groundnut in SA (Van Zyl 2013). Nematode management strategies investigated to reduce *M. incognita* populations on Bambara groundnut showed that biofumigation and solarisation were effective (Kwerepe and Labuschagne 2003). Under glasshouse conditions the effect of biofumigation, using soil-amended-aerial parts of *Brassica oleracea capitata* (cvs Drumhead and Glory of Enkhuizen) and *Brassica napus* (cv. Forage Star) at rates of 2–6 kg m^{-2}, resulted in significant ($P \leq 0.05$) reduction in root-knot nematode galling. Galling decreased by 60–95 % in roots of landrace DIPC compared to that of the untreated control. However, phytotoxic effects were evident on Bambara groundnut plants at such high application rates and resulted in a significant reduction in dry weight of plant shoots. Soil amendment with low rates of cruciferous residues (2 kg m^{-2} and 20 MT ha^{-1}) was not successful in reducing *M. incognita*-galling on Bambara groundnut roots in glasshouse experiments. A combination of biofumigation, at an application rate of 4 kg m^{-2} of aerial parts of cabbage (cv. Drumhead) followed by solarisation, resulted in better control of *M. incognita* than the individual treatments alone. This combined treatment reduced galling by up to 100 % compared to the untreated control and had similar efficacy as the application of a synthetic nematicide standard (aldicarb). Individual treatments of solarisation and biofumigation resulted in 62 and 78 % reductions, respectively, in *M. incognita* root galling. Ultimately results from this study showed that the amendment of soil with cruciferous crops at an application rate of 40 MT ha^{-1} proved to be effective in substantially reducing *M. incognita* population levels, while higher rates were phytotoxic to plants.

Information on the host suitability of this crop to *M. incognita* race 2 (Kwerepe and Labuschagne 2004) and *M. javanica* (Mc Donald and De Waele 1989) has been reported for genotypes from Botswana and SA, respectively. Host status investigations of 50 Bambara groundnut landraces from Botswana and SA showed that only landrace HVA 38–3 was resistant to *M. incognita*. Three landraces, namely, CLDRE, M4 and A12, were classified as tolerant since they produced higher yields in the presence of *M. incognita* than the others screened (Kwerepe and Labuschagne 2004). For *M. javanica*, Mc Donald and De Waele (1989) identified 15 genotypes from a glasshouse experiment as being susceptible to this pest. Variable pod production among the genotypes infected with *M. javanica* suggested that some of them may exhibit tolerance to this pest.

9.7 Cowpea

Cowpea is an important rotation crop and used as food or fodder in commercial and smallholding farming systems in SA (Riekert and Henshaw 1998). Smallholdings are generally the largest producers of cowpea under rain-fed conditions. Cowpea production areas in SA are situated within the Limpopo, Mpumalanga, North-West and KwaZulu-Natal provinces. Crop performance is optimal during summer seasons where rainfall ranges between 400 and 700 mm annum^{-1}.

9.7.1 Plant-Parasitic Nematodes Associated with Cowpea

Numerous plant-parasitic nematodes are listed in association with cowpea in SA, namely, *Criconema*, *Criconemoides*, *Longidorus*, *Meloidogyne*, *Pratylenchus*, *Rotylenchulus*, *Rotylenchus*, *Tylenchorhynchus* and *Xiphinema* spp. (Keetch and Buckley 1984; Kleynhans et al. 1996; SAPPNS).

Cowpea is a particularly good host for root-knot nematodes and may increase numbers of these pests considerably, which may adversely affect the growth and yield of follow-up or intercropped plants (Riekert and Henshaw 1998).

9.7.1.1 Symptoms, Damage Potential and Management Strategies

Typical symptoms of root-knot nematode infection occur on cowpea plants. Roots of infected plants are galled, while above-ground symptoms include stunting of plants and leaf chlorosis.

Limited information is available on the impact of nematode pests on cowpea in SA. However, Riekert and Henshaw (1998) investigated the effect of a mixed root-knot nematode population of *M. incognita* and *M. javanica* in a 4-year crop rotation cycle in a field experiment in the northern Free State Province. One rotation with cowpea (cv. Glenda) resulted in a 20-fold increase in root-knot nematode numbers in roots of follow-up maize plants (cv. PAN6043). Two rotations of cowpea resulted in final root-knot nematode population levels that ranged from 15,690 to 73,600 eggs and J2 g^{-1} roots in follow-up maize plants. A substantial decrease in maize yield, from 4,8 MT ha^{-1} (monoculture maize) to 1,9 MT ha^{-1} (after cowpea rotation), was recorded. This was due to a substantial increase in root-knot nematode population levels in maize roots during the duration of the study. This study demonstrated that a root-knot nematode susceptible crop such as cowpea (cv. Glenda in this case) should be avoided in cropping cycles where root-knot nematodes pose problems.

No nematicide or biological agent is registered for use on cowpea in SA (Van Zyl 2013). Of the 21 genotypes that were screened in glasshouse experiments for their host status to *M. javanica*, 12 had Rf>1, indicating susceptibility to this pest (Riekert 1999). Two genotypes (lines T182b-889 and R6A) were, however, identified as resistant to *M. javanica*, with Rf<1. For *M. incognita*, 11 genotypes were also identified with resistance.

9.8 Lupin

Lupin production in SA is very low. The first official figures for ha planted (16,300) and production (9,000 MT) were reported for the 1998/1999 growing season (PRF 2016). Although production figures for this crop are generally lacking, it was reported that 71,000 ha were planted during 2009 with 20,654 MT being harvested (FAO 2016). The lack of a stable market and fair prices for sweet lupins seems to be the biggest concern for the industry, resulting in bitter lupins being the favoured cvs to be produced. High meat prices may, however, lead to an increase in lupin production in the future because of the important role that this protein source plays in sheep farming (Aggenbach 2007).

9.8.1 Plant-Parasitic Nematodes Associated with Lupin

Only the root-knot nematode species *M. arenaria*, *M. hapla*, *M. incognita* and *M. javanica* have been listed in association with lupin in SA (Keetch and Buckley 1984; Kleynhans et al. 1996, SAPPNS). Large populations of *Meloidogyne* spp. occurred in crop rotation experiments in which lupin was included as reported by Schoeman and Riekert (1999).

9.8.2 Symptoms and Management Strategies

Below-ground symptoms caused by root-knot nematodes on lupin include typical gall formation on the roots.

No nematicide or biological agent is registered on lupin in SA (Van Zyl 2013). In terms of the host status of 29 locally adapted lupin cvs, SAL 34 supported the lowest average *M. javanica* population levels (8 eggs and J2 g roots^{-1}), and Esta the highest (667 eggs and J2 g roots^{-1}) in a microplot experiment (Schoeman and Riekert 1999).

9.9 Conclusions

Research on the impact of plant-parasitic nematodes on oilseed and leguminous crops in SA and the development of management strategies to limit damage inflicted by such pests have to be pursued. Various gaps in terms of the generation of knowledge need to be addressed concerning the distribution and impact of nematode pests on several oilseed crops, as has been outlined in this chapter. Initiatives to reduce population levels of plant-parasitic nematodes in leguminous and oilseed crops have to supply producers with practical and cost-effective strategies to ensure sustainable crop production. The current approach towards environmentaly-friendly strategies to combat diseases and pests, particularly plant-parasitic nematodes, increases the pressure on scientists and related industries to coordinate related research initiatives.

Acknowledgements Funding institutions, the Protein Research Foundation (PRF), Oil and Protein Seed Development Trust (OPDT) and the Oilseeds Advisory Committee (OAC), are acknowledged for their continual financial contributions towards nematode research on various leguminous and oilseed crops. The ARC and NWU also funded research and infrastructure.

References

Agenbag M (2016) Identification and reproduction potential of South African *Meloidogyne* species. MSc Dissertation, North-West University, Potchefstroom

Aggenbach H (2007) Lupin production in the Western and Southern Cape. http://www.proteinresearch.net. Accessed 12 Aug 2013

Anonymous (1979) Pulses: tropical legumes, resources for the future. National Academy of Sciences, Washington, pp 47–53

Anonymous (1997) Agricultural Product Standards Act No. 119 of 1990 of the Republic of South Africa. Government Gazette 18056. Juta and Co, Cape Town

Anonymous (2001) Report on the South African variety list as maintained by the registrar of Plant Improvement. Department of Agriculture – Directorate of Genetic Resources, Gezina

Anonymous (2012) Endosulfan ban – South Africa. http://www.cgcsa.co. Accessed 23 May 2013

Aryal SK, Davis RF, Stevenson KL et al (2011) Induction of systemic acquired resistance by *Rotylenchulus reniformis* and *Meloidogyne incognita* in cotton following separate and concomitant inoculations. J Nematol 43:160–165

Basson S, De Waele D, Meyer AJ (1990) An evaluation of crop plants as hosts for *Ditylenchus destructor* isolated from peanut. Nematropica 20:23–29

Basson S, De Waele D, Meyer AJ (1991) Population dynamics of *Ditylenchus destructor* on peanut. J Nematol 23:485–490

Basson S, De Waele D, Meyer AJ (1992) Effect of host plant age on population development and pathogenicity of *Ditylenchus destructor* on peanut. J Nematol 24:310–314

Bekker S, Venter GA, Fourie H (2007) Nematode diagnostic services at the ARC-Grain Crops Institute. S Afr J Plant Soil 24:258–259

Bolton C, De Waele D (1989) Host suitability of commercial sunflower hybrids to *Pratylenchus zeae*. J Nematol 4:682–685

Bolton C, De Waele D, Loots GC (1989) Plant-parasitic nematodes on field crops in South Africa. 3. Sunflower. Rev Nématol 12:69–76

Bos WS (1977) *Aphelenchoides arachidis* n. sp. (Nematoda: Aphelenchoidea), an endoparasite of the testa of groundnuts in Nigeria. Z Pflanzenkr Pflanzenschutz 84:95–99

Bridge J, Hunt DJ (1985) *Aphelenchoides arachidis*. C.I.H. Descriptions of Plant-parasitic Nematodes, Set 8, No 116. Commonwealth Agricultural Bureau, London

Bridge J, Starr JL (2007) Plant nematodes of agricultural importance. Academic, Boston

Bridge J, Bos WS, Page LJ et al (1977) The biology and possible importance of *Aphelenchoides arachidis*, a seed-borne endoparasitic nematode of groundnuts from northern Nigeria. Nematologica 23:253–259

Coetzee V (1968) The distribution of the family Heteroderidae (Filipjev, 1934) in South Africa and some host records of *Meloidogyne* species. S Afr J Agric Sci 11:775–788

DAFF (2010a) Groundnut. Production Guideline. Government Printers, Pretoria

DAFF (2010b) Sunflower. Production Guideline. Government Printers, Pretoria

DAFF (2011) Production guidelines for Bambara groundnut. Government Printers, Pretoria

De Beer AS, Bronkhorst L (2015) Soybean cultivar recommendations for 2015/2016. ARC-Grain Crops Institute, Potchefstroom

De Beer A, Fourie H, Venter C et al (2014) Evaluation of root-knot nematode (*Meloidogyne incognita*) resistant soybean genotypes for their host status and yield potential. In: Abstracts of the combined congress. Rhodes University, Grahamstown, 20–23 Jan

De Waele D, Jordaan EM (1988) Plant-parasitic nematodes on field crops in South Africa. 1. Maize. Rev Nématol 11:65–74

De Waele D, Wilken R (1990) Effect of temperature on the *in vitro* reproduction of *Ditylenchus destructor* isolated from groundnut. Rev Nématol 13:171–174

De Waele D, Jones BL, Bolton C et al (1989) *Ditylenchus destructor* in hulls and seeds of peanut. J Nematol 21:10–15

De Waele D, Jordaan EM, Basson S (1990) Host status of seven weed species and their effects on *Ditylenchus destructor* infestation of peanut. J Nématol 22:292–293

De Waele D, Wilken R, Lindeque J (1991) Response of potato cultivars to *Ditylenchus destructor* isolated from groundnut. Rev Nématol 14:123–126

De Waele D, Venter C, Mc Donald AH (1997) The peanut pod nematode *Ditylenchus africanus*. Nematology Circular No. 218. Agricultural Research Council–Grain Crops Institute, Potchefstroom

Dickson DW, De Waele D (2005) Nematode parasites of peanut. In: Luc M, Sikora RA, Bridge J (eds) Plant parasitic nematodes in subtropical and tropical agriculture, 2nd edn. CAB International, Wallingford, pp 393–436

FAO (2016) Statistics. http://faostat.fao.org/. Accessed 1 Feb 2016

Fourie H, Mc Donald AH (2001) Chemical control options for plant-parasitic nematodes associated with soybean in South Africa. Report on project M151/60. Agricultural Research Council–Grain Crops Institute (ARC-GCI), Potchefstroom

Fourie H, Mc Donald AH (2007) Chemical control options for plant-parasitic nematodes associated with soybean in South Africa. Report on project M151/60. Agricultural Research Council–Grain Crops Institute (ARC-GCI), Potchefstroom

Fourie H, Zijlstra C, Mc Donald AH (1998) ITS-PCR sequence based identification of *M. chitwoodi* and screening of crops for host suitability. Afr Plant Prot 4:107–111

Fourie H, Mc Donald AH, Loots GC (1999) Host suitability of South African commercial soybean cultivars to two root-knot nematode species. Afr Plant Prot 2:119–124

Fourie H, Mc Donald AH, Loots GC (2001) Plant-parasitic nematodes in field crops in South Africa. 6. Soybean. Nematol 3:447–454

Fourie H, Schoeman MJ, Mtshali MB et al (2002) The challenge towards integrated control of root-knot nematodes for developing agriculture in South Africa. In: Abstracts of the 4th international congress of nematology, Tenerife, 8–13 June 2002

Fourie H, Mc Donald AH, De Waele D (2006) Host suitability of South African and foreign soybean cultivars to *Meloidogyne incognita* race 2. S Afr J Plant Soil 23:132–137

Fourie H, Mienie CMS, Mc Donald AH et al (2008) Identification of genetic markers associated with *Meloidogyne incognita* race 2 resistance in soybean (*Glycine max* L. Merr). Nematol 10:651–661

Fourie H, Mc Donald AH, De Waele D (2010) Relationships between initial population densities of *Meloidogyne incognita* race 2 and nematode population development in terms of variable soybean resistance. J Nematol 42:55–61

Fourie H, Bekker S, Mc Donald AH et al (2011) Prospective extended benefits of plant-nematode diagnostic and advisory services should closer cooperation between laboratories be attainable. S Afr J Plant Soil 28:265

Fourie H, Mc Donald AH, De Waele D et al (2013a) Comparative cellular responses in resistant and susceptible soybean cultivars infected with *Meloidogyne incognita*. Nematol 15:695–708

Fourie H, Mc Donald AH, De Waele D (2013b) *In vivo* host and yield responses of *Meloidogyne incognita*-resistant and susceptible soybean genotypes. Int J Pest Manag 59:111–121. doi:10.1080/09670874.2013.77226

Fourie H, De Waele D, Mc Donald AH et al (2015) Nematode pests threatening soybean production in South Africa, with reference to *Meloidogyne*. S Afr J Sci 111. doi:10.17159/sajs.2015/20140212

Grain SA (2016) Grain market overview. http://www.grainsa.co.za. Accessed 2 Jan 2016

Haydock PJ, Woods SR, Grove IG et al (2013) Chemical control of nematodes. In: Perry RN, Moens M (eds) Plant nematology. CAB International, Wallingford, pp 459–479

Hussey RS, Janssen GJW (2002) Root-knot nematodes. In: Starr JL, Cook R, Bridge J (eds) Plant resistance to parasitic nematodes. CAB International, Wallingford, pp 43–70

Jones BL, De Waele D (1988) First report of *Ditylenchus destructor* in pods and seeds of peanut. Plant Dis 72:453

Jones BL, De Waele D (1990) Histopathology of *Ditylenchus destructor* on peanut. J Nematol 22:268–272

Karssen G, Wesemael W, Moens M (2013) Root-knot nematodes. In: Perry RN, Moens M (eds) Plant nematology. CAB International, Wallingford, pp 73–108

Keetch DP (1989) A perspective of plant nematology in South Africa. S Afr J Sci 85:506–508

Keetch DP, Buckley NH (1984) A check-list of the plant-parasitic nematodes of Southern Africa. Technical communication. Department of Agriculture, Government Printer, Pretoria

Kleynhans KPN, Van den Berg E, Swart A et al (1996) Plant nematodes in South Africa. Plant Protection Research Institute Handbook No 8. Agricultural Research Council–Plant Protection Research Institute, Pretoria

Kwerepe BC, Labuschagne N (2003) Biofumigation and solarisation as integrated pest management (IPM) components for the control of root-knot nematode (*Meloidogyne incognita* (Kofoid and White) Chitwood) on Bambara groundnut (*Vigna subterranean*) L. (Verdc). Uniswa J Agric 11:56–63

Kwerepe BC, Labuschagne N (2004) Screening of Bambara groundnut landraces for resistance/tolerance to *Meloidogyne incognita* race 2. Afr Plant Prot 10:35–41

Labuschagne H, Kotze JM, Wehmer FC (1980) *Thielaviopsis basicola* infection of groundnuts in South Africa. Phytophylactica 12:177–180

Lesufi MM, Swart A, Mc Donald AH et al (2015) Morphological and molecular studies on *Aphelenchoides arachidis* Bos, 1977 (Tylenchina: Aphelenchoididae) from groundnuts in South Africa. Nematol 17:433–445. doi:10.1163/15685411-00002879

Liebenberg AJ (2002) Dry bean production. Department of Agriculture, Pretoria

Liebenberg AJ (2012) Soybean production manual: Your guide to successful soybean production. Agricultural Research Council–Grain Crops Institute (ARC–GCI), Potchefstroom

Linneman AR (1987) Bambara groundnut (*Vigna subterranean* (L.) Verde.) – a review. Abstr Trop Agric 12:9–25

Mc Donald AH (1998) Management of *Ditylenchus africanus* (Nematoda: Tylenchidae) on groundnut (*Arachis hypogaea*). PhD Thesis, Potchefstroom University for Christian Higher Education, Potchefstroom

Mc Donald AH, De Waele D (1989) Effect of *Meloidogyne javanica* on Bambara groundnut in South Africa. Phytophylactica 21:429–431

Mc Donald AH, Loots GC, Fourie H et al (2005) A microplot study on *Ditylenchus africanus* population densities and damage symptoms on groundnut in relation to commercial yields. Nematol 7:647–653

Mienie CMS, Fourie H, Smit MA et al (2002) Identification of AFLP markers in soybean linked to resistance to *Meloidogyne javanica* and conversion to Sequence Characterised Amplified Regions (SCARs). Plant Growth Regul 37:157–166

Ntidi N, Fourie H, Mc Donald AH et al (2012) Plant-parasitic nematodes associated with weeds in developing agriculture in South Africa. Nematol 14:875–887

Ogbuji RO (1979) Effects of two *Meloidogyne* spp. on growth and reproduction of Bambara groundnut (*Voandzeia subterranea*) in Nigeria. Tropenlandwirt 80:47–51

Pretorius M, Steenkamp S, Mc Donald AH et al (2014) Host status of commercial wheat and sunflower cultivars to *Meloidogyne* spp. J Nematol 46(2):223

PRF (2016) Statistics and estimates. http://www.proteinresearch.net. Accessed 1 Feb 2016

Prinsloo GC (1980) *Thielaviopsis basicola* associated with a pod disease of groundnuts in South Africa. Phytophylactica 12:177–180

Riekert HF (1996) Economic feasibility of nematode control in dryland maize in South Africa. Afr Crop Sci J 4:477–481

Riekert HF (1999) Screening of local cowpea cultivars and lines for resistance to *Meloidogyne javanica*. Afr Plant Prot 5:11

Riekert HF, Henshaw GE (1998) Effect of soybean, cowpea and groundnut rotations on root-knot nematode build-up and infestation in dryland maize. Afr Crop Sci J 6:377–383

Roberts PA (2002) Concepts and consequences of resistance. In: Starr JL, Bridge J et al (eds) Plant resistance to parasitic nematodes. CAB International, Wallingford, pp 23–41

Schoeman MJ, Riekert HF (1999) Microplot screening of local lupin cultivars for resistance to *Meloidogyne javanica*. Afr Plant Prot 5:17

Seinhorst JW (1957) *Ditylenchus* races, pathogenicity, and ecology. Proceedings S-19 workshop on Phytonematodes, Nashville, pp 1–7

Shurtleff W, Aoyagi A (2009) History of soybeans and soyfoods in Africa (1857–2009): Extensively Annotated Bibliography and Sourcebook. Soyinfo Centre, Lafayette

Sikora RA, Bridge J, Starr JL (2005a) Management practices: an overview of integrated nematode management technologies. In: Luc M, Sikora RA, Bridge J (eds) Plant parasitic nematodes in subtropical and tropical agriculture, 2nd edn. CAB International, Wallingford, pp 793–825

Sikora RA, Greco N, Veloso Silva JF (2005b) Nematode parasites of food legumes. In: Luc M, Sikora RA, Bridge J (eds) Plant parasitic nematodes in subtropical and tropical agriculture, 2nd edn. CAB International, Wallingford, pp 259–318

Smit MA, De Beer GP (1998) Report of the national soybean cultivar trials 1998/99. Oil and Protein Seed Centre, Agricultural Research Council–Grain Crops Institute (ARC–GCI), Potchefstroom

Starr JL, Mc Donald AH, Claudius-Cole A (2013) Nematode resistance in crops. In: Perry RN, Moens M (eds) Plant nematology, 2nd edn. CAB International, Wallingford, pp 411–436

Steenkamp S, Mc Donald AH, De Waele D (2010a) Resistance to *Ditylenchus africanus* present in peanut breeding lines. J Nematol 42:159–165

Steenkamp S, Jordaan A, Mc Donald AH et al (2010b) Differential responses of resistant and susceptible groundnut genotypes at cellular level to *Ditylenchus africanus*. Nematol 13. doi:10.1163/13885541OX508723

Svamv PM, Narasimhareddy S (1977) Changes in the leakage of electrolytes from groundnut seeds (*Arachis hypogaea*) during after-ripening. Seed Sci Technol 5:654–648

Swanevelder CJ (1997) Grondbone – Altyd 'n wenner. Landbounavorsingsraad-Instituut vir Graangewasse, Potchefstroom

Van Biljon ER (2004) Integrated pest management programme for the control of plant parasitic nematodes. In: Abstracts of the Fourteenth Soilborne Plant Diseases Interest Group Symposium, Stellenbosch, 15–16 Sept

Van den Berg E, Mc Donald AH (1991) Screening of soybean cultivars for resistance against *M. javanica* and *M. incognita* race 4. In: Proceedings of the tenth symposium of the Nematological Society of Southern Africa, Wildernis, 7–10 Apr 1991

Van der Linde WJ, Clemitson JG, Crous ME (1959) Host-parasite relationships of South African root-knot eelworm (*Meloidogyne spp.*). Dep Agric Technol Serv Repub S Afr Ent Ser 44:3–16

Van Zyl K (2013) A guide to crop pest management in South Africa, 1st edn. VR Print, Pinetown

Venter C (2014) Exploitation and characterisation of resistance to the root-knot nematode *Meloidogyne incognita* in soybean. MSc Dissertation, North-West University, Potchefstroom

Venter C, De Waele D, Meyer AJ (1991) Reproductive and damage potential of *Ditylenchus destructor* on peanut. J Nematol 23:12–19

Venter C, De Waele D, Van Eeden CF (1992) Plant-parasitic nematodes on field crops in South Africa. 4. Groundnut. Fundam Appl Nematol 15:7–14

Venter C, De Waele D, Meyer AJ (1993) Reproductive and damage potential of *Ditylenchus destructor* on six peanut cultivars. J Nematol 25:59–62

Venter C, Van Aswegen G, Meyer A et al (1995) Histological studies of *Ditylenchus africanus* within peanut pods. J Nematol 27:284–291

Verdoorn G (2012) Endosuflan dies a sudden death. http://www.pcsib.org.za/wp-content. Accessed 13 Feb 2013

Viaene N, Coyne DL, Davies KG (2013) Biological and cultural management. In: Perry RN, Moens M (eds) Plant nematology. CAB International, Wallingford, pp 383–410

Wendt KR (1992) Taxonomic differentiation of three species of *Ditylenchus* (Nematoda) and their races using ribosomal DNA analysis. MSc thesis, Simon Fraser University, Burnaby

Wendt KR, Vrain TC, Webster JM (1993) Separation of three species of *Ditylenchus* and some host races of *D. dipsaci* using RFLP's. J Nematol 25:555–563

Wendt KR, Swart A, Vrain TC et al (1995) *Ditylenchus africanus* sp. n. from South Africa; a morphological and molecular characterization. Fundam Appl Nematol 18:241–250

Windham GL, Williams WP (1988) Reproduction of *Meloidogyne javanica* on corn hybrids. J Nematol 20(Annals 2):25–28

Chapter 10
Nematode Pests of Potato and Other Vegetable Crops

Robin K. Jones, Sheila G. Storey, Rinus Knoetze, and Hendrika Fourie

10.1 Introduction

South Africa (SA) is largely self-sufficient in the production of potato (*Solanum tuberosum*) and other vegetables and has a significant export market for both fresh and processed produce. Production is predominantly undertaken by commercial farmers, utilising land that is intensively farmed. As a result, when certain soil conditions prevail and specific cultural conditions are practised, nematode problems specific to such crops will arise. The major vegetable crops that experience nematode-induced problems are potato and tomato (*Solanum lycopersicon*), while a variety of other vegetable crops are also parasitised by these pests and are reported on.

10.2 Potato

South Africa is the sixth largest by area and the largest by yield per hectare (ha) of producers of potato in Africa. In 2014, nearly 55,000 ha of potato were planted in SA (Anonymous 2016), with plantings and harvest occurring throughout the year.

R.K. Jones (✉)
Robin Jones Nematological Services, 24, Victor Street, Murrayfield, South Africa
e-mail: robin.rsa@gmail.com

S.G. Storey
Nemlab, Nematode Advisory Service, PO Box 2825, Durbanville 7551, South Africa

R. Knoetze
Department of Agriculture, Forestry and Fisheries, Plant Quarantine Station, Polkadraai Road, Stellenbosch 7600, South Africa

H. Fourie
North-West University, Unit for Environmental Sciences and Management, Private Bag X6001, Potchefstroom 2520, South Africa

H. Fourie et al. (eds.), *Nematology in South Africa: A View from the 21st Century*, DOI 10.1007/978-3-319-44210-5_10

Table 10.1 Potato production figures for South Africa (Anonymous 2016)

Market distribution of potato production 2014		Regional distribution of potato production 2014		
Market	Percentage of harvest	Region	Percentage of harvest	Percentage of ha
Seed	8	Limpopo	20	21
Processing	20	Sandveld	15	14
Export	8	Western Free State	17	13
Formal market	35	Eastern Free State	14	19
Informal market	29	KwaZulu-Natal	7	7
		Others (11)	27	26

Eighteen percent by area and 8 % by production volume were registered for seed production. A total of 2,240,000 metric tonnes (MT) of potato were harvested in the major production regions of SA, which was split between the various markets as listed in Table 10.1.

Both the targeted market and the region influence the production systems used (e.g. planting dates, cultivars (cvs.), control practices and others) and consequently also the nature of the resultant nematode pest incidence.

Keetch and Buckley (1984) listed a wide range of plant-parasitic nematode genera/species that parasitise potato in SA and other southern African countries. Kleynhans et al. (1996) listed 39 species of plant-parasitic nematodes from 19 genera as infecting potato in SA. Data obtained from the SAPPNS[1] in 2015 lists 95 species associated with potato. However, only a few species from three genera, *Meloidogyne* (root-knot nematodes; six species), *Pratylenchus* (lesion nematodes; 12 species) and *Globodera* (cyst nematodes; one species), are recognised amongst the principle production constraints of potato in SA. The annual losses caused by plant-parasitic nematodes on potato in SA were assessed in 1989 as being nearly 17 % of production (Keetch 1989), then representing a loss in current value of ZAR1.05 billion. These losses result from reduced plant vigour, on-farm sorting in the packhouse of damaged tubers, further rejection by seed inspectors relating to seed quality, rejection at processing plants and rejection at the fresh produce markets. In 2014, nematode damage caused 9 % of losses due to downgrading from class 1 of the tubers (Vermeulen 2015).

Louw (1982) published a review of the nematode problems of potato in southern Africa under the title 'Nematode Pests of Potatoes'. Since then the major developments that have impacted on the nematode problems in potato have been:

---

[1] Dr Mariette Marais of the Nematology Unit, Biosystematics Division, Agricultural Research Council–Plant Protection Research Institute is thanked for the use of data from the South African Plant-Parasitic Nematode Survey (SAPPNS) database; E-mail: maraism@arc.agric.za

(i) The identification of new sites where the golden cyst nematode, Globodera *rostochiensis* (Wollenweber, 1923) Behrens, 1975, occurs (Knoetze et al. 2004)
(ii) The identification of new root-knot nematode species (Fourie et al. 2001; Onkendi and Moleleki 2013a)
(iii) The first molecular-based report on the distribution and genetic diversity of root-knot nematodes in potatoes in SA (Onkendi and Moleleki 2013b)
(iv) The increase in the area of potato grown under irrigation from <50 % in 1987 to 83 % in 2014 (Vermeulen 2015)
(v) The increase of production in warmer, higher-yielding areas of the Limpopo Province and Sandveld (Western Cape Province), which are dependent on irrigation
(vi) The associated increase in yield from approximately 21 MT ha^{-1} in 1990 to 44 MT ha^{-1} in 2014 (Vermeulen 2015)
(vii) The increase in intensity of land use and the resultant shorter rotation cycles between potato plantings
(viii) The increase in the use of nematicides on potato with most potato plantings receiving at least a single and often multiple applications of nematicides during one growing season
(ix) The tightening of regulations controlling the use of nematicides

The economically most important nematode pests that damage local potato crops, viz. *Meloidogyne* and *Pratylenchus* spp. and *G. rostochiensis*, are elaborated on below.

10.2.1 Root-Knot Nematodes

10.2.1.1 Biology and Distribution

Six root-knot nematode species are reported by Kleynhans et al. (1996) to infect and parasitise potato in SA. *Meloidogyne incognita* (Kofoid & White, 1919) Chitwood, 1949, and *Meloidogyne javanica* (Treub, 1885) Chitwood, 1949, are the most widespread and damaging and are present in 27 and 41 % of populations, respectively, that occur in potato-producing areas (Coetzee 1968; Steyn 1997). Onkendi and Moleleki (2013b) reported on the distribution and genetic diversity of *Meloidogyne* spp. using a molecular-based assay. Samples were collected from 78 sites from all provinces excluding the Eastern and Western Cape. *Meloidogyne* were recovered from 81 % of the samples, with *M. javanica* occurring in 24 %, *M. incognita* in 23 %; *Meloidogyne arenaria* (Neal, 1889) Chitwood, 1949, in 17 %; *Meloidogyne enterolobii* Yang and Eisenback, 1983 (a new identification on potato in SA), in 14 %; *Meloidogyne chitwoodi* Golden, O'Bannon, Santo and Finley, 1980, in 3 %; and *Meloidogyne hapla* Chitwood, 1949, in 1 %. Recently, Agenbag (2016) also identified *M. enterolobii* from potato in the Delmas region (Mpumalanga Province).

Table 10.2 Relationship between precipitation and soil temperatures and the occurrence of *Meloidogyne incognita* and *Meloidogyne javanica* (Taylor et al. 1982)

Meloidogyne spp.	Percentage occurrence in soils with 500 mm or less annual precipitation	Percentage occurrence in soils with average temperature of coolest month of 12 °C or below	Percentage occurrence in soils with average temperature of warmest month at or above 30 °C
M. javanica	42	53	9
M. incognita	17	18	6

The rate of development of *M. incognita* and *M. javanica* is very temperature dependent. Milne and du Plessis (1964) reported that the life cycle of *M. javanica* was completed in 56 days at 12 °C in tobacco, but in only 21 days at 26 °C. Often the invasion of the root system is non-damaging, but as soon as tuber initiation occurs, tubers are invaded by the infective second-stage juveniles (J2) and rapid development and spread occur. Thus, if potato roots and/or tubers become infected early in the growth of the plant, several generations of the pest will occur before tuber lifting, which typically is about 110 days after planting. However, this interval is cv. dependent. The expansion of potato production to the warmer regions of the Limpopo Province and the Sandveld area (Western Cape Province), where the soils are very sandy, required that irrigation be used at planting. The resultant mix of favourable temperatures, soil structure and moisture status has led to a significant increase in the severity of root-knot nematode infections and the resultant increased dependence on the use of chemical control practices.

The global distribution of *M. incognita* and *M. javanica,* as influenced by soil temperature and rainfall, has been shown to differ (Taylor et al. 1982; Evans and Perry 2013) with *M. javanica* occurring in cooler and drier areas (Table 10.2). A need exists to survey thoroughly the geographical distribution of these two species in the potato-production areas to determine more accurately their relative abundance. Onkendi and Moleleki (2013b) reported that *M. javanica* and *M. incognita* were found in all the seven provinces sampled, whereas *M. enterolobii* was found only in KwaZulu-Natal, *M. chitwoodi* in the Free State and *M. hapla* in Mpumalanga. In addition, populations were found to be highly homogeneous.

Kleynhans (1991) first reported *M. chitwoodi* in SA from severely damaged tubers from Ugie in Eastern Cape Province. This author also identified the species from infected seed potato tubers (cv. Sebago) imported from Australia. According to the SAPPNS, *M. chitwoodi* has been recorded from nine locations in five provinces of SA. Six of these reports were on potato. In temperate regions of North America, Australia and Europe, *M. chitwoodi* is a widespread and important pest of potato where it causes serious economic losses due to direct damage to the plant and loss of marketable yield due to reduced tuber quality. This root-knot nematode species tolerates much lower temperatures than *M. incognita* and *M. javanica* and is reported to cause damage to tubers down to 6 °C. In addition, the approved rates in the USA of some nematicides for controlling *M. chitwoodi* are higher than for other *Meloidogyne* spp.; thus, any spread of this nematode will further complicate control problems.

Fourie et al. (2001) also reported the first record of *Meloidogyne fallax* Karssen, 1996, in SA, from groundnut in the Vaalharts area (Northern Cape Province), an important potato production area. This root-knot nematode species is recognised as a serious pest of potato in Europe, New Zealand and Australia where it is also a cold-tolerant species. Thus, as with *M. chitwoodi*, it is important that the distribution and host specificity of *M. fallax* are studied and that the spread is controlled. *Meloidogyne acronea* Coetzee, 1956; *M. arenaria*; and *M. hapla*, have also been recorded as infecting potato but are rare in commercial plantings and little is known of their distribution (Kleynhans 1991; Kleynhans et al. 1996; SAPPNS).

The identification of the highly damaging and resistance-breaking *M. enterolobii* (Onkendi and Moleleki 2013b) in the KwaZulu-Natal and Mpumalanga (Agenbag 2016) provinces, notably on seed producer units in KwaZulu-Natal, is of critical concern, and measures to prevent its spread need be implemented.

10.2.1.2 Symptoms and Economic Damage

Key factors influencing the extent of root-knot nematode damage to potato tubers are the initial abundance of the species and the length of time the crop is in the soil. If a large root-knot nematode population is present at planting, or if infected seed is planted, potato roots will be invaded rapidly. Although not always visible, the above-ground symptoms are typical of a plant with an unhealthy root system. Ultimately stunted growth, yellow foliage, premature wilting and dieback of root-knot nematode-infected potato plants occur. Such symptoms are less common where basic pre-plant control measures are implemented, e.g. the use of certified seed, extended crop rotations and the standard use of chemical control measures in areas where root-knot nematodes pose severe problems to potato production. The latter particularly includes areas where sandy soils occur. However, losses due to root-knot nematode parasitism and subsequent damage of tubers are common. By the time tubers are initiated, J2 from any existing root infection and the soil population that survived any pre- or at plant chemical or biological treatment will readily invade the developing tubers. If this infection occurs early during the growing season or if the mature tubers are stored in the soil after they should have been lifted, a high percentage of tubers will be invaded and severely blemished (Fig. 10.1a). This will result in reduced marketable yield and the downgrading and possible market rejection of the crop due to an adverse impact on the quality (Fig. 10.1a, b). Soil storage of tubers increases the chances and severity of root-knot nematode damage. At the site where the infected tubers were monitored, the percentage of tubers infected with root-knot nematodes in the untreated control increased by 21 % over 18 days (Jones 2008). Similar increases in damage of infected tubers, albeit from a lower existing root-knot nematode population level, occurred in all treatments even though pre- or at-plant nematicides had been applied.

The specific target market influences root-knot nematode damage tolerance levels. Seed producers must meet the criteria of the SA Seed Certification Scheme (DAFF 2010a), which are implemented by inspectors visiting all registered seed

Fig. 10.1 (**a**) Galled and blemished tubers caused by root-knot nematodes and (**b**) cut potato with root-knot damage inside (Diedrich Visser, Agricultural Research Council–Vegetable and Ornamental Research Institute, Pretoria, South Africa)

production units. Table 10.3 lists the standards applied for the specific nematode pests. Potato destined for the processing industry, constituting about 20 % of the market, must also meet a virtual zero tolerance for damage based on visual symptoms. Severely damaged tubers will be discarded during the on-farm packing process, an increased input into the packing process, but as indicated by the 9 % downgrading of tubers at the fresh produce markets, on-farm sorting frequently only removes the worst of the root-knot nematode damaged tubers. Further losses occur at markets before the produce is sold.

10.2.1.3 Disease Interactions

Root-knot nematode species interact with other disease-causing organisms including bacterial wilt caused by *Pseudomonas solanacearum* and blackleg caused by *Erwinia* spp. Pathogenic fungi such as *Verticillium* spp., *Fusarium* spp. and *Rhizoctonia solani* (Brodie et al. 1993; Scurrah et al. 2005; Manzanilla-Lopez and Starr 2009) also concomitantly occur in potato-producing areas.

10.2.1.4 Damage Threshold Levels

Population thresholds and resultant damage for *Meloidogyne* spp. on potato in SA have not been quantified. However, thresholds (Table 10.4) have been determined for *M. incognita and M. chitwoodi* infecting the crop in the USA on the basis of sampling soil directly after the harvest of the previous crop (Ferris and Roberts 2005). It must be noted that these thresholds are higher than would be expected for local soils. Studies to establish such threshold populations in SA remain a research priority.

Table 10.3 Maximum percentages of nematode-infected seed potato permissible for South Africa (DAFF 2010a)

Plant-parasitic nematode genus	Generation (G) of potato plants								
	G 1–3			G 4–6			G 7–8		
	Class category of potato tubers according to quality (elite class is the superior)								
	Elite	Class1	Standard	Elite	Class1	Standard	Elite	Class1	Standard
Meloidogyne spp.	0.1	0.2	1.0	0.1	0.5	1.0	0.2	0.5	1.0
Pratylenchus spp.	0.1	0.5	5.0	0.5	1.0	5.0	1.0	2.0	5.0

Table 10.4 Expected percentage yield loss at pre-plant densities of *Meloidogyne incognita* and *M. chitwoodi* second-stage juveniles (J2) and yield losses for potato in the USA (Ferris and Roberts 2005)

J2 numbers 250 cm soil^{-3}	10		20		50		100		200	
Root-knot nematode species	*Mi*	*Mc*	*Mi*	*Mc*	*Mi*	*Mc*	*Mi*	*Mc*	*Mi*	*Mc*
Percentage yield loss	4	7	15	8	34	12	47	15	51	18

Mi = *M. incognita*, *Mc* = *M. chitwoodi*

10.2.1.5 Management Strategies

Due to the combination of factors outlined above and set against the potential financial losses that can occur, input costs on potato in SA can be up to ZAR120,000–200,000 ha^{-1} under irrigation and 60,000–80,000 under rain-fed conditions (F Niederwieser, Potato South Africa, December 2015, Pretoria, personal communication). Producers must invest heavily on control strategies to minimise both quality and yield losses due to root-knot nematode damage. The use of registered synthetic nematicides is the principal method of nematode control (see Sect. 6.2.2, Table 6.1, Chap. 6). A crop rotation cycle of non- or poor hosts of root-knot nematodes, recommended to be at least 4 years, together with effective weed control, is the foundation to limit the extent of damage by these parasites in potato.

Both *M. incognita* and *M. javanica* infect a wide range of crops, weeds and indigenous plants (Coetzee 1968; Keetch and Buckley 1984; Kleynhans 1991; Kleynhans et al. 1996; SAPPNS). Crops that should be avoided prior to planting potato include lucerne (*Medicago sativa*); maize (*Zea mays*); solanaceous crops such as tomato, paprika (*Capsicum annuum*) and tobacco (*Nicotiana tabacum*); as well as most vegetable crops. Certain cvs. of oat (*Avena sativa*), cotton (*Gossypium hirsutum*) and grasses that exhibit resistance to root-knot nematodes may be used in potato-based cropping systems to minimise damage. Crops that may be included in rotation systems are *Crotalaria* spp., Bloubuffels grass (*Cenchrus ciliaris*), Borseltjie grass (*Anthephora pubescens*), millets, red millet (*Panicum miliaceum*), Rhodes grass (*Chloris gayana*), smuts finger grass (*Digitaria eriantha*), sorghum

(*Sorghum bicolor*), oulandsgras (*Eragrostis curvula*) as well as Brassicaceae crops (mustard and *Raphanus* spp.). Regrettably, due to land pressure and market forces, this cycle is often compromised by the need to plant maize, root-knot nematode susceptible cotton cvs. and other host crops.

Other recommended practices are the regular and timely cultivation and drying of the soil, immediate destruction of volunteer potato plants and tubers, planting certified seed, selecting planting dates to avoid high root-knot nematode infection potential during tuber development, shortening the growing period of the potato crop by using early maturing cvs. and, finally just before planting, at planting and post-planting, the application of registered nematicides.

10.2.2 Lesion Nematodes

10.2.2.1 Biology and Distribution

Twelve lesion nematode species have been identified from potato roots and tubers in SA with *Pratylenchus brachyurus* (Godfrey, 1929) Filipjev and Schuurmans Stekhoven, 1941, being the most damaging and predominant (Van den Berg 1971; Kleynhans et al. 1996; SAPPNS). It is reported from most potato-producing areas.

10.2.2.2 Economic Damage

Large numbers of lesion nematodes damage roots and tubers such that growth of potato plants above-ground is stunted. Plants wilt on hot sunny afternoons and premature plant death occurs. When the plant dies and the roots decay, large numbers of lesion nematodes migrate to and infect tubers. Initially symptoms exhibited are small brownish, blackish or greyish lesions on the surface of tubers. The centre of a lesion is usually raised and is surrounded by a slight depression. Where heavy lesion nematode infections occur, the lesions coalesce to form a crust that covers a large section of the tuber. Lesion nematodes generally invade the tubers to a depth of 0.5 mm, with the underlying tissue not being affected. During storage lesion nematode-infected tubers wither, lose weight and become hard. Lesion development is most rapid between 20 and 30 °C and is halted at 5 °C (Koen and Hogewind 1967).

10.2.2.3 Disease Interactions

Lesion nematodes are known to interact with a number of other soilborne pathogens, notably *Fusarium, Phytophthora, Pythium* and *Verticillium* spp. (Bridge and Starr 2007). Although no such observations have been recorded locally on potato, it is likely that lesion nematodes influence the incidence of a number of soilborne pathogens that are associated with local potato production.

10.2.2.4 Damage Threshold Levels

Lesion nematodes are only likely to cause severe crop losses where the previous crop sustained a very high population, and the follow-up potato crop was invaded early during its development. This is most likely to occur in the warmer production areas on sandy soils if potato plantings follow a susceptible crop host, e.g. maize, soybean (*Glycine max*), sunflower (*Helianthus annuus*) and others. No damage threshold population levels for lesion nematodes have been specifically developed for potato in SA, but typically pre-plant populations of 200 l^{-1} soil are considered severe. Potter and Olthof (1993) indicated damage levels to range from 50–1,800 in 100 g^{-1} soil. According to Bridge and Starr (2007), damage threshold levels of as little as 1–2 individual g^{-1} soil are known for *Pratylenchus penetrans* (Cobb, 1917) Filipjev and Schuurmans Stekhoven, 1941, and *Pratylenchus scribneri* Steiner, 1943, in potato fields. For *Pratylenchus neglectus* (Rensch, 1924) Filipjev and Schuurmans Stekhoven, 1941, and *Pratylenchus crenatus* Loof, 1960, however, damage threshold levels are considered higher. According to the latter authors, damage levels vary with species, climate, soil type and host crop.

10.2.2.5 Management Strategies

The principles for the control of lesion nematodes are the same as for root-knot nematodes. In both instances, a rotation involving non- or poor hosts is critical. However, the wide range of susceptible hosts of these nematodes complicates effective control. Fortunately, if effective pre-plant chemical control is applied to lower population levels of root-knot nematodes, lesion nematodes will also be controlled. No cvs with resistance to lesion nematodes are available (Bridge and Starr 2007).

10.2.3 Golden Cyst Nematodes

10.2.3.1 Distribution

It has been reported that *G. rostochiensis* originated from the South American Andes along with the potato plant (Stone 1979). Along with the sibling species, *Globodera pallida* Stone, 1973, it is one of the most specialised and successful plant-parasitic nematodes of agricultural crops, including potato.

During the 1950s, viable cysts of *G. rostochiensis* were intercepted in soil adhering to seed potato imported from Scotland. Golden cyst nematodes were reported for the first time in SA in 1971 from an irrigated farm north of Pretoria and subsequently on smallholdings around Johannesburg and Bon Accord (Gauteng Province) (Kleynhans 1998). Very strict quarantine measures were imposed to prevent the spread of this nematode pest to other potato-producing areas. These measures were successful, allowing the quarantine restrictions to be lifted at that time. During 1999, almost 28 years later, high population levels of *G. rostochiensis* were reported for the first time in the Ceres area in the Western Cape Province where potatoes are produced (Knoetze et al. 2004). In both the Plant Improvement Act (Act No. 53 of 1976) (NDA 2015a) and Agricultural Pest Act (Act No.36 of 1983) (NDA 2015b), this nematode is listed as a prohibited pest in SA. Local distribution of golden cyst nematodes by means of seed potato is prevented in the South African Seed Potato Certification Scheme of 15 May 1998 (Anonymous 2016), with no tolerance for infection being permissible.

More than 15,000 ha of local potato fields have been tested for the presence of golden cyst nematodes since 1999 (Knoetze et al. 2006; Knoetze 2014). This includes samples from a number of surveys and for certification and export purposes. To date, approximately 500 ha countrywide have been placed under quarantine because of the presence of *G. rostochiensis*. These plots are situated in the Ceres and Sandveld areas (Western Cape Province), Hankey (Eastern Cape Province) as well as in some parts of Gauteng Province. No records of *G. pallida* have been reported from SA.

10.2.3.2 Biology and Symptoms

Globodera rostochiensis has a very narrow host range, restricted to potato, tomato and aubergine (*Solanum melongena*). Solanaceous weeds are also infected by J2 of this species, but detailed host range studies in this regard are lacking for SA (Kleynhans 1975; Knoetze 2014).

Nijboer and Parlevliet (1990) proposed that within *G. rostochiensis*, only pathotypes Ro1 (old Ro1 and Ro4), Ro3 (old Ro2 and Ro3) and Ro5 should be recognised. Therefore, Ro1/Ro4 is distinguishable by reproduction on susceptible potato cvs but not on those containing the *G. rostochiensis*-resistant H1 gene (derived from *S. tuberosum* ssp. *andigena*). On the other hand, pathotypes Ro2/Ro3 will reproduce on both susceptible and resistant cvs. Research with three andigena-derived cvs. showed that the Bon Accord population *G. rostochiensis* was unable to produce a new generation of cysts (Kleynhans 1975), indicating that it represented the Ro1 pathotype. Recently, reproductive tests on golden cyst nematode populations from the Ceres, Sandveld, Hankey and Gauteng areas showed that these populations clearly all belong to pathotype Ro1 (old Ro1 and Ro4), showing no multiplication on a resistant cv. containing the H1 gene (Knoetze 2014).

The optimum temperature range for hatching and population build-up of golden cyst nematodes is 15–27 °C (Kaczmarek et al. 2014). Multiplication rates depend partly on the initial population density (Pi) because of competition for root space and its influence on the sex ratio. With few eggs per g soil multiplication may be 60-fold. With more than 100 eggs per g soil, the post-harvest population may be smaller than the initial population because root systems become too severely damaged (Winslow and McKenna 1972). In cool soils of Northern Europe, there is usually one major generation of golden cyst nematodes each year (Jones 1950).

In moist soils a partial second generation may occur, but activity declines sharply at temperatures above 25 °C (Berry et al. 1977). At Bon Accord in Gauteng, an area that experiences cold winter periods, the population density in one field increased from 2 to only 4 eggs g^{-1} soil over 6 years of intermittent potato cropping (Kleynhans 1998). Le Roux (2000) conducted field experiments in the Ceres area to study the life cycle, epidemiology and control of *G. rostochiensis* under hotter SA conditions. This study showed that the life cycle was completed within only 6 weeks after plant emergence (8 weeks after planting) (Fig. 10.2). This is considerably less than the 90 days required under European conditions and the 80 days in New Zealand and Australia (Stanton and Sartori 1990; Van Riel and Mulder 1998). The minimum and maximum ambient temperatures at the Ceres experiment were recorded as 4–17 and 17–34 °C, respectively, while the maximum soil temperature was 21–39 °C. Glasshouse studies revealed that a small second generation is possible within one potato-growing season. The small, second generation may cause a faster increase in population density if the root system of the potato plant is not already overcrowded by J2.

The eggs of the golden cyst nematode are retained in the body of the old female that forms a protective cyst (Fig. 10.3a, b). Once the J1 have developed inside the eggs, they enter diapause during which time they cannot be stimulated to hatch into J2. Diapause generally terminates before the next crop, and most J1 only enter diapause during their first year (Turner and Evans 1998; Turner and Subbotin 2013).

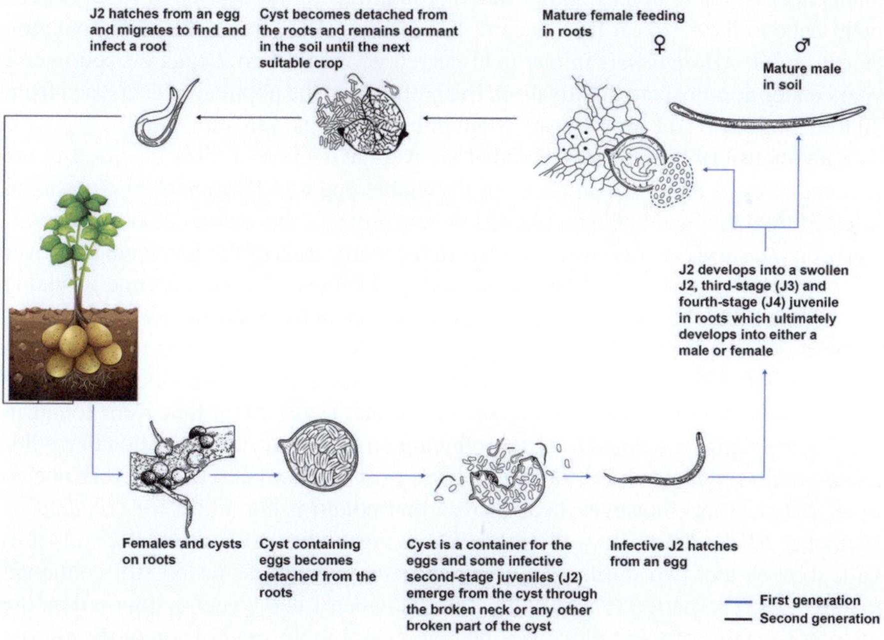

Fig. 10.2 The life cycle of the golden cyst nematode (Hannes Visagie, North-West University, Potchefstroom, South Africa)

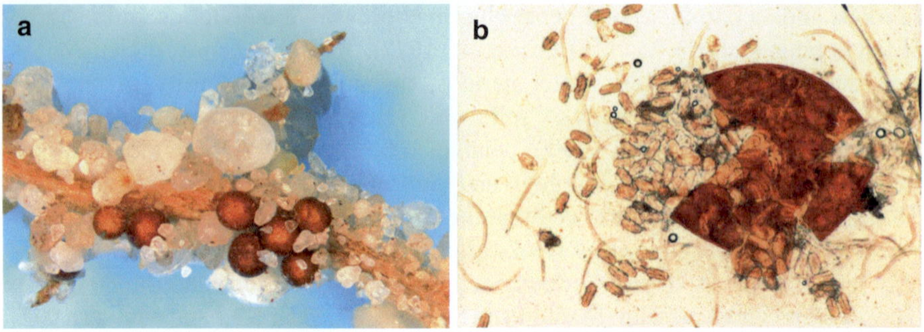

Fig. 10.3 (**a**, **b**) Brown, roundish cysts of the golden cyst nematode on potato roots (**a**) and a cyst filled with eggs and second-stage juveniles (**a**) Caroline Mouton, Department of Agriculture, Forestry and Fisheries, Stellenbosch, South Africa; (**b**) (Antoinette Malan, University of Stellenbosch, Stellenbosch, South Africa)

The hatching of golden cyst nematode J2 is stimulated by root diffusates of the host crop and can be as high as 80 %. Conversely, without the presence of host root diffusates, spontaneous hatching of J2 will be less than 30 %.

The ability of potato cyst nematodes to multiply on a susceptible potato cv. after the prolonged absence of a host plant decreases with time. Depending on the age and state of the cysts, the number of eggs and J2 they contain can vary considerably. The decline in the number of viable eggs and J2 within the cysts in the absence of host plants appears to be at least partly due to conditions in the soil itself. In an experiment done in the 1970s at Bon Accord, Kleynhans (1975) found that the population density of *G. rostochiensis* in one field decreased from 16 to 2 eggs g^{-1} soil over 2 years under non-host crop cultivation. In another field the population decreased from 30 to 0.9 eggs g^{-1} soil over 3 years when non-host crops were cultivated.

Kleynhans (1975) also reported that an average decline of 49 % occurred in the number of cysts and 97 % in terms of the viable egg and J2 content of cysts in an infested field that had not been planted to host crops of this nematode for 6 years. A study on the survival of *G. rostochiensis* in the warm soils of the Sandveld area over a 2-year period concluded that leaving the soil fallow caused a decline in viable eggs in the cysts, but suggests that the cysts will be able to survive for much longer in these soils. This observation was confirmed where cysts remained viable in soils that were left fallow for more than 3 years (Knoetze 2014). Individuals of a new species, *Globodera capensis* Knoetze, Swart and Tiedt, 2013, that were found in most of the samples seemed to be reproducing on opportunistic vegetation that colonised potato fields when left fallow, but this host has not been identified (Knoetze et al. 2012). It has, however, been proven that potato is not a host for *G. capensis* (Knoetze 2014). A similar study in the Ceres area, where slightly cooler soils prevail, showed that two thirds of *G. rostochiensis* populations tested still contained viable cysts. The period of non-cultivation (between 1 and 9 years), imposed by the quarantine status of these plots, was not successful in the eradication of the *G. rostochiensis* population (R Knoetze, unpublished data). These results indicate that *G. rostochiensis* might be able to survive without a host for longer periods in the extreme climates of the Sandveld and Ceres areas than was previously believed. Suggestions that the quarantine period for infested plots in these areas can be shortened were therefore premature and unsubstantiated.

10.2.3.3 Damage

Potato cyst nematodes, *G. rostochiensis* and *G. pallida*, are known globally to be the most important pests of potato in cool temperate regions (Scurrah et al. 2005). Although they also attack and multiply on aubergine (*Solanum melongena*) and tomato, they are especially important as pests of potato crops. Severe infections of potato by potato cyst nematodes can result in serious losses in yield and quality of harvested tubers. Infected potato cvs that are susceptible to these pests grow slowly because their roots are stunted, which limits uptake of water and nutrients from the soil. Haulms emerge slowly from the soil, remain small and may wilt in dry weather and show symptoms of nutrient deficiencies (Whitehead 1998; Scurrah et al. 2005; Bridge and Starr 2007; Subbotin et al. 2010). Tubers remain small. Potato cyst nematodes reduce the size of root systems but never invade the tubers. As the invasion of the root system increases, the plant is unable to compensate for the damage to the roots and ultimately exhibits a range of symptoms resulting from a poor and inefficient root system. In extreme cases the weight of tubers harvested can be less than the weight of the seed tubers planted. Because of the effects of temperature on the survival of cyst nematodes, it is speculated that conditions in much of SA are unsuitable for potato cyst nematodes to thrive. This could apply in certain of the hotter production areas. If the infestation of cyst nematodes is expanded to the cooler areas, severe recurrent losses can be expected for potato crops. The presence of the golden cyst nematode in an area will eventually lead to severe losses for potato crops, either directly by yield loss or indirectly by restriction of management options.

Once present in a field, the golden cyst nematode is extremely difficult to eradicate. Long rotations, the need to restrict cv. use to resistant ones (not available locally) and the use of nematicides can affect farming practices. Smallholding farmers are especially vulnerable to this nematode pest because of poor resources, limited finances to buy expensive nematicides and limitations on available land. The presence of potato cyst nematodes could also severely affect the industry through quarantine restrictions. Potato tubers intended for export can only be from a field that tested free of these pests. It is essential to be able to convince importers that SA does have effective control measures to combat potato cyst nematodes when they occur in agricultural soils.

10.2.3.4 Management Strategies

These nematodes have been controlled by crop rotations, use of resistant cvs, trap cropping, use of nematicides and, most successfully, by the combined use of two or more of these measures (Scurrah et al. 2005; Bridge and Starr 2007). Control in SA is, however, achieved mainly by quarantine measures since no plantings are permitted on fields infested with cyst nematodes (Knoetze 2014). Existing policy objectives concerning the control and reduction of golden cyst nematodes to non-detectable levels seek to avoid large-scale future problems. The protocol developed to manage the spread of this pest has the same objective as occurred in the 1970s, namely:

(i) To identify precisely the infested areas by means of countrywide surveys
(ii) To ensure that traded seed in SA remains free of golden cyst nematodes. Trade should not occur with uncertified seed due to the risk of contaminating additional lands. A zero tolerance exists for potato seed due to the threat of spread of cysts in soil adhering to seed
(iii) To isolate infested lands to ensure further spread of golden cyst nematodes is prevented
(iv) To plant only non-host crops of golden cyst nematodes in infested fields
(v) To contain, reduce and finally eliminate golden cyst nematode pests

These measures are essentially aimed at the long term, but have been adopted successfully before in several countries including New Zealand and Australia (Marshall 1998). For an individual grower whose fields are currently free of golden cyst nematodes, the objective should be to maintain a barrier to the introduction of the pest by using nematode-free certified seed. If production units are in a region where the pest has been identified, strict access must be implemented and irrigation sources must be separated.

On fields that are currently infested with golden cyst nematodes, the guidelines in the protocol for the 'Regulatory Control and Management of the golden cyst nematode in South Africa' should be followed. The purpose of this protocol is to provide specific guidelines for the management of fields tested positive for the presence of this nematode species. The protocol states that seed potato are only to be planted in fields not under a served order and tested free of *G. rostochiensis*. A served order will only be lifted after 8 years of non-host cultivation and only if an official test shows the field to be free from *G. rostochiensis*. Final certification will be subject to a negative test result at harvest of the crop. Plantings for table potato on previously infested plots may take place whenever an official test shows the unit to be free of viable cysts. The 8-year period of non-host cultivation does not apply in the latter case.

10.2.4 Other Plant-Parasitic Nematodes Associated with Potato

Although nematode pests other than *Meloidogyne* spp., *Pratylenchus* spp. and *G. rostochiensis* have been recorded on potato in SA (Keetch and Buckley 1984; Kleynhans et al. 1996; Marais et al. 2015), they have not been associated with crop losses. Internationally, the ectoparasitic root-tip feeders *Nanidorus*, *Paratrichodorus* and *Trichodorus* are considered as important pests of potato. These genera can cause severe damage to the root system of young developing plants and have the ability to transmit viral diseases (Marshall 1998; Scurrah et al. 2005). Although *Nanidorus* and *Paratrichodorus* spp. occur widely in local agricultural-producing areas (SAPPNS), particularly in sandy soils planted to potato, problems encountered have not been reported. However, the importance of these nematodes as potential pests that could damage potato plantings should always be considered.

10.2.4.1 Management Strategies

Nematode control on potato is complex with certain specific crop-related problems. Firstly, the need exists for control of both root invasion directly after planting and tuber invasion from flowering until lifting. This requires very effective pre-plant control using a range of strategies as well as the well-planned and responsible use of registered nematicides up until as close to harvest as the specific chemical permits. Secondly, regional differences in approach to plant-parasitic nematode control exist, primarily due to climatic and soil factors, but also due to land use patterns. Where fields can be rotated effectively with poor root-knot nematode hosts over an extended period, e.g. planting of pasture crops, control by cultural practices may be sufficient. As the pressure on land use increases, so will the root-knot and lesion nematode burden and chemical options will be needed. In areas of intense utilisation, for example, areas with warm and sandy soil, an integrated programme of cultural control, pre-plant fumigation and at plant and post-plant application of registered nematicides are generally applied by producers. Thirdly, despite the use of such programmes, damage still occurs such that the crop may be downgraded or totally rejected as seed, processed or fresh produce. Usually such situations have resulted from the producer ignoring some of the basic cultural and chemical control recommendations, such as delaying lifting because of poor market prices. The control of root-knot nematodes is the major focus of nematode control practices for potato in SA. Lesion nematodes cause losses, but in the nematode control planning process, producers will not adopt specific strategies that address lesion nematode control in isolation of those for root-knot nematodes. The control strategies in place for *G. rostochiensis* are based on very specific regulations as outlined above. The recommendations outlined below cannot be used to provide control of *G. rostochiensis* but are applicable to other nematode pests such as *Meloidogyne* and *Pratylenchus* spp. in particular.

Seed Certification

The production of a reliably clean ('nematode-free') source of potato seed forms the foundation of effective control against the predominant plant-parasitic nematodes associated with the crop as referred to earlier. The certification of seed potato is under the auspices of the Plant Improvement Act 53 of 1976 and the South African Seed Potato Certification Scheme. Registered plantings are inspected by trained officers during and after lifting and categorised as listed in Table 10.3. Many producers are tempted to plant farm-retained seed where infection levels of plant-parasitic nematodes are poorly known and can easily exceed the regulated standards. The use of such seed will result in immediate infection of roots and the subsequent downgrading of the tubers of the harvested crop.

Chemical Control

Potato crops generally form the largest market for nematicides in SA. All products to control nematodes must be registered with the Department of Agriculture in terms of Act 36 of 1947. The general characteristics of the products are not specific to potato, and thus these details are outlined in Chap. 6 (see Sect. 6.2.2, Table 6.1).

Due to the extended period of nematode invasion in potato roots and tubers, nematicides are applied either pre-plant, at plant or post-plant. In most cases a combination of all three is used. Furfural, with a very short pre-harvest interval of only 14 days, is a versatile product that can be used nearly throughout the production cycle (Van Zyl 2013).

Cultural Control

The fundamental components of cultural control are based on the manipulation of the soil environment to reduce the plant-parasitic nematode populations between potato crops. This is not easy to achieve in the intensely farmed areas due to the wide and differing host range of *M. incognita* and *M. javanica,* as well as that for lesion nematodes. Recommendations include:

(i) The removal of unharvested tubers, destruction of crop residues and volunteer plants, though currently no herbicidal remedies are registered for this process
(ii) Planting potato after a long rotation with poor host crops; a minimum of 4 years is currently recommended
(iii) Examine the roots of the previous crop prior to root dieback to determine the extent of root-knot nematode galling, lesion formation due to *Pratylenchus* spp. or the presence of cyst nematodes. If the crop, for example, has extensive root-knot nematode damage, extended rotation with poor or non-host crops is recommended until the root-knot nematode population has declined substantially
(iv) Before planting the next crop, the soil surface must be left to dry out thoroughly immediately after the preceding crop has been harvested. Dry hot soil with no crop cover is an unfavourable environment, particularly for root-knot nematodes. Cultivation of the soil to enhance this process is also recommended as an additional strategy
(v) Planting of certified seed only should be practised by producers
(vi) Planting of seed should be done at such a stage that tuber initiation occurs during the cooler months of the growing season. Such periods are unfavourable for the population build-ups of the thermophilic root-knot nematode

spp., e.g. *M. enterolobii, M. incognita* and *M. javanica*. This period of planting will differ between regions. Producers in cooler potato-production areas should also be aware of the potential of cryophilic spp., viz. *M. hapla* and *M. chitwoodi*, recorded to infect potato locally. Such root-knot nematode species optimally develop and reproduce under low temperatures and can cause big quality losses to potato crops (Kleynhans 1991)

(vii) The crop should be harvested as early as possible. In areas where root-knot nematodes are known to cause problems to potato, quick-maturing cvs should be planted, e.g. BP13, Koos Smit, Van der Plank, Sandvelder and King George

(viii) Tubers should not be stored in the soil

10.3 Tomato

Tomato is, after potato, the second most important vegetable produced in SA (FAO 2016). During 2013, the crop was planted to approximately 7,900 ha, with 566,180 MT being produced (FAO 2016). In SA, tomato production constituted near to 19 % of the total gross value of vegetable production that excludes potato (DAFF 2014a). Although tomato is planted throughout the country, the major tomato-producing areas are situated in the Limpopo Province, followed by Mpumalanga and the Eastern Cape Provinces (DAFF 2014a). Production is limited to frost-free areas or under semi-controlled conditions, such as tunnels or glasshouses. The crop is cultivated by commercial and smallholding producers as well as home and community gardeners. During 2010, the commercial sector contributed 95 % and the emerging sector 5 % of the total produce (DAFF 2014a).

10.3.1 Plant-Parasitic Nematodes Associated with Tomato

Numerous plant-parasitic nematode genera and species parasitise tomato in SA (Keetch and Buckley 1984; Kleynhans et al. 1996; SAPPNS). Keetch (1989) reported local yield losses as a result of nematode parasitism of approximately 21 %, which at that time equalled monetary losses of ZAR35.3 million. Root-knot (Fig. 10.4) and lesion nematodes are, economically, the most important on tomato crops in both the commercial and smallholding sectors.

10.3.1.1 Root-Knot Nematodes

Root-knot nematode species that infect tomato in SA are *M. arenaria*; *M. chitwoodi*; *M. enterolobii*; *M. fallax*; *M. hapla*; *Meloidogyne hispanica* Hirschmann, 1986; *M. incognita*; and *M. javanica* (Keetch and Buckley 1984; Kleynhans et al. 1996; Mtshali et al. 2002; SAPPNS).

Fig. 10.4 A root-knot nematode-infested field of a smallholding farmer in Matsulu near White River (Mpumalanga Province) with patches where plants are stunted or dead as a result of high population levels (Mieke Daneel, Agricultural Research Council–Institute for Tropical and Subtropical Crops, Mbombela, South Africa)

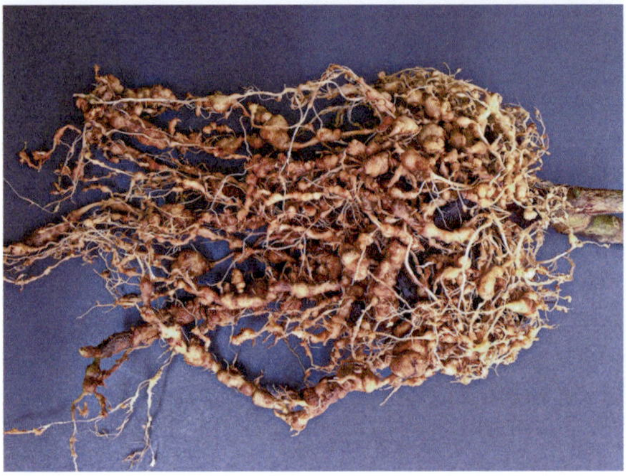

Fig. 10.5 A severely galled root system of a tomato plant infected with root-knot nematodes (cv. Floradade) (Kirk West, Port Elizabeth, South Africa)

Symptoms and Damage

Above-ground symptoms in root-knot nematode infested fields usually include patches of poor-growing plants (Fig. 10.4) that may be wilted, stunted and/or chlorotic. Feeding of root-knot nematode individuals in roots of tomato plants causes the typical formation of galls or knots (Fig. 10.5), which limit water and nutrient uptake (Abad et al. 2009). Where severe root-knot nematode infections occur, the whole root system may be visible as thickened structures due to excessive galling. The life cycle of root-knot nematodes in tomato roots is influenced by factors such as temperature, initial population density (Pi), cv. and soil type, as well as other biotic and abiotic factors. These play an important role in the duration of the life cycle.

Damage Potential

No extensive research has been done on tomato in SA. Daiber (1989a) reported cumulative galling and yield losses of tomato (cv. Rodade) in a micro-plot study following inoculation with eggs of a mixed root-knot nematode population (70:30

ratio of *M. incognita* and *M. javanica*). The highest root-knot nematode population levels and yield losses were recorded in tomato roots 84 days after transplanting. Also, infected tomato seedlings were significantly smaller than those that were not inoculated with root-knot nematodes. In the same study the effect of pre- (10 days before transplanting) and post-inoculation (21 days after transplanting) of soil with the same mixed *M. incognita* and *M. javanica* population was also investigated. Although root-knot nematode population development for the two treatments showed similar trends, tomato growth differed. Plants that grew in fumigated plots were visibly healthy 112 days after transplanting, while 50 % of such plants showed the same tendency 133 days after transplanting. Also, no galls were visible on roots of the latter plants. It was concluded that big second-generation root-knot nematode populations developed and were maintained in tomato roots whether nematode inoculation was done pre- or post-seedling transplanting. The presence of root-knot nematode males in samples from both treatments further indicated that conditions were adverse for optimal development of these pests due to the presence of their abnormal high populations in tomato roots.

Daiber (1990a) investigated the effect of various root-knot nematode Pi, viz. 6,000, 30,000 and 180,000 eggs plot^{-1} 10 days after transplanting. From 35 days after transplanting, population levels of J2 and females increased substantially with significant, linear relationships being evident with plant height, yield and Pi. This research showed that root-knot nematode population densities that exist during the early seedling stage of tomato are decisive in terms of plant vigour and yield. The data show that tomato seedlings should be protected from root-knot nematode infections at seedling transplanting and during the early growth stages of the plants.

Population development of root-knot nematodes was also investigated in roots of susceptible and resistant cvs using various Pi levels in micro-plot experiments (Daiber 1990b) and glasshouse studies (Fourie et al. 2012). Results showed that root-knot nematode J2 infected and developed in roots of susceptible cv. Rodade with mature females and eggs present 21 days after transplanting. Severity of root galling increased with increasing Pi, and roots were heavily galled at the termination of the trial (154 days after seedling transplanting). Conversely, in roots of the resistant cv. Rotam 4, only a few J2 were recorded with none of them developing into females. Cultivar Rodade tolerated root-knot nematode infection up to Pi levels of 9,000 eggs plot^{-1} without yield loss being recorded. Only at higher Pi levels (ca 27,000, 81,000 and 243,000 eggs plot^{-1}) did this cv. suffer significant yield losses. By contrast, yields of cv. Rotam 4 were not affected by increasing Pi levels.

Fourie et al. (2012) studied the multiplication of a local *M. incognita* population in roots of both a susceptible (Moneymaker) and resistant (Rhapsody) cv. using a range of Pi levels, ca 0, 100, 500, 1,000, 5,000, 10,000 and 20,000 seedling^{-1}. Populations of *M. incognita* increased and were significantly higher in roots of cv. Moneymaker compared to those in roots of cv. Rhapsody at most of the Pi levels. A significant, non-linear relationship existed between Pi and the final population (Pf) density in the roots of both cvs. The regression line for cv. Moneymaker generally levelled off between Pi = 5,000 and Pi = 10,000, showing that Pf in roots did not increase substantially at these Pi levels. On the other hand, the regression line for cv.

Rhapsody levelled off after a slight increase in eggs and J2 at the highest Pi = 20,000. Although yield data for the two cvs was adversely affected by excessive rains throughout the growing season, percentage yield loss was substantially higher for cv. Moneymaker compared to that for cv. Rhapsody. At the highest Pi = 20,000, yield losses of up 19 % were recorded for cv. Moneymaker in contrast to 0.2 % for cv. Rhapsody.

Management Strategies

The availability of resistant vegetable cvs remains one of the most viable and environmentally friendly options for limiting crop yield and quality losses due to plant-parasitic nematodes (Williamson and Roberts 2009). Although a number of root-knot nematode-resistant tomato cvs are available worldwide (Roberts 1992; Starr et al. 2013; Williamson and Roberts 2009), the host status of genotypes to the two most common *Meloidogyne* spp. (*M. incognita* and *M. javanica*) in many, particularly Third World countries, is generally unknown (Fourie et al. 2012). Twenty-one tomato cvs that were commercially available in SA during the early 2000s were evaluated for their host suitability to *M. incognita* and *M. javanica* populations in separate, concurrent glasshouse experiments (Fourie et al. 2012). Various nematode parameters, e.g. Pf, reproduction factor (Rf) and percentage resistance, showed that the cvs differed significantly in their host status for both species. Cultivars Rhapsody, MFH 9324, F1454 and FA 593 were identified with resistance to *M. incognita,* whereas none was resistant to *M. javanica*.

The control options for plant-parasitic nematodes in tomato and other vegetable crops are in general the same as those used to protect potato crops (see Sect. "Chemical Control" and 10.2.4.1.2). Various nematicides, synthetically and plant-derived as well as a biological product that contains the fungus *Purpureocillium lilacinum*, are currently registered for use on tomato (Van Zyl 2013). As with potato, synthetic nematicides are used as the major control strategy in tomato plantings. Daiber (1990c), for example, reported the efficacy of both pre- and post-plant fenamiphos (a.s. 400 g l^{-1}) treatments in reducing a mixed population of *M. javanica* and *M. incognita* in tomato.

As with potato, nematode-free planting material should also be used to grow tomato crops. This could be achieved by obtaining seedlings from accredited nurseries or by growing seedlings in nematode-free soil in seedbeds or containers.

A variety of cultural control methods, e.g. soil solarisation, planting of antagonistic crops in soils/substrates where seedlings will be grown, well-planned crop rotation using crops that are poor hosts of *Meloidogyne* spp., amendment of soil with manures, permacultures and resistant cvs., are also used to protect tomato plants against nematode pests (Sikora and Fernandez 2005).

More detail on eco-friendly, alternative nematode management strategies that are employed against plant-parasitic nematodes in the smallholding farming sector in SA are elaborated on in Chap. 7.

10.4 Other Vegetable Crops

Although a substantial amount of work has been done by local nematologists on crops other than potato and tomato, a dearth of published information exists for a wide range of vegetable crops. For this reason only an outline, including major cornerstones achieved to date with regard to selected vegetable crops, is presented.

10.4.1 Onion

Onion (*Allium cepa*) is the most consumed vegetable in SA after potato and tomato (DAFF 2014b). A total of 23,300 ha was planted to onion during 2013, with 518,062 MT being produced (FAO 2016). Onion crops are produced in almost all provinces but mainly in the Western Cape (Ceres area), Northern Cape, North-West and Limpopo (DAFF 2014b).

Although most producers and agricultural consultants generally perceive that onion is not a good host for plant-parasitic nematodes, a variety of genera and species have been reported to parasitise the crop in SA (Keetch and Buckley 1984; Meyer 1984; Kleynhans et al. 1996; SAPPNS). These include *Meloidogyne* spp., *Pratylenchus* spp. (S Storey, Nemlab, Durbanville, 2015, personal communication), *Paratrichodorus* spp. and *Nanidorus* spp. (Meyer 1984). The latter author reported that *Nanidorus minor* (Colbran, 1956) Siddiqi, 1974, caused substantial yield losses to onion on a farm in the Koue Bokkeveld near Ceres (Western Cape Province). Above-ground symptoms were represented by patches of stunted and poorly growing onion plants. Below-ground, the roots were severely stunted with some plants having hardly any. A follow-up experiment, in which a synthetic, fumigant nematicide was applied, resulted in a substantial increase (6 to more than 45 MT ha^{-1}) in the yield of the onion crop. Fenamiphos and furfural are registered for use on onion in SA (Van Zyl 2013).

10.4.2 Cabbage and Other Brassica Crops

Production of 132,600 MT of cabbage (*Brassica oleracea* var. *capitata*), broccoli (*Brassica oleracea* var. *italica)*, cauliflower (*Brassica oleracea* var. *botrytis)*, radish (*Raphanus sativus*) and Swiss chard (*Beta vulgaris* subsp. *vulgaris*) from 2,314 ha was recorded for SA during 2013 (FAO 2016). Although cabbage is generally produced throughout the country, its production is mainly concentrated within the Mpumalanga Province as well as in the Camperdown and Greytown areas of KwaZulu-Natal (DAFF 2010b).

Heterodera schachtii A. Schmidt, 1871, and *Meloidogyne* spp. (Kleynhans et al. 1996) are regarded as the predominant nematodes associated with cabbage and related crops in SA (Keetch and Buckley 1984; Kleynhans et al. 1996; Mtshali et al.

2002). Although no recent data are available, plant-parasitic nematodes were estimated to cause losses of close to 14 % in brassica crops during the 1980s, which was then equivalent to a monetary yield loss of ZAR2.4 billion (Keetch 1989).

Daiber (1989b) reported a mean of 119 *H. schachtii* eggs and J2 cyst^{-1} from potted cabbage plants. According to this author, almost 50 % of these cysts contained 100 eggs and J2 or more, with 1–7 % of the cysts being empty. Daiber (1991) also reported that *H. schachtii* population levels decreased from a mean of 37 to below 4 eggs and J2 g^{-1} soil 18 months after cabbage was harvested from a field in the Johannesburg area (Gauteng Province). Second-stage juveniles emerged continuously from cysts in moist soil and infected beetroot tubers 17 months after cabbage was harvested at this field site. In terms of chemical control, only products with ethoprophos as the a.s. are registered for use on cabbage in SA (Van Zyl 2013).

10.4.3 Carrot

Carrot (*Daucus carota*) is considered one of the major vegetables consumed in SA. During 2013, local carrot production was approximately 180,000 MT. Although carrot grows best in cooler regions or during winter seasons, the crop can be planted during summer or in warmer areas of the country. Production is concentrated in the Western Cape, Gauteng, Free State, North-West, KwaZulu-Natal and Mpumalanga provinces (DAFF 2014c).

A wide range of plant-parasitic nematodes parasitise carrot in SA (Keetch and Buckley (1984); Kleynhans et al. 1996; SAPPNS). The predominant plant-parasitic nematodes that damage carrot are the root-knot nematodes, *M. arenaria*; *Meloidogyne ethiopica* Whitehead, 1968; *M. hapla*; *M. incognita*; and *M. javanica* (Fig. 10.6) (Kleynhans et al. 1996; Mtshali et al. 2002; Bekker et al. 2007; Fourie et al. 2011). In addition, *Heterodera carotae* Jones, 1950 is an important pest (Kleynhans et al. 1996; SAPPNS). In the Western Cape Province in particular, *Paratrichodorus* spp., *Nanidorus* spp. and *Hemicycliophora* spp. are regarded as important pests that infect carrot crops (SAPPNS).

Although no recent figure is available, Keetch in 1989 estimated local yield losses to carrot as a result of plant-parasitic nematodes to be approximately 9.3 %, with monetary losses of ZAR1.3 billion. The only product that is registered locally for use on carrot is furfural (Van Zyl 2013). Although no synthetically-derived nematicide is registered for use on carrot locally, application of several nematicides in soils of seed beds/fields before the planting of vegetable crops, such as carrot, can reduce the potential of plant-parasitic nematodes that may infect the crop.

10.4.4 Sweet Potato

Sweet potato (*Ipomoea batatas*) was planted on 19,000 ha in SA during 2013, with a production volume of 56,000 MT (FAO 2016). The main production areas are situated in the Northern Cape, Western Cape, Limpopo, Free State, Eastern Cape

Fig. 10.6 Galls visible on root systems of carrot plants infected with root-knot nematodes (Willem Steyn, Agricultural Research Council–Institute for Tropical and Subtropical Crops, Mbombela, South Africa)

and Gauteng provinces (DAFF 2014d). A range of plant-parasitic nematodes are associated with sweet potato in SA (Keetch and Buckley 1984; Kleynhans et al. 1996; SAPPNS) of which root-knot nematodes are probably the most damaging. *Paratrichodorus lobatus* (Colbran, 1965) Siddiqi, 1974, was found infecting sweet potato tubers (cv. Ribbok) in the Sandveld area (Western Cape Province) (Malan et al. 2002). Above-ground symptoms of infection were conspicuous circular patches of stunted plants with chlorotic leaves. Although visual symptoms on the tubers resembled insect-feeding damage, investigations revealed that hollows in tubers were filled with *P. lobatus* individuals. As part of the same study, a glass-house experiment with potted sweet potato resulted in a 30-fold increase in *P. lobatus* numbers within 30 days. At present no nematicides are registered for use on sweet potato in SA (Van Zyl 2013).

10.4.5 Green Pea

During 2013, green pea (*Pisum sativum*) production was recorded as 9,458 MT from an area of 5,306 ha (FAO 2016). Pea crops are grown in almost all parts of the cooler areas of SA, particularly in KwaZulu-Natal, the Brits and Rustenburg areas in the North-West Province and in the Mpumalanga Lowveld (DAFF 2011). Although a wide variety of plant-parasitic nematodes are associated with green pea in SA (Keetch and Buckley 1984; Kleynhans et al. 1996; SAPPNS), *Meloidogyne* followed by *Pratylenchus* are generally the predominant genera that cause damage to this crop. Damage to green pea by *Meloidogyne* spp. and *P. penetrans*, in particular, resulted in suboptimal yields during the early 1980s (Van der Vegte and Daiber 1985). Evaluation of dichloropropene, ethylene dibromide and fenamiphos at reduced rates resulted in an increase in plant growth in soils where *Meloidogyne* spp. and *P. penetrans* occurred. A yield increase of up to 100 % was reported from one experiment, following nematicide application. Currently only ethoprophos and fenamiphos are registered locally for use on pea (Van Zyl 2013).

10.4.6 Beetroot

The gross value for local beetroot (*Beta vulgaris*) production was recorded as approximately ZAR200 million during 2013, with 65,000 MT being produced (DAFF 2014e). Beetroot is widely grown both commercially and in home and community gardens throughout the country. The crop grows best during spring and autumn seasons, but also performs well when planted during the summer season in the Highveld and in winter season in the Lowveld of the Mpumalanga Province. The main production areas of beetroot are situated in the North-West, Gauteng, Mpumalanga, KwaZulu-Natal and Western Cape provinces.

Beet cyst (*H. schachtii*) and root-knot nematodes are generally regarded as the predominant plant-parasitic nematodes that adversely affect the crop in SA. Symptoms of root-knot nematode damage on beetroot appear as galled areas on the tap and secondary roots. In many cases the tubers are deformed, particularly where high infection levels occur (Fig. 10.7).

Cysts of the beet cyst nematode were reported from vegetable fields where beetroot was planted in rotation with cabbage, cauliflower, carrot, lettuce (*Lactuca sativa*) and onion in the Rondebult area near Johannesburg (Gauteng Province) (Daiber 1987). An average of 37 *H. schachtii* eggs and J2 g^{-1} soil was recorded in beetroot plantings at this site, resulting in a yield reduction of more than 66 %.

Treatment with fenamiphos (EC; 40 %) at a dosage of 8 l ha^{-1} on *H. schachtii*-infested soil during the 1986/87 growing season resulted in a significantly higher yield than for crops grown on untreated plots. In another experiment, application of metam sodium and ethylene dibromide resulted in 92 and 90 % yield increases, respectively, of a beetroot crop grown in cyst nematode-infested soils (Daiber 1990d). No nematicides are registered for use on beetroot in SA (Van Zyl 2013).

10.4.7 Green Pepper

Green pepper (*Capsicum annuum*) is grown on farms in both the high- and lowveld areas of the Gauteng Province and in the Northern Cape, Eastern Cape, Western Cape, Limpopo and KwaZulu-Natal provinces (DAFF 2013). The crop is

Fig. 10.7 The deformed root systems of beetroot plants infected with root-knot nematodes (Mieke Daneel, Agricultural Research Council–Institute for Tropical and Subtropical Crops, Mbombela, South Africa)

parasitised by a range of plant-parasitic nematodes (Keetch and Buckley 1984; Kleynhans et al. 1996; SAPPNS). *Meloidogyne javanica* and *M. incognita* are the most abundant nematode pests of green pepper. Spiral *(Helicotylenchus* spp.), ring (*Criconema* spp.) and dagger (*Xiphinema* spp.) nematodes have also been associated with the crop (Mtshali et al. 2002; SAPPNS).

Aluvilu et al. (2010) reported resistance to *M. incognita* in *Capsicum frutescens* cv. Capistrano under micro-plot conditions in saline soils near Polokwane (Limpopo Province). In the same experiment, the green pepper cv. Serrano was identified as being susceptible. *Capsicum* sp. cv. Tabasco was identified as resistant in glasshouse and micro-plot experiments (Steyn et al. 2015). However, at Pi levels >30,000 eggs and J2 seedling^{-1}, the resistance could not be verified in the micro-plot experiment. Furfural is the only nematicide registered for use on pepper (Van Zyl 2013). However, some nematicides (e.g. fumigants) can be applied to seed beds/fields prior to the planting of vegetable crops (e.g. beetroot, carrot and pepper) should it be an economically viable option for producers. Nematode control should, however, commence at nursery level where cuttings should only be rooted and transplanted into nematode-free soil.

Meloidogyne spp. are generally the most important nematodes on most vegetable crops grown in SA. The species associated with the various crops are listed in Table 10.5.

Table 10.5 Root-knot nematode species that are known to parasitise vegetable crops in South Africa (Kleynhans et al. 1996; Mtshali et al. 2002; S Storey, unpublished data; SAPPNS)

Crop	*Meloidogyne* species										Crop	*Meloidogyne* species									
	Ma	*Mac*	*Mc*	*Me*	*Met*	*Mf*	*Mh*	*Mhi*	*Mi*	*Mj*		*Ma*	*Mac*	*Mc*	*Me*	*Met*	*Mf*	*Mh*	*Mhi*	*Mi*	*Mj*
Allium cepa (onion)	▲						▲		▲	▲	*Daucus carota* (carrot)	▲				▲		▲		▲	▲
Allium porrum (leek)										▲	*Ipomoea batatas* (sweet potato)							▲		▲	▲
Apium graveolens (celery)									▲	▲	*Lactuca sativa* (lettuce)	▲						▲		▲	▲
Beta vulgaris (beetroot)	▲								▲	▲	*Solanum lycopersicon* (tomato)	▲		▲	▲		▲	▲	▲	▲	▲
Brassica oleracea (cabbage)									▲	▲	*Pisum sativum* (green pea)	▲						▲		▲	▲
Capsicum sp. (pepper)				▲			▲		▲	▲	*Phaseolus* spp. (beans)	▲	▲					▲		▲	▲
Cichorium endivia (endive)									▲		*Raphanus sativus* (radish)									▲	▲
Cynara scolymus (artichoke)									▲	▲	*Solanum melongena* (eggplant)									▲	▲
Cucumus sativus (cucumber)									▲	▲	*Solanum tuberosum* (potato)	▲		▲	▲			▲		▲	▲
Cucurbita pepo (pumpkin)							▲		▲	▲	*Spinacia oleracea* (spinach)									▲	▲

▲ *Meloidogyne* spp. associated with each crop; *Ma* = *M. arenaria*, *Mac* = *M. acronea*, *Mc* = *M. chitwoodi*, *Me* = *M. enterolobii*, *Met* = *M. ethiopica*, *Mf* = *M. fallax*, *Mh* = *M. hapla*, *Mhi* = *M. hispanica*, *Mi* = *M. incognita*, *Mj* = *M. javanica*

References

Abad P, Castagnone-Sereno P, Rosso M-N et al (2009) Invasion, feeding and development. In: Perry RN, Moens M, Starr JL (eds) Root-knot nematodes. CAB International, Wallingford, pp 163–176

Agenbag (2016) Identification and reproduction potential of South African *Meloidogyne* species. MSc Dissertion, North-West University, Potchefstroom

Aluvilu A, Mashela PW, Mokgalong NM et al (2010) Influence of *Meloidogyne incognita* race 2 and salinity on the productivity of two *Capsicum* cultivars. S Afr J Plant Soil 27(3):273

Anonymous (2016) Potato Certification Service. http://www.potatocertification.co.za/home.aspx. Accessed 30 Jan 2016

Bekker S, Venter E, Fourie H (2007) Nematode diagnostic service at the ARC-Grain Crops Institute. S Afr J Plant Soil 24:258–259

Berry MM, Stone AR, Parrott DM (1977) Nematology Department. Report of Rothamsted experimental station: 1976: Part 1. Rothamsted

Bridge J, Starr JL (2007) Plant nematodes of agricultural importance. Academic, Boston

Brodie BB, Evans K, Franco J (1993) Nematode parasites of potato. In: Evans K, Trudgill DL, Webster JM (eds) Plant parasitic nematodes in temperate agriculture. CAB International, Wallingford, pp 87–132

Coetzee V (1968) The distribution of the family Heteroderidae (Filipjev, 1934) in South Africa and some host records of *Meloidogyne* species. S Afr J Agr Sci 11:775–788

DAFF (2010a) South African seed potato certification scheme. Department of Agriculture, Forestry and Fisheries, Pretoria

DAFF (2010b) Production guideline for cabbage. Department of Agriculture, Forestry and Fisheries, Pretoria

DAFF (2011) Garden peas (*Pisum sativum*) production. Department of Agriculture, Forestry and Fisheries, Pretoria

DAFF (2013) Production Guideline: sweet peppers (*Capsicum annuum*). Department of Agriculture, Forestry and Fisheries, Pretoria

DAFF (2014a) A profile of the South African tomato market value chain. Department of Agriculture, Forestry and Fisheries, Pretoria

DAFF (2014b) A profile of the South African onion market value chain. Department of Agriculture, Forestry and Fisheries, Pretoria

DAFF (2014c) A profile of the South African carrot market value chain. Department of Agriculture, Forestry and Fisheries, Pretoria

DAFF (2014d) A profile of the South African sweet potato market value chain. Department of Agriculture, Forestry and Fisheries, Pretoria

DAFF (2014e) A profile of the South African beetroot market value chain. Department of Agriculture, Forestry and Fisheries, Pretoria

Daiber KC (1987) The beet cyst nematode, *Heterodera schachtii*, Schmidt. of some vegetable farms near Johannesburg. In: Proceedings of the eighth symposium of the Nematological Society of Southern Africa, Potchefstroom University for Christian Higher Education, Potchefstroom, 7–8 Apr

Daiber KC (1989a) Population development of *Meloidogyne incognita* and *M. javanica* in tomato roots and their effects on plant growth and yield. Phytophylactica 21:159–163

Daiber KC (1989b) Observations on *Heterodera schachtii*, the beet cyst nematode. In: Proceedings of the ninth symposium of the Nematological Society of Southern Africa, Golden Gate Highlands National Park, Clarens, 25–28 Apr

Daiber KC (1990a) Effect of egg density on the population dynamics of *Meloidogyne incognita* and *M. javanica* on growth and yield of tomatoes. Phytophylactica 22:303–306

Daiber KC (1990b) Effects of mixed populations of *Meloidogyne incognita* and *M. javanica* originating from different numbers of eggs on growth and yield of a susceptible and resistant tomato cultivar. Phytophylactica 22:453–456

Daiber KC (1990c) Effect of fenamiphos on population dynamics of *Meloidogyne incognita* and *M. javanica* and on yield of tomatoes. Phytophylactica 22:307–310

Daiber KC (1990d) *Heterodera schachtii* in South African vegetable fields. Phytophylactica 22:133–135

Daiber KC (1991) Decrease of *Heterodera schachtii* populations in vegetable soil. In: Proceedings of the tenth symposium of the Nematological Society of Southern Africa, Wilderness, 7–10 Apr

Evans AAF, Perry RN (2013) Survival mechanisms. In: Perry RN, Moens M, Starr JL (eds) Root-knot nematodes. CAB International, Wallingford, pp 201–222

FAO (2016) Statistics. http://faostat.fao.org/. Accessed 3 Jan 2016

Ferris H, Roberts P (2005) Nematode thresholds and damage levels for Californian crops. Nematological work group, March 2005. http://www.plpnemweb.ucdavis.edu/nemaplex/powerpoint/Thresholds.ppt. Accessed 10 Feb 2015

Fourie H, Zijlstra C, Mc Donald AH (2001) Identification of root-knot nematode species occurring in South Africa using the SCAR – PCR technique. Nematol 3:675–680

Fourie H, Bekker S, Mc Donald AH et al (2011) Prospective extended benefits of plant-nematode diagnostic and advisory services should closer cooperation between laboratories be attainable. Afr J Plant Soil 28:265

Fourie H, Mc Donald AH, Mothata TS et al (2012) Indications of variation in host suitability to root-knot nematode populations in commercial tomato cultivars. Afr J Agr Res 7:2344–2352

Jones FGW (1950) Observations on the beet eelworm and other cyst-forming species of *Heterodera*. Ann Appl Biol 37:407–440

Jones RK (2008) Should nematode control be expected to last till lifting? Chips 19:39–41

Kaczmarek A, MacKenzie K, Kettle H et al (2014) Influence of soil temperature on *Globodera rostochiensis* and *Globodera pallida*. Phytopathol Mediterr 53:376–405

Keetch DP (1989) A perspective of plant nematology in South Africa. S Afr J Sci 85:506–508

Keetch DP, Buckley NH (1984) A check-list of the plant-parasitic nematodes of Southern Africa, Technical communication. Department of Agriculture, Government Printer, Pretoria

Kleynhans KPN (1975) The potato cyst eelworm in South Africa. Nematol Soc S Afr Newsletter 7:10–12

Kleynhans KPN (1991) The root-knot nematodes of South Africa, Technical Communication No 231. Department of Agricultural Development, Plant Protection Research Institute, Pretoria

Kleynhans KPN (1998) Potato cyst nematodes (*Globodera* species) in Africa. In: Marks RJ, Brodie BB (eds) Potato cyst nematodes. Biology, distribution and control. CAB International, Wallingford, pp 347–351

Kleynhans KPN, Van den Berg E, Swart A et al (1996) Plant nematodes in South Africa, Plant Protection Research Institute Handbook No 8. ARC-Plant Protection Research Institute, Pretoria

Knoetze R (2014) A molecular, morphological and biological characterisation of the genus *Globodera* (Nematoda: Heteroderidae) in South Africa. PhD thesis, University of Stellenbosch, Stellenbosch

Knoetze R, Malan AP, Swart A et al (2004) Present status of potato cyst nematode in South Africa. Afr Plant Prot 10:129

Knoetze R, Malan AP, Mouton C (2006) Differentiation of South African potato cyst nematodes (PCN) by analysis of the rDNA internal transcribed spacer region. Afr Plant Prot 12:103–110

Knoetze R, Swart A, Tiedt LR (2012) Description of *Globodera capensis* n. sp. (Nematoda: Heteroderidae) from South Africa. Nematol 15:233–250

Koen H, Hogewind WL (1967) Symptoms and characteristics of *Pratylenchus brachyurus* infestation on stored potato. S Afr J Agr Sci 10:543–550

Le Roux S (2000) Life cycle, epidemiology and control of the golden cyst nematode (race: Ro1) under South African conditions at Ceres, Cape. Report 1999–2000, Project No. 103402, Agricultural Research Council–Vegetable and Ornamental Plants, Roodeplaat

Louw IW (1982) Nematode pests of potato. In: Keetch DP, Heyns J (eds) Nematology in Southern Africa, Science Bulletin No. 400. Department of Agriculture and Fisheries, Pretoria, pp 80–87

Malan AP, Storey S, Meyer AJ (2002) Field observations of *Paratrichodorus lobatus* on sweet potato (*Ipomoea batatas*). Afr Plant Prot 8:14

Marais M, Swart A, Knoetze R et al (2015) Plant nematodes in South Africa. 13. A checklist of nematodes associated with potatoes. Chips 2015:38–43

Manzanilla-Lopez RH, Starr JL (2009) Interactions with other pathogens. In: Perry RN, Moens M, Starr JL (eds) Root-knot nematodes. CAB International, Wallingford, pp 223–245

Marshall JW (1998) Potato cyst nematodes (*Globodera* species) in New Zealand and Australia. In: Marks RJ, Brodie BB (eds) Potato cyst nematodes. Biology, distribution and control. CAB International, Wallingford, pp 353–394

Meyer AJ (1984) Damage to onion by *Nanidorus minor* in the Koue Bokkeveld. Phytophylactica 16:331

Milne DL, Du Plessis DP (1964) Development of *Meloidogyne javanica* (Treub) Chitwood on tobacco under fluctuating soil temperature. S Afr J Agr Sci 7:673–680

Mtshali MB, Fourie H, Mc Donald AH (2002) Incidence of root-knot nematodes in resource-poor agriculture in South Africa. Afr Plant Prot 8:88–89

NDA (2015a) http://www.nda.agric.za/doaDev/topMenu/services/doc/Plant%20Improvement%20 Act%205 3%200f%201976.pdf. Accessed 12 Sep 2013

NDA (2015b) http://www.nda.agric.za/docs/NPPOZA/Agricultural%20Pests%20Act.pdf. Accessed 12 Sep 2013

Nijboer H, Parlevliet JE (1990) Pathotype-specificity in potato cyst nematodes, a reconsideration. Euphytica 49:39–47

Onkendi EM, Moleleki LN (2013a) Detection of *Meloidogyne enterolobii* in potatoes in South Africa and phylogenetic analysis based on intergenic region and the mitochondrial DNA sequences. Eur J Plant Pathol 136:1–5

Onkendi EM, Moleleki LN (2013b) Distribution and genetic diversity of root-knot nematodes (*Meloidogyne* spp.) in potatoes in South Africa. Plant Pathol 62:1184–1192

Potter JW, Olthof Th HA (1993) Nematode pests of vegetable crops. In: Evans K, Trudgill DL, Webster JM (eds) Plant parasitic nematodes in temperate agriculture. CAB International, Wallingford, pp 171–207

Roberts PA (1992) Current status of the availability, development and use of host plant resistance to nematodes. J Nematol 24:213–227

Scurrah ML, Niere B, Bridge J (2005) Nematode parasites of solanum and sweet potatoes. In: Luc M, Sikora RA, Bridge J (eds) Plant parasitic nematodes in subtropical and tropical agriculture, 2nd edn. CAB International, Wallingford, pp 193–220

Sikora R, Fernandez E (2005) Nematode parasites of vegetables. In: Luc M, Sikora RA, Bridge J (eds) Plant parasitic nematodes in subtropical and tropical agriculture, 2nd edn. CAB International, Wallingford, pp 319–392

Stanton JM, Sartori M (1990) Hatching and reproduction of the potato cyst nematode, *Globodera rostochiensis*, from potato fields in Western Australia as influenced by soil temperature. Nematologica 36:457–464

Starr JL, Mc Donald AH, Claudius-Cole AO (2013) Nematode resistance in crops. In: Perry R, Moens M (eds) Plant nematology, 2nd edn. CAB International, Wallingford, pp 411–436

Steyn P (1997) The control of nematodes on potatoes. In: Urquhart L (ed) Potato short course: potato production in SA with the emphasis on KwaZulu-Natal. Agricultural Research Council–Vegetable and Ornamental Plant Institute, Pretoria, pp 52–65

Steyn WP, Daneel MS, Slabbert MM (2015) Host suitability and response of different vegetable genotypes to *Meloidogyne incognita* race 2 and *Meloidogyne javanica* in South Africa. Int J Pest Manag 60:59–66. doi:10.1080/09670874.2014.900587

Stone AR (1979) Co-evolution of nematodes and plants. Symbolae Botanicae Uppsala 22:46–61

Subbotin SA, Mundo-Ocampo M, Baldwin JG (2010) Systematics of the cyst nematodes (Nematoda: Heteroderinae). Nematology monographs and perspectives 8A. Hunt DJ, Perry RN (Series eds) Systematics of cyst nematodes (Nematoda: Heteroderinae). Brill, Leiden

Taylor AL, Sasser JN, Nelson LA (1982) Relationship of climate and soil characteristics to geographical distribution of *Meloidogyne* species in agricultural soils. North Carolina State University Graphics, Raleigh

Turner SJ, Evans K (1998) The origins, global distribution and biology of potato cyst nematodes (*Globodera rostochiensis* (Woll.) and *Globodera pallida* Stone). In: Marks RJ, Brodie BB (eds) Potato cyst nematodes. Biology, distribution and control. CAB International, Wallingford, pp 7–26

Turner SJ, Subbotin SA (2013) Cyst nematodes. In: Perry RN, Moens M (eds) Plant nematology, 2nd edn. CAB International, Wallingford, pp 109–141

Van der Vegte FA, Daiber KC (1985) A preliminary report on experiments with nematicides in green pea fields. In: Proceedings of the seventh symposium of the Nematological Society of Southern Africa, University of Stellenbosch, Stellenbosch, 2–3 Apr

Van Riel HR, Mulder A (1998) Potato cyst nematodes (*Globodera* species) in western Europe. In: Marks RJ, Brodie BB (eds) Potato cyst nematodes. Biology, distribution and control. CAB International, Wallingford, pp 271–298

Van den Berg E (1971) The root-lesion nematodes of South Africa. Department of Agricultural Technical Services, Pretoria

Van Zyl K (2013) A guide to crop pest management in South Africa. A compendium of acaricides, insecticides, nematicides, molluscicides, avicides and rodenticides. A CropLife South African Compendium, 1st edn. VR Print, Pinetown

Vermeulen H (2015) The South African potato industry in perspective. Potato South Africa. Chips Maart/April, Pretoria

Whitehead AG (1998) Plant nematode control. CAB International, Wallingford

Williamson VM, Roberts PA (2009) Mechanisms of resistance. In: Perry RN, Moens M, Starr JL (eds) Root-knot nematodes. CAB International, Wallingford, pp 301–325

Winslow RD, McKenna LA (1972) Comparative rates of development of strains of potato cyst nematode, *Heterodera rostochiensis*, in outdoor pot experiments. Rec Agr Res Minist Agr No Irel 20:17–20

Chapter 11
Nematode Pests of Sugarcane

Shaun D. Berry, Patrice Cadet, and Vaughan W. Spaull

11.1 Introduction

Sugarcane is believed to have originated in the Melanesian archipelago in New Guinea from where it was carried to other parts of the world (Rosenfeld 1956). It is thought to have been introduced into southern Africa by wandering tribesmen from the north or perhaps by Arab traders (Herbert 1991).

Modern-day sugarcane cultivars (cvs) are complex hybrids between *Saccharum officinarum* and *Saccharum spontaneum* and are grown throughout the tropical and subtropical regions of the world (Butterfield et al. 2001). In most countries the crop is grown primarily for the extraction and production of sugar. However, in Brazil, in particular, and in several other countries, it is also used to produce ethanol as a biofuel (Gosnell 2013).

There are a number of useful by-products in the sugar milling process, including molasses, bagasse and filter cake. Molasses is the remaining liquor in the manufacturing process, after the crystallised sucrose has been separated out. It is used as an animal feed, in the production of yeast and as a fertiliser. It is also used in the production of various alcohols and organic acids (Paturau 1982; Blume 1985). Bagasse is the fibrous residue from the crushed sugarcane. Most is used as fuel for the boiler furnaces of the sugar mill. It is also used in the manufacture of various types of

S.D. Berry
BASF Corporation, 26 Davis Drive, Research Triangle Park, Durham, NC 27709, USA
e-mail: shaun.berry@basf.com

P. Cadet
Les Martins, 42640 St Forgeux Lespinasse, France
e-mail: cadetpatrice@orange.fr

V.W. Spaull (✉)
170 Edmonds Road, Glenwood, Durban 4001, South Africa
e-mail: vaughanspaull@gmail.com

H. Fourie et al. (eds.), *Nematology in South Africa: A View from the 21st Century*, DOI 10.1007/978-3-319-44210-5_11

board and paper, for the production of furfuraldehyde and the generation of electricity (Paturau 1982; Gosnell 2013). Filter cake is the sediment produced after clarifying the juice expressed from the crushed cane. It contains relatively large amounts of phosphorus, and about 60 % of the solids is organic matter. It is used mainly as a fertiliser and as a soil ameliorant (Moberly and Meyer 1978; Rein and Purchase 2013). One metric tonne (MT) of sugarcane produces about 120 kg sugar, 42 kg molasses, 300 kg bagasse and 40 kg filter cake (Blume 1985; Cadet and Spaull 2005). Sugarcane is probably unique among the plant kingdom in providing commercial quantities of food, fibre, fuel, feed and fertiliser.

11.2 Sugar Industries in southern Africa

11.2.1 South Africa

Reports of sugarcane being grown in South Africa (SA) date back to the 1600s, but commercial production only began in 1848 (Osborn 1964). By the beginning of the twentieth century, sugar output had reached 16,000 MT annum^{-1}, and by the end of the century, it had increased to an average of more than two million MT (Anonymous 2009). South Africa is the 15th largest producer of sugar worldwide (Anonymous 2014a) and the top producer in southern Africa (Table 11.1). About 24 % of the sugar is exported overseas; the remaining 76 % is sold within the southern African Customs Union (Anonymous 2014a).

Sugarcane is grown in the eastern part of SA, stretching about 1600 km from Pondoland in the south to Komatipoort in the north and 200 km from the coast inland to the Midlands region of KwaZulu-Natal. Most (84 %) of the cane is grown by about 1400 large-scale growers on farms of more than 100 hectares (ha) in size. The rest of the crop is grown by about 22,500 small-scale growers and by miller-cum-planters (Anonymous 2014a). Over the past 14 years, yield of cane per ha harvested averaged about 65 MT. Unlike the rest of southern Africa, most of the cane in SA is rain fed, apart from the areas north of Mtubatuba, where rainfall averages less than 900 mm annum^{-1}, and full or supplementary irrigation is required (Anonymous 2009).

Table 11.1 Area under cane, metric tonnes (MT) of sugarcane harvested and actual and relative sugar production in southern Africa during 2012/2013 (Anonymous 2014b)

Country	Area under cane (hectares)	Sugarcane harvested (metric tonnes)	Sugar production (metric tonnes)	Percentage sugar production relative to the total for the region
Malawi	24,732	2,460,000	299,494	7
Mozambique	50,105	3,393,000	396,719	9
South Africa	371,662	17,826,000	2,038,389	47
Swaziland	57,103	5,662,000	676,527	16
Zambia	26,590	3,254,000	406,000	10
Zimbabwe	53,600	3,300,000	479,127	11

There are 14 sugar mills in SA and the milling season is typically 9 months long, extending from April to December. Each year about three quarters of the crop is harvested. About 85 % of the crop is burnt at harvest and practically all is cut manually. Burning the dead leaves facilitates the cutting and handling of the stalks, resulting in increased efficiency of harvesting. However, environmental and soil health concerns, as well as the encroachment of urban areas closer to farm land and the associated nuisance caused by the soot fallout from the burnt leaves, have led to more producers adopting 'green-cane' harvesting practices. These include the retention of the cut tops and leaves as a thick mulch on the soil surface. This practice conserves soil moisture, improves soil health and can suppress plant-parasitic nematodes (Stirling 2008; van Antwerpen et al. 2009).

11.2.2 Malawi, Mozambique, Swaziland, Zambia and Zimbabwe

In contrast to the situation in SA, in Malawi, Mozambique, Swaziland, Zambia and Zimbabwe, sugarcane is mostly grown on a few, large miller-cum-planter estates, with some contribution from smallholdings.

Swaziland is the second largest producer of sugar in the region (Table 11.1) (Anonymous 2014b). The industry dates back to the 1950s, when a mill was first built at Big Bend on the Great Usutu River. There are now three large estates and three mills. The milling season runs from May to November. The production of sugar is an important source of revenue for the country.

Zimbabwe's first sugar estate and mill was established in the 1930s on the Triangle ranch in the south-east Lowveld. A second mill was built some 40 years later on the nearby Hippo Valley Estate (Anonymous 2014b). In Malawi, there are two sugar estates, one at Dwangwa and the other at Nchalo, each with a mill. Sugarcane is one of the major crops for the country.

Sugarcane in Mozambique is grown mainly on four large miller-cum-planter estates served by four mills. In Zambia, there are two mills on two estates near the Kafue River, one of which produces 90 % of the country's sugar (Anonymous 2014b). Throughout the region, where the climate is suitable, sugarcane is widely grown in small patches on smallholdings as 'chewing cane'.

11.3 Cultivation

Sugarcane is a tall, thick-stemmed, perennial grass. It is propagated vegetatively by planting setts (stalk cuttings) with two or more nodes. Within a few days, fine sett roots develop from primordia around the nodes. These sustain the initial growth and development of the primary shoots that arise from the axillary buds on the setts. Additional shoots develop by tillering, and these and the primary shoots develop shoot roots which, about 6–12 weeks later, replace the sett roots. The shoot roots are initially thick, fleshy and white and less branched than the sett roots; they become suberised with age (van Dillewijn 1952). The sugarcane plants grow in tufts or stools comprising several stalks of varying length. As the young stalks grow, they compete for light and space and a large proportion die. Those that survive increase in diameter and length. After about 12–24 months, depending on soil moisture and temperature, the sucrose content of the stalks approaches its peak, and the crop is harvested.

Shortly after the first harvest, new shoots develop from axillary buds on the stubble of the previous crop and give rise to the first ratoon crop. Initially, the young shoots are reliant upon the roots of the previous crop, the stool roots, but after 4–8 weeks these are replaced by new shoot roots. The developing stalks compete with one another and those that survive are subsequently harvested, as in the plant crop. This cycle of growth and harvesting is continued for a number of ratoon crops. The actual number depends on the yields as, over time, these tend to decline due to damage caused by pests and diseases, as well as physical damage to the stools caused by infield vehicles. Nematodes are an important component of this yield decline, especially on the lighter textured soils (Cadet and Spaull 2003). Replanting takes place after the stubble of the previous crop has been killed. This is done either by ploughing out the stools or by killing the regrowth with the herbicide, glyphosate. Usually there is only a 4–12-week fallow period between crop cycles.

11.4 Plant-Parasitic Nematodes Associated with Sugarcane

11.4.1 Distribution

Most data concerning the occurrence and distribution of plant-parasitic nematodes associated with sugarcane in the region are from SA. These show that the nematode communities are remarkably diverse with more than 90 species of 28 genera having been identified (Spaull and Heyns 1991; Kleynhans et al. 1996; SAPPNS[1]). The greatest diversity occurs within the genus *Xiphinema*, with more than 20 species recorded. Unusually, a large number of the species associated with sugarcane in SA are sedentary parasites, viz. *Meloidogyne hispanica* Hirschmann, 1986; *Meloidogyne incognita* (Kofoid and White, 1919) Chitwood, 1949; *Meloidogyne javanica* (Treub, 1885) Chitwood, 1949; *Meloidogyne kikuyensis* De Grisse, 1960; *Rotylenchulus macrosomoides* Van den Berg, Palomares-Rius, Vovlas, Tiedt, Castillo and Subbotin, 2016; *Rotylenchulus parvus* (Williams, 1960) Sher, 1961; *Rotylenchulus sacchari* Van den Berg and Spaull, 1981; *Rotylenchulus clavicaudatus* Dasgupta, Raski and Sher, 1968; *Meloidoderita safrica* Van den Berg and Spaull, 1982; *Sphaeronema*

---

[1]Dr Mariette Marais of the Nematology Unit, Biosystematics Division, Agricultural Research Council–Plant Protection Research Institute is thanked for the use of data from the South African Plant-Parasitic Nematode Survey (SAPPNS) database; E-mail: maraism@arc.agric.za.

cornubiensis Van den Berg and Spaull, 1982; *Tylenchulus furcus* Van den Berg and Spaull, 1982; and an unidentified *Heterodera* sp. (Spaull 1981a; Kleynhans et al. 1996; Van den Berg et al. 2016).

The number of plant-parasitic nematode genera present in a single soil sample from sugarcane fields in SA ranged from three to nine with an average of almost six (Cadet and Spaull 2005). Surveys conducted in 1978 and 2005 show that such a soil sample would usually include *Helicotylenchus, Paratrichodorus, Pratylenchus, Scutellonema* and *Xiphinema*, as species of these genera are among the more abundant and frequently occurring nematodes in sugarcane fields in SA (Table 11.2). If the soil was sandy, *Meloidogyne* would usually also be present. Most of these genera were frequently recovered from sugarcane soils in the other southern African countries (Table 11.2). In SA the most common species are *Helicotylenchus dihystera* (Cobb 1893) Sher, 1961; *Nanidorus minor* (Colbran, 1956) Siddiqi, 1974; *Pratylenchus zeae* Graham, 1951; *Scutellonema brachyurus* (Steiner, 1938) Andrássy, 1958; *Xiphinema elongatum* Schuurmans, Stekhoven and Teunissen, 1938; *Xiphinema mampara* Heyns, 1979; and *M. javanica* (South African Sugarcane Research Institute, unpublished data). Curiously, *Pratylenchus* was the only nematode that was commonly found associated with sugarcane in Malawi; all the other nematodes were recovered from fewer than half the samples, albeit from a few fields (Table 11.2).

The data collected in the 1978 and 2005 surveys in SA indicate that several changes in the frequency of the common genera have occurred over the 27-year period. Most notable was a reduction in the frequency of occurrence of *Rotylenchulus*, particularly *R. parvus*, from almost 100 % in 1978 to 9 % in 2005, and of species of *Longidorus*, from 63 to 18 % over the same period (Table 11.2). Species of *Scutellonema, Meloidogyne* and various criconematid genera were also less frequently recovered in the more recent survey (Berry 2006). This was thought to be linked to changes in various agronomic practices, including the adoption of new cvs, a shorter cropping cycle and reduced tillage at planting. A recent study suggests that changing cane cvs may not necessarily have a significant effect on the incidence of certain plant-parasitic nematode species as local soil conditions have a greater effect on the composition of the nematode fauna than does plant cv. (Berry et al. 2008). Possibly the differences are related to the different extraction techniques used in the two surveys. Despite the adoption of new agronomic practices and with the exception of *Rotylenchulus*, the list of dominant nematodes remained the same.

A study of the distribution of 37 species of the longidorid genera, *Longidorus, Paralongidorus* and *Xiphinema* in SA showed *X. elongatum* to be the most common, occurring in 60 % of 124 fields (Spaull and Heyns 1991). *Xiphinema mampara* was also quite common, occurring in 46 % of the fields, followed by *Longidorus pisi* Edwards, Misra and Singh, 1964, in 35 % and *Longidorus laevicapitatus* Williams, 1959, in 25 %. The remaining species were found in fewer than 10 % of the fields. The distribution of some of the species appeared to be related to soil texture; thus, *Paralongidorus deborae* (Jacobs and Heyns, 1987) Escuer and Arias, 1997, and *Paralongidorus paramaximus* (Heyns, 1965) Escuer and Arias, 1997, are among the largest of the nematodes in sugarcane fields and both show a preference for the

Table 11.2 Frequency of occurrence (%) of the more common plant-parasitic nematodes associated with sugarcane in southern Africa (Cadet and Spaull 2005; Berry 2006; SD Berry et al. 2007, unpublished data; VW Spaull 2013, unpublished data)

	Pratylenchus	*Helicotylenchus*	*Tylenchorhynchus*	*Meloidogyne*	Trichodorids[a]	*Xiphinema*	*Hoplolaimus*	Criconematids[b]	Longidorids[c]	*Paratylenchus*	*Rotylenchulus*	*Hemicycliophora*	*Scutellonema*
Malawi seven fields (2003–2005)	87	44	40	44	34	34	11	36	12	4	16	11	37
Mozambique eight fields (2003 and 2013)	87	62	75	87	50	50	0	25	0	12	12	62	75
South Africa 124 fields (1978)	96	95	30	71	93	94	8	75	72	9	99	11	97
South Africa 39 fields (2005)	83	87	12	41	69	72	8	32	18	<1	9	7	63
Swaziland 21 fields (2007)	100	84	43	82	95	82	0	74	11	3	5	87	71
Zimbabwe 27 sites (1966)	85	41	67	41	81	78	19	22	22	0	15	11	59

[a]Trichodorids = *Nanidorus* and *Paratrichodorus*
[b]Criconematids = *Criconemella* and related genera
[c]Longidorids = *Longidorus* and *Paralongidorus*

coarser-textured soils where the greater diameter of the pores between soil particles would facilitate their movement. However, this logic does not apply to the distribution of the two common species of *Xiphinema*; the notably larger, more obese *X. mampara* occurred much more frequently in clay soils than the more slender *X. elongatum*.

A number of studies have been conducted to identify edaphic factors associated with the distribution of nematode species and nematode communities in sugarcane fields, but with variable and contrasting results. In a survey of large- and small-scale farms in KwaZulu-Natal, a number of combinations of nematodes and soil factors were strongly associated, but the nature of the relationships was not consistent in both farming systems. The combinations that were the same in both farm types were an inverse relationship between the abundance of *Pratylenchus, Helicotylenchus* and *Paratrichodorus* with soil pH and a positive relationship with iron (Fe). Numbers of *Meloidogyne* showed the opposite relationship with pH and Fe. Numbers of *Xiphinema* showed no association with any of the soil characteristics in either the small- or large-scale farms (Spaull and Cadet 2001). In two similar investigations, but in smaller areas within single fields, manganese (Mn) levels in one field were negatively correlated with populations of *H. dihystera* and *X. elongatum* as well as with all nematodes combined (Cadet et al. 2004). In the other field, there was a strong positive association between numbers of *H. dihystera* and most of the soil cations, especially calcium, magnesium and Mn. The reverse was true for *X. elongatum* (Dana et al. 2002).

Factors affecting the distribution of nematodes in sugarcane in SA were studied on a broader scale using multivariate analysis of the nematode data collected from the 1978 survey, combined with edaphic factors, climate and topography (Spaull et al. 2003). The analysis showed that according to the nematode data, two contrasting areas could be described, one corresponding to the entire coastal region plus the irrigated area of Pongola and the other corresponding to the inland, higher altitude area. The main determining abiotic factors were higher temperatures, radiation and soil pH in the former regions and these were associated with greater numbers of *Pratylenchus* and *Xiphinema* and fewer *Helicotylenchus* and *Rotylenchus*. The distribution of *Meloidogyne* depended more on soil type than climatic or topographic factors.

Besides the abiotic component of the soil environment, certain bacteria were associated with the distribution of nematodes in sugarcane. Thus, species of the bacterium *Burkholderia*, which are common in the roots and rhizoplane of sugarcane in SA, have a spatial association with plant-parasitic nematodes. Omarjee et al. (2008) found that, in parts of a cane row where *X. elongatum* was the dominant plant-parasitic nematode, the *Burkholderia* community was dominated by *Burkholderia fungorum, Burkholderia gladioli, Burkholderia graminis* and *Burkholderia silvatlantica*. By contrast, in adjacent parts of the same row, where *H. dihystera* and *P. zeae* predominated, *Burkholderia tropica* was the most numerous. Coinertia analysis confirmed a positive correlation between this species and *H. dihystera*, but a negative one with *X. elongatum*. The significance of this observation is that *X. elongatum* is among the more pathogenic of the nematodes associated with sugarcane. By contrast, nematode communities dominated by *H. dihystera* have been associated with higher-yielding cane (see Sect. 11.5.4.5, Box 11.2).

11.4.2 Interaction Between Nematodes and Sugarcane

Yield of sugarcane is a function of the number, length and diameter of the stalks. Sucrose yield is a function of the sucrose content of the stalks. Only occasionally does root damage by plant-parasitic nematodes affect stalk diameter or sucrose content. Data from numerous field experiments show that SA plant-parasitic nematodes reduce both the number and the length of stalks in the plant crop and the ratoons (Moberly et al. 1974; Rostron 1976). This contrasts with the situation in Burkina Faso where, in the plant crop, nematodes reduce the number of stalks but have little effect on the length of the stalks. Also, in the ratoon crops, they have little effect on either the number or length of stalks (Cadet 1985). These differences were investigated in a plant crop and a ratoon crop, in both SA and Burkina Faso (Cadet and Spaull 1985; Spaull and Cadet 1991). The deductions made from these studies are discussed below.

11.4.2.1 Plant Crop

In both plant cane experiments, during the early period of growth when shoot and tiller development was greatest, damage to the sett roots by the endoparasites *M. javanica* and *P. zeae* delayed the emergence and slowed the development of the primary shoots. As a result they either produced fewer tillers or were unable to compete with adjacent shoots. This, in turn, lead to fewer stalks surviving through to harvest. The effect was greater when larger numbers of endoparasites invaded the sett roots, as occurred in Burkina Faso. Where the ectoparasites *X. elongatum* and species of *Paratrichodorus* were numerous during the early stages of growth, as in SA, the damage they caused to the shoot roots was also associated with suppressed tillering and increased competition between shoots, resulting in fewer stalks at harvest. In SA, large numbers of ectoparasites persisted, and they appeared to be the principal cause of the reduction in stalk length. This followed from extensive damage they caused to the shoot roots, which would have restricted the uptake of water. Since sugarcane is particularly sensitive to drought stress, this affected stalk elongation and thus stalk length. In Burkina Faso, where the ectoparasitic fauna was dominated by *H. dihystera*, rather than more pathogenic species, there was little or no effect on stalk length (Cadet and Spaull 1985).

11.4.2.2 Ratoon Cane

In SA, ratoon cane is as susceptible to plant-parasitic nematodes as plant cane. In the ratoon cane trial in SA, there was a notable reduction in the length of stalks. This could be attributed to the considerable damage to the shoot roots caused by *X. elongatum* and possibly the *Paratrichodorus* spp. These ectoparasites were also thought to be responsible for the reduction in the number of stalks, since large numbers were present in the soil during the early, critical period of shoot and tiller development. The endoparasites, *M. javanica* and *P. zeae,* were also implicated in reducing the number of stalks, but to a lesser extent as very few were recovered from the roots during the first month when tiller suppression was greatest. However, they did appear to affect stalk length as their numbers increased considerably in the following 3 months when stalk elongation was first affected. Thus both ecto- and endoparasites contribute to the restricted growth of ratoon cane in SA.

By contrast to this situation, ratoon crops in Burkina Faso are much less vulnerable, and the effect of nematodes on yield is usually not significant (Cadet 1985). In the ratoon cane trial, populations of endoparasites remained low and seemingly had no effect on the cane. Populations of ectoparasites were much greater and were dominated (87 %) by *H. dihystera*, which at that time was believed to be no more than a weak pathogen of sugarcane. This species is now considered to be beneficial to cane growth as it appears to mitigate the pathogenicity of other nematodes (Cadet et al. 2002; Cadet and Spaull 2005; Berry et al. 2007). The predominance of *H. dihystera* and the lack of a significant response to nematicide treatment in ratoon cane in Burkina Faso may be linked to the way the cane is harvested (Berry et al. 2007). In this country, just prior to harvest, the leaves and leaf sheaths are removed from the stalks and left scattered on the ground. At harvest the stalks are cut above the thick mulch of leaf material, several centimetres above-ground level, creating somewhat elongated stubble from which the shoots of the next crop emerge. The new shoot roots develop and can grow within the mulch layer for some weeks before reaching the soil. By contrast, in SA, the leaves and leaf sheaths are mostly burnt at harvest, and the stalks are cut close to ground level. The new shoots emerge from the stool below-ground, and the roots of the new ratoon are at once exposed to plant-feeding nematodes. These two harvesting scenarios were simulated in a field trial in SA in a plant crop and the following two ratoon crops. It indicated that the dominance of *H. dihystera* in the sugarcane fields in Burkina Faso and the lack of a response in ratoon crops to treatment with a nematicide may be linked, in particular, to the size of the stubble from which the subsequent ratoons emerged (Berry et al. 2007).

11.4.3 Symptoms of Nematode Damage

There are some above-ground symptoms that, although not diagnostic, are often associated with the damage caused by plant-parasitic nematodes (Fig. 11.1a, b). These include a reduction in the number and length of the stalks and slower development of a full canopy of leaves over the inter-row, giving the cane rows a more open appearance, and the young leaves curl longitudinally and appear spiky. These are also symptoms of drought-stressed cane. Below-ground the symptoms are somewhat more diagnostic, with the proviso that, in field-grown cane, the roots show the combined symptoms of all the nematodes that have been feeding on them and that other organisms also feed on the roots. Since several plant-parasitic

Fig. 11.1 (**a**, **b**) Four-month-old sugarcane treated with a nematicide (**a**) in a field trial on a sandy soil adjacent to an untreated plot (**b**) showing the reduction in plant growth of sugarcane plants infected with plant-parasitic nematodes (Kirk West, Port Elizabeth, South Africa)

nematode species cause similar damage, it is usually not possible to use the symptoms to identify the nematodes responsible. However, the sparse root system and root stunting (Fig. 11.2), associated with large numbers of *Xiphinema* and *Paratrichodorus,* appear to be diagnostic, as are the galls produced by *Meloidogyne* (Cadet and Spaull 2005).

11.4.4 Association with Other Pathogens

Spaull and Bailey (1993) investigated the combined effect of plant-parasitic nematodes and the bacterium, *Leifsonia xyli* subsp. *xyli,* on sugarcane. This bacterium is the causal agent of ratoon stunt, an important and widespread disease of sugarcane in southern Africa. The bacterium resides in the xylem vessels and is spread by planting infected stalks or on cane cutters' knives, which carry bacteria from diseased to healthy stalks. Spread may also occur when healthy cane is planted in fields that previously carried an infected crop, indicating infection via the roots from propagules that survive in the soil. Since nematodes feed on the roots, they may facilitate entry of the bacteria. However, results from experiments on a plant and a ratoon crop indicated that the combined effect of *L. xyli* subsp. *xyli* and plant-parasitic nematodes on the yield of cane was additive. The loss in yield from nematode pests was about four times that from the bacterium.

Fig. 11.2 Composite symptoms of damage to sugarcane roots caused by plant-parasitic nematodes (Kirk West, Port Elizabeth, South Africa)

11.5 Management Practices

Studies in the 1960s and 1970s showed that plant-parasitic nematodes were a serious constraint to cane production on sandy soils in SA (Dick 1961; Moberly et al. 1974; Dick and Harris 1975; Rostron 1976). Since that time various management practices have been investigated to reduce the impact of nematodes on sugarcane in problem fields.

11.5.1 Chemical Control

Information concerning nematicide use on sugarcane is available only from SA. This is mainly due to the fact that most of the cane in the other countries in southern Africa is grown on soils with a relatively high clay or silt content, where nematodes are not generally a serious problem. In SA, however, two thirds of the soils are sandy and nematodes are a significant yield constraint.

In the 1970s and 1980s, a number of nematicides were tested on sugarcane. Only three of these, all carbamates, were registered for use on cane, viz. aldicarb and carbofuran, as 15 % and 10 % granular formulations, respectively, as well as a 12 % liquid formulation of oxamyl. During that period a great deal of research was conducted to determine their optimum use (Donaldson 1988). These investigations included work on split applications (Rostron 1976; Richardson and Watt 1980), application to mulched vs burnt fields (Donaldson 1983), rates of application (Harris 1973; Rostron 1976; Moberly and Clowes 1981), residual responses (Rau and Moberly 1975; Donaldson 1987), placement of granules (Rostron 1976; Donaldson 1983), interaction with herbicides (Donaldson and Turner 1984); time of application (Rostron 1976; Spaull and Donaldson 1983; Donaldson 1988) and effect of soil pH, clay content and rainfall on response to treatment (Donaldson 1985). With the withdrawal of aldicarb from the market, the nematicides now registered for use on sugarcane in SA are carbofuran and oxamyl (granular formulations) and furfural (2-furfuraldehyde), produced from sugarcane bagasse. In a number of field experiments under rain-fed and irrigated conditions, furfural gave inconsistent results (Spaull 1997; Berry and Spaull 2008). However, when applied at 100 l ha^{-1}, which is twice the registered rate, cane yields were improved (Berry et al. 2009).

Treatment with a nematicide may not only increase the yield of the treated crop but also that of the following ratoon. It is assumed that this residual response results from the better root system of the treated crop which favours the initial growth of the following crop. This assumption is supported by the observed significant correlation in field trials between the yield of ratoon crops with that of the plant crop (Cadet and Spaull 2003).

11.5.1.1 Factors Affecting Response to Nematicides

Greater responses to nematicide treatment occur in sandy, coarse-textured soils than in finer-textured soils (Cadet and Spaull 2005). Also, responses are greater for nematode-susceptible cvs (Spaull and Cadet 2003). The magnitude of the response to treatment varies considerably from season to season, and this appears to be related to rainfall, with smaller responses occurring in drier years (Donaldson 1985; Spaull 1996). Contrary to this observation, Donaldson and Turner (1988) found that the response to treatment was greater in cane growing under rain-fed conditions than when irrigated.

The size of the response to a nematicide varies according to which nematode species are present in the soil. Thus, in two field trials on similar sandy soils, a few hundred metres apart, the response to treatment with a nematicide was greater in the site where *M. javanica* occurred. The difference in response between the two sites amounted to an additional annual loss of 15 MT cane ha^{-1} over a 4-year period, equivalent to 30 % of production (Cadet and Spaull 2003).

Greater responses to nematicides follow application at higher rates or at the conventional rate repeated over time. This indicates that crop loss from plant-parasitic nematodes is greater than the estimates based on treatment at 'economic' rates (Berry et al. 2004; Cadet and Spaull 2005).

11.5.1.2 Economics of Chemical Control

In most field experiments done on sandy soils, the response to aldicarb (previously the most widely used nematicide in the sugar industry) more than justified the cost. Moreover, the benefits derived from using a nematicide included not just an increase in yield per crop but also (i) an increase in the number of high-yielding ratoon crops, which means that the need to replant the cane is delayed (Fig. 11.3) (Cadet and Spaull 2003); (ii) a reduction in the number of herbicide treatments, due to the more rapid formation of a full leaf canopy that shades out the inter-row; (iii) where the cane is cut without burning, the provision of a thicker and more effective mulch of

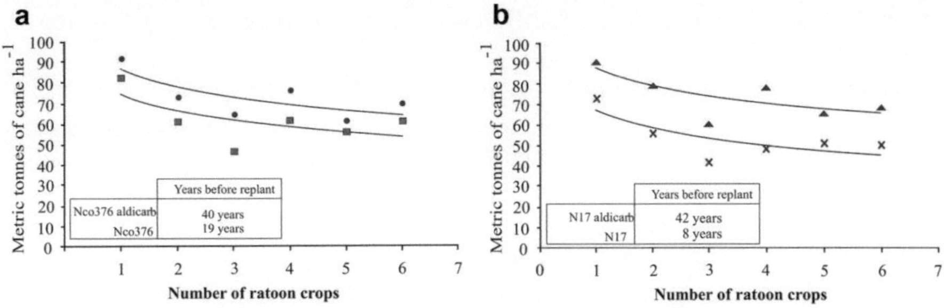

Figs. 11.3 (**a**, **b**) Logarithmic regression curves fitted to the ratoon yield decline for sugarcane cvs NCo376 (**a**) and N17 (**b**), untreated (*lower curves*) and treated with aldicarb (*upper curves*), observed over six ratoon crops. Expected yield of untreated N17 falls below a plough-out threshold fixed at 40 t cane ha^{-1} before that of NCo376. Rate of yield decline was similar in crops treated with aldicarb and much slower than in untreated controls (South African Sugarcane Research Institute, Mount Edgecombe, South Africa, unpublished data)

cane tops and leaves for the following crop; and (iv) an improved root system able to access water and nutrients over a greater soil volume, resulting in increased plant health and increased resistance to drought.

Estimates from more than 170 field experiments, conducted on a range of mostly sandy soils, indicate that in SA, plant-parasitic nematodes are responsible for an annual loss in yield of 1.6 million MT cane (Spaull 1995; Spaull and Cadet 2003). Currently this is equivalent to a 9.2 % loss of production (Anonymous 2014a). Data from 8 plant crops and 22 ratoon crops suggest that on sandy soils, losses from plant-parasitic nematodes are four times greater than those from the sugarcane stalk borer, *Eldana saccharina*, an important insect pest in SA (Berry et al. 2010). The severe damage caused by the stalk borer in the drought of 2015 (Baker 2015) indicates that such a difference may not always occur.

11.5.2 Biological Control

Natural control of nematodes is widespread. Thus, in a nematode survey conducted in more than 100 sugarcane fields in SA, spores of *Pasteuria* spp. were found attached to *Pratylenchus* in 30 %, to *Helicotylenchus* in 58 % and to *Scutellonema* in 72 % of the fields (Spaull 1981b). The highest level of infection of individual infested fields was 48 % for *Helicotylenchus*. Altogether 13 plant-parasitic species of 9 genera were infected by these bacteria. The widespread occurrence of the bacterial spores suggested that they may regulate populations of some species. However, this appeared not to be the case with *Pasteuria penetrans* infecting *Meloidogyne*. Although infected females were recovered from a third of 81 fields sampled, *Meloidogyne* populations were generally larger in fields with *P. penetrans* than in those without. This phenomenon suggests that the bacterium had no regulatory effect (Spaull 1984). Judging from differences in infectivity on *M. incognita* and *M. javanica,* there were two isolates of *P. penetrans*.

Other studies to find naturally occurring nematode pathogens showed that of more than 100 isolates of the *Burkholderia cepacia* complex (Bcc), recovered from sugarcane roots and rhizosphere, two thirds paralysed *M. javanica* second-stage juveniles (J2) in vitro (Omarjee 2006). Species of *Burkholderia* are relatively common in cane fields, representing up to 25 % of the cultivable bacteria in the roots (Vogel et al. 2002). Research on *B. tropica* found that 13 isolates of this species were also able to paralyse *Meloidogyne* J2 in vitro. In addition, some isolates were able to fix nitrogen, and some were antagonistic to the fungus, *Ustilago scitaminea,* which is an important pathogen of sugarcane (Guyon et al. 2003). Five of nine isolates of the Bcc, three of four isolates of *B. tropica* and single isolates of *Burkholderia caribensis* and *Burkholderia vietnamiensis* reduced the number of root-knot galls on sett roots of cane in pots in a glasshouse. Half of the isolates, encompassing all four species, were non-paralysing strains, indicating a second means of interfering with the normal development of *Meloidogyne*. Other plant-parasitic nematode genera in the soil were not affected by these strains (Omarjee 2006).

A few commercial biocontrol agents have been tested in the field. These included formulations of three *Bacillus* spp. (bacteria), the kd strain of the fungus *Trichoderma harzianum* as well as a commercial strain of the fungus *Purpureocillium lilacinum.* None had a significant effect on the plant-parasitic nematodes present, and none affected cane yield (Berry et al. 2009).

11.5.3 Physical Control

Damage to sugarcane due to plant-parasitic nematodes is greatest on sandy soils (Cadet and Spaull 2005). One solution to improving sugarcane growth on such soils is to increase the clay content by inverting and mixing the sandy topsoil with a clay subsoil layer. This practice, known as 'marling', was attempted in SA in the 1980s, and although it provides a permanent solution, it is expensive and is only possible in areas where a suitable subsoil layer exists (Anonymous 1982). Tillage can reduce the number of nematodes, presumably as result of a combination of physical damage and drying out the soil and root pieces (Stirling 2008), but such activity is detrimental to soil health (Prammanee et al. 2001; Pankhurst et al. 2003; Govaerts et al. 2007).

11.5.4 Cultural Control

Several cultural practices have been investigated with the aim of reducing yield losses due to parasitism by nematodes. These included planting tolerant cvs, using organic amendments, growing green manure crops, selecting the optimum time of planting and nematode community management.

11.5.4.1 Cultivar Selection

The roots of sugarcane are fed upon by many species of nematodes. The more common and those causing the most damage in SA are *M. javanica, P. zeae, X. elongatum* and other longidorids and two or three *Paratrichodorus* spp. Sustained production of sugarcane on sandy soils can only be achieved if these nematode species are controlled. Nematicides are effective and cost-efficient on the light sandy soils but less so on the sandy loams and sandy clay loams (Cadet and Spaull 2005). Cultivating plants that are resistant to nematodes is a very effective way of sustaining yields (Cook and Evans 1987). However, sugarcane is attacked by several plant-parasitic species that interact with the host plant in different ways. Breeding for combined resistance, even to the more important components of such a community, is likely to be difficult (Luc and Reversat 1985). Breeding and selecting cvs for tolerance to plant-parasitic nematodes would be a better option. Although no attempts have been made by the South African sugar industry to breed for tolerance to nematodes, by selecting the higher-yielding cvs, some random selection has occurred (Berry et al. 2008). This is shown by data from a number of cultivar-x-nematicide field trials, conducted on sandy soils over the past 35 years. The trials identified certain cvs more tolerant than others, including N12, N14, N25 and NCo376. Tolerance was deduced by the smaller response to treatment with aldicarb compared with other cvs (Moberly and Clowes 1981; McArthur and Spaull 1995; Cadet and Spaull 2003; Spaull and Cadet 2003; Cadet et al. 2005; Spaull et al. 2005).

The contribution that tolerant cvs make in minimising losses from nematodes was quantified by comparing their yield with susceptible cvs in six field trials, with three or more crops per trial, conducted on sandy soil in three regions of the South African sugar industry (Spaull and Cadet 2003). Where a nematicide was not used, growing a tolerant cv. increased yields by between 25 and 124 % (average 83 %) over that of a susceptible cv. The benefit was broadly equivalent to the increased yield achieved by treating the susceptible cv. with a nematicide. In the 23 crops of the 6 trials, this represented an average increase of 34 MT cane ha^{-1}. Additionally, it was shown that nematode-tolerant cvs can sustain production over a longer period through a greater number of high-yielding ratoon crops (Cadet and Spaull 2003; Cadet et al. 2005; Fig. 11.3).

11.5.4.2 Organic Amendments

Filter cake produced in the sugar milling process is often returned to the field as a soil ameliorant (Rein and Purchase 2013). It contains plant nutrients and organic matter and almost always improves cane growth when incorporated into the soil (Martin 1967; Moberly and Meyer 1978). In most instances fewer plant-parasitic nematodes were recovered from soil to which filter cake had been added (Cadet and Spaull 2005). In two field trials, a reduction in the number of nematodes was found in the sett and shoot roots of sugarcane planted in a furrow in an envelope of bagasse (Berry et al. 2005). By contrast, kraal manure and fly ash, which is the residue from the burning of bagasse and filter cake to produce energy for the mill, had no noticeable depressive effect on the nematode community. In these trials the yield of cane and sucrose was not affected by the three amendments. Evidence shows that harvesting the cane without burning and leaving the tops and leaves on the soil surface, in addition to having soil health benefits, such as increasing labile carbon in the soil and increasing microbial activity, also results in reduced numbers of plant-parasitic nematodes (Stirling 2008; Van Antwerpen et al. 2009).

11.5.4.3 Rotation/Green Manure Crops

Growing leguminous rotation crops between sugarcane cycles is used by cane growers in Australia as part of a revised farming system to relieve the deleterious effects of continuous monocropping of sugarcane (Stirling 2008). Currently only a few growers in SA include rotation crops with sugarcane, although this practice is promoted (Ramouthar 2015). Previously several studies had been conducted to monitor the effect of green manure crops on nematode populations (Berry and Wiseman 2003; Berry and Rhodes 2006; Berry et al. 2013). The crops tested included buckwheat (*Fagopyrum esculentum*), cabbage (*Brassica oleracea* var. *capitata*), cowpea (*Vigna unguiculata*), dolichos bean (*Lablab purpureus*), forage peanut (*Arachis glabrata*), forage sorghum (*Sorghum bicolor*), giant English rape (*Brassica* sp.), grazing vetch (*Vicia sativa*), hairy vetch (*Vicia villosa*), Japanese millet (*Echinochloa esculenta*), pearl millet (*Pennisetum glaucum*), lucerne (*Medicago sativa*), lupin (*Lupinus* sp.), oat (*Avena sativa*), marigold (*Tagetes* sp.), red clover (*Trifolium pratense*), Rhodes grass (*Chloris gayana*), serradella (*Ornithopus sativus*), sunn hemp (*Crotalaria juncea*), velvet bean (*Mucuna pruriens*) and wheat (*Triticum aestivum*). They were planted in nematode-infested sandy soil in 25-l capacity pots and their effects on nematode populations measured. Pearl millet, cowpea and serradella were associated with an increase in the numbers of *M. javanica,* whereas giant English rape, grazing vetch, lucerne, lupin and red clover were associated with a decrease. Most of the crops decreased the numbers of *P. zeae* significantly, although numbers of *X. elongatum* were not affected. Forage peanut, marigold and sunn hemp increased the number of beneficial/nonparasitic nematodes.

11.5.4.4 Manipulating Planting and Ratooning Dates

Manipulating planting dates to avoid periods of peak plant-parasitic nematode activity has been used in a variety of crops in several countries (Sikora et al. 2005a, b). In Taiwan, sugarcane planted in spring was more susceptible to plant-parasitic nematodes than when planted in autumn (Cadet and Spaull 2005). Similar data from field trials in the northern irrigated areas of SA showed that, without a nematicide, both planting and ratooning in autumn (early in the milling season), rather than in spring (late season), conferred an average 1.5 MT sucrose ha^{-1} yield advantage for eight sugarcane cvs averaged over the plant crop and the following four ratoon crops (Spaull et al. 2005). With a nematicide this advantage did not occur, though yields were greater compared with untreated cane (+1.5 MT sucrose ha^{-1} early season vs + 3.1 MT sucrose ha^{-1} late season). The early development of cane planted or ratooned during the early part of the season takes place during a period of decreasing rainfall, decreasing evapotranspiration and lower air and soil temperatures. The lower temperatures not only slow down the activity of the sugarcane plant but also that of its parasitic nematodes (Webster 1987). Thus, if a nematicide is not used, confining the time of planting and ratooning of sugarcane grown on sandy soils to the early part of the season helps reduce the yield loss caused by nematodes (Spaull et al. 2005).

11.5.4.5 Nematode Community Management

Cadet and Spaull (2005) suggested that in the long term, instead of the current reliance on the use of nematicides, a more sustainable way of reducing the effect of nematodes would be to select for sugarcane cvs that favour less pathogenic nematode communities. The concept of manipulating the composition of plant-parasitic nematode communities arose from observations that growth of sugarcane and millet was greater in situations where *H. dihystera* was the dominant, or at least an important, component of the community (Cadet 1986; Villenave and Cadet 1998) (Boxes 11.1 and 11.2). The beneficial effect of *H. dihystera* has since been observed in a number of trials where it seemingly mitigated the pathogenicity of other plant-parasitic nematodes (Cadet et al. 2002; Rimé et al. 2003; van Antwerpen et al. 2007; Cadet unpublished). Thus, an option for nematode management is to create a soil environment that favours populations of *H. dihystera,* and/or other species, that moderate the damage caused by the more pathogenic species within the community. Sugarcane growers already do this when they plant cv. N12 on their sandy soils, as this cv. supports greater populations of *H. dihystera* (Cadet and Spaull 2005; Ramouthar et al. 2012). Other ways to promote this, and other moderating species, would be through:

(i) The use of new cvs with gene/s that promote the multiplication of selected plant-parasitic nematode species. It was argued that, unlike genes for resistance, those that promote the wellbeing of a particular nematode species would not be broken down. Cultivars with these genes would grow well in the presence of nematode pests and would not increase the number of pathogenic species (Cadet and Spaull 2005)

(ii) Planting cvs that host micro-organisms that favour *H. dihystera*, as may be the case with *B. tropica* and with *H. dihystera*-helper bacteria (Omarjee et al. 2008)

(iii) Mimicking cultural practices from other sugar industries where *H. dihystera* is particularly abundant (e.g. in Burkina Faso) (Berry et al. 2007)

Box. 11.1 How to Study the Effect of a Nematode Community on Crop Yield (Part 1)

The numbers of each of the nematode species collected from soil and roots (r) from 30 plots in a field trial are organised as a matrix (only eight plots shown), with species in columns and plots in rows. The yield from each plot is arranged in a single column. A correlation matrix principal component analysis (PCA) performed on the nematode matrix calculates factorial values for each plot (row), which describe the entire community structure according to all other plots. Once the nematode community is translated into a single column, a regression is calculated between the factorial values (nematodes) and the yields of the corresponding plots. The yield data were also normed $(x - \dot{x})/\sigma$ to ease the reading of the factorial plan.

Nematode community

Plot	Melo/g	Prat/g	Melo	Prat	Heli	Xiphi	...
1	17	138	84	160	72	408	
2	333	69	4968	80	12	10	
3	51	1116	108	360	360	108	
4	2473	54	132	200	12	72	
5	1536	196	36	80	684	468	
6	184	63	1740	20	2172	900	
7	968	1903	480	140	1548	96	
8	212	303	84	140	1836	12	
...							

PCA

F1	F2	Yield
-4,7719	-1,0255	87.6
-2,5426	-1,6882	65.2
-5,0189	-2,9311	105.1
-1,1184	0,7053	89.6
-0,8169	2,0405	54.8
0,3087	1,6563	62.3
-1,4749	2,1039	78.5
0,2079	0,5785	118.6
...	...	...

Regression

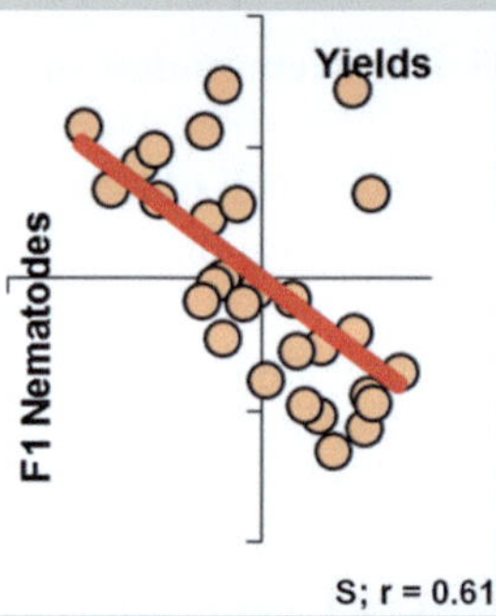

The linear regression between the factorial value describing the nematode community and the yields was significant, with a correlation coefficient of 0.61. A similar regression done with edaphic factors and yield was not significant. It can therefore be assumed that variations in the nematode community composition explain the variations in yield. Identification of the effect of each particular plant-parasitic nematode species on yield is demonstrated in Box 11.2.

Attempts to increase the numbers of *H. dihystera* by combining a mixture of four sugarcane cvs in the planting furrow were not successful (Cadet et al. 2006). However, there was a slight change in the nematode community structure which might suggest that combining the right cv. mix could promote a less pathogenic nematode community.

Box. 11.2 How to Study the Effect of a Nematode Community on Crop Yield (Part 2)

To identify which species could explain the result in Box 11.1, the normed yield values are projected directly on the factorial plan drawn from the principal component analyses of the nematode community. Along F1, the first axis in abscissa, *Meloidogyne* is opposed to *Helicotylenchus*, as shown by the correlation circle of the variables corresponding to the species in columns in the matrix. On the factorial plan of the rows, each plot is positioned according to the structure of its nematode community. Circles, representing yields above average, are mainly located on the left of the factorial plan, corresponding to plots with greater populations of *Helicotylenchus*. Squares, corresponding to yields below average, are located mainly on the right of the plan, corresponding to plots with larger populations of *Meloidogyne*. The size of the circles and squares is proportional to their absolute value.

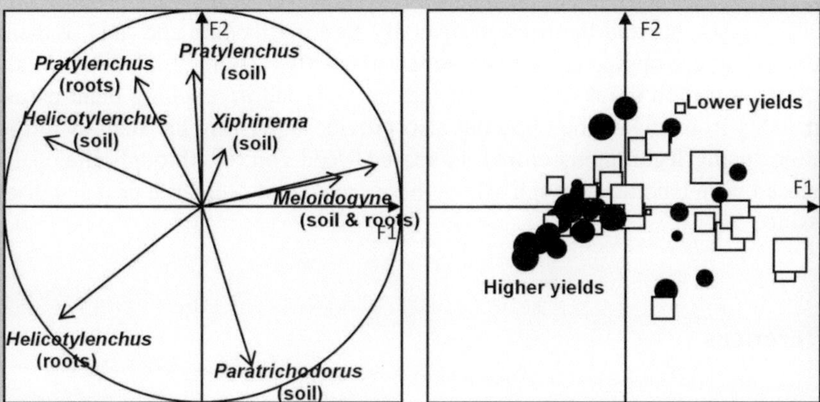

11.6 Outlook

In many countries around the world, sugarcane is grown not only for its sugar but also for the production of ethanol, as a biofuel for vehicles, and the production of biomass for electricity generation (Gosnell 2013; van Antwerpen et al. 2013). As the need for clean alternatives to fossil fuels increases, the area under sugarcane in these and other countries will likely expand to meet the energy demand.

The future use of carbamate nematicides will probably become more restricted. As a result greater emphasis will be placed on adopting alternative nematode control methods with lower hazard and risk levels (Spaull 2013). However, these methods may not be sufficient for sugarcane on very sandy soils infested with *M. javanica* and *X. elongatum*, where, currently, sustained economic yields are usually only possible if carbamate nematicides are used. The problem will be compounded if marginal land with light-textured soils is used for the expansion of sugarcane cultivation. However, on other, less sandy soils, crop loss from nematode parasitism can be minimised through the adoption of existing management options, together with the restoration of soil health by adopting a green-cane harvesting system, with the retention of a thick mulch of leaves (Stirling 2008; van Antwerpen et al. 2009).

Future research should include studies on inducing resistance to nematode damage through the use of specific elicitor molecules (e.g. salicylic acid, jasmonate). Such resistance has been shown to minimise nematode damage in wheat and barley (*Hordeum vulgare*) (Oka and Cohen 2004), tomatoes (Cooper et al. 2005) and rice (Oostendorp et al. 2004; Nahar et al. 2011). Preliminary work has begun on sugarcane (Guimaraes et al. 2008; Edmonds 2014); however, more work is needed. Due to land constraints and the need to diversify farming income and land usage, the practice of intercropping is receiving renewed attention (Berry et al. 2005; Mathias 2013). Companion crops are required that not only aid in repelling plant-parasitic nematodes from sugarcane roots but also provide other benefits such as nitrogen fixation (using leguminous crops), increased weed control (through allelopathy), increased farm income (through sale of these crops) and decreased pest and disease pressure.

References

Anonymous (1982) Growing sugarcane on the recent sands. S Afr Sugar J 66:369–370

Anonymous (2009) South African sugar industry directory 2008/2009. South African Sugar Association, Mount Edgecombe

Anonymous (2014a) South African sugar industry directory. http://www.sasa.org.za/Files/Industry%20Directory%202013%20-%202014.pdf. Accessed 12 Nov 2015

Anonymous (2014b) Annual SADC digest 2014: your guide to the sugar industry. http://sadcsugardigest.com/digest_flip_book/. Accessed 14 Nov 2015

Baker C (2015) Message from the Director. The Link. South Afr Sugarcane Res Inst 24:2–3

Berry SD (2006) Diversité et dynamique des communautés de nématodes phytoparasites en culture de canne à sucre en Afrique du Sud; répercussion sur la plante. PhD Thesis, University of Montpellier II, Montpellier

Berry S, Rhodes R (2006) Green manure crops: agronomic characteristics and effect on nematodes. Proc S Afr Sugar Techn Ass 80:269–273

Berry S, Spaull VW (2008) Use of cropguard to improve sugarcane yield on nematode infested soils of South Africa. Proc S Afr Sugar Techn Ass 81:310–318

Berry S, Wiseman EJ (2003) The effect of green manure crops on plant parasitic nematodes in the South African Sugar Industry. Proc S Afr Sugar Techn Ass 77:114–117

Berry SD, Cadet P, Spaull VW (2004) Estimation of nematode damage to sugarcane in sandy soil in KwaZulu Natal. Afr Plant Prot 10:125

Berry S, Cadet P, Spaull VW (2005) Effect of certain cultural practices on nematode management in a small-scale farming system. Proc S Afr Sugar Techn Ass 79:149–164

Berry S, Spaull VW, Cadet P (2007) Impact of harvesting practices on nematode communities and yield of sugarcane. Crop Prot 26:1239–1250

Berry S, Cadet P, Parfitt R et al (2008) Nematological environment and selection pressure on six SASRI research farms. Proc S Afr Sugar Techn Ass 81:274–286

Berry SD, Spaull VW, Cadet P (2009) Field assessment of biologically-based control products against nematodes on sugarcane in South Africa. Afr Plant Prot 15:1–12

Berry SD, Leslie GW, Spaull VW et al (2010) Within-field damage and distribution patterns of the stalk borer, *Eldana saccharina,* (Lepidoptera: Pyralidae) in sugarcane and a comparison with nematode damage. Bull Entomol Res 100:373–385

Berry SD, Rhodes R, Foster J et al (2013) The effect of cover crops on plant parasitic nematodes. Intern J Pest Manag 54:363–375

Blume H (1985) Geography of sugar cane: environmental, structural and economical aspects of cane sugar production. Verlag Dr. Albert Bartens, Berlin

Butterfield MK, D'Hont A, Berding N (2001) The sugarcane genome. A synthesis of current understanding, and lessons for breeding and biotechnology. Proc S Afr Sugar Techn Ass 75: 1–5

Cadet P (1985) Incidence des nématodes sur les repousses de canne à sucre au Burkina Faso et en Côte d'Ivoire. Rev Nématol 8:277–284

Cadet P (1986) Evolution des nématodes ectoparasites dans la rhizosphère de la canne à sucre au Burkina Faso. Rev Ecol Biol du Sol 23:205–213

Cadet P, Spaull VW (1985) Studies on the relationship between nematodes and sugarcane in South and West Africa: plant cane. Rev Nématol 8:131–142

Cadet P, Spaull VW (2003) Effect of nematodes on sustainability of sugarcane production in South Africa. Field Crops Res 83:91–100

Cadet P, Spaull VW (2005) Nematode parasites of sugarcane. In: Luc M, Sikora RA, Bridge J (eds) Plant parasitic nematodes in subtropical and tropical agriculture, 2nd edn. CAB International, Wallingford, pp 645–674

Cadet P, Spaull VW, McArthur DG (2002) Role of plant parasitic nematodes and abiotic soil factors in growth heterogeneity of sugarcane on a sandy soil in South Africa. Plant and Soil 246:259–271

Cadet P, Berry S, Spaull VW (2004) Mapping of interactions between soil factors and nematodes. Europ J Soil Biol 40:77–86

Cadet P, Berry S, Spaull VW (2005) Long term ratooning of sugarcane on sandy soils in KwaZulu-Natal. Proc S Afr Sugar Techn Ass 79:175–178

Cadet P, Berry S, Leslie G et al (2006) Management of nematodes and a stalk borer by increasing within-field sugarcane cultivar diversity. Plant Pathol 56:526–535

Cook R, Evans K (1987) Resistance and tolerance. In: Brown RH, Kerry BR (eds) Principles and practice of nematode control in crops. Academic, Sydney, pp 179–231

Cooper WR, Jia L, Goggin L (2005) Effects of jasmonate-induced defences on root-knot nematode infection of resistant and susceptible tomato cultivars. J Chem Ecol 31:1953–1967

Dana P, Cadet P, Spaull VW (2002) Field variation of abiotic factors and their relationship with nematode communities in sugarcane. Proc S Afr Sugar Techn Ass 76:597–600

Dick J (1961) Eelworms and sugarcane. Proc S Afr Sugar Techn Ass 35:110–113

Dick J, Harris RHG (1975) Nematodes and sugarcane. S Afr Sugar J 59:397–410

Donaldson RA (1983) The effect of placement on the efficacy of granular nematicides. Proc S Afr Sugar Techn Ass 57:120–122

Donaldson RA (1985) The effects of soil pH, clay content, rainfall and age at harvest on the yield response of sugarcane to Temik. Proc S Afr Sugar Techn Ass 59:164–167

Donaldson RA (1987) Some aspects related to the use of nematicides on sugarcane in South Africa. Proc S Afr Sugar Techn Ass 61:117–120

Donaldson RA (1988) Nematicides for sugarcane. S Afr Sugar J 72:57–61, 92–97

Donaldson RA, Turner PET (1984) The interaction of herbicides and nematicides on plant cane grown in weak sandy soils. Proc S Afr Sugar Techn Ass 58:137–142

Donaldson RA, Turner PET (1988) A preliminary report of the effect of soil moisture level on responses to Temik and Curaterr. Proc S Afr Sugar Techn Ass 62:164–168

Edmonds GJ (2014) Investigating induced resistance in sugarcane. MSc Dissertation, University of KwaZulu-Natal, Pietermaritzburg

Gosnell J (2013) Biomass management: ethanol and electricity generation. In: Meyer J, Rein P, Turner P et al (eds) Good management practices for the cane sugar industry. Verlag Dr. Albert Bartens, Berlin, pp 395–420

Govaerts B, Mezzalam M, Unno Y et al (2007) Influence of tillage, residue management, and crop rotation on soil microbial biomass and catabolic diversity. Appl Soil Ecol 37:18–30

Guimaraes LMP, Pedrosa EMR, Coelho RSB et al (2008) Efeito de Metil Jasmonato e Silicato de Potássio no Parasitismo de *Meloidogyne incognita* e *Pratylenchus zeae* em Cana-de-Açúcar. Nematol Brasil 32:50–55

Guyon S, Vogel JL, Omarjee J et al (2003) *Burkholderia tropicalis*, a potential bacterial inoculant to control nematodes and improve sugarcane growth. Proc S Afr Sugar Techn Ass 77: 118–123

Harris RHG (1973) Nematicide experiments in the South African Sugar Industry. Proc S Afr Sugar Techn Ass 47:185–190

Herbert A (1991) CG Smith: One hundred years. Fishwick Printers, Durban

Kleynhans KPN, Van den Berg E, Swart A et al (1996) Plant nematodes in South Africa. Handbook no.8. Plant Protection Research Institute, Pretoria

Luc M, Reversat G (1985) Possibilités et limites des solutions génétiques aux affections provoquées par les nématodes sur les cultures tropicales. Comptes rendus des Séances de l'Académie d'Agriculture de France 71:781–791

Martin GC (1967) Plant parasitic nematodes associated with sugarcane production in Rhodesia. FAO Plant Protection Bulletin 15:45–58

Mathias K (2013) Outgrowers. In: Meyer J, Rein P, Turner P et al (eds) Good management practices for the cane sugar industry. Verlag Dr. Albert Bartens, Berlin, pp 549–584

McArthur DG, Spaull VW (1995) A note on the effect of aldicarb on the yield of four sugarcane varieties on a sandy soil in the Natal Midlands. Proc S Afr Sugar Techn Ass 69:25–27

Moberly PK, Meyer JH (1978) Filter cake – a field and glasshouse evaluation. Proc S Afr Sugar Techn Ass 52:131–135

Moberly PK, Clowes MSJ (1981) Trials with nematicides registered for use on sugarcane in South Africa. Proc S Afr Sugar Techn Ass 55:92–98

Moberly PK, Harris RHG, Millard E (1974) An investigation into the problem of poor sugarcane growth on some sandy soils of the Natal Sugar belt. Proc Internat Soc Sugar Cane Technol 15:923–931

Nahar K, Kyndt T, De Vleesschauwer D et al (2011) The jasmonate pathway is a key player in systemically induced defense against root knot nematodes in rice. Plant Physiol 157:305–316

Oka Y, Cohen Y (2004) Induced resistance to cyst and root-knot nematodes in cereals by DL-β-amino-n-butyric acid. Europ J Plant Pathol 107:219–227

Omarjee J (2006) Effets des peuplement de *Burkholderia* et de nématodes sur la coissance de la canne à sucre. PhD thesis, University of Claude Bernard-Lyon, Lyon

Omarjee J, Balandreau J, Spaull VW et al (2008) Relationship between *Burkholderia* populations and plant parasitic nematodes in sugarcane. Appl Soil Ecol 39:1–14

Oostendorp M, Kunz W, Dietrich B et al (2004) Induced disease resistance in plants by chemicals. Europ J Plant Pathol 107:19–28

Osborn RF (1964) Valiant harvest: the founding of the South African Sugar Industry 1848–1926. Brown Davis and Platt Ltd, Durban

Pankhurst CE, Magarey RC, Stirling GR et al (2003) Management practices to improve soil health and reduce the effects of detrimental soil biota associated with yield decline of sugarcane in Queensland, Australia. Soil and Tillage Res 72:125–137

Paturau JM (1982) By-products of the cane sugar industry: an introduction to their industrial utilization, 2nd edn. Elsevier Scientific Publishing, Amsterdam

Prammanee P, Grange I, Prasertsak P et al (2001) Minimum tillage of sugarcane in Thailand: its effect on yields, soil properties and cost effectiveness. Proc Internat Soc Sugar Cane Technol 24:202–204

Ramouthar P (2015) Impact of green manures on nematode control. The Link. South Afr Sugarcane Res Inst 24:24–25
Ramouthar PV, De Haas O, Berry SD (2012) Long term effects of nematodes on six South African sugarcane varieties. Proc S Afr Sugar Techn Ass 85:155–170
Rau S, Moberly PK (1975) Nematicide application to ratoon crops of sugarcane grown in some sandy soils of the Natal sugarbelt. Proc S Afr Sugar Techn Ass 49:171–173
Rein P, Purchase B (2013) Management of wastes and effluents. In: Meyer J, Rein P, Turner P et al (eds) Good management practices for the cane sugar industry. Verlag Dr. Albert Bartens, Berlin, pp 477–499
Richardson FE, Watt EJ (1980) Foliar applications of oxamyl to ratoon sugarcane for nematode control. Proc Internat Soc Sugar Cane Technol 17:1865–1872
Rimé D, Nazaret S, Gourbiere F et al (2003) Comparison of sandy soils suppressive or conducive to ectoparasitic nematode damage on sugarcane. Phytopathol 93:1437–1444
Rosenfeld AH (1956) Sugarcane around the world. University of Chicago Press, Chicago
Rostron H (1976) Rate, time and methods of Temik application in ratoon sugarcane. Proc S Afr Sugar Techn Ass 50:29–33
Sikora RA, Bridge J, Starr JL (2005a) Management practices: an overview of integrated nematode management technologies. In: Luc M, Sikora RA, Bridge J (eds) Plant parasitic nematodes in subtropical and tropical agriculture. 2nd edn. CAB International, Wallingford, pp 793–825
Sikora RA, Greco N, Silva JFV (2005b) Nematode parasites of food legumes. In: Luc M, Sikora RA, Bridge J (eds) Plant parasitic nematodes in subtropical and tropical agriculture. 2nd edn. CAB International, Wallingford, pp 259–318
Spaull VW (1981a) Nematodes associated with sugarcane in South Africa. Phytophylactica 13:175–179
Spaull VW (1981b) *Bacillus penetrans* in South African plant parasitic nematodes. Nematologica 27:244–245
Spaull VW (1984) Observations on *Bacillus penetrans* infecting *Meloidogyne* in sugarcane fields in South Africa. Rev Nématol 7:277–282
Spaull VW (1995) A revised estimate of crop loss in sugarcane caused by nematodes. Proc S Afr Sugar Techn Ass 69:28–34
Spaull VW (1996) Factors affecting the response of sugarcane to treatment with nematicide. Proc S Afr Sugar Techn Ass 70:14
Spaull VW (1997) On the use of furfural to control nematodes in sugarcane. Proc S Afr Sugar Techn Ass 71:96
Spaull V (2013) Pest control. In: Meyer J, Rein P, Turner P et al (eds) Good management practices for the cane sugar industry. Verlag Dr. Albert Bartens, Berlin, pp 301–330
Spaull VW, Bailey RA (1993) Combined effect of nematodes and ratoon stunting disease on sugarcane. Proc S Afr Sugar Techn Ass 67:29–33
Spaull VW, Cadet P (1991) Studies on the relationship between nematodes and sugarcane in South and West Africa: ratoon cane. Rev Nématol 14:183–186
Spaull VW, Cadet P (2001) Nematodes and nutrients: association between plant-parasitic nematodes and soil chemicals. Proc S Afr Sugar Techn Ass 75:116–117
Spaull VW, Cadet P (2003) Impact of nematodes on sugarcane and the benefit of tolerant varieties. Proc S Afr Sugar Techn Ass 77:230–238
Spaull VW, Donaldson RA (1983) Relationship between time of nematicide application, numbers of nematodes and response to treatment in ratoon sugarcane. Proc S Afr Sugar Techn Ass 57:123–127
Spaull VW, Heyns J (1991) Distribution of longidorid nematodes in sugarcane fields in South Africa. Phytophylactica 23:167–172
Spaull VW, Cadet P, Gers C (2003) Factors affecting the distribution of plant feeding nematodes in KwaZulu-Natal. Proc S Afr Sugar Techn Ass 77:220–222
Spaull VW, Cadet P, Berry S (2005) Sugarcane varieties suitable for sandy soils in Mpumalanga. Proc S Afr Sugar Techn Ass 79:165–174

Stirling GR (2008) The impact of farming systems on soil biology and soilborne diseases: examples from the Australian sugar and vegetable industries – the case for better integration of sugarcane and vegetable production and implications for future research. Australasian Plant Pathol 37:1–18

van Antwerpen T, van Antwerpen R, Meyer JH et al (2007) Factors associated with a healthy soil in sugarcane production in KwaZulu-Natal. Proc Internat Soc Sugar Cane Technol 26: 273–281

van Antwerpen R, Berry SD, van Antwerpen T et al (2009) Indicators of soil health for use in the South African Sugar Industry: a work in progress. Proc S Afr Sugar Techn Ass 82:551–563

van Antwerpen R, Berry SD, van Antwerpen T et al (2013) Sugarcane as an energy crop: its role in biomass economy. In: Singh BP (ed) Biofuel crop sustainability. Wiley, Ames, pp 53–108

Van den Berg E, Palomares-Rius JE, Vovlas N et al (2016) Morphological and molecular characterisation of one new and several known species of the reniform nematode genus *Rotylenchulus* Linford & Oliveira, 1940 (Rotylenchulinae, Hoplolaimidae) and a phylogeny of the genus. Nematol 18:67–107

van Dillewijn C (1952) Botany of sugarcane. Chronica Botanica Co., Waltham

Villenave C, Cadet P (1998) Interaction of *Helicotylenchus dihystera, Pratylenchus pseudopratensis* and *Tylenchorhynchus gladiolatus* on two plants from the soudano-sahelian zone of West Africa. Nematropica 28:31–39

Vogel JL, Balandreau J, van Antwerpen T et al (2002) Isolation and characterisation of sugarcane rhizobacteria and their effect on nematodes. Proc S Afr Sugar Techn Ass 76:305–309

Webster JM (1987) Introduction. In: Brown RH, Kerry BR (eds) Principles and practice of nematode control in crops. Academic Press, Sydney, pp 1–11

Chapter 12
Nematode Pests of Tobacco and Fibre Crops

Elizabeth R. van Biljon

12.1 Introduction

Tobacco (*Nicotiana tabacum*) and cotton (*Gossypium hirsutum*) historically were major crops in South Africa (SA) and were subject to considerable research of their nematode pest problems. Recent production has declined markedly such that they are now relatively minor crops, but a considerable volume of research data has been published. The production of other fibre crops, hemp (*Cannabis sativa*), kenaf (*Hibiscus cannabinus*) and flax (*Linum usitatissimum*), has increased, notably by smallholding farmers, and has been studied to a limited extent to determine the impact of plant-parasitic nematodes on production.

12.2 Tobacco

Tobacco is a member of the Solanaceae or potato family, also frequently called the nightshade family. Tobacco belongs to the genus *Nicotiana* (Garner 1947), with *N. tabacum* representing the cultivated form. Tobacco was native to the Americans and was grown and used by the Mayans. The popularity of the crop spread throughout the north and south as other tribes began to cultivate it. The written history of tobacco started on October 12 in 1492 when Christopher Columbus reached the beaches of San Salvador in the West Indies (Collins and Hawks 1993). Tobacco was introduced into southern Africa by the early Portuguese explorers, and its use spread rapidly among the indigenous people (Taylor 1927). The early Dutch settlers were growing

E.R. van Biljon
Agricultural Research Council–Institute for Industrial Crops,
Private Bag X 82075, Rustenburg 0300, South Africa
e-mail: vanbiljon.elizabeth@gmail.com

H. Fourie et al. (eds.), *Nematology in South Africa: A View from the 21st Century*, DOI 10.1007/978-3-319-44210-5_12

tobacco in 1657, and by 1875 the Cape Colony was producing more than 1,360 metric tonnes (MT) year^{-1}. Further expansion became possible when the Zuid-Afrikaansche Republiek was established and developed in the late 1800s (Akehurst 1968).

Despite the negativity around tobacco and tobacco-related products, this crop is still a cash entity that generates the highest income when managed correctly. Local commercial flue-cured tobacco farmers produced 17,010 MT on 5,139 hectares (ha). Despite high revenues tobacco production is rapidly declining in SA, with approximately 67 % fewer ha planted and 43 % less tobacco produced since 2000 (FAO 2016).

The Agricultural Research Council–Institute for Industrial Crops (ARC-IIC) in Rustenburg (North-West Province) supplies 80 % of all commercial flue-cured tobacco seed and 100 % of all commercial air-cured tobacco seed planted locally. Being such an important role player, this research institution needs constantly to keep in touch with international market trends regarding the style and quality of tobacco. All registered cultivars (cvs) must also be resistant or tolerant to the major diseases and plant-parasitic nematodes that attack the crop. A sustainable cv. development programme and a viable tobacco germplasm bank are thus of utmost importance. Projects currently running include root-knot nematode resistance evaluation on tobacco as well as other crops that are grown with it in rotation.

12.2.1 Plant-Parasitic Nematodes Associated with Tobacco

Various nematode species are associated with tobacco, but only a few are known to cause damage to the crop and are represented below.

12.2.1.1 Root-Knot Nematodes

Tobacco is highly susceptible to root-knot nematode damage, and as such, high priority must be given to nematode management where the crop is grown. Galling of root systems is a typical below-ground symptom of *Meloidogyne* spp. infection (Fig. 12.1). Within 2 months after transplanting, the leaves of tobacco plants will go into a process of early maturation (Fig. 12.2). In severe cases of infection, a condition known as 'rimfiring' occurs that includes necrosis of leaf tips and margins (Rich and Kinloch 2005). Susceptible tobacco cvs are severely damaged by *Meloidogyne* spp. (Lamberti 1979; Di Vito et al. 1983). According to the literature, *Meloidogyne javanica* (Treub, 1885) Chitwood, 1949 is more damaging than *Meloidogyne incognita* (Kofoid and White, 1919) Chitwood, 1949 (Arens and Rich 1981) and causes two to four times greater yield losses (Rich and Dunn 1985). In SA, *M. javanica* occurred in more than 37 % of all flue-cured tobacco fields, while *M. incognita* was present in only 10 % of the fields sampled (SAPPNS).[1]

---

[1] Dr Mariette Marais of the Nematology Unit, Biosystematics Division, Agricultural Research Council–Plant Protection Research Institute is thanked for the use of data from the South African Plant-Parasitic Nematode Survey (SAPPNS) database; E-mail: maraism@arc.agric.za.

Fig. 12.1 A root-knot nematode-infected tobacco root system (Jeannie van Biljon, Agricultural Research Council–Institute for Industrial Crops, Rustenburg, South Africa)

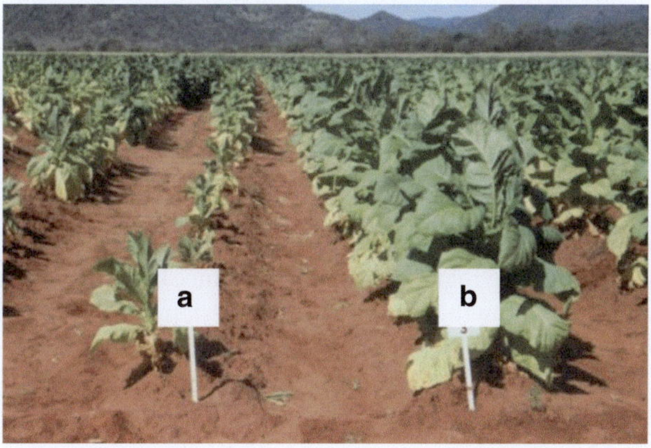

Fig. 12.2 Chlorotic and stunted tobacco plants (**a**), which are symptomatic of root-knot nematode damage compared to (**b**) bigger and healthier plants that were treated with a nematicide (Jeannie van Biljon, Agricultural Research Council–Institute for Industrial Crops, Rustenburg, South Africa)

Meloidogyne enterolobii Yang and Eisenback, 1983, previously identified as *Meloidogyne mayaguensis* Rammah and Hirschmann 1988 and only associated with guava trees in SA prior to 1997 (SAPPNS), was also identified from tobacco roots in the Mpumalanga Province in the late 1990s (Willers 1997). Since then it has been reported from roots of tomato (*Solanum lycopersicon*) in the Limpopo Province and green pepper (*Capsicum* sp.) near Barberton (Mpumalanga Province) (M Marais, Agricultural Research Council–Plant Protection Research Institute, Pretoria, 2016, personal communication) as well as recently from potato (*Solanum tuberosum*) tubers in the KwaZulu-Natal (Onkendi et al. 2014) and Mpumalanga provinces (Agenbag 2016). This root-knot nematode species is regarded as an emerging threat (see Sect. 17.5.1, Box 17.1, Chap. 17); it has a variety of hosts and is morphologically similar to *M. incognita* (Brito et al. 2004, 2005; Jones et al. 2013).

12.2.1.2 Lesion Nematodes

Following root-knot nematodes, *Pratylenchus* spp. are among the most important pests that infect tobacco in SA. Graham (1951) showed that the root systems of certain tobacco cvs were severely damaged when tobacco was grown after maize (*Zea mays*). Several lesion nematode species were found in association with tobacco, viz. *Pratylenchus brachyurus* (Godfrey, 1929) Filipjev and Schuurmans Stekhoven, 1941; *Pratylenchus delattrei* Luc, 1958; *Pratylenchus flakkensis* Seinhorst, 1968; *Pratylenchus hexincisus* Taylor and Jenkins, 1957; *Pratylenchus neglectus* (Rensch, 1924) Filipjev and Schuurmans Stekhoven, 1941; *Pratylenchus scribneri* Steiner, 1943; *Pratylenchus thornei* Sher and Allen, 1953; *Pratylenchus vulnus* Allen and Jensen, 1951; and *Pratylenchus zeae* Graham 1951 (SAPPNS). Infected roots may show different degrees of decay with lesions varying in colour. The lesions may completely girdle the root so that the cortex tissues slough off, leaving only the vascular cylinder (Collins and Hawks 1993). The latter was observed with *P. flakkensis* damage on flue-cured tobacco (Jeannie van Biljon, Agricultural Research Council–Institute for Industrial Crops, Rustenburg, 2015, personal communication). *Pratylenchus hexincisus* is also associated with the development of 'Black root rot' (*Thielaviopsis basicola*) in tobacco that is grown in the black turf soils that are wet at planting (Milne 1972).

12.2.1.3 Stubby-Root Nematodes

Paratrichodorus meyeri De Waele and Killian, 1992 was recorded in 1991 in the Vaalwater tobacco-growing region (Limpopo Province) in association with tobacco plants displaying poor growth and stubby-root systems (Van Biljon 1992). Crop losses due to parasitism by this nematode pest amounted to more than a 25 % reduction in income for tobacco producers (Jeannie van Biljon, Agricultural Research Council–Institute for Industrial Crops, Rustenburg, 2015, personal communication). Stubby-root nematodes are especially damaging to young seedlings. When the seed germinates, the nematode damages the root tip and root growth declines and finally ceases. Newly developed lateral roots are in turn attacked, resulting in a root system consisting of stubby and stunted roots on a shallow much-branched root system (Fig. 12.3) (Christie and Perry 1951; Rohde and Jenkins 1957). Abnormal stimulation of newly developed lateral roots can sometimes occur behind damaged root tips (van Biljon 1992).

12.2.1.4 Stunt Nematodes

Tylenchorhynchus claytoni Steiner, 1937, can cause moderate stunting of infected tobacco plants (Deubert et al. 1967; MacGowan 1980). In SA, *Tylenchorhynchus brevilineatus* Williams, 1960, and *Tylenchorhynchus ventralis* (Loof, 1963) Fortuner

Fig. 12.3 Typical stunted roots of a stubby-root nematode-infected tobacco plant (Jeannie van Biljon, Agricultural Research Council–Institute for Industrial Crops, Rustenburg, South Africa)

and Luc, 1987, were associated with tobacco on an experimental farm near Mbombela (Mpumalanga Province) (SAPPNS).

12.2.2 Management Strategies

12.2.2.1 Chemical Control

Various nematicides have been used over the years for the control of nematode pests on tobacco (Van Zyl 2013), but typically their use is under threat due to: (i) effective chemical control methods being relatively expensive, (ii) the repeated use of such products that leads to accelerated microbial degradation (AMD) (see Box 6.4, Sect. 6.4, Chap. 6), and (iii) key products that have been withdrawn (e.g. aldicarb and methyl bromide) that have been used as general-purpose pesticides on tobacco fields. As a result new products are being used, e.g. the fumigant 1,3-D, increasingly by local farmers for pre-plant fumigation of nematode-infested soils. Also, metam sodium is commonly used as an effective replacement for methyl bromide in seedbed sterilisation (Miller 2007).

Another relatively new acitve substance (a.s) with nematicidal properties is furfural, which was registered as a nematicide on tobacco in SA during 2009. This chemical has the potential to damage the outer protective protein-lipid layers of

Meloidogyne spp. egg masses. Higher concentrations (10,000–50,000 ppm) of the product can kill the eggs (Jansen van Vuuren et al. 2005). Recently fluopyram was registered for use on tobacco as a single or a split application (Anonymous 2015). In field trials, the application of this product resulted in a significant increase in yield and quality in a broad spectrum of crops *inter alia* tobacco. Fluopyram is one of the first nematicides acting via Complex ll inhibition, thus selectively inhibiting the generation of cellular energy by the nematode individual. Compared to existing chemical products, it has a favourable safety and environmental profile and offers long-lasting control at very low application rates. An additional benefit is that it is also effective against fungal diseases such as Alternaria leaf spot, powdery mildew and Sclerotinia rot.

12.2.2.2 Biological Control

Organisms, e.g. bacteria and fungi, have been tested widely under local environmental conditions. Some of these products are already available and registered in SA for general use. One of them contains the nematode-egg fungal parasite *Purpureocillium lilacinum*. It has been registered on a number of crops in SA, including tobacco (Neethling 2004; Van Zyl 2013). In a field trial, the efficacy of *P. lilacinum*, with and without a growth stimulant, was investigated at various dosage rates and application times. The combined treatment with *P. lilacinum* and a growth stimulant reduced the number of *M. javanica* females in tobacco roots by 50 %. All the *P. lilacinum* treatments performed better than the untreated control (Van Biljon and Botha 1998).

12.2.2.3 Genetic Host Plant Resistance

Although resistance in tobacco cvs to the various root-knot nematode species exists (Ng'ambi et al. 1999), it is being improved for potential introgression in local commercially available genotypes (collectively referring to breeding lines and cvs) by nematologists and breeders of the ARC-IIC. Various selections of genotype OD 86 have shown to be either susceptible or tolerant to *M. javanica*, but resistant to *M. incognita* races 2 and 4 (Van Biljon 2010).

12.2.2.4 Crop Rotation

When selecting rotation crops for tobacco, all the important nematode pest species that they might host should be considered and identified (Van Biljon and Botha 1999; Van Biljon and Meyer 2000; Van Biljon et al. 2015). It is usual for a single field to have several plant-parasitic species present. For example, *P. meyeri*, *M. javanica* and *M. incognita* races 2 as well as 4 might inhabit the same tobacco field.

Root-knot nematode species are usually the key nematode problem since they usually have a devastating effect on tobacco. The development of a management system should, therefore, focus on root-knot nematodes.

Paratrichodorus meyeri might, however, emerge as a problem when the root-knot nematodes are no longer a threat and thus should not be ignored.

Pratylenchus spp. are present in many tobacco fields. Since most lesion nematode species have a wide host range, this causes problems in the selection of rotation crops. Fortunately, tobacco is not a good host to many of the *Pratylenchus* spp., and depending on the dominant species present, suitable crops can be selected for inclusion in a rotation system (Shepherd and Barker 1990). Rotating nematode suppressive cover crops, such as sunn hemp (*Crotalaria juncea*) with field or cash crops may suppress root-knot nematode populations and benefit the following crop. However, certain crops (including sunn hemp) can increase the numbers of certain lesion nematode species (Van Biljon et al. 2015). Therefore, the identification of nematode pests to species level is important when crop rotation is employed as a nematode control strategy.

12.3 Cotton

Cotton plants originated from tropical shrubs, which have been developed to produce the most important textile fibre of tropical regions. Cotton is a member of the genus *Gossypium* and belongs to the Malvaceae family, which also includes the flowering shrubs hibiscus (*Hibiscus* spp.) and okra (*Abelmoschus esculentus*). More than 90 % of the commercial crop is upland cotton, while long-staple cotton *Gossypium barbadense* occupies a small production area of approximately 8 % globally (Buchanan 1999). Cotton is grown in tropical, subtropical and warm-temperature climates (Brubaker et al. 1999), and the major nematode pests of cotton are well adapted to warm environments (Robinson 2007). The total ha planted to cotton in SA declined from 22,574 in 2003/2004 to 6,827 in 2013/2014 (Cotton 2014). A mere 17,000 MT was produced during the 2013 growing season (FAO 2016).

Cotton is the world's most important natural textile fibre. The seed is an important source of feed, foodstuff and oil (Chen et al. 2007). Cotton oil is refined to remove gossypol (Adams et al. 1960), which is toxic to humans and monogastric animals (Anonymous 2008; Scheffler and Romano 2008). Understanding cotton's growth and development pattern is very important for timely management of plant-parasitic nematodes. The demand for carbohydrates and nutrients increases with the growth of the cotton plant, while the plant's ability to supply carbohydrates and nutrients becomes limited during boll growth. This is because the plant shifts its priorities from building the suppliers (leaves and roots) to feeding the consumers (seed and lint) (Albers 1993). Root growth dominates the growth of the cotton plant during germination and seedling establishment, which is a critical time for the development of the root system (Ritchie et al. 2007).

12.3.1 *Plant-Parasitic Nematodes Associated with Cotton*

Root-knot nematodes (*Meloidogyne* spp.) have been reported as a limiting factor of cotton production in some of the irrigation schemes in SA as early as the 1970s (Louw 1982; SAPPNS). In two surveys a total of 167 sites, representing 24 areas, in the two major cotton growing areas (western and eastern parts) of the Northern Cape Province were sampled. Sixteen plant-parasitic nematode genera and 23 species were recorded with *M. incognita*, race 4 dominating and present in 79 % of the root samples. Individuals of *Paratrichodorus* spp. were present in 63 % of the samples, while 49 % of the samples contained *Pratylenchus* spp. (Van Biljon and Swanepoel 2007).

Fusarium oxysporum f sp. *vasinfectum*, a pathogen that forms a synergistic disease complex with *M. incognita* race 4, was isolated from a number of nematode samples obtained from local cotton crops (Van Biljon and Swanepoel 2007). Associations between plant-parasitic nematodes and soil microbes exist (Box 12.1), with the synergistic association between the *M. incognita* race 4 and *F. oxysporum* f. sp. *vasinfectum* that destroyed cotton plants in the Northern Cape Province (see Sect. 12.3.3.5) being illustrated in Fig. 12.4.

Box. 12.1 Three basic types of interactions are known to occur between plant-parasitic nematodes and bacteria and/or fungi

Synergistic: the primary pathogen (e.g. a plant-parasitic nematode) causes mechanical/physical/anatomical/biochemical/physiological damage and/or modifications in the host plant and facilitates interaction by a secondary pathogen (e.g. a bacterium or fungus). Damage to the host plant caused by two such pathogens is greater than the sum of damage caused by both pathogens if they should have infected the host plant separately (1 + 1 > 2).

Antagonistic: infection by one pathogen (e.g. the plant-parasitic nematode) modifies the host or its response in such a way that it becomes less suitable and/or unfavourable for infection by a secondary pathogen (e.g. a bacterium or fungus). Damage to the host plant is hence smaller than the sum of the damage that would have been caused by the two organisms had they been alone (1 + 1 < 2).

Additive: two or more organisms that infect a host plant interact and cause damage that equals the sum of individual damage that would be caused by the two organisms (1 + 1 = 2).

Fig. 12.4 (**a–c**) A synergistic interaction between *Meloidogyne incognita* race 4 (**a**) and the fungus *Fusarium oxysporum* f. sp. *vasinfectum* (**b**), resulting in severe damage to infected cotton plants (**c**) in a field near Kakamas in the Northern Cape Province (**a**), Ebrahim Shokoohi, North-West University, Potchefstroom, South Africa; (**b**), Maryke Craven, Agricultural Research Council–Grain Crops Institute, Potchefstroom, South Africa; (**c**), Jeannie van Biljon, Agricultural Research Council–Institute for Industrial Crops, Rustenburg, South Africa)

12.3.1.1 Root-Knot Nematodes

Meloidogyne incognita is a major pest of cotton and is considered to be the most important pest of the crop worldwide (Starr et al. 2005). This root-knot nematode species causes significant yield losses, both directly (Kirkpatrick and Sasser 1984) and indirectly through interactions with soilborne fungal pathogens (Powell 1971; Starr et al. 2005). Only races 3 and 4 of this species parasitise cotton (Hartman and

Sasser 1985; Starr and Veech 1986; Maqbool 1992; Robinson 2007). Race 4 is widespread and the cause of severe yield losses in SA (Botha and Van Biljon 1996).

Another species, *Meloidogyne acronea* Coetzee 1956, also parasitises cotton but it has not been identified as a pest of this crop in SA. However, this species has been found only near Vryburg (Northern Cape Province) (Coetzee 1956; SAPPNS) and in two isolated areas in the lower Shire valley of Malawi (Bridge et al. 1976). This nematode is unique to Africa and is known as the cotton root-knot nematode (Bridge and Muller 1984). The biology of *M. acronea* in cotton is similar to that of *M. incognita*, except that the mature female is semi-endoparasitic (Coetzee and Botha 1965). The slight root swellings and growth distortions resulting from *M. acronea* parasitism on cotton are also markedly different from the typical root galling caused by *M. incognita* (Fig. 12.5) (Bridge et al. 1976). The symptoms of root damage caused by *M. acronea* are termed the so-called turned-aside tap-root effect, including proliferation of the lateral roots and distortion of the tap-root (leading to stunting), delayed flowering and significant yield loss. These symptoms are exacerbated when cotton is grown under water-stressed conditions (Page and Bridge 1994).

12.3.1.2 Lesion Nematodes

Pratylenchus brachyurus is another nematode pest of cotton. Large numbers of *P. brachyurus* were recorded from local cotton fields where crops performed poorly (SAPPNS). In Zimbabwe, control of *P. brachyurus* resulted in a 15 % yield increase in cotton (Anonymous 1971). Another lesion nematode species, *Pratylenchus teres* Khan and Singh, 1974, was first reported on cotton in the Jan Kempdorp area (Vaalharts Irrigation Scheme, Northern Cape Province) of SA during the 1995/1996 season (Carta et al. 2002).

Fig. 12.5 Characteristic galls on the roots of a cotton plant as a result of parasitism by *Meloidogyne incognita* race 4 (Jeannie van Biljon, Agricultural Research Council–Institute for Industrial Crops, Rustenburg, South Africa)

12.3.1.3 Rotylenchulus *spp.*

Two species that belong to the genus *Rotylenchulus* are pests of cotton. The reniform nematode *Rotylenchulus reniformis* Linford and Oliveira, 1940, is distributed throughout the subtropical and tropical regions of the world. Limited information on associations between *R. parvus* Sher 1961 and cotton is available (Starr et al. 2005). No problems have been experienced with these nematode pests on cotton in SA.

12.3.1.4 Lance Nematodes

Four lance nematode species are considered pests of cotton worldwide, viz. *Hoplolaimus aegypt* Shafiee and Koura, 1969; *Hoplolaimus indicus* Sher, 1963; *Hoplolaimus columbus* Sher, 1963; and *Hoplolaimus seinhorsti* Luc, 1958 (Starr et al. 2005). However, only *Hoplolaimus pararobustus* (Schuurmans Stekhoven and Teunissen, 1938) Sher, 1963, has been recorded on cotton in SA (Kleynhans et al. 1996; SAPPNS).

12.3.1.5 Stubby-root Nematodes

Cotton was proven to be a host for the stubby-root nematode *Paratrichodorus meyeri* De Waele and Kilian, 1992, in pot experiments (Van Biljon 1992). Stunting of plants was visible when infected with different population levels of this pest compared to a untreated control. On some farms in the Mpumalanga Province, large numbers of an undisclosed stubby-root nematode species were found in association with cotton plants (Louw 1982). Such infected plants exhibited poor growth and stubby necrotic root systems.

12.3.1.6 Needle and Dagger Nematodes

Longidorus pisi Edward, Misra and Singh, 1964 (needle nematode); *Xiphinema louisi* Heyns, 1979; *Xiphinema vanderlindei* Heyns, 1962; and *Xiphinema coomansi* Kruger and Heyns, 1986 (dagger nematodes), were present in rhizosphere soil samples from cotton obtained near Upington (Northern Cape Province) (Swanepoel 2005; SAPPNS). *Longidorus pisi* has been associated with pre-emergence death of cotton in Egypt (Aboul-Eid 1970), but this condition never presented itself in SA. Other dagger nematode species recorded on cotton in SA include *Xiphinema elongatum* Schuurmans Stekhoven and Teunissen, 1938; *Xiphinema dimorphicaudatum* Heyns, 1966; *Xiphinema imitator* Heyns, 1965; and *Xiphinema vitis* Heyns, 1974 (Kleynhans et al. 1996; SAPPNS).

12.3.2 Symptoms and Damage

No adverse effects were detected on fibre quality of cotton infected with plant-parasitic nematodes in local production areas. Thomas and Smith (1993) also reported that *M. incognita* had no effect on cotton fibre quality in New Mexico (USA). It is reported, however, that plant-parasitic nematodes can hamper the quality of the crop, depending on the developmental stage during which infection occurs (Anonymous 2009a). Infected roots may show discolourations, brown lesions, development of lots of short truncated roots and abnormal development at the nematode feeding spot (Kleynhans et al. 1996). In addition to direct damage, such as a reduction in the number and size of bolls, nematode infection may also delay cotton plants reaching maturity (Walker et al. 1998).

12.3.3 Management Strategies

In SA, reducing the damage caused by nematodes on cotton is imperative because the profitability of cotton is marginal. Effective chemical control methods are relatively expensive. Root-knot nematode-resistant cultivars (cvs) are not commercially available anymore due to the market being dominated by genetically modified (GM) cotton cvs. The various control measures that are available for use by cotton producers are referred to below.

12.3.3.1 Chemical Control

Chemical control remains the most important means of reducing the adverse impact of nematode pests on cotton in SA (Botha-Greeff and van Biljon 1999). Substantial increases in yield can be obtained with the use of nematicides to control *M. incognita* (Kinloch and Rich 1998; Baird et al. 2000). The management of plant-parasitic nematodes relied mostly on the use of aldicarb which, since its withdrawal from the market, has left producers with few control options (Anonymous 2012a; Verdoorn 2012).

Trial results from the Northern Cape Province suggested that oxamyl granules (125 g 100 m row^{-1}) reduced root-knot nematode population levels with a subsequent increase in yield. A slight phytotoxic effect was, however, present on the lower leaves of the plants (Van Biljon 2005a). A product containing both terbufos and phorate was also included in trials from 2005 to 2008, and in most instances, yield increases of more than 14 % were achieved (Van Biljon 2005a). In 2005, abamectin was launched as a seed treatment for early-season control of damage to seedlings (Anonymous 2005). Variable results have been achieved in suppressing *M. incognita* when using this product at 150-μg seed^{-1} (Faske and Starr 2006; Lawrence et al. 2006; Monfort et al. 2006; Phipps et al. 2006; De Beer et al. 2007; Van Biljon 2007a). The limited protection of early-stage root development suggests that only a small portion of abamectin applied to the seed is transferred to the developing root system of plants. This could possibly explain the variability in results. Seed treatment alone may also not be sufficient to provide protection to cotton plants in fields where high population densities of economically important nematode species occur (Faske and Starr 2007; Kirkpatrick 2007). During the 2006/2007 season furfural, which is a by-product of sugarcane, was evaluated on cotton. Application of this product reduced root-knot nematode numbers by 38 % 12 weeks after application and resulted in a 17 % increase in yield when compared to the untreated control (Van Biljon 2007a). However, furfural is not registered on cotton in SA (Van Zyl 2013).

Nematicides are effective but do not provide season-long protection. Their future availability is uncertain due to environmental concerns (Colyer et al. 2000). The use of nematicides is also being limited as a result of inconsistency in efficacy or economic constraints in production. Factors affecting nematicide efficacy, especially granular nematicides, are product degradation, field selection, application error, warm and moist soil conditions, soil biology, species population, unusual crop growth and variety choice (Anonymous 2013; Noling 2013). The release of nematode-resistant cotton cvs with superior yield potential and high fibre quality (Starr et al. 2007) remains a high priority with the decrease in nematode control options currently available.

12.3.3.2 Biological and Other Agents

Since 1992 various applications of biological control agents, alone or in combination with synthetic nematicides, have been evaluated. These included products that contained: (i) the fungus *P. lilacinum*, (ii) *Trichoderma* spp. combined with a growth stimulant and (iii) three *Bacillus* spp. (viz. *Bacillus chitinosporus*, *Bacillus laterosporus* and *Bacillus licheniformis*). Also, other alternative and ecofriendly products such as chicken manure, an earthworm derivative, chitin and mycorrhizae were included in research trials. Synthetics such as aldicarb, oxamyl, furfural, a phorate-terbufos mixture, fenamiphos as well as abamectin were used either alone or in combination with the biological products (Van Biljon 2007a). None of the biological agents or organic amendments performed satisfactorily on their own.

12.3.3.3 Genetic Host Plant Resistance

Germplasm with genetic resistance to *M. incognita* race 4 has been identified, incorporated and released into local conventional cotton breeding lines. A nematode-tolerant line was selected, and yield quality characteristics were monitored at different localities (Botha-Greeff and Van Biljon 1999). According to Van der Walt (2006), GM cotton covers approximately 92 % of the total area planted in SA. Since GM cotton cvs steadily increased in market share, no further development of the conventional nematode-tolerant lines occurred. Colyer et al. (2000) showed that some transgenic cvs could be more susceptible to root-knot nematodes than their non-transgenic parents. None of the transgenic cvs available in SA have resistance to root-knot nematodes (De Beer 2010).

12.3.3.4 Crop Rotation

A long-term crop rotation trial was done at Jan Kempdorp (Northern Cape Province) to determine whether selected wheat (*Triticum aestivum*) and oat (*Avena sativa*) cvs could be used to effectively reduce damage caused by plant-parasitic nematodes. Although the oat and wheat cvs initially suppressed the *Meloidogyne* spp. populations, they increased again towards the third season. The lowest *Meloidogyne* spp. population density was maintained in roots of the cotton cv. Tetra, followed by the wheat cv. Marico (Joubert and Botha 1999). Although small-grain winter cover crops are hosts to many nematode pests, their reproduction is in most cases suppressed by the prevailing low soil temperatures during winter (Starr et al. 2007).

Table 12.1 Crop cultivars identified with resistance to *Meloidogyne incognita* race 4 (Van Biljon 2004a)

Crop		
Common name	Scientific name	Cultivar
Cabbage	*Brassica oleracea* var. *capitata*	Prize drumhead
Cauliflower	*Brassica oleracea* var. *botrytis*	Snowball, Snowcap
Sorghum (annual fodder)	*Sorghum bicolor*	Superdon
Grass (weeping love)	*Eragrostis curvula*	Ermelo
Grass (blue buffalo)	*Cenchrus ciliaris*	Gayndah
Grass (buffalo)	*Panicum maximum*	Green panic, Mutali
Grass (rye)	*Lolium* sp.	Caversham, Agri-hilton, Nui
Grass (smutsfinger)	*Digitaria eriantha*	Irene
Lucerne	*Medicago sativa*	Beryl, Granada, Topaz
Maize	*Zea mays*	Pantera
Oats	*Avena sativa*	Drakensberg, Heros, Kompasberg, Maluti, Overberg, Perdeberg, Swartberg, Witteberg
Soybean	*Glycine max*	A 7119, Bakgat, Bamboes, Columbus, Crawford, Forrest. PAN494, PAN812
Triticale	Triticosecale	SSKR-1
Wheat	*Triticum aestivum*	Adam Tas, BSP89/24, H1-2 Duzi, Gamtoos, Harts, Inia, SSKR-1, H1-4 Krokodil, Palmiet, UP2H1-Baviaans, UP2H1-3 Kariega, UP2H1-1 Olifants, UP2H1-9, UP2H1-15 SST 876, PAN3478, UPSST 33, SST 55, SST 66, SST 86, UP2H1-6 Steenbras, SST 806, SST 822, SST 825, T4, W84/17

Extensive glasshouse screenings have been conducted to identify suitable rotation crops for cotton producers (Van Biljon 2004a). Root galls, formed by feeding root-knot nematodes, were rated according to an index of 0–5 (Sasser et al. 1982). The degree of resistance of *M. incognita* race 4 was assigned for each cv. according to Canto-Saenz's quantitative scheme (Canto-Saenz and Brodie 1982, 1986). Cultivars of crops with resistance to *M. incognita* race 4 are indicated in Table 12.1. Undisclosed cvs of buckwheat (*Fagopyrum esculentum*), lupin (*Lupinus albus*) and sunn hemp (*Crotalaria juncea*) were also identified with resistance to *M. incognita* race 4 during this study.

The development of effective crop rotation systems is difficult when crop choices increase and fields are infested with many plant-parasitic nematode species. Crops with resistance or tolerance to the root-knot nematode, *M. incognita* race 4, were evaluated for management of plant-parasitic nematodes in organic cotton production for five consecutive seasons. During the first two summers, sesame (*Sesamum indicum*), marigold (*Tagetes erecta*), sunn hemp, oat and cotton were grown. During the winter, oat followed sesame, marigold, sunn hemp, and one of the cotton regimes, while the summer oat was followed by Abyssinian crambe (*Crambe abyssinica*) during spring. The other cotton regime was left fallow during the winter. During the third summer, cotton was grown in all the rotations. The main nematode pest species present at the trial site were *M. incognita* race 4 and *P. zeae*. Sesame-oat, marigold-oat, and sunn hemp-oat rotations reduced *M. incognita* race 4 population densities in this study. *Pratylenchus zeae* population densities, however, increased following sunn hemp in summer. Fortunately, cotton is not a good host for this lesion nematode species. Cotton in the sesame-oat rotation resulted in a higher yield than other rotations (Van Biljon et al. 2015).

12.3.3.5 Disease Complexes

Disease complexes involving nematode pests and fungal pathogens are known to cause significant yield losses in a range of crops (Noling 1999). Disease management strategies are being modified to reduce the amount of damage caused by nematode-fungal complexes. Controlling the nematode component of such an interaction has proven effective in reducing the incidence and severity of Fusarium wilt in some trials in the USA (Davis et al. 2006).

During the 2006/2007 growing season, a field experiment was conducted on a farm in the Kakamas district (Northern Cape Province) to determine whether Fusarium wilt could be controlled by reducing the numbers of *M. incognita* race 4, which were abundant at the site. These two pathogens caused serious damage to the cotton crop previously cultivated on this particular site (see Box 12.1 and Fig. 12.4c). During the 12-week sampling period, an aldicarb treatment (100 g 100 m row^{-1}) resulted in the lowest *M. incognita* race 4 population density followed by a mycorrhiza and fenamiphos combination treatment (150 g seed coat ha^{-1} + 150 g 100 m row^{-1}). The latter gave the highest yield increase of approximately 13 % relative to the untreated control (Van Biljon 2007a).

Fig. 12.6 Mature hemp plants (Jeannie van Biljon, Agricultural Research Council–Institute for Industrial Crops, Rustenburg, South Africa)

12.4 Hemp

Hemp is an annual, herbaceous plant belonging to the family Cannabaceae. Botanists commonly consider it as the only species in the genus *Cannabis* (Ehrensing 1998). Hemp as a fibre crop (Fig. 12.6) is unknown to most local farmers but has been grown in southern Africa for medicinal purposes for centuries. During the past half century, it has also been cultivated as an illegal drug crop, known as marijuana (*Cannabis sativa*) (Dippenaar et al. 1996). In SA, cannabis production was prohibited by the law under the Medical, Dental and Pharmacy Act No. 13 of 1928 (Wynn 1998). Industrial hemp and marijuana, however, are different varieties of the same species. They differ in their psychoactive ingredient, delta-9-tetrahydrocannabinol (THC). Marijuana contains 3–15 % THC on a dry-weight basis, whereas fibre hemp contains less than 1 % of the substance (Vantreese 1997). Hence, the term hemp generally refers to the fibre-producing strain of cannabis. Hemp has been grown for many centuries for the strong fibre produced in its stems. Hemp produces both long, coarse fibres, which extend nearly the entire length of the stalk (primary bast fibres), and short, fine fibres that tend to adhere to the woody core (secondary bast fibres) (Ehrensing 1998). There is renewed interest in hemp as a source of cellulose fibre and seed oil in the western European countries, Australia, the USA and Canada (De Meijer 1995). Since 1994, agricultural trials were initiated in the Rustenburg area that are commissioned and financed by members of the South African Bast Crop Consortium (SABCC) (Wynn 1998).

12.4.1 Plant-Parasitic Nematodes Associated with Hemp

Since 2003, various localities where hemp trials were conducted in the Eastern Cape (Addo, Bathurst, Döhne, Fort Cox, Libode, Mthiza, Qamata and Tsolo) and in the Western Cape (Robertson and Elsenburg) provinces of SA were surveyed for the presence of plant-parasitic nematodes. *Meloidogyne javanica* and *M. incognita* races 2 and 4 were the root-knot nematode species commonly associated with hemp at such sites. Noteworthy also was that roots of cv. Kompolti maintained high population levels of *Pratylenchus* spp. (Dippenaar et al. 1996). To date *P. teres*, *P. zeae*, *P. scribneri* and *P. brachyurus* were recorded from some of the hemp plantings. Ectoparasitic nematode species recorded from hemp plantings in the same surveys included: *Criconema mutabile* (Taylor, 1936) Raski and Luc, 1985; *Geocenamus brevidens* (Allen, 1955) Siddiqi, 1970; *Helicotylenchus dihystera* (Cobb, 1893) Sher, 1961; *Helicotylenchus pseudorobustus* (Steiner, 1941) Golden, 1956; *Helicotylenchus paraplatyurus* Siddiqi, 1972; *Helicotylenchus serenus* Siddiqi, 1963; *Hemicycliophora labiata* Colbran, 1960; *Longidorus pisi*; *Criconemella xenoplax* (Raski 1952) Loof and de Grisse, 1989, *Nanidorus minor* (Colbran, 1956) Siddiqi, 1974; *Paratrichodorus renifer* Siddiqi, 1974; *Paratylenchus projectus* Jenkins, 1956; *R. parvus*; *Rotylenchulus unisexus* Sher, 1965; *Scutellonema brachyurus* (Steiner, 1938) Andrassy, 1958; *Scutellonema truncatum* Sher, 1963; *Telotylenchus ventralis* Loof, 1963; *Quinisulcius capitatus* Allen, 1955; and a *Xiphinema* sp. (Van Biljon 2003, 2004b, 2005b; SAPPNS).

12.4.2 Control Measures

12.4.2.1 Genetic Host Plant Resistance

Various hemp cvs (Diana, F-17, F-75, Futura 75, Kompolti, Kubanskaia Rannaja, Multiseed Novosadska, Unico-B, VIR-140, USO 31) were screened for resistance to the root-knot nematode species *M. javanica* and *M. incognita* races 2 and 4 in glasshouse experiments. The host suitability of most cvs varied between tolerant and susceptible, while three (Futura 75, Diana and Kubanskaia Rannaja) were resistant to *M. incognita* race 2. Results from another study suggested that cvs Kompolti, Futura 75, Felina and Ferrimon were tolerant to *M. javanica* (Pofu et al. 2010). Although some of the hemp cvs best adapted for local conditions are those with tolerance against these root-knot nematode species and races, the nematode reproduction factor is still too high to use the tolerant cvs as a tool to ensure sustainable crop production. Besides planting a poor-host hemp cv., additional nematode control measures will still have to be implemented to optimise the reduction in numbers of root-knot nematodes in infected plantings (Van Biljon 2005c; Pofu et al. 2010).

12.4.2.2 Chemical Control

No nematicides are specifically registered for use on hemp. A nematicide treatment tested on hemp, a phorate and terbufos combination as well as individual applications of aldicarb and oxamyl decreased root-knot and lesion nematode numbers in roots and increased yields (Van Biljon 2005b, 2006, 2007b). In another study, fenamiphos and the phorate and terbufos combination decreased lesion nematode populations (Van Biljon 2009). Hemp treated with aldicarb and oxamyl produced higher yields than when treated with other nematicides (Van Biljon 2007b).

12.4.2.3 Organic Amendments

Among the various organic amendments tested, chicken manure at a dosage rate of 6 MT ha^{-1} increased the yield of hemp cv. Kompolti in a field where high infestation levels of *Pratylenchus* spp. occurred (15 individuals 250 ml $soil^{-1}$ before planting) (Van Biljon 2007b).

12.5 Kenaf

Kenaf is a fibre plant native to east-central Africa (LeMahieu et al. 1991) and is related to okra, jute (*Corchorus capsularis* or *C. olitorius*) and cotton, which all belong to the Malvaceae family. The bark of kenaf, which contains long, soft bast fibres, constitutes 30–40 % (dry weight) of the stem. The kenaf plant has an ideal blend of long and short fibres (LeMahieu et al. 1991), is grown throughout the tropics and subtropics (McSorley and Parrado 1986) and is well adapted to a wide range of soil types (El Bassam 1998).

Kenaf has two distinct leaf shapes, viz. palmatified and entire. The palmatified-shaped leaf closely resembles marijuana and can be mistaken for the illegal weed as is the case with hemp (Baldwin et al. 2006). Sustainable Fibre Solutions (SFS), a joint venture between The Seardel Investment Corporation Ltd and the Industrial Development Corporation (IDC), was the first to successfully cultivate kenaf in SA (Anonymous 2009b). The crop is largely produced and processed near Winterton (KwaZulu-Natal Province) (Anonymous 2012b).

12.5.1 Plant-Parasitic Nematodes Associated with Kenaf

Meloidogyne is a common constraint in kenaf production worldwide (Schieber et al. 1961; Wilson and Summers 1966; Minton et al. 1970; Tu and Cheng 1971; Minton and Adamson 1979; Lawrence 1994; Botha-Greeff 2002a). *Meloidogyne incognita* is a severe pest of kenaf and is capable of reducing the growth, yield and harvesting efficiency of the crop. In some areas, this species is known to predispose kenaf to pathogenic soilborne fungi (Cook and Mullin 1994). Root-knot nematodes were recorded in association with kenaf at Rustenburg (North-West Province), Addo (Eastern Cape Province), in the El Salvador Seed Production trials (Bloemfontein, Free State Province) (Botha-Greeff 2002b) and on three farms in the Winterton area (KwaZulu-Natal Province) (Botha-Greeff et al. 2004).

In cotton growing areas worldwide, the Meloidogyne-Fusarium wilt complex is also expected to limit yield potential of both cotton and kenaf (Coetzee 2004). Susceptible root-knot nematode kenaf cvs permit population densities of *M. javanica* and *M. incognita* races 2 and 4 to rapidly increase to levels that are disastrous for the production of other root-knot nematode susceptible crops.

Other plant-parasitic nematode species identified from soil and root samples collected from kenaf crops in the Winterton area are: *S. brachyurus*; *P. zeae*; *H. dihystera*; *H. paraplatyurus*; *Helicotylenchus microcephalus* Sher, 1966; *N. minor*; *R. parvus*; *R. unisexus*; and *Mesocriconema sphaerocephalum* (Taylor, 1936) Loof and De Grisse, 1989 (Botha-Greeff et al. 2004).

12.5.2 Management Strategies

A population density of 100 root-knot nematode J2 cm^{-3} soil during planting can reduce the yield of susceptible kenaf cvs by 32 %. At population densities of 500 J2 cm^{-3}, reductions in yield as high as 67 % can be expected. It is therefore recommended that a nematode analysis be conducted on all fields where it is planned to grow kenaf. Options for nematode management include resistant cvs, crop rotation and the use of nematicides (Lawrence 1994; Yu 1994).

12.5.2.1 Genetic Host Plant Resistance

Various kenaf cvs (Cuba 108, Tainung 2, Everglades 41, El Salvador, SF 459, Gregg, Dowling, Endora, Whitten and Everglades 71) were identified as susceptible to *M. incognita* race 4 in glasshouse trials. However, cvs Cuba 108, Tainung 2, Everglades 41, El Salvador and SF 459 varied in their host suitability between being tolerant or susceptible to monoculture populations of *M. javanica* and *M. incognita* race 2 (Botha-Greeff 2002b).

12.6 Flax

Flax belongs to the genus *Linum*, which is one of the 10 genera that comprise the family Linaceae (Rowland 1996). Cultivated flax, *Linum usitatissimum*, is one of the oldest fibre crops that were extensively cultivated in ancient Egypt (Anonymous 2009c).

Dyed flax fibres found in a cave in ancient Georgia (USA) dated back 30,000 years (Anonymous 2009c). Flax remnants were also found in Stone Age dwellings in Switzerland (Berglund and Zollinger 2007). Flax is the source of industrial fibre and, as currently processed, results in long-line and short fibres (Van Sumere 1992), which are used to produce linen while seeds yield linseed oil (Duke 1983).

The linseed plant has a bushy nature and is approximately 80 cm high, whereas fibre flax varieties are almost unbranched and can reach a height of 1.2 m (Jacobsz and Van der Merwe 2007). Fibre flax grows best in a cool, moist climate (Sharma and van Sumere 1992; Elhaak et al. 1999; Jacobsz and Van der Merwe 2007).

12.6.1 Plant-Parasitic Nematodes Associated with Flax

Although various plant-parasitic nematode species have been found in association with flax, no above- or below-ground symptoms have ever been reported (Rashid 2003). Interestingly, some parts of the flax plants could possibly be used for nematode control. A substance toxic to eggs and J2 of *Meloidogyne* was extracted from mature, dried oat straw and leaves and stems of flax (Johnson 1974).

12.6.2 Management Strategies

12.6.2.1 Host Plant Resistance

Glasshouse screenings to determine the host status of flax cvs to *M. javanica* showed that Ariane, Diane, Elise and Hermes were all susceptible. Cultivars G7 and G8 were identified as tolerant (Van Biljon 2004a). No data are available as to the host status of these cvs to *M. incognita* races 2 and 4.

References

Aboul-Eid HZ (1970) Systematic notes on *Longidorus* and *Paralongidorus*. Nematologica 16:159–179

Adams R, Geissman TA, Edwards JD (1960) Gossypol, a pigment of cottonseed. Chem Rev 60:555–574

Agenbag M (2016) Identification and reproduction potential of South African *Meloidogyne* species. MSc Dissertion, North-West University, Potchefstroom

Akehurst BC (1968) Tobacco. Longmans Green & Co, London

Albers D (1993) Cotton plant development and plant mapping. Agricultural publication G4268 – University of Missouri Extension, United States of America. http://extension.missouri.edu/p/G4268. Accessed 15 Aug 2012

Anonymous (1971) Report of the Secretary of Agriculture, Rhodesia, for the period 1st Oct 1970 to the 30th Sept 1971. Department of Research and Specialist Services, Rhodesian Ministry of Agriculture, Salisbury

Anonymous (2005) Syngenta to launch breakthrough seed treatment in the United States of America. Media release. In: AgProfessional Magazine. Basel, 7 Mar 2005

Anonymous (2008) Gossypol as undesirable substance in animal feed – scientific opinion of the panel on contaminants in the food chain. Eur Food Saf Auth J 908:1–55

Anonymous (2009a) Market information in the commodities area. Cotton pests and diseases. UNCTAD. http://www.unctad.org/infocomm/anglais/cotton/crop.htm. Accessed 10 Sept 2015

Anonymous (2009b) Sustainable Fibre Solutions (Pty.) Ltd. http://www.kenaf.co.za. Accessed 2 Sept 2015

Anonymous (2009c) Flax. http://en.wikipedia.org/wiki/Flax. Accessed 10 Sept 2015

Anonymous (2012a) Endosulfan ban – South Africa. Online. http://www.cgcsa.co.za. Accessed 15 Oct 2014

Anonymous (2012b) Kenaf. Department of Agriculture, Forestries and Fisheries. Directorate Communication Services, Pretoria

Anonymous (2013) Factors affecting nematicide efficacy http://www.syngenta-crop.co.uk.pdfs. Accessed 28 Jan 2016

Anonymous (2015) Velum® Prime. http://www.cropscience.bayer.co.za/en/Products/Fungicides/Velum-Prime.aspx. Accessed 18 Dec 2015

Arens ML, Rich JR (1981) Yield response and injury levels of *Meloidogyne incognita* and *M. javanica* on the susceptible tobacco cultivar 'McNair 944'. J Nematol 13:196–201

Baird RE, Rich JR, Herzog GA et al (2000) Management of *Meloidogyne incognita* in cotton with nematicides. Nematol Mediter 28:255–259

Baldwin BS, Hollowell JE, Mosley JW et al (2006) Registration of "Whitten" kenaf. Crop Sci 46:988–989

Berglund DR, Zollinger RK (2007) Flax production in North Dakota. North Dakota State University A-1038 (revised). http://www.ag.ndsu.edu/pubs/plantsci/crops/a1038.pdf. Accessed 10 Sept 2015

Botha MS, Van Biljon ER (1996) Identification by the Agricultural Research Council–Tobacco and Cotton Research Institute. Cotton SA Publication, Rustenburg, pp 84–99

Botha-Greeff MS (2002a) Nematodes on kenaf (*Hibiscus cannabinus*) in South Africa. In: Proceedings of the 3rd SANCRA Symposium, Nelspruit, 26–28 June 2002

Botha-Greeff MS (2002b) Evaluation of nematode interaction with kenaf cultivars. Annual ARC-IIC Kenaf project report 2001/2002, Agricultural Research Council–Institute for Industrial Crops, Rustenburg

Botha-Greeff MS, Van Biljon ER (1999) Integrated nematode control on cotton in South Africa: present status. Proceedings of the Beltwide Cotton conference, San Antonio, 5–8 Jan 1999

Botha-Greeff MS, Swart WJ, Labuschagne M (2004) Nematode survey on kenaf (*Hibiscus cannabinus*) planted on field conditions in KZN. Project Report, Plant Production, UFS 2003–2004 season

Bridge J, Muller RLJ (1984) Role of nematodes as constraints to agricultural development in Africa. Advancing agricultural production in Africa. In: Proceedings of CAB's 1st scientific conference, Arusha, Tanzania, Commonwealth Agricultural Bureax, Farnham Royal, Slough, 12–18 Feb 1984

Bridge J, Jones E, Page SLJ (1976) *Meloidogyne acronea* associated with reduced growth of cotton in Malawi. Plant Dis Rep 60:5–7

Brito JA, Stanley JD, Cetintas R et al (2004) Identification and host preference of *Meloidogyne mayaguensis*, and other root-knot nematodes from Florida, and their susceptibility to *Pasteuria penetrans*. J Nematol 36:308–309

Brito JA, Stanley JD, Cetintas R et al (2005) A new root-knot nematode infesting vegetables in Florida. J Nematol 37:359

Brubaker CL, Bourland FM, Wendel JF (1999) The origin and domestication of cotton. In: Smith CW, Cothren JT (eds) Cotton: origin, history, technology and production. Wiley, New York

Buchanan R (1999) A weaver's garden: growing plants for natural dyes and fibres. Dover Publications, Mineola, New York

Canto-Saenz M, Brodie BB (1982) Factors affecting host efficiency of potatoes to *Meloidogyne incognita*. J Nematol 14:433–434

Canto-Saenz M, Brodie BB (1986) Host efficiency of potato to *Meloidogyne incognita* and damage threshold densities on potato. Nematropica 16:109–116

Carta LK, Handoo ZA, Skantar AM et al (2002) Redescription of *Pratylenchus teres* Khan & Singh, 1974 (Nemata: Pratylenchidae), with the description of a new subspecies from South Africa, and a phylogenetic analysis of related species. Afr Plant Prot 8:13–24

Chen Z, Scheffler B, Dennis E (2007) Towards sequencing cotton (*Gossypium*) genomes. Plant Physiol 145:1303–1310

Christie JR, Perry VG (1951) A root disease of plants caused by a nematode of the genus *Trichodorus*. Science 113:491–493

Coetzee V (1956) *Meloidogyne acronea*, a new species of root-knot nematode. Nature 177:899–900

Coetzee R (2004) Characterization of kenaf (*Hibiscus cannabinus* L.) cultivars in South Africa. MSc dissertation, Faculty of Natural and Agricultural Sciences, Department of Plant Science, University of the Free State, Bloemfontein

Coetzee V, Botha HJ (1965) A rediscription of *Hypsoperine acronea* (Coetzee, 1956) Sledge and Golden, 1964 (Nematoda: Heteroderidae) with a note on its biology and host specificity. Nematologica 11:480–484

Collins WK, Hawks SN Jr (1993) Principles of flue-cured tobacco production. North Carolina State University, Raleigh

Colyer PD, Kirkpatrick TL, Caldwell WD et al (2000) Root-knot nematode reproduction and root galling severity on related conventional and transgenic cotton cultivars. J Cott Sci 4:232–236

Cook CG, Mullin BA (1994) Growth responses of kenaf cultivars in root-knot nematode/soilborne fungi infested soil. Crop Sci 34:1455–1457

Cotton SA (2014) Hectares planted and yields for the Republic of South Africa. http://www.cottonsa.org.za. Accessed 22 Oct 2014

Davis RM, Colyer PD, Rothrock CS et al (2006) Fusarium wilt of cotton: population diversity and implications for management. Plant Dis 90:692–703

De Beer EFL (2010) A study of the effect of abamectin and different cultivars on root-knot nematode infection of cotton. MSc Dissertation, North West University, Potchefstroom Campus, Potchefstroom

De Beer EFL, Fourie H, Mc Donald AH et al (2007) Evaluation of abamectin seed treatment as a potential nematicide for plant-parasitic nematodes in cotton. S Afr J Plant Soil 24:233–260

De Meijer EPM (1995) Fibre hemp cultivars: a survey of origin, ancestry, availability and brief characteristics. J Int Hemp Assoc 2:66–73

Deubert KH, Norgren RL, Paracer SM et al (1967) The influence of *Tylenchorhynchus agricola* and *Tylenchorhynchus claytoni* on corn roots under gnotobiotic conditions. Nematologica 13:56–62

Di Vito M, Greco N, Carella A (1983) The effect of population densities of *Meloidogyne incognita* on yield of cantaloupe and tobacco. Nematol Medit 11:169–174

Dippenaar MC, Du Toit CLN, Botha-Greeff MS (1996) Response of hemp (*Cannabis sativa* L.) varieties to conditions in Northwest province, South Africa. J Int Hemp Ass 3:63–66

Duke JA (1983) Handbook of energy crops. Electronic publication. www.hort.purdue.edu/newcrop/duke_enery/dukeindex.html. Accessed 21 Nov 2015

Ehrensing DT (1998) Feasibility of industrial hemp production in the United States Pacific Northwest. Oregon State University Extension Service Station Bulletin 681. http://cropandsoil.oregonstate.edu/bioenergy/sites/default/files/Hemp/body.html. Accessed 15 Aug 2012

El Bassam N (1998) Energy plant species: their use and impact on environment and development. James & James Science Publishers Ltd, London

Elhaak M, El-Shourbagy N, El-Nagar R (1999) Effect of edaphic factors on technical properties of flax fibre. J Agron Crop Sci 182:113–120

FAO (2016) Statistics. http://faostat.fao.org/. Accessed 5 Jan 2016

Faske TR, Starr JL (2006) Sensitivity of *Meloidogyne incognita* and *Rotylenchulus reniformis* to abamectin. J Nematol 38:240–244

Faske TR, Starr JL (2007) Cotton root protection from plant-parasitic nematodes by abamectin-treated Seed. J Nematol 39:27–30

Garner WW (1947) The production of tobacco. The Blakiston Company, Philadelphia

Graham TW (1951) Nematode root rot of tobacco and other plants. South Carolina Agriculture Experiment Station Bulletin, South Carolina

Hartman KM, Sasser JN (1985) Identification of *Meloidogyne* species on the basis of differential hosts test and perineal pattern morphology. In: Barker KR, Carter CC, Sasser JN (eds) An advanced treatise on *Meloidogyne*. North Carolina State University Graphics, Raleigh, pp 69–77

Jacobsz MJ, Van Der Merwe WJC (2007) Production guidelines for flax (*Linum usitatissimum* L.). Directorate Communication Services, Department of Agriculture, Forestry and Fisheries, Pretoria

Jansen Van Vuuren R, Bunting E, Steyn A (2005) A study to determine the virulence of root-knot nematode eggs after treatment with various concentrations of Crop Guard® (furfural). Afr Plant Prot 12:34–35

Johnson LF (1974) Extraction of oat straw, flax, and amended soil to detect substances toxic to the root-knot nematode. Phytopathol 64:1471–1473

Jones JT, Haegeman A, Danchin EGJ et al (2013) Top 10 plant-parasitic nematodes in molecular plant pathology. Mol Plant Pathol 14:946–961

Joubert AJ, Botha MS (1999) Rotational crops for the cotton root-knot nematode in South Africa. Afr Plant Prot 5:15

Kinloch RA, Rich JR (1998) Response of cotton yield and *Meloidogyne incognita* soil populations to soil applications of aldicarb and 1,3-D in Florida. J Nematol 30:639–642

Kirkpatrick TL (2007) Innovations in chemical control. In: Proceedings of the World Cotton Research Conference – 4, Lubbock Memorial Civic Centre, Lubbock, 10–14 Sept 2007

Kirkpatrick TL, Sasser JN (1984) Crop rotation and races of *Meloidogyne incognita* in cotton root-knot management. J Nematol 16:323–328

Kleynhans KPN, Van den Berg E, Swart A et al (1996) Plant nematodes in South Africa. Plant Protection Research Institute handbook No. 8. ARC-Plant Protection Research Institute, Pretoria

Lamberti F (1979) Economic importance of *Meloidogyne* spp. in subtropical and mediterranean climates. In: Lamberti F, Taylor CE (eds) Root-knot nematodes (*Meloidogyne* species) systematics, biology and control. Academic Press, London, pp 341–357

Lawrence GW (1994) Plant parasitic nematodes – Pests of kenaf. In: Fowler MJ (ed) A summary of Kenaf production and product development research 1989–1993. Bulletin 1011. Agricultural and Forestry Experiment Station, Mississippi, pp 13–14

Lawrence KS, Burmester CH, Lawrence GW et al (2006) Evaluation of Avicta formulations as compared to Temik 15G for reniform nematode management in cotton in north Alabama. Fungicide and nematicide tests. Report 61: No.15

LeMahieu PJ, Oplinger ES, Putnam DH (1991) Alternative field crops manual. University of Wisconsin-Extension, Cooperative Extension; University of Minnesota: Centre for alternative Plant and Animal Products and the Minnesota Extension Service, Minnesota. http://onlinebooks.library.upenn.edu/webbin/book/lookupid?key=olbp10537. Accessed 2 Sept 2015

Louw I (1982) Nematode pests of cotton. In: Keetch DP, Heyns J (eds) Nematology in southern Africa, Science Bulletin No 400. Department of Agriculture and Fisheries, Government Printer, Pretoria

MacGowan JB (1980) *Tylenchorhynchus claytoni* Steiner 1937 the tobacco stunt nematode. Nematology circular no. 67. Florida Department of Agriculture and consumer Services, Division of Plant Industry, Florida

Maqbool MA (1992) Distribution and host association of plant-parasitic nematodes in Pakistan. National Nematological Research Centre, University of Karachi, Karachi

McSorley R, Parrado JL (1986) Relationship between height of kenaf and root galling by *Meloidogyne incognita*. Nematrop 16:205–211

Miller C (2007) Herbifume (metam sodium), a suitable replacement for methyl bromide in tobacco seedbed sterilisation in Malawi. Afr J Plant Soil 24:233–260

Milne DL (1972) Nematodes of tobacco. In: Webster JM (ed) Economic Nematology. Academic Press, London, pp 159–186

Minton NA, Adamson WC (1979) Control of *Meloidogyne javanica* and *M. arenaria* on Kenaf and roselle with genetic resistance and nematicides. J Nematol 11:37–41

Minton NA, Adamson WC, White GA (1970) Reaction of Kenaf and roselle to three root-knot nematode species. Phytopathol 12:1844–1845

Monfort WS, Kirkpatrick TL, Long DL et al (2006) Efficacy of a novel nematicidal seed treatment against *Meloidogyne incognita* on cotton. J Nematol 38:245–249

Neethling D (2004) The commercialisation of *Paecilomyces lilacinus* as an agent for the control of plant-parasitic nematodes. In: Proceedings of the 4th International Congress of Nematology, Tenerife, 8–13 June 2002

Ng'ambi BS, Rufty RC, Barker KR, Melton TA (1999) Identification of sources of resistance to four species of root-knot nematodes in tobacco. J Nematol 31:272–282

Noling JW (1999) Nematode management in commercial vegetable production. University of Florida, IFAS extension ENY-014. http://edis.ifas.ufl.edu/pdffiles/NG/NG00400.pdf. Accessed 13 May 2015

Noling JW (2013) Movement and toxicity of nematicides in the plant root zone. University of Florida, IFAS extension ENY-041. http://edis.ifas.ufl.edu. Accessed 30 Jan 2016

Onkendi EM, Kariuki GM, Marais M et al (2014) The threat of root-knot nematodes (*Meloidogyne* spp.) in Africa: a review. Plant Pathol 68:727–737

Page SLJ, Bridge J (1994) The African cotton-root nematode, *Meloidogyne acronea*; its pathogenicity and intra-generic infectivity within *Gossypium*. Fund Appl Nematol 17:67–73

Phipps PM, Partridge DE, Eisenback JD (2006) Efficacy of abamectin (A16006) on seed and Temik 15G in-furrow for root-knot nematode control in cotton, 2005. Fungicide Nematicide Tests 61:N003

Pofu MK, Van Biljon ER, Mashela WP et al (2010) Responses of selected fibre hemp cultivars to *Meloidogyne javanica* under greenhouse conditions. Am Eurasian J Agric Environ Sci 9:509–513

Powell NT (1971) Interactions between nematodes and fungi in disease complexes. Annu Rev Phytopathol 9:253–273

Rashid KR (2003) Principal diseases of flax. In: Muir A, Westcot D (eds) Flax: the genus Linum. CRC Online Press, Taylor and Francis Group, Boca Raton, Oxford

Rich JR, Dunn RA (1985) Root-knot disease in Florida tobacco. Nematology Circular No. 122. Department of Agriculture & Consumer Services, Division of Plant Industry, Florida. http://edis.ifas.ufl.edu/pdffiles/NG/NG01900.pdf. Accessed 21 Nov 2015

Rich JR, Kinloch RA (2005) Tobacco nematode management. Fact sheet ENY-005 (NG019), (Revised December 2005). Department of Entomology and Nematology, Florida Cooperative Extension Service, Institute of Food and Agricultural Sciences, University of Florida, Gainesville

Ritchie GL, Bednarz CW, Jost PH et al (2007) Cotton growth and development. Bulletin 1253 (Revised 2007). The University of Georgia Cooperative Extension. Colleges of Agricultural and Environmental Sciences and Family and Consumer Services, Georgia

Robinson AF (2007) Nematode management in cotton. In: Ciancio A, Mukerji KG (eds) Integrated Management of plant pests and diseases. Integrated management and biocontrol of vegetables and grain crops nematodes. Springer, Dordrecht, pp 149–184

Rohde RA, Jenkins WR (1957) Host range of a species of *Trichodorus* and its host-parasite relationship on tomato. Phytopathol 47:295–298

Rowland G (1996) Growing flax-production, management and diagnostic guide. Flax Council of Canada and the Saskatchewan Flax Development Commission, 4th edn. Laird Books, Saskatchewan

Sasser JN, Carter CC, Taylor AL (1982) A guide to the development of a plant nematology program. Cooperative Publication, Department Plant Pathology, North Carolina State University and the United States Agency of Interior Development, Raleigh

Scheffler JA, Romano GB (2008) Modifying gossypol in cotton (*Gossypium hirsutum* L.): a cost effective method for small seed samples. J Cott Sci 12:202–209

Schieber E, Sosa ON, Escobar P (1961) Root-knot nematode on kenaf in Guatemala. Plant Dis Rep 45:119

Sharma H, Van Sumere C (1992) The biology and processing of flax. M Publications, Belfast

Shepherd JA, Barker KR (1990) Nematode Parasites of Tobacco. In: Luc M, Sikora RA, Bridge J (eds) Plant parasitic nematodes in subtropical and tropical agriculture. CAB International, Wallingford, pp 493–517

Starr JL, Veech JA (1986) Comparison of development, reproduction and aggressiveness of *Meloidogyne incognita* races 3 and 4 on cotton. J Nematol 18:413–415

Starr JL, Carneiro RG, Ruano O (2005) Nematode parasites of cotton and other tropical crops. In: Luc M, Sikora RA, Bridge J (eds) Plant parasitic nematodes in subtropical and tropical agriculture. CAB International, Wallingford, pp 733–750

Starr JL, Koenning SR, Kirkpatrick TL et al (2007) The future of nematode management in cotton. J Nematol 39:283–294

Swanepoel A (2005) Survey of Fusarium wilt in cotton in the cotton producing areas of the Northern Cape Province. Annual ARC-IIC Project Report 2004–2005

Taylor HW (1927) Tobacco culture with special reference to South African conditions. Central News Agency, Hugh Wilbur Taylor Publisher, Johannesburg

Thomas SH, Smith DW (1993) Effects of 1,3-dichloropropene for *Meloidogyne incognita* management on cotton produced under furrow irrigation. J Nematol 25:752–757

Tu CC, Cheng YH (1971) Interaction with *Meloidogyne javanica* and *Macrophomina phaseoli* in kenaf root rot. J Nematol 3:39–42

Van Biljon ER (1992) Ontwikkeling van 'n tegniek vir die massateelt van *Paratrichodorus*, en bepaling van die invloed van hierdie plaag op tabak en katoen. MSc Thesis, Rand Afrikaans University, Johannesburg

Van Biljon ER (2003) Survey of the various trial sites in the Eastern Cape to determine the nematodes species found in association with hemp. Annual ARC-IIC Hemp project report 2002–2003, Agricultural Research Council–Institute for Industrial Crops, Rustenburg

Van Biljon ER (2004a) Crop rotation as part of an integrated pest management programme for the control of plant parasitic nematodes. In: Proceedings of the Soilborne Plant Diseases Interest Group symposium, Agricultural Research Council–Institute for Industrial Crops, Stellenbosch, 14–16 Sept 2004

Van Biljon ER (2004b) Survey of the various trial sites in the Eastern Cape to determine the nematodes species found in association with hemp. Annual ARC-IIC Hemp project report 2003–2004, Agricultural Research Council–Institute for Industrial Crops, Rustenburg

Van Biljon ER (2005a) Integrated nematode control strategy. The chemical and/or biological control of nematodes in the various cotton production areas. Annual ARC-IIC Cotton project report 2004–2005, Agricultural Research Council–Institute for Industrial Crops, Rustenburg

Van Biljon ER (2005b) Determination of the effect of various nematicides and other control measures for nematodes found in association with hemp under field conditions and a survey of the various trial sites in the Eastern and Western Cape to determine the nematodes species found in association with hemp. Annual ARC-IIC Hemp project report 2004–2005, Agricultural Research Council–Institute for Industrial Crops, Rustenburg

Van Biljon ER (2005c) Evaluation of the resistance of various hemp cultivars against *Meloidogyne javanica* and *M. incognita* races 2 and 4 under greenhouse conditions. Annual ARC-IIC Hemp project report 2004–2005, Agricultural Research Council–Institute for Industrial Crops, Rustenburg

Van Biljon ER (2006) Determination of the effect of various nematicides and other control measures for nematodes found in association with hemp under field conditions and a survey of the various trial sites in the Eastern and Western Cape to determine the nematodes species found in association with hemp. Annual ARC-IIC Hemp project report 2005–2006, Agricultural Research Council–Institute for Industrial Crops, Rustenburg

Van Biljon ER (2007a) Integrated nematode control strategy. The chemical and/or biological control of nematodes in the various cotton production areas. Annual ARC-IIC Cotton project report 2006–2007, Agricultural Research Council–Institute for Industrial Crops, Rustenburg

Van Biljon ER (2007b) Nematode management for hemp small scale farmers. Annual ARC-IIC Hemp project report 2006–2007, Agricultural Research Council–Institute for Industrial Crops, Rustenburg

Van Biljon ER (2009) Nematode management for hemp small scale farmers. Annual ARC-IIC Hemp project report, 2008–2009. Agricultural Research Council–Institute for Industrial Crops, Rustenburg

Van Biljon ER (2010) Root-knot nematode resistance evaluation of different air/flue cured tobacco breeding lines and cultivars to the principal root-knot nematode species and races. Annual ARC-IIC Tobacco project report 2009–2010, Agricultural Research Council–Institute for Industrial Crops, Rustenburg

Van Biljon ER, Botha MS (1998) Evaluation of *Paecilomyces lilacinus* with and without a growth stimulant for the control of *Meloidogyne javanica* in flue-cured tobacco within an integrated control system. In: Proceedings of the 36th South African Society for Plant Pathology Congress – Champagne Sports Resort, KwaZulu-Natal, 16–18 Jan 1998. J Plant Sci 94:xviii

Van Biljon ER, Botha MS (1999) The role of crop rotation in an integrated control system against nematodes. Afr Crop Sci Proc 4:541–545

Van Biljon ER, Meyer AJ (2000) Reproduction of *Pratylenchus zeae* and *P. delattrei* on crops resistant to *Meloidogyne javanica* and *M. incognita* races 2 and 4. Afr Plant Prot 6:43–45

Van Biljon ER, Swanepoel A (2007) Plant-parasitic nematode species associated with cotton in the two major cotton-growing regions of the Northern Cape Province of South Africa. S Afr J Plant Soil 24:233–260

Van Biljon ER, Mc Donald AH, Fourie H (2015) Population responses of plant-parasitic nematodes in selected crop rotations over five seasons in organic cotton production. Nematropica 45:102–112

Van der Walt WJ (2006). Adoption of GM crops in South Africa in 2006. http://www.seedquest.com/News/releases/2006/may/15694.htm. Accessed 3 Oct 2016

Van Sumere C (1992) Retting flax with special reference to enzyme retting. In: Sharma H, van Sumere C (eds) The biology and processing of flax. M publications, Belfast, pp 157–198

Vantreese VL (1997) Industrial hemp: global markets and prices. University of Kentucky, Department of Agricultural Economics, Lexington. http://www.uky.edu/Classes/GEN/101/Hemp/HEMP98.PDF. Accessed 19 Oct 2013

Van Zyl K (2013) A guide to crop pest management in South Africa. A compendium of acaracides, insecticides, nematicides, molluscicides, avicides and rodenticides, 1st edn. VR Print, Pinetown

Verdoorn G (2012) Endosulfan dies a sudden death. Online http://www.pcsib.org.za/wp-content. Accessed 15 Oct 2015

Walker NR, Kirkpatrick TL, Rothrock CS (1998) Interaction between *Meloidogyne incognita* and *Thielaviopsis basicola* on cotton (*Gossypium hirsutum*). J Nematol 30:415–422

Willers P (1997) First record of *Meloidogyne mayaguensis* Rammah & Hirschmann, 1988: Heteroderidae on commercial crops in the Mpumalanga Province. Inligting bulletin, Agricultural Research Council–Institute for Tropical and Subtropical Crops 294:19–29

Wilson FD, Summers TE (1966) Reaction of kenaf, roselle and related species of *Hibiscus* to root-knot nematodes. Phytopathol 56:687–690

Wynn J (1998) South African Hemp Feasibility Report. Commissioned by Interim Task Team on Bast Fibre Crops, South African Plant Fibre Cluster. http://www.sahc@icon.co.za. Accessed 19 Oct 2009

Yu Y (1994) The application of integrated control technique of kenaf root-knot nematode. Guangxi Agric Sci 1:24–25

Chapter 13
Nematode Pests of Citrus

Mathys C. Pretorius and Hendrik F. Le Roux†

13.1 Introduction

Citrus is the most important fruit crop in the world in terms of production, with 240,780 million metric tons (MT) produced in 2013 (FAO 2016). Citrus is grown in more than 125 countries located within the 40° latitude lines north and south of the equator (Davies and Albrigo 1994; Spiegel-Roy and Goldschmidt 1996). The major world citrus-producing countries include China, Brazil, the United States of America (USA), Mexico, India, Spain, Italy, Iran, Egypt and Turkey. In Africa, the major citrus-producing countries are Algeria, Egypt, Morocco, South Africa (SA) (the 12th largest producer worldwide), Swaziland, Tunisia and Zimbabwe. In tropical Africa, citrus is mainly grown by smallholding farmers for local consumption. Although in relatively small volumes, Swaziland and Zimbabwe are the only southern African countries besides SA that produce citrus for export (Pretorius 2005). Botswana produces citrus for local consumption on fewer than 1,000 hectares (ha) along the Limpopo River, and Namibia has citrus orchards (<500 ha) in the Tsumeb/Grootfontein area. Angola and Mozambique used to have large citrus groves while they were Portuguese colonies. Most of these orchards were neglected during their respective revolutions, and there is now only one farm left in Mozambique that is exporting citrus. Angola, however, has started to re-establish some citrus orchards. Zambia has very few citrus orchards and only produces fruit for local consumption. Citrus grown closer to the equator does not colour properly and is hence only produced for local consumption.

Citrus, originally imported from St. Helena Island in 1654, is grown in SA between the 22.5 and 33.8 °S latitudes (Cartwright 1977). Local citrus production is

†Author was deceased at the time of publication

M.C. Pretorius (✉) • H.F. Le Roux
Citrus Research International, PO Box 28, Nelspruit 1200, South Africa
e-mail: mc@cri.co.za

H. Fourie et al. (eds.), *Nematology in South Africa: A View from the 21st Century*, DOI 10.1007/978-3-319-44210-5_13

limited to irrigated areas. Production occurs in the following provinces: Limpopo (16,255 ha), Mpumalanga (11,681 ha), Eastern Cape (12,923 ha), KwaZulu-Natal (4,004 ha), Western Cape (9,524 ha) and Northern Cape (639 ha) (Burger 2009). The SA citrus industry is export orientated, with total exports averaging 65 % of the total production, while processing and local consumption constitute 25 and 10 %, respectively (Siphugu 2009). In 2013, SA was the second largest exporter of fresh citrus, exporting 1.54 million MT of citrus compared to the world's number one exporter, Spain, that exported 3.86 million MT of citrus. Although production is relatively small compared to other countries, the citrus industry contributes significantly to the local economy. In 2015, 113 million 16-kg cartons of citrus were shipped from SA to more than 50 countries worldwide. This earned an income of more than ZAR8 billion for the country.

The citrus cultivars (cvs.) grown in southern Africa include sweet orange (*Citrus sinensis*), grapefruit (*C. paradisi*), lemon (*C. limon*), lime (*C. aurantifolia*) and mandarin (*C. reticulata*). They also include hybrids such as tangelo (a cross between grapefruit and tangerine) and rootstocks such as rough lemon (*C. jambhiri*), Volkamer lemon (*C. volkameriana*) and trifoliate orange (*Poncirus trifoliata*), as well as their hybrids Troyer citrange, Carrizo citrange, C35 citrange and Swingle citrumelo (Saunt 2000). The SA citrus industry was established on rough lemon (*Citrus jambhiri*) rootstock, a vigorous grower but highly susceptible to the citrus nematode, *Tylenchulus semipenetrans* Cobb 1913 (Castle et al. 1993). This orchard practice has changed considerably in recent years. Whereas almost 90 % of trees older than 30 years were budded on rough lemon rootstock, only 15 % of those planted in 1991 and 10 % of those planted in 2015 were budded on rough lemon rootstock.

Citrus nursery trees were produced in open soil for centuries, often between old nematode-infected citrus trees used for commercial production. Symptoms of citrus nematode infection are not apparent in young nursery trees, even when severe infections occur, with the result that infected nursery stock was often planted in new orchards (Cohn 1965). The citrus nematode, which is not indigenous to SA, has consequently spread to all citrus-growing areas in the country (De Villiers and Milne 1976).

13.2 Plant-Parasitic Nematodes Associated with Citrus

Milne (1982) gives an extended overview of the citrus nematode and other nematodes associated with citrus, while Keetch and Buckley (1984), Keetch and Kleynhans et al. (1996) and the SAPPNS[1] database also list nematode pests associated with the crop. Economically, the most important nematode pest associated with citrus in SA is *T. semipenetrans*. The sheath nematode, *Hemicycliophora*, occurs

[1] Dr Mariette Marais of the Nematology Unit, Biosystematics Division, Agricultural Research Council–Plant Protection Research Institute is thanked for the use of data from the South African Plant-Parasitic Nematode Survey (SAPPNS) database; E-mail: maraism@arc.agric.za.

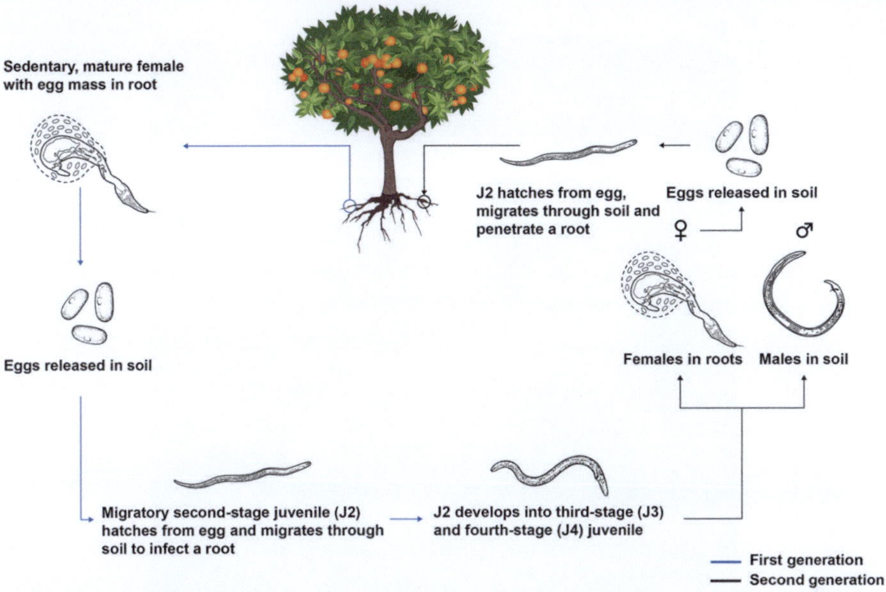

Fig. 13.1 The life cycle of the citrus nematode (Hannes Visagie, North-West University, Potchefstroom, South Africa)

together with the citrus nematode in certain citrus-producing countries (Van Gundy 1959), but the effect of this nematode on citrus yield is unknown. Sheath nematodes and certain *Xiphinema* spp. have been identified from a few citrus-producing regions in SA (Pretorius et al. 2002). The burrowing nematode, *Radopholus citrophilus* (Cobb 1983) Thorne, 1949, formerly known as the citrus race of *R. similis*, has not been found in SA (L Huisman, Citrus Research International Diagnostic Centre, Nelspruit, 2012, personal communication).

13.2.1 Citrus Nematode

Tylenchulus semipenetrans is an obligate plant parasite of citrus and has few other host plants. The life cycle of *T. semipenetrans* consists of an egg, four juvenile stages and male and female adults (Fig. 13.1). The female is attached to the root with the head imbedded inside the root, while the posterior part protrudes from the root (Fig. 13.2a, b). Eggs are deposited in a gelatinous layer covering the posterior area of the female.

As with other plant-parasitic nematodes, *T. semipenetrans* has a very limited ability to move actively over long distances under its own power and does not readily spread from tree to tree in established orchards. Infection of new orchards occurs mainly through infected planting material and contaminated irrigation water (Tarjan 1971; Baines 1974).

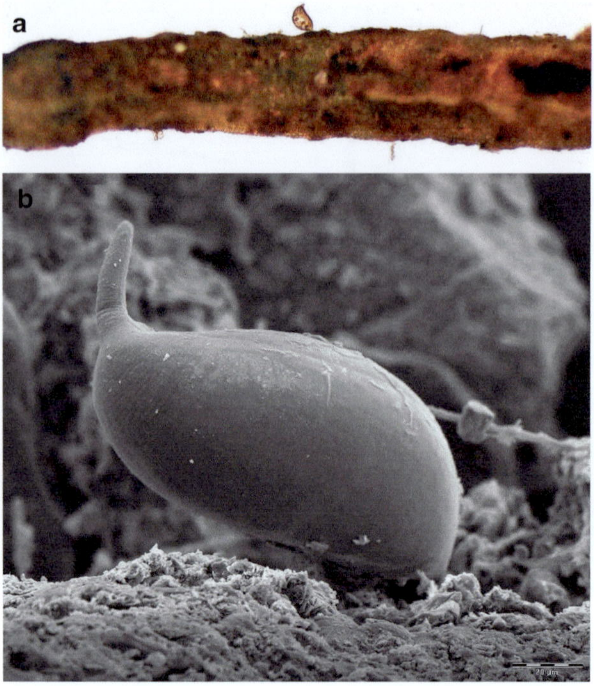

Fig. 13.2 (**a, b**) Posterior parts of the bodies of citrus nematode females outside the roots of citrus in which they feed as visible under a dissection microscope (**a**) and a scanning electron microscope (**b**) (**a**: MC Pretorius, Citrus Research International, South Africa; (**b**) Louwrens Tiedt, North-West University, South Africa)

Irrigation water was implicated as a source of *T. semipenetrans* infection by Cohn (1976), who recommended decontamination procedures such as settling ponds and filtration systems. In SA, the citrus nematode has been detected in major sources of irrigation water such as the Crocodile, Sabie and Letaba rivers in the Mpumalanga and Limpopo provinces. However, despite the presence of *T. semipenetrans* in these rivers, orchards established with nematode-free Citrus Improvement Scheme (CIS)-certified trees on virgin soils predominantly remained free of the citrus nematodes for up to 14 years after planting. Research on the control of *T. semipenetrans* has, therefore, mainly been directed at existing orchards infected with the nematodes (HF Le Roux, unpublished data). *Tylenchulus semipenetrans* is able to survive as eggs in the absence of its host and has been detected in citrus soil for up to 9 years after removal of trees (Van Gundy et al. 1967). The apparent absence of *T. semipenetrans* in soils during the first 2 years after replanting is possibly due to the high soil temperatures resulting from the lack of tree shade, which deplete the energy reserves of infective second-stage juveniles (J2) after hatching and prevent them from establishing successfully on the feeder roots (Le Roux 1995).

The South African CIS was initiated in 1973 to establish a more productive citrus industry to compete internationally. Budwood is rendered virus-free through shoot-tip grafting, after which the material is inoculated with a mild strain of the citrus tristeza virus (CTV). Irrigation water used in nurseries is kept free of nematodes and pathogens such as *Phytophthora* by flocculation, filtration and/or chlorination. Trees are grown in containers raised above the soil surface to prevent cross-contamination. All the nurseries participating in the CIS changed to nematode-free substrates consisting of composted bark, coarse sand or a mixture of these. These changes resulted in the exclusion of the citrus nematode in CIS-certified planting material. This was a considerable improvement of the situation that existed up until the early 1980s when trees were produced in open soil nurseries where nematode populations were as high as 40,000 J2 250 ml^{-1} soil (MNN du Toit, Citrus Research International, Uitenhage, 2015, personal communication). The main source of citrus nematode spread, namely, new planting material (Van Gundy and Meagher 1977), was thus eliminated successfully by the CIS.

13.3 Seasonal Occurrence

The seasonal fluctuation of citrus nematode populations differs between the summer and winter rainfall areas of southern Africa. In the summer rainfall areas including the northern citrus-producing areas in the Mpumalanga, Limpopo, Gauteng, North West and KwaZulu-Natal provinces, as well as Swaziland and Zimbabwe, the J2 populations in the soil and roots tend to peak after each root flush. These root flushes occur during spring (September and October), summer (November and December) and autumn (March and April) (Le Roux 1995). In the winter rainfall areas including the southern citrus-producing areas of the Western Cape, seasonal changes in the soil environment appear to be more important than the timing of root flushes. Populations of J2 start to increase with the commencement of the rainy season during autumn (March and April) and reach a peak in late winter (August). Second-stage juveniles are more sensitive to extreme moisture conditions than females that have their anterior ends embedded in root tissue. Female populations in the roots, therefore, fluctuate less than J2 populations in the soil during the season (Le Roux 1995).

13.4 Biotypes

Three biotypes of the citrus nematode are commonly recognised on citrus, viz. the Citrus, Mediterranean and Poncirus types. The Poncirus biotype, which reproduces on citrus and grapes, occurs only in California (the USA). The Mediterranean biotype reproduces on grapes, persimmon and most citrus varieties and is found throughout the Mediterranean region, southern Africa and India (Duncan 2005). The Citrus biotype has a similar host range but it also reproduces on olive. It was introduced into SA during the early 2000s, through the importation of persimmon plants from Israel (Storey and Malan 2007). The biotypes vary by geographic region as do suitably resistant cvs.

Fig. 13.3 A poorly growing citrus tree infected with the citrus nematodes alongside a healthy tree (Kirk West, Port Elizabeth, South Africa)

13.5 Symptoms

Symptom development depends on the overall orchard condition. Infected trees growing under optimum conditions may yield smaller fruit, while the tree may still appear healthy. As the local citrus industry is focussed on fresh fruit exports, fruit size is of utmost importance. The largest impact of citrus nematodes to the industry is a reduction in fruit size, resulting in lower prices. When conditions become less suitable for tree growth, the effects of citrus nematode parasitism are more apparent. In new citrus plantings in nematode-infested replant soils, symptom development progresses slowly as nematode populations develop to high levels. Leaves are smaller and may become chlorotic, often showing zinc deficiency symptoms (Fig. 13.3). Under saline conditions, excessive sodium may accumulate in leaves. Heavily infected feeder roots are slightly thicker than healthy roots and have a dirty appearance due to soil particles that adhere to gelatinous egg masses of female nematodes on the root surface. There are no visible galls on citrus roots as a result of citrus nematode infection (Duncan 2005).

13.6 Citrus Replant Problem

This is a condition where new trees planted in soils from which deteriorating citrus trees were removed do not grow as vigorously as they would in virgin soils (Tsao et al. 1989). The citrus replant problem was observed in the USA in California as early as the 1920s and later in Florida (Martin and Batchelor 1952),

Texas (Sleeth 1953) as well as in SA (Marloth 1954; Martin 1960) and Australia (Barkley 1981). The most common theories explaining the problem include the accumulation of toxic organic substances, nutrient deficiencies or excesses and deterioration of soil physical properties. Available data indicate, however, that none of these factors are a primary cause of the problem (Tsao et al. 1989). The majority of evidence points to the development of root parasites in the soil during prolonged cropping of citrus. *Tylenchulus semipenetrans* is the most commonly encountered pest associated with the replant problem (Baines and Clarke 1951), although a number of soilborne fungi could also be involved, e.g. *Phytophthora nicotianae*, *P. citrophthora* (Klotz et al. 1958), *Fusarium* spp. and *Pythium* spp. (Martin 1960). Labuschagne et al. (1989) demonstrated an interaction between *F. solani* and *T. semipenetrans.*

Reynolds and O'Bannon (1963) and Le Roux (1995) investigated the build-up of citrus nematode population in a citrus replant situation and reported a slow increase in their numbers during the first 3 years after planting. Young trees provide little shade on the soil beneath the trees, resulting in soil temperatures unsuitable for nematode establishment. In midsummer, soil temperatures of up to 43 °C have been measured at a depth of 10 cm below the soil surface in the Limpopo Province (Le Roux 1995). Once trees become large enough to cast a larger shadow, a more suitable environment is provided for nematode propagation, and a rapid build-up of nematode populations occur. A steep increase in the nematode population between years three and five can thus be expected on replant soils infected with citrus nematode.

13.7 Management Strategies

Damage thresholds are influenced by several factors, including aggressiveness of the nematode population, soil type, rootstock, other diseases and grove management practices (Garabedian et al. 1984). Treatment threshold values for commercial orchards in SA used to be 10,000 J2 250 cm^{-3} soil. However, as a result of the huge seasonal fluctuation in J2 counts, the damage threshold level for the citrus nematode has since been set at 1,000 females 10 g^{-1} of fresh roots. Considerably less fluctuation occurs in the female counts in root samples (Le Roux 1995). Methods to control *T. semipenetrans* normally include reducing the nematode population levels using an integrated pest management (IPM) approach.

Nematicide treatments should be considered only after citrus roots have been tested for the presence of citrus nematodes and numbers found to exceed the threshold levels. In most cases, nematicide treatments should not be considered until all other potential causes of tree decline have been evaluated and corrected. Cultural practices such as irrigation, fertilisation and pruning should be optimal before attention is given to the nematode status of an orchard.

13.7.1 Non-fruit-Bearing Trees

The success of the CIS programme was such that no citrus nematodes could be detected in any of the samples collected from accredited nurseries over a period of more than 30 years. The use of CIS-certified trees (referring to nematode-free planting material) eliminates the possibility of a citrus nematode problem in new groves planted on virgin soils and irrigated with nematode-free water; this is normally the case where borehole water is used. The use of certified trees also reduces damage during the early years of growth after establishment in old, previously infected soil (Meagher 1967; Tarjan 1971). As discussed before, the high soil temperatures under young citrus trees normally prevent the hatching of J2, and the nematode population is thus unable to establish successfully. Even when orchards are irrigated with water in which citrus nematodes occur, the development of populations on young citrus trees is slow. This is in strong contrast with other crops and their associated nematode pests, where considerable numbers can build up within one season.

Organic mulches or persistent weed growth can reduce soil temperatures and result in the re-establishment of citrus nematode populations on replant soils within 2 years after replanting. Sampling of trees on replant soils on an annual basis is therefore necessary. Trees planted in virgin soils need to be monitored for nematodes only every third year because of the low risk of the soils becoming infested with citrus nematodes.

13.7.2 Bearing Trees

The effect of nematodes on yield differs from orchard to orchard, depending on the overall orchard condition. General thresholds based on adult female counts from root samples have been developed to determine when nematicide applications are necessary. The samples can be taken at any time of the year as female nematode counts, which do not fluctuate as much as J2 counts throughout the season, are used as criteria. The samples must be taken from trees that are in an apparently healthy condition. Research has shown that healthy-looking trees, with nematode counts exceeding 1,000 females 10 g^{-1} roots, will respond to nematicide treatments much quicker than trees with the same counts but in a poorer condition visually. The first response will be an increase in fruit size, resulting in a yield increase of between 15 and 20 %.

As stated previously, the citrus nematode has spread throughout SA through infected planting material. This mode of dissemination has ceased since the implementation of the CIS and the supply of citrus nematode-free planting material. New plantings with CIS-certified trees could thus remain nematode-free for many years or for the tree's whole lifespan if the citrus nematode is not introduced through irrigation or run-off water infested with citrus nematode.

13.7.3 *Cultural Practices*

13.7.3.1 Citrus Improvement Scheme (CIS)-Certified Trees

Nematode damage in newly established orchards could be minimised by planting trees certified free of citrus nematodes by the CIS. Proper grove management is critical to mitigate damage caused by plant-parasitic nematodes. There is no net profit in managing nematode pests when other, more critical problems such as poor soil drainage, poor soil preparation, insufficient irrigation, *Phytophthora* collar and fibrous root rot, improper fertilisation or poor disease control limit root function and reduce tree quality.

13.7.3.2 Rootstock Resistance

The ideal control method to deal with the citrus nematode problem is to utilise resistant rootstocks. It is a safe and relatively inexpensive means of control (Cohn 1965). Resistance mechanisms employed against the citrus nematode include either a hypersensitive cell reaction or formation of wound periderm in the cortex in response to nematode feeding or a toxic factor in the root sap (Van Gundy and Kirkpatrick 1964). *Poncirus trifoliata* is resistant to most populations of *T. semipenetrans*. Resistant hybrids of *P. trifoliata* also provide acceptable rootstocks in some regions. Swingle citrumelo *(C. paradisi* x *P. trifoliata)* is a commercially acceptable rootstock with a high degree of resistance to most populations of *T. semipenetrans*. It is also resistant to tristeza virus and tolerant to *Phytophthora nicotianae* var. *parasitica* (Duncan 2005). Two replant orchards in the Sundays River Valley (Eastern Cape Province) on Swingle rootstock were found to be susceptible to citrus nematodes. These are an exception.

Forty different rootstocks were evaluated in a pot trial by Citrus Research International (CRI) in Nelspruit to determine their host sustainability towards the citrus nematode. Table 13.1 (CRI 2003) lists 12 of the most commonly used rootstocks in SA. Unfortunately, it is not always feasible to use resistant rootstocks because of factors such as unsuitable soil properties, incompatibility between certain rootstocks and scions or horticultural disadvantages such as internal fruit quality and yields.

13.7.4 *Chemical Control*

13.7.4.1 Pre-plant Treatments

The most effective strategy followed to eliminate the citrus nematode and pathogenic fungi after the removal of old infected trees is by means of a fumigant such as methyl bromide. However, this product is no longer available. Research into alternative methods of control is currently being conducted by CRI.

Table 13.1 Host suitability of the most commonly used rootstocks in southern Africa to the citrus nematode

Rootstock	Relative tolerance to *Tylenchulus semipenetrans*
Rough lemon	Highly susceptible
Volkameriana	Highly susceptible
Empress mandarin	Highly susceptible
Cleopatra mandarin	Highly susceptible
Carrizo citrange	Susceptible
Troyer citrange	Susceptible
Yuma citrange	Susceptible
X639 citrange	Susceptible
Minneola x trifoliate	Susceptible
Swingle citrumelo	Tolerant[a]
C35	Tolerant
Trifoliate	Resistant

[a]The exception is a few replant orchards in the Eastern Cape Province where the citrus nematode population overcame the normal tolerance

13.7.4.2 Post-plant Treatments

From the recovery rate reported by O'Bannon et al. (1967), it is clear that the pre-plant fumigants not only killed soil fungi and the J2 and adult stages of the citrus nematode but also the eggs. This activity on eggs is the most important difference between the pre-plant soil fumigants and the non-fumigant post-plant chemicals.

The following post-plant nematicides are registered on citrus in SA: cadusafos, ethoprophos, fenamiphos, fosthiazate, furfural and terbufos (Van Zyl 2013). After these nematicides have degraded to non-toxic levels, the nematode's life cycle can continue (Le Roux 1995). Yield and fruit size responses to nematicides are often the result of the additional insecticidal/acaricidal effects and not of nematicidal activity only (Milne and Willers 1979; Duncan et al. 2011). Determining the effectiveness of a nematicide in citrus should thus be based on nematode counts rather than on yield data of a single season. Such counts are now included in orchard management procedures in SA.

When the concept of multiple nematicide applications was introduced on a commercial scale to citrus orchards in SA, situations occurred where growers were not successful in disrupting the nematode's life cycle, despite adhering strictly to prescribed procedures. An investigation was initiated to determine the efficacy of cadusafos in soil where aldicarb and fenamiphos had failed as a result of accelerated microbial degradation (AMD) (see Box 6.4, Chap. 6). It was found that, in the absence of sufficient irrigation water, none of the nematicides had distributed effectively through the soil profile and consequently failed to eliminate the citrus nematode (Le Roux et al. 1998). With sufficient rain and irrigation, the nematicides were able effectively to reduce nematode populations.

In trials conducted by CRI, it was shown that three consecutive applications of cadusafos were effective in breaking the life cycle of the citrus nematode. In these trials aldicarb, fenamiphos and terbufos failed to reduce *T. semipenetrans* populations, in spite of adequate soil moisture. The possibility of AMD was thus investigated. It is well known that once a population of micro-organisms has adapted to utilise a chemical as a nutrient source, its numbers start to increase (Racke and Coats 1988). Consequently, the chemical is degraded at a more rapid rate and fails to control the target pest effectively (Suett and Walker 1988) (see Sect. 6.4, Box 6.4, Chap. 6). Samples sent to Bayer Crop Science in Germany indicated that accelerated degradation of fenamiphos was caused by micro-organisms since soils fumigated 6 years previously without subsequent applications of a nematicide did not support AMD of the compound. Investigations confirmed that both aldicarb and fenamiphos were degraded at such rapid rates that prolonged control of *T. semipenetrans* was unattainable. Most citrus orchards are sprayed frequently with organophosphates to control other pests and substantial quantities of these compounds end up in the soil under the trees. This could predispose soil to AMD of soil-applied organophosphate nematicides. There were, however, cases where it was established that even if AMD of a specific organophosphate was present, it might not necessarily mean that there would be accelerated degradation of another organophosphate (Le Roux et al 1996). Nematicides should not be applied through drip irrigation systems as it will predispose the soil to AMD (Le Roux et al. 2001). They should rather be applied in a 1-m-wide band along the drip line during the rainy season.

13.7.4.3 Timing of Nematicide Applications for Different Rainfall Regions

In the summer rainfall areas, timing of nematicide applications is aimed to coincide with the three root flushes mentioned earlier in order to create an opportunity to disrupt the life cycle of the citrus nematode to such an extent that surviving nematodes never reach damage threshold levels. Therefore, CRI has established that applications should commence after the first good summer rain (at least 30 mm) and thereafter twice every 2 months. In the winter rainfall areas, the first treatment should be applied with the commencement of the rainy season in March/April, followed by a second treatment 2 months later in May/June and a final treatment 2 months later (July/August). If the nematicides are applied further apart than 2 months, it will allow J2 to hatch and to develop to adult egg-laying females. The nematicide application will have no effect on the eggs from which the J2 will hatch and continue with its life cycle once the nematicide has broken down. The timing between applications is thus as important as sufficient irrigation to move the nematicides through the root profile and the even spread of the nematicide over the root surface.

13.8 Conclusions

Effective nematode control is only successful when all factors affecting the root system of the plant are taken into consideration and an IPM approach is followed. When establising new orchards on virgin soils it is essential to make use of nematode-free planting material supplied by CIS-accredited nurseries, and to prevent nematode-infested run-off water into virgin orchards. Only utilise nematode-free irrigation water including the use of settling ponds. For replant or existing orchards infested with citrus nematodes, IPM consists of the use of tolerant rootstocks when compatible with scion. Multiple applications of post-plant nematicides within specified application windows within the same season to ensure interruption of the nematode's life cycle is advised together with raising awareness of AMD and the use of sufficient water to disperse nematicides through the soil profile.

References

Baines RC (1974) The effect of soil type on movement and infection rate of larvae of *Tylenchulus semipenetrans*. J Nematol 6:60–62

Baines RC, Clarke OF (1951) Citrus-root nematode. Calif. Citrogr. 37:60. 62, 86

Barkley P (1981) Replanting citrus. Farmers' Newsl 150:22–27

Burger D (2009) South African yearbook 2008/09. Government Communication and Information System, Pretoria

Cartwright AP (1977) A history of the South African citrus industry. Purnell, Cape Town

Castle WS, Tucker DPH, Krezdorn AH et al (1993) Rootstocks for Florida citrus, 2nd edn. University of Florida, Florida

Cohn E (1965) On the feeding and histopathology of the citrus nematode. Nematologica 11:47–54

Cohn E (1976) Report of investigations on nematodes of citrus and subtropical fruit crops in South Africa. Citrus and Subtropical Fruit Research Institute, Nelspruit

CRI (2003) Citrus Research International Group Annual Research Report. CRI, Nelspruit, pp 278–284

Davies FS, Albrigo LG (1994) Citrus. CAB International, Wallingford

De Villiers EA, Milne DL (1976) Sitrusaalwurm, *Tylenchulus semipenetrans* Cobb. Boerdery in Suid-Afrika H5:1976

Duncan LW (2005) Nematode parasites of citrus. In: Luc M, Sikora RA, Bridge J (eds) Plant parasitic nematodes in subtropical and tropical agriculture, 2nd edn. CAB International, Wallingford, pp 437–466

Duncan LW, Noling JW, Inserra RN (2011) Florida citrus pest management guide: Nematodes. Institute of Food and Agricultural Sciences, University of Florida, Florida, Publication SP-43, ENY- 606

FAO (2016) Statistics at FAO. http://faostat.fao.org/site/567/default.aspx#ancor. Accessed 2 Jan 2016

Garabedian S, Van Gundy SD, Mankau R et al (1984) Nematodes. In: Integrated pest management for citrus. Division of Agriculture and Natural Resources Publications, University of California, Berkeley, pp 129–131

Keetch DP, Buckley NH (1984) A check-list of the plant-parasitic nematodes of Southern Africa. Technical Communication, Department of Agriculture. Government Printer, Pretoria

Kleynhans KPN, Van den Berg E, Swart A et al (1996) Plant nematodes in South Africa. Business Print, Pretoria

Klotz LJ, De Wolfe TA, Wong P (1958) Decay of fibrous roots of citrus. Phytopathol 48:616–622

Labuschagne N, Van der Vegte FA, Kotzé JM (1989) Interaction between *Fusarium solani* and *Tylenchulus semipenetrans* on citrus roots. Phytophylactica 21:29–33

Le Roux HF (1995) Control of the citrus nematode in South Africa. PhD Thesis, University of Pretoria, Pretoria

Le Roux H, Ware AB, Suett DL et al (1996) Accelerated degradation of some soil applied nematicides in a South African orchard. Inter Soc Citricul 1:593–596

Le Roux HF, Ware AB, Pretorius MC (1998) Comparative efficacy of preplant fumigation and postplant chemical treatment of replant citrus trees in an orchard infested with *Tylenchulus semipenetrans*. Plant Dis 82:1323–1327

Le Roux HF, Pretorius MC, Huisman L (2001) Comparison of accelerated degradation rates of nematicides applied through drip- and micro-irrigation. Afr Plant Prot 8:6

Marloth RH (1954) Replanting old citrus orchard sites. Farming S Afr 29:319–323

Martin JP (1960) Fungi and nematodes in South African citrus orchard soils in relation to the citrus replant problem. Proc Soil Sci Soc Am 24:469–472

Martin JP, Batchelor LD (1952) The difficulties of replanting lands to the same species of orchard trees. Proc Rio Grande Valley Hort Inst 6:1–10

Meagher JW (1967) Observations on the transport of nematodes in subsoil drainage and irrigation water. Aust J Exp Agric Anim Husb 7:577–579

Milne DL (1982) Nematode pests of citrus. In: Keetch DP, Heyns J (eds) Nematology in southern Africa, Science Bulletin No. 400. Department of Agriculture and Fisheries, Pretoria, pp 12–18

Milne DL, Willers P (1979) Yield and nutritional responses to fenamiphos treatment of citrus infested with citrus nematodes. Subtropica 1:11–14

O'Bannon JH, Leather CR, Reynolds HW (1967) Interactions of *Tylenchulus semipenetrans* and *Fusarium* species on rough lemon *(Citrus limon)*. Phytopathol 57:414–417

Pretorius MC (2005) Epidemiology and control of *Pseudocercospora angolensis* fruit and leaf spot disease on citrus in Zimbabwe. MSc Dissertation, University of Stellenbosch, Stellenbosch

Pretorius MC, Huisman L, Le Roux HF et al (2002) Detection of a *Hemicycliophora* species on citrus in the Gamtoos River Valley. Afr Plant Protect 8:77

Racke KD, Coats JR (1988) Enhanced degradation and the comparative fate of carbamate insecticides in soil. J Agric Food Chem 36:1067–1072

Reynolds HW, O'Bannon JH (1963) Factors influencing the citrus replants in Arizona. Nematologica 9:337–340

Saunt J (2000) Citrus varieties of the world, 2nd edn. An illustrated guide. Sinclair International Ltd, Norwich, p 130–133

Siphugu L (2009) Republic of South Africa, citrus annual 2009. Global Agriculture Information Network Report SF 1-8

Sleeth B (1953) An old citrus soil problem in the lower Rio Grande Valley. Proc Rio Grande Valley Hortic Inst 7:31–32

Spiegel-Roy R, Goldschmidt EE (1996) Biology of Citrus. Cambridge University Press, Cambridge

Storey SG, Malan AP (2007) First report of *Tylenchulus semipenetrans* on persimmon (*Diospyros kaki*) in South Africa. S Afr J Plant Soil 24:233–260

Suett DL, Walker A (1988) Accelerated degradation of soil applied pesticides – implications for UK horticulture. Ann Appl Biol 17:213–222

Tarjan AC (1971) Migration of three pathogenic citrus nematodes through two Florida soils. Soil Crop Sci Soc Florida Proc 31:253–255

Tsao PH, Martin JP, Davis M (1989) The citrus replant problem. In: Reuther W, Calavan EC, Carman GE (eds) The citrus industry, vol 5. Division of Agricultural Sciences, University of California, Berkeley, pp 261–279

Van Gundy SD (1959) The life history of *Hemicycliophora arenaria* Raski (Nematoda: Criconematidae). P Helm Soc Wash 26:67–72

Van Gundy SD, Kirkpatrick JD (1964) Nature of resistance in certain citrus rootstocks to citrus nematodes. Phytopathol 54:419–427

Van Gundy SD, Meagher JW (1977) Citrus nematode *(Tylenchulus semipenetrans)* problems worldwide. Proc Int Soc Citric 3:823–826

Van Gundy SD, Bird AF, Wallace HR (1967) Ageing and starvation of larvae of *Meloidogyne javanica* and *Tylenchulus semipenetrans*. Phytopathol 57:571–599

Van Zyl K (2013) A guide to crop pest management in South Africa. A compendium of acaracides, insecticides, nematicides, molluscicides, avicides and rodenticides. A CropLife South African Compendium, 1st edn, VR Print, Pinetown

Chapter 14
Nematode Pests of Grapevine

Sheila G. Storey, Antoinette P. Malan, and Hans J. Hugo

14.1 Introduction

Viticulture in South Africa (SA) dates back to the first grapevines (*Vitis vinifera*) that were planted by Jan van Riebeeck in Cape Town in 1655. In 1659, Van Riebeeck reported that the first wine had been pressed from Cape grapes (Smith 1982). The second major influence on viticulture was the arrival of the French Huguenots at the end of the 17th century.

14.2 Major Production Areas

In SA, grapevines are grown both for wine production and for table grapes, with a small percentage producing grapes for the dried fruit market (Table 14.1). Of the 117,675 hectares (ha) under grape production in 2014/2015 (J Lombardt, SA Wine Industry Information and Systems NPC, Stellenbosch, 2016, personal communication; C Whitehead, South African Table Grape Industry, Stellenbosch, 2016,

S.G. Storey (✉)
Nemlab, Nematode Advisory Service, PO Box 2825, Durbanville 7551, South Africa
e-mail: sheila@nemlab.co.za

A.P. Malan
Department of Conservation Ecology and Entomology, Faculty of AgriSciences, Stellenbosch University, Private Bag X1, Matieland, 7602 Stellenbosch, South Africa
e-mail: apm@sun.ac.za

H.J. Hugo
Agricultural Research Council–Infruitec/Nietvoorbij, Private Bag X5026, Stellenbosch 7599, South Africa
e-mail: hans.hugo1887@gmail.com

H. Fourie et al. (eds.), *Nematology in South Africa: A View from the 21st Century*, DOI 10.1007/978-3-319-44210-5_14

Table 14.1 Grape production figures for the 2014/2015 growing season

	Wine grapes[a]	Table and dried grapes[b]
Planted area (hectares)	99,463	18,212
Production (metric tonnes)	1,519,708	291,442
Rand value (ZAR in billions)	4.7	5
Export percentage	44	90
Rand value of export (ZAR in billions)	8	7.4

[a]J Lombardt, SA Wine Industry Information and Systems NPC, Stellenbosch, 2016, personal communication

[b]C Whitehead, South African Table Grape Industry, Stellenbosch, 2016, personal communication

Table 14.2 Geographic distribution of South African vineyards during the 2014/2015 growing season

Regions	Wine grapes[a]	Table and dry grapes[b]
	Area (hectares)	Area (hectares)
Worcester	8,858	–
Hex River	–	6,419
Berg River	15,835	4,053
Stellenbosch	16,037	–
Swartland	13,591	–
Robertson	14,652	–
Olifants River	10,149	1,210
Orange River	4,659	5,081
Little Karoo	2,660	–
Limpopo	–	1,449
Total	99,463	18,212

[a]J Lombardt, SA Wine Industry Information and Systems NPC, Stellenbosch, 2016, personal communication
[b]C Whitehead, South African Table Grape Industry, Stellenbosch, 2016, personal communication

personal communication), the majority (85 %) were wine grapes. During the 2014/2015 season, 1,811,150 metric tonnes (MT) of grapes were produced in SA. The export market is by far the most important market, earning SA around ZAR15 billion (Table 14.1).

South Africa's viticulture is located in ten regions, with almost 90 % of the production occurring in the Western Cape Province (Table 14.2). The production has spread as far afield as the Olifants River (Vredendal and Klawer), the lower Orange River and the Overberg as well as to the Free State and Limpopo provinces. In addition, isolated pockets of wine grapes exist in the KwaZulu-Natal Province and the southern Cape. The wide-ranging distribution of grapevine cultivation covers a variety of soils and climates. In the hotter, inland regions, table grapes are grown for export, while the cooler regions in the south specialise in premium wine grapes.

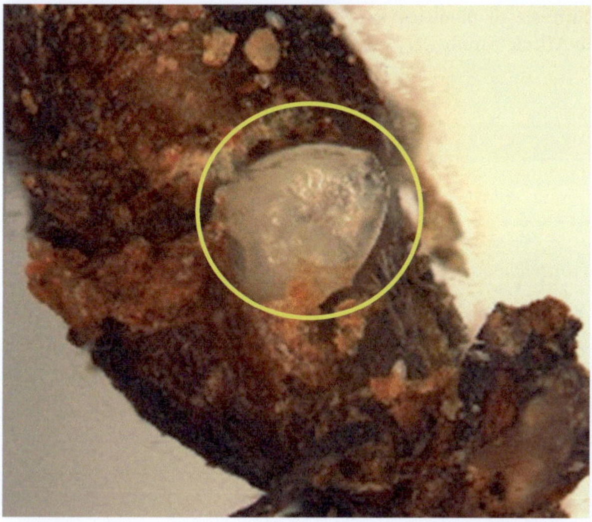

Fig. 14.1 The posterior part of a root-knot nematode female (*yellow circle*) visible in a vine root in which it is feeding (Welma Pieterse, Department of Agriculture Forestry and Fisheries, Stellenbosch, South Africa)

14.3 Plant-Parasitic Nematodes Associated with Grapevine

Worldwide, the economically most important plant-parasitic nematodes that parasitise grapevine are root-knot nematodes (*Meloidogyne* spp.), especially *Meloidogyne javanica* (Treub, 1885) Chitwood, 1949 followed by *Meloidogyne incognita* (Kofoid & White, 1919) Chitwood, 1949 and *Meloidogyne arenaria* (Neal, 1889) Chitwood, 1890 (Fig. 14.1). Other nematode pests include several lesion nematode species (primarily *Pratylenchus vulnus* Allen & Jensen, 1951), ring nematode (*Criconemoides xenoplax* Raski, 1952), dagger nematodes (*Xiphinema americanum* Cobb, 1913, *Xiphinema index* Thorne & Allen, 1950 and several other species) and stubby-root nematodes (*Paratrichodorus* and *Nanidorus* spp.) (Addison and Fourie 2008; Walker and Stirling 2008; SAPPNS[1]).

In SA, the earliest report of nematode damage on grapevine was published in 1915 and estimated that about two-thirds of the Cape nurseries were affected by *Meloidogyne* spp. (Lounsbury 1915). The first survey to assess the importance of nematode pests was undertaken several decades later by Smith (1977). As elsewhere in the world, the prevalent genera identified in local vineyards are *Meloidogyne*, *Pratylenchus* and *Xiphinema* (Table 14.3). Concise summaries of the plant-parasitic nematode species that prevail in local vineyards are presented below. Most of these nematode pests are listed in Kleynhans et al. (1996) if listed before this date, but others are added to the SAPPNS databases as they are found.

---

[1]Dr Mariette Marais of the Nematology Unit, Biosystematics Division, Agricultural Research Council–Plant Protection Research Institute is thanked for the use of data from the South African Plant-Parasitic Nematode Survey (SAPPNS) database; E-mail: maraism@arc.agric.za.

Table 14.3 Occurrence of plant-parasitic nematode genera in vineyards in the Western Cape Province of South Africa (Smith 1977)

Nematode genus	Frequency of occurrence (percentage) of genera in 100 soil samples
Pratylenchus	86
Meloidogyne	77
Xiphinema	70
Criconemoides	48
Nanidorus and *Paratrichodorus*	41
Paratylenchus	30
Helicotylenchus	27
Longidorus	27
Scutellonema	20
Tylenchorhynchus sensu lato	13
Rotylenchus	12
Tylenchulus	10
Hoplolaimus	9
Hemicycliophora	6
Criconema	3

14.3.1 Root-Knot Nematodes

Root-knot nematodes are common in SA vineyards, with *M. javanica* being the most common species, followed by *M. incognita*, *M. arenaria* and *Meloidogyne hapla* Chitwood, 1949. Collectively, these species were found in 77 % of sites sampled in the Western Cape (Smith 1977). Loubser and Meyer (1987a) found that the presence of the two most important species was determined by the vastly different climatic regions under which grapevine are grown in SA. These authors reported that *M. incognita* dominated in the Northern Cape Province and *M. javanica* in the Western Cape Province.

The life span of root-knot nematodes in grapevine ranges from one to several months. The egg population is greatest in March when soils are driest. Second-stage juveniles (J2) in soil are generally two to five times more numerous during autumn and winter than they are in spring and summer. Depending on soil conditions and tillage practices, the bulk of the population is located at a depth of between 150 and 900 mm beneath the vine row (Loubser and Meyer 1987a), with almost half of the J2 being recovered from the 200–400 mm layer. Smith (1982) noted that *Meloidogyne* was the only genus that was influenced by soil type, with the proportion of infested sites decreasing as the soil became heavier (higher clay content). Root-knot nematodes have a wide host range, including many broadleaf weed species and cover crops that are present in vineyards.

Meloidogyne spp. are the only nematodes that cause diagnostic symptoms on grapevine in the form of root galls (Fig. 14.2). These are clearly seen when young, infected roots are rinsed in water. Such galls are distinguished from the crown gall,

Fig. 14.2 Typical root-knot nematode galling on infected vine roots (Kirk West, Port Elizabeth, South Africa)

caused by the bacterium *Agrobacterium vitis,* as root-knot nematode galls are visible as thickenings of the entire root, typically on young feeder roots. Crown gall symptoms are seen on the side of older, thicker roots and can be removed or rubbed off easily.

The interference of root-knot nematodes with plant growth and nutrient uptake is reflected in the potassium (K) deficiency symptoms seen on vines (Fig. 14.3). Severely galled roots are often found on own-rooted vines in the sandy and sandy loam soils of the lower Orange River area. The problems due to *Meloidogyne* spp. in the warm, sandy soils led in the 1980s, to research programmes that concentrated mainly on the introduction of resistant rootstocks and the use of nematicides. The adoption of such practices resulted in a substantial reduction in root-knot nematode infections of vines. However, losses due to root-knot nematodes still occur, with the degree of damage depending on the rootstock involved, the *Meloidogyne* sp. present and its aggressiveness on grapevine and its population density.

14.3.2 Lesion Nematodes

Several *Pratylenchus* spp. have been reported from grapevine worldwide (Quader et al. 2003), including SA. These are *Pratylenchus flakkensis* Seinhorst, 1968, *Pratylenchus minyus* Sher & Allen, 1953, *P. penetrans* (Cobb, 1917) Filipjev &

Fig. 14.3 Potassium deficiency symptoms (*white circle*) visible in a vineyard due to root-knot nematode infection (Kirk West, Port Elizabeth, South Africa)

Schuurmans Stekhoven, 1941, *Pratylenchus pratensis* (De Man, 1880) Filipjev, 1936, *Pratylenchus scribneri* Steiner, 1943 and *P. vulnus* (Smith 1982; Kleynhans et al. 1996). Since relatively large populations of lesion nematodes are recorded in some vineyards and since such nematodes are widespread across all soil types, they should be considered as being among the more important nematode pests of vines.

14.3.3 Ring Nematodes

Grapevine is a highly suitable host plant for the ring nematode, *C. xenoplax*. It is the most common and abundant nematode species recorded in vineyards in Spain, Germany, France, Switzerland and in the states of Oregon and California in the United States of America (USA) (Pinkerton et al. 1999). *Criconemoides xenoplax* is especially prevalent in vineyards throughout the Western Cape Province (Smith 1977). Since *C. xenoplax* is not easily recovered by many of the commonly used extraction methods, its incidence and population densities are often higher than previously reported in early surveys. The use of the centrifugal flotation extraction technique (Jenkins 1964), suited for the extraction of ring nematodes, resulted in the recovery of *C. xenoplax* from almost 90 % of grapevine soil samples (S Storey, unpublished data; H Hugo, Agricultural Research Council–Infruitec/Nietvoorbij, unpublished data). In Oregon, the presence of *C. xenoplax* was associated with unhealthy vine plants and extensive root damage (Pinkerton et al. 1999, 2005), and in California, McKenry (1992) estimated that *C. xenoplax* numbers >500 kg soil^{-1} reduced grape yields by 10–25 %. In SA, *C. xenoplax* is the nematode pest species that is considered to be responsible for most of the damage on grapevine.

The feeding of *C. xenoplax* on grapevine causes rapid local darkening and destruction of root tissue, resulting in a stunted root system with relatively few feeder roots. Plants parasitised by ring nematodes often tend to sprout more side roots, which are usually short, discoloured and often dead (a symptom known as 'witches' broom').

14.3.4 Stubby-Root Nematodes

Paratrichodorus and *Nanidorus* spp. are present on a wide variety of crops that are cultivated across SA, although in most cases in low population levels (Smith 1977; Kleynhans et al. 1996). *Paratrichodorus porosus* (Allen, 1957) Siddiqi, 1974, *Paratrichodorus lobatus* (Colbran, 1965) Siddiqi, 1974 and the related species, *Nanidorus minor* (Colbran, 1956) Siddiqi, 1974, have been associated with grapevine (Marais and Swart 2002). A pathogenicity experiment showed that the population density of *N. minor* was inversely related to the shoot and root growth of Thompson seedless grapevine (Smith 1982). *Nanidorus* spp. feed as ectoparasites on the epidermal cells near the root tip, causing severe stunting of vine roots.

14.3.5 Dagger and Needle Nematodes

Xiphinema spp. (dagger nematodes) reduce the yield of grapevine, especially when the plants are under stress. In Oregon, *X. americanum* was reported to be pathogenic to grapevine, causing darkening and excessive branching of the root system (Pinkerton et al. 1999). In Australian vineyards, *X. americanum* is probably the most widely distributed nematode-pest species (Walker and Stirling 2008).

Individuals of *Xiphinema* spp. were recovered from 70 of the 100 vineyards surveyed in the Western Cape Province by Smith (1977). According to Smith (1982) and Kleynhans et al. (1996), the *Xiphinema* spp. present in SA vineyards include *Xiphinema diffusum* Lamberti & Bleve-Zacheo, 1979, *X. americanum*; *Xiphinema elongatum* Schuurmans Stekhoven & Teunissen, 1938, *Xiphinema meridianum* Heyns 1971, *Xiphinema italiae* Meyl, 1953 and *X. index*. The most common were *X. diffusum, X. elongatum* and *X. americanum*. Worldwide, the most important dagger nematode on grapevine is *X. index*, the vector of *Grapevine fanleaf virus* (GFLV). The virus causes shoot, leaf and berry abnormalities, poor growth and yield loss. The nematodes associated with grapevine viruses are discussed in Sect. 14.4. Dagger nematodes prefer the young roots of woody plants. Weakened grapevine plants have been shown to yield correspondingly less as the population levels of these nematode pests increase. The feeding of the dagger nematodes near the root tip (Fig. 14.4) causes an enlargement or swelling of the root. An investigation of the root distribution of various grape rootstocks in a deep sandy soil along the Olifants River in the north Western Cape showed *Xiphinema* spp. and *Longidorus* spp. to be numerous at depths of 100–300 mm (Smith 1982).

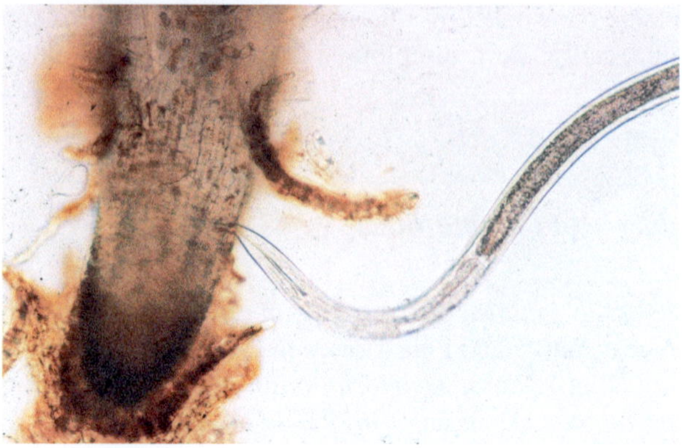

Fig. 14.4 *Xiphinema index* feeding on a vine root (Antoinette Malan, Stellenbosch University, Stellenbosch, South Africa)

Two needle nematode species, *Paralongidorus costatus* (Jacobs & Heyns, 1987) Escuer & Arias, 1997 and *Paralongidorus maximus* (Bütchli, 1874) Escuer & Arias, 1997 have been identified from grapevine in SA.

14.3.6 The Citrus Nematode

The citrus nematode, *Tylenchulus semipenetrans* Cobb, 1913, was first recorded as a pest of grapevine in Australia. It was subsequently recorded on grapes in a number of other countries, including SA (Smith 1982). When present in high population densities, it can cause considerable damage that according to McKenry (1992) can amount to as much as 13 % loss in yield. Although only one species is involved, three biotypes exist (see Sect. 13.4, Chap. 13). Although they all feed on citrus and grapevine, they can be distinguished according to whether or not they reproduce on persimmon and olive. The Mediterranean and citrus biotypes have been reported in SA (Inserra et al. 1980). The citrus nematode, which favours loam-type soils, reduces the vigour and yield of its host plants.

14.3.7 Pin Nematodes

Pin nematode species occur in vineyards in many parts of the world. In SA, they are reasonably common (Table 14.3), with *Paratylenchus projectus* Jenkins, 1956, and *Paratylenchus obtusicaudatus* Raski, 1975, identified from grapevine (Smith 1982). Pin nematodes are probably not pathogenic on grapevine (Ferris and McKenry 1975).

14.3.8 Spiral and Lance Nematodes

Various spiral nematode genera, viz. *Helicotylenchus*, *Rotylenchus* and *Scutellonema*, are relatively common in local vineyard soils (Kleynhans et al. 1996). The lance nematode, *Hoplolaimus pararobustus* (Schuurmans Stekhoven & Teunissen, 1938) Sher, 1963, was also recorded on grapevine in SA. Although population densities of these genera are sometimes high, they do not seem to be of any economic importance.

14.4 General Above-Ground Symptoms and Yield Losses

Plant-parasitic nematodes, which are commonly found in vineyards in all regions of the world, are often associated with areas of poor growth (Ferris and McKenry 1975; Smith 1982; Pinkerton et al. 1999; Walker and Stirling 2008). Estimated losses range from 5–15 % in most grape-growing districts of Australia (Walker and Stirling 2008), with as much as 60 % loss reported from South Australia (Nicol et al. 1999; Quader et al. 2001). Yield losses in SA amount to approximately 15 % (Hugo 2003).

Above-ground symptoms of nematode damage are poor growth, short internodes and comparatively smaller berries and bunches with a resultant decrease in yield. Affected vines may show potassium (K) deficiency symptoms in the leaves despite sufficient amounts being available in the soil. Grapevines infected with high population levels of plant-parasitic nematodes will also show drought symptoms, especially under rain-fed conditions, more quickly due to their poorly developed root systems (Hugo 2003). An aerial view of a vineyard shows typical 'poor-growing' areas as a result of plant-parasitic nematode infection (Fig. 14.5).

14.5 Nematodes Associated with Grapevine Viruses

Certain nematodes cause crop losses through the role that they play as virus vectors and through their interactions with other root pathogens. In 1958, Hewitt et al. (1958) reported the first experimental evidence of a plant virus being transmitted by a nematode. By that time, the grapevine fanleaf virus (GFLV) and its nematode vector, *X. index*, had already been distributed by means of soil and infected plant material to all viticulture regions around the world (including SA). In the 1960s, it was also demonstrated that members of the family Trichodoridae are responsible for the transmission of several plant viruses (Sol et al. 2009). Since then, many different virus-nematode associations have been described. The position of the virus particles in the oesophageal region of the nematode body is elaborated in Box 14.1 and Fig. 14.6.

Box 14.1 Longidoridae Virus Vectors

Virus particles are associated with the lining of the lumen of the oesophagus and, depending on the genus involved, with different parts of the oesophagus (Fig. 14.6a–c). Electron microscopy studies of thin sections through the feeding apparatus of virus-vectoring nematodes have shown that, in the case of *Longidorus* and *Paralongidorus,* isometrical virus particles (nepoviruses) were present in the lining of the stoma between the odontostyle and the guiding ring.

In the genus *Xiphinema*, nepovirus particles were observed in a monolayer attached to the cuticular lining of the stylet extension (odontophore) (Fig. 14.6c). These nepovirus particles were also observed throughout the rest of the oesophagus, although not in the odontostyle (Fig. 14.6b). In *Nanidorus*, *Paratrichodorus* and *Trichodorus* spp., tubular viruses (tobraviruses) are associated with the cuticular lining of the oesophagus. Nematode individuals tend to lose the virus during each moult and thus only retain the virus for transmission to a host plant for a few weeks until they become adults.

The presence of weeds in orchards and the use of infected planting material are the predominant sources of the dissemination of nematode-vectored viruses. A certification scheme for disease-free planting material (as stated in the Plant Improvement Act of South Africa, Act No. 53 of 1976), pre-planting soil tests to establish whether nematode pests are vectors of viruses and strict weed control are of utmost importance to limit and/or prevent the occurrence of such diseases in vineyards.

Fig. 14.5 Aerial view of nematode damage in a vineyard in the Upington area in the Northern Cape Province of South Africa (Caroline Mouton, Nemconsult, Upington, South Africa)

In SA, information regarding the occurrence and distribution of longidorid species associated with grapevine is limited (Barbercheck et al. 1985; Barbercheck and Heyns 1986; Van Reenen and Heyns 1986; Van Mieghem and Pieterse 1989; Malan and Meyer 1994). The only virus vector in the genus *Paralongidorus*, *P. maximus*, was reported from a vineyard in the Western Cape Province (Swart et al. 1996). However,

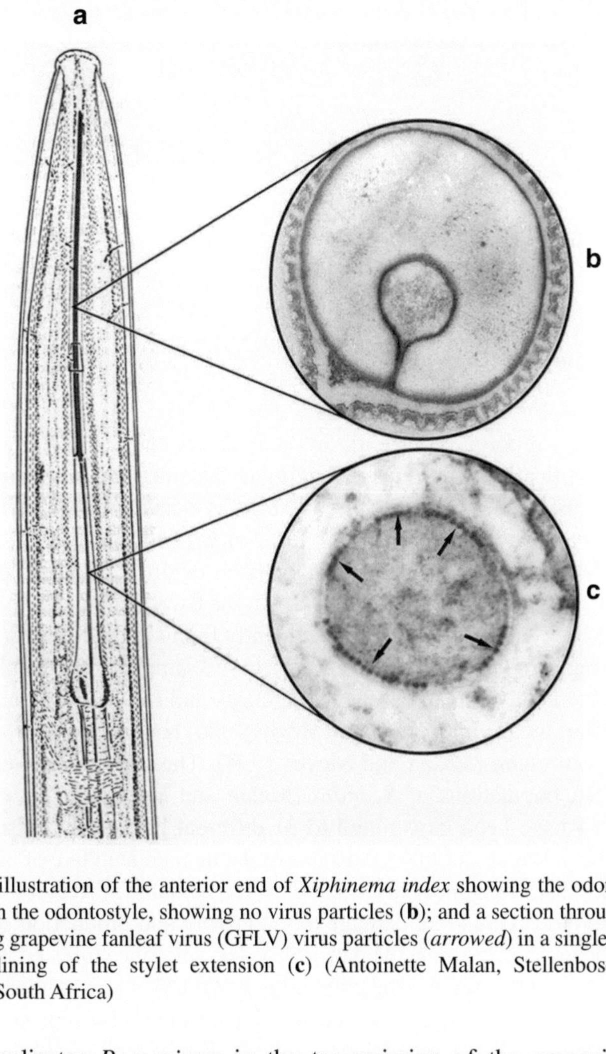

Fig. 14.6 An illustration of the anterior end of *Xiphinema index* showing the odontostylet (**a**); a section through the odontostyle, showing no virus particles (**b**); and a section through the odontophore, showing grapevine fanleaf virus (GFLV) virus particles (*arrowed*) in a single layer attached to the inner lining of the stylet extension (**c**) (Antoinette Malan, Stellenbosch University, Stellenbosch, South Africa)

no report implicates *P. maximus* in the transmission of the grapevine strain of *Raspberry ringspot virus*. Of the *Xiphinema* spp. that are implicated in the transmission of grapevine viruses, four have been reported from SA vineyards (Barbercheck and Heyns 1986; Van Mieghem and Pieterse 1989; Malan 1995) (Table 14.4). Of the two trichodorid species considered pests of grapevine, only *N. minor* is a vector of tobraviruses, but again no transmission has been reported in SA (Brown et al. 2004).

Xiphinema americanum sensu Heyns, which might be *X. americanum* sensu stricto according to Loots and Heyns (1984), is one of the most common species in vineyards in the Western Cape Province (Malan and Meyer 1994). After a reappraisal of all populations identified in the last century as *X. americanum*, it was concluded that the *X. americanum* group is comprised of a complex of several species (Lamberti et al. 2000). The virus-vector relationship of this nematode pest, as

Table 14.4 Species of the family Longidoridae reported from South Africa and implicated internationally as vectors of grapevine viruses

Species	Virus	Reference
Paralongidorus maximus[a]	Raspberry ringspot (grapevine strain)	Jones et al. (1994)
Xiphinema americanum[a]	Peach rosette mosaic	Kloss (2009)
	Tobacco ringspot	Brown et al. (2004)
	Tomato ringspot	Brown and Halbrendt (1992)
Xiphinema index	Grapevine fanleaf	Hewitt et al. (1958)
Xiphinema italiae[a]	Grapevine fanleaf	Cohn et al. (1970)

[a]Not a vector of viruses in South Africa

indicated in many reports, is currently in doubt due to the confusion with regard to the correct identification of species belonging to the americanum group. *Xiphinema americanum* sensu stricto transmits three viruses associated with grapevine, namely, *Peach rosette mosaic*, *Tobacco ringspot* and *Tomato ringspot viruses* (Table 14.4). None of the viruses concerned has, however, been reported from grapevine in SA.

Xiphinema index was first recorded in SA in the Swellendam area of the Western Cape Province (Heyns 1971). It was subsequently found in the Robertson area of the Breede River region (Barbercheck et al. 1985), in soil samples from the plant improvement vineyards in the Worcester area (Barbercheck and Heyns 1986), along the Berg River in the Paarl area (Van Reenen and Heyns 1986) and in Calitzdorp vineyards in the Klein Karoo region (Malan and Meyer 1994). The transmission of GFLV was reported for SA populations of *X. index* (Malan and Meyer 1992), with the virus being shown to have been transmitted to 31 different grapevine rootstocks (Malan and Meyer 1993). Reports of the occurrence of the nematode in five of the viticultural areas in SA are disturbing, not only because of the nematodes' vector ability but also because of the direct economic damage to the growth and production of susceptible grapevine rootstocks (Fig. 14.7) (Malan 1995). The soil, once infested with viruliferous *X. index*, cannot be used for the cultivation of GFLV-free grapevine.

Xiphinema italiae has been reported in all the viticultural regions of the Cape provinces, except for the Orange River and Klein Karoo regions (Malan and Meyer 1994). Cohn et al. (1970) showed *X. italiae* to be a vector of GFLV. However, SA specimens of *X. italiae* were found not to transmit GFLV under laboratory conditions (Malan 1995).

Of the 12 nepoviruses of grapevine transmitted by nematode pests (Brown et al. 2004), only GFLV has, so far, been reported from SA. However, four nematode vectors associated with grapevine virus transmission have been reported (Table 14.4). Importing virus-free plant material from countries in which these viruses occur is therefore vital. More surveys to determine the occurrence and distribution of virus-vector nematodes and the nepoviruses that they transmit are required. Conducting such surveys should give an indication of the risks involved due to the natural spread of virus-vector nematodes where nepoviruses occur in SA vineyards. Such action

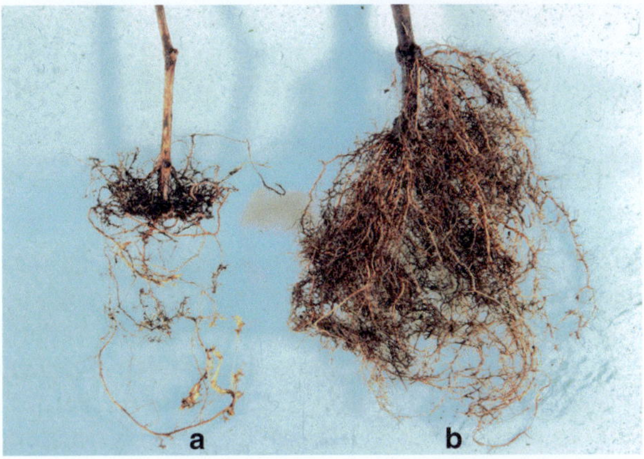

Fig. 14.7 Grapevine root systems (rootstock Paulsen 779) from a pot experiment 6 months after inoculation with 20 *Xiphinema index* individuals showing a damaged root system as a result of concomitant nematode and virus infection (**a**), compared to a healthy, non-infected grapevine root system (**b**) (Antoinette Malan, Stellenbosch University, Stellenbosch, South Africa)

would also enable the immediate setup and implementation of contingency plans for the eradication of these exotic viruses, should it be required.

14.6 Management Strategies

In the local grape industry, table and dried grapes as well as wine are subjected to stringent export regulations, which make the management of the nematode problem a challenging one. Vineyard production is moving towards an integrated pest management (IPM) approach, in line with the Scheme for Integrated Production of Wine (promulgated under the Act on Liquor Products, No. 60 of 1998). Cultural practices and the biological control of pests form the basis of an IPM programme.

When grapevines are planted in nematode-infested soil, e.g. when replanting old vineyards or planting grapevine on old stone fruit orchards, early growth of plants and establishment will likely be adversely affected. When replanting such vineyards, growers should consider using clean plant material, tolerant or resistant rootstocks and the application of a pre-plant nematicide.

14.6.1 Certified Plant Material

The Plant Improvement Act requires that registered vine nurseries abide by the related regulations. In the case of nematodes, the root material must be visually free of any nematode damage. In addition, all registered nursery plots must test free of the presence of *X. index*. Vineyards that are infected with GFLV can neither supply plant material for propagation nor be used as mother blocks.

14.6.2 Hot Water Treatment

As early as the 1900s, hot water treatment was recommended for control of the grape phylloxera on rooted grapevine in SA (Smith 1982). Similar treatment is recommended for the control of plant-parasitic nematodes and a variety of other pests in other countries (Smith 1982). In SA, Barbercheck (1986) recommended that dormant grapevine is immersed in water at 50 °C for 15 min to eradicate individuals of *M. javanica*. This is a practice that is still used by nurseries when compliance with the regulations of the Plant Improvement Act is required.

14.6.3 Resistant Rootstocks

Resistance of vines has mainly been studied with regard to *Meloidogyne* spp. In SA, excellent results have been obtained with several table grape varieties on Ramsey rootstock on sandy soils in the Hex River Valley (Smith 1982), and resistant rootstocks have been extensively used to limit root-knot nematode damage. The resistance of grapevine rootstocks to *M. incognita* under field conditions was tested by Loubser and Meyer (1987b). Only rootstock Ramsey was found to be resistant, while others ranged from moderately resistant to susceptible (Table 14.5). However, resistance to root-knot nematodes can be overcome, should soil temperatures rise above 30 °C for extended periods. None of the locally available rootstocks seems to be completely resistant to lesion, ring or dagger nematodes.

The use of the highly vigorous rootstocks (Ramsey and 99 Richter), which dominated the industry for the past 20 years, has declined in recent years in favour of less vigorous and often less resistant rootstocks. Such rootstocks tend to be more compatible with the production of high-quality wines on the varied soils that are present in the wine-growing regions. The reintroduction of the moderately susceptible rootstock, US 8–7, is a case in point. Critically, the highly vigorous rootstocks, Ramsey and 99 Richter, would appear to be susceptible to *C. xenoplax*. Thus rootstock resistance must be evaluated for all the major nematode pests that parasitise grapevine.

The development of nematode-resistant rootstocks, with better and lasting resistance, against a greater range of nematode pests will provide a more cost-effective and environmentally sustainable solution than is currently available. The rootstocks concerned must be tested under local conditions. The reason for this is to eliminate possible differences in the geographical populations of the nematode pests, as was shown by Malan and Meyer (1993) for *X. index* in SA and by Pinkerton et al. (2005) for *C. xenoplax* in Oregon and California.

Table 14.5 Resistance of grapevine rootstocks to *Meloidogyne incognita* under field conditions (Loubser and Meyer 1987b)

Rootstock	Resistance level	Rootstock	Resistance level
Ramsey	Resistant	110 Richter	Moderately susceptible
99 Richter	Moderately resistant	US 1–9	Moderately susceptible
101–14 Mgt	Moderately resistant	US 8–7	Moderately susceptible
143 B Mgt	Moderately resistant	Jacquez	Susceptible
1103 Paulsen	Moderately resistant	140 Ruggeri	Susceptible
US 2–1	Moderately resistant	C Metallica	Susceptible
3306 C	Moderately resistant	1202 C	Susceptible

14.6.4 Cultural Practices

Cover crops are planted in winter to prevent water erosion and to improve soil structure. A study by Addison and Fourie (2008) to determine the effect of cover crops on plant-parasitic nematode populations showed variable results. Grazing vetch (*Vicia sativa*) supported higher numbers of root-knot nematodes than did the other cover crops tested. Perennial ryegrass (*Lolium perenne*) did not support significantly higher numbers of lesion nematodes. In the case of ring nematodes, oat (*Avena sativa*) cv. Overberg maintained significantly lower nematode population levels during 2 of the 4 years that the cover crops were tested. However, no significant differences were found in cumulative counts over a period of 4 years.

The use of biofumigants for the control of soilborne diseases, including plant-parasitic nematodes, has been investigated under local conditions (Kruger 2013). These cover crops form isothiocyanate during soil incorporation. Laboratory bioassays, with macerated cover crops, showed a significant suppression of *M. javanica* in the case of white mustard (*Sinapis alba*) cv. Braco, Indian mustard (*Brassica juncea*) cv. Calienté 199 and rocket salad (*Eruca sativa*) cv. Nemat. However, there was no significant effect on *C. xenoplax* with any of the cover crops tested (Kruger et al. 2015a).

In glasshouse experiments, using several cover crops with biofumigation properties, *M. javanica* was able to complete its life cycle on all crops tested. The severity of gall formation was, however, less in the case of *E. sativa* cv. Nemat (Kruger et al. 2015a). There was some evidence that numbers of *C. xenoplax* were reduced following treatment with canola (*Brassica napus*) and *B. juncea* (cv. Calienté 199).

In general, the use of annual cover crops in local vineyards is an established cultural practice. In a study by Kruger et al. (2015b), cover crops were evaluated in a vineyard, using different management practices for their biofumigation impact on the suppression of economically important plant-parasitic nematodes over three growing seasons. Although no dramatic results were obtained, a steady decrease in root-knot nematode and ring nematode population densities was observed. Results from the study confirmed the positive effects of cover crops as part of a crop rotation programme since it prevented the population density build-ups of economically important nematodes in the vine row over time. Given the more integrated approach to pest management that is currently being adopted, additional research is required regarding the host status of the various cover crops in relation to the various plant-parasitic nematodes.

14.6.5 Chemical Control

By far the most common method of controlling nematodes in local vineyards is by the application of nematicides.

14.6.5.1 Pre-plant Fumigation

It is strongly recommended that vines should not be replanted until at least a year after the removal of the previous crop. This provides a time for the old roots to break down and so prevent the survival of nematodes in the old roots. Only after this interval should pre-plant fumigants be used. However, since fumigants are both expensive and have a negative impact on the environment, they are seldom used. An exception is where table grapes are replanted on old soils. When correctly done, fumigation is still the most cost-effective method.

14.6.5.2 Established Vineyards

The control of nematodes in established vineyards includes the use of the nematicides, cadusafos, fenamiphos or furfural. Immediately after being applied to the soil surface, the products are washed into the soil profile by means of irrigation. The berm ('bankie') should be weed and mulch-free. According to Loubser and Meyer (1987a), chemical control in established vineyards should commence immediately after harvest and/or during early spring. The new root flushes that appear during such periods should especially be protected against nematode infection. A noticeable improvement in berry and bunch size, as well as in vine growth, was recorded 6 months after nematicide treatment (Smith 1982). In some cases, the value of the yield increased by an average of 58 % in the third season after treatment. A progressive improvement in the growth and yield of grapevine following treatment was recorded over successive years. However, subsequent studies have shown that the response to treatment can be extremely variable (Loubser and De Klerk 1985; Loubser and Meyer 1986; Schneider et al. 2006), making the continual use of such chemicals difficult to justify.

14.6.6 Biological Control

The established fumigants and nematicides are under both consumer and regulatory pressures, and the need for alternative control measures is urgent. Only one biological product, DiTera® DF, derived from the soil fungus *Myrothecium verrucaria* strain AARC-0255, is currently registered for the treatment of grapevine. Given the limitations of biological products with regard to the mode of action and application, long-term control of nematodes in a perennial crop requires the application of repeated treatments. This makes the cost of DiTera® DF often prohibitive.

Awareness of the key role that is played by soil organic matter in maintaining soil health, a term that refers to a soil's capacity to function in its broadest sense, is steadily increasing (Walker and Stirling 2008). The use of mulches, cover crops, composts, compost teas, formulations containing various biological control agents and organic products (e.g. plant growth-promoting rhizobacteria) are important future options and need to be subjected to rigorous and independent testing.

14.7 Conclusions

The viticulture industry is under constant pressure both from within and from its end users in terms of implementing long-term environmentally sustainable production systems. Vineyard management will have to include the control of traffic, reduction of tillage and use of herbicides and pesticides, improvement of irrigation scheduling and paying more attention to improving soil health. Climate change (Sohlenius and Boström 1999), the use of shallower and shortened irrigation cycles and shallower rootstocks (McCarthy et al. 1997) as well as the use of mulches and cover crops have impacted on the presence of nematode pests (Van Huyssteen et al. 1984; Ingels 1998; Nicol et al. 1999; Guerra and Steenwerth 2012). This was demonstrated by either an increase (due to stress or increased temperatures) or a reduction (improved soil health) in nematode-pest population levels. Although such changes are likely to alter the available nematode management options, new challenges will arise, and new opportunities will be created to reduce the impact of nematode pests in grapevine.

References

Addison P, Fourie JC (2008) Cover crop management in vineyards of the Lower Orange River region, South Africa: 2. Effect on plant parasitic nematodes. S Afr J Enol Vitic 29:26–32

Barbercheck M (1986) Control of *Meloidogyne javanica* in dormant grapevine nursery stock. Phytophylactica 18:39–40

Barbercheck M, Heyns J (1986) Occurrence of *Xiphinema* in South African plant improvement vineyards. Phytophylactica 18:59–61

Barbercheck M, Smith P, Heyns J (1985) Occurrence and distribution of *Xiphinema* in vineyards of the Breede River Valley. Phytophylactica 17:27–30

Brown DJF, Halbrendt JM (1992) The virus vector potential of *Xiphinema americanum* and related species. J Nematol 24:584

Brown JFB, Zheng J, Zhou X (2004) Virus vectors. In: Chen ZX, Chen SY, Dickson DW (eds) Nematology advances and perspectives, vol 2. CAB International, Wallingford, pp 717–770

Cohn E, Tanne E, Nitzany FE (1970) *Xiphinema italiae*, a new vector of grapevine fanleaf virus. Phytopathol 60:181–182

Ferris H, McKenry MV (1975) Relationship of grapevine yield and growth to nematode densities. J Nematol 7:295–304
Guerra B, Steenwerth K (2012) Influence of floor management technique on grapevine growth, disease pressure, and juice and wine composition: a review. Am J Enol Vitic 63:149–164
Hewitt WB, Raski DJ, Goheen AC (1958) Nematode vector of soilborne fanleaf virus of grapevine. Phytophylactica 48:586–595
Heyns J (1971) A guide to the plant and soil nematodes of South Africa. AA Balkema, Cape Town
Hugo HJ (2003) Nematodes of Grapevine. Agricultural Research Council–Infruitec-Nietvoorbij. http://www.dfptresearch.co.za/admin/ProductsList.asp. Accessed 3 Jan 2015
Ingels CA (ed) (1998) Cover cropping in vineyards: a Grower's handbook: practical implications of interactions between soil biota and cover crops for Vineyards. University of California, Division of Agriculture and Natural Resources. Publication 3338. Davis, p 77–79
Inserra RN, Vovlas N, O'Bannon JH (1980) A classification of *Tylenchulus semipenetrans* biotypes. J Nematol 12:283–287
Jenkins WR (1964) A rapid centrifugal-flotation technique for separating nematodes from soil. Plant Dis Rep 48:692
Jones AT, Brown DJF, McGravin WJRM et al (1994) Properties of an unusual isolate of raspberry ringspot virus from grapevine in Germany and evidence for its possible transmission by *Paralongidorus maximus*. Ann Appl Biol 124:283–300
Kleynhans KPN, van den Berg E, Swart A et al (1996) Plant nematodes in South Africa. Plant Protection Research Institute. Handbook No. 8. Agricultural Research Council–Plant Protection Research Institute, Pretoria
Kloss EJ (2009) Rosette Mosaic. Virus diseases and non-infectious disorders of stone fruits in North America. USDA Agricultural Handbook No 437. USDA, Washington
Kruger DHM (2013) The role of cover crops with biofumigation potential for the suppression of plant-parasitic nematodes in vineyards. MSc Agric Thesis. Department of Conservation Ecology and Entomology, Stellenbosch University, Stellenbosch
Kruger DHM, Fourie JC, Malan AP (2015a) The effect of cover crops and the management thereof on plant-parasitic nematodes in vineyards. S Afr J Enol Vitic 36:195–209
Kruger DHM, Fourie JC, Malan AP (2015b) The control potential of Brassicaceae cover crops as green manure and their host status for *Meloidogyne javanica* and *Criconemoides xenoplax*. S Afr J Enol Vitic 36:165–174
Lamberti F, Molinari S, Moens M et al (2000) The *Xiphinema americanum* group. I. Putative species, their geographical occurrence and distribution and regional polytomous identification keys for the group. Russ J Nematol 8:65–84
Loots GC, Heyns J (1984) A study of *Xiphinema americanum* sensu Heyns, 1974 (Nematoda). Phytophylactica 16:313–320
Loubser JT, De Klerk C (1985) Chemical control of root-knot nematodes in established vineyards. S Afr J Enol Vitic 6:31–33
Loubser JT, Meyer AJ (1986) Strategies for chemical control of root-knot nematodes (*Meloidogyne* spp.) in established vineyards. S Afr J Enol Vitic 7:84–89
Loubser JT, Meyer AJ (1987a) Population dynamics of the root-knot nematodes *Meloidogyne incognita* Chitwood and *Meloidogyne javanica* Chitwood on grapevine in two different regions of South Africa. S Afr J Enol Vitic 8:36–40
Loubser JT, Meyer AJ (1987b) Resistance of grapevine rootstocks to *Meloidogyne incognita* under field conditions. S Afr J Enol Vitic 8:70–74
Lounsbury CP (1915) Division of Entomology: Annual Report 1913–1914. Rep Dep Agric S Afr 199–216
Malan AP (1995) The distribution, identity and virus vector potential of *Xiphinema* spp. in the Western Cape Province. S Afr J Enol Vitic 15:12–16
Malan AP, Meyer AJ (1992) Transmission of grapevine fanleaf virus by South African populations of *Xiphinema index*. Phytophylactica 24:217–219
Malan AP, Meyer AJ (1993) Interaction between a South African population of *Xiphinema index* and different grapevine rootstocks. S Afr J Enol Vitic 14:11–15

Malan AP, Meyer AJ (1994) Distribution of Longidoridae in the viticultural regions of the Cape Province. S Afr J Enol Vitic 15:12–16

Marais M, Swart A (2002) Plant nematodes in South Africa. 4. Modimolle area, Limpopo Province. Afr Plant Prot 8:25–32

McCarthy MG, Cirami RM, Furkaliev DG (1997) Rootstock response of Shiraz (*Vitis vinifera*) grapevine to dry and drip-irrigated conditions. Aust J Grape Wine Res 3:95–98

McKenry MV (1992) Nematodes. In: Flaherty DL, Christensen LP, Lenin WT et al (eds) Grape pest management, vol 3343, 2nd edn, Publication No. Division of Agricultural Science, University of California, Berkeley, pp 279–293

Nicol JM, Stirling GR, Rose BJ et al (1999) Impact of nematodes on grapevine growth and productivity: current knowledge and future directions, with special reference to Australian viticulture. Aust J Grape Wine Res 5:109–127

Pinkerton JN, Forge TA, Ivors KL et al (1999) Plant-parasitic nematodes associated with grapevine, *Vitis vinifera*, in Oregon vineyards. J Nematol 31:624–634

Pinkerton JN, Vasconcelos MC, Sampaio TL et al (2005) Reaction of grape rootstocks to ring nematode *Mesocriconema xenoplax*. Am J Enol Vitic 56:377–385

Quader M, Riley IT, Walker GE (2001) Distribution pattern of root-knot nematode (*Meloidogyne* spp.) in South Australian vineyards. Australas Plant Pathol 30:357–360

Quader M, Riley IT, Walker GE (2003) Spatial and temporal distribution patterns of dagger (*Xiphinema* spp.) and root lesion (*Pratylenchus* spp.) nematodes in a South Australian vineyard. Australas Plant Pathol 32:81–86

Schneider SM, Ajwa H, Trout TJ (2006) Chemical alternatives to methyl bromide for nematode control under vineyard replant conditions. Am J Enol Vitic Ame 57:183–193

Smith PC (1977) Distribution of plant-parasitic nematodes in vineyards in the Western Cape Province. Phytophylactica 9:27–28

Smith PC (1982) Nematode pests of grapevine. In: Keetch DP, Heyns J (eds) Nematology in Southern Africa, vol 400, Science Bulletin No. Department of Agriculture and Fisheries, Pretoria, pp 88–95

Sohlenius B, Boström S (1999) Effects of climate change on soil factors and metazoan microfauna (nematodes, tardigrades and rotifers) in a Swedish tundra soil – a soil transplantation experiment. Appl Soil Ecol 12:113–128

Sol HH, Van Heuven JC, Seinhorst JW (2009) Transmission of rattle virus and *Atropa belladonna* virus by nematodes. Tijdschr PlZiekten 66:228–231

Swart A, Malan AP, Heyns J (1996) *Paralongidorus maximus* (Bütschli) Siddiqi (Dorylaimida: Longidoridae) from South Africa. Phytopathol 2:31–33

Van Huyssteen L, Van Zyl JL, Koen AP (1984) The effect of cover crop management on soil conditions and weed control in a colombar vineyard in Oudtshoorn. S Afr J Enol Vitic 5:7–17

Van Mieghem AP, Pieterse W (1989) The distribution of Longidoridae in vineyards of the Western Cape. Phytophylactica 21:17–19

Van Reenen E, Heyns J (1986) A survey of Longidoridae in vineyards along the Berg River system. Phytophylactica 18:203–207

Walker GE, Stirling GR (2008) Plant-parasitic nematodes in Australian viticulture: key pests, current management practices and opportunities for future improvements. Australas Plant Pathol 37:268–278

Chapter 15
Nematode Pests of Deciduous Fruit

Hans J. Hugo and Sheila G. Storey

15.1 Introduction

Deciduous fruit production is a major contributor to the agricultural economy in South Africa (SA) and includes apple (*Malus domestica*), pear (*Pyrus* sp.), apricot (*Prunus armeniaca*), nectarine and peach (*Prunus persica*), and plum (*Prunus domestica*). These fruit crops are one of the most important commodities of the Western Cape Province. The Mediterranean-type climate, with its hot summers, the absence of rain and hail during the harvesting period, combined with cold winters to put trees in dormancy, makes this area highly suited for deciduous fruit production.

Deciduous fruit became a major farming activity during the early 20th century when improved storage management techniques allowed fruit to be exported to the United Kingdom (UK) and European countries in large quantities. Presently, SA, with an annual production of *ca* 3.26 million MT, is the fourth largest producer of deciduous fruit in the southern hemisphere, following Argentina, Brazil, and Chile (Hortgro 2015). South Africa is also the world's seventh largest exporter of apples, with approximately 413,292 MT being exported in 2014 (Hortgro 2015).

With deciduous fruit being a major export crop, most research over the past 100 years focused on insect and disease control to ensure that unblemished fruit reached the market. Below-ground problems caused by various diseases and pests, other than insects, received very little attention until 1964. Occasional popular articles appeared from time to time in local agricultural magazines such as "*Boerdery in*

H.J. Hugo (✉)
Agricultural Research Council–Infruitec/Nietvoorbij,
Private Bag X5026, Stellenbosch 7599, South Africa
e-mail: hans.hugoh1887@gmail.com

S.G. Storey
Nemlab, PO Box 2825, Durbanville 7551, South Africa

H. Fourie et al. (eds.), *Nematology in South Africa: A View from the 21st Century*, DOI 10.1007/978-3-319-44210-5_15

Suid-Afrika" by Koorts in 1961 as well as in the "*Landbouweekblad*" and "*Farmer's Weekly*. However, the first in-depth study of nematode damage to peach was done during the 1960s (Koorts 1964). Giliomee and Strydom (1968) investigated nematode pests as part of the replant problem in fruit orchards, but it was Meyer (1973, 1976) who demonstrated the importance of plant-parasitic nematodes as a limiting factor in peach production.

The association between particular nematode pests and fungi and/or bacteria in local apple, apricot, nectarine, peach and plum orchards poses another big problem for producers and is highlighted in Box 15.1. However, our knowledge of the interaction and damage caused by nematodes, fungi, and bacteria to deciduous fruit trees still lags far behind information on major annual crops such as maize (*Zea mays*), potato (*Solanum tuberosum*), various vegetable crops, and soybean (*Glycine max*).

Plant-parasitic nematodes that are associated with different deciduous crops locally (Kleynhans et al. 1996; SAPPNS[1]) are elaborated on below, with emphasis on the economically important species, their damage potential, symptoms, and strategies to manage them.

Box 15.1 Disease complexes of deciduous fruit crops in South African orchards as a result of the association between economically important plant-parasitic nematodes and other soilborne organisms (bacteria and fungi)

Fruit Crop	Decease	Causal organisms (Nematode pest + micro-organism)
Apricot	Die-back of branches and tree death	*Criconemoides xenoplax* + *Pseudomonas syringae* pv *syringae* (bacterial canker)
Nectarine	Peach tree short life (PTSL)	*Criconemoides xenoplax* + *Pseudomonas syringae* pv *syringae* (bacterial canker)
Plum	Plum tree death	*Criconemoides xenoplax* + *Pseudomonas syringae* pv *syringae* (bacterial canker)
Apple	Apple re-plant disease (ARD)	*Pratylenchus* spp. and/or complex of unknown micro-organism(s)

15.2 Pome Fruit

With 23,625 hectares (ha) harvested, apples are the largest component of the deciduous fruit crop, followed by the 12,697 ha of pears harvested in 2015 (Hortgro 2015). In total, pome fruit contribute *ca.* 1.34 million tons (920,406 MT for apples and 410,840 MT for pears) of the total deciduous fruit yield in SA during 2014 (Hortgro 2015).

[1]Dr Mariette Marais of the Nematology Unit, Biosystematics Division, Agricultural Research Council – Plant Protection Research Institute is thanked for the use of data from the South African Plant-Parasitic Nematode Survey (SAPPNS) database; E-mail: maraism@arc.agric.za

Table 15.1 The incidence of the major nematode pests, expressed as their frequency of occurrence (%), in apple orchards in three major apple-production areas of South Africa (Hugo 1994a)

Nematode pest genera	Grabouw-Villiersdorp ($n=43$)[a]	Ceres ($n=24$)	Langkloof ($n=33$)	Total ($n=100$)
Pratylenchus	95	92	100	96
Nanidorus/Paratrichodorus	44	54	80	58
Xiphinema	51	29	33	41

[a]Number of orchards sampled in a region

Table 15.2 The frequency of occurrence (%) of *Pratylenchus* spp. in root samples from 43 apple orchards sampled in the Grabouw-Villiersdorp production area (Hugo 1984)

Praylenchus spp.	No. of samples with >100 *Pratylenchus* spp. individuals 1 g root^{-1}	Percentage of orchards surveyed ($n=43$)
P. flakkensis	14	33
P. scribneri	6	14
P. vulnus	6	14
P. penetrans	5	12
P. zeae	3	7
P. pratensis	2	5

15.2.1 Apple

The most common plant-parasitic nematodes present in SA apple orchards are lesion (*Pratylenchus* spp.), stubby-root (*Nanidorus/Paratrichodorus* spp.), and dagger nematodes (*Xiphinema* spp.). Different genera and species often occur together in one orchard, making it difficult to relate nematode damage to a single species or even a single genus.

Results from a survey, during which 100 apple orchards were sampled in three production areas, showed that six species of lesion nematodes were encountered frequently and occurred in 96 % of the orchards (Hugo 1994a) (Tables 15.1 and 15.2). In the Grabouw-Villiersdorp area, the numbers of *Pratylenchus* spp. individuals in most of the 43 orchards were regarded as high, exceeding 100 individuals g roots^{-1} (Table 15.2). *Pratylenchus flakkensis* (Seinhorst, 1968) was the most abundant species, occurring in 14 of the 43 orchards. *Pratylenchus penetrans* (Cobb, 1917) Filipjev & Schuurmans Stekhoven, 1941, associated with apple replant disease, was identified in 5 of the 43 orchards sampled (Hugo 1984) (Table 15.2). Other lesion nematode species identified from this study were *Pratylenchus pratensis* (De Man, 1880) Filipjev, 1936, *Pratylenchus scribneri* Steiner, 1943, *Pratylenchus vulnus* Allen & Jensen, 1951 and *Pratylenchus zeae* Graham, 1951. In nine of these orchards, various combinations of two to three *Pratylenchus* spp. occurred in mixed populations. This indicates that *Pratylenchus* damage is not limited to one species, as is the case in The Netherlands or the United States of America (USA), where *P. penetrans* is considered the only lesion nematode species that causes damage on apple roots (Hoestra 1968; Mai and Abawi 1981).

Stubby-root nematodes (Trichodoridae) were present in 58 % of orchards in the main apple-growing areas of the Grabouw-Villiersdorp, Ceres, and the Langkloof areas (Table 15.1). At the time of the survey, their numbers were relatively low (Hugo 1994a). During subsequent investigations by the Agricultural Research Council–Infruitec/Nietvoorbij's Nematode Diagnostic Services and Nemlab Diagnostic Laboratory in Durbanville, high numbers (>250 individuals in 250 cm^3 $soil^{-1}$) of stubby-root nematodes were recorded from roots of apple trees. These nematodes were hence associated with poor tree growth and severe stubbiness of roots (H Hugo and S Storey, unpublished data).

Dagger nematodes (*Xiphinema* spp.) are major pests that parasitize apple roots in SA. They occurred in over 40 % of apple orchards (Table 15.1), often in numbers exceeding 250 individuals in 250 cm^3 $soil^{-1}$ with >500 individuals in 250 cm^3 $soil^{-1}$ not being uncommon (H Hugo Agricultural Research Council–Infruitec/Nietvoorbij 2015, personal observation). *Xiphinema americanum* Cobb, 1913, *Xiphinema diffusum* Lamberti & Bleve-Zacheo, 1979 and *Xiphinema elongatum* Schuurmans Stekhoven & Teunissen, 1938 were the species most frequently encountered. *Xiphinema diffusum* was the most common and occurred as a single population or in combination with either *X. americanum* or *X. elongatum*. High numbers of dagger nematodes are often associated with apple trees that exhibit poor growth and reduced yield (Hugo 1984). Although *X. americanum* is the vector of tomato ringspot virus (TmRSV), causing apple union necrosis or brown-line disease (Rosenberger et al. 1989), this virus has not yet been recorded in SA.

Helicotylenchus spp. are widespread in apple orchards, but their numbers are seldom very high, and they are therefore not regarded as major nematode pests of this fruit crop. Other parasitic nematodes that are infrequently observed are species of *Hemicycliophora*, *Longidorus*, *Paratylenchus*, and *Scutellonema* (S Storey and H Hugo, personal observation). However, their potential to damage apple crops is unknown.

A number of nematicide trials on apple trees demonstrated that controlling lesion and dagger nematodes significantly increased tree growth and yield (Hugo 1984, 1994b). In a glasshouse experiment where young apple trees were planted in soil containing a mixed population of *ca.* 1000 *X. americanum* and *X. diffusum* 250 cm^3 $soil^{-1}$, treatment with aldicarb, fenamiphos, and oxamyl significantly increased shoot length (Hugo 1984) (Table 15.3).

In a field trial on 14-year-old Granny Smith trees (M793 rootstock) that were cultivated in soil where high numbers of both lesion and dagger nematodes were present, fenamiphos (4 ml a.s. m^{-2}) increased shoot growth by up to 27 % (Hugo 1984). In another orchard where lesion nematodes occurred in excess of 100 individuals g^{-1} roots, treatment with aldicarb, fenamiphos, terbufos, and oxamyl reduced the lesion nematode numbers to nearly zero and increased yield between 53 and 89 % over a 2-year period (Hugo 1994b) (Tables 15.3 and 15.4).

One of the biggest concerns SA apple growers have when establishing a new orchard is apple replant disease (ARD). This disease presents itself as poor initial growth of young trees when old orchards are replanted. Unlike the situation in the USA and Europe where the concomitant occurrence of *P. penetrans* and/or fungi

Table 15.3 Effects of nematicide treatment on shoot growth of apple trees grown in soils infested with *Xiphinema* (Hugo 1984)

Treatment and active substance	Increase in shoot length (cm)
Autoclaved, sterile soil	240 a[a]
Aldicarb soil application (4.5 g a.s. m^{-2})	207 ab
Fenamiphos root dip (600 mg a.s. l^{-1})	194 ab
Fenamiphos soil application (4 ml a.s. m^{-2})	186 b
Oxamyl root dip (1200 mg a.s. l^{-1})	182 b
Oxamyl soil application (50 mg l^{-1} a.s. m^{-2})	168 bc
Untreated control	134 c
LSD ($P=0.01$)	46

[a]Means followed by the same letter do not differ significantly at $P<0.05$

Table 15.4 Effects of nematicides on the yield of apple trees grown in soils infested with *Pratylenchus* spp. (Hugo 1994b)

Treatment and dosage (active substance)	*Pratylenchus* spp. numbers (g $roots^{-1}$)		Yield (kg $tree^{-1}$)	Improvement (%)
	Before treatment	After treatment		
Fenamiphos EC (4 ml a.s. m^{-2})	962	1	24 a	89
Aldicarb (4.5 g a.s. m^{-2})	2,680	0	23 ab	78
Terbufos (0.45 g a.s. m^{-2})	1,147	2	23 ab	78
Fenamiphos GR (4 g a.s. m^{-2})	1,260	2	22 ab	70
Oxamyl SL (50 mg a.s. m^{-2})	660	171	20 ab	53
Control	1,309	329	13 c	-
LSD ($P=0.05$)			6.4	

[a]Means within a column followed by the same letter do not differ significantly at $P<0.05$

such as *Rhizoctonia* and *Pythium* spp. were shown to be the cause in some areas, no definite cause has yet been established in SA (Van Schoor et al. 2009). To control ARD, soil sterilization in the form of soil fumigation is done. This treatment also controls nematodes during the crucial first few years of apple tree establishment and is recommended as part of an integrated nematode control strategy.

At present, producers only have nematicides as an option to control nematode pests in apple orchards. *Pratylenchus* spp. pests are controlled by applications of carbamate and organophosphate nematicides, but these chemicals do not always result in a marked reduction in *Xiphinema* spp. numbers. Nematicide treatment does, however, consistently result in increased tree growth and yield (Hugo 1994b). Currently, cadusaphos and furfural are registered for use on pome fruit in SA (Van Zyl 2013).

As yet, there are no effective biological control measures available, although the application of compost at planting seems to stimulate plant growth to such an extent that satisfactory tree growth is obtained during the crucial establishment period (Van Schoor et al. 2009). Various biological products that claim to control nematode pests are available, but none can yet be recommended.

Although Parker and Mai (1974) reported that the growth response of different rootstocks varied in reaction to *P. penetrans* infestation, none of the current commercially available apple rootstocks are known to be tolerant or resistant to nematodes.

15.2.2 Pear

Stubby-root nematodes, belonging to the family Trichodoridae, are the most important nematodes that parasitize pear crops in SA. In a survey, these nematodes were recorded from 50 % of 101 pear orchards in the Eastern and Western Cape provinces and were associated with poor growth (Swart and Hugo 1984). *Nanidorus minor* (Colbran, 1956) Siddiqi, 1974 is the most common species, with a presence of 32 %, followed by *Paratrichodorus porosus* (Allen, 1957) Siddiqi, 1974 (28 %) and *Paratrichodorus lobatus* (Colbran, 1965) Siddiqi, 1974 (8 %).

Dagger nematodes (*Xiphinema* spp.) occurred in 48 % of pear orchards sampled during the survey, with one or more species present per orchard (Swart and Hugo 1984). *Xiphinema diffusum* was the most widespread species, having been recorded from 29 %, and *Xiphinema parvistillus* Heyns, 1971 from 8 % of the orchards sampled. *Xiphinema diversicaudatum* (Micoletzky, 1927) Thorne, 1934, *X. americanum, X. elongatum*, and *Xiphinema mluci* Heyns, 1976 were present in 3 % or less of the orchards. Eighteen percent of farmers interviewed during the survey reported poor tree growth and/or below-average pear yields. All these orchards sampled were associated with higher population densities of *Nanidorus*, *Paratrichodorus*, or *Xiphinema* spp. than orchards rated with growth and yield being normal (Swart and Hugo 1984).

Lesion nematodes occurred in 40 % of SA pear orchards (Swart and Hugo 1984). *Pratylenchus vulnus*, present in 13 % of the orchards, was the most widespread species. *Pratylenchus flakkensis*, *Pratylenchus neglectus* (Rensch, 1924) Filipjev & Schuurmans Stekhoven, 1941, *P. penetrans*, *P. pratensis*, *P. scribneri*, and *P. zeae* all occurred in 5 % or less of the orchards.

Helicotylenchus spp. were present in 55 % of all pear orchards surveyed in the Eastern and Western Cape provinces (Swart and Hugo 1984), with *Helicotylenchus dihystera* (Cobb 1893) Sher, 1961 being the only species present in all the production areas. Although *Helicotylenchus* spp. are widespread, their numbers are seldom high, and they are not regarded by the authors as serious pests of pear trees, based on numerous samples analyzed over three decades.

15.3 Stone Fruit

Stone fruit are fleshy fruit with a hard endocarp (stone pip) such as apricot, peach, and plum. These fruit are well-suited to the Mediterranean type of climate of the Western Cape. Since 1980, stone fruit production, notably peach and plum, has expanded rapidly to the summer rainfall areas of the North-West, Limpopo, and Mpumalanga provinces. However, the Western Cape province still delivers approximately 90 % of the total SA stone fruit production (Hortgro 2015).

Based on the results of ARC-Infruitec/Nietvoorbij's Nematode Diagnostic Services and Nemlab Diagnostic Laboratory, lesion, ring, dagger, and stubby-root nematodes are the most important nematode groups causing damage to SA stone fruit trees. However, it is important to determine the specific root stock used before allocating nematode-pest species to a stone fruit crop (Box 15.2).

Box 15.2 Important information to be obtained for samples designated for diagnostic analyses and/or research purposes

Care should be taken as the rootstock may differ from the scion. It is thus not correct to ascribe a nematode as a pest of a specific crop without establishing which rootstock is used. For example, the plum rootstock Marianna is commonly used for plums and apricot trees, whereas peach seedling rootstocks, although mainly used for peaches and nectarines, are also sometimes used for apricots and plum trees.

The different rootstocks differ in their suitability as a nematode host, e.g., peach rootstocks are parasitised by root-knot and ring nematodes, while Marianna is not a host to root-knot nematodes, and apricot is a poor host of these nematodes. The presence of a specific nematode pest in an orchard sample can, therefore, lead to wrong assumptions, since the nematode could have been hosted by a weed and not by the rootstock. This can lead to incorrect nematode management recommendations being supplied to growers.

Important questions to ask when samples are sent in for diagnostic analyses and/or research purposes are:

(i) What is the nematode complex that causes problems on the specific rootstock (especially when budding was done on another stone fruit rootstock)?
(ii) What nematode species are present in the soil and root samples?
(iii) Were the samples taken correctly (see Chap. 4, Sect. 4.2), and do they represent the rootstock and not weeds growing in the vicinity?

15.3.1 Apricot

A survey of plant-parasitic nematodes in apricot orchards by Kleynhans and Hugo (1983) indicated that the incidence and population densities of plant-parasitic nematodes were substantially lower in this crop than in peach orchards. This phenomenon

Fig. 15.1 Die-back of an apricot tree due to infection by high ring nematode population densities. The white mass is batting used to control insects, not fungal growth (Hans Hugo, Stellenbosch, South Africa)

was documented by Meyer (1976). This could possibly be due to apricot often being planted in more marginal soils under dry-land conditions, which is not conducive to the expansion of nematode populations. Based on the results of the survey, the ring nematode *Criconemoides xenoplax* Raski, 1952 can be regarded as the most important nematode pest of apricot in SA. This species was present in 40 % of the 89 orchards included in the survey. According to ARC-Infruitec/Nietvoorbij's Nematode Diagnostic Services' records, *C. xenoplax* occurred in numbers higher than 500 individuals 250 cm^3 $soil^{-1}$ in approximately 50 % of the apricot samples received from growers. High *C. xenoplax* numbers have been associated with branch die-back and tree death caused by bacterial canker in the Ceres and Ladismith areas (Fig 15.1).

Meloidogyne spp. were found in just five (6 %) of the 89 apricot orchards sampled. This could be attributed to apricot rootstocks being less suitable hosts for root-knot nematodes than are peach rootstocks. Casual observations by the authors have shown that Royal apricot seedling, a rootstock commonly used for apricots, may be a poor host to *Meloidogyne* spp., but its host status needs to be confirmed.

Seven species of the dagger nematode, *Xiphinema*, were recorded from 29 (33 %) of the 89 orchards sampled. *Xiphinema diffusum* was the most common species, occurring in 11 % of the orchards. *Pratylenchus* spp., typically the most common genus present in deciduous fruit orchards, was present in 26 % of the apricot orchards.

15.3.2 Peach and Nectarine

Peaches and nectarines are produced in all nine provinces of SA on approximately 9,690 ha (during 2014), of which 91 % are in the Western Cape (Hortgro 2015). Production of peaches and nectarines amounted to approximately 174,230 MT in 2014 (Hortgro 2015)

Fig. 15.2 Typical symptoms as a result of plant-parasitic nematode infection of a peach tree are underdevelopment of shoots and a rosette of leaves, which are rolled up and often reddish along the edges resembling glyphosate damage (Hans Hugo, Stellenbosch, South Africa)

A survey by Meyer (1976) of 300 peach orchards showed that Criconematinae, Trichodoridae, and species of *Helicotylenchus*, *Meloidogyne*, *Paratylenchus*, *Pratylenchus*, *Scutellonema* and *Xiphinema* occurred in at least 40 % of the sites sampled. Numerous field observations by the authors between 1982 and 2010 showed that ring, root-knot, and dagger nematodes are the most important plant-parasitic nematodes parasitising peach and nectarine roots in SA. Typical symptoms of nematode infection on peach trees are poor or no development of shoots, while a rosette of leaves is formed. Leaves are rolled up and are often reddish along the edges resembling glyphosate damage (Fig. 15.2).

Meyer (1973) identified two ring nematode species, viz., *C. xenoplax* and *Criconema mutabile* (Taylor, 1936) De Grisse & Loof, 1965. The authors regard *C. xenoplax* as the more important species based on long-term diagnostic experience. In almost all samples, this is the only ring nematode species present, often in numbers exceeding 1,000 individuals 250 cm^3 $soil^{-1}$. *Criconemoides xenoplax* is also often associated with poor tree growth. In 2011, peach tree short-life (PTSL) was confirmed in a nectarine orchard in Tulbagh, Western Cape Province (Fig. 15.3). This disease represents a complex of *C. xenoplax* and bacterial canker (*Pseudomonas syringae* pv *syringae*), which causes branch die-back and tree death (Hugo and Storey 2011).

Coetzee (1968) reported the presence of four *Meloidogyne* spp. from the Western Cape Province, namely *Meloidogyne arenaria* (Neal, 1889) Chitwood, 1949, *Meloidogyne hapla* Chitwood, 1949, *Meloidogyne incognita* (Kofoid and White 1919) Chitwood, 1949 and *Meloidogyne javanica* (Treub, 1885) Chitwood, 1949. Although root-knot nematodes occurred in only about half of all the peach orchards surveyed by Meyer (1973), they cause severe damage. This nematode genus is also

Fig. 15.3 A severe case of peach tree short-life, showing dying trees in the middle of the row, in a nectarine orchard due to plant-parasitic nematode infection (Hans Hugo, Stellenbosch, South Africa)

the only one that has been observed by the authors to cause the death of young peach trees within 1 year after planting. Data indicate that only *M. incognita* race 2 and *M. javanica* are present in peach orchards in the Western Cape Province (Kleynhans et al. 1996). *Meloidogyne javanica* is the species most often encountered during diagnostics. However, *M. incognita*, although present in some orchards in the Western Cape Province, is more prevalent in peach and nectarine orchards in the cooler areas of the Highveld (Kleynhans 1991). Symptoms of damage on older trees include premature leaf fall during January and die-back of small shoots. Leaves sometimes show deficiency symptoms for macro-elements, despite adequate levels present in the soil.

Another very common genus, especially in the Western Cape Province, is *Xiphinema,* which is reported to parasitise peaches and nectarines. Populations often consist of more than one species, with *X. diffusum* present in almost all localities, often in high numbers. Meyer (1976) reported *Xiphinema* from 16 peach orchards (5 %) in numbers in excess of 500 individuals 250 cm^3 $soil^{-1}$, which are regarded as very high for this genus. Other species identified by Meyer were *Xiphinema meridianum* Heyns, 1971, *Xiphinema parvistilus* Heyns, 1971 and *X. elongatum,* as confirmed by Kleynhans et al. (1996). *Xiphinema elongatum* has been identified from various deciduous fruit and vine crops throughout the Western Cape. The study by Meyer and Hugo (1994) reported that *Xiphinema* can cause serious damage to peach trees.

Stubby-root nematodes are also sometimes encountered during diagnostic analyses and cause damage to peaches and nectarines, especially on sandy soils. Kleynhans et al. (1996) identified *N. minor, P. lobatus,* and *P. porosus* from deciduous fruit tree roots sampled in the Western Cape Province.

Lesion nematodes, although very common and present in 88 % of the orchards sampled, are usually only present in relatively low numbers (Meyer 1973). Species identified were *Pratylenchus coffeae* (Zimmerman, 1898) Filipjev & Schuurmans Stekhoven, 1949, *P. neglectus*, *P. penetrans*, *P. pratensis*. *P. scribneri*, and *P. vulnus*.

15.3.3 Plum

Plum is a high-value crop in SA, with *ca.* 60,000 MT (75 %) of the total crop of 81,250 MT exported in 2014 (Hortgro 2015). However, it is also very susceptible to what is locally known as "plum tree death," a disease complex consisting of ring nematodes, bacterial canker, as well as environmental stress factors. In some years, more than 50 % of the trees in an orchard can die. The symptoms are similar to PTSL, with gummosis, die-back of branches, and a sour-sap smell of affected branches shortly after blossom.

The rootstock most often used for plum trees, Marianna, is regarded as immune to root-knot nematodes (Nyczepir and Halbrendt 1993). However, it is a very good host for the ring nematode *C. xenoplax*, which causes serious damage to roots of plum trees. According to the Agricultural Research Council–Infruitec/Nietvoorbij's Nematode Diagnostic Services' records, *C. xenoplax* has been identified from 68 % of Marianna rootstock samples examined.

Dagger nematodes, mostly *X. diffusum*, are also often associated with Marianna rootstock. According to the Agricultural Research Council–Infruitec/Nietvoorbij's Nematode Diagnostic Services, records of *Xiphinema* spp. occur in approximately 33 % of samples taken from soil around Marianna roots.

Stubby-root nematodes are also present in soils associated with Marianna rootstock (11 % of samples analyzed by Agricultural Research Council–Infruitec/Nietvoorbij's Nematode Diagnostic Services). Severe root damage and poor growth have been observed in orchards planted on coarse sandy soils (H Hugo, Agricultural Research Council–Infruitec/Nietvoorbij's, Stellenbosch, 2015, personal communication).

Lesion nematodes are occasionally present in plum orchards but seldom in high numbers. *Pratylenchus vulnus* is the only species of this group that has been identified from plum orchards (Kleynhans et al. 1996), but according to the authors, *Pratylenchus* spp. could be more frequently associated with plum than is documented. This possible underestimation of the association of *Pratylenchus* spp. with plum trees could be due to the large numbers of *Criconemoides* spp. (often exceeding 1,000 individuals 250 cm^3 $soil^{-1}$), which masks the presence of other nematodes in diagnostic samples.

15.4 Conclusions

Deciduous fruit has been a major commodity for over a century in the Western Cape. Insect pests and plant diseases have received wide attention, but nematode pests only came to the fore with the work by Meyer (1973) on peach and Hugo

(1984) on apple. Due to the scarcity of virgin soil available for new plantings, more and more deciduous fruit orchards are replanted every year, causing a build-up of nematode pest population densities. At the same time, growers only have a limited choice of fruit crops and rootstocks. With dagger, lesion, and stubby-root nematodes attacking the roots of all deciduous fruit crops and ring nematodes parasitising all stone fruit rootstocks, nematode problems are bound to increase. With major gaps in our knowledge of the interactions between nematode pests and deciduous fruit trees and the damage that nematodes can cause, more research is urgently needed.

References

Coetzee V (1968) The distribution of the family Heteroderidae (Filipjev, 1934) in South Africa and some host records of *Meloidogyne* species. S Afr J Agric Sc 11:775–788

Giliomee JH, Strydom DK (1968) Replant diseases of deciduous fruit trees. Deciduous Fruit Grower 18:187–189

Hoestra H (1968) Replant disease of apple in the Netherlands. Mededelingen Landbouwhogeschool Wageningen 68:1–105

Hortgro (2015) Key deciduous fruit statistics 2015. Hortgro, Paarl

Hugo HJ (1984) Introductory investigations of some nematodes associated with apple trees in the Grabouw area. MSc Agric Dissertation, Stellenbosch University, Stellenbosch

Hugo HJ (1994a) Survey of the different nematode species occurring in Western Cape apple orchards. Project 2300/03 final report, ARC – Infruitec-Nietvoorbij, Stellenbosch

Hugo HJ (1994b) Chemical control of nematodes on apple. Project 2300/04: final report, ARC – Infruitec-Nietvoorbij, Stellenbosch

Hugo HJ, Storey SG (2011) Confirmation of the presence of PTSL in South Africa. S Afr J Plant Soi 28:252

Kleynhans KPN (1991) The root-knot nematodes of South Africa. Technical Communication No. 231. Directorate of Agricultural Information. Department of Agricultural Development, Pretoria

Kleynhans KPN, Hugo HJ (1983) Plant-parasitic nematodes in apricot orchards. Deciduous Fruit Grower 33:209–212

Kleynhans KPN, Van den Berg E, Swart A et al (1996) Plant nematodes in South Africa. Plant Protection Research Institute Handbook No. 8. Agricultural Research Council, Pretoria

Koorts LJ (1961) Aalwurm op die aanval. Boerdery in Suid-Afrika. September 1961:57–59

Koorts LJ (1964) Die aalwurmprobleem in die Westelike Kaapprovinsie. In: Proceedings of a Symposium on Entomological Problems, 17–21 July 1961. Technical Communication No. 12. Department of Agricultural Technical Services, Pretoria, p 146–150

Mai WF, Abawi GS (1981) Controlling replant diseases of pome and stone fruits in northeastern United States by preplant fumigation. Plant Dis 5:859–864

Meyer AJ (1973) 'n Studie van die verhouding tussen plant-parasitiese nematodes en die perske in Suidwes-Kaapland. PhD Thesis, Stellenbosch University, Stellenbosch

Meyer AJ (1976) Plant-parasitic nematodes associated with peach orchards in the Western Cape Province. Phytophylactica 8:21–22

Meyer AJ, Hugo HJ (1994) *Xiphinema americanum* damaging peach trees in South Africa. J Nematol 26:111

Nyczepir AP, Halbrendt JM (1993) Nematode pests of deciduous fruit and nut trees. In: Evans K, Trudgill DL, Webster JM (eds) Plant parasitic nematodes in temperate agriculture. CAB International, Wallingford, pp 381–425

Parker KG, Mai WF (1974) Root diseases of fruit trees in New York State. VI. Damage caused by *Pratylenchus penetrans* to apple trees in the orchard growing on different rootstocks. Plant Dis Rep 58:1007–1011

Rosenberger DA, Cummins JN, Gonsalves D (1989) Evidence that tomato ringspot virus causes apple union necrosis and decline: Symptom development in inoculated apple trees. Plant Dis 73:262–265

Swart A, Hugo HJ (1984) Plant-parasitic nematodes in pear orchards in the Eastern and Western Cape. Phytophylactica 16:49–52

Van Schoor L, Denman S, Cook NC (2009) Characterisation of apple replant disease under South African conditions and potential management strategies. Sci Hortic 119:153–162

Van Zyl K (2013) A guide to crop pest management in South Africa. A compendium of acaracides, insecticides, nematicides, molluscicides, avicides and rodenticides. A CropLife South African Compendium, 1st edn. VR Print, Pinetown

Chapter 16
Nematode Pests of Banana

Mieke S. Daneel and Dirk De Waele

16.1 Introduction

Banana (*Musa*) is believed to have originated in the South-East Asian and West Pacific regions where their edible, seed-bearing, diploid ancestors are still found in natural forests. The earliest records dealing with the cultivation of these crops are from India, approximately 2,500 years ago. Although the exact route and time frame of the distribution of *Musa* outside Asia is uncertain, banana were known during the 15th century on the west coast of Africa, from where the Portuguese introduced them into the Canary Islands. In the 16th century banana were found in Haiti, the Caribbean and tropical America, which is where most export dessert banana are now produced. There are widely varying views on the establishment of banana in East Africa, although it is accepted that they were introduced into Mozambique in the 16th century. Introduction of banana into South Africa (SA) was apparently much later (Jones and Milne 1982).

The banana industry in SA is based on the production of the Cavendish subgroup of cultivars (*Musa acuminata* Colla; AAA group). The most popular cultivars (cvs) are Grand Nain, Williams, Chinese Cavendish and Dwarf Cavendish (Robinson 1993). Lowland, tropical regions with rainfall in excess of 1250 mm year^{-1} and minimum temperatures above 15 °C are preferred for banana production (Simmonds 1966). Despite suboptimal conditions with regard to minimum temperatures and rainfall, a banana-growing industry is well established in low-lying and frost-free

M.S. Daneel (✉)
Agricultural Research Council–Institute for Tropical and Subtropical Crops,
Private Bag X11208, Mbombela 1200, South Africa
e-mail: mieke@arc.agric.za

D. De Waele
Laboratory of Tropical Crop Improvement, Department of Biosystems, University of Leuven,
Willem de Croylaan 42, 3001 Heverlee, Belgium
e-mail: dirk.dewaele@agr.kuleuven.ac.be

H. Fourie et al. (eds.), *Nematology in South Africa: A View from the 21st Century*, DOI 10.1007/978-3-319-44210-5_16

areas of SA on an overall area of 11,360 hectares (ha) (DAFF 2011). These regions are Levubu (Limpopo Province), Hazyview and Onderberg (both in Mpumalanga Province), and the North and South Coast of KwaZulu-Natal Province. Lately banana has been grown more extensively in Mozambique in the Maputo and Nampula areas. Swaziland also has a small area under banana.

Plant-parasitic nematodes are a serious problem of banana worldwide (Gowen et al. 2005), including SA. Based on 1996 production figures, Willers (1998) estimated that these pests caused a direct loss of 19 % in total production of the crop in SA. Thirty-four plant-parasitic nematode species have been associated with banana in SA (Kleynhans et al. 1996; SAPPNS[1]). However, economic damage can only be ascribed to a limited number of these species. Plant-parasitic nematodes usually occur in banana roots or root zones and represent concomitant infestations of three or more species. Quantification of the damage by individual species is, therefore, problematic. Above-ground symptoms of nematode infection on banana can be confused with those associated with damaged or diseased root systems.

The economically important plant-parasitic nematodes associated with banana in SA are root-knot (*Meloidogyne* spp.), spiral (*Helicotylenchus* spp.), burrowing (*Radopholus similis*) (Cobb, 1893) Thorne, 1949, and lesion (*Pratylenchus* spp.) nematodes. The major species that parasitise banana crops are summarised below with special reference to the symptoms and damage they cause to banana as well as their host range. All nematode pests associated with banana in SA are recorded in Kleynhans et al. (1996), whilst those identified after this publication are being added on a continuous basis to the SAPPNS Database[1] (see Chap. 21).

16.2 Root-Knot Nematodes

The root-knot nematode species commonly reported in association with local banana plantations are *Meloidogyne incognita* (Kofoid and White, 1919) Chitwood, 1949, and *Meloidogyne javanica* (Treub, 1885) Chitwood, 1949. *Meloidogyne arenaria* (Neal, 1889) Chitwood, 1949, was found on banana in Zimbabwe (Jones and Milne 1982).

A survey of commercial plantations of banana revealed that root-knot nematodes were present in 93 % of the samples (Table 16.1). Root-knot nematodes were the most abundant and together with spiral nematodes constituted 72 % of the total plant-parasitic nematode complex (De Jager et al. 1999). A similar survey of banana in smallholding plantations in the rural areas of SA and Swaziland showed that root-knot nematodes were present in 93 % of the root samples (Table 16.2) (Daneel et al. 2003).

[1] Dr Mariette Marais of the Nematology Unit, Biosystematics Division, Agricultural Research Council–Plant Protection Research Institute is thanked for the use of data from the South African Plant-Parasitic Nematode Survey (SAPPNS) database; E-mail: maraism@arc.agric.za.

Table 16.1 Mean number of nematodes recorded from soil and roots in a survey of plantations within the three major commercial banana-producing areas of South Africa (Daneel et al. 2015)

	Soil samples (250 ml)			Root samples (30 g)			
Area[a]	O[a]	H[b]	S KZN[c]	O	H	S KZN	Percentage occurrence
Number of samples	163	124	117	166	119	117	
Helicotylenchus spp.	474	1142	1953	1059	1292	1532	95
Meloidogyne spp.	730	749	835	1571	571	750	93
Pratylenchus coffeae	17	55	15	37	76	26	26
Radopholus similis	3	14	11	79	45	41	19
Paratylenchus sp. (84 %)[d]	0	0	185	0	0	266	22
Rotylenchulus spp.	279	1	0	7	0	0	2
Nanidorus spp.	3	31	30	0	0	0	4

[a]*O = Onderberg*
[b]*H = Hazyview*
[c]*S = KZN* South Coast of KwaZulu-Natal
[d]Percentage for KZN as this genus was not recorded from banana in any of the other areas

Table 16.2 Percentage frequency of occurrence of plant-parasitic nematodes associated with banana grown in informal gardens in South Africa and Swaziland (Daneel et al. 2003)

	Soil		Roots		
Nematode genera/species	*n*[a]	Percentage occurrence[b]	*n*	Percentage occurrence	Locations[c]
Meloidogyne spp.	152	96.8	147	93.6	All regions
Helicotylenchus spp.	152	96.8	147	93.6	All regions
Radopholus similis	0 0 6	3.8	1 1 12	8.9	BBR K N KZN
Pratylenchus coffeae	5 0 2 5	7.6	0 1 1 3	3.2	BBR V S EC, S KZN
Rotylenchulus spp.	4 1 2	4.4	0 0 2	1.3	K S BBR
Nanidorus/Paratrichodorus spp.	11	7.0	–	–	EC, S KZN
Criconemoides spp.	4 3 2	5.7	–	–	K N KZN EC, S KZN
Paratylenchus spp.	1	0.6	–	–	EC, S KZN

[a]*n* number of localities where nematode species were found out of 157 sampled
[b]Percentage of localities where nematodes were present in soil or root samples
[c]Localities where nematode taxa were present: *BBR* = Bushbuck Ridge (*n*=29); *EC and S KZN* = Eastern Cape and South KwaZulu-Natal (*n*=39); *K* = Komatipoort (*n*=18); *N KZN* = North KwaZulu-Natal (*n*=29); *S* = Swaziland (*n*=7); *V* = Venda (*n*=35)

16.2.1 Damage

Heavily infected plants are stunted and have thin pseudostems, whilst leaves are yellowish or show discoloured, greenish-yellow bands along the leaf blades (Milne and Kuhne 1968). Infected plants are also prone to wilting during moderately hot days.

Rabie (1991) reported that the concomitant occurrence of root-knot nematodes and various stress factors, including drought and cold temperatures, are involved in inflicting symptoms of 'False Panama Disease', which resembles 'Panama Disease'. In the former case transverse sections of rhizomes of infected plants showed reddish-brown to brownish-purple discoloured vascular tissue. Leaf symptoms include progressive dying back of older leaves, starting at the tips. Galls occur on the primary and secondary roots, whilst distortion of roots and sometimes bifurcation occurs after heavy nematode infections.

Since root-knot nematodes are present in almost all banana plantations (Daneel et al. 2015) (Tables 16.1 and 16.2), it is particularly important to monitor their population levels in young crops. The general perception is that these nematode pests can cause severe damage to young plants, resulting in suboptimal growth and yield.

16.2.2 Host Range

In the subtropical areas of SA, the host range of both *M. incognita* and *M. javanica* includes most broadleaf weed species and cultivated crops. Special attention should, therefore, be given to weeding during fallowing or crop selection for rotation to control root-knot nematodes in banana (Willers et al. 2001).

16.3 Hoplolaimidae

Helicotylenchus multicinctus (Cobb, 1893) Golden, 1956, *Helicotylenchus dihystera* (Cobb, 1893) Sher, 1961, and *Helicotylenchus erythrinae* (Zimmermann, 1904) Golden, 1956, are the most abundant spiral nematodes that parasitise banana in SA (Jones and Milne 1982; Daneel et al. 2015). A survey of commercial plantations by the latter authors showed that species of *Helicotylenchus* (mainly *H. multicinctus*) were present in 95 % of all the samples (Table 16.1). These nematodes had the highest average numbers overall. Spiral nematodes (mainly *H. multicinctus*) were also recorded in most of the soil and root samples collected in a survey by Daneel et al. (2003) in the rural areas of SA and Swaziland (Table 16.2). These results agree with those of Gowen et al. (2005), who suggested that *H. multicinctus* is often the major parasitic nematode on banana where temperature and rainfall conditions are suboptimal for the crop.

Fig. 16.1 Necrotic areas on a banana root showing damage caused by feeding and migrating burrowing nematodes (Kirk West, Port Elizabeth, South Africa)

Fig. 16.2 (**a**, **b**) A damaged banana plant infected with burrowing nematodes (**a**) resulting in toppling of plants (**b**) (Kirk West, Port Elizabeth, South Africa)

16.3.1 Damage

The above-ground symptoms caused by *H. multicinctus* are similar to those caused by other nematode pests of banana. Toppling may occur when infection levels of this nematode pest are very high (Gowen and Quénéhervé 1990). Symptoms of damage inflicted by *H. multicinctus* are blackening of the root epidermis, small reddish lesions in the superficial cortex and a reduction in the number of lateral roots (Jones and Milne 1982). When heavy infections of this spiral nematode occur, lesions coalesce and cause extensive necrosis in the outer root cortex and even root dieback. Lesions may also be present in rhizome tissue of infected plants (Quénéhervé and Cadet 1985). *Helicotylenchus multicinctus* has no distinct biotypes or races (Gowen and Quénéhervé 2005).

16.3.2 Host Range

Most edible banana and plantain cvs and a range of alterative host plants, such as pigweed (*Amaranthus* spp.), purslane (*Portulaca oleracea*) and ornamentals (Gowen et al. 2005) are hosts to *H. multicinctus*.

16.4 Burrowing Nematode

The burrowing nematode, *R. similis* is a notoriously destructive pest of banana crops throughout the world. However, it is not widespread in SA or in the southern African region. The nematode was present in only 19 % of samples collected during a survey of the three main banana-producing areas, viz. the KwaZulu-Natal Province and Hazyview and Komatipoort (Mpumalanga Province) (Table 16.1) (Daneel et al. 2015). The Levubu and Tzaneen areas in the Limpopo Province are free of *R. similis*. In a survey conducted in banana in the rural areas of SA and Swaziland, *R. similis* was found in fewer than 9 % of the root samples examined (Daneel et al. 2003) (Table 16.2). A study on 57 burrowing nematode isolates collected from Australia, Cameroon, Central America, Cuba, Dominican Republic, Florida, Guadeloupe, Hawaii, Nigeria, Honduras, Indonesia, Ivory Coast, Puerto Rico, SA and Uganda showed that 55 of them were morphologically similar to *R. similis*. Seven of these isolates, all obtained from material collected in Florida (USA), were parasites of citrus (*Citrus* spp.) (Kaplan et al. 2000). The SA isolates were not parasitic to citrus, and there is no record that a citrophylic isolate is present in SA.

16.4.1 Damage

Radopholus similis infection causes mass destruction of primary roots and poor anchorage of banana plants (Fig. 16.1). The nematode could be present throughout the entire root system, including the rhizome (De Villiers et al. 2007). Toppling of plants often occurs during windy conditions when bunches are present, hence the general name of 'toppling disease' (Fig. 16.2). Jones and Milne (1982) explained that the lateral movement of *R. similis* in the root cortex, as well as secondary colonisation of the vacated cavities by parasitic and saprophytic fungi, enlarges lesions so that the stele may be disrupted. When this occurs, the entire root beyond the initial nematode entry site becomes functionless. The stele is never entered by this nematode (Jones and Milne 1982).

After the identification of *R. similis* in SA in 1969 (Jones and Milne 1982), a strict quarantine system was introduced. The practice of using rhizomes as propagating material was stopped in SA and many other parts of the world since *R. similis* could be spread this way. A permit needs to be obtained for the transport of suckers and rhizomes between local production areas. In quarantined regions, notably the KwaZulu-Natal Province, a permit also has to be obtained even to move material

Fig. 16.3 Tissue culture banana plants reared in vitro in glass vials in the laboratory (Mieke Daneel, Agricultural Research Council–Institute for Tropical and Subtropcial Crops, Mbombela, South Africa)

Fig. 16.4 Banana plants, derived from in vitro tissue culturing in the laboratory, growing after being transplanted to fields (Mieke Daneel, Agricultural Research Council–Institute for Tropical and Subtropcial Crops, Mbombela, South Africa)

between districts or farms. With the introduction of tissue-cultured propagating material, the spread of this nematode pest between commercial plantations is no longer an issue (Box 16.1 and Figs. 16.3 and 16.4). However, smallholding producers that are still planting suckers need to follow alternative measures to prevent the

Box. 16.1 Prevention of the spread of *Radopholus similis* and other plant-parasitic nematodes in and between banana plantations

Originally banana plantations were developed using rhizomes, which are often infected with plant-parasitic nematodes. Planting such nematode-infected rhizomes leads to nematode infections in new plantations. Burrowing nematodes were spread in this way to many banana-growing areas, including SA.

Although legislation, developed in SA in 1983, was very effective in limiting the spread of *R. similis* in local banana-growing areas, it was the combination of this legislation with the development of tissue culture plants (as a means of propagating nematode-free plants) that really limited the spread of *R. similis*. Tissue culture plants are developed in the laboratory from clean planting material and are free of pests and diseases. Before the tissue-cultured plants are distributed to the producers, the plantations have to be tested and declared virus-free.

spread of *R. similis* (M. Daneel, Agricultural Research Council–Institute for Tropical and Subtropical Crops, Mbombela and D. De Waele, University of Leuven, Leuven, 2016, personal communication).

16.4.2 Host Range

After Duchame and Birchfield (1956) established the existence of a *R. similis* biotype that also attacked citrus, it was confirmed that the *R. similis* that occurs in SA did not attack citrus (Keetch 1972; Milne and Keetch 1976). An experiment was conducted by Keetch (1972) and Milne and Keetch (1976) in which plants were planted in soil infested with *R. similis*, and the reaction of these plants was noted to determine the host status of these plants. Only 20 out of 100 plants tested were found to be able to act as host for *R. similis*. However, in both the National Collection of Nematodes (NCN) and South African Plant-Parasitic Nematode Survey (SAPPNS), any record of *R. similis* in the field is only associated with bananas.

16.5 Lesion Nematodes

The pandemic lesion nematodes are also well represented in the southern African region, but their distribution in banana plantations is limited. Two species of *Pratylenchus* are regularly found in SA plantations, viz. *Pratylenchus coffeae* (Zimmerman, 1898) Filipjev and Schuurmans Stekhoven, 1941, and *Pratylenchus brachyurus* (Godfrey, 1929) Filipjev and Schuurmans Stekhoven, 1941, with the former being more frequently found (Daneel et al. 2003). *Pratylenchus coffeae* was found in 25.9 % of root and soil samples in commercial plantations (Table 16.1) (Daneel et al. 2015). Samples from several individual plantations had no lesion nematodes, but these pests were widely distributed throughout the banana-producing areas. In commercial banana plantations, lesion nematode numbers varied from 0 to 1,400 nematodes 30 g roots^{-1}. The survey of Daneel et al. (2003) in rural banana-producing areas showed that, although lesion nematodes were present in all areas, *P. coffeae* constituted only 3.2 % of the nematode pest complex present in banana roots and 7.6 % in soil samples (Table 16.2).

16.5.1 Damage

The symptoms of damage caused by *P. coffeae* are very similar to those caused by *R. similis*. They include stunting of the plants, slowing of the vegetative phase, reduction in the number of leaves, lower bunch mass and reduced lifespan of plantations. Willers et al. (2001) reported that smallholding banana plantations in the Giyani area (Limpopo Province) were rendered totally unproductive as a result of *P. coffeae* infections during the crop cycle. This was despite the fact that good management practices were maintained by the producers.

16.5.2 Host Range

Pratylenchus coffeae has a wide host range, including many broadleaf weed species (Gowen et al. 2005). The known host range of *P. coffeae* in SA is grapevine, banana, citrus and veld (dune thicket, grasses) (Kleynhans et al. 1996 and information from both NCN and SAPPNS databases).

16.6 Reniform Nematodes

The occurrence of *Rotylenchulus* spp. in bananas has not been studied in detail in SA since it is not regarded as a widespread pest in the banana industry. *Rotylenchulus* spp. individuals were present in 2 % of the samples in a survey conducted in commercial banana plantations (Table 16.1). Reniform nematodes were only found in the Onderberg area (Mpumalanga Province) in numbers ranging from 0 to 2,985 250 ml soil^{-1} (Table 16.2).

16.7 Other

Numerous other nematode species have been associated with banana but are not considered economically important. In a survey of 52 farms, Jones (1979) recorded *Nanidorus renifer* Siddiqi, 1974, in 53 % of samples, *Paratylenchus* sp. in 35 %, *Scutellonema brachyurus* (Steiner 1938) Andrássy, 1958, and *Scutellonema truncatum* Sher, 1964, in 8 %, *Tylenchorhynchus ventrosignatus* Tobar Jiménez, 1969, in 7 %, *Hemicycliophora* in 2 % and *Xiphinema* in 1 %. The survey that was done in commercial banana plantations by the ARC – ITSC showed that *Paratylenchus* was present only in the KwaZulu-Natal Province where it occurred in 84 % of the samples. *Nanidorus*/*Paratrichodorus* spp. were found in only 4 % of all the samples.

16.8 Management Strategies

16.8.1 Legislation

Following the discovery of *R. similis* in SA, legislation was passed to prevent further distribution of the pest. The Agricultural Pest Act 36 of 1983 (NDA 2015) requires that a permit must be obtained for the transportation of plant propagation material between individual farms as well as between magisterial districts. The act specifies traditional propagation material such as suckers, rhizomes and setts. Although tissue cultured plants are exempted by the legislation, a permit is also required once they have been established in a nursery.

16.8.2 Preparation of Plant Material

Almost all commercial banana crops are established exclusively from tissue culture material that is free of plant-parasitic nematodes. However, suckers and rhizomes are still used to establish new plantings in rural areas, particularly when a shortage of tissue culture material exists. These suckers are infected with plant-parasitic nematodes present in the soil where such planting material was sourced. Paring, to remove visible lesions caused by nematode pests and banana weevil so that only white rhizome tissue remains, is recommended in such cases. In many African countries the rhizome is dipped in hot water for a specific period of time to kill remaining nematode pests, but in SA this technique is seldom or never used. Once cleaned of nematode-infected material, the rhizomes must be planted in soil where nematode pest numbers have been substantially reduced. This could be obtained by using one or combinations of the following strategies (see also Sect. 16.8.3), viz. planting in virgin soil, leaving soil fallow for an extended period, addition of organic amendments or heat sterilisation using a transparent plastic cover for several weeks. The latter technique can only be done when planting is done in a very small area.

16.8.3 Cultural Control

Fallowing for at least 6 months (Loos 1961; Tarjan 1961) and rotation with selected cover crops can reduce populations of certain nematodes prior to planting banana. Milne and Keetch (1976) tested several cover crops and reported that radish (*Raphanus sativus*) and *Tagetes patula* reduced populations of *R. similis* after 5 months to a level comparable to that of ethylene dibromide (EDB) fumigation. Rotation with Buffalo grass (*Megathyrsus maximus* var. *trichoglum*; syn *Panicum maximum*) and purple bean (*Phaseolus atropurpureus*) supported neither *R. similis* nor *Meloidogyne* spp., whilst growing sugarcane (*Saccharum* hybrid) eliminated *R. similis* after 10 weeks (De Villiers et al. 2007). Intercropping with crops such as coffee (*Coffea arabica*), vegetables, maize (*Zea mays*) and cassava (*Manihot esculenta*) often used in West and East Africa is not practiced in SA.

Chicken and cattle manure amendments at high volumes (40 MT ha^{-1} or more) in combination with chemical fertilisers are frequently used in banana production in SA to suppress nematode pest populations over the long term. Before planting banana, application of 15 MT chicken manure ha^{-1} or 30 MT cattle manure ha^{-1} is generally recommended (De Villiers et al. 2007). Stirling (1991) reported that the effect of organic amendments is a complex process that enhances plant growth by improving soil structure and fertility, enhanced plant resistance and the stimulation of micro-organisms, which act as natural enemies of nematodes. Microbial degradation of organic amendments affects the release of metabolites that have direct nematicidal effects (Stirling 1991). In cases of severe nematode infections, treatment of banana plants with a nematicide is recommended (see Sect. 16.8.5).

16.8.4 Biological Control

Daneel et al. (1998) demonstrated the efficacy of the soil fungus *Purpureocillium lilacinum* (syn *Paecilomyces lilacinus*) for the control of banana nematode pests including *R. similis* and *Meloidogyne* spp. Aside from reducing numbers of nematodes, the product shortened the period from flowering to harvest of banana. The product was registered for use in SA on banana at a dosage rate of 2×10^9 spores g^{-1} in suspension at a rate of 2–4 g mat^{-1}, depending on severity of nematode infestation (Van Zyl 2013).

16.8.5 Chemical Control

Conventional, synthetically derived nematicides have been widely used in SA for nematode control on banana. Although fumigants have been shown to be effective (Keetch et al. 1976), such products are not used in banana production mainly due to high input costs. Pre- and post-plant carbamate and organophosphate nematicide applications are more regularly used for nematode control in banana. These products are applied as granular or liquid formulations around the base of the pseudostems or suckers. Furfuraldehyde is also registered on banana in SA and can be used when nematode population pressure is not severe (Van Zyl 2013).

In the local banana industry, plant-parasitic nematodes are seldom regarded as the primary yield-limiting factor. Therefore, it is recommended that nematode samples are taken annually and that nematicides are only applied to reduce nematode-pest populations likely to limit yield or cause long-term yield decline.

16.9 Conclusions

Although plant-parasitic nematodes are not the major limiting factor in banana production in SA, they can still cause severe damage to the crop and reduce yields considerably. Because tissue culture material is used in more than 90 % of the plantations replanted, nematode pests seldom become problematic during the first years after establishment of commercial plantations. Besides, many producers send in soil and root samples annually for nematode analysis to monitor plant-parasitic nematode populations and follow recommendations. This way, the effect of nematode pests in commercial banana production is effectively limited. However, there will always be a need to investigate alternative strategies to keep nematodes under control. For SA and surrounding countries, the need for nematode management strategies in the smallholding farming sector is significant and requires more attention.

References

DAFF (2011) A profile of the South African banana market value chain. Department of Agriculture, Fisheries and Forestry, Pretoria. www.nda.agric.za/docs/AMCP/Bananamvcp2011–12

Daneel MS, De Jager K, Dreyer S et al (1998) PL Plus is an environmentally friendly nematicide for banana nematodes. Neltropica Bull 300:32–34

Daneel MS, De Jager K, Van den Bergh I et al (2015) Occurrence and pathogenicity of plant-parasitic nematodes on commonly grown banana cultivars in South Africa. Nematropica 45:118–127

Daneel MS, Dillen N, Husselman J et al (2003) Results of a survey on nematodes of *Musa* in household gardens in South Africa and Swaziland. InfoMusa 12:8–11

De Jager K, Daneel MS, Desmet M et al (1999) Pathogenicity and distribution studies to determine threshold levels for nematodes on banana. Banana Grow Assoc S Afr Yearb 3:72–77
De Villiers EA, Daneel MS, Schoeman PS (2007) Pests. In: Robinson JC, De Villiers EA (eds) The cultivation of banana. Ingwe Print, Nelspruit, pp 194–219
Duchame EP, Birchfield W (1956) Physiologic races of the burrowing nematode. Phytopathol 46:615–616
Gowen S, Quénéhervé P (2005) Nematode parasites of bananas and plantains. In: Luc M, Sikora RA, Bridge J (eds) Plant parasitic nematodes in subtropical and tropical agriculture. CAB International, Wallingford, pp 611–643
Jones RK (1979) Nematodes associated with bananas in South Africa. Phytophylactica 11:79–81
Jones RK, Milne DL (1982) Nematode pests of bananas. In: Keetch DP, Heyns J (eds) Nematology in southern Africa. Science Bulletin No. 400. Department of Agriculture and Fisheries, Pretoria, pp 30–37
Kaplan DT, Thomas WK, Frisse LM et al (2000) Phylogenetic analysis of geographically diverse *Radopholus similis* via rDNA sequence reveals a monomorphic motif. J Nematol 32:134–142
Keetch DP (1972) Some host plants of the burrowing eelworm, *Radopholus similis* (Cobb) in Natal. Phytophylactica 4:51–58
Keetch DP, Reynolds RE, Mitchell JA (1976) An evaluation of pre- and post–plant nematicides for the control of plant parasitic nematodes on bananas. Citrus Subtrop Fruit J 506:5–7
Kleynhans KPN, Van den Berg E, Swart A et al (1996) Plant nematodes in South Africa. Plant Protection Research Institute Handbook No 8. Agricultural Research Council–Plant Protection Research Institute, Pretoria
Loos CA (1961) Eradication of the burrowing nematode, *Radopholus similis*, from bananas. Plant Dis Rep 45:457–461
Milne DL, Keetch DP (1976) Some observations on the host plant relationships of *Radopholus similis* in Natal. Nematropica 6:13–17
Milne DL, Kuhne FA (1968) Nematodes attack bananas. Farming S Afr 44:5–9
NDA (2015) http://www.nda.agric.za/docs/NPPOZA/Agricultural%20Pests%20Act.pdf. Accessed 12 Sep 2013
Quénéhervé P, Cadet P (1985) Localisation des nematodes dans les rhizomes du bananier cv Poyo. Rev Nématol 8:3–8
Rabie EC (1991) Die invloed van *Meloidogyne javanica* en *M. incognita* op die voorkoms van Valspanamasiekte by piesangs. MSc Dissertation, University of Pretoria, Pretoria
Robinson JC (1993) Handbook of banana growing in South Africa. Dynamic Ad, Nelspruit
Simmonds NW (1966) Bananas, 2nd edn. Longmans, London
Stirling GR (1991) Biological control of plant parasitic nematodes. Progress, problems and prospects. CAB International, Wallingford
Tarjan AC (1961) Longevity of *Radopholus similis* (Cobb) in host free soil. Nematologica 6:170–175
Willers P (1998) Nematologiese navorsing in subtropiese bedrywe. Neltropica Bull 299:12–13
Willers P, Daneel MS, De Jager K (2001) Banana. In: Van den Berg MA, De Villiers EA, Joubert PH (eds) Pest and beneficial arthropods of tropical and non-citrus subtropical crops in South Africa. Agricultural Research Council–Institute for Tropical and Subtropical Crops, Nelspruit, pp 34–43
Van Zyl K (2013) A guide to crop pest management in South Africa. A compendium of acaricides, insecticides, nematicides, molluscicides, avicides and rodenticides. A CropLife Compendium, 1st edn. VR Print, Pinetown

Chapter 17
Nematode Pests of Minor Tropical and Subtropical Crops

Mieke S. Daneel

17.1 Introduction

Besides banana (see Chap. 16), various subtropical and tropical crops are grown in South Africa (SA). These include avocado (*Persea americanum*), which is the biggest crop in terms of production, followed by mango (*Mangifera indica*), macadamia (*Macadamia integrifolia*), guava (*Psidium guajava*), pecan (*Carya illinoinensis*), papaya (*Carica papaya*), litchi (*Litchi chinensis*), ginger (*Zingiber officinale*), granadilla (*Passiflora edulis* and *P. edulis* f. *flavicarpa*), coffee (*Coffea arabica*), tea (*Camellia sinensis*) and black pepper (*Piper nigrum*).

Although a wide range of plant-parasitic nematodes is associated with the different subtropical and tropical crops, only those that are economically important are presented in this chapter.

17.2 Avocado

Avocado originates from Central America and is consumed primarily as fresh fruit. The local avocado industry consists of 18,000 hectares (ha) of commercial orchards. The majority of these orchards are situated in the north-eastern part of the country in the Limpopo and Mpumalanga provinces. Avocado is also grown commercially in certain areas of the KwaZulu-Natal Province. Annual production is in the region of 100,000 metric tons (MT), of which approximately 46,000 MT is exported to Europe and the United Kingdom (UK) (SAAGA 2014).

M.S. Daneel
Agricultural Research Council–Institute for Tropical and Subtropical Crops, Private Bag X11208, Mbombela 1200, South Africa
e-mail: mieke@arc.agric.za

H. Fourie et al. (eds.), *Nematology in South Africa: A View from the 21st Century*, DOI 10.1007/978-3-319-44210-5_17

17.2.1 Plant-Parasitic Nematodes Associated with Avocado

Thirty-seven plant-parasitic nematode species have been recorded on avocado in SA (Kleynhans et al. 1996, SAPPNS[1]) of which only *Meloidogyne javanica* (Treub, 1885) Chitwood, 1949, *Criconemoides xenoplax* (Raski, 1952) de Grisse and Loof, 1989, *Helicotylenchus dihystera* (Cobb, 1893) Sher, 1961 and *Xiphinema elongatum* Schuurmans Stekhoven and Teunissen, 1938 are potential pests.

17.2.2 Damage, Symptoms, Interactions with Other Pathogens and Management Strategies

Although large populations of plant-parasitic nematodes can occur, there are no reports of nematode damage to avocado in SA (Willers 2001). It is, therefore, difficult to determine the economic importance of nematode pests on avocado production, as also noted by El-Borai and Duncan (2005). However, the root rot pathogen *Phytophthora cinnamomi* is responsible for severe decline on avocado orchards in SA and it might be possible that the nematode pest effect is overshadowed by this fungus, as suggested by Milne (1982a).

Willers (1999, 2001) reported that a large avocado producer in the Nelspruit (now known as Mbombela) area of Mpumalanga reversed decline in tree performance in orchards by using cadusaphos (10 % granules). Before this intervention, the application of potassium phosphate for *Phytophthora* root rot control was not successful. Ultimately, soil and root samples taken from the orchards revealed the presence of *C. xenoplax* and *H. dihystera* in high and *X. elongatum* in moderate numbers.

In Israel, *Xiphinema diffusum* Lamberti and Bleve-Zacheo, 1979 is often recovered from avocado orchards in high numbers. Cohn (1968) reported reduced seedling growth in pot trials, but post-plant treatment with 1,2-Dibromo-3-chloropropane (DBCP) treatments in orchards did not consistently improve tree performance. Sher (1955) attributed reduced tree growth in California (USA) to the presence of *Pratylenchus vulnus* Allen and Jensen, 1951. This was confirmed in glasshouse inoculation experiments and pre-plant fumigation trials with 1,2-Dichoropropane; 1,3-Dichloropropene (DD) (Sher et al. 1959). Locally the coffee lesion nematode, *Pratylenchus coffeae* (Zimmerman, 1898) Filipjev and Schuurmans Stekhoven, 1941, was often found in high numbers in avocado root and soil samples from the Levubu area (Limpopo Province) (Willers 2001).

No nematicides are registered for use on avocado in SA. Orchards with declined tree performance often react negatively to *Phytophthora* root rot treatments in the presence of high plant-parasitic nematode numbers. For such situations, Willers

[1]Dr Mariette Marais of the Nematology Unit, Biosystematics Division, Agricultural Research Council–Plant Protection Research Institute is thanked for the use of data from the South African Plant-Parasitic Nematode Survey (SAPPNS) database; E-mail: maraism@arc.agric.za.

(2001) recommended the use of organic amendments, including kraal manure, to supplement the fertilizer program. The possible interaction between nematode pests and *Phytophthora* root rot on avocado still has to be investigated locally.

17.3 Mango

In SA, the mango industry expanded rapidly in the 1980s and 1990s but subsequently declined. Currently the commercial production area is about 7,500 ha and is situated mainly in the Limpopo Province, followed by Mpumalanga Province and a small area in the KwaZulu-Natal Province.

17.3.1 Plant-Parasitic Nematodes Associated with Mango

Although 40 plant-parasitic nematode species have been associated with mango in SA (Kleynhans et al. 1996, SAPPNS), none seems to be damaging. Milne (1982a) observed the ring nematode *Hemicriconemoides strictathecatus* Esser, 1960, feeding on mango roots together with *X. diffusum*. Later surveys, however, indicated that when litchi and mango orchards were adjacent to one another, *H. strictathecatus* was more prevalent in litchi than mango roots (Willers 1998). According to McSorley (1981), *H. strictathecatus* is the most widely distributed nematode species associated with mango. No endoparasitic nematode species have been recorded on mango in SA and according to McSorley and Parrado (1983), *Rotylenchulus reniformis* Linford and Oliveira, 1940, is the only sedentary nematode affecting mango. In the Onderberg region (Mpumalanga Province), this nematode occurs on papaya and banana but no records of infections of mango exist.

17.3.2 Damage Potential and Management Strategies

In inoculation trials, mango seedlings were damaged at a population level of 6 *H. strictathecatus* individuals cm^{-3} soil (Saeed 1974). However, in trials with fenamiphos in a 5-year-old orchard (cv. Sensation) Willers (1998) found no response to treatment, indicating that plant-parasitic nematodes had no significant effect on yield. He suggested that the local populations of *H. strictathecatus* might belong to a different biotype with a different host range than the potentially damaging species reported by Saeed (1974). Badra and Khattab (1982) found that the growth regulator, ethephon, reduced *R. reniformis* population levels in soil and roots of mango. This response should be evaluated on other subtropical crops where nematode problems occur and where ethephon is used.

17.4 Macadamia

Macadamia belongs to the Proteaceae family and originates from Queensland, Australia, from where it has been introduced to SA. About 21,500 ha are planted in the Mpumalanga, Limpopo, KwaZulu-Natal and the Eastern Cape provinces.

17.4.1 Plant-Parasitic Nematodes Associated with Macadamia

Twenty-eight plant-parasitic nematode species have been associated with macadamia in SA (Kleynhans et al. 1996; SAPPNS). These include potential pests such as *Meloidogyne ethiopica* Whitehead, 1968, *Meloidogyne incognita* (Kofoid and White, 1919) Chitwood, 1949, *Criconema mutabile* (Taylor, 1936) Raski and Luc, 1985 and *H. dihystera*. In the Lowveld of Mpumalanga, very high numbers of *H. dihystera* have been associated with trees in macadamia orchards. However, no negative effects on crop production or growth of the infected trees were observed. Willers (2001) suggests that Proteaceae might have an intrinsic tolerance to plant-parasitic nematodes since no damage has been reported from the cut flower industry in the Cape where many *Protea* spp. are cultivated intensively for export.

17.5 Guava

Guava is indigenous to tropical America and is consumed as fresh fruit and in processed forms as juice, canned fruit and jelly. There are about 1,200 ha of commercial plantings in the Western Cape, Limpopo and Mpumalanga provinces.

17.5.1 Plant-Parasitic Nematodes Associated with Guava

Thirty-nine plant parasitic nematode species have been associated with guava in local orchards (Kleynhans et al. 1996, SAPPNS). Nonetheless, economic damage is limited to trees infected with *Meloidogyne* and *H. dihystera* (Willers 1997a). According to Willers (2001), economic nematode damage was limited to two species, namely, *Meloidogyne enterolobii* Yang and Eisenback, 1983 (previously known as *Meloidogyne mayaguensis* Rammah and Hirschmann, 1988), which is regarded as an emerging threat to crop production worldwide (Jones et al. 2013) (Box 17.1) and *H. dihystera*. Severe root-knot nematode damage on commercial guava plantings was observed for the first time in 1991 and the species initially identified was *M. mayaguensis* (Willers 1997b). More recently *M. incognita* and *M. javanica* have been recorded on guava trees in the Limpopo and Mpumalanga provinces in combination with *M. enterolobii*. In more than 50 % of the orchards sampled, a combination of at least two *Meloidogyne* spp. was found (Agenbag 2016). More studies are underway to better understand the *Meloidogyne* spp. complex on guava and to determine the life cycle of *M. enterolobii* under local conditions.

Box. 17.1 *Meloidogyne enterolobii*

Meloidogyne enterolobii is a very aggressive root-knot nematode that is able to overcome the root-knot resistance genes in a number of crops, including tomato, bell pepper, cotton, cowpea, potato, sweet pepper, sweet potato and soybean (Fargette et al. 1996; Brito et al. 2004; Anonymous 2011; EPPO 2011; Castagnone-Sereno 2012; Jones et al. 2013).

Since its discovery in roots of the Pacara Earpod tree (*Enterolobium contortisiliquum*) in China in 1983, this thermophilic root-knot nematode species has also been reported from various tropical and subtropical countries in Asia, Africa, Central America and the Caribbean, North and South America as well as Europe (Blok et al. 2002; Brito et al. 2004; Kiewnick et al. 2009; Tigano et al. 2010; Anonymous 2011; Ramirez-Suáres et al. 2014). *Meloidogyne enterolobii* was first recorded from SA, from roots of guava (*Psidium guajava*) and black-jack (*Bidens pilosa*), just prior to 1997 in the Mbombela area (Mpumalanga Province) (M. Marais, unpublished data; Onkendi and Moleleki 2013a). In 2011 and 2013, individuals of *M. enterolobii* were recorded from glasshouses near Barberton (Mpumalanga Province) in roots of green pepper (Marais 2014) and in Letsitele (Limpopo Province) in guava roots (Marais 2014). Moreover, after 2012 *M. enterolobii* was identified from potato tubers in the KwaZulu-Natal Province (Onkendi and Moleleki 2013b) and the Mpumalanga Province (Agenbag 2016). It appears that this species is more widespread in South Africa than previously anticipated.

It is important to note that the identity of *M. enterolobii* is often confused with that of *M. incognita*, which is widely distributed and a serious pest of numerous crops in SA. The reason for this is that to distinguish these two species from one another, using only morphological/morphometrical characters, identification is challenging. Rather, molecular techniques should be used in combination with the standard measurements to ensure accurate identification of these two cryptic species.

17.5.2 *Symptoms, Interactions with Other Pathogens and Management Strategies*

Root-knot nematodes can have a devastating effect on the more recently available guava cvs. Root systems of guava vines are severely galled (Fig. 17.1a) and rotting takes place due to infection by other soil pathogens. Above-ground symptoms are

Fig. 17.1 (**a**, **b**) Galled roots of a guava tree infected with *Meloidogyne enterolobii* (**a**), resulting in a poorly growing and dying tree on the left (**b**) in the Mbombela area (Mpumalanga Province) (Kirk West, Port Elizabeth, South Africa)

reduced growth and vigor, leaf yellowing, smaller leaves, reduced fruit set, smaller fruit and, in severe cases, trees die (Fig. 17.1b).

Fan Retief, a highly resistant cv. to root-knot nematodes, was used historically as the only germplasm source for production and is still widely grown in the Western Cape Province. Root stock TSG2, resistant to the crippling guava wilt disease caused by *Nalanthamala psidii*, was reported as highly susceptible to plant-parasitic nematodes in the late 1970s (Willers 1997a) (Box 17.2).

Box. 17.2 Breeding for Resistance Against Pests and Diseases. An Example from the Guava Industry

During the early 1990s, a new disease, guava wilt disease (GWD) caused by *Nalanthamala psidii*, broke out in SA and almost crippled the guava industry in the Limpopo and Mpumalanga provinces. At that time one cultivar (cv.), Fan Retief, was used in practically all the guava orchards. Plant material from local and international seed was screened for resistance to this disease and eventually, during the late 1990s, a cv., TSG2, was developed that was highly resistant to GWD. However, it soon became obvious that this cv. was much more susceptible to plant-parasitic nematodes than Fan Retief. Before the advent of TSG2, plant-parasitic nematodes were not perceived to be a real problem on guava. However, this scenario changed and nematode pests proved to be a major problem with damage so severe that adult trees were dying.

It is important, when evaluating plant material, to include the major pests and diseases in the screening programs. As part of the guava breeding program at the Agricultural Research Council–Institute for Tropical and Subtropical Crops (ARC–ITSC), plant material is now routinely tested for resistance against GWD and the more damaging nematode pest species, to identify resistance or tolerance before the plant material is released.

When planting a new orchard, it is very important to establish the presence or absence of root-knot nematodes in the field. When root-knot nematodes are present, nematode control is compulsory prior to planting or at planting with one of the below-mentioned products.

Screening for resistance or tolerance of guava cvs to *M. incognita*, *M. javanica* and *M. enterolobii* (Willemse et al. 2014) and guava wilt disease is conducted on an ongoing basis.

Nematicides that contain the active substances (a.s.) fenamiphos and cadusaphos are registered on guava in SA. Application should be done at the correct time to ensure that no residues are found in the fruit at harvesting. The pre-harvesting interval (PHI) for fenamiphos is 120 days. Nematodes should be monitored annually to prevent the build-up of high nematode population densities. Application of organic material (e.g., cattle or chicken manure) at 40 MT ha^{-1} can help to suppress nematode pest densities (Willers 2001). Any organic amendments (mulches, compost, manure) and root growth stimulants (e.g., *Trichoderma*) can be used to improve root growth, especially in the initial 6 months after planting.

17.6 Pecan

Pecan is indigenous to the central southern states of the USA. It is well adapted to large areas of SA where short, cold winters and long, very hot summers occur. Trees have been successfully established in valleys and along rivers where the winter temperatures are low and frost occurs. The main production area is in the Northern Cape Province, with small plantings in other provinces.

17.6.1 Plant-Parasitic Nematodes Associated with Pecan

In SA, 28 plant-parasitic nematode species have been associated with the roots of pecan trees (Kleynhans et al. 1996, SAPPNS). Of these, the pecan root-knot nematode *Meloidogyne partityla* Kleynhans, 1986, as well as *Xiphinema vitis* Heyns, 1974, and *C. xenoplax* are considered the major pests. *Meloidogyne partityla* only parasitises pecan trees and was probably imported with plant material from the USA (Kleynhans et al. 1996).

17.6.2 Symptoms and Management Strategies

The root symptoms caused by *M. partityla* are not the typical galling as presented on other crops by *Meloidogyne* spp., but rather a general absence of feeder roots in the top soil. When high numbers of *M. partityla* eggs and second-stage juveniles (J2) are present, newly formed roots are often swollen at the tip and club-shaped. Damage by the plant-parasitic nematode complex associated with pecan is of a subtle nature and symptoms manifest as a general decline in the overall growth of infected trees over a prolonged period. This may become serious if left untreated as both pecan kernel size and yield are influenced adversely (Willers 2001).

Products containing the a.s. fenamiphos (GR) are the only nematicides registered on pecan. It is applied in the irrigation zone underneath the tree canopy during spring each year with a 120 day PHI.

To prevent the spreading of *M. partityla* to other possible pecan-cultivating areas in SA, control measures were enforced on nurseries in Mpumalanga to ensure that they provide producers with trees that are free of nematode pests (Willers and Daneel 1993).

Recommendations to achieve this included:

(i) Soil fumigation with methyl bromide (hot-gas method) at 175 g m^{-3} for 48 h under gastight tarpaulins
(ii) Use of a primary water source for irrigation
(iii) Location of nurseries in areas where re-infestation by run-off water from infested orchards can be prevented

17.7 Papaya

Papaya originates from tropical America but is widely distributed today. Commercial production of the crop locally is limited to about 800 ha in the Limpopo and Mpumalanga provinces.

17.7.1 Plant-Parasitic Nematodes Associated with Papaya

In SA, 25 plant-parasitic nematode species have been associated with papaya (Kleynhans et al. 1996, SAPPNS), but only two genera (*Meloidogyne* and *Rotylenchulus*) seem to be linked with economic damage (Milne 1982a; Willers and Neethling 1997; El-Borai and Duncan 2005). *Rotylenchulus reniformis* Linford and Oliveira, 1940, has been reported on papaya from around the world (El-Borai and Duncan 2005). Locally it has been recorded from the Onderberg region in the Mpumalanga Province by Willers and Neethling (1994). Mean population levels were 4,100 second-stage juveniles (J2) 250 ml $soil^{-1}$, with numbers reaching 12,900 J2 250 ml $soil^{-1}$.

17.7.2 *Symptoms, Interaction with Other Soil-borne Pathogens and Management Strategies*

Root damage caused by root-knot nematode parasitism is often severe and galls can reach the size of golf balls (Milne 1982a). Additionally root infections enhance the chances of root rot, which reduces yield and life expectancy (Milne 1982a). Seedling growth in pot trials was greatly retarded as a result of root-knot nematode parasitism (Lamberti et al. 1980), with symptoms including severely chlorotic leaves, damage to the tap root and proliferation of lateral roots (Lamberti et al. 1980; Darekar and Mhase 1986).

Severe yield losses, toppling and plant death have been attributed to the *R. reniformis* infection of papaya trees (Ayala et al. 1971; Singh and Farrell 1972). The combined occurrence *of R. reniformis* and *Phytophthora nicotianae* var. *parasitica* killed papaya trees in Brunei in Asia (El-Borai and Duncan 2005). Damage by *R. reniformis* has not been determined in SA, but it is believed that this nematode pest might be of economic importance in papaya orchards in the Onderberg region (Mpumalanga Province) where it occurs in combination with *Meloidogyne* spp. (Willers and Neethling 1997). Root rot caused by *Phytophthora* might be enhanced by the presence of these nematode pests.

Fenamiphos is registered for control of plant-parasitic nematodes on papaya and producers have reported considerable yield and growth improvements using the product (Willers 2001).

17.8 Litchi

Litchi is endemic to southern China and is marketed as fresh, canned and dried fruit. Local production of litchi covers about 1,800 ha and is situated in the Limpopo, Mpumalanga and KwaZulu-Natal provinces.

17.8.1 *Plant-Parasitic Nematodes Associated with Litchi*

Nine plant-parasitic nematodes have been associated with litchi in SA (Kleynhans et al. 1996, SAPPNS). During a nematode survey from three provinces, Daneel et al. (2010) recorded that many plant-parasitic nematode species were limited to one area only (Table 17.1). For example, *Pratylenchus* and *Xiphinema* spp. were recorded only in the Tzaneen area (Limpopo Province),

Table 17.1 Plant-parasitic nematodes recorded from soil samples (250 g) obtained during a survey of litchi trees with the minimum and maximum numbers of nematodes, their frequency of occurrence (%) in the samples and the areas where the species occurred (Daneel et al. 2010)

Plant-parasitic nematode species	Minimum and maximum population levels	Percentage occurrence in samples[a]	Presence of nematode species listed province		
			Limpopo	Mpumalanga	KwaZulu-Natal
Hemicriconemoides strictathecatus	0–4,950	100	+[b]	+	+
Meloidogyne sp.	0–900	43	–[c]	–	+
Helicotylenchus sp.	0–2,050	88	–	+	–
Pratylenchus sp.	0–50	7	+	–	–
Xiphinema spp.	0–900	100	+	–	–
Hemicycliophora sp.	0–8,350	17	–	–	+

[a]Percentages are valid for the provinces in which the nematode species were found. Except for ring nematodes all species were limited to one area only
[b]Presence of nematode species
[c]Absence of nematode species

Fig. 17.2 (**a**, **b**) Typical symptoms of nematode damage in a litchi tree (**a**: *tree on right side*) with bare twigs and branches visible. This contrasts with symptoms of litchi dieback (**b**), which results in branches or the whole tree dying off and which is caused by a fungal complex (**a** Kirk West, Port Elizabeth, South Africa; **b** Mieke Daneel, Agricultural Research Council – Institute for Tropical and Subtropical Crops, Mbombela, South Africa)

whereas *Hemicycliophora* spp. (sheath nematode) was only observed in KwaZulu-Natal Province (Table 17.1). *Criconema* sp. (ring nematodes), *H. strictathecatus* and *Hemicycliophora* sp. are responsible for most of the damage associated with litchi-tree decline (Daneel et al. 2010). Milne (1982b) recorded *X. diffusum* also as a major pathogen, but from the survey by Daneel et al. (2010), *Xiphinema* spp. were only found in one production area and did not seem to be linked with severe damage to litchi trees.

17.8.2 Symptoms, Interaction with Other Pathogens and Management Strategies

Symptoms associated with nematode damage include the occurrence of stubby, darkened roots, with limited development of feeder roots resulting in low root mass. The loss of feeder roots culminates in poor uptake of nutrients and water (Daneel 2010).

Typical above-ground symptoms are the presence of bare twigs and branches, leaf chlorosis, leaf tip burn, poor flowering and excessive fruit drop (Fig. 17.2a). However, litchi dieback (when branches or the whole tree dies back: Fig. 17.2b) is caused by a series of fungal pathogens initiated by environmental stress, including soil compaction and water stress. In such situations, nematode control will not have a significant effect on tree vigor as the fungi need to be controlled (Daneel 2010; Steyn et al. 2013).

Recovery of trees can be facilitated by effective mulching, using any organic material such as manures throughout the year, while nematicide treatments should be done in spring or late summer after harvest. In poor performing orchards, all other cultivation practices such as fertilization and irrigation should be optimal to limit nematode-induced root damage.

Fenamiphos (Nemacur® GR) and cadusaphos (Rugby® GR and ME) are registered in SA for nematode control on litchi. The products should be applied evenly under the canopy after the area has been cleared from leaves and other organic material. Fenamiphos can only be applied in early spring because of a 70-day PHI. Cadusaphos has a zero day PHI and can be used throughout the season, but should preferably be applied after the first good rains have fallen. To minimize the possibility of accelerated microbial degradation (AMD) (see Sect. 6.4, Chap. 6), repeated applications of the same product or group of compounds with the same active substance should be avoided.

17.9 Granadilla

Granadilla is a vine and is mostly found in the tropics and subtropics. It is believed to have originated in a region ranging from southern Brazil to northern Argentina. In SA, a yellow form, *P. edulis* f. *flavicarpa*, and a purple form, *P. edulis*, of granadilla are grown. The yellow form is more tolerant to nematode damage (Milne 1982a) and was therefore commonly used as a root stock for the purple form as the latter is preferred in taste and colour. Both are produced in SA. The granadilla is mainly planted for juice although some fruit is sold for the fresh market, both locally and internationally. Granadilla is grown in most provinces in SA but the total production area does not exceed 500 ha in total.

17.9.1 Plant-Parasitic Nematodes Associated with Granadilla

Thirty-eight plant parasitic nematode species have been associated with granadilla (Kleynhans et al. 1996, SAPPNS), but only *M. incognita* and *M. javanica* are of economic importance in SA. *Rotylenchulus reniformis* is also described as causing economic damage to granadilla (El-Borai and Duncan 2005), but this nematode has not been recorded infecting granadilla in SA.

17.9.2 Symptoms, Interactions with Other Pathogens and Management Strategies

Root damage as a result of nematode infection manifests as typical galling with a reduced root system. Above-ground symptoms are leaf chlorosis, excessive leaf and fruit drop, reduced growth and faster (premature) wilting on hot days.

The life span of the vine is drastically reduced by either the nematode infection alone or by the concomitant occurrence of nematode pests and root and soil fungal pathogens, including *Phytophthora*, *Pythium* and *Fusarium* spp. Daneel and Garber (2009) reported the frequent concurrent occurrence of root rot pathogens with root-knot nematodes in granadilla, with devastating effect on the vines.

Management practices such as optimal irrigation schedule, good fertilization and proper soil preparation can reduce the effects of pests and pathogens and extend the life span of the crop. No post-plant nematicides are registered for use on granadilla in SA, but more environmentally-friendly products were tested locally. Results showed that CropGuard® (a.s. furfuraldehyde), Bioneem (a neem-based product) and cadusaphos provided excellent nematode control in all areas tested, while other products only gave good control in some of the areas (Daneel and Garber 2009). However, until now no nematicide, synthetically-derived or biological agent, is registered on granadilla in SA (Van Zyl 2013).

For new granadilla plantings, it is important to ensure that nematode-free nursery material is obtained. Field trials with fumigants showed yield increases of 40 % over a 3-year period (De Villiers and Milne 1973). This practice is, however, not used in the granadilla industry. New plantings can be started with the application of a nematicide prior to, or at, planting. Milne (1982a) suggested that crop rotation should be useful for nematode control as well as the use of rootstocks that are tolerant to root-knot nematodes, but limited research has been done on this aspect. Research conducted recently on resistance screening showed that cv. Edulis maintains the highest nematode numbers followed by cv. Ester. By contrast, *Passiflora alata x* (*P. alata x P. caerulea*) and *P. edulis* f. *flavicarpa* selections seemed to exhibit resistance (Husselman et al. 2014). However, much more research is needed before a resistant cv. will be available for release.

17.10 Ginger

Ginger is the rhizome of a herbaceous perennial belonging to the Zingiberaceae family. The country from where ginger originated is unknown but is believed to be China or India where more than 50 % of the world's dried ginger is produced. Ginger plantings are propagated by seed rhizomes or setts, which are cut into small pieces of 2.5–5 cm length with each having one or two good buds (Koshy et al. 2005). In SA, the main ginger-producing areas are Burgershall and Kiepersol in the Mpumalanga Province. Production is limited to areas with high temperatures and humidity in summer.

17.10.1 Plant-Parasitic Nematodes Associated with Ginger

A large number of plant-parasitic nematodes have been recorded from ginger worldwide, but the most important are *Meloidogyne* spp., *Radopholus similis* (Cobb, 1893) Thorne, 1949, and *P. coffeae* (Koshy et al. 2005). All these nematode pests occur in SA, but severe damage is only caused by *M. incognita* and *M. javanica*. Although *R. similis* is present in banana plantations in the main ginger-producing areas, no natural infection of ginger by this nematode pest has been recorded (Willers 2001) since old banana soils are seldom used for ginger cultivation. High numbers of *H. dihystera* have often been associated with ginger but the pathogenicity has not been evaluated. In the Levubu area of the Limpopo Province, isolated cases of severe *P. coffeae* infections have been recorded on ginger.

17.10.2 Symptoms, Damage Potential and Management Strategies

Root-knot nematodes cause galling and rotting of the roots and rhizomes. When ginger is exported as a fresh market commodity, nematode-infected rhizomes adversely affect the appearance and quality of the produce and limit its shelf life drastically by decay, rapid water loss and shrinkage of the rhizomes (Willers 2001). Second-stage juveniles of *Meloidogyne* spp. invade the rhizome through the axils of the leaf sheaths in the shoot apex during the early growth stages. In fleshy roots, the entire length is invaded while in fibrous roots, invasion takes place in the differentiation zone behind the root tip. In the roots, the life cycle takes about 21 days, whereas in the rhizome the life cycle is completed in 40 days at 30 °C (Cheng and Tu 1979). Galls are limited to the fibrous roots whereas internal necrosis is visible in fleshy roots and rhizomes. Infected rhizomes have brown, water soaked areas in the outer tissues, particularly in the angles between shoots. In these areas, root-knot females

Fig. 17.3 Stunted growth of root-knot nematode infected ginger plants (*front*) (Kirk West, Port Elizabeth, South Africa)

can be observed. Nematodes continue to develop in the rhizome even after harvest and induce seed breakdown. Infected rhizomes, used as seed pieces, serve as a source of infection and dissemination. Root-knot nematode infestations in the field are observed as patchy areas where plants are stunted and appear chlorotic, and premature wilting occurs even where soil moisture levels are adequate (Fig. 17.3). In addition, despite proper fertilization nematode-infected plants often show various degrees of macro- and micronutrient deficiencies. Potassium uptake, in particular, is reduced when plants are infected by root-knot nematodes (Willers 2001).

In SA, severe root-knot infections of ginger plants may result in complete crop failure. Infected ginger is downgraded and can only be dried and milled for the spice market, at great financial loss. Root-knot nematode damage is aggravated by various stress factors experienced during ginger cultivation. A fertilizer programme based on leaf analysis, sufficient water and cooling irrigation during hot days (>28 °C) are prerequisites to achieve quality ginger rhizomes for export purposes (Willers 2001). When harvesting ginger during autumn and early winter, lifting of the rhizomes should happen as quickly as possible to prevent root-knot nematode infections in the rhizomes, especially when soil temperatures have not dropped during that period.

Effective nematode control on ginger starts with producing nematode-free planting material by selecting nematode-free rhizomes. In Australia (Pegg et al. 1974) and Fiji (Anonymous 1971), hot water treatment for 10–20 min, dependent on temperatures, is recommended to kill nematodes inside rhizomes. The rhizomes should be planted within one week after treatment. When severe nematode problems have been experienced in a field, it might be necessary to select an area where no ginger was planted previously.

Willers (2001) recommended soil fumigation as a basic requirement for the production of ginger for the export markets. Post-plant applications of fenamiphos granules at 1 kg 100 m^{-1} row should only be used as a supplementary control measure. Residues of systemic post-plant nematicides tend to accumulate in ginger rhizomes at the end of the season when reserves from the foliage are stored in rhizomes. Thus on ginger, a PHI of 250 days following application of fenamiphos must be strictly applied. Composted cattle manure applied at 40 MT ha^{-1} is recommended as a cultural control measure to suppress nematode numbers in ginger plantings. In Australia, a high level of root-knot nematode control was obtained by sawdust mulching to a depth of 5.0–7.5 mm, combined with post-plant applications of fenamiphos (Koshy et al. 2005).

In Queensland, Australia, Pegg et al. (1974) were able to increase yields by 80 % after fumigation of soils with DD before planting nematode-free seed rhizomes. Significant weight loss was recorded by Sukumaran and Sundararaju (1986) 6 months after inoculation of 10,000 root-knot nematodes $plant^{-1}$. They also reported significant yield losses with a soil population of 1 J2 30 g $soil^{-1}$.

17.11 Coffee

After a growing interest in local production from the 1970s to the 1990s, coffee plantings are almost non-existent in SA. At present, only about 50 ha is under coffee production. However, local communities have recently revived the interest in coffee and some new commercial plantings are planned (Naudé 2016). Two species in the Erythrocoffea group, namely, *Coffea arabica* (Arabica coffee) and *Coffea canephora* (Robusta coffee) are grown locally.

17.11.1 Plant-Parasitic Nematodes Associated With Coffee

Twenty-four plant-parasitic nematode species from 13 genera have been associated with coffee in SA (Cohn 1976; Anonymous 1989; Kleynhans et al. 1996, SAPPNS) but *P. coffeae*, which is an economically important pest of coffee worldwide (Campos and Villain 2005), has not been recorded (M Marais, Agricultural Research Council–Plant Protection Research Institute, Pretoria, 2016, personal communication). In a survey in the Hazyview area, undisclosed species of the three important nematode genera associated with coffee, namely, *Meloidogyne*, *Rotylenchulus* and *Pratylenchus* were recorded (Anonymous 1989). In 1999, *M. incognita* was identified for the first time on coffee plantings in the Bush Buck Ridge area of the Limpopo Province (Willers 2001).

17.11.2 Symptoms and Management Strategies

Root symptoms of nematode-infected plants appear as typical galling, peeling and cracking of infected roots. Above-ground symptoms include a general decline in tree growth, chlorosis, leaf fall and eventually plant death. Willers (2001) stated that only 0.6 MT ha^{-1} of green coffee beans was harvested in nematode-infested plantings. Leaf analysis indicated deficiencies of nitrogen (N), magnesium (Mg), manganese (Mn) and zinc (Zn).

For the control of *M. incognita*, clean planting material should be used at all times and when replanting, the land should be treated with a registered fumigant. No post-plant nematicides are registered on coffee and the use of organic fertilizer or organic amendments, together with foliar applications of macro and micronutrients, to compensate for root damage, is recommended.

17.12 Tea

Tea is grown in various areas where a wide variety of climatic conditions prevail. It requires well drained acid soil with a pH range of 4.5–5.5 and well distributed rainfall, totaling not less than 1000 mm $annum^{-1}$ (Campos and Villain 2005). Previously in SA, there were some 3,000 ha of tea plantations, mainly in the subtropical area of the Limpopo Province, with small areas established in the Mpumalanga, KwaZulu-Natal and the Eastern Cape provinces. Currently hardly any tea (less than 50 ha) is cultivated in SA, but projects are underway to establish plantations in the Limpopo Province.

17.12.1 Plant-Parasitic Nematodes Associated with Tea

Twenty-three plant-parasitic nematode species have been associated with tea in SA (SAPPNS). However, records of nematode damage on tea in SA are absent (Willers 2001). According to Campos and Villain (2005), *Meloidogyne* is the most commonly encountered nematode pest in tea plantations in the different tea-growing areas of the world. *Meloidogyne brevicauda* Loos, 1953, was identified in a few limited areas in India and Sri Lanka. *Pratylenchus brachyurus* (Godfrey, 1929) Filipjev and Schuurmans Stekhoven, 1941, has also been recorded as a major pathogen of tea especially in young (1–3-year-old) plants (Campos and Villain 2005). In SA, this species has a wide distribution on a variety of crops but has not been recorded on tea, while *Pratylenchus zeae* Graham, 1951, has been recorded locally on tea.

17.12.2 Symptoms and Management Strategies

Meloidogyne spp. cause losses to mature tea with symptoms visible as stunting of bushes and yellow, small and dull colored leaves (Loos 1953). However, *Meloidogyne* spp. seem to mostly attack roots of young plants; mature plants seem to become less susceptible to this pest. Cohn (1976) recorded above-ground stunting and yellowing of tea seedlings in a nursery near Tzaneen (Limpopo Province) due to root-knot nematode parasitism. It is therefore important to ensure that potting soil and water that are free of nematode pests are used in tea nurseries. No chemicals have been registered for nematode control on tea.

17.13 Black Pepper

Black pepper is a perennial climber belonging to the Piperaceae family and is cultivated in the hot and humid parts of the world. It originated from the hills in the south-west of India where it is known as the 'king of spices' (Koshy et al. 2005). In SA, pepper production is restricted to the Levubu (Limpopo Province) and Hazyview (Mpumalanga Province) areas. Pepper can be propagated by cuttings or seed but the former is universally adopted.

17.13.1 Plant-Parasitic Nematodes Associated with Black Pepper

Meloidogyne incognita, *Meloidogyne hapla* Chitwood, 1949, *X. diffusum* and *Xiphinema mampara* f. *major* Hutsebaut, Heyns and Coomans, 1989, are associated with pepper locally (Kleynhans et al. 1996). Although *R. similis* is present in several banana-producing areas in SA, it has not been recorded from pepper roots, probably due to the small pepper production area and the use of virgin land to establish such crops.

17.13.2 Symptoms, Damage Potential and Management Strategies

Damage caused by *Meloidogyne* spp. is characterized by a gradual decline in the growth of black pepper vines and is visible as unthrifty growth and yellowing of leaves. The leaves exhibit a yellow discoloration of the interveinal areas, making the leaf veins quite prominent with a deep green color. On the other hand, leaves of vines attacked by *R. similis*, a major pathogen on pepper worldwide, have a uniform pale yellow or whitish discoloration and typical drooping (Koshy et al. 2005).

Root-knot nematode infected pepper roots can become heavily galled (Fig. 17.4). Although no damage studies have been done on pepper in SA, studies in India have shown that a 16 % growth reduction was observed over a period of 1 year in sterile soil in pots, with an initial *M. incognita* population of 10 J2 rooted cutting^{-1}.

Fig. 17.4 Root-knot nematode damage on pepper roots (Kirk West, Port Elizabeth, South Africa)

However, up to a 50 % growth reduction was recorded when 100,000 J2 were inoculated in the same experiment (Koshy et al. 1979). Freire and Bridge (1985) found *M. incognita* to be highly pathogenic when inoculated at 100–10,000 J2 seedling^{-1}.

No post-plant nematicides have been registered on pepper in SA, but the use of kraal manure at 5 kg vine^{-1} to replace standard fertilization programs would, in part, be useful to suppress nematode pest levels over the long term and provide sustainable production of the crop. Other amendments like sugarcane tops, grain straw and grass could also contribute toward higher carbon contents in soil in pepper production areas and contribute toward effective management of nematode pests. Effective irrigation scheduling and fertilization based on the exact requirements of plants, including micronutrients, would enhance pepper production and compensate for root damage caused by root-knot nematodes (Willers 2001). Additionally, nursery practices such as steam sterilization or fumigation should be such that potting soil and water are nematode free.

17.14 Conclusions

Plant-parasitic nematodes are not considered a serious problem of minor subtropical crops in many countries. However, nematode pests may adversely affect the production of coffee, ginger, granadilla, guava and litchi in certain countries, including SA. Furthermore, some plant-parasitic nematodes can cause severe damage when in combination with other soil-borne pathogens (e.g., bacteria and fungi). Due to the small size of the crops, in area planted and production quantities, limited funds are available for research. Also, chemical companies do not seem to be particularly interested in obtaining registrations for nematicides on these relatively minor crops. Therefore, it is important that alternative strategies are developed to control the nematode pests.

References

Agenbag M (2016) Identification and reproduction potential of South African *Meloidogyne* species. MSc Dissertation, North West University, Potchefstroom

Anonymous (1971) Annual report, 1971. Department of Agriculture, Fiji

Anonymous (1989) Nematodes on coffee plants. Citrus and Subtropical Fruit Research Institute. Inform Bull 207:6

Anonymous (2011) *Meloidogyne enterolobii*. European and Mediterranean Plant Protection Organization Bull 41:329–339

Ayala A, Acosta N, Adsuar JA (1971) A preliminary report on the response of *Carica papaya* to foliar applications of two systemic nematicides. Nematropica 1:10

Badra T, Khattab MM (1982) Chemically-induced resistance to *Rotylenchulus reniformis* by ethephon growth regulant and relevant pathometabolites in mango seedlings. Nematol Medit 10:41–56

Blok VC, Wishart J, Fargette M et al (2002) Mitochondrial DNA differences distinguishing *Meloidogyne mayaguensis* from the major species of tropical root-knot nematodes. Nematol 4:773–781

Brito J, Powers TO, Mullin PG et al (2004) Morphological and molecular characterisation of *Meloidogyne mayaguensis* isolates from Florida. J Nematol 36:232–240

Campos VP, Villain L (2005) Nematode parasites of coffee and cocoa. In: Luc M, Sikora RA, Bridge J (eds) Plant parasitic nematodes in subtropical and tropical agriculture, 2nd edn. CAB International, Wallingford, pp 529–579

Castagnone-Sereno P (2012) *Meloidogyne enterolobii* (=*M. mayaguensis*): profile of an emerging, highly pathogenic, root-knot nematode species. Nematol 14:133–138

Cheng YH, Tu CC (1979) Pathogenesis of *Meloidogyne incognita* to edible ginger. J Agric Res China 28:91–99

Cohn E (1968) Nematodes as possible vectors of virus diseases of citrus and other fruit crops. Second annual report to USDA, project no. A10-CR-72. Bet Dagan

Cohn E (1976) Report on investigations on nematodes of citrus and subtropical fruit crops in South Africa. Citrus and Subtropical Fruit Research Institute, Nelspruit

Daneel MS (2010) Nematodes. In: De Villiers EA, Joubert PH (eds) The cultivation of litchi. ARC-Institute for Tropical and Sub-tropical Crops, Mbombela, pp 196–198

Daneel M, Garber H (2009) Efficacy, residue and phytotoxicity studies for the control of nematodes on granadilla conducted in the RSA during 2006–2009 for support in the acquisition of EU import tolerances as well as local registration under Act 36 of 1947. Project B 1287 – Field Trials 1, 2 and 3. Agricultural Research Council–Institute for Tropical and Subtropical Crops, Nelspruit

Daneel MS, Serfontein S, Steyn W et al (2010) A study into the causes of litchi tree dieback and control strategies for the disease. SA Litchi Growers' Association Yearbook, pp 22–25

Darekar KS, Mhase NL (1986) Effect of initial inoculum levels of *Meloidogyne incognita* on growth parameters of papaya seedlings. Int Nematol Network Newslett 3:4–5

De Villiers EA, Milne DL (1973) Die bestryding van aalwurms by granadillas. Citrus Subtrop Fruit J 480:7–13

El-Borai FE, Duncan LW (2005) Nematode parasites of subtropical and tropical fruit tree crops. In: Luc M, Sikora RA, Bridge J (eds) Plant parasitic nematodes in subtropical and tropical agriculture, 2nd edn. CAB International, Wallingford, pp 467–492

EPPO (2011) PQR – EPPO database on quarantine pests. http://www.eppo.int. Accessed 22 Sep 2015

Fargette M, Phillips MS, Blok VC et al (1996) An RFLP study of relationships between species, isolates and resistance-breaking lines of tropical species of *Meloidogyne*. Fundam Appl Nematol 19:193–200

Freire FCO, Bridge J (1985) Influence of different inoculum levels of *Meloidogyne incognita*, *Nectria haematococca* f. sp. *piperis* and *Phytophthora palmivora* on black pepper plants. Fitopatol Bras 10:559–575

Husselman J, Daneel MS, Steyn WP (2014) Susceptibility of passion fruit selections from the ARC-ITSC breeding programme to root-knot nematode (*Meloidogyne incognita* and *Meloidogyne javanica*). J Nematol 46:177

Jones JT, Haegeman A, Danchin EGJ et al (2013) Top 10 plant-parasitic nematodes in molecular plant pathology. Mol Plant Pathol 14:946–961

Kiewnick S, Dessimoz M, Franck L (2009) Effects of the Mi-1 and the N root-knot nematode resistance gene on infection and reproduction of *Meloidogyne enterolobii* on tomato and pepper cultivars. J Nematol 41:134–139

Kleynhans KPN, Van den Berg E, Swart A et al (1996) Plant nematodes in South Africa. Plant Protection Research Institute, Handbook 8. Business Print, Pretoria

Koshy PK, Premachandran D, Sosamma VK et al (1979) Effect of *Meloidogyne incognita* population on black pepper (*Piper nigrum* L.). Ind Phytopathol 32:115–121

Koshy PK, Eapen SJ, Pandey R (2005) Nematode parasites of spices, condiments and medicinal plants. In: Luc M, Sikora RA, Bridge J (eds) Plant parasitic nematodes in subtropical and tropical agriculture, 2nd edn. CAB International, Wallingford, pp 751–791

Lamberti F, Rohini HM, Ekananyake V et al (1980) Effects of two *Meloidogyne* species on the growth of papaya seedlings. Ind J Nematol 10:225–230

Loos CA (1953) *Meloidogyne brevicauda* n. sp. a cause of rootknot of mature tea in Ceylon. P Helm Soc Wash 20:83–91

Marais M (2014) *Meloidogyne enterolobii (=Meloidogyne mayaguensis)*,'n gevallestudie van'n plantparasitiese nematode. Suid-Afrikaanse Tydskrif vir Natuurwetenskap en Tegnologie 33:Art. #1237, 1 page. http://dx.doi.org/10.4102/satnt.v33i1.1237

McSorley R (1981) Plant parasitic nematodes associated with tropical and subtropical fruits. Agric. Exp. Station Institute of Food and Agricultural Science, Bull. 823. University of Florida, Gainsville

McSorley R, Parrado JL (1983) Effects of Nemacur on phytoparasitic nematodes of mango, 1982. Fungicide and Nematicide tests. APS 39:92–93

Milne DL (1982a) Nematode pests of miscellaneous subtropical crops. In: Keetch DP, Heyns J (eds) Nematology in Southern Africa. Science Bulletin Department of Agriculture and Fisheries, No. 400. Pretoria, pp 42–46

Milne DL (1982b) Nematode pests of litchi. In: Keetch DP, Heyns J (eds) Nematology in Southern Africa. Science Bulletin Department of Agriculture and Fisheries, No. 400, Pretoria, pp 38–41

Naudé L (2016). New committees for coffee study group. http://gpsnews.co.za/?p=15769

Onkendi EM, Moleleki LN (2013a) Detection of *Meloidogyne enterolobii* in potatoes in South Africa and phylogenetic analysis based on intergenic region and the mitochondrial DNA sequences. Eur J Plant Pathol 136:1–5

Onkendi EM, Moleleki LN (2013b) Distribution and genetic diversity of root-knot nematodes (*Meloidogyne* spp.) in potatoes from South Africa. Plant Pathol 62:1184–1192

Pegg KG, Moffett MC, Colbran RC (1974) Disease of ginger in Queensland. Qld Agricult J 100:611–618

Ramirez-Suáres A, Rosas-Herhandez L, Alcasio S et al (2014) First report of the root-knot nematode *Meloidogyne enterolobii*, parasitizing watermelon from Veracruz, Mexico. Plant Dis 98:428

SAAGA (2014) Avocado production in South Africa. http://www.avocado.co.za/local-marketing/south-african-avocado-production/. Accessed 15 Nov 2015

Saeed M (1974) Studies on some stylet-bearing nematodes associated with sapodilla (*Achras zapota* L.) with special reference to *Hemicriconemoides mangiferae* Siddiqi, 1961. PhD thesis, University of Karachi, Karachi

Sher SA (1955) Nematodes attacking avocado. California Citrograph 40:198

Sher SA, Foote FJ, Boswell SB (1959) A root lesion nematode disease of avocados. Plant Dis Rep 43:797–800

Singh ND, Farrell KM (1972) Occurrence of *Rotylenchulus reniformis* in Trinidad, West Indies. Plant Dis Rep 56:551

Steyn WP, Daneel MS, Serfontein S et al (2013) Investigations into the control of litchi tree dieback in South Africa. Acta Horticult 1029:439–444

Sukumaran S, Sundararaju P (1986) Pathogenicity of *Meloidogyne incognita* on ginger (*Zingiber officinale* Rosc.). Ind. J Nematol 16:258

Tigano M, De Siqueira K, Castagnone-Sereno P et al (2010) Genetic diversity of the root-knot nematode *Meloidogyne enterolobii* and development of a SCAR marker for this guava-damaging species. Plant Pathol 59:1054–1061

Van Zyl K (2013) A guide to crop pest management in South Africa. A compendium of acaricides, insecticides, nematicides, molluscicides, avicides and rodenticides. A CropLife Compendium, 1st edn. VR Print, Pinetown

Willemse S, Daneel MS, De Jager K et al (2014) Nematode tolerance enhanced using soil amendments on TSG2-guava trees in South Africa. J Nematol 46:257

Willers P (1997a) The nematode problem of guavas and possible use of cadusafos nematicide. Institute for Tropical and Subtropical Crops. Inform Bull 293:10–12

Willers P (1997b) First record of *Meloidogyne mayaguensis* Rammah and Hirschmann, 1988: Heteroderidae on commercial crops in the Mpumalanga Province, South Africa. Institute for Tropical and Subtropical Crops. Subtropica 18:19–20

Willers P (1998) Nematologiese navorsing in subtropiese bedrywe. Institute for Tropical and Subtropical Crops. Neltropika Bull 299:12–13

Willers P (1999) Do nematodes play a role in avocados? Neltropika Bull 306:7–8

Willers P (2001) Plant nematodes. In: Van den Berg MA, De Villiers EA, Joubert P (eds) Pest and beneficial arthropods of tropical and non-citrus subtropical crops in South Africa. Dynamic Add Mbombela, pp 50–52

Willers P, Daneel MS (1993) Pecan root knot nematode-prohibited. Institute for Tropical and Subtropical Crops. Inform Bull 245:10–11

Willers P, Neethling C (1994) Reniform nematode – a potential papaya pathogen? Institute for Tropical and Subtropical Crops. Inform Bull 267:13–14

Willers P, Neethling C (1997) Reniform nematode on papaya in the Onderberg area – possible pest status? Institute for Tropical and Subtropical Crops, Nelspruit. Inform Bull 293:9

Chapter 18
Nematode Pests of Pineapple

Elmarie C. Rabie

18.1 Introduction

Pineapple, *Ananas comosus* var *comosus*, originated in southern Brazil and Paraguay where wild relatives still occur. It was spread by the indigenous people through South and Central America. Pineapple reached the Caribbean, after centuries of Indian migration and commerce, where it was eventually discovered by Christopher Columbus on the island of Guadeloupe. From there it was introduced to tropical and subtropical areas throughout the world (Rohrbach et al. 2003) and first grown in South Africa (SA) in 1655. Nonetheless, the start of the local pineapple industry only took place around 1860 when it was planted in the KwaZulu-Natal and Eastern Cape (Bathurst district) provinces (Keetch 1982; Anonymous 2003).

Pineapple is mainly cultivated in northern KwaZulu-Natal (Zululand) and in the coastal belt of the Eastern Cape from East London to Port Alfred (Rabie 2001). Small plantations occur in southern KwaZulu-Natal, the Levubu district in the Limpopo Province as well as in the Lowveld of the Mpumalanga Province between Malelane and Komatipoort. Pineapples are produced in most of the countries included under the Southern African Development Community (SADC).

Three pineapple cultivars (cvs) are grown in SA, with Queen and MD2 (Honey Gold, Del Monte Gold®) being produced in northern KwaZulu-Natal mainly for the fresh market. Cultivar Smooth Cayenne is grown in the Eastern Cape Province, mainly for canning, with a small portion also supplied to the fresh market.

The cultivated pineapple belongs to the Bromeliaceae family (Anonymous 2003). It is a perennial monocotyledonous plant with a terminal inflorescence, which produces multiple fruit. After harvesting the fruit, plant growth continues through the

---

E.C. Rabie
Agricultural Research Council–Institute for Tropical and Subtropical Crops, Hluhluwe Research Station, P.O.Box 194, Hluhluwe 3960, South Africa
e-mail: erabie@mtuba.co.za

H. Fourie et al. (eds.), *Nematology in South Africa: A View from the 21st Century*, DOI 10.1007/978-3-319-44210-5_18

development of suckers from auxiliary buds on the mother stem. These suckers, as well as slips produced on the peduncle or the crown, can be used for planting material. Alternatively, the suckers can mature on the mother plant and a second crop (first ratoon) or even a third crop (second ratoon) can be harvested. Primary roots are found on the very young shoots, even while it is still attached to the plant. The initial growth and vigour of the plant depend on this first root system, but the roots soon die and are replaced by a second adventitious root system (Godfrey 1936). Pineapple plants retain their epiphytic ability to absorb water and nutrients through the leaves while supported by a fragile root system (Py et al. 1984). The root system is generally short and compact with numerous strong roots and little branching, but well supplied with rootlets and root hairs. It can develop extensively (1–2 m laterally and 1 m deep) under optimum conditions (Coppens d'Eeckenbrugge and Leal 2003).

Commercial pineapples are normally planted at 60,000 (cvs Cayenne and MD2) to 120,000 (cv. Queen) plants per hectare (ha). They are planted on ridges with two (for cvs Cayenne and MD2) or 4–6 plant rows (cv. Queen) ridge^{-1}. Ridges can be covered with plastic mulch. The growth cycle depends on the number of ratoons, but for the plant crop it varies between 12 and 18 months for cv. Queen in northern KwaZulu-Natal and 19–27 months for cv. Cayenne in the Eastern Cape Province. Only one ratoon crop of Queen is produced on heavy soils, while for Cayenne it is standard practice to grow one or two ratoons. A two-crop cycle can therefore take up to 4 years in the Eastern Cape Province. The phenology of variety MD2 in SA is still under investigation since it was only introduced in 2006. At present only a plant crop is grown.

18.2 Plant-Parasitic Nematodes Associated with Pineapple

Nematodes are considered a major pest of pineapple worldwide and are responsible for considerable losses. Although more than 100 nematode species have been found in association with pineapple roots, most of them are not considered as pests of the crop and/or their pathogenicity is unknown. Only four species are considered to be of economic importance, namely, the root-knot nematode species *Meloidogyne incognita* (Kofoid and White, 1919) Chitwood, 1949, *Meloidogyne javanica* (Treub, 1885) Chitwood, 1949, the reniform nematode *Rotylenchulus reniformis* Linford and Oliviera, 1940 and the lesion nematode *Pratylenchus brachyurus* (Godfrey, 1929) Filipjev and Schuurmans Stekhoven, 1941 (Py et al. 1984; Sipes et al. 2005). In SA, Keetch (1976) recorded 19 plant-parasitic nematode genera and 20 species in association with pineapple crops. More nematode pest genera and species were added thereafter (Keetch and Buckley 1984; Kleynhans et al. 1996), while records of plant-parasitic nematodes associated with pineapple are updated continuously in the SAPPNS Database[1]. However, only *M. javanica* and *P. brachyurus* are considered to be of economic importance for local pineapple producers.

[1] Dr Mariette Marais of the Nematology Unit, Biosystematics Division, Agricultural Research Council–Plant Protection Research Institute is thanked for the use of data from the South African Plant-Parasitic Nematode Survey (SAPPNS) database; E-mail: maraism@arc.agric.za.

The most frequently found plant-parasitic nematode species that have been recorded from pineapple roots in the Eastern Cape Province are *M. javanica* and spiral nematode genera *Helicotylenchus*, *Rotylenchus* and *Scutellonema*. *Pratylenchus brachyurus* is the predominant species in northern KwaZulu-Natal, especially on sandy soils (Rabie 2001), while *M. javanica* is found to a lesser extent. The latter nematode pest can, however, be the dominant species in areas with heavy soils. Keetch (1982) also listed *Helicotylenchus dihystera* (Cobb, 1893) Sher, 1961, *Meloidogyne* spp., *P. brachyurus* and *Scutellonema unum* Sher, 1964 from roots of natural vegetation in Zululand. Controlling the indigenous grass species such as *Panicum maximum*, a host for *P. brachyurus*, is therefore important in pineapple cultivation in Zululand. Nematodes reported from roots of natural vegetation in the Eastern Cape Province were *Meloidogyne* spp., *Rotylenchus unisexus* Sher, 1965 and *Scutellonema* spp. (Keetch 1982).

From Swaziland, eight nematode genera (*Meloidogyne*, *Pratylenchus*, *Paratylenchus*, *Rotylenchus*, *Hoplolaimus*, *Helicotylenchus*, *Rotylenchulus* and *Tylenchulus*) were recorded in root and soil samples (Keetch 1982). Results obtained through the diagnostic service of the Agricultural Research Council–Institute for Tropical and Subtropical Crops (ARC – ITSC) during 2005–2010 indicated that 67 % of the root and soil samples of cv. Cayenne were infected with root-knot nematodes. Spiral nematodes were the second most abundant (52 %), followed by lesion nematodes (20 %). Pin nematodes, *Paratylenchus* spp., occurred only occasionally in soil samples from Swaziland (less than 1 %). Concise summaries of the economically most important nematode pests of pineapple in SA is presented below.

18.2.1 Root-Knot Nematodes

A number of *Meloidogyne* spp. are associated with pineapple roots in various parts of the world, with *M. incognita* and *M. javanica* being considered as the most important (Guerout 1975; Sipes et al. 2005). Locally, *Meloidogyne arenaria* (Neal, 1889) Chitwood, 1949, *Meloidogyne ethiopica* Whitehead, 1968, *M. incognita* and *M. javanica* are associated with the root systems of pineapple (Keetch 1982; Keetch and Buckley 1984; Kleynhans et al. 1996; SAPPNS). However, only *M. javanica* has been associated with crop losses in local pineapple plantings and is considered a severe pest of such crops.

18.2.1.1 Symptoms and Damage Potential

The life cycle of *Meloidogyne* spp. is illustrated in Chap. 7 (Sect. 7.3.1, Fig. 7.3). The galls that develop on the roots as a result of feeding by second-stage juveniles (J2) and females are not always conspicuous. A high incidence of terminal galls and branching are an indication of an early, but severe root-knot nematode infection. This results in a stunted root system with poor anchorage and plants are then more susceptible to nutrient and moisture stress. Inadequate or ineffective treatment of soils with nematicides allows a relatively gall-free pineapple root system to develop, with a severe root-knot nematode infection generally being reached after 6–9 months.

A sufficiently long rotation crop or fallow period, irrespective of an effective pre-plant nematicide treatment, generally results in a root-knot nematode symptom-free root system for the plant crop. However, the ratoon crop can be severely infected. This is experienced in most of the plantings in local production areas. Root-knot nematode galls on pineapple roots may often be invaded by various fungi that cause the galls to turn black and rot internally.

According to Keetch (1982), the presence of a single *M. javanica* J2 in a root or soil sample is indicative of potential yield losses in pineapple crops to be planted in such infested fields.

18.2.2 Lesion Nematodes

Pratylenchus brachyurus is widespread in the subtropical northern KwaZulu-Natal area. Although this lesion nematode species is also present in the Eastern Cape Province where pineapple is produced, it is suggested to be of lesser importance compared to *M. javanica* that dominate in this area (Keetch 1982).

Pineapple planted on virgin soils can suffer substantial yield losses where a severe infestation of *P. brachyurus* occurs (Willers and Moolman 1990). This is especially applicable for cv. Queen planted in sandy soils where this species is the dominant pest. Although a mixed population of lesion and root-knot nematodes often occurs early in the growing season, lesion nematodes usually dominate (Rabie and Tustin 2002). This phenomenon was also recorded in pineapple-producing area in the Ivory Coast was attributed to the higher latitude (Sipes et al. 2005). Despite the presence of very high numbers of lesion nematodes in roots of cv. MD2, it does not necessarily correlate with a reduction in plant growth and yield. It is hence suggested that this variety may be tolerant to *P. brachyurus* (Rabie 2014).

18.2.2.1 Symptoms, Damage Potential and Factors Impacting on Population Development

Pratylenchus spp. are migratory endoparasites (Duncan and Moens 2013) that penetrate and move through the cortical parenchyma of pineapple roots during feeding (Sipes et al. 2005). Symptoms of root damage caused by lesion nematodes are generally difficult to detect. Black, necrotic lesions may be visible where the nematodes penetrate but can extend across the entire root surface. Although root growth continues after lesion nematode infection, it stops after penetration of the secondary roots by these pests (Py et al. 1984). Root systems of pineapple plants infected with numerous lesion nematodes can be devoid of secondary roots and consist only of poorly developed primary roots (Sipes et al. 2005). Infection by lesion nematodes ultimately destroys the parenchyma tissues of roots, while the cortex separates from the central cylinder (Guerout 1975).

Above-ground symptoms of nematode infected pineapple plants resemble those caused by a variety of factors such as moisture and nutrient stress. Zones of stunted plant growth is generally indicative of nematode-pest infestations in a pineapple field. In clay soils these zones can be ellipsoid, but in sandy soils they are more diffuse (Py et al. 1984). Emergence of leaves as well as leaf weight and width are reduced as a result of plant-parasitic nematode infection. Leaves of infected plants can also turn pinkish-yellow to red (Fig. 18.1a), loose turgidity and in severe cases leaf tips can die back whereas these symptoms are absent in non-infected plants (Fig. 18.1b). Such foliar symptoms are the result of deficient water and mineral supply to the plant since water and fertilizer absorption is suppressed by nematode damage to the roots. The extent of nematode damage to pineapple roots and the subsequent yield losses can be also attributed to the non-generative nature of the pineapple root system.

While population dynamics of lesion nematodes are directly related to climatic factors (especially rainfall) and plant growth (root emergence and elongation), survival of these pests between crops is dependent on the presence of living roots fragments in the soil (Guerout 1975). *Pratylenchus brachyurus* survived under laboratory conditions for more than 22 months in viable pineapple roots in fallow soils in the Ivory Coast. Conversely, survival of this lesion nematode species was only 7 months in soils in which such roots were absent (Sipes et al. 2005). Other hosts, such as weeds, also enhance the survival of *P. brachyurus* between crops. Keetch (1977) reported that temperature also influenced the survival of lesion nematodes, with only 25–50 % of a population surviving after 35 days at 44 °C. The optimum temperature for *P. brachyurus* development is 25–30 °C, while nematode movement is restricted at soil temperatures higher than 40 °C (Olowe and Corbett 1976).

When pineapple is planted in Zululand from November to February, representing the biggest part of the summer rainy season (from September to March), lesion nematode numbers generally peak approximately 2 months into the dry season (May/June) (Rabie and Tustin 2002). When planting occurs just before March/April or after the dry season (August to October), nematode populations generally stay low and peak in January/February, followed by another peak in June. The highest peak in nematode numbers occurs in May/June, while it decreases in September/October. Similar tendencies were found in the Ivory Coast (Sarah and Hugon 1991).

Soil temperature, soil structure and radiation affecting plant growth as well as the level of lesion nematode inoculum in soils at planting are other factors influencing the population dynamics of lesion nematodes (Sarah 1983). Results of nematode control trials as well as nematode analyses from diagnostic services indicated that *P. brachyurus* numbers increased sharply as soon as an infection level of ±450 nematodes 10 g roots^{-1} is reached. This density does not serve as an economic threshold but is a useful guide to initiate nematode control measures in pineapple plantings. The efficacy of pre-plant nematode control is in particular important in preventing such high lesion nematode population build-ups. Soil pH can also influence lesion-nematode population densities. Sarah et al. (1991) reported that population densities of *P. brachyurus* were higher in acid soils and remained low in soils with a pH exceeding 5–5.5. Pineapple is best cultivated in acid soils with a pH ranging from 4.5–6.5 (Py et al. 1984; Morton 1987). Most soils in Zululand are acid (as low as pH 3.4) and soil pH is increased by adding lime before planting.

Fig. 18.1 (**a**, **b**) Areas in a pineapple field where high population densities of *Pratylenchus* infection caused stunting of plants and *pinkish-yellow* to *red* discoloration of leaves (**a**) compared to healthy, non-infected pineapple plants (**b**) (Elmarie Rabie, Agricultural Research Council–Institute for Tropical and Subtropical Crops, Hluhluwe, South Africa)

18.3 Economic Importance of Nematode Pests and Estimated Yield Loss

Keetch (1989) estimated a yield loss of approximately 15 % in pineapple production in SA. In a pot trial with cv. Smooth Cayenne, the individual effects of spiral, root-knot and lesion nematodes were reported to be approximately 1, 10 and 25 % reduction in plant mass after 10 months (Keetch 1982). The latter author also reported an

Table 18.1 Average yield increases (%) obtained from 10 trials that were done in Zululand in pineapple fields infested with *Pratylenchus brachyurus* and *Meloidogyne javanica* (E Rabie, 2011, unpublished data)

Nematode control strategy	Average yield increase: metric tonnes per hectare	Percentage yield increase
Pre-plant only	12	34
Pre- plus post-plant	27	40
Post-plant only	10	19

average yield increase of 34 % in eight fumigation trials on cv. Smooth Cayenne. Milne et al. (1976) recorded a 367 % increase in yield of pineapple in Zululand after the application of nematicides. The treatments consisted of pre-plant fumigation with ethylenedibromide (EDB), dipping of planting material in oxamyl and two post-plant sprays with fenamiphos. Table 18.1 shows average yield increases from 10 nematode control trials that were done in Zululand on cv. Queen over a period of 16 years using registered nematicides. Treatments were divided into three nematode control strategies, namely, pre-plant only, pre-plant plus post-plant as well as the additive effect of post-plant control to a standard pre-plant control practice (E Rabie, 2011 unpublished data). Results showed that pre-plant control is the most important strategy for protecting pineapple crops against plant-parasitic nematodes. Willers and Smart (1990) also reported that lesion nematode control using pre-plant application of either EDB or methyl bromide increased fruit size, resulting in a higher percentage of Queen fruits in the popular sizes.

18.4 Management Strategies

Several factors necessitate the need for nematode control in pineapple in SA, namely, (i) monoculture cropping, (ii) crop cycles with cv. Cayenne (in the Eastern Cape) that can be as long as 8 years, (iii) the polyphagous nature of nematode-pest species that parasitise pineapple roots, (iv) growth in a plantation of cv. Queen must extend for 6–10 months after harvesting for the production of healthy planting material and (v) pineapple fields are often replanted within a year after harvest.

Control strategies against nematodes are initiated prior to planting and include practices such as fallowing, crop rotation, planting of cover crops, application of soil amendments as well as chemical control. Since pineapple is cultivated under rain-fed conditions, the application of post-plant control strategies presents a challenge for local producers.

18.4.1 Cultural Control

Aiming all cultural practices towards the reduction of the nematode populations is an integral part of nematode management. The practice of burning plant residues in the Eastern Cape is being phased out and replaced by mulching and incorporating plant residues into the soil. By adopting these practices, pre-plant soil fumigation can be eliminated. In Zululand, however, very little organic material is available in pineapple fields as suckers are used for planting material. Therefore, a longer intercrop cycle is advised (3 years plus). The addition of any organic matter can lead to a reduction in plant-parasitic nematode populations by increasing the number of nematode antagonists and predators in the soil and can have a positive effect on soil fertility and structure.

Keetch (1982) observed a reduction in nematode numbers in the Eastern Cape with frequent ploughing. In the Ivory Coast, a 6-week clean fallow period before planting reduced *P. brachyurus* numbers in plant crop roots after 12 months by 50 % (Guerout 1975). The effect of different cultivation practices, prior to planting in a short intercrop cycle, on plant-parasitic nematode populations and pineapple yield was investigated in Zululand (Rabie 2003). The effect of slashing plant residues prior to rotavating or discing was compared to rotavating, discing, burning of plant residues and clean fallow (all plant residues removed). Numbers of *P. brachyurus* were the lowest when all plant residues were removed from pineapple fields or burned. Slashing of residues prior to rotavating or discing soils also reduced lesion nematode numbers and suggested that incorporating smaller pieces of plant residue into the soil resulted in a lower nematode inoculum levels than when incorporating larger pieces. Ultimately, higher yields were correlated with lower nematode numbers as a result of this study.

Application of lime prior to planting is a near standard practice in Zululand due to the high acidity of soils. Sarah et al. (1991) reported that soil pH influences the ability of plant-parasitic nematodes to enter plant roots, with low pH enhancing the infection rate. The effects of the time of lime application as well as dosage applied (with or without nematode control) on lesion nematode populations were thus investigated. It was found that nematode-pest populations decreased with higher amounts of lime added to the soil, resulting in an increase in pH. However, time of lime application did not have an effect on nematode-pest populations but the earlier application of lime resulted in increased pineapple yields. It is therefore recommended that lime application is done well in advance (e.g. 3 months) before planting (Rabie 2003).

The polyphagous nature of the plant-parasitic nematodes that parasitize pineapple roots, as well as rain-fed cultivation of pineapples, makes the selection of alternative crops for crop rotation or cover crop purposes difficult. Sunn hemp (*Crotalaria* spp.) planted in Zululand as a cover crop in the intercrop cycle reduced nematode-pest populations in the pineapple crop more than did velvetbean (*Mucuna* sp.), which was claimed to be tolerant to *P. brachyurus* (Rabie and Tustin 2009; Rabie 2010). Pineapple yield was also higher following sunn hemp (*Crotalaria juncea*) than velvetbean. Wang et al. (2002) also reported that *Crotalaria* can increase populations of bacterivore nematodes and nematode-trapping fungi, while it is a poor host for *Meloidogyne* spp. Keetch and Dalldorf (1980) found that Rhodes grass (*Chloris gayana* variety Katambora) or silver leaf desmodium (*Desmodium uncinatum*) can reduce spiral and root-knot nematode numbers after a 3-year intercropping cycle.

18.4.2 Chemical Control

An effective nematode management strategy is based on the length of the crop cycle and number of ratoons to be grown. Options include pre-plant and post-plant nematicide applications.

Between 1970 and 1988, research on plant-parasitic nematodes that were associated with pineapple concentrated mainly on understanding the species complex, while the registration of nematicides for the control of such pests led to numerous publications (Marr 1972; Keetch 1975, 1976, 1977, 1982; Milne et al. 1976, 1977). In this period, the soil fumigants were registered and widely used as were fenamiphos (EC) and oxamyl (EC) as post-plant leaf applications. Granular formulations with cadusafos and aldicarb as active substances (a.s.) were later registered as pre-plant applications. Aldicarb was also registered for post-plant control of nematode pests of pineapple, but was withdrawn from the market in 2011 (see Sect. 6.2.2, Chap. 6).

18.4.2.1 Pre- and At-Plant Chemical Control

Pre- or at plant treatment is essential when a nematode-pest population is already present. Such a treatment generally represents fumigation, but can also include the incorporation of a nematicide into the soil during or before preparation of the field or during ridging (Fig. 18.2). In Zululand, pre-plant nematode control is essential, especially in a re-plant situation on the sandy soils. Typically 1,3 dichloropropene is applied at 225 l ha^{-1} followed by application of a registered post-plant nematicide to ensure optimal yield. In the Eastern Cape, only pre-plant control is applied using mainly 1,3 dichloropropene applied at 100 l ha^{-1}.

Keetch (1975) reported that temperatures were too low for the use of EDB in the months June to September in the Eastern Cape. Rabie and Tustin (1996) also reported that DD was more effective than EDB when applied in June in Zululand, while EDB was more effective in October. Current practice in Zululand is to apply the fumigant during ridging and subsequently cover the ridges with plastic mulch to retain the fumigant and to suppress weed growth. Since the planting of suckers takes place within a few days after fumigation, concern was raised about the possibility of a phytotoxic effect that can be caused by the fumigant. Measuring root and plant growth has, however, proven that no phytotoxic effect is experienced when fumigation and planting are done under optimum conditions (Rabie 2010). This can be due to the fact that root formation and elongation take place 2–3 weeks after planting, which corresponds with the recommended waiting period of 2 weeks after fumigation. Fumigation with EDB can be effective and reduce nematode populations for 3–6 months in the plant crop in Zululand (Rabie and Tustin 1996), while root-knot nematode numbers stayed low until 12 months after treatment in fumigation trials with 1,3 dichloropropene in the Eastern Cape (Petty 2009). Fumigation in the Eastern Cape resulted in lower yields on clay-loam soils than on

Fig. 18.2 Pineapple plants that are stunted and have discoloured leaves as a result of the soil not being fumigated before planting for nematode control (**a**) compared to healthy-looking pineapple plants planted in soil treated with a fumigant before planting (**b**) (Elmarie Rabie, Agricultural Research Council–Institute for Tropical and Subtropical Crops, Hluhluwe, South Africa)

sandy loam soils. Higher dosages (125–300 l ha^{-1}) than the standards treatment (100 l ha^{-1}) increased pineapple yield in the plant and ratoon crops but was not economically justifiable (Petty 2009).

Depending on planting time, application of cadusafos controlled nematodes for 3–4 months, whereas aldicarb was effective for 6–8 weeks. Root formation must therefore take place within a limited period after nematicide application, and optimum soil (moisture) and growth conditions (quality of planting material and rainfall) must therefore prevail. Post-plant applications of nematicides should be timeous to prevent rapid build-up of plant-parasitic nematode populations (Rabie and Tustin 1996).

Various other nematicides which contain a.s. ethoprophos, carbofuran, carbosulfan and fenamiphos (GR and EC) were evaluated for pre-plant control in two trials. The three registered nematicides, EDB, DD and cadusafos, resulted in the best nematode control and the highest pineapple yields (Rabie 2003).

18.4.2.2 Post-plant Chemical Control

The systemic properties of fenamiphos and oxamyl, as well as the fact that oxamyl can translocate basipetally in the plant, enable foliar application of such products using a boom sprayer during the growth cycle of pineapple plants. Though fenamiphos and oxamyl did not reduce high plant-parasitic nematode population densities, they delayed the increase in population levels of such pests. Time of application is therefor critical and will be determined by the nematode control practice used before or at planting. Aldicarb, applied to the soil between the rows before the plant canopy forms, was the only nematicide reported to reduce high population densities of the total plant-parasitic nematode complex.

Sarah (1980, 1981a, b) reported that fenamiphos caused heart and leaf burn of young plant tissue of pineapple plants as well as disturbance of flowering. Application must therefore cease 12 weeks prior to flower induction. In Zululand, the application of fenamiphos resulted in the production of up to 5 % non-flowering plants when applied close to flower induction (Rabie 2000). Milne et al. (1977) reported that fenamiphos inhibited oxidase activity, resulting in increased levels of the plant hormone indole acetic acid and consequent stimulation of growth in the absence of nematode pests.

18.5 Conclusions

Pineapple is an important commodity for SA since it is a high value crop in monetary terms. The adverse effect by root-knot and lesion nematodes are, however, devastating in some production areas and limit pineapple production. Although various nematode management strategies are employed to combat nematode pests of pineapple, chemical control is still regarded as the most effective. Future nematode research on pineapple is, however, crucial and alternative options should be exploited to protect crops against such pests in intensive production systems. The withdrawal of effective, synthetically derived nematicides is a challenge for producers and researchers and implies that finding alternative nematicides and/or strategies that holds promise in ensuring sustainable crop production be identified and investigated.

References

Anonymous (2003) The Biology and Ecology of Pineapple (*Ananas comosus* var *comosus*) in Australia. Australian Government. Office of the Gene Technology Regulator. Department of Health and Ageing, Canberra

Coppens d'Eeckenbrugge G, Leal F (2003) Morphology, anatomy and taxonomy. In: Bartholomew DP, Paull RE, Rohrbach KG (eds) The Pineapple: Botany, production and uses. CAB International, Wallingford, pp 13–32

Duncan LW, Moens M (2013) Migratory endoparasitic nematodes. In: Perry RN, Moens M (eds) Plant nematology, 2nd edn. CAB International, Wallingford, pp 144–178

Godfrey GH (1936) The pineapple root system as affected by the root-knot nematode. Phytopathol 26:408–428

Guerout R (1975) Nematodes of pineapples. A review. PANS 21:123–140

Keetch DP (1975) Soil temperatures and the use of the fumigants D-D, EDB and DBCP in pineapple lands. Citrus Subtropical Fruit J 500:5–8

Keetch DP (1976) Nematodes of pineapples in South Africa. Crop Prod 5:51–54
Keetch DP (1977) Nematodes in pineapple. Farming in South Africa. Pineapple Series H 18
Keetch DP (1982) Nematode pests of Pineapple. In: Keetch DP, Heyns J (eds) Nematology in Southern Africa. Science Bulletin No. 400. Department of Agriculture and Fisheries, Pretoria, pp 19–29
Keetch DP (1989) A perspective of plant nematology in South Africa. S Afr J Sci 85:506–508
Keetch DP, Buckley NH (1984). A check-list of the plant-parasitic nematodes of southern Africa. Technical Communication, Department of Agriculture, Plant Protection Research Institute, Pretoria
Keetch DP, Dalldorf ER (1980) The use of grass and legume rotations for Pineapple nematode control in the Eastern Cape. Citrus Subtropical Fruit J 557:10–12
Kleynhans KPN, Van den Berg E, Swart A et al (1996) Plant nematodes in South Africa, Plant Protection Research Institute Handbook No 8. Agricultural Research Council–Plant Protection Research Institute, Pretoria
Marr GS (1972) The effect of soil fumigation, applied fertiliser and climate on the growth and nutient levels of Cayenne pineapples under field conditions. PhD. Thesis, Rhodes University, Grahamstown
Milne DL, Reynolds R, Webster G (1976) We told you so. Citrus and Subtropical Fruit Research Institute, Nelspruit. Hluhluwe Inform Bull 45:6–7
Milne DL, Van Lelyveld LJ, De Villiers EA (1977) The response to peroxidase and indolacetic acid oxidase in pineapple roots to foliar applications of fenamiphos. Agrochemophysica 9:65–70
Morton JF (1987) Pineapple. In: Fruits of warm climates. http://www.hort.purdue.edu/newcrop/morton/pineapple.html. Accessed 23 Oct 2014
Olowe T, Corbett DCM (1976) Aspects of the biology of *Pratylenchus brachyurus* and *P. zea*. Nematologica 22:202–211
Petty GJ (2009) Evaluation of insecticides and nematicides for control of nematode and beetle soil pests of pineapples in the Eastern Cape. Annual Pineapple Project Report 2008–2009. Agricultural Research Council–Institute for Tropical and Subtropical Crops, Nelspruit
Py C, Lacoeuilhe JJ, Teisson C (1984) The pineapple: cultivation and uses. G.P. Maisonneuve et Larose
Rabie EC (2000) Nematode control in Queen pineapple – To establish a post-plant control program for preventative and correctional control. Annual Pineapple Project Report 1999–2000. Agricultural Research Council–Institute for Tropical and Subtropical Crops, Nelspruit
Rabie EC (2001) Biology and control of pests. Nematoda. Pineapples. In: Van den Berg MA, De Villiers EA, Joubert PH (eds) Pests and beneficial arthropods of tropical and non-citrus subtropical crops in South Africa. Ad Dynamics, Nelspruit, pp 56–63
Rabie EC (2003) Nematode control in Queen pineapple – Preplant control of nematodes in Queen pineapple – alternatives for EDB. Annual Pineapple Project Report 2002–2003. Agricultural Research Council–Institute for Tropical and Subtropical Crops, Nelspruit
Rabie EC (2010) Nematode control in Queen pineapple - To determine the effect of Telone II for nematode control on root development of the Queen pineapple. Annual Pineapple Project Report 2009–2010. Agricultural Research Council–Institute for Tropical and Subtropical Crops, Nelspruit
Rabie EC (2014) Phenological studies to improve growth and fruit quality of Queen pineapple. Annual Pineapple Project Report 2013–2014. Agricultural Research Council–Institute for Tropical and Subtropical Crops, Nelspruit
Rabie EC, Tustin HA (1996) Preplant control of nematodes in pineapple. Acta Horticul 425:435–442
Rabie EC, Tustin HA (2002) An integrated approach to nematode control in queen pineapple in northern KwaZulu-Natal, South Africa. Acta Horticul 666:229–236
Rabie EC, Tustin HA (2009) The effect of different cover crops on nematode populations and yield in 'queen" pineapple cultivation. Acta Horticul 822:205–213

Rohrbach KG, Leal F, Coppens d'Eeckenbrugge G (2003) History, distribution and world production. In: Bartholomew DP, Paull RE, Rohrbach KG (eds) The pineapple: Botany, production and uses. CAB International, Wallingford, pp 1–12

Sarah JL (1980) Use of systemic nematicides in the control of *Pratylenchus brachyurus* (Godfrey) in pineapple growing. I Preventive and curative activity on root infestations by foliar application. Fruits 35:745–757

Sarah JL (1981a) Use of systemic nematicides for the control of *Pratylenchus brachyurus* in pineapples. II Secondary effects of foliar applications on the vegetative phase of the pineapple development cycle. Fruits 36:275–283

Sarah JL (1981b) Use of systemic nematicides for the control of *Pratylenchus brachyurus* in pineapples. III Secondary effects of foliar applications on response to flower induction treatment and flowering. Fruits 36:491–500

Sarah JL (1983) The use of systemic nematicides in the control of *Pratylenchus brachyurus* on pineapple plantations. IV Secondary effects of foliar applications on plant development and fruit maturation after flower induction treatment. Fruits 38:523–540

Sarah JL, Hugon R (1991) Population dynamics of *Pratylenchus brachyurus* in pineapple growing in Cote D'Ivoire. Fruits 46:241–250

Sarah JL, Osseni B, Hugon R (1991) Effect of soil pH on development of *Pratylenchus brachyurus* population in pineapple roots. Nematropica 21:211–216

Sipes BS, Caswell-Chen EP, Sarah J-L et al (2005) Nematode Parasites of Pineapple. In: Luc M, Sikora RA, Bridge J (eds) Plant parasitic nematodes in subtropical and tropical agricultural, 2nd edn. CAB International, Wallingford, pp 709–731

Wang KW, Sipes BS, Schmitt DP (2002) Management of *Rotylenchus reniformis* in pineapple, *Ananas comosus*, by intercycle cover crops. J Nematol 34:106–114

Willers P, Moolman M (1990) Nuwe grond: nie altyd sonder aalwurmprobleme nie. Citrus and subtropical Fruit Research Institute, Nelspruit, South Africa. Information Bull 213:7

Willers P, Smart G (1990) Lesion nematode: an underestimated problem for pineapple growers in Hluhluwe. Subtropica 11:11–13At exceprerchil molo eos illum quod etum vellenditium velitatia iurerum eseque vit dolupta net idusa dolenis nesent re eaquatusae ra si solupta vidunt, quam,

Chapter 19
Nematodes of Grasses and Weeds

Keikantsemang N. Ntidi, Suria Bekker, and Hendrika Fourie

19.1 Introduction

Grasses and weeds did not receive priority in terms of nematode research in South Africa (SA). Nonetheless, the limited and fragmented information concerning the associations between plant-parasitic nematodes and grasses and weeds that has been generated over the past 60 years is reported on. Defining the term "weeds" is important (Box 19.1) to clarify its use in this chapter and put it in perspective.

Box 19.1: Weeds

Weeds are generally defined as plants or vegetation that grow in undesirable places (Bromilow 2001), specifically referring to agricultural systems, for the purpose of this chapter. Referring to agriculture, a weed may hence be defined as any plant or vegetation that interferes with and adversely impacts on the objectives of farming such as growing crops and grazing of cattle. Hence, weeds can be referred to as plants that grow where they are not wanted. For example, grasses that grow in fields where farmers cultivate crops are regarded as weeds. When grasses are, however, grown as a fodder crop for cattle or for their seed, they are not regarded as weeds.

K.N. Ntidi (✉)
Agricultural Research Council–Grain Crops Institute,
Private Bag X1251, Potchefstroom 2520, South Africa
e-mail: NtidiN@arc.agric.za

S. Bekker • H. Fourie
North-West University, Unit for Environmental Sciences and Management,
Private Bag X6001, Potchefstroom 2520, South Africa
e-mail: suriabekker2@gmail.com; driekie.fourie@nwu.ac.za

H. Fourie et al. (eds.), *Nematology in South Africa: A View from the 21st Century*, DOI 10.1007/978-3-319-44210-5_19

Fig. 19.1 Seed-galls of *Eragrostis curvula*, representing blackish structures (in *white circles*) infected with *Subanguina wevelli* (Suria Bekker, Agricultural Research Council—Grain Crops Institute, Potchefstroom, South Africa)

19.2 Seed- and Leaf-Gall Nematodes

Seed- and leaf-gall nematode (SLGN) species, belonging to the genera *Anguina* and *Subanguina*, parasitize seeds, leaves, and other above-ground parts of grasses and crops such as wheat (*Triticum aestivum*) (Agrios 1997; Mc Donald and Nicol 2005; Duncan and Moens 2013; SAPPNS[1]). The most noticeable symptom in grass seeds infected by these nematodes is the presence of characteristic, dark galls (Fig. 19.1) (Inserra et al. 2003; Bekker 2010). In terms of survival, SLGN are some of the most successful anhydrobiotic organisms (Womersley et al. 1982). Their survival is greatest when the desiccation process is slow, since most individuals are killed when it happens too quickly (Barrett 1991).

Eight SLGN species are important in terms of regulatory aspects and are thus recorded as quarantine organisms (Duncan and Moens 2013). The problem caused by these nematodes in graminaceous hosts is aggravated due to their association with both fungal and bacterial diseases (Riley and McKay 1990; Mc Donald and Nicol 2005). For example, *Anguina funesta* Price, Fisher & Kerr, 1979, in association with a coryneform bacterium (*Clavibacter* sp.), causes a disease known as annual ryegrass toxicity (Riley et al. 1988), which can result in the death of livestock. This disease was first recorded in 1956 in South Australia (McKay 1993) and has resulted in livestock deaths in South Australia since 1979 (Anonymous 1981). Similarly, the concomitant occurrence of *Anguina agrostis* (Steinbuch, 1799) Filipjev, 1936 and either *Clavibacter toxi-*

[1]Dr Mariette Marais of the Nematology Unit, Biosystematics Division, Agricultural Research Council-Plant Protection Research Institute is thanked for the use of data from the South African Plant-Parasitic Nematode Survey (SAPPNS) database; E-mail: maraism@arc.agric.za

Table 19.1 Seed and leaf-gall nematodes of *Anguina* and *Subanguina* spp. associated worldwide with grasses (Family Gramineae) (Cid Del Prado Vera and Maggenti 1984; Subbotin et al. 2004; Fleming et al. 2015)

Seed and leaf-gall nematode species	Grass host plant	Infected plant part/other medium	Country
Anguina agropyri Kirjanova, 1955	*Elymus repens*	Basal stem	Estonia
Anguina agrostis (Steinbuch, 1799) Filipjev, 1936	*Agrostis capillaris* *Lolium perenne*	Seed	Australia, Belgium, New Zealand, Russia, South Africa, The United States of America
Anguina askenasyi Bütschli, 1873	*Calliergon cuspidata*	Terminal	Estonia
Anguina australis Steiner, 1940	*Ehrharta longiflora*	Leaf	Australia
Anguina funesta Price, Fisher & Kerr, 1979	*Lolium rigidum*	Seed	Australia
Anguina graminis (Hardy, 1850) Filipjev, 1936	*Festuca rubra*	Leaf	Russia
Anguina microlaenae (Fawcett, 1938) Steiner, 1940	*Microlaena stipoides*	Leaf	Australia
Anguina pacificae Cid Del Prado Vera & Maggenti, 1984	*Poa annua*	Stem Root	California Ireland
Anguina phalaridis Steinbuch, 1799	*Phleum phleoides*	Seed	Estonia
Anguina tritici (Steinbuch, 1799) Chitwood, 1935	*Triticum aestivum*	Seed	Australia, South Africa
Anguina woodi Swart, Subbotin, Tiedt & Riley, 2004	*Ehrharta villosa*	Stem	South Africa
Subanguina radicicola (Greef, 1872) Paramonov, 1967	*Phalaris arundinacea* *Poa annua* *Poa* sp.	Root Root Root	Estonia Russia Belgium
Subanguina tumefaciens (Cobb, 1932) Fortuner & Maggenti, 1987	*Cynodon transvaalensis*	Stem, leaf, and flower head	South Africa
Subanguina wevelli (Van den Berg, 1985) Ebsary, 1991	*Eragrostis curvula*	Seed	South Africa

cus or *Corynebacterium rathayi* caused annual ryegrass toxicity and the death of livestock in the Western Cape, in particular where ryegrass was used for pasture (Anonymous 1981; Morton and Meyer 1989). A number of SLGN species have been reported from grasses worldwide (Table 19.1), with some undisclosed species also being reported (Subbotin et al. 2004).

Fig. 19.2 *Anguina woodi* male and female individuals in seed galls of dune grass (*Ehrharta villosa* var. *villosa*) (Antoinette Swart, Agricultural Research Council-Plant Protection Research, Pretoria, South Africa)

Five SLGN species have been reported for SA and include *A. agrostis*, *Anguina woodi* Swart, Subbotin, Tiedt & Riley, 2004 (Fig. 19.2), *Anguina tritici* (Steinbuch, 1799), Chitwood, 1935, *Subangiuna wevelli* (Van den Berg 1985) Fortuner & Maggenti, 1987 and *Subanguina tumefaciens* (Cobb, 1932) Fortuner & Maggenti, 1987.

Anguina agrostis was recorded from the seeds of *Agrostis capillaris*, *Lolium perenne*, and *Lolium* spp. in SA (Schneider 1981). It also occurred in volunteer grasses (*Lolium* spp.) in wheat fields in the Ceres, Caledon, and Bredasdorp areas, and resulted in numerous livestock deaths (Anonymous 1981). *Anguina agrostis* was also recorded from pasture grass in SA by Kleynhans et al. (1996). *Anguina woodi* was identified from galled seeds of *Ehrharta villosa* f. var. *villosa* that grew on dunes near Milnerton beach, Western Cape (Swart et al. 2004). The SLGN *A. tritici*, was reported to parasitise wheat crops (Kleynhans et al. 1996; SAPPNS).

Subanguina wevelli was first identified in SA by Van den Berg (1985) from seed samples of *Eragrostis curvula*, collected near Harrysmith and Reitz (Free State Province) and near Bethal (Mpumalanga Province). The presence of *S. wevelli* was also reported from pasture grasses (Kleynhans et al. 1996), and in 2000, in galled seeds from *E. curvula* that was cultivated in the Mpumalanga and the Free State provinces (A. Swart, Agricultural Research Council-Plant Protection Research, Pretoria, 2016, personal communication). Also, seed samples obtained from Amersfoort (Mpumalanga Province) and Potchefstroom (North-West Province) during 2008 were found to be infected with *S. wevelli* (Bekker 2010). Individuals of the other *Subanguina* sp., viz., *S. tumefaciens*, were isolated from galled seeds of *Cynodon transvaalensis* and from lawns in SA (Keetch and Buckley 1984; Kleynhans et al. 1996). The life cycle of SLGN is illustrated in Fig. 19.3.

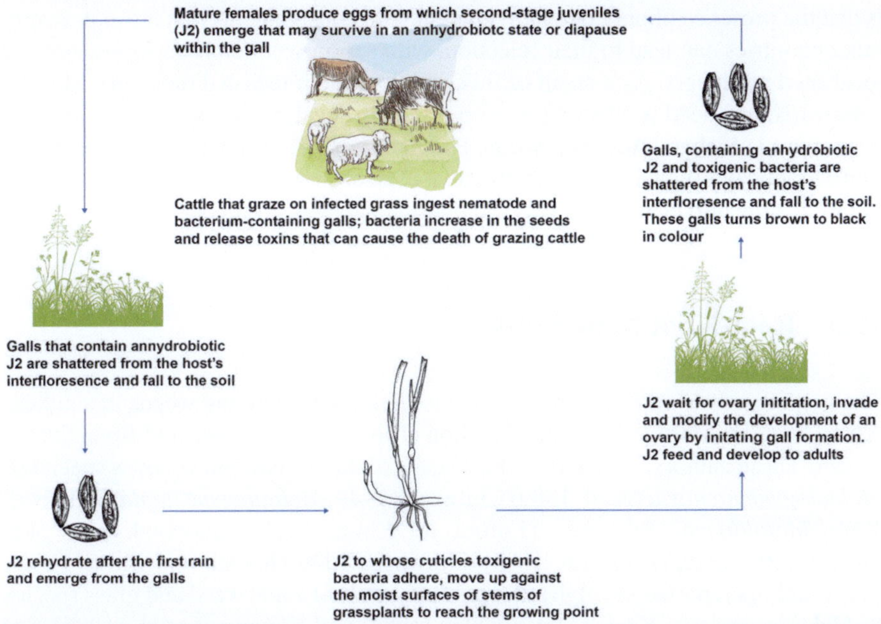

Fig. 19.3 The life cycle of *Anguina* species that infect grasses and can cause the death of livestock due to their concomitant occurrence with pathogenic bacteria (Hannes Visagie, North-West University, Potchefstroom, South Africa)

Due to the quarantine status of SLGN, the presence of single individuals of *Anguina* species in *E. curvula* seed in shipments exported to the United States of America (USA) resulted in the rejection of consignments on several occasions (T. Siebert, Advance Seeds, Potchefstroom, 2015, personal communication). This emphasizes the negative impact that SLGN have on the local grass seed industry. It resulted in a survey being conducted in fields where *E. curvula* and *Lolium* spp. were grown in the major grass-producing areas of SA. *Subangiuna wevelli* was recovered in large numbers from 92 % of the localities sampled (Bekker 2010).

The evaluation of commercializable sieving and flotation techniques was investigated as a means of reducing population levels of *S. wevelli* in grass seeds. Substantial reductions were achieved, but the mere presence of a few infected seeds in the samples still poses a problem to the local grass-seed industry. Subsequently, the use of nematicides to reduce *S. wevelli* numbers was investigated in a microplot study (Bekker 2010). Treatment of *E. curvula* seed with 0.25 mg abamectin per seed and soil application of carbofuran (1.5 g m^{-1}) and terbufos (0.66 g m^{-1}), resulted in a significant reduction of *S. wevelli* numbers. Treatment with aldicarb, cadusafos, ethylene dibromide, ethoprophos, and oxamyl was less effective. Results from this research suggested that a single control strategy is not sufficient to ensure that grass-seed consignments are free of SLGN. The bottom line of the local SLGN problem is that the presence of only one individual in seed consignments that are exported to other countries can lead to their rejection, with serious economic repercussions for local seed producers. As a result of this situation, the forecasted monetary value of *Eragrostis* spp. seed produced for export and the local market since 2001 has not been met (T. Siebert, Advance Seeds, Potchefstroom, 2015, personal communication). Practical, efficient, and cost-effective control strategies for SLGN hence have to be developed to enable producers to optimize grass-seed production.

19.3 Root-Knot Nematodes

The first records of root-knot nematode research on grasses and weeds in southern Africa were by Martin (1956) and Daulton (1963) for Zimbabwe and Koen (1965) for SA. These authors tabulated the host status of three weed and 11 grass species to *Meloidogyne arenaria* (Neal, 1889) Chitwood, 1949, *Meloidogyne hapla* Chitwood, 1949, *Meloidogyne incognita* (Kofoid & White, 1919) Chitwood, 1949 and *Meloidogyne javanica* (Treub, 1885) Chitwood, 1949 (Keetch and Milne 1982). These authors reported the high susceptibility levels of some weed and grass species to *Meloidogyne* spp. Keetch and Buckley (1984) and Kleynhans et al. (1996) also reported *Meloidogyne* spp. associated with various weeds and grasses. Additional information is contained in the database of the South African Plant Parasitic Survey (SAPPNS) to which listings are continuously added. Roots of weeds infected with root-knot nematodes generally show galling (Fig. 19.4).

A study on the inclusion of *E. curvula* (cv. Ermelo) grass as a suitable cover crop to reduce population levels of *M. chitwoodi* Golden, O' Bannon, Santo & Finley, 1980 in a potato field in the Mooi river area (KwaZulu-Natal Province) was done by Fourie et al. (1998). This grass species had a reproduction factor (Rf) of 0.07, whereas that of the susceptible tomato standard (cv. Rutgers) was approximately 20-fold higher. The use of this poor host grass variety is, however, not recommended for planting in fields where SLGN are present, since it is a good host of these pests (see Table 19.1). In the same study, *Eragrostis tef* (cv. SA Bruin) and *Lolium multiflorum* (cv. Midmar) were reported as highly susceptible to the Mooi River population of *M. chitwoodi*, with Rf values of 24 and 7, respectively. These two grass species are thus not suitable cover or rotation crops in fields infested with this root-knot nematode species.

In a pot trial, Berry and Rhodes (2006) found that babala (*Pennisetum glaucum*), which is normally used as a summer cover crop, increased *M. javanica* population densities. In the same study, Rhodes grass (*Chloris gayana*) allowed multiplication of *M. javanica*, but maintained relatively low numbers of the pest.

Ntidi et al. (2012) conducted a nematode survey of weeds (including various grasses) in 67 fields of smallholding farmers in SA. The predominant root-knot species in both soil and root samples were *M. hapla*, *M. incognita*, and *M. javanica*. These species were associated with 30 weed species. They were particularly abundant in roots of feather finger grass (*Chloris virgata*), common couch grass (*Cynodon*

Fig. 19.4 Galled, root-knot nematode infected roots of *Amaranthus greacizans* (Bosbok) infected with *Meloidogyne incognita* (Nancy Ntidi, Agricultural Research Council - Grain Crops Institute, Potchefstroom, South Africa)

dactylon), and common pigweed (*Amaranthus hybridus*). Interestingly, the presence of *Meloidogyne* spp. was significantly and positively correlated with soils with high clay and silt percentages and also with high rainfall. A negative correlation was observed for pH. An important observation from this research is that numbers of *Meloidogyne* spp. may build up and reach damaging levels in weeds growing in soils with a high clay content where these pests are often not considered a problem.

Furthermore, *Meloidogyne* spp. occurred more frequently at localities where maize was monocropped, and where maize was rotated with vegetable crops. Relatively high prominence values for *Meloidogyne* spp. were recorded for marigold (*Tagetes minuta*), which is generally perceived as a poor host of plant-parasitic nematodes (Bromilow 2001). This signals a definite warning that this weed species should not be underestimated as a host of root-knot nematodes.

Ntidi et al. (2015) also screened 20 weed species, commonly occurring in fields of smallholding farmers, for their host suitability to *M. incognita* and *M. javanica* in glasshouse and field experiments. Using Rf as the parameter in glasshouse experiments, seven weed species were classified as susceptible and 13 as resistant to both nematode species. Bladderweed (*Hibiscus trionum*) and Joseph's coat (*Amaranthus tricolor*) were the most susceptible, while green goosefoot (*Chenopodium carinatum*) and large thorn apple (*Datura ferox*) were the poorest hosts for *M. incognita* and *M. javanica*, respectively. In field experiments at Kuruman, nightshade

(*Solanum retroflexum*) was the most susceptible weed species to a *M. javanica* population, while *H. trionum* was the most susceptible weed at Mbombela where a mixed population of *M. incognita* and *M. javanica* occurred. *Hibiscus trionum* was also the most susceptible weed at Potchefstroom where a population of *M. incognita* was present.

19.4 Other Plant-Parasitic Nematodes Associated with Grasses and Weeds

A wide range of nematode pests, other than SLGN and root-knot nematodes, have been reported to parasitize numerous weed and grass species that occur in SA (Keetch and Buckley 1984; Fourie et al. 1998; Berry and Rhodes 2006; Ntidi et al. 2012, 2015; Marais 2015). Also a large number of both ectoparasitic and endoparasitic nematodes have been recorded from pasture grasses, lawns, golf courses, and bowling-greens by Kleynhans et al. (1996) and Swart et al. (2000). Berry and Rhodes (2006) reported that babala maintained high numbers of *Pratylenchus zeae* Graham, 1951, but low numbers of *Helicotylenchus dihystera* (Cobb 1893) Sher, 1961, *Paratrichodorus* sp., and *Xiphinema elongatum* Schuurmans Stekhoven & Teunissen, 1938. In the same study, Rhodes grass was also recorded as maintaining relatively low numbers of *H. dihystera*, a *Paratrichodorus* sp. and *X. elongatum*, but intermediate numbers of *P. zeae*. The study by Ntidi et al. (2012) reported that *P. zeae*, *H. dihystera*, and *Rotylenchus unisexus* Sher, 1965 followed in terms of dominance after *Meloidogyne* (see Sect. 19.3). The groundnut pod nematode, *Ditylenchus africanus* Wendt, Swart, Vrain & Webster, 1995 (previously identified as *D. destructor*), has also been reported from various weeds sampled in the study by Ntidi et al. (2012). This may have serious consequences should groundnut be included in gardens and fields of smallholding farmers where this nematode pest species occur. Earlier, De Waele et al. (1990) evaluated the host status of seven weed species that commonly occur in groundnut fields in SA, against *D. africanus*. These authors also studied the effect of growing groundnut in pots where *D. africanus*-infected weeds were cultivated. Results showed that white goosefoot (*Chenopodium album*), feather finger grass, purple nutsedge (*Cyperus rotundus*), jimson weed (*Datura stramonium*), goose grass (*Eleusine indica*), khaki weed (*T. minuta*), and cocklebur (*Xanthium strumarium*) were poor hosts for the groundnut-pod nematode. However, it survived in roots of all weed species screened and hence demonstrated that these weeds can play an important role in the survival of this pest in the absence of groundnut crops. Such a scenario may result in an increase in population densities of the groundnut pod nematode in hulls and seeds of groundnut that is planted as a follow-up crop where these weeds occur. Roots of purple nutsedge that were left in the soil and in which a groundnut crop was planted suppressed populations of the groundnut pod nematode that developed in roots and pods of groundnut plants. However, nematode populations in hulls and seeds of groundnut were not suppressed.

The effects of weeds on the infection of maize roots by the lesion nematode, *P. zeae*, were investigated in a glasshouse study (Jordaan and De Waele 1987). Five weed species (*E. indica*, *Crotalaria sphaerocarpa*, *A. hybridus*, *D. stramonium*, and *T. minuta*) that commonly occur in maize-producing areas were investigated for their host status and the effects that their root exudates have on nematode infection. All of the weed species, as well as maize, were hosts to *P. zeae*, with *E. indica* supporting the highest and *T. minuta* the lowest population densities. Still as part of this study, maize was planted in pots where the weed species had previously been cultivated and then removed. Interestingly, *P. zeae* numbers were only significantly reduced in roots of maize plants grown in pots in which *T. minuta* had been grown.

19.5 Conclusions

It is evident from the nematode research done locally for grasses and weeds that various species serve as hosts of plant-parasitic nematodes and in this way adversely affect crop production. Timely and effective removal of weeds is thus crucial to optimize nematode management strategies and contribute to sustainable crop production.

References

Agrios GN (1997) Plant pathology, 4th ed. Academic Press, San Diego

Anonymous (1981) G1. Giftige plante tussen koring in die somerreëngebied. Boerdery in Suid-Africa. Departement van Landbou en Visserye, Pretoria

Barrett J (1991) Anhydrobiotic nematodes. Agric Zool Rev 4:161–176

Bekker S (2010) Assessment of the identity, distribution and control options for seed- and leaf-gall nematodes in grass in South Africa. MSc Dissertation, North-West University, Potchefstroom

Berry S, Rhodes R (2006) Green manure crops: agronomic characteristics and effect on nematodes. Proc S Afr Sugar Technol Assoc 80:270

Bromilow C (2001) Problem plants of South Africa. Briza Publications, Pretoria

Cid Del Prado Vera I, Maggenti A (1984) A new gall-forming species of *Anguina* Scopoli, 1977 (Nemata: Anguinidae) on bluegrass, *Poa annua* L., from the coast of California. J Nematol 16:386–392

Daulton RAC (1963) Controlling *M. javanica* in Southern Rhodesia. Rhodesia Agricult J 60:150–152

De Waele D, Jordaan EM, Basson S (1990) The host status of seven weed species and their effects on *Ditylenchus destructor* infestation of peanut. J Nematol 22:292–296

Duncan LW, Moens M (2013) Migratory endoparasitic nematodes. In: Perry RN, Moens M (eds) Plant nematology. CAB International, Wallingford, pp 144–178

Fleming TR, Maule AG, Martin T et al (2015) A first report of *Anguina pacificae* in Ireland. J Nematol 47:97–104

Fourie H, Zijlstra C, Mc Donald AH (1998) ITS-PCR sequence based identification of *M. chitwoodi* and screening of crops for host suitability. Afr Plant Prot 4:107–111

Inserra R, Brito JA, Dong K et al (2003) List of exotic nematode plant pests of agricultural and environmental significance to the United States. http://nematode.unl.edu/pest33.htm. Accessed 8 Jan 2014

Jordaan EM, De Waele D (1987) Effects of weeds on the infestation of *Pratylenchus zeae* in maize. Phytophylactica 19:254

Keetch, DP, Buckley NH (1984) A checklist of the plant-parasitic nematodes of southern Africa., Technical Communication No.195. Department of Agriculture, Pretoria

Keetch DP, Milne DL (1982) The control of plant-parasitic nematodes. In: Keetch DP, Heyns J (eds) Nematolgy in southern Africa. Science Bulletin No 400, Department of Agriculture and Fisheries, Pretoria

Kleynhans KPN, Van den Berg E, Swart A et al (1996) Plant nematodes in South Africa., Plant Protection Research Institute Handbook No. 8. Agricultural Research Council – Plant Protection Research Institute, Pretoria

Koen H (1965) The influence of a number of crops on root-knot nematodes (*Meloidogyne* spp.) over a period of six years. Meded Landbouwhogeschool Wageningen 30:2087–2092

Marais M (2015) Dataset from South African Plant-Parasitic Nematode Survey (SAPPNS) database. Identification Job Sheet N03655, Nematology Unit, Biosystematics Division. Agricultural Research Council - Plant Protection Research Institute, Pretoria

Martin GC (1956) The common root-knot nematode (*Meloidogyne javanica*) and crop rotation. No. 20. Summary. Rhodesian Farmer 27:24–25

Mc Donald AH, Nicol JM (2005) Nematode parasites of cereals. In: Luc M, Sikora RA, Bridge J (eds) Plant-parasitic nematodes in subtropical and tropical agriculture, 2nd edn. CAB International, Wallingford, pp 131–191

McKay AC (1993) Livestock deaths associated with *Clavibacter/Anguina* sp. infection in seedheads of *Agrostis avenacea* and *Polypogon monspeliensis*. Plant Dis 77:635–641

Morton AJ, Meyer AJ (1989) Observations on the life cycle, toxicity and spread of *Anguina agrostis* in the Western Cape. Proceedings of the Ninth Symposium of the Nematological Society of Southern Africa. Phytophylactica 21:214

Ntidi KN, Fourie H, Daneel M (2015) Greenhouse and field evaluations of commonly occurring weed species for their host suitability to *Meloidogyne* species. Int J Pest Manag. doi:10.1080/09670874.2015.1087602

Ntidi KN, Fourie H, Mc Donald AH et al (2012) Plant-parasitic nematodes associated with weeds in developing agriculture in South Africa. Nematol 14:875–887

Riley IT, McKay AC (1990) Specificity of the adhesion of some plant pathogenic micro-organisms to the cuticle of nematodes in the genus *Anguina* (Nematoda: Anguinidae). Nematologica 36:90–103

Riley IT, Reardon TB, McKay AC (1988) Electrophoretic resolution of species boundaries in seedgall nematodes, *Anguina* spp. (Nematoda: Anguinidae), from some Graminaceous hosts in Australia and New Zealand. Nematologica 34:401–411

Schneider DJ (1981) First report of annual ryegrass toxicity in the Republic of South Africa. Onderstepoort J Veter Res 48:251–255

Subbotin SA, Krall EL, Riley IT et al (2004) Evolution of the gall-forming plant parasitic nematodes (Tylenchida:Anguinidae) and their relationships with hosts as inferred from Internal Transcribed Spacer sequences of nuclear ribosomal DNA. Mol Phylogenet Evol 30:226–235

Swart A, Marais M, Schoeman AS (2000) Plant nematodes in South Africa. 2. Golf course putting greens. Afr Plant Prot 6:35–39

Swart A, Subbotin SA, Tiedt LR et al (2004) *Anguina woodi* sp. n. (Tylenchida: Anguinidae) from dune grass, *Ehrharta villosa*, in South Africa. Nematol 6:129–144

Van den Berg E (1985) Notes on the genus *Afrina* Brzeski, 1981 (Anguinidae: Nematoda) with descriptions of new and known species. Phytophylactica 17:67–79

Womersley C, Thompson SN, Smith L (1982) Anhydrobiosis in nematodes II: carbohydrate and lipid analysis in undesiccated and desiccated nematodes. J Nematol 14:145–153

Chapter 20
Non-parasitic, Terrestrial and Aquatic Nematodes

Antoinette Swart, Mariette Marais, Caroline Mouton, and Gerhard C. du Preez

20.1 Introduction

Conventionally, all nematodes that are not animal parasites are called non-parasitic, including the herbivores or plant-parasitic nematodes (Poinar 1983). However, in this chapter the term non-parasitic will be used in a more restricted sense to refer only to non-plant-parasitic, terrestrial, freshwater and, to a certain extent, estuarine nematodes (see Chap. 24). As plant-parasitic nematodes (see Chap. 3) constitute an important feeding group in the soil, their contribution to ecological processes will briefly be discussed under non-parasitic, terrestrial nematodes (see Sect. 20.2).

For a clearer picture of the extent of the work being done on the non-parasitic nematodes of South Africa (SA), this chapter will be divided into five parts, viz. non-parasitic terrestrial nematodes, non-parasitic freshwater nematodes, non-parasitic nematodes from caves and caverns, non-parasitic nematodes from estuaries and a checklist of non-parasitic nematodes described from or found in SA up to 2015. An interesting phenomenon, namely, the discovery of the first multicellular organisms deep under the surface of the earth, that represented non-parasitic nematodes, is discussed in Box 20.1.

Dr Mariette Marais of the Nematology Unit, Biosystematics Division, Agricultural Research Council – Plant Protection Research Institute is thanked for the use of data from the South African Plant-Parasitic Nematode Survey (SAPPNS) database; E-mail: MaraisM@arc.agric.za

A. Swart (✉) • M. Marais
Nematology Unit, Biosystematics, Agricultural Research Council – Plant Protection Research Institute, Private Bag X134, Queenswood 0121, South Africa
e-mail: SwartA@arc.agric.za; maraism@arc.agric.za

C. Mouton
Nemconsult, P.O. Box 877, Upington 8800, South Africa

G.C. du Preez
Unit for Environmental Sciences and Management, North-West University, Potchefstroom, Private Bag X6001, Potchefstroom 2520, South Africa

H. Fourie et al. (eds.), *Nematology in South Africa: A View from the 21st Century*, DOI 10.1007/978-3-319-44210-5_20

Box 20.1. The Deepest Living, Multicellular Organism Discovered Recently: A Non-parasitic Nematode

The subterranean biosphere extends more than 3 km below the surface of the Earth and presents near inhospitable conditions (extreme temperature, limited space and lack of oxygen) for life. Although prokaryotic organisms have been known to survive at these depths, it was only in 2011 that an international team of scientists provided evidence on the existence of multicellular organisms as deep as 3.6 km below the surface (Borgonie et al. 2011). These organisms were nematodes, sampled from ancient (2,900–12,100 years old) borehole water associated with mineshafts in South Africa. While deoxyribonucleic acid (DNA) evidence of a monhysterid species was found at the latter depth (with a recorded temperature of 48 °C), a new species, namely, *Halicephalobus mephisto* Borgonie, Garcia-Moyano, Litthauer, Bert, Bester, Van Heerden, Möller, Erasmus and Onstott 2011, was discovered 1.3 km below the surface. Temperatures at this depth were recorded at 37 °C. Two other species (*Plectus aquatilis* Andrássy, 1985) and a monhysterid sp. were also collected from borehole water in another mine at a depth of nearly 1 km below the surface. Evidence was provided by Borgonie et al. (2011) that these nematodes fed on a mixture of aerobic and anaerobic bacteria associated with the paleometeoric water. The discovery that multicellular life, in this case nematodes, can survive in extreme conditions deep below the surface of the Earth renewed hopes for finding life on Mars.

20.2 Non-parasitic, Terrestrial Nematodes

Non-parasitic nematodes are common in natural veld with some species being particularly prevalent within SA in cultivated fields (Heyns 1971). These include members of the orders Araeolaimida, Cephalobida, Dorylaimida, Enoplida, Mononchida, Panagrolaimida, Rhabditida, Triplonchida and Tylenchida. Although many taxonomical studies have been conducted, only three ecological works focusing on non-parasitic nematodes discovered locally have thus far been published (Durand et al. 2012; Du Preez et al. 2013, 2015b). However, quite a few dissertations and theses on this topic are in the process of being concluded. Furthermore, presentations with nematode ecology as the topic have also been presented at various nematology congresses/symposia (Du Preez et al. 2013, 2014, 2015a; Marais et al. 2009, 2010; Swart et al. 2014).

Scientists within SA have become increasingly concerned about the health of agricultural soils and are using nematodes as biological indicators. For this purpose, an automated web-based tool known as the Nematode Indicator Joint Analysis (NINJA) (available at http://spark.rstudio.com/bsierieb/ninja) is very useful for the calculation of several ecological indices (Sieriebriennikov et al. 2014). Nematodes may well be the most suitable organisms for environmental diagnosis, especially as more information exists on their taxonomy and feeding roles (Gupta and Yeates 1997) than that exists about other mesofauna. For the producer, soil-inhabiting nematodes are of

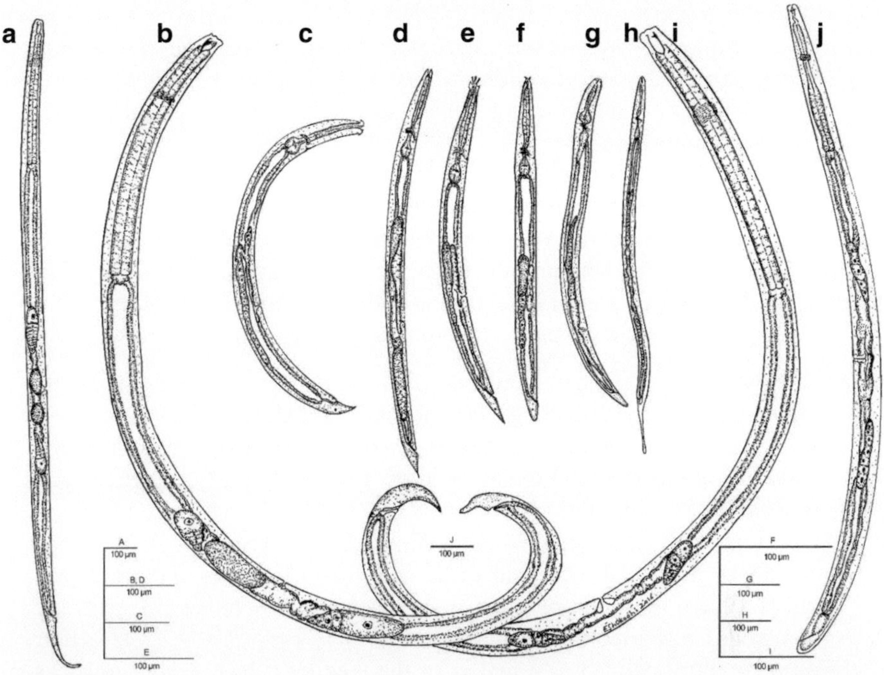

Fig. 20.1 *Mononchus truncatus* Bastian, 1865 (**a** predator), *Mylonchulus sigmaturus* Cobb, 1917 (**b** predator), *Elaphonema* sp. (**c** bacterivore), *Panagrolaimus rigidus* (Schneider, 1866) Thorne, 1937 (**d** bacterivore), *Acrobeles iranicus* Shokoohi, Abolafia and Zad, 2007 (**e** bacterivore), *Acrobeloides nanus* (de Man, 1880) Anderson, 1968 (**f** bacterivore), *Aphelenchus avenae* Bastian, 1965 (**g** fungivore), *Psilenchus aestuarius* Andrássy, 1962 (**h** fungivore), *Clarkus papillatus* (Bastian, 1865) Jairajpuri, 1970 (**i** predator), *Labronema vulvapapillatum* (Meyl, 1954) Loof and Grootaert, 1981((**j** predator) (Ebrahim Shokoohi, North West University, Potchefstroom, South Africa)

particular interest. They are small, generally between 0.3 and 5.0 mm long and can be abundant (in their millions) but also diverse (commonly more than 30 taxa) in all soils (Yeates 1979). Although the body form of soil nematodes is similar in all life stages, their greatest apparent morphological diversity can be seen in the head and mouth structures, which are closely related to their feeding habits (Fig. 20.1). Some important and useful terms referred to in the literature are listed and defined in Box 20.2.

Box 20.2. Glossary
Trophic groups of nematodes.

cp (coloniser-persister) scale: assignment of soil and freshwater nematode taxa to a 1–5 linear scale according to their *r* and *K* characteristics (Ferris et al. 2001):

- cp-1: Short generation time, small eggs, high fecundity, mainly bacterivores, feed continuously in enriched media, form dauer juveniles as microbial blooms subside

- cp-2: Longer generation time and lower fecundity than the cp-1 group, very tolerant of adverse conditions and may become cryptobiotic. Feed more deliberately and continue feeding as resources decline. Mainly, bacteri- and fungivores
- cp-3: Longer generation time, greater sensitivity to adverse conditions. Fungi-, bacteri- and carnivores
- cp-4: Longer generation time, lower fecundity, greater sensitivity to disturbance. Besides the other trophic roles, smaller omnivore species
- cp-5: Longest generation time, largest body sizes, lowest fecundity, greatest sensitivity to disturbance. Predominantly carni- and omnivores

r strategists: relatively small nematodes with short life cycles and potentially higher reproductive rates (e.g. non-parasitic nematodes with cp-values of 1 and 2, parasitic nematodes such as *Meloidogyne* spp. and *Ditylenchus africanus* Wendt, Swart, Vrain and Webster, 1995 (previously reported as *D. destructor* Thorne, 1945) (see Sect. 9.3, Chap. 9). Some grow large and have long life cycles with low rates of population increase.

K strategists: relatively large nematodes with long life cycles and low reproduction rates (e.g. non-parasitic nematodes with cp-values >3, parasitic nematodes such as *Longidorus* spp.).

Functional guild: defined as a matrix of the feeding habits of nematodes as well as incorporating their biological, ecological and life history characteristics (which are all incorporated in the cp classification) (Neher et al. 2004).

Faunal profile/simplified food web: an indicator of the state of a given food web where soils are categorised into four quadrants according to the presence, abundance and diversity of non-parasitic, soil-inhabiting nematodes as affected by stressor disturbance(s) (Ferris et al. 2001; Neher et al. 2004).

Enrichment index: a measure of opportunistic bacteri- and fungivore nematodes (cp 1 and 2 values) present in a given soil substrate (Ferris et al. 2001; Neher et al. 2004).

Structure index: using non-parasitic nematodes with higher cp-values (three and more) as an indicator of soil health (Ferris et al. 2001; Neher et al. 2004).

Channel index (CI): an indicator of predominant decomposition pathways (bacterial or fungal) that occur in soil food webs where nematodes are used as indicators of soil health (Ferris et al. 2001; Neher et al. 2004).

20.2.1 Trophic Groups

According to Yeates et al. (1993), the following nematode-feeding/trophic groups are recognised.

20.2.1.1 Herbivores or Plant-Feeders (Plant-Parasitic Nematodes)

These nematode pests feed on vascular plants. To accomplish this, a stomatostylet (Tylenchida and some Aphelenchida) or an onchio- (Triplonchida) or odontostylet (Dorylaimida) is always present in such individuals.

20.2.1.2 Fungivores or Hyphal Feeders

These nematodes penetrate fungal hyphae by using a small and delicate stomato- or odontostylet. In addition to obligate hyphal feeders, this group includes the alternative life cycle of some invertebrate parasites (e.g. *Deladenus* spp.).

20.2.1.3 Bacterivores or Bacterial Feeders

Such nematodes feed on any prokaryotic food source present in the soil substrate. These organisms ingest their food through either a narrow (e.g. *Rhabditis* spp., *Alaimus* spp.) or broad (e.g. *Diplogaster* spp.) mouth.

20.2.1.4 Substrate Feeders

Ingestion by substrate feeders may be incidental to bacterivores, predator and unicellular, eukaryotic-feeding nematodes because more than one food source is ingested by such nematode individuals. The mouth form ranges from being short and broad to long and narrow. Also, teeth may be present in the mouth, which suggests a more predatory lifestyle. The expression 'nonselective, deposit feeding' used in reference to aquatic nematodes refers to ingestion of more than one food source.

20.2.1.5 Predators or Animal Feeders

Nematodes referred to under this category ingest invertebrates such as protozoa, nematodes and rotifers either as 'ingesters' (e.g. *Diplogaster* spp., *Mononchus* spp., *Nygolaimus* spp.) or as 'piercers' (e.g. *Seinura* spp., *Labronema* spp., *Laimaphelenchus* spp.), sucking body fluids through a narrow stylet.

20.2.1.6 Feeders on Eukaryotes

A wide range of nematodes feed on diatoms or other algae, as well as fungal spores and yeast cells. Examples of this trophic group are *Achromadora* spp., *Diplogaster* spp. and *Fictor* spp.

20.2.1.7 Dispersal or Infective Stages of Animal Parasites

Stages of animal-parasitic nematodes occur in the soil as invertebrate (e.g. *Deladenus* spp., *Heterorhabditis* spp.) or vertebrate (e.g. *Strongyloides* spp.) parasites. When these stages feed and contribute to soil processes, they should be included in other appropriate categories, such as fungi- or bacterivores. Furthermore, when they die in the soil they contribute to the nutrient pool. Individuals from, for example, Rhabditida and Diplogasteridae that use animals as phoretic (transport) hosts are, however, not included in this group.

20.2.1.8 Omnivores

Some nematode species appear normally to feed on a wide range of foods (particularly combining feeding types two to six referred to above). These species are restricted to a few members of the Dorylaimida. Examples include *Actinolaimus*, *Aporcelaimellus* and *Kochinema* spp.

Given this range of feeding types, the soil nematode fauna interacts with many other groups of soil organisms. As the soil biota play critical roles in controlling the mineralisation of nutrients for plant growth, studies were conducted on various functional groups of soil organisms in an effort to understand soil processes. According to Magdorf (2001), soil health is a term used by farmers to refer to the condition of their agricultural soil as it relates to growing viable crops. He stated that high-quality soil has the following characteristics:

(i) Sufficient, but not too high in supply of nutrients
(ii) Good structure or tilth
(iii) Sufficient depth for rooting and drainage
(iv) Good internal drainage
(v) Low populations of plant disease and parasitic organisms
(vi) High populations of organisms that promote plant health
(vii) Low weed pressure
(viii) No chemicals harmful to plants
(ix) Resistance to being degraded
(x) Resilience following an episode of degradation

Only a few studies, aimed at studying non-parasitic nematode assemblages, have been conducted in SA. From 2006 to 2008, a nematode survey was conducted in mixed agricultural gardens in three rural villages (Vhembe, Limpopo Province) by the Nematology Unit of the Agricultural Research Council–Plant Protection Research (ARC-PPR, Roodeplaat, Pretoria). The survey was part of a project titled 'Legumes and protein for resource-poor farmers in Limpopo Province'. The focus of the nematology part of the project rested mainly on the influence of plant-parasitic nematodes on the yield of the different crops, such as bambara groundnut (*Vigna subterranea*), bean (*Phaseolus* sp.), cabbage (*Brassica oleracea*), groundnut (*Arachis hypogaea*), maize (*Zea mays*), okra (*Abelmoschus esculentus*), onion (*Allium cepa*), pumpkin (*Cucurbita pepo*), spinach (*Spinacia oleracea*), sweet potato (*Ipomoea batatas*) and tomato (*Solanum lycopersicum*), while also monitoring the incidence of non-parasitic nematodes. Probably as a result of the lack of agrochemical usage, the abundance and incidence of non-parasitic nematodes was high in the gardens of all three communities. During the drought of 2007, the numbers of all nematodes were reduced, but of great interest was the high population numbers of especially endoparasitic lesion nematodes (*Pratylenchus* spp.) in the roots of plants.

Berry and Rhodes (2006) and Engelbrecht (2012) conducted studies aimed at, amongst others, determining the effects of different green manure cover crops on nematode populations in soils where sugarcane and potato were grown. These authors chose green manure crops as they are known to improve soil health and, when chosen correctly, reduce pest and disease problems. The main benefits of green manure crops are their ability to:

(i) Produce a large amount of biomass, which can be returned to the soil as sufficient organic matter
(ii) Allow crops to develop a large and deep root system to alleviate the effects of soil compaction
(iii) Provide good ground cover so that weed problems are alleviated

As a result of this study, no significant changes in population levels of plant-parasitic (both in roots and soil samples) and non-parasitic nematodes for the summer and winter crops in the sugarcane production system were recorded.

Results from the sugarcane study, however, showed that none of the green manure cover crops resulted in a significant increase in non-parasitic nematode numbers. By contrast, results from the potato study showed pronounced shifts in non-parasitic nematode assemblages in plots where four Brassicaceae cvs, viz. Nemat (*Brassica juncea*), Calienté (*Eruca sativa*), Doublet and Terranova (*Raphanus sativus*), were planted and their aerial parts subsequently incorporated into the soil, compared to nematode assemblages in control plots. Such soils were regarded as 'stress depleted' (degraded) and/or 'stressed enriched' (disturbed) before incorporation of the green manure crops, but as 'stable enrich' (maturing) afterwards and demonstrated the positive effect of these crops in terms of soil health.

A study by Jansen (2014) furthermore showed that no differences in non-parasitic nematode assemblages were recorded from soils in which conventional soybean (*Glycine max*) cvs were grown (with no application of glyphosate for 5 years prior to the study) compared with those where Roundup® Ready cvs were cultivated. Moreover, non-parasitic nematode assemblages from these two systems also did not differ from those collected from adjacent natural veld sites.

Of great interest to local nematologists studying the role of nematodes in soil are the findings of Moens et al. (2004). They commented that feeding-type classification of nematodes, which usually relies strongly on buccal morphology, may create problems especially as there are many unresolved questions on nematode food sources and feeding rates. For instance, the food sources of the delicately speared Tylenchidae and Psilenchidae remain largely unknown. Although these nematode groups often occur in high population densities in the rhizosphere of plants, direct feeding on plant roots has not been documented. Furthermore, no apparent negative effects on plants are evident. They have been classified as fungal feeders, root hair

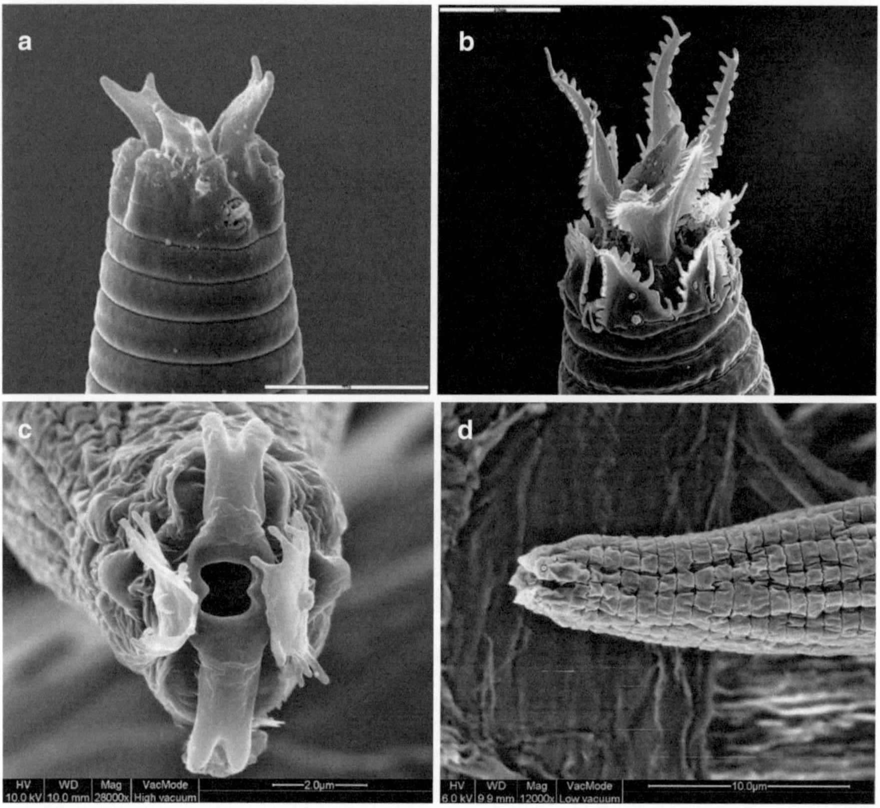

Fig. 20.2 (**a**–**d**) Bacterivore nematodes *Chiloplacus* (**a**), *Acrobeles* (**b**), *Diploscapter* (**c**), *Panagrolaimus* (**d**) (**a** and **b** Ebrahim Shokoohi and **c** and **d** Louwrens Tiedt, North-West University, South Africa)

feeders and plant-associated nematodes (Yeates et al. 1993; Yeates and Bongers 1999). Also, soil bacterial feeders may ingest whole bacterial cells, but the range of differently sized and shaped buccal cavities, lips and appendages exhibited by bacterivores suggests various strategies for obtaining this resource. This phenomenon is illustrated by Fig. 20.2a–d that represents bacterivores that have been recorded from the Bakwena Cave, Gauteng Province, SA, by Durand et al. (2012). Likewise, Venette and Ferris (1998) demonstrated that not all bacteria are suitable food for bacterivores and linked this, amongst other things, to cell size. Moens et al. (1999) showed that four coexisting species of Monhysteridae all respond differently to the presence of different bacteria. Such response differences relate not only to the bacterial strains offered, but also to density, age, growth conditions and activity of the bacteria. Also of great interest is the work by Mikola and Setälä (1998), who demonstrated that bacterial-feeding nematodes perform species-specific rather than guild-specific functions in the soil food web (Fig. 20.3). These studies have far-reaching implications, e.g. the effectiveness of the enrichment of soil by adding bacteria and fungi.

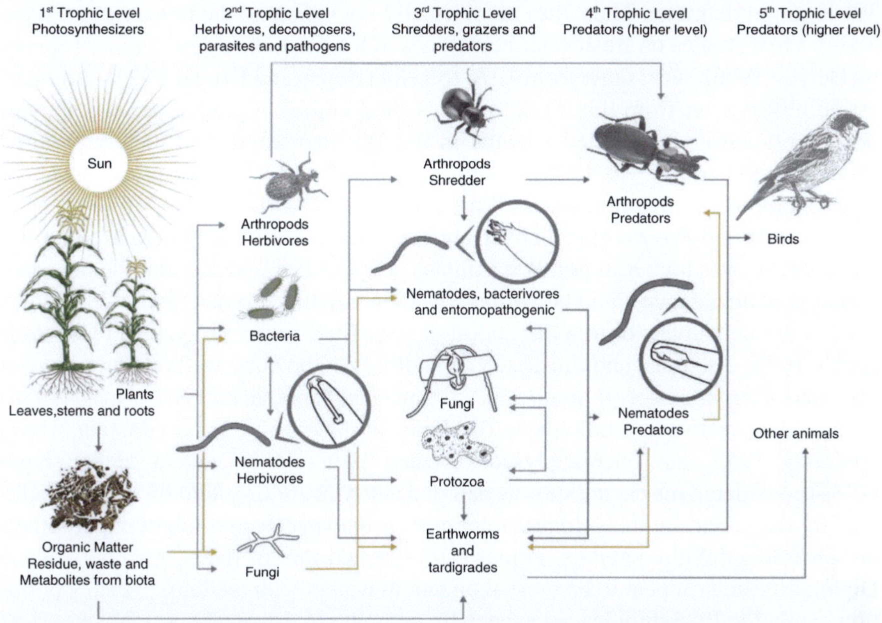

Fig. 20.3 A food web with special reference to nematode individuals from various trophic groups that form an intrinsic part of this process in soils (Hannes Visagie, North West University, Potchefstroom)

Nematodes normally face two options during their journey in soils, namely, (i) maximising energy gains either by moving to a better feeding location or (ii) foraging within a given patch.

The choice between these options will depend on the nematode's functional response to a given type of food and on the presence, suitability and detectability of alternative food sources. Current work suggests that the relative abundance of fungi- and bacterivores is sensitive to management activities and may thus be a good indicator of underlying changes in the composition of the nematode faunal structure. The decrease in diversity of nematode fauna with increasing level of management reflects not only physical disturbance and change in quality of organic matter being returned to the soil but also possible increases in specific herbivores associated with crops (Yeates and Bongers 1999). Generally, soils with annual arable crops contain fewer nematode species, whereas up to 154 species have been recorded in grasslands (Hodda and Wanless 1994). Should permanent grassland be regarded as sustainable, it may provide a baseline for the composition/diversity of the nematode fauna in a given soil.

20.3 Non-parasitic, Freshwater Nematodes

Nematodes usually encountered in freshwater habitats (e.g. dams, karst systems, lakes and streams) are all microscopic in size, ranging from about 0.25 to 10 mm in length. Although some of these are several mm long, they are seldom more than

30–40 µm in diameter. Thus, they are normally not visible to the human eye (Heyns 2002). Most studies on freshwater nematodes in SA, including a preliminary survey by Heyns (1976), were done from 1976 to 1984 (Heyns and Kruger 1983). An interesting observation from this work was that the number of genera per sample was perceptibly lower in polluted streams in the Witwatersrand than in less polluted streams elsewhere (Heyns 1982). Groups particularly affected in this way were the Actinolaimoidea, Belondiroidea, Chromadorida, Dorylaimoidea, *Ironus* spp., Mononchida and *Plectus* spp. Conversely, the abundance of the genera *Monhystera* and *Tobrilus* was higher in polluted samples. Heyns (1982) stated that the observed variation in abundance could be attributed either to a direct toxic effect of the pollutants on the nematodes or to a disruption of their food source. Heyns and Coomans (1980, 1983) discussed and illustrated crystalloids in the body wall of *Chronogaster* spp. and *Chronogaster rotundicauda* Heyns and Coomans, 1984. Nuss (1984) described refractive crystalloids in the body of *Tobrilus gracilis* (Bastian, 1865) Andrássy, 1959, that, once analysed, revealed high sulphur content. Hence, Nuss (1984) considered the crystalloids as part of a detoxification system for sulphide ions that are expressed by the sulphide tolerance of the species and might explain their presence in sulphidic habitats. Poinar (1983) stated that in rivers, members of the Diplogasteroidea appear to be most abundant in waters with medium to heavy pollution, while the Rhabditoidea are especially common in extremely polluted water.

Relatively little taxonomic work has been done on terrestrial non-parasitic nematodes and even less on freshwater nematodes. However, a series of papers on the nematodes from the rivers of the Kruger National Park were published between 1991 and 1993 (Botha and Heyns 1990, 1991, 1992a, b, c, 1993a, b) and included a compendium of the *Mesodorylaimus* spp. and a checklist of the 31 nematode species of the orders Araeolaimida, Chromadorida, Dorylaimida, Enoplida, Monhysterida, Mononchida and Tylenchida from SA. These nematodes were extracted from sediments collected from several rivers in the park. Although aquatic nematodes were found, several species that are usually found in terrestrial habitats were also collected from these rivers. This is not unusual since terrestrial nematodes are commonly found in rivers and other freshwater bodies after having been washed into them with runoff from irrigation and/or torrential rains.

These studies indicate the complexity of the freshwater habitat and also that a good understanding of the benthic nematode community remains of great importance for determining pollution and other types of aquatic disturbances. Heyns (2002) gave valuable information on the extraction of nematodes from aquatic sediments, their morphology and biology as well as keys designed specifically for the identification of freshwater nematodes. He also constructed a key to the orders of the Nematoda that may occur in the freshwaters of southern Africa and stated that the process was essentially complicated by two factors: Firstly, the lack of information since little research has been done on freshwater nematodes in this region and, secondly, the fact that not all nematodes collected from the mud or sand at the bottom of a dam or stream can arbitrarily be regarded as aquatic organisms. Terrestrial nematodes, both non-parasitic and plant-parasitic, sourced from other water bodies are frequently found in large numbers in the bottom sediment where they can survive for long periods. Moreover, many nematode families and even genera contain both terrestrial and aquatic species. Thus, when such a species is found in a water sample, it is particularly difficult to discern whether or not it is truly an aquatic form. *Chronogaster* appears to exhibit an exceptional amount of genetic plasticity regarding habitat selection and adaptation. In fact, *Chronogaster africana* Heyns and Coomans 1980 is cited as occurring in multiple aquatic and terrestrial habitats (Heyns and Coomans 1980).

Of special interest is the report by Hugo and Malan (2010) on nematodes found during a survey of irrigation water in SA. Plant-parasitic nematodes were targeted, and although the nematode numbers seemed to be relatively low, more research is needed to establish the presence and extent of nematodes in local freshwater systems.

20.4 Nematodes from Caves and Caverns

Karst system research has only recently started in southern Africa, and therefore relatively little is known about the unique ecology characterising these systems (Durand et al. 2012). Delicate ecosystems are housed in karst environments containing unique and complex faunal assemblages, some of which are endemic (Culver and Pipan 2010). Furthermore, solution cavities, conduits, fissures and aquifers are predominant features of such landscapes resulting in complex, dynamic and sometimes highly interconnected networks (Williams 2008). This may lead to the formation of an interface between the surface and groundwater. However, as a result of its interconnectivity, these subterranean environments are extremely vulnerable to pollution and degradation by humans (Bonacci et al. 2009). Polluted aquifers pose a severe risk to human and livestock health as abstracted groundwater is commonly used for agricultural and domestic use. Dropping groundwater levels may also lead to sinkhole formation, which has claimed many lives in the past. South Africa is famous for its old and extensive karst system, which is dominated by dolomite deposits covering a large area stretching from the North-West Province, through Gauteng, into Mpumalanga and Limpopo Province (Durand 2008; Du Preez 2014).

When karst systems are damaged by pollutants, including sewage and agrochemicals, to such a degree that their ability to sustain life is compromised, it is very likely that many species of organisms will disappear before they have been studied and documented. For documentation of a karst system, the Bakwena Cave near Irene (Gauteng) has been studied by the SA Karst Ecology Study Group (SAKES) from May 2008 to May 2010. Durand et al. (2012) stated that this cave system is inhabited by bats, fungi, bacteria, nematodes and several arthropod groups in a complex, intricate and interdependent food web. Six localities from the cave were sampled, viz. (i) under ferns and mosses against the walls of the entrance in the daylight zone, (ii) the floor of the main chamber in the twilight zone, (iii) dry bat guano from the floor of the main chamber in the twilight zone, (iv) fresh guano from a side chamber in total darkness, (v) the sediment from a groundwater pool approximately 30 m below the surface and in total darkness and (vi) the floor of the side chamber in total darkness (Durand et al. 2012). Results from this study showed that nematodes were the most common and widely diffused organisms in the Bakwena Cave. In total they represented 11 orders, 23 families and 27 genera. Two localities, viz. the entrance of the cave and underground pool, represented subterranean sampling areas where the most nematode species were recorded.

Another study by Du Preez et al. (2013) reported five nematode genera, namely, *Aphelenchoides*, *Criconema*, *Helicotylenchus*, *Panagrolaimus* and *Rhabditis*, from Knocking Shop and Gatkop, two caves located near Krugersdorp (Gauteng Province) and Thabazimbi (Northern Province), respectively. More recently, Du Preez et al. (2015b) reported a total of 53 nematode genera from sediment, soil and guano samples collected from the Wonderfontein Cave (Gauteng Province). This cave system developed in the dolomitic river banks of the Wonderfontein Spruit (Witwatersrand Basin) and is subjected to the influx of water from the same water body. Of the 53 identified nematode genera, which represented most trophic groups, 22 have never before been reported from a cave environment. Du Preez et al. (2015b) also provided an updated list of cavernicolous nematodes from southern Africa (Table 20.1).

Although a number of nematode taxa have been reported from local subterranean environments, especially in cave ecosystems, information on the functioning of,

Table 20.1 Cavernicolous nematodes from southern Africa

Genus	Trophic group (cp-value)	Habitat	Country	Reference
Achromadora	Eu (3)	Cave pool	SA	Jansen Van Rensburg (2010); Durand et al. (2012)
Acrobeles	Ba (2)	Soil (daylight zone)	SA	Jansen Van Rensburg (2010); Durand et al. (2012)
Acrobeloides	Ba (2)	Cave pool, soil (daylight zone)	SA	Jansen Van Rensburg (2010); Durand et al. (2012)
Alaimus	Ba (4)	Cave pool, soil (daylight zone)	SA	Jansen Van Rensburg (2010); Durand et al. (2012)
Anaplectus	Ba (2)	Soil (daylight zone)	SA	Jansen Van Rensburg (2010); Durand et al. (2012)
Aphelenchoides	Fu/Her	Soil (daylight zone)	SA , BS	Jansen Van Rensburg (2010); Durand et al. (2012); Du Preez et al. (2013)
Aphelenchus	Fu (2); Her	Soil (twilight zone), Cave pool	SA	Jansen Van Rensburg (2010); Durand et al. (2012)
Aporcelaimus	Om (5)	Cave pool, soil (daylight zone)	SA	Jansen Van Rensburg (2010); Durand et al. (2012)
Boleodorus	Her	Cave pool, soil (daylight zone)	SA	Jansen Van Rensburg (2010); Durand et al. (2012)
Cephalobus	Ba (2)	Soil	BS	Du Preez et al. (2013)
Chiloplectus	Ba (2)	Soil (total darkness)	SA	Jansen Van Rensburg (2010); Durand et al. (2012)
Criconema	Her	Soil	BS	Du Preez et al. (2013)
Cylindrolaimus	Ba (3)	Cave pool	SA	Jansen Van Rensburg (2010); Durand et al. (2012)

Table 20.1 (continued)

Genus	Trophic group (cp-value)	Habitat	Country	Reference
Diplogasteroides	Ba (1)	Fresh guano (total darkness)	SA	Jansen Van Rensburg (2010); Durand et al. (2012)
Diploscapter	Ba (1)	Cave pool	SA	Jansen Van Rensburg (2010); Durand et al. (2012)
Discolaimus	Ca (5)	Cave pool, soil (daylight zone)	SA	Jansen Van Rensburg (2010); Durand et al. (2012)
Ditylenchus	Fu/Her (2)	Cave pool, soil (daylight zone)	SA	Jansen Van Rensburg (2010); Durand et al. (2012)
Eumonhysterida	Ba (1)	Soil (total darkness)	SA	Jansen Van Rensburg (2010); Durand et al. (2012)
Helicotylenchus	Her	Cave pool, soil	SA, BS	Jansen Van Rensburg (2010); Durand et al. (2012); Du Preez et al. (2013)
Meloidogyne	Her	Soil	BS	Du Preez et al. (2013)
Mesorhabditis	Ba (1)	Cave pool, soil (daylight zone)	SA	Jansen Van Rensburg (2010), Durand et al. (2012)
Monhystrella	Ba (1)	Cave pool, dry guano (twilight zone)	SA	Jansen Van Rensburg (2010); Durand et al. (2012)
Mononchus	Ca (4)	Cave pool	SA	Jansen Van Rensburg (2010); Durand et al. (2012)
Mylonchulus	Ca (4)	Cave pool, soil (twilight zone)	SA	Jansen Van Rensburg (2010); Durand et al. (2012)
Neoactinolaimus	Om (4)	Soil	BS	Du Preez et al. (2013)
Panagrolaimus	Ba (1)	Cave pool, dry guano (twilight zone), fresh guano (total darkness)	SA, BS	Jansen Van Rensburg (2010); Durand et al. (2012); Du Preez et al. (2013)
Paracrobeles	Ba (2)	Soil	BS	Du Preez et al. (2013)
Plectus	Ba (2)	Cave pool, soil (daylight zone)	SA	Jansen Van Rensburg (2010); Durand et al. (2012)
Pratylenchus	Her	Cave pool	SA	Jansen Van Rensburg (2010); Durand et al. (2012)
Prismatolaimus	Ba (3)	Cave pool, soil	SA, BS	Jansen Van Rensburg (2010); Durand et al. (2012); Du Preez et al. (2013)
Rhabditis	Ba (1)	Soil	BS	Du Preez et al. (2013)
Trichodorus	Her	Cave pool	SA	Jansen Van Rensburg (2010); Durand et al. (2012)
Tylenchorhynchus	Her	Cave pool, soil (twilight zone)	SA	Jansen Van Rensburg (2010); Durand et al. (2012)
Xiphinema	Her	Soil	BS	Du Preez et al. (2013)
Zeldia	Ba (2)	Soil	BS	Du Preez et al. (2013)

SA = South Africa, *BS* = Botswana
Du Preez et al. (2015b)

services provided by and trophic interactions of such nematode assemblages remains extremely limited. Furthermore, Hodda et al. (2006) stated that true cavernicolous (cave-dwelling) (Box 20.3) nematodes seem to consist of only a few species restricted to such or very similar habitats, together with accidental occupants and species with wide environmental tolerances. However, in order to understand fully the functioning and structure of nematodes occurring in cave environments, further research efforts should be undertaken.

Box 20.3. How Nematodes Survive in Caves: A World View

Similar to the deep-subsurface nematodes found in South Africa, caves also present unique challenges for the survival of life. The lack of sunlight and thus autotrophic production (Hodda et al. 2006), as well as the desolate nature of some cave environments, sometimes only allow the survival of highly adapted species (Romero 2009; Culver and Pipan 2010). Nonetheless, energy is still required to sustain subterranean ecosystems. This is mostly available in the form of organic matter that enters the subterranean environment via percolating and flowing water (dissolved and/or suspended), wind, gravity, animal movement and roots (Culver and Pipan 2010). Once energy becomes available to the subterranean environment, it forms part of the associated ecosystem, which may include various vertebrate (e.g. bats, rodents, fish and salamanders) and invertebrate (e.g. nematodes, insects, arachnids and crustaceans) species (Romero 2009). Biodiversity hotspots may also be found in caves and include bat guano heaps, freshwater bodies and areas subject to substantial energy flow from the surface.

However, some cave ecosystems, such as that associated with Movile Cave (Romania), are chemoautotrophically based (Poinar and Sarbu 1994). Within the latter system, the presence of hydrogen sulphide-rich thermal water facilitates the growth of floating fungal mats, which ultimately sustain an ecosystem of 33 endemic species (Sarbu et al. 1996). Nematodes also form part of this extreme, isolated cave ecosystem and include five highly adapted bacterivores, namely, *Poikilolaimus* sp., *Monhystrella* sp., *Panagrolaimus c.f. thienemanni*, *Udonchus tenuicaudatus* Cobb, 1913 and *Chronogaster troglodytes* Poinar and Sarbu 1994 (Poinar and Sarbu 1994; Muschiol and Traunspurger 2007; Muschiol et al. 2015). The latter species was described as the first true cavernicolous (cave-dwelling) nematode.

Although some nematode species clearly play an intricate role in the functioning of subterranean ecosystems, it is hypothesised that most cavernicolous nematodes are only accidental inhabitants (temporary residents). Such nematodes likely enter the subterranean environment via similar pathways to that of organic matter and other invertebrates (Hodda et al. 2006; Du Preez et al. 2015b). Ultimately, the role that nematodes play in the functioning of subterranean ecosystems is poorly understood and requires further study.

20.5 Non-parasitic Nematodes from Estuaries

Estuaries and estuarine sediments are commonly defined as having more than 1 % salinity (Hodda et al. 2006). Furstenberg and De Wet (1983) recorded two plant-parasitic nematode species, *Helicotylenchus californicus* Sher, 1966, and an unknown *Tylenchus* sp., as being predominant at all sites of salt marsh vegetation sampled at the Swartkops Estuary, Port Elizabeth (Eastern Cape Province). Whether the sediment was also sampled is not clear from their report. According to Hodda et al. (2006), some freshwater nematodes in estuaries are indeed plant-feeders associated with vascular plants (e.g. *Tylenchus* spp., *Helicotylenchus* spp., *Meloidogyne* spp., *Heterodera* spp. and *Hirschmanniella* spp.).

Since 2012, the Nematology Unit of the ARC–PPR has been involved in a study of the nematodes from the Isipingo Estuary near Durban Harbour, KwaZulu-Natal Province. The sediment from the rhizosphere of mangroves (*Rhizophora mucronata*) was sampled as part of a PhD study of Krishni Naidoo at the University of KwaZulu-Natal. The nematode genera found during this survey are listed according to their trophic levels in Table 20.2. This is an ongoing study and other research initiatives on the nematodes are underway. For a more in depth description of the nematodes in estuaries, see Chap. 24.

20.6 Checklist of Non-parasitic Nematodes in South Africa

This list (Table 20.3) contains the non-parasitic nematodes identified in SA from aquatic and terrestrial ecosystems. Classification is according to Andrássy (2005, 2007, 2009), except for the Mononchida, which was classified according to Ahmad and Jairajpuri (2010).

Table 20.2 Nematodes identified from the Isipingo Mangrove Estuary (A. Swart, Agricultural Research Council–Plant Protection Research Institute, Pretoria, 2016, personal communication)

Bacterivore	Predator	Bacterivore/predator	Herbivore (plant parasite)
Areaolaimida, *Camacolaimus* sp., *Deontolaimus* sp., *Desmodora* sp., *Ethmolaimus* sp., *Monhystera* sp., *Monhystrella* sp., *Panagrolaimus* sp., *Plectus* sp., *Prodesmodora* sp., *Rhabditis* sp., *Theristus* sp.	*Tobrilus* sp.	*Koerneria* sp., *Fictor* sp.	*Hemicycliophora ripa* Van den Berg, 1981, *Hemicycliophora typica* de Man, 1921, *Rotylenchus* sp.

Table 20.3 The classification of non-parasitic nematodes that were identified from aquatic (A) and terrestrial (T) habitats in arboreal areas (Ar), caves (C) and estuaries (E) according to Andrássy (2005, 2009), except for the Mononchida, which was classified according to Ahmad and Jairajpuri (2010)

Class Secernentea von Linstow, 1905
Order Rhabditida Chitwood, 1933
Suborder Teratocephalina Andrássy, 1974
Family Teratocephalidae Andrássy, 1958
Genus *Teratocephalus* de Man, 1876
Teratocephalus diversiannulatus Swart and Heyns, 1989 (T)
Teratocephalus lirellus Anderson, 1969 (T)
Suborder Cephalobina Andrássy, 1974
Superfamily Cephaloboidea Filipjev, 1934
Family Cephalobidae Filipjev, 1934
Subfamily Acrobelinae Thorne, 1937
Genus *Acrobeles* von Linstow, 1877
Acrobeles annulatus Heyns and Hogewind, 1969 (T)
Acrobeles bushmanicus Heyns, 1969 (T)
Acrobeles farzanae Heyns, 1995 (T)
Acrobeles sheasbyi Heyns and Hogewind, 1969 (T)
Acrobeles singulus Heyns, 1969 (T)
Acrobeles sparsus Heyns, 1969 (T)
Acrobeles thornei Heyns, 1962 (T)
Genus *Acrobeloides* Cobb, 1924
Acrobeloides butschlii (de Man, 1884) Thorne, 1925 (T)
Genus *Acrobelophis* Andrássy, 1984
Acrobelophis lentus (Maupas, 1900) Vinciguerra and Clausi, 1996 (T)
Genus *Paracrobeles* Heyns, 1968
Paracrobeles laterellus Heyns, 1968 (T)
Genus *Penjatinema* Heyns and Swart, 1998
Penjatinema natalense Heyns and Swart, 1998 (T)
Genus *Seleborca* Andrássy, 1985
Seleborca complexa (Thorne, 1925) Andrássy, 1985 (T)
Seleborca dimorpha (Heyns and Hogewind, 1969) Andrássy, 1958 (T)
Seleborca mariannae (Andrássy, 1968) Andrássy, 1985 (T)
Seleborca recurva (Heyns, 1969) Andrássy, 1985 (T)
Genus *Stegelleta* Thorne, 1983
Stegelleta incasa (Thorne, 1937) Thorne, 1938 (T)
Genus *Zeldia* Thorne, 1937
Zeldia punctata (Thorne, 1925) Thorne, 1937 (T)
Subfamily Cephalobinae Filipjev, 1934
Genus *Cephalobus* Bastian, 1865
Cephalobus persegnis Bastian, 1865 (T)
Genus *Eucephalobus* Steiner, 1936

Table 20.3 (continued)

Eucephalobus oxyuroides (de Man, 1876) Steiner, 1936
Eucephalobus tribei Swart and Heyns, 1997 (Ar, insect galleries, dead pine tree)
Family Elaphonematidae Heyns, 1962
Subfamily Elaphonematinae Heyns, 1962
Genus *Elaphonema* Heyns, 1962
Elaphonema mirabile Heyns, 1962
Elaphonema messinae Van den Berg, Swart and Heyns, 1984 (T)
Family Osstellidae Heyns, 1962
Subfamily Osstellinae Heyns, 1962
Genus *Osstella* Heyns, 1962
Osstella hamata Heyns, 1962
Family Panagrolaimidae Thorne, 1937
Subfamily Panagrolaiminae Thorne, 1937
Genus *Panagrolaimus* Fuchs, 1930
Panagrolaimus magnivulvatus Boström, 1995 (T, nesting material, Antarctica)
Subfamily Tricephalobinae Andrássy, 1976
Genus *Halicephalobus* Timm, 1956
Halicephalobus mephisto Borgonie, García-Moyano, Litthauer, Bert, Bester, Van Heerden, Möller, Erasmus and Onstott, 2011 (A)
Subfamily Turbatricinae Goodey, 1943
Genus *Turbatrix* Peters, 1927
Turbatrix aceti (Mueller, 1783) Peters, 1927 (A)
Superfamily Chambersielloidea Thorne, 1937
Family Chambersiellidae Thorne, 1937
Subfamily Macrolaiminae Sanwal, 1971
Genus *Macrolaimus* Maupas, 1900
Macrolaimus richteri Swart and Heyns, 1992 (T)
Suborder Rhabditina Chitwood, 1933
Superfamily Rhabditoidea Örley, 1880
Family Mesorhabditidae Andrássy, 1976
Subfamily Mesorhabditinae Andrássy, 1976
Genus *Mesorhabditis* Osche, 1952
Mesorhabditis spiculigera (Steiner, 1936) Dougherty, 1953 (T)
Mesorhabditis striatica Dassonville and Heyns, 1984 (A)
Subfamily Parasitorhabditinae Lazarevskaya, 1965
Genus *Parasitorhabditis* Fuchs, 1937
Parasitorhabditis obtusa (Fuchs, 1915) Chitwood and Chitwood, 1950 (Ar, insect galleries, dead pine tree)
Family Diploscapteridae Micoletzky, 1922
Genus *Diploscapter* Cobb, 1913
Diploscapter coronatus (Cobb, 1893) Cobb, 1913 (C, bat guano)
Suborder Diplogastrina Micoletzky, 1922

(continued)

Table 20.3 (continued)

Superfamily Diplogastroidea Micoletzky, 1922
Family Diplogastridae Filipjev and Schuurmans Stekhoven, 1941
Genus *Acrostichus* Rahm, 1928
Acrostichus secundus (Bovien, 1837) Andrássy, 2005 (A)
Genus *Diplogasteritus* Paramonov, 1952
Diplogasteritus nudicapitatus (Steiner, 1914) Paramonov, 1952 (A)
Genus *Paroigolaimella* Paramonov, 1952
Paroigolaimella bernensis (Steiner, 1914) Andrássy, 1958 (A)
Family Neodiplogastridae Paramonov, 1952
Subfamily Neodiplogastrinae Paramonov, 1952
Genus *Mononchoides* Rahm, 1928
Mononchoides gracilis Dassonville and Heyns, 1984 (A)
Genus *Mononchoides* Rahm, 1928
Mononchoides gracilis Dassonville and Heyns, 1984 (A)
Order Aphelenchida Siddiqi, 1980
Suborder Aphelenchina Geraert, 1966
Superfamily Aphelenchoidea Fuchs, 1937
Family Aphelenchidae Fuchs, 1937
Genus *Aphelenchus* Bastian, 1865
Aphelenchus avenae Bastian, 1865 (T)
Family Paraphelenchidae Goodey, 1951
Genus *Paraphelenchus* Micoletzky, 1922
Paraphelenchus pseudoparietinus (Micoletzky, 1922) Micoletzky, 1925 (T)
Paraphelenchus amblyurus Steiner, 1934 (T)
Superfamily Aphelenchoidoidea Skarbilovich, 1947
Family Aphelenchoididae Skarbilovich, 1947
Genus *Laimaphelenchus* Fuchs, 1938
Laimaphelenchus patulus Swart, 1997 (Ar)
Genus *Aphelenchoides* Fischer, 1894
Aphelenchoides africanus Dassonville and Heyns, 1984 (A)
Aphelenchoides chamelocephalus (Steiner, 1926) Filipjev, 1934 (T)
Aphelenchoides composticola Franklin, 1957 (T)
Aphelenchoides haguei Maslen, 1979 (Ar)
Aphelenchoides helicus Heyns, 1964 (T)
Order Tylenchida Thorne, 1949
Suborder Tylenchina Chitwood in Chitwood and Chitwood, 1950
Superfamily Tylenchoidea Örley, 1880
Family Anguinidae Nicoll, 1935
Genus *Ditylenchus* Filipjev, 1936
Ditylenchus equalis Heyns, 1964 (T)
Class Penetrantia Andrássy, 1974
Subclass Enoplia Pearse, 1942
Order Enoplida Filipjev, 1929

Table 20.3 (continued)

Suborder Oncholaimina De Coninck, 1965
Superfamily Oncholaimoidea Filipjev, 1916
Family Oncholaimidae Filipjev, 1916
Genus *Oncholaimus* Dujardin, 1845
Oncholaimus deconincki Heyns and Coomans, 1977 (A)
Oncholaimus jessicae Coomans and Heyns, 1986 (A)
Suborder Ironina Siddiqi, 1983
Superfamily Ironoidea de Man, 1876
Family Ironidae de Man, 1876
Genus *Ironus* Bastian, 1865
Ironus ignavus Bastian, 1865 (A)
Ironus crassatus Argo and Heyns, 1972 (A)
Ironus dentifurcatus Argo and Heyns, 1972 (A)
Ironus ernsti Argo and Heyns, 1972 (A)
Ironus lautus Argo and Heyns, 1972 (A)
Ironus longicaudatus de Man, 1884 (A)
Ironus tenuicaudatus de Man, 1876 (A)
Suborder Tripylina Andrássy, 1974
Superfamily Prismatolaimoidea Micoletzky, 1922
Family Prismatolaimidae Micoletzky, 1922
Genus *Prismatolaimus* de Man, 1880
Prismatolaimus parvus Milne, 1963 (A)
Family Onchulidae Andrássy, 1964
Genus *Limonchulus* Andrássy, 1963
Limonchulus heynsi Swart and Furstenberg, 1993 (A)
Superfamily Tripyloidea de Man, 1876
Family Tobrilidae De Coninck, 1965
Subfamily Tobrilinae De Coninck, 1965
Genus *Neotobrilus* Tsalolikhin, 1981
Neotobrilus longus (Leidy, 1852) Tsalolikhin, 1981 (A)
Genus *Eutobrilus* Tsalolikhin, 1981
Eutobrilus ampiei Joubert and Heyns, 1979 (A)
Eutobrilus annetteae Joubert and Heyns, 1979 (A)
Eutobrilus diversipapillatus (Daday, 1905) Andrássy, 1959 (A)
Eutobrilus floridensis Joubert and Heyns, 1979 (A)
Eutobrilus heptapapillatus (Joubert and Heyns, 1979) Swart and Heyns, 1988 (A)
Genus *Epitobrilus* Tsalolikhin, 1981
Epitobrilus stefanskii (Micoletzky, 1925) Andrássy 2007 (A)
Genus *Macrotobrilus* Tsalolikhin, 1981
Macrotobrilus elephas (Andrássy, 1964) Tsalolikhin, 1981 (A)
Subfamily Tobriloidinae Tsalolikhin, 1976
Genus *Tobriloides* Loof, 1973

(continued)

Table 20.3 (continued)

Tobriloides loofi Swart and Heyns, 1990 (T)
Family Tripylidae de Man, 1876
Genus *Trischistoma* Cobb, 1913
Trischistoma ursulae Argo and Heyns, 1973 (A)
Subclass Dorylaimia Inglis, 1983
Order Alaimida Siddiqi, 1983
Suborder Alaimina Clark, 1961
Superfamily Alaimoidea Micoletzky, 1922
Family Amphidelidae Andrássy, 2002
Genus *Paramphidelus* Andrássy, 1977
Paramphidelus monohystera (Heyns, 1962) Andrássy, 1977
Paramphidelus trichurus (Siddiqi and Brown, 1965) Andrássy, 1977
Order Mononchida Jairajpuri, 1969
Suborder Mononchina Kirjanova and Krall, 1969
Superfamily Mononchoidea Filipjev, 1934
Family Mononchidae Filipjev, 1934
Subfamily Mononchinae
Genus *Mononchus* Bastian, 1865
Mononchus truncatus Bastian, 1865 (T)
Mononchus aquaticus Coetzee, 1968 (A, T)
Mononchus tunbridgensis (T)
Genus *Prionchulus* Cobb, 1916
Prionchulus muscorum (Dujardin, 1845) Wu and Hoeppli, 1929 (T)
Genus *Clarkus* Jairajpuri, 1970a
Clarkus papillatus (Bastian, 1865) Jairajpuri, 1970 (T)
Clarkus sheri (Mulvey, 1967) Jairajpuri, 1970 (T)
Genus *Coomansus* Jairajpuri and Khan, 1977
Coomansus parvus (de Man, 1880) Jairajpuri and Khan, 1977 (T)
Family Mylonchulidae Jairajpuri, 1969
Subfamily Mylonchulinae Jairajpuri, 1969
Genus *Mylonchulus* Cobb, 1916
Mylonchulus minor (Cobb, 1893) Cobb, 1916 (A, T)
Mylonchulus brachyuris (Bütschli, 1873) Cobb, 1917 (T)
Mylonchulus brevicaudatus (Cobb, 1917) Altherr, 1954 (T)
Mylonchulus hawaiiensis (Cassidy, 1931) Andrássy, 1958 (T)
Mylonchulus incurvus (Cobb, 1917) Andrássy, 1958 (T)
Mylonchulus lacustris (Cobb N.A. in Cobb, M.V., 1915) Andrássy, 1958 (T)
Mylonchulus sigmaturus (Cobb, 1917) Altherr, 1953 (T)
Mylonchulus striatus (Thorne, 1924) Andrássy, 1958 (T)
Subfamily Sporonchulinae Jairajpuri, 1969
Genus *Granonchulus* Andrássy, 1958
Granonchulus decurrens (Cobb, 1917) Andrássy, 1958 (T)

Table 20.3 (continued)

Granonchulus subdecurrens Coetzee, 1966 (T)
Family Cobbonchidae Jairajpuri, 1969
Genus *Cobbonchus* Andrássy, 1958
Cobbonchus artemisiae Coetzee, 1968 (T)
Cobbonchus charlesi Coetzee, 1966 (T)
Cobbonchus diannae Coetzee, 1965 (T)
Cobbonchus eurystoma Coetzee, 1965 (T)
Cobbonchus heynsi Coetzee, 1965 (T)
Cobbonchus mauritianus (Williams, 1958) Clark, 1960 (T)
Cobbonchus megalus Coetzee, 1966 (T)
Cobbonchus ockerti Coetzee, 1965 (T)
Cobbonchus rotundicaudatus Coetzee, 1968 (T)
Cobbonchus thesigeri Coetzee, 1968 (T)
Family Iotonchidae Jairajpuri, 1969
Subfamily Iotonchinae Jairajpuri, 1969
Genus *Iotonchus* Cobb, 1916
Iotonchus acutus Heyns and Lagerwey, 1965 (T)
Iotonchus geminus Heyns and Lagerwey, 1965 (T)
Iotonchus litoralis Coetzee, 1967 (T)
Iotonchus loteniae de Bruin and Heyns, 1992 (T)
Iotonchus monhystera (Cobb, 1917) Jairajpuri, 1970 (T)
Iotonchus pauli Heyns and Lagerwey, 1965 (T)
Iotonchus rinae Coetzee, 1967 (T)
Iotonchus risoceiae (Carvalho, 1955) Andrássy, 1858 (T)
Iotonchus spinacaudatus Coetzee, 1967 (T)
Iotonchus transkeiensis Heyns and Lagerwey, 1965 (T)
Genus *Jensenonchus* Jairajpuri and Khan, 1982
Jensenonchus antedontoides (Coetzee, 1967) Andrássy, 1993
Order Dorylaimida Pearse, 1942
Suborder Nygolaimina Ahmad and Jairajpuri, 1979
Superfamily Nygolaimoidea Thorne, 1935
Family Nygolaimidae Thorne, 1935
Subfamily Nygolaiminae Thorne, 1935
Genus *Aquatides* Heyns, 1968
Aquatides thornei (Schneider, 1937) Thorne, 1974 (A)
Genus *Solididens* Heyns, 1968
Solididens bisexualis (Thorne, 1930) Heyns, 1968 (T)
Solididens capensis Heyns, 1967 (T)
Solididens spiralis Loos, 1946 (T)
Solididens vulgaris (Thorne, 1930) Thorne, 1974 (T)
Solididens xosorum Heyns, 1967 (T)
Genus *Nygolaimus* Cobb, 1913
Nygolaimus brachyuris (de Man, 1880) Thorne, 1930 (T)

(continued)

Table 20.3 (continued)

Nygolaimus anneckei Heyns, 1967 (T)
Nygolaimus directus Heyns, 1967 (T)
Nygolaimus dorotheae Heyns, 1967 (T)
Nygolaimus elainnae Botha and Heyns, 1990b (T)
Genus *Laevides* Heyns, 1968
Laevides laevis (Thorne, 1939) Heyns, 1968 (T)
Genus *Paravulvus* Heyns, 1968
Paravulvus andrassyi Heyns, 1967 (T)
Paravulvus hartingii (de Man, 1880) Thorne, 1929 (T)
Family Nygolaimidae Thorne, 1935
Subfamily Nygolaimellinae Clark, 1961
Genus Nygolaimellus Loos, 1949
Nygolaimellus macmacus Heyns, 1967
Nygolaimellus rectalus Heyns, 1967
Suborder Dorylaimina Pearse, 1936
Superfamily Dorylaimoidea de Man, 1876
Family Actinolaimidae Thorne, 1939
Subfamily Actinolaiminae Thorne, 1939
Genus *Actinolaimus* Cobb, 1913
Actinolaimus perplexus Heyns and Argo, 1970
Genus *Neoactinolaimus* Thorne, 1967
Neoactinolaimus crassidens Heyns and Argo, 1970 (T)
Genus Paractinolaimus Meyl, 1957
Paractinolaimus microdentatus (Thorne, 1939) Meyl, 1957 (T)
Paractinolaimus prodenticulatus Heyns and Argo, 1970 (T)
Paractinolaimus vigor Thorne, 1967 (T)
Paractinolaimus xosorum Heyns and Argo, 1970 (T)
Family Crateronematidae Siddiqi, 1969
Subfamily Lordellonematinae Siddiqi, 1969
Genus *Lordellonema* Andrássy, 1959
Lordellonema porosum (Heyns, 1963) Heyns, 1963 (T)
Family Thornenematidae Siddiqi, 1969
Subfamily Thornenematinae Siddiqi, 1969
Genus *Thornenema* Andrássy, 1959
Thornenema baldum (Thorne, 1939) Andrássy, 1959 (A, T)
Thornenema cavalcantii Lordello, 1955 (T)
Family Nordiidae Jairajpuri and Siddiqi, 1964
Subfamily Pungentinae Siddiqi, 1964
Genus *Lenonchium* Siddiqi, 1965
Lenonchium fimbricaudatum Swart and Heyns, 1991 (A)
Genus *Kochinema* Heyns, 1963
Kochinema proamphidum Heyns, 1963 (T)
Subfamily Nordiinae Jairajpuri and Siddiqi, 1964

Table 20.3 (continued)

Genus *Longidorella* Thorne, 1939
Longidorella microdorus (de Man, 1880) Goodey, 1963 (T)
Family Dorylaimidae de Man, 1876
Subfamily Dorylaiminae de Man, 1876
Genus *Laimydorus* Siddiqi, 1969
Laimydorus africanus Botha and Heyns, 1993 (A)
Laimydorus olifanti Botha and Heyns, 1991 (A)
Genus *Dorylaimus* Dujardin, 1845
Dorylaimus asymphydorus Andrássy, 1969 (T)
Subfamily Mesodorylaiminae Andrássy, 1969
Genus *Calcaridorylaimus* Andrássy, 1986
Calcaridorylaimus sirgeli Heyns and Meyer, 1995 (A, T)
Genus *Mesodorylaimus* Andrássy, 1959
Mesodorylaimus mesonyctius (Kreis, 1930) Andrássy, 1959 (T)
Mesodorylaimus aegypticus (Andrássy, 1958) Andrássy, 1959 (A)
Mesodorylaimus arvensis (Cobb in Thorne and Swanger, 1936) Andrássy, 1959 (A, T)
Mesodorylaimus bainsi Basson and Heyns, 1974 (T)
Mesodorylaimus importunes Basson and Heyns, 1974 (A, T)
Mesodorylaimus intermedius Dassonville and Heyns, 1984 (A)
Mesodorylaimus johanni Basson and Heyns, 1974 (T)
Mesodorylaimus kowyni Basson and Heyns, 1974 (T)
Mesodorylaimus margaritus Basson and Heyns, 1974 (T)
Mesodorylaimus mesonyctius (Kreis, 1930) Andrássy, 1959 (A, T)
Mesodorylaimus paralitoralis Basson and Heyns, 1974 (T)
Mesodorylaimus potus Heyns, 1963 (T)
Mesodorylaimus pseudosubtilis Basson and Heyns, 1974 (T)
Mesodorylaimus rotundolabiatus Basson and Heyns, 1974 (T)
Mesodorylaimus sanctus Basson and Heyns, 1974 (T)
Mesodorylaimus transkeiensis Basson and Heyns, 1974 (T)
Mesodorylaimus usitatus Basson and Heyns, 1974 (A)
Mesod*orylaimus vaalensis* Heyns and Kruger, 1983 (A, T)
Genus *Namaquanema* Heyns and Swart, 1993
Namaquanema hanki Heyns and Swart, 1993 (T)
Subfamily Prodorylaiminae Andrássy, 1959
Genus *Prodorylaimus* Andrássy, 1959
Prodorylaimus paralongicaudatus (Micoletzky, 1925) Andrássy, 1959 (T)
Prodorylaimus rionensis (Gerlach, 1954) Andrássy, 1959 (T)
Family Aporcelaimidae Heyns, 1965
Subfamily Aporcelaiminae Heyns, 1965
Genus *Aporcelaimellus* Heyns, 1965
Aporcelaimellus obtusicaudatus (Bastian, 1865) Altherr, 1968 (A)

(continued)

Table 20.3 (continued)

Aporcelaimellus adriani Botha and Heyns, 1990 (T)
Aporcelaimellus amylovorus Thorne and Swanger (1936), Heyns (1965)
Aporcelaimellus glandus Botha and Heyns, 1991 (A, T)
Aporcelaimellus micropunctatus Botha and Heyns, 1990 (A, T)
Aporcelaimellus parapapillatus Botha and Heyns, 1990 (T)
Genus *Aporcelaimus* Thorne and Swanger, 1936
Aporcelaimus pseudospiralis Botha and Heyns, 1990 (T)
Genus *Makatinus* Heyns, 1965
Makatinus punctatus Heyns, 1965 (T)
Makatinus capensis Heyns, 1965 (T)
Makatinus macropunctatus Heyns, 1967 (T)
Genus *Tubixaba* Monteiro and Lordello, 1980
Tubixaba minima Botha and Heyns, 1990 (T)
Subfamily Sectonematinae Siddiqi, 1969
Genus *Sectonema* Thorne, 1930
Sectonema brevicauda Heyns, 1965 (T)
Sectonema probulbum (Heyns, 1965) Siddiqi, 1995 (T)
Sectonema pseudoventrale Heyns, 1965 (T)
Family Qudsianematidae Jairajpuri, 1965
Subfamily Qudsianematinae Jairajpuri, 1965
Genus *Allodorylaimus* Andrássy, 1986
Allodorylaimus diadematus (Cobb in Thorne and Swanger, 1936) Andrássy, 1986 (T)
Genus *Crassolabium* Yeates, 1967
Crassolabium annae Van Reenen and Heyns, 1986 (T)
Crassolabium christiani Van Reenen and Heyns, 1986 (T)
Crassolabium nothus (Thorne and Swanger, 1936) Peña-Santiago and Ciobanu, 2008 (T)
Crassolabium surikae Van Reenen and Heyns, 1986 (T)
Genus *Skibbenema* Van Reenen and Heyns, 1986
Skibbenema constrictum Van Reenen and Heyns, 1986 (T)
Genus *Talanema* Andrássy, 1991
Talanema mauritiense Williams, 1959 (T)
Genus *Labronema* Thorne, 1939
Labronema pygmaeum Heyns, 1963 (T)
Subfamily Lordellonematinae Siddiqi, 1969
Genus *Lordellonema* Andrássy, 1959
Lordellonema porosum (Heyns, 1963) Heyns, 1963 (T)
Subfamily Discolaiminae Siddiqi, 1969
Genus *Discolaimus* Cobb, 1913
Discolaimus acuticapitus Furstenberg and Heyns, 1965
Discolaimus bicorticus Furstenberg and Heyns, 1965

Table 20.3 (continued)

Discolaimus constrictus Heyns, 2001 (T)
Discolaimus deaconi Botha and Heyns, 1991 (A)
Discolaimus intermedius Heyns and Lagerwey, 1965
Discolaimus krugeri Furstenberg and Heyns, 1965
Discolaimus levinae Furstenberg and Heyns, 1965
Discolaimus major Thorne, 1939 (T)
Discolaimus monoplanes Heyns, 1963 (A, T)
Discolaimus similis Thorne, 1939
Genus *Discolaimium* Thorne, 1939
Discolaimium sublatum Heyns, 1963 (T)
Genus *Eudorylaimus* Andrássy, 1959
Eudorylaimus fransus Heyns, 1963 (T)
Eudorylaimus nudicaudatus Heyns, 1993 (T, Antarctica)
Genus *Microdorylaimus* Andrássy, 1986
Microdorylaimus rapsus (Heyns, 1963) Andrássy, 1986 (T)
Subfamily Carcharolaiminae Thorne, 1967
Genus *Carcharolaimus* Thorne, 1939
Carcharolaimus crassicostatus Heyns and Argo, 1969 (T)
Superfamily Belondiroidea Thorne, 1939
Family Dorylaimellidae Jairajpuri, 1964
Genus *Dorylaimellus* Cobb, 1913
Dorylaimellus aferoides Jordaan and Heyns, 1984 (T)
Dorylaimellus andrassyi Heyns, 1963 (T)
Dorylaimellus imitator Heyns, 1963 (T)
Dorylaimellus jonsoni (Jordaan and Heyns, 1984) Jairajpuri and Ahmad, 1992 (T)
Dorylaimellus meridionalis Jordaan and Heyns, 1984 (T)
Dorylaimellus monticolus Clark, 1963 (T)
Dorylaimellus projectus Heyns, 1962 (T)
Dorylaimellus tenuidens Thorne, 1939 (T)
Dorylaimellus vexator Heyns, 1963 (T)
Genus *Axodorylaimellus* Jairajpuri and Ahmad, 1980
Axodorylaimellus parvulus (Thorne, 1939) Jairajpuri and Ahmad, 1980 (T)
Family Swangeriidae Jairajpuri, 1964
Subfamily Swangeriinae Jairajpuri, 1964
Genus *Oxydirus* Thorne, 1939
Oxydirus gangeticus Siddiqi, 1966 (A, T)
Superfamily Tylencholaimoidea Filipjev, 1934
Family Tylencholaimidae Filipjev, 1934
Subfamily Tylencholaiminae Filipjev, 1934
Genus *Chitwoodius* Jiménez-Guirado and Peña-Santiago, 1992
Chitwoodius transvaalensis Furstenberg and Heyns, 1966 (T)

(continued)

Table 20.3 (continued)

Genus *Tylencholaimus* de Man, 1876
Tylencholaimus gertii Kruger, 1956 (T)
Tylencholaimus obscurus Jairajpuri, 1965 (T)
Tylencholaimus proximus Thorne, 1939 (T)
Family Mydonomidae Thorne, 1964
Subfamily Mydonominae Thorne, 1964
Genus *Dorylaimoides* Thorne and Swanger, 1936
Dorylaimoides dactylurus Heyns, 1963 (T)
Dorylaimoides paraconurus Heyns, 1963 (T)
Dorylaimoides pretoriensis Heyns, 1963 (T)
Dorylaimoides thecolaimus Heyns, 1963 (T)
Subfamily Vanderlindiinae Siddiqi, 1969
Genus *Vanderlindia* Heyns, 1964
Vanderlindia duplopapillata Heyns, 1964 (T)
Family Leptonchidae Thorne, 1935
Subfamily Leptonchinae Thorne, 1935
Genus *Leptonchus* Cobb, 1920
Leptonchus transvaalensis Heyns, 1963 (A, T)
Genus *Proleptonchus* Lordello, 1955
Proleptonchus krugeri Botha and Heyns, 1992 (A)
Subfamily Tyleptinae Jairajpuri, 1964
Genus *Tyleptus* Thorne, 1939
Tyleptus striatus Heyns, 1963 (T)
Subfamily Xiphinemellinae Jairajpuri, 1964
Genus *Xiphinemella* Loos, 1950
Xiphinemella christiae de Bruin and Heyns, 1991 (T)
Xiphinemella eversa (Heyns, 1963) Siddiqi, 1966 (T)
Xiphinemella marindae de Bruin and Heyns, 1991 (T)
Order Mermithida Hyman, 1951
Suborder Isolaimina Inglis, 1983
Superfamily Isolaimoidea Timm, 1969
Family Isolaimidae Timm, 1969
Genus *Isolaimium* Cobb, 1920
Isolaimium africanum (Hogewind and Heyns, 1967) Heyns and Swart, 1988 (T)
Isolaimium incus Hogewind and Heyns, 1967 (T)
Class Torquentia Andrássy, 1974
Order Monhysterida De Coninck and Schuurmans Stekhoven, 1933
Suborder Monhysterina De Coninck and Schuurmans Stekhoven, 1933
Superfamily Monhysteroidea de Man, 1876
Family Monhysteridae de Man, 1876
Subfamily Monhysterinae de Man, 1876
Genus *Monhystera* Bastian, 1965

Table 20.3 (continued)

Monhystera stagnalis Bastian, 1965 (A)
Monhystera gabaza (Joubert and Heyns, 1980) Jacobs and Heyns, 1994 (A)
Monhystera magnacephala (Joubert and Heyns, 1980) Jacobs and Heyns, 1994 (A)
Monhystera somereni Allgén, 1952 (A)
Monhystera wangi Wu and Hoeppli, 1929 (T)
Genus *Geomonhystera* Andrássy, 1981
Geomonhystera pervaga (Argo and Heyns, 1973) Andrássy, 1981 (T)
Family Xyalidae Chitwood, 1951
Order Araeolaimida De Coninck and Schuurmans Stekhoven, 1933
Suborder Leptolaimina Lorenzen, 1979
Superfamily Haliplectoidea Chitwood, 1951
Family Haliplectidae Chitwood, 1951
Genus *Haliplectus* Cobb, 1913
Haliplectus algoensis Swart and Heyns, 1992 (T)
Haliplectus bickneri (Chitwood, 1956) Swart, Heyns and Coomans, 1993 (T)
Family Plectidae Örley, 1880
Subfamily Plectinae Örley, 1880
Genus *Plectus* Bastian, 1865
Plectus parietinus Bastian, 1865 (T)
Plectus antarcticus de Man, 1904 (T, Antarctica)
Plectus aquatilis Andrássy, 1985 (A)
Plectus cirratus Bastian, 1865 (A)
Subfamily Anaplectinae Zell, 1993
Genus *Anaplectus* De Coninck and Schuurmans Stekhoven, 1933
Anaplectus granulosus (Bastian, 1865) De Coninck and Schuurmans Stekhoven, 1933 (T)
Subfamily Wilsonematinae Chitwood, 1951
Genus *Wilsonema* Cobb, 1913
Wilsonema otophorum (de Man, 1880) Cobb, 1913 (T)
Genus *Tylocephalus* Crossman, 1933
Tylocephalus auriculatus (Bütschli, 1873) Anderson, 1966 (T)
Family Rhabdolaimidae Chitwood, 1951
Subfamily Rhabdolaiminae Chitwood, 1951
Genus *Rhabdolaimus* de Man, 1880
Rhabdolaimus terrestris de Man, 1880 (A)
Superfamily Metateratocephaloidea Eroshenko, 1973
Family Metateratocephalidae Eroshenko, 1973
Genus *Euteratocephalus* Andrássy, 1958
Euteratocephalus palustris (de Man, 1880) Andrássy, 1958 (T)
Euteratocephalus spiraloides (Micoletzky, 1913) Heyns, 1977 (A)

(continued)

Table 20.3 (continued)

Genus *Metateratocephalus* Eroshenko, 1973
Metateratocephalus crassidens (de Man, 1880) Eroshenko, 1973 (T)
Superfamily Plectoidea Örley, 1880
Family Chronogastridae Gagarin, 1975
Genus *Chronogaster* Cobb, 1913
Chronogaster africana Heyns and Coomans, 1980 (T and A)
Chronogaster glandifera Heyns and Coomans, 1980 (A)
Chronogaster longicauda Heyns and Coomans, 1980 (A)
Chronogaster multispinata Heyns and Coomans, 1980 (A)
Order Chromadorida Chitwood, 1933
Suborder Chromadorina Chitwood and Chitwood, 1937
Superfamily Chromadoroidea Filipjev, 1917
Family Selachinematidae Cobb, 1915
Genus *Cobbionema* Filipjev, 1922
Cobbionema capense Furstenberg and Heyns, 1987 (E)
Superfamily Cyatholaimoidea Filipjev, 1918
Family Cyatholaimidae Filipjev, 1918
Genus *Achromadora* Cobb, 1913
Achromadora ruricola (de Man, 1880) Micoletzky, 1925 (A, T)
Genus *Synonchium* Cobb, 1920
Synonchium capense Heyns and Swart, 1995 (T)
Order Desmodorida De Coninck, 1965
Suborder Desmodorina De Coninck, 1965
Superfamily Desmodoroidea Filipjev, 1922
Family Desmodoridae
Genus *Sibayinema* Swart and Heyns, 1991
Sibayinema natalensis Swart and Heyns, 1991 (A)

20.7 Conclusions

Non-parasitic nematodes are an integral part of the interlocking chain of nutrient conversions that occur in terrestrial and aquatic environments. They function in the recycling of carbon-containing substances, mineral nutrients and nitrogenous components. Likewise, they control explosions of microflora and microfauna and maintain the stability of life forms that constitute the delicate balance of nature. Viglierchio (1991) duly emphasised the fact that although non-parasitic nematodes are considered benign by mankind, they constitute one of the vital components in the preservation of the balance of life processes of our world.

References

Ahmad W, Jairajpuri MS (2010) Order mononchida jairajpuri 1969. In: Ahmad W, Jairajpuri MS (eds) *Mononchida;* the predatory soil nematodes. Nematology Monographs and Perspectives 7 (Series editors: Hunt DJ, Perry RN). Brill, Leiden, p 45–50

Andrássy I (2005) Non-parasitic nematodes of Hungary I. (Nematoda errantia) Pedozoologica Hungarica No. 3. Hungarian Natural History Museum and Systematic Zoology Research Group of the Hungarian Academy of Sciences, Budapest

Andrássy I (2007) Non-parasitic nematodes of Hungary II. (Nematoda errantia) Pedozoologica Hungarica No. 4. Hungarian Natural History Museum and Systematic Zoology Research Group of the Hungarian Academy of Sciences, Budapest

Andrássy I (2009) Non-parasitic nematodes of Hungary III. (Nematoda errantia) Pedozoologica Hungarica No. 5. Hungarian Natural History Museum and Systematic Zoology Research Group of the Hungarian Academy of Sciences, Budapest

Berry S, Rhodes R (2006) Green manure crops: agronomic characteristics and effect on nematodes. Proc S Afr Sugar Techn Ass 80:269

Bonacci O, Pipan T, Culver DC (2009) A framework for karst ecohydrology. Environ Geol 56:891–900

Borgonie G, García-Moyano A, Litthauer D et al (2011) Nematoda from the terrestrial deep surface of South Africa. Nature 474:4. doi:10.1038/nature09974

Botha A, Heyns J (1990) A preliminary check list of the non-parasitic and plant-parasitic nematodes of the Kruger National Park. Koedoe 33:73–76

Botha A, Heyns J (1991) Dorylaimoidea (Nematoda) from rivers in the Kruger National Park. Koedoe 34:1–24

Botha A, Heyns J (1992a) Freshwater nematodes from the genera *Thornenema* and *Mesodorylaimus* from the Kruger National Park with a diagnostic species compendium for South African species of the genus *Mesodorylaimus* (Nematoda: Dorylaimida). Koedoe 35:25–42

Botha A, Heyns J (1992b) Species of *Tyleptus*, *Proleptonchus*, *Aquatides* and *Afractinolaimus* from rivers in the Kruger National Park (Nematode: Dorylaimida). Koedoe 35:43–54

Botha A, Heyns J (1992c) Further records and descriptions of nematodes from rivers in the Kruger National Park (Orders Enoplida, Chromadorida, Monhysterida, Mononchida and Araeolaimida). Koedoe 35:11–25

Botha A, Heyns J (1993a) Species of the genera *Oxidirus*, *Dorylaimellus* (*Axodorylaimellus*), *Laimydorus* and *Rhabdolaimus* from rivers of the Kruger National Park (Nematode: Dorylaimida and Araeolaimida). Koedoe 36:49–60

Botha A, Heyns J (1993b) New records of Tylenchida, Araeolaimida and Enoplida from the Kruger National Park, with an addendum to the check list of nematode species in the park. Koedoe 36:61–65

Culver D, Pipan T (2010) The biology of caves and other subterranean habitats. University Press, Oxford

Du Preez GC (2014) Determining the effect of polluted mine water on the ecosystem health of a karstic cave environment in the Witwatersrand Basin. MSc Dissertation, North-West University, Potchefstroom

Du Preez GC, Theron P, Fourie D (2013) Terrestrial mesofauna biodiversity in unique karst environments in southern Africa. In: Filippi M, Bosak P (eds) Proceedings of the 16th International Congress of Speleology, Brno

Du Preez GC, Fourie H, Theron P et al (2014) Nematodes: indicators of metal pollution in the Wonderfontein Cave (South Africa). Proceedings of the 6th International Congress of Nematology. J Nematol 46:130–260

Du Preez GC, Fourie H, Swart A (2015a) Using nematodes as bioindicators of anthropogenic influence in the Wonderfontein Cave (Witwatersrand basin, South Africa). Proceedings of the 9th Symposium for European Freshwater Sciences, 5–10 July, Geneva, p 542

Du Preez GC, Swart A, Fourie H (2015b) Nematodes of the wonderfontein cave (Witwatersrand Basin, South Africa). Nematology 17:967–980

Durand JF (2008) Die karstekologie van Suid-Afrika met spesiale verwysing na die Wieg van die Mensdom Wêrelderfenisgebied. S-Afr Tydskr Natuurwet Tegnol 4:5–15

Durand F, Swart A, Marais W et al (2012) Die karst-ekologie van die Bakwenagrot, Gauteng. S-Afr Tydskr Natuurwet Tegnol 31. doi:org/10.4102/satnt.v31i1.275

Engelbrecht E (2012) Nematode (Phylum Nematoda) Community Assemblages: a tool to implement environmentally-sound management strategies for root-knot nematodes (*Meloidogyne* spp.) in Potato-based Cropping Systems. MS Dissertation North-West University, Potchefstroom

Ferris H, Bonger T, De Goede RGM (2001) A framework for soil food web diagnostics: extension of nematode faunal analysis concept. Appl Soil Ecol 18:13–29

Furstenberg JP, De Wet AG (1983) Effect of salt marsh vegetation on *Helicotylenchus* and *Tylenchus* species of Swartkops Estuary. Nematological Soc S Afr Newsl 16:5–6

Gupta VVR, Yeates GW (1997) Soil microfauna as bioindicators of soil health. In: Doube BM, Gupta VVR (eds) Biological indicators of soil health. CAB International, Wallingford, pp 201–234

Heyns J (1971) A guide to the plant and soil nematodes of South Africa. AA Balkema, Cape Town

Heyns J (1976) Preliminary results of a survey of freshwater nematodes in South Africa. J Limnol Soc S Afr 2:43–45

Heyns J (1982) Non-parasitic nematodes. In: Keetch DP, Heyns J (eds) Nematology in southern Africa. Science Bulletin of the Department of Agriculture and Fisheries Division of Agricultural Information, Pretoria, pp 149–150

Heyns J (2002) Nematoda. In: Day JA, De Moor IJ (eds) Guides to the freshwater invertebrates of Southern Africa, vol 5. Non-Arthropods. Water Research Commission Report No TT167/02, Pretoria, p 136–161

Heyns J, Coomans A (1980) Freshwater nematodes from South Africa. 5. *Chronogaster* Cobb, 1913. Nematol 26:187–208

Heyns J, Coomans A (1983) New and little known species of *Chronogaster* Cobb, 1913 (Nematoda). Nematol 29:245–265

Heyns J, Kruger JC de W (1983) Freshwater nematodes from South Africa. 6. *Mesodorylaimus* Andrássy, 1959. Phys Chem Phys 15:47–53

Hodda M, Wanless FR (1994) Nematodes from an English chalk grassland: species distributions. Nematol 40:116–132

Hodda M, Ocaña A, Traunspurger W (2006) Nematodes from extreme freshwater habitats. In: Abebe E, Traunspurger W, Andrássy I (eds) Freshwater nematodes: ecology and taxonomy. CAB International, Wallingford, pp 179–210

Hugo HJ, Malan AP (2010) Occurrence and control of plant parasitic nematodes in irrigation water – a review. S Afr J Enol Vitic 31:169–180

Jansen C (2014) Microbial community structure and nematode diversity in soybean-based cropping systems. MSc Dissertation. North-West University, Potchefstroom

Jansen Van Rensburg C (2010) Nematodes from the Bakwena Cave in Irene, South Africa. MSc Dissertation. Ghent University, Ghent

Magdorf F (2001) Concepts, components, and strategies of soil health in agroecosystems. J Nematol 33:169–172

Marais M, Swart A, Lamprecht SC et al (2009) Die uitwerking van grondbewerking op nematodes. S-Afr Tydskr Natuurwet Tegnol 28:165

Marais M, Swart A, Lamprecht SC et al (2010) The effect of tillage on nematodes. S Afr J Plant Soil 27:280

Mikola J, Setälä H (1998) Relating species diversity to ecosystem functioning: mechanistic backgrounds and experimental approach with a decomposer food web. Oikos 83:180–194

Moens T, Verbeeck L, De Maeyer A et al (1999) Selective attraction of marine bacterivorous nematodes to their bacterial food. Mar Ecol Prog Ser 176:165–178

Moens T, Yeates GW, De Ley P (2004) Use of carbon and energy sources by nematodes. Nematol Monogr Perspect 2:529–545

Muschiol D, Traunspurger W (2007) Life cycle and calculation of the intrinsic rate of natural increase of two bacterivorous nematodes, *Panagrolaimus* sp. and *Poikilolaimus* sp. from chemoautotrophic Movile Cave, Romania. Nematol 9:271–284

Muschiol D, Giere O, Traunspurger W (2015) Population dynamics of a cavernicolous nematode community in a chemoautotrophic groundwater system. Limnol Oceanogr 60:127–135

Neher D, Bongers T, Ferris H (2004) Computation of nematode community indices. Society of Nematologists Estes Park, Colorado

Nuss B (1984) Ultrastructurelle und ökophysiologische Untersuchungen an kristalloiden Einschlüssen der Muskeln eines sulfid-toleranten limnischen Nematoden (*Tobrilus gracilis*). Veröffentlichungen Institut für Meeresforschung in Bremerhaven 20:3–15

Poinar GO Jr, Sarbu S (1994) *Chronogaster troglodytes* sp. n. (Nemata: Chronogasteridae) from Movile Cave, with a review of cavernicolous nematodes. Fundam Appl Nematol 17:231–237

Poinar GO Jr (1983) The natural history of nematodes. Prentice-Hall, Englewood Cliffs

Romero A (2009) Cave biology: life in darkness. Cambridge University Press, Cambridge

Sarbu SM, Kane T, Kinkle BK (1996) A chemoautotrophically based cave ecosystem. Science 272:1953–1955

Sieriebriennikov B, Ferris H, de Goede R (2014) NINJA: an automated calculation system for nematode-based biological monitoring. Eur J Soil Biol 61:90–93

Swart A, Marais M, Habig J (2014) The effect of conservation agriculture on nematode populations and soil microbial community dynamics. Proceedings of the 6th International Congress of Nematology. J Nematol 46:243

Venette RC, Ferris H (1998) Influence of bacterial type and density on population growth of bacterial-feeding nematodes. Soil Biol Biochem 30:949–960

Viglierchio DR (1991) The world of nematodes. Department of Nematology, University of California, Davis

Williams P (2008) World heritage caves and karst. A thematic study. IUCN Programme on Protected Areas. World Heritage Studies no.2. p 1–34

Yeates GW (1979) Soil nematodes in terrestrial ecosystems. J Nematol 11:213–229

Yeates GW, Bongers T (1999) Nematode diversity in agroecosystems. Agric Ecosyst Environ 74:113–135

Yeates GW, Bongers T, De Goede RGM et al (1993) Feeding habits in soil nematode families and genera – an outline for soil ecologists. J Nematol 25:315–331

Chapter 21
Overview of the South African Plant-Parasitic Nematode Survey (SAPPNS)

Mariette Marais, Antoinette Swart, and Naomi Buckley

21.1 Introduction

In the mid-1980s Prof Dirk de Waele, then stationed at the Grain Crops Institute in Potchefstroom, proposed a development programme based on his experience with the European Plant-Parasitic Nematode Survey (De Waele 1980; Alphey and Taylor 1986). A facet of the programme was the focus on the countrywide occurrence and control of plant-parasitic nematodes of grain, viz. maize (*Zea mays*), sorghum (*Sorghum bicolor*), wheat (*Triticum aestivum*) and leguminous and oilseed crops (groundnut (*Arachis hypogaea*), soybean (*Glycine max*), sunflower (*Helianthus annuus*), dry bean (*Phaseolus vulgaris*) in South Africa (SA); after deliberation, David Keetch, Esther van den Berg and Dirk de Waele agreed that a South African Plant-Parasitic Nematode Survey (SAPPNS) be launched and that it would be co-ordinated by the Nematology Unit of Plant Protection Research of the Agricultural Research Council (ARC-PPR). The SAPPNS was launched in 1987 (Van Vuuren 1992; Marais 2006).

21.2 Main Objectives of the SAPPNS

The four main aims of the SAPPNS are:

(i) Making an inventory of all the plant-parasitic nematodes found in SA
(ii) Studying the biogeography of plant-parasitic nematodes
(iii) Establishing an electronic database at the ARC-PPR of all plant-parasitic nematodes listed
(iv) Drawing maps that represent the distribution of all listed nematodes

M. Marais (✉) • A. Swart • N. Buckley
Nematology Unit, Biosystematics Division, Agricultural Research Council-Plant Protection Research Institute, Private Bag X134, Queenswood 0121, South Africa
e-mail: MaraisM@arc.agric.za

H. Fourie et al. (eds.), *Nematology in South Africa: A View from the 21st Century*, DOI 10.1007/978-3-319-44210-5_21

Table 21.1 Nematologists contributing data in the mid-1980s during the first drive to establish a South African Plant-Parasitic Nematode Survey (SAPPNS) database

Contributor	Associated crop/plant	Locality/province
A.J. Meyer	*Prunus persica* (peach)	Western Cape Province
D. de Waele C. Venter	*Sorghum bicolor* (sorghum), *Zea mays* (maize), *Arachis hypogaea* (groundnut)	Highveld Region (Gauteng and Mpumalanga provinces)
E. Rabie	*Ananas comosus* (pineapple)	KwaZulu-Natal and Eastern Cape provinces
E. Jordaan	*Triticum aestivum* (wheat)	Free State, Western Cape provinces
F. Steyn	*Solanum tuberosum* (potato)	Gauteng, Mpumalanga, Northern Cape, Western Cape provinces
H. Hugo	Deciduous and stone fruit	Western Cape Province
J. Heyns	Various crops and plants	Material deposited in the Nematology Collection of the Rand Afrikaans University
J. Loubser	*Vitis vinifera* (grapevine)	Northern and Western Cape provinces
K. Daiber	Various vegetables	Eastern Cape, Gauteng, KwaZulu-Natal and Limpopo provinces
M. Botha-Greeff	*Nicotiana tabacum* (tobacco) and *Gossypium hirsutum* (cotton)	Loskop Dam Irrigation Scheme (Mpumalanga Province)
P. Willers	*Coffea arabica* (coffee)	Lowveld of the Mpumalanga Province
R. van Wyk	Various crops	Vaalharts Irrigation Scheme (Northern Cape Province)
V. Spaull	*Saccharum officinarum* (sugarcane)	KwaZulu-Natal and Mpumalanga provinces

Hercula Coetzee was appointed as co-coordinator of the initial phase of SAPPNS, and her first task was to visit the different research institutes of the Department of Agriculture, Fisheries and Forestry (DAFF) and the universities in SA to introduce the initiative to other nematologists. Her second task was to establish what data to record. In consultation with Dirk de Waele and the taxonomists of PPR, it was decided to include information about the collector, sampling date, locality (country, province, magisterial district, altitude, geo-reference), associated plant (biome, vegetation type, host type, plant species), soil type (percentage clay, silt, sand and soil pH), mean rainfall and the plant-parasitic nematodes found (with genus name being the minimum requirement). The primary sources of data were the information contained in the National Collection of Nematodes (NCN), data from the plant-parasitic nematodes collected in SA since the 1950s, published data and unpublished historical data contributed by local nematologists (Table 21.1).

The second tier of information was that of systematic surveys undertaken by the personnel of the NCN (Table 21.2) and data obtained from the diagnostic services rendered by the different institutes of the then Department of Agriculture (now part of the ARC). Contributions of data by various local nematologists (e.g. from persimmon donated by Ms S Storey, Nemlab, Durbanville) were also included. Sampling during a national survey was based on the 25-km grid scheme of the SA

Table 21.2 Papers published containing data from the South African Plant-Parasitic Nematode Survey (SAPPNS)

Survey area or crop plant	Province	References
Barberton and Mbombela areas	Mpumalanga	Swart (1994), Van den Berg (1994a), Marais (1998)
Caledon area	Western Cape	Marais and Swart (1998), Van den Berg (1998), Decraemer and Marais (2000), Van den Berg and Tiedt (2001b)
Carolina area	Mpumalanga	Kleynhans (1992b), Decraemer and Kilian (1992), Van den Berg (1992b), Decraemer and Marais (1993), Marais (1993), Swart (1994), Van den Berg (1994a), Marais (1998)
Douglas area	Northern Cape	Marais and Swart (2001)
Garden route	Western Cape	Marais and Van den Berg (1996), Marais et al. (2004)
Goegap and Witsand Nature Reserves	Northern Cape	Van den Berg et al. (2003)
Lower Orange River	Northern Cape	Marais and Swart (1996), Van den Berg (1996a); Marais (1998)
Lusikisiki-Bizana area	Eastern Cape	Marais and Swart (2007)
Modimolle area	Limpopo	Van den Berg and Tiedt (2001a), Marais and Swart (2002)
Northern KwaZulu-Natal	KwaZulu-Natal	Van den Berg (1991, 1992a), Van den Berg and Tiedt (2001a)
Plantations	Eastern Cape, KwaZulu-Natal, Limpopo Mpumalanga	Kruger et al. (1992), Decraemer and Marais (1993), Marais and Buckley (1993), Marais and Swart (1999), Swart (2000), Van den Berg and Tiedt (2000, 2001a, c, 2002)
Protected areas	Eastern Cape, KwaZulu-Natal	Marais and Swart (2013, 2014)
Swartberg Nature Reserve	Western Cape	Heyns and Van den Berg (1996), Van den Berg (1998), Van den Berg et al. (2003)
Tsitsikamma National Park	Eastern Cape	Kleynhans (1988), De Waele and Kilian (1992), Van den Berg (1996b), Marais (1998)
Tzaneen area	Limpopo	Van den Berg and Tiedt (2001c, 2002), Marais and Swart (2003)
Venda	Limpopo	Subbotin et al. (2011)
Wetlands	Kwazulu-Natal	Heyns and Swart (2002), Van den Berg et al. (2007a, b; 2009)
Wilderness National Park	Western Cape	Marais (1998), Van den Berg (1994b)

(continued)

Table 21.2 (continued)

Survey area or crop plant	Province	References
Glycine max (soybean) in various areas	Eastern Cape, Free State, Gauteng, KwaZulu-Natal, Mpumalanga, Northern Cape, North West	Fourie et al. (2015)
Medicago sativa (lucerne) fields in various areas	Eastern Cape, Western Cape, Northern Cape, Free State, North West, Mpumalanga	Van den Berg (1989); Marais (1990), Kleynhans (1992)
Solanum tuberosum (potato) in various areas	Eastern Cape, Northern Cape, Western Cape, Free State, Gauteng, KwaZulu-Natal, Limpopo, Mpumalanga, North West	Marais et al. (2015)

Coordinate System (Zaca 2015). In each of these grids, all cultivated and uncultivated areas were sampled. Sampling and extraction of nematodes were done as described in the 'Plant nematodes in South Africa series' (Marais and Swart 1996, 1998, 1999, 2001, 2002, 2003, 2007).

In 1996, 10 years into the SAPPNS programme, personnel of the Nematology Unit of the ARC-PPR were able to publish a book entitled 'Plant nematodes in South Africa' (Kleynhans et al. 1996), drawing together the knowledge on the occurrence of plant-parasitic nematode species occurring in the country.

In 2006 the SAPPNS database was transferred into a module of the web-based relational Nematode System. This represented the second phase of the SAPPNS being launched when digitising of the specimens deposited in the NCN commenced. More than 200,000 specimens collected since the 1950s are now housed in the NCN at the ARC-PPR's Biosystematics Programme in Pretoria. The digitising of all the specimens deposited into the NCN was completed in early 2014. Being part of a biosystematics programme, researchers and technicians are dedicated to the NCN and Nematode System. All taxonomic or geographic name changes are immediately implemented and principles of bioinformation management adhered to. The datasets read into the SAPPNS database also contribute to the fulfilment of South Africa's obligations as part of international agreements, viz. the Convention on Biodiversity, the Nagoya protocol and the National Environmental Management: Biodiversity Act (Act 10 of 2004) (Anonymous 2014). The change in technology also makes it possible to draw maps of the distribution of any nematode reported from South Africa, in contrast with the initial aim to draw only maps of nematode pests.

Anyone interested in the SAPPNS programme can contact Mariette Marais electronically at 'MaraisM@arc.agric.za'. Datasets are usually made available to bona fide students, researchers or local producers, but because of the sensitive nature of some of the data, potential users will be asked to sign an agreement that spells out the conditions under which datasets will be made available.

Other data logged into the Nematode System include records of nematodes found in fresh water, from dams, streams and rivers and subterranean systems such as karst. Because of the wealth of information contained in the NCN (now more easily available because of digitising) and the nearly 9,000 locality records of the SAPPNS module, we now know, for example, that:

- *Helicotylenchus brevis* is typically found in the Forest Biome of SA.
- A number of the plant-parasitic nematodes found in the Fynbos and Forest biomes are endemic to SA.
- Eight of the nine *Trichodorus* spp. found in SA are endemic, namely, *Trichodorus iuventus* Decraemer & Marais, 2000; *Trichodorus kilianae* Decraemer & Marais, 1993, *Trichodorus magnus* Decraemer & Marais, 1993, *Trichodorus parorientalis* Decraemer & Kilian, 1992, *Trichodorus petrusalberti* De Waele, 1988, *Trichodorus philipi* De Waele, Meyer & Van Mieghem, 1990, *Trichodorus sanniae* Vermeulen & Heyns, 1984 and *Trichodorus vandenbergae* Decraemer & Kilian, 1992. Interestingly, one record exists of more than one *Trichodorus* sp. occurring together in the same locality. These species are *T. iuventus* and *T. vandenbergae* and are associated with fynbos in the Western Cape Province.
- A number of nematodes including *Discocriconemella* spp.; *Helicotylenchus* spp.; *Longidorus laevicapitatus* Williams, 1959; *Meloidogyne* spp.; *Paralongidorus* spp.; *Nanidorus* spp.; *Pratylenchus* spp.; *Pratylenchus zeae*; *Rotylenchulus clavicaudatus* Raski and Sher, 1968; *Rotylenchus* spp.; *Rotylenchus unisexus* Sher, 1965; *Scutellonema* spp.; *Scutellonema truncatum* Sher, 1964; *Xiphinema spp.*; *Xiphinema elongatum* Schuurmans Stekhoven and Teunissen, 1938; *Xiphinema mampara* Heyns, 1979; and *Xiphinema umobae* Heyns and Spaull, 1979 have been found in soils with clay contents higher than 60 %.
- *Meloidogyne arenaria* (Neal, 1889) Chitwood, 1949; *Meloidogyne hapla* Chitwood, 1949; *Meloidogyne incognita* (Kofoid & White, 1919) Chitwood, 1949; and *Meloidogyne javanica* (Treub, 1885) Chitwood, 1949 are the most common root-knot nematodes species and were present in 0.4, 1.7, 5.0 and 7.4 % of the records, respectively. These four species are found from sea level up to an altitude of 1,800 m.
- Huge gaps in our knowledge of the Thicket, Grassland, Savanna, Nama- and Succulent Karoo biomes still exist.

References

Alphey TJW, Taylor CE (1986) European atlas of the longidoridae and trichodoridae. Scottish crop research institute with a grant-in-aid from the European science foundation. Invergowrie

Anonymous (2014) No. 10 of 2004: National environmental management: biodiversity Act of 2004. Government Gazette, Cape Town, South Africa, Vol 467, No. 26436. https://www.environment.gov.za/sites/default/files/legislations/nema_amendment_act10.pdf. Accessed 30 Apr 2016

Decraemer W, Kilian S (1992) Description of *Trichodorus parorientalis* n. sp. (Nemata: Diphtherophoroidea) from South Africa. Fundam Appl Nematol 15:539–543

Decraemer W, Marais M (1993) Two new species of *Trichodorus* from South Africa, with a note on *T. petrusalberti* (Nemata: Trichodoridae). Fundam Appl Nematol 16:211–214

Decraemer W, Marais M (2000) A new *Trichodorus* species from South Africa with notes on *T. vandenbergae* (Triplonchida: Diphtherophorina). Ann Zool 50:193–203

De Waele D (1980) Objectives, methods and preliminary results concerning a survey of virus-vector nematodes in Belgium. Meded Fac Landbouwwet, Rijksuniv Gent 45:807–813

De Waele D, Kilian S (1992) *Trichodorus vandenbergae* n. sp. and *Paratrichodorus meyeri* n. sp. (Nematoda: Trichodoridae) from South Africa, with notes on *P. teres*, *P. catharinae* and *P. sacchari*. Fundam Appl Nematol 15:397–406

Fourie H, De Waele D, Mc Donald AH et al (2015) Nematode pests threatening soybean production in South Africa, with reference to *Meloidogyne*. S Afr J Sci 2015:111(9/10): Art. #2014–0212, 9 pages

Heyns J, Swart A (2002) The genus *Xiphinema* in South Africa. XXVII. *Xiphinema zyzy* sp.n. and *X. louisi* Heyns, 1979 (Nematoda: Longidoridae). Nematol Mediterr 30:73–77

Heyns J, Van den Berg E (1996) *Neothada hades* n. sp. from South Africa, with notes on the genus and a key to the species (Nematoda: Tylenchidae). South Afr J Zool 31:165–169

Kleynhans KPN (1988) *Meloinema silvicola* sp. nov. (Nemata: Heteroderidae) from an indigenous forest in the southern Cape. Phytophylactica 20:257–261

Kleynhans KPN (1992) Cyst-forming nematodes in South Africa. *Afenestrat africana* (Luc, Germani and Netscher, 1973) Baldwin & Bell, 1985 (Nemata: Heteroderidae). Phytophylactica 24:307–311

Kleynhans KPN, Van den Berg E, Swart A et al (1996) Plant nematodes in South Africa. Plant Protection Research Institute Handbook No. 8. Agricultural Research Council – Plant Protection Research Institute, Pretoria

Kruger JC de W, Kilian SM, Heyns J (1992) The genus *Xiphinema* in southern Africa. XXIII. X. *parasimplex* spec. nov. and notes on *X. machoni* Hunt, 1980 and *X. heynsi* Siddiqi, 1979 (Nematoda: Longidoridae). Phytophylactica 24:271–278

Marais M (1990) Plant-parasitic nematodes in lucerne fields in South Africa. Phytophylactica 22:449–452

Marais M (1993) On some *Helicotylenchus* Steiner, 1945 from South Africa (Hoplolaimidae: Nemata). Phytophylactica 25:21–38

Marais M (1998) Some species of *Helicotylenchus* Steiner, 1945 from South Africa (Nematoda: Hoplolaimidae). Fundam Applied Nematol 21:327–352

Marais M (2006) South African plant-parasitic nematode survey (SAPPNS). Plant Prot News 67:6

Marais M, Buckley NH (1993) Plant-parasitic nematodes in forest soils in the Transvaal and Natal, South Africa. South Afr For J 166:9–16

Marais M, Swart A (1996) Plant-parasitic nematodes of the lower orange river irrigation area, South Africa. Afr Plant Prot 2:25–30

Marais M, Swart A (1998) Plant nematodes in South Africa. 1. Caledon area, Western Cape Province. Afr Plant Prot 4:27–33

Marais M, Swart A (1999) Plant nematodes in the Lottering Plantation, Eastern Cape Province. South Afr For J 186:77–79

Marais M, Swart A (2001) Plant nematodes in South Africa. 3. Douglas area, Northern Cape Province. Afr Plant Prot 7:33–38

Marais M, Swart A (2002) Plant nematodes in South Africa. 4. Modimolle area, Limpopo Province. Afr Plant Prot 8:25–32

Marais M, Swart A (2003) Plant nematodes in South Africa. 6. Tzaneen area, Limpopo Province. Afr Plant Prot 9:99–107

Marais M, Swart A (2007) Plant nematodes in South Africa. 8. Bizana, Lusikisiki and Port St. Johns area, Eastern Cape Province. Afr Plant Prot 13:16–27

Marais M, Swart A (2013) Plant nematodes in South Africa. 11. Checklist of plant nematodes of the protected areas of KwaZulu-Natal. Koedoe 55(1): Art. #1086, 2 pages plus online Appendix

Marais M, Swart A (2014) Plant nematodes in South Africa. 12. Checklist of plant nematodes of the protected areas of the Eastern Cape Province. Koedoe 56, Art. #1220: 2 pages plus online Appendix

Marais M, Swart A, Knoetze R et al (2015) Plant nematodes in South Africa. 13. A checklist of nematodes associated with potatoes. Chips 2015:38–43

Marais M, Van den Berg E (1996) Plant nematodes in the Bergplaas Plantation, Western Cape province with a description of a new *Criconema* species and notes on two known species of Criconematidae. South Afr For J 177:1–6

Marais M, Van den Berg E, Swart A et al (2004) Plant nematodes in South Africa. 7. A check list of plant nematodes from the Fynbos Biome, with a description of *Helicotylenchus curatus* sp. n. Koedoe 47:67–78

Subbotin SA, Inserra RN, Marais M et al (2011) Diversity and phylogenetic relationships within the spiral nematodes of *Helicotylenchus* Steiner, 1945 (Tylenchida: Hoplolaimidae) as inferred from analysis of the D2-D3 expansion segments of 28S rRNA gene sequences. Nematol 13:333–345

Swart A (1994) Description of three new *Xiphinema* species from South Africa (Nematoda: Dorylaimida). Fundam Appl Nematol 17:455–467

Swart A (2000) A survey of the plant parasitic nematodes associated with *Pinus patula* in selected plantations in the Eastern Cape Province. South Afr For J 188:13–16

Van den Berg E (1989) A new and a known nematode from the little Karoo, South Africa (Tylenchoidea: Nemata). Phytophylactica 21:21–27

Van den Berg E (1991) New and known species from northern Natal coastal sandy soils (Criconematinae: Nemata). Phytophylactica 23:157–165

Van den Berg E (1992a) Redescriptions and new records of six known *Criconema* species from Natal, South Africa (Criconematinae: Nemata). Phytophylactica 24:29–38

Van den Berg E (1992b) New criconematinae (nemata) from the Carolina area of the eastern Transvaal, South Africa. Phytophylactica 24:253–269

Van den Berg E (1994a) New and known Criconematoidea (Nemata) from forests in South Africa with a redescription of *Mesocriconema teres* (Raski, 1952) Loof and De Grisse, 1989. Fundam Appl Nematol 17:551–562

Van den Berg E (1994b) A first list of plant-parasitic nematodes from the Wilderness National Park, with a description of *Ogma sekgwaum* spec. nov. Koedoe 36:61–76

Van den Berg E (1996a) Two new species of the subfamily Criconematinae (Tylenchida: Criconematidae) from South Africa. Afr Plant Prot 2:35–44

Van den Berg E (1996b) A first list of plant-parasitic nematodes from the Tsitsikamma National Park, with descriptions of two new species of the subfamily Criconematinae. Koedoe 39:43–54

Van den Berg E (1998) New records and notes on known species of Hoplolaimidae (Nemata) in South Africa. J Nematode Morphol Syst 1:29–46

Van den Berg E, Tiedt LR (2000) *Hemicycliophora stiaani* sp. n. with notes on three tylenchid (Nemata) species from South Africa. J Nematode Morphol System 2:137–148

Van den Berg E, Tiedt LR (2001a) Description of four new *Criconema* species with notes on *Criconema lantanum* (Van den Berg, 1984) Raski & Luc, 1987 (Nemata: Tylenchina) from South Africa. J Nematode Morphol Syst 3:119–137

Van den Berg E, Tiedt LR (2001b) One new and some known species of *Hemicycliophora* De Man, 1921 (Nemata: Hemicycliophorinae) from South Africa. J Nematode Morphol Syst 3:175–190

Van den Berg E, Tiedt LR (2001c) New records of Criconematoidea (Nematoda) from South Africa, with description of *Criconema zantene* n. sp. Nematol 3:797–815

Van den Berg E, Tiedt LR (2002) Two known criconematids from South Africa with a description of *Ogma chrisbarnardi* (Heyns, 1970) Andrássy, 1979 (Nemata: Criconematinae). J Nematode Morphol Syst 4:119–135

Van den Berg E, Marais M, Tiedt LR (2003) Plant nematodes in South Africa. 5. Plant nematodes from the Gouekrans area of the Swartberg Nature Reserve, with descriptions of one new and one known Hemicycliophorinae (Nemata). Afr Plant Prot 9:43–52

Van den Berg E, Marais M, Tiedt LR (2007a) Plant nematodes in South Africa. 9. Check-list of plant nematodes from the Goegap and Witsand Nature Reserves, Northern Cape Province, with a description of a new *Rotylenchus* species (Hoplolaimidae: Nematoda). Afr Plant Prot 13:28–35

Van den Berg E, Schroeder S, Tiedt LR (2007b) Two new species from KwaZulu-Natal wetlands, South Africa (Nematoda: Criconematidae). J Nematode Morphol Syst 10:63–74

Van den Berg E, Subbotin SA, Handoo ZA et al (2009) *Hirschmanniella kwazuna* sp. n. from South Africa with notes on a new record of *H. spinicaudata* (Schuurmans Stekhoven, 1944) Luc & Goodey, 1964 (Nematoda: Pratylenchidae) and on the molecular phylogeny of *Hirschmanniella* Luc & Goodey, 1964. Nematol 11:523–540

Van Vuuren C (1992) Instituut spog met databank. Landbounuus 34:5–6

Zaca MN (2015) South African co-ordinate systems. http://www.durban.gov.za/City_Services/engineering%20unit/Surveying_Land_Information/Documents/SouthAfricanCoordinateSystems.pdf. Accessed 10 Jan 2015

Chapter 22
Entomopathogenic Nematodes

Antoinette P. Malan and Tiarin Ferreira

22.1 Introduction

Many nematodes are associated with insects, and the relationship varies from simple phoresis to facultative or obligate parasitism with pathogenicity (Koppenhöfer 2002). Entomopathogenic nematodes (EPN) are an assemblage of obligate insect-parasitic nematodes, defined by Onstad et al. (2006) as nematodes which are mutualistically associated with bacterial symbionts and in which all life stages of the nematode, except for the free-living third-stage infective juvenile (IJ), are found inside the insect host.

The first EPN was discovered in 1923 and described as *Aplectana kraussei* Steiner 1923. In 1929, Glaser and Fox retrieved a parasitic nematode from larvae of the Japanese beetle, *Popillia japonica*, at the Tavistock Golf Course near Haddonfield (New Jersey; USA), which was described as *Neoaplectana glaseri* Steiner 1929. Glaser was also, coincidently, the first person to realise the potential of these nematodes as biological control agents of insects and managed to culture them *in vitro,* although unaware of the associated bacterial symbiont at the time (Glaser 1931). Bovien (1937) was responsible for linking the interrelationship between bacteria, nematodes and insects. In 1955, the parasitic nematode *Steinernema carpocapsae* (Weiser 1955) Wouts, Mráček, Gerdin and Bedding, 1955, was isolated from overwintering codling moth larvae in a Virginia apple orchard in the United States of the America (USA) (Dutky and Hough 1955). Another isolate of the same species was recovered in the same year from codling moth in the then Czech Republic (Weiser 1955), and the presence of an associated bacterium was confirmed. From this group's first discovery until the early 1960s, research surrounding EPN was neglected. At that stage, chemical control was a lucrative and effective method of

A.P. Malan (✉) • T. Ferreira
Department of Conservation Ecology and Entomology, Faculty of AgriSciences, Stellenbosch University, Private Bag X1, 7602 Matieland, South Africa
e-mail: apm@sun.ac.za

H. Fourie et al. (eds.), *Nematology in South Africa: A View from the 21st Century*, DOI 10.1007/978-3-319-44210-5_22

insect control in the absence of strict regulation (Adams and Nguyen 2002). Research on the use of biological control agents for the management of insect pests was renewed, as the negative effects on the environment associated with the use of pesticides became known. At the same time, the use of chemicals became more restricted, unsatisfactory and costly. During the 1980s, research on EPN in the USA expanded rapidly, with the backing of government and industry (Adams and Nguyen 2002). Such a trend expanded to several other countries, leading to the current situation where these bioinsecticides are used worldwide.

The first record of EPN in SA was made by Harington (1953). He found such nematodes in larval, pupal and adult stages of the maize beetle *Heteronychus arator* that was collected from a maize field near Grahamstown (Eastern Cape Province). In 1997 *Parasitorhabditis obtusa* (Fuchs, 1915) Chitwood and Chitwood, 1950, was found for the first time in SA in wood containing insect galleries, cut from the stem of a dead *Pinus radiata* tree, Tokai Plantation, Western Cape Province (Swart and Heyns 1997). Hitherto the species has been recorded from Europe and North America where it is associated with Ips beetles (Andrássy 1984).

Research on EPN in SA is still in its infancy. Up until now it has included research on the use of these nematodes to control wood wasps (*Sirex noctilio*), codling moth (*Cydia pomonella*), false codling moth (*Thaumatotibia leucotreta*), the banded fruit weevil (*Phlyctinus callosus*), mealybugs (*Planococcus ficus*, *Planococcus citri* and *Pseudococcus viburni*), fruit flies (*Ceratitis capitata* and *Ceratitis rosa*) as well as the sugarcane stalk borer (*Eldana saccharina*).

22.2 Taxonomy

In recent years, several developments have impacted on the systematics of EPN. These include:

(i) The standardisation of criteria for species description and suggestions regarding name changes (Hominick et al. 1997)
(ii) A proposal regarding the phylogenetic relationships within the phylum Nematoda based on molecular evidence (Blaxter et al. 1998)
(iii) A proposed phylogenetic species concept (Adams et al. 1998)
(iv) A proposal for a new classification of Nematoda (De Ley and Blaxter 2002)

As a result of such developments, the systematics of EPN is regarded as being relatively stable (Hominick 2002). Currently, the family Steinernematidae contains two genera, viz. *Steinernema,* with currently more than 100 species, and *Neosteinernema* with only one species, *Neosteinernema longicurvicauda* Nguyen and Smart, 1994, a parasite of termites (Nguyen and Smart 1994). In 1976, Poinar erected the family Heterorhabditidae when he described the genus and species *Heterorhabditis bacteriophora* Poinar 1976. The family contains one genus, *Heterorhabditis*, with currently more than 20 species (Nguyen 2007). Both the families Steinernematidae and Heterorhabditidae belong to the order Rhabditida, and although they do not share a common ancestry, both have evolved a similar life

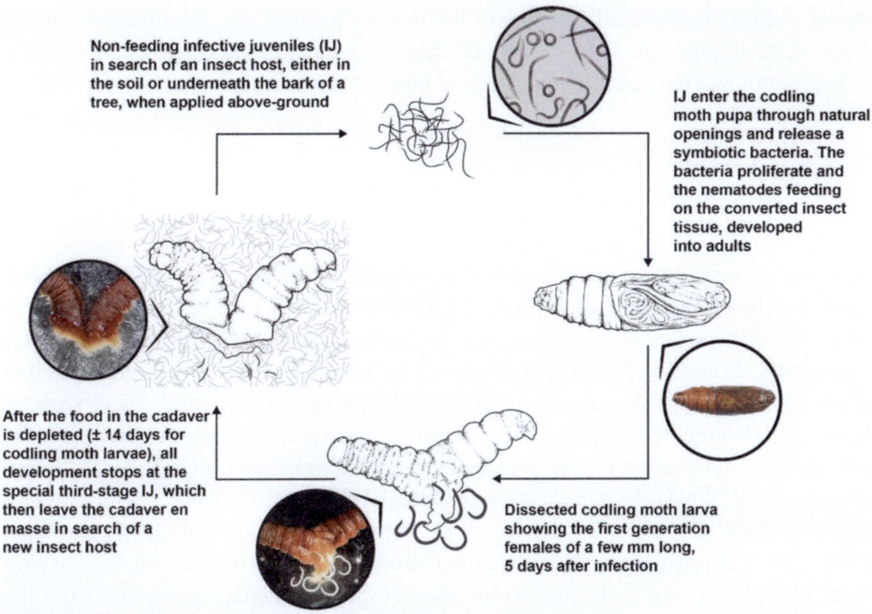

Fig. 22.1 A simplified illustration of the life cycle of entomopathogenic nematodes (Photos: Antoinette Malan, University of Stellenbosch, Stellenbosch, South Africa; Illustration: Hannes Visagie, North-West University, South Africa)

cycle (Blaxter et al. 1998). The number of newly discovered and subsequently described species and isolates of both genera is increasing since more elaborate and directed surveys are conducted worldwide.

The IJ stage, also referred to as the invasive stage or dauer larva, is the only free-living stage of EPN that occurs in the soil (Forst and Clarke 2002) (Fig. 22.1). This life stage of the nematode is a specially adapted third-stage development, during which the mouth of the IJ is closed. These juveniles locate their host, either by following cues, such as carbon dioxide, excreted by the host, or by attaching to a passing insect (Griffin et al. 2005). The host is entered by way of natural openings, viz. the mouth, the spiracles or the anus. *Heterorhabditis* is also equipped with a dorsal tooth, which can facilitate entry into the host by abrasion of thin parts of the cuticle (Kaya and Gaugler 1993). After the IJ enters the haemocoel of the insect host, the symbiotic bacteria within the intestine are released, either by way of the mouth or anus. The bacteria multiply exponentially, turning the inside of the host into a bacterial paste. The IJ transforms into a feeding stage, using the bacterial paste as a food source. The mutualistic bacteria of the genus *Steinernema* belong to *Xenorhabdus* and those of *Heterorhabditis* to *Photorhabdus* (Forst et al. 1997). The host is usually killed within 24–48 h after infection. Depending on the size of the host, the nematode completes one generation or more. In the case of *Steinernema,* the first generation develops into males and females, whilst in the case of *Heterorhabditis* this generation turns into hermaphrodites (Hazir et al. 2003; Griffin et al. 2005). For both genera, the second and following generations produce males and females. As soon as the food in the insect cadaver is depleted, it triggers the development of the pre-infective and IJ stage, retaining the cuticle of the second stage as a sheath. The IJ leaves the cadaver in large numbers in search of a new insect host to infect (Adams and Nguyen 2002).

22.3 Biogeography

Reviews of surveys for the worldwide occurrence of EPN, up to 2002, have been published by Hominick et al. (1996), (2002). Before 1990, the identification of EPN species was regarded as confusing, due to the lack of suitable keys and molecular techniques. Poinar's (1990) contribution was generally regarded as the baseline study for the geographical origins of EPN (Hominick 2002). Since then, surveys have been conducted in a number of countries, mostly in Europe and the USA. On the African continent, surveys have been carried out in Egypt (Atwa 2004), Ethiopia (Mekete et al. 2005), Kenya (Waturu and Reid 1997; Waturu 1998; Mwaniki et al. 2008) and SA (Spaull 1991; Malan et al. 2006; Molotsane et al. 2007; De Waal 2008, 2011; Hatting et al. 2009). All nematodes isolated from recent SA surveys were identified using molecular techniques, and the sequences generated from the different species and isolates are available in GenBank.

The first survey of EPN in SA was conducted in the KwaZulu-Natal Province. The purpose of the survey was to find virulent isolates for the control of the sugarcane stalk borer. From a total of 68 sites, seven isolates of *Heterorhabditis* (at 10 % of the sites) and 15 isolates of *Steinernema* (at 22 % of the sites) were recovered (Spaull 1991). At three of the sites, isolates of the two genera occurred simultaneously. In a previous study, a virulent isolate of *Heterorhabditis* and three *Steinernema* isolates had been collected on an ad hoc basis from the Province (Spaull 1988). Recently *Steinernema sacchari* Nthenga, Knoetze, Berry, Tiedt and Malan, 2014, has been described from a sugarcane field in Gingindlovu, KwaZulu-Natal (Nthenga et al. 2014); *Steinernema fabii* Abate, Malan, Tiedt, Wingfield, Slippers and Hurley 2016, from a black wattle plantation in Piet Retief, Mpumalanga Province (Abate et al. 2016); and *Steinernema nguyeni* Malan, Knoetze and Tiedt, 2016, from fynbos in the Western Cape Province (Malan et al. 2016).

During the period from 1993 to 1994, a total of 57 soil samples were collected from deciduous fruit orchards in the Western Cape Province. In 28 % (16) of the samples, EPN were present (Basson 1993). All the nematode isolates were identified as heterorhabditids. Specimens sent to France were identified as *H. bacteriophora,* using species-specific satellite deoxyribonucleic acid (DNA) as diagnostic probes. This was the first EPN to be identified to species level from SA (Grenier et al. 1996).

At the Agriculture Research Council-Small Grain Institute (ARC-SGI), near Bethlehem (Free State Province), 1,056 samples were processed from 2003 to 2005, incorporating collections from seven geographic regions throughout the country (Hatting et al. 2009). Of these samples that were found to be positive for EPN, 44 (4 %) contained steinernematids and 35 (3 %) heterorhabditids. Species recovered included *H. bacteriophora*; *Steinernema innovationi* Çimen, Lee, Hatting, Hazir and Stock, 2014; *Steinernema* sp.; and *Steinernema tophus* Çimen,

Lee, Hatting, Hazir and Stock, 2014, (Çimen et al. 2014a, b). A *Steinernema* sp. was isolated as a dual infection from the cadaver of a white grub (Scarabaeidae). In addition, *H. bacteriophora* was also recovered from the black vine weevil, *Otiorhynchus sulcatus*. Additional specimens have since been added to the collection (Hatting et al. 2009).

During 2004 and 2005, a total of 498 randomly collected soil samples were taken in the south-western parts of SA (Malan et al. 2006). Entomopathogenic nematodes were isolated from 36 of the samples (7 %). *Heterorhabditis* was found to be the dominant genus, whilst *Steinernema* was rare. The most common species was *H. bacteriophora*. The isolation of *Heterorhabditis zealandica* Poinar, 1990, in this survey represented a new record for SA (Malan et al. 2006). Two new species, *Steinernema khoisanae* Nguyen, Malan and Gozel, 2006, and *Heterorhabditis safricana* Malan, Nguyen, De Waal and Tied, 2008, were identified during the survey (Nguyen et al. 2006; Malan et al. 2008). In another survey, conducted during 2007 and 2008, De Waal (2008) collected soil samples from several different habitats and regions in the Western Cape Province. Nematodes were recovered from 20 of the 200 soil samples. Eight of the samples were found to include *Steinernema* spp., which were identified as three isolates of *S. khoisanae* and five undescribed *Steinernema* spp. The other 12 samples contained *Heterorhabditis* spp., comprising six isolates of *H. bacteriophora*, five *H. zealandica* and one *H. safricana.* In all the surveys described, the *Galleria* baiting technique was used for trapping nematodes. The main objective of these surveys was to obtain EPN for the biological control of codling moth, which is a serious pest of deciduous fruit in SA. Samples were collected from both cultivated and non-cultivated habitats, including deciduous fruit orchards, vineyards and natural vegetation (Table 22.1).

Similarly, work on the biological control of the false codling moth, a citrus pest of considerable economic importance, started with a targeted survey to determine the occurrence and distribution of EPN in citrus orchards (Malan et al. 2011). From a total of 202 samples, 35 (17 %) contained nematodes. Of the nematodes recovered, four isolates (11 %) were *Steinernema* spp. and 31 isolates represented *Heterorhabditis* spp. The *Steinernema* spp. were *S. khoisanae* and *Steinernema yirgalemense* Nguyen, Tesfamariam, Gozel, Gaugler and Adams, 2004, and two new species, *Steinernema citrae* Stokwe, Malan, Nguyen, Knoetze and Tiedt, 2011, (Stokwe et al. 2011) and *Steinernema jeffreyense* Malan, Knoetze and Tiedt, 2015, (Malan et al. 2015). *Steinernema citrae* was found in the Piketberg and Porterville regions of the Western Cape Province. The *Heterorhabditis* species included *H. bacteriophora*, *H. zealandica* and an unknown species from the Mbombela area (Mpumalanga Province). It was later identified as *Heterorhabditis noenieputensis* Malan, Knoetze and Tiedt, 2014, a new species found from a garden fig in the settlement, Noenieput (Northern Cape Province), close to the Namibian border of SA (Malan et al. 2014). In all the aforementioned surveys, *H. bacteriophora* was the dominant EPN species (except in KwaZulu-Natal). *Heterorhabditis zealandica* had only once been reported as occurring in natural vegetation in the Eastern Cape Province, with the conclusion being drawn that it was endemic to SA (Malan et al. 2011). In the survey conducted by Malan et al. (2011), *H. zealandica* was obtained from two citrus orchards in the Western Cape Province, as well as in one orchard in

Table 22.1 Occurrence of *Heterorhabditis* and *Steinernema* spp. in different provinces and habitats throughout South Africa

Species	Authority	Province	Habitat	Reference
H. bacteriophora	Poinar, 1976	Eastern Cape, KwaZulu-Natal, Mpumalanga, Western Cape	Various habitats	Grenier et al. (1996); Malan et al. (2006, 2011); De Waal (2008); Hatting et al. (2009)
H. noenieputensis[a]	Malan, Knoetze & Tiedt, 2012	Eastern Cape, Northern Cape	Citrus, fig	Malan et al. (2014)
H. safricana[a]	Malan, Nguyen, De Waal & Tiedt, 2008	Western Cape	Natural vegetation, peach	Malan et al. (2006); De Waal (2008)
H. zealandica	Poinar, 1990	Eastern Cape, Northern Cape, North West, Mpumalanga, Western Cape	Citrus, natural vegetation	Malan et al. (2006, 2011); Molotsane et al. (2007); De Waal (2008)
S. citrae[a]	Stokwe, Malan, Nguyen, Knoetze & Tiedt, 2011	Western Cape	Citrus	Stokwe et al. (2011)
S. fabii[a]	Abate, Malan, Tiedt, Wingfield, Slippers & Hurley, 2016	Mpumalanga	Black wattle	Abate et al. (2016)
S. innovationi[a]	Çimen, Lee, Hatting, Hazir & Stock, 2014	Free State	Grain field	Çimen et al. (2014a)
S. jeffreyense[a]	Malan, Knoetze, Tiedt 2015	Eastern Cape	Guava	Malan et al. (2015)
S. khoisanae[a]	Nguyen, Malan & Gozel, 2006	Western Cape	Apple, citrus, grapevine, grass, grassland, natural vegetation, rooibos	Malan et al. (2006, 2011); Molotsane et al. (2007); De Waal (2008); Hatting et al. (2009)
S. nguyeni[a]	Malan, Knoetze & Tiedt, 2016	Western Cape	Fynbos	Malan et al. (2016)
S. sacchari[a]	Ntengha, Knoetze, Berry & Tiedt, 2014	KwaZulu-Natal	Sugarcane	Nthenga et al. (2014)
S. tophus[a]	Çimen, Lee, Hatting, Hazir & Stock, 2014	Western Cape	Grapevine	Çimen et al. (2014b)
S. yirgalemense	Nguyen, Tesfamariam, Gozel, Gaugler & Adams, 2004	Mpumalanga	Citrus	Malan et al. (2011)

[a]Type specimen

Mpumalanga. It was also reported from Gauteng and the North West Provinces (Molotsane et al. 2007). De Waal (2008) obtained several isolates of *H. zealandica* from disturbed and non-disturbed soil in the Western Cape Province, indicating the wider occurrence of this species throughout SA.

The finding of *S. yirgalemense* in a survey of citrus orchards was the first such report for SA and just the third report for the African continent (Malan et al. 2011). It was described by Nguyen et al. (2004) from specimens collected near the town of Yirgalem in Ethiopia. It was also reported from the Central Rift Valley region of Kenya (Mwaniki et al. 2008). *Steinernema yirgalemense* belongs to the *bicornutum*-group, consisting of 11 species, of which the exsheathed IJ has two horn-like structures on the cephalic region. This nematode has not yet been reported from outside the African continent. In Ethiopia, it was the dominant species and occurred in 6.3 % of the samples collected, whilst only two isolates of *H. bacteriophora* were recorded from 0.7 % of the samples (Mekete et al. 2005).

22.4 Bacterial Symbionts

Entomopathogenic nematodes share a mutualistic symbiotic relationship with bacteria belonging to the family Entobacteriaceae. Each species of nematode is specifically associated with a certain species or subspecies of bacterium. Whereas *Heterorhabditis* is associated with species of *Photorhabdus*, *Steinernema* is associated with species of *Xenorhabdus* (Boermare and Akhurst 1988; Boermare 2002). Although both partners benefit from such an association, the bacteria are regarded as playing the more important role, being mainly responsible for the rapid death of the insect, normally within the space of 48 h. Whereas *Xenorhabdus* occurs generally in a special intestinal vesicle of *Steinernema* (Bird and Akhurst 1983), *Photorhabdus* is mainly contained in the anterior part of the intestine in *Heterorhabditis* (Ciche and Ensign 2003). The bacteria play an important role in the commercial production of EPN, during which process they can undergo a phenotypic or phase variation. The original bacterial isolate, which was obtained from nematodes found in the wild, has been termed the Phase I variant. After repeated *in vitro* subculturing, such bacteria can change into a Phase II variant, which subsequently can display a range of different phenotypic characters (Boermare and Akhurst 1988). Currently, some 20 species have been proposed for the genus *Xenorhabdus*, with three for *Photorhabdus,* but many bacterial symbionts from recently described nematode species still require characterisation (Koppenhöfer 2007). Symbiotic bacteria associated with EPN that occur in SA are referred to in Box 22.1.

Box 22.1: Status of the Symbiotic Bacteria Associated with Entomopathogenic Nematodes in South Africa

Not all symbiotic bacteria associated with locally occurring entomopathogenic nematodes have been described (Ferreira and Malan 2014b):

(i) Two *Xenorhabdus* spp., *Xenorhabdus khoisanae* associated with *Steinernema khoisanae* and *Xenorhabdus indica* associated with *Steinernema yirgalemense* have been identified.
(ii) *Xenorhabdus khoisanae* was described as a new species (Ferreira et al. 2013b), whilst *X. indica* was a new nematode bacteria association (Ferreira et al. 2016b).
(iii) Two new *Photorhabdus* spp. were identified, namely, *Photorhabdus luminescence* subsp. *noenieputensis* in association with *Heterorhabditis noenieputensis* (Ferreira et al. 2013a) and *Photorhabdus heterorhabditis* associated with *Heterorhabditis zealandica* (Ferreira et al. 2014a). The latter association is interesting, since the *H. zealandica* described from New Zealand is associated with *Photorhabdus temperata*.

22.5 Interaction of Entomopathogenic Nematodes with Plant-Parasitic Nematodes

An interaction between EPN and plant-parasitic nematode populations in the soil was originally documented in 1986 in two separate and unrelated studies (Bird and Bird 1986; Ishibashi and Kondo 1986). However, no obvious or direct ecological links exist between the two different nematode groups.

Laboratory (Perez and Lewis 2002), glasshouse (Perez and Lewis 2004) and field trials (Grewal et al. 1997; Somasekhar et al. 2002) were conducted to determine whether the presence of EPN has the capacity to reduce plant-parasitic nematode populations. Several different trials that were conducted with the same plant-parasitic nematode species and at the same locality produced variable results. Only in the case of turf grass using *Steinernema riobrave* Cabanillas, Poinar and Raulston, 1994 was some level of consistent interaction noted (Grewal et al. 1997). Three different mechanisms were proposed to explain the reduction in numbers of plant-parasitic nematodes. More than one factor might have the capacity to act together (Lewis and Grewal 2005). The first explanation for such an interaction was suggested by Bird and Bird (1986), who deduced that the accumulation of a layer of carbon dioxide along the roots of plants serves to attract EPN to the plants, subsequently driving the plant-parasitic nematodes away. A second possible explanation for such an interaction is that when exceptionally high numbers (2.5 billion ha^{-1}) of EPN are applied to the soil, the result is the build-up of antagonists and the eventual development of a suppressive soil (Ishibashi and Kondo 1986; Ishibashi and Choi 1991). Lastly, Grewal et al. (1999) proposed that allelochemicals, which were found to have been produced by EPN and their associated symbiotic bacteria, served to intoxicate the plant-parasitic nematodes and as a result cause a decline in their numbers. For all three theories, both positive and negative results have been documented relating to each of the mechanisms concerned.

Interestingly, two previous studies showed that, although the number of plant-parasitic nematodes was found to decrease after an application of EPN, no detectable effect was found on the free-living nematode populations concerned (Jagdale et al. 2002; Somasekhar et al. 2002). Such a finding highlights how little is known of the impact of EPN on soil ecology, in general. However, most studies showed at least some level of interaction between the two groups of nematodes. The question as to whether EPN can lower plant-parasitic nematode populations to such a degree as to make their use viable for producers remains unanswered.

22.6 Regulation in South Africa

Two EPN species, viz. *Steinernema carpocapsae* (Weiser 1955) Wouts, Mracek, Gerdin and Bedding, 1982, and *Steinernema feltiae* (Filipjev) Wouts, Mráček, Gerdin and Bedding, 1882, are produced commercially, mainly in North America and Western Europe. Neither species has yet been reported as occurring in SA. In many different countries, such as Austria, Germany, the United Kingdom (UK) and the USA, the use of EPN as biopesticides has been exempted from registration (Ehlers 2005). However, the introduction of nonindigenous species is strictly controlled. Nevertheless, governing regulations still vary between various European countries. In 1987, the regulatory actions regarding the import of exotic species were reviewed for the USA. Currently, complex regulatory procedures and safeguards are in place for the introduction of exotic EPN into the USA (Rizvi et al. 1996).

According to the amended Act 18 of 1989 (South African Agricultural Pests Act No. 36 of 1947), the introduction of exotic animals such as non-endemic EPN is only allowed under permit, which has to be accompanied by a full impact study. In 1993, such a permit was issued to allow for the importation of *Heterorhabditis megidis* Poinar, Jackson and Klein, 1987 and *S. carpocapsae* from Biosys in California (USA), with another being issued in 1995 to allow for the importation of both species from Koppert Biological Systems, for research into the control of the banded fruit weevil (Basson 1993). The nematodes were applied to apple trees in an orchard in the Elgin area of the Western Cape Province under strict quarantine measures (digging of trenches around treatment blocks and access control indicated with chevron banding). Unfortunately the exact location of application is unknown; however, should these species be isolated in the Elgin area, their endemicity should be scrutinised. In the 2000s the parasitic nematode *Beddingia siricidicola* (Bedding 1968) Blinova and Korenchenko, 1986, was introduced in South Africa against the wood wasp *S. noctilio* (Tribe 2006). It was sourced from Australia.

During 2005, a permit was obtained to import IJ of *Steinernema arenarium* (Artyukhovsky, 1967) Wouts, Mráček, Gerdin and Bedding, 1982, (Russian strain); *Steinernema diaprepesi* Nguyen and Duncan, 2002, (Polk strain); and *Steinernema glaseri* (Steiner, 1929) Wouts, Mracek, Gerdin and Bedding, 1982, (NJ strain) from Florida in the USA for cross-breeding tests directed at the description of *S. khoisanae*. In 2008, *S. feltiae* were imported from the Czech Republic for a cross-breeding test for the description of *S. citrae* (Stokwe et al. 2011). All cross-breeding tests were conducted at the Plant Quarantine Station (Stellenbosch, Western Cape Province) with the imported nematodes being destroyed after use by quarantine officers.

Currently, permits have been issued for the import of *H. bacteriophora* and *S. feltiae* from e-nema in Germany and BASF in England. *Heterorhabditis bacteriophora* is one of the most common EPNs found in agricultural soil in SA, whilst

S. feltiae is an exotic nematode for SA and has only once being reported as naturally occurring on the African continent, in Algeria (Tarasco et al. 2009).

22.7 Research in South Africa

22.7.1 Control of Sugarcane Stalk Borer

A chronological timeline of EPN research conducted in SA is given in Box 22.2, whilst extensive information on such research done locally is published in Malan and Hatting (2015).

Box 22.2 A Timeline of Entomopathogenic Nematode (EPN) (Steinernematidae and Heterorhabditidae) Research Conducted in South Africa

Year	Nematode/bacteria	Target insect pest	Nematodes tested
1953	First mention of EPN from the maize beetle (*Heteronychus arator*) in the Eastern Cape Province	-	-
1988-1992	First survey and isolation of EPN from KwaZulu-Natal Province in sugarcane	Sugarcane borer (*Eldana saccharina*)	*Heterorhabditis* spp.; *Steinernema* spp.
1993-1994	-	Use of imported EPN against the banded fruit weevil (*Phlyctinus callosus*)	*H. megidis* (imported); *S. carpocapsae* (imported)
2006	First new *Steinernema* described from SA: *S. khoisanae*	-	-
2008	First new *Heterorhabditis* described from SA: *H: safricana*	-	-
2009	-	Mediterranean fruit fly (*Ceratitis capitata*); Natal fruit fly (*Ceratitis rosa*)	*H. bacteriophora, S. khoisanae, S. yirgalemense*
2010	-	Coding moth	*H. zealandica*
2011	*S. citrae*	Codling moth; false codling moth (*Thaumatotibia leucotreta*)	*H. bacteriophora, H. zealandica, S. citrae, S. jeffreyense, S. khoisanae, S. yirgalemense*
2012	-	Citrus mealybug (*Planococcus citri*)	*H. bacteriophora, H. safricana, H. zealandica, S. khoisanae, S. citrae, S. yirgalemense*
2013	First description of *Xenorhabdus* and *Photorhabdus* for SA: *P. luminescens* subsp. *Noenieputensis, X. khoisanae*	Codling moth; banded fruit weevil, vine mealybug (*Planococcus ficus*), citrus mealybug	*H. bacteriophora, H. noenieputensis, H. zealandica, S. citrae, S. khoisanae, S. yirgalemense*
2014	*H. noenieputensis, S. innovation, S. sacchari, P. zealandica*	Citrus mealybug, mealybug ladybird (*Cryptolaemus montrouzieri*)	*H. zealandica, S. yirgalemense*
2015	*S. jeffreyense, S. tophus, P. zealandica*	Obscure (*P. viburni*), citrus and vine mealybug	*H. bacteriophora, H. zealandica, S. yirgalemense, S. feltiae* (imported)
2016	*S. fabii, S. khuyeni, X. indica*	Coding moth	*H. bacteriophora, H. zealandica,S. feltiae* (imported), *S. yirgalemense*

Apart from the observations made by Harington (1953) on the EPN that he found in black maize beetle near Grahamstown, no further mention of these nematodes was made until 35 years later when Spaull (1988) reported on a study aimed at the control of the sugarcane stalk borer, *E. saccharina*. The work was initiated after obtaining a *Heterorhabditis* isolate, designated Hsp1, and three *Steinernema* isolates, collected from the KwaZulu-Natal Province. The nematodes used in the study were reared on the larvae of the stalk borer. The study showed the borer larvae to be more susceptible to *Heterorhabditis* than to the *Steinernema* species and that the pupae were less susceptible than the larvae. Although additional isolates of both genera were collected, none was as pathogenic as was the *Heterorhabditis* isolate, Hsp1. In a series of 21 field tests, Spaull (1988, 1990, 1991, 1992) evaluated the performance of this isolate against the stalk borer. Better control was obtained when nematodes were sprayed onto the cane in the late afternoon, rather than when they were applied before sunrise or at midday or when the cane was pre-wetted. Some other factors that served to improve control included the addition of a methylcellulose water thickener to the nematode suspension and applying the IJ in high concentrations in large volumes of water. The need for large concentrations is reasonable, in view of the observation that only about 12 % of the nematodes survived on the stalks or leaf sheaths, 30 min after the spraying had taken place. It was concluded that the cost of rearing sufficient nematodes to achieve a high level of control in standing cane would far exceed the benefit to be gained from such an exercise (Spaull 1992).

22.7.2 Control of Banded Fruit Weevil

During 1993, imported *S. carpocapsae* and *H. megidis* were used in laboratory and field trials under quarantine conditions in apple orchards in the Elgin area of the Western Cape Province, for control of the banded fruit weevil (Basson 1993). In 1995, *S. carpocapsae* and *H. megidis* were imported for a second time and used to determine the vertical migration of nematodes in two orchards, one with a clay soil and the other with a sandy soil. The soil stages of the banded fruit weevil, including the larvae and pupae, can be up to 30 cm deep into the soil. It was found that the nematodes penetrated the soil for at least 10 cm and that *H. megidis* survived longer in the soil than did *S. carpocapsae*.

22.7.3 Control of Pests of Deciduous and Citrus Fruit

Research on EPN at Stellenbosch University started in 2004, in the form of surveys to obtain local isolates for the control of codling moth and other pests of deciduous fruit, grapevine (*Vitis vinifera*) and citrus (*Citrus* spp.). For codling moth and mealybugs, an aerial application approach was followed to control the above-ground insect stages with EPN. For false codling moth, the vine mealybug and the banded fruit weevil, the below-ground insect stages were targeted with a soil application. Initially, EPN isolates obtained through surveys were screened against target insects, followed by various bioassays under controlled conditions, and finally nematode isolates were tested in field trials. This research was followed with two industry-funded projects to initiate the *in vivo* and *in vitro* mass production and formulation of selected EPN species, such as *S. yirgalemense*, *H. bacteriophora* and *H. zealandica*. Results from these (Ferreira and Malan 2014a, 2016a; Van Zyl and Malan 2014a, b, 2015) led to a government-funded project in 2015 conducted by a private company to investigate the mass culture and formulation of promising endemic EPN species.

22.7.3.1 Codling Moth Research

Results from research on EPN in SA indicate that they have potential for incorporation in the current integrated pest management (IPM) programme against codling moth in apple and pear orchards. That is despite the fact that the environmental conditions in the Western Cape are not conducive to aerial application of the nematodes (Odendaal et al. 2015a, b). Different aspects relating to the efficacy of EPN against the codling moth have been investigated in SA (Fig. 22.2). These studies included the control of diapausing overwintering codling moth larvae using six endemic EPN species. The results indicated that *S. yirgalemense* and *S. jeffreyense* were the most promising (De Waal et al. 2011a; Odendaal et al. 2015b, 2016). Key factors contributing to the success in the control of codling moth in wooden fruit bins were investigated; pre-wetting and the maintenance of high humidity for at least 3 days post-treatment were found to be essential (De Waal et al. 2010). Studies showed that the addition of a superabsorbent polymer formulation to nematode suspensions assisted in maintaining adequate moisture levels required for nematode survival and efficacy (De Waal et al. 2013).

Field trials to evaluate the effect of mulches, such as pine chips, wheat straw, pine wood shavings, black wood and apple wood chips, on codling moth control using EPN were also conducted. High codling moth mortality was found in pine wood, with high humidity maintained for at least 3 days. An interesting result from this study was that IJ was able to move 10 cm from the soil upwards into a mulch to infect codling moth larvae (De Waal et al. 2011b).

22.7.3.2 False Codling Moth Research

False codling moth is endemic to SA where it is a key pest of citrus. No previous research on the use of EPN to control the soil-borne stages of false codling moth has been undertaken (Fig. 22.3). In contrast to codling moth, most of the life stages of false codling moth are found in the soil, viz. the last instar larvae, pre-pupae, pupae and the emerging moth. As soil is the natural habitat for EPN, these organisms are ideally suited to control this insect. Results from local studies showed all endemic EPN isolates can control the soil stages of false codling moth, with the added

Fig. 22.2 A codling moth larva infected with entomopathogenic nematodes (Antoinette Malan, University of Stellenbosch, Stellenbosch, South Africa)

potential of persistence in the soil environment. An important finding in these studies is the susceptibility of the emerging moth to EPN infection and the potential of long distance dispersal of the nematodes with the infected adults (Malan et al. 2011).

22.7.3.3 Mealybug Research

In general mealybugs are difficult to control because of their cryptic lifestyle, hiding in crevices in the bark and sometimes below ground on the roots. They are covered by a hydrophobic waxy secretion making chemical control difficult and are prone to the rapid development of resistance to chemicals. An urgent need exists for improved control measures against mealybugs, which potentially includes the use of EPN (Le Vieux and Malan 2013a). Laboratory bioassays were conducted to determine the effectiveness of EPN isolates to control mealybugs. *Heterorhabditis zealandica* was found to be the most effective against *P. viburni*, whilst *S. yirgalemense* was most effective against *P. citri* and *P. ficus* (De Waal et al. 2007; Van Niekerk and Malan 2013; Le Vieux and Malan 2013a) (Fig. 22.4). Laboratory bioassays showed that the nematodes were able to reproduce successfully in the three mealybug species (Stokwe 2009; Van Niekerk and Malan 2012; Le Vieux and Malan 2013b; Stokwe and Malan 2016). When apples naturally infested with *P. viburni* were treated with EPN in the laboratory, they were able to locate and infect mealybugs on the surface, in the calyx and in the core of the fruits (Stokwe 2009).

To determine the potential of EPN to control *P. citri*, it was shown that *S. yirgalemense* was able to locate and infect *P. citri* more quickly than *H. zealandica* and was able to infect mealybugs in an exposure time as short as 30 min (Van Niekerk and Malan 2012). The addition of adjuvants to nematode application suspensions was found significantly to increase mortality. Adding both Nu–Film–P® and Zeba® retarded sedimentation significantly and increased the average number of nematodes deposited on 2-cm^2 leaf discs by ten (Van Niekerk and Malan 2014a).

Fig. 22.3 A false codling moth larva infected with *Heterorhabditis bacteriophora,* in a partially opened cocoon (Antoinette Malan, University of Stellenbosch, Stellenbosch, South Africa)

The compatibility of EPN with larvae of the predatory beetle, *Cryptolaemus montrouzieri* (Coleoptera, Coccinellidae), a commercially produced biological control agent of mealybugs, was investigated. Results showed the insect to be highly susceptible to *H. zealandica* and *S. yirgalemense*. Adult beetles were found to be twice as susceptible to *S. yirgalemense* as they were to *H. zealandica* (Van Niekerk and Malan 2014b). In an IPM system where these beetle larvae are used against the citrus mealybug, nematodes should not be applied in orchards where these beetle larvae are released.

It was also shown that EPN could be used within an IPM programme to control *P. ficus* on grapevine roots. In a field trial, 50 % mortality of *P. ficus* females, buried 15 cm into the soil, was achieved after 48 h. The persistence of *S. yirgalemense,* measured using codling moth larval mortality, was found to be zero in one vineyard, whereas in another it was 70 %, 12 weeks after application (Le Vieux and Malan 2015). Olfactometry tests indicated a significant difference between the numbers of *S. yirgalemense* that were attracted to damaged grapevine roots and to *P. ficus*, indicating the active movement of the IJ and the attractiveness of organic compounds produced by the roots. These studies showed that *S. yirgalemense* has potential as a biological control agent for the control of *P. ficus* soil populations (Le Vieux and Malan 2015), but more research is needed.

22.7.3.4 Banded Fruit Weevil Research

In 2008, further trials were conducted to establish whether EPN could be used to control the banded fruit weevil, *P. callosus* (Fig. 22.5a, b). The screening of several nematode isolates for their potential to control the soil stage of the weevil pest showed that extensive variation of infectivity existed. The weevil was less susceptible to the isolates of *H. zealandica* that were tested against codling moth and false codling moth. Doubling the concentration and the amount of time was needed to produce comparable results. Banded fruit weevil larvae mainly occur in the top 10 cm of orchard soil. Excellent results were obtained with bioassays in sandy loam soil and sand. Mortality of weevil larvae in soil bioassays in the laboratory was more

Fig. 22.4 (**a**, **b**) An orange-coloured noninfected mealybug (*Pseudococcus viburni*) (**a**) compared to a reddish-coloured infected mealybug with entomopathogenic nematodes (*Heterorhabditis bacteriophora*) (**b**) (Antoinette Malan, University of Stellenbosch, Stellenbosch, South Africa)

than 75 %, which shows that the EPNs were able to move downwards 15 cm through the soil within 7 days after inoculation (Ferreira 2012; Ferreira and Malan 2014a).

22.7.3.5 Fruit Fly Research

Laboratory bioassays were undertaken to assess the susceptibility of two fruit fly species, *C. capitata* and *C. rosa,* to *H. bacteriophora, H. zealandica* and *S. khoisanae* (Malan and Manrakhan 2009). The different life stages of the two fruit fly species were susceptible to nematode infection (Fig. 22.6), except that no infection was recorded for fully developed pupae. Pupating larvae of *C. capitata* generally proved to be more susceptible to nematode infection than did those of *C. rosa*. A significantly greater number of larvae of *C. capitata* were infected by *H. bacteriophora*. For *C. rosa,* the highest infection of larvae was obtained with *H. zealandica*. The adults of both species were highly susceptible to *S. khoisanae* (Malan and Manrakhan 2009).

22.8 Conclusions

Entomopathogenic nematodes are produced and sold in several countries in the northern hemisphere. Researchers are investigating the use of these nematodes as biological control agents for the management of several insect pests. In SA,

Fig. 22.5 (**a**, **b**) A banded fruit weevil (*Phlyctinus callosus*) larva infected with *Heterorhabditis bacteriophora* (**a**) compared to noninfected larvae (**b**) (Tiarin Ferreira, University of Stellenbosch, Stellenbosch, South Africa)

Fig. 22.6 A fruit fly larva infected with entomopathogenic nematodes (Antoinette Malan, University of Stellenbosch, Stellenbosch, South Africa)

the opportunities for using EPN against certain pest species are promising. Even though species such as *H. bacteriophora* and *H. zealandica* have been mass cultured by Plant Health Products Pty (Ltd), mainly for research purposes, they have not yet been produced and sold in SA. Current research is aimed at controlling the most important insect pests as well as the commercial production of selected species. An important, though often neglected, aspect that is currently being addressed is that of research into practical methods for the successful application of EPN.

Despite progress having been made in SA in the field of entomopathogenic nematology, many future challenges await. These include the isolation of native nematode species with commercial potential, as has been undertaken in the case of *S. carpocapsae* and *S. feltiae* in other countries. The practical application of EPN on a commercial scale and the investigation of mass rearing and formulation methodology best suited to local isolates also require attention.

In SA, a need to clarify government registration requirements for the commercialisation of EPN formulations also warrants investigation. The general awareness levels of consultants and producers regarding the use of EPN also need to bc increased through technology transfer opportunities. Even though the use of EPN is likely to cost more than the use of conventional insecticides, the greater positive effect on the environment and the human population in general cannot be measured purely in monetary terms.

References

Abate BA, Malan AP, Tiedt LR et al (2016) *Steinernema fabii* n. sp. (Rhabditidae: Steinernematidae), a new species of entomopathogenic nematodes from South Africa. Nematol 18:235–255

Adams BJ, Nguyen KB (2002) Taxonomy and systematics. In: Gaugler R (ed) Entomopathogenic nematology. CAB International, Wallingford, pp 1–33

Adams BJ, Burnell AM, Powers TO (1998) A phylogenetic analysis of *Heterorhabditis* (Nemata: Rhabditidae) based on internal transcribed spacer 1 DNA sequence data. J Nematol 30:22–39

Andrássy I (1984) Klasse nematoda ordnungen monhysterida, desmoscolecida, araeolaimida, chromadorida, rhabditida. Gustav Fischer Verlag, Stuttgart, pp 261–270

Atwa AA (2004) Isolation and identification of entomopathogenic nematodes in Egypt. Int J Nematol 14:40–43

Basson S (1993) Biological control of *Phlyctinus callosus* by entomogenous nematodes. Proceedings of the 11th Symposium of the Nematological Society of Southern Africa, Rustenburg. Phytophylactica 25:303

Bird AF, Akhurst RJ (1983) The nature of the intestinal vesicle in nematodes of the family Steinernematidae. Int J Parasitol 13:599–606

Bird AF, Bird J (1986) Observations on the use of insect parasitic nematodes as a means of biological control of root-knot nematodes. Int J Parasitol 16:511–516

Blaxter ML, Garey JR, Liu LX et al (1998) A molecular evolutionary framework for the phylum Nematoda. Nature 392:71–75

Boermare NE (2002) Biology, taxonomy and systematics of *Photorhabdus* and *Xenorhabdus*. In: Gaugler R (ed) Entomopathogenic nematology. CAB International, Wallingford, pp 35–56

Boermare NE, Akhurst RJ (1988) Biochemical and physiological characterization of colony form variants in *Xenorhabdus* spp. (Enterobacteriaceae). J Gen Microbiol 134:751–761

Bovien P (1937) Some types of association between nematodes. Videnskabelige Meddelelser Dansk Naturhistorisk Forening 101:1–114

Ciche TA, Ensign JC (2003) For the insect pathogen *Photorhabdus luminescens*, which end of a nematode is out? Appl Environ Microbiol 69:1890–1897

Çimen HC, Lee M–M, Hatting J et al (2014a) *Steinernema innovationi* n. sp. (Panagrolaimomorpha: Steinernematidae), a new entomopathogenic nematode species from South Africa. J Helminthol. doi:10.1017/S0022149X14000182

Çimen HC, Lee M–M, Hatting J (2014b) *Steinernema tophus* sp. n. (Nematoda: Steinernematidae), a new entomopathogenic nematode from South Africa. Zootaxa 3821:337–353

De Ley P, Blaxter M (2002) Systematic position and phylogeny. In: Lee DL (ed) The biology of nematodes. Taylor and Francis, New York, pp 1–30

De Waal JY (2008) Entomopathogenic nematodes (Rhabditida: Steinernematidae and Heterorhabditidae) for the control of codling moth, *Cydia pomonella* (L.) under South African conditions. MSc Agric Thesis, Department of Conservation Ecology and Entomology/ University of Stellenbosch, Stellenbosch

De Waal JY (2011) Biological and ecological factors contributing to the successful use of entomopathogenic nematodes (Rhabditida: Steinernematidae and Heterorhabditidae) for the control of codling moth, *Cydia pomonella* (L.) (Lepidoptera: Tortricidae) under South African conditions. PhD Agric Dissertation, Department of Conservation Ecology and Entomology/University of Stellenbosch, Stellenbosch

De Waal JY, Wolhlfarter M, Malan AP (2007) Laboratory bioassays for the differential susceptibility of *Planococcus ficus* and *Pseudococcus viburni* (Hemiptera: Pseudococcidae) to entomopathogenic nematodes (Rhabditida: Heterorhabditidae and Steinernematidae). S Afr Plant Soil 24:243–244

De Waal JY, Malan AP, Addison MF (2010) Key elements in the successful control of diapausing codling moth, *Cydia pomonella* (L.) (Lepidoptera: Tortricidae) in wooden fruit bins with a South African isolate of *Heterorhabditis zealandica* (Rhabditida: Heterorhabditidae) Biological. Sci Technol 20:489–502

De Waal JY, Malan AP, Addison MF (2011a) Efficacy of entomopathogenic nematodes (Rhabditida: Heterorhabditidae and Steinernematidae) against codling moth, *Cydia pomonella* (Lepidoptera: Tortricidae) in temperate regions. Biocontrol Sci Tech 21:1161–1176

De Waal JY, Malan AP, Addison MF (2011b) Evaluating mulches together with *Heterorhabditis zealandica* (Rhabditida: Heterorhabditidae) for the control of diapausing codling moth larvae, *Cydia pomonella* (L.) (Lepidoptera: Tortricidae). Biocontrol Sci Tech 21:255–270

De Waal JY, Addison MF, Malan AP (2013) A superabsorbent polymer formulation for improved efficacy of *Heterorhabditis zealandica* (Rhabditida: Heterorhabditidae) control of codling moth larvae, *Cydia pomonella* (L.) (Lepidoptera: Tortricidae). Biocontrol Sci Tech 23:62–78

Dutky SR, Hough WS (1955) Note on a parasitic nematode from codling moth larvae, *Carpocapsa pomonella* (Lepidoptera, Olethreutidae). Proc Entomol Soc Wash 57:244

Ehlers R-U (2005) Forum on safety and regulation. In: Grewal PS, Ehlers R-U, Shapiro-Ilan DI (eds) Nematodes as biocontrol agents. CAB International, Wallingford, pp 107–114

Ferreira T (2012). Characterisation of nematode symbiotic bacteria and the *in vitro* liquid culture of *Heterorhabditis zealandica* and *Steinernema yirgalemense*. PhD dissertation, Department of Conservation Ecology and Entomology, Stellenbosch University

Ferreira T, Malan AP (2014a) Potential of entomopathogenic nematodes for the control of the banded fruit weevil, *Phlyctinus callosus* (Schönherr) (Coleoptera: Curculionidae). J Helminthol 88:293–301

Ferreira T, Malan AP (2014b) *Xenorhabdus* and *Photorhabdus*, bacterial symbionts of the entomopathogenic nematodes *Steinernema* and *Heterorhabditis* and their *in vitro* liquid mass culture: a review. Afr Entomol 22:1–14

Ferreira T, Van Reenen C, Pagès S et al (2013a) Description of *Photorhabdus luminescens* subsp. *noenieputensis* subsp. nov., a symbiotic bacterium associated with a new *Heterorhabditis* species related to *Heterorhabditis indica*. Int J Syst Evol Microbiol 63:1853–1858

Ferreira T, Van Reenen CA, Endo A et al (2013b) Description of *Xenorhabdus khoisanae* sp. nov., a symbiont of the entomopathogenic nematode *Steinernema khoisanae*. Int J Syst Evol Microbiol 63:3220–3224

Ferreira T, Van Reenen CA, Endo A et al (2014a) Description of *Photorhabdus heterorhabditis* sp. nov., a symbiont of the entomopathogenic nematode *Heterorhabditis zealandica*. Int J Syst Evol Microbiol 64:1540–1545

Ferreira T, Addison MF, Malan AP (2014b) *In vitro* liquid culture of a South African isolate of *Heterorhabditis zealandica* for the control of insect pests. Afr Entomol 22:80–92

Ferreira T, Addison MF, Malan AP (2016a) Development and population dynamics of *Steinernema yirgalemense* and growth characteristics of its associated *Xenorhabdus* symbiont in liquid culture. J Helminthol 90:364–371

Ferreira T, Van Reenen CA, Tailliez P et al (2016b) First report of the symbiotic bacteria, *Xenorhabdus indica*, associated with the entomopathogenic nematode *Steinernema yirgalemense*. J Helminthol 90:108–112

Forst S, Clarke D (2002) Bacteria–nematode symbiosis. In: Gaugler R (ed) Entomopathogenic nematology. CAB International, Wallingford, pp 57–77

Forst S, Dowds B, Boemare N et al (1997) *Xenorhabdus* and *Photorhabdus* spp.: bugs that kill bugs. Ann Rev Microbiol 51:47–72

Glaser RW (1931) The cultivation of a nematode parasite of an insect. Science 73:614–615

Glaser RW, Fox H (1929) A nematode parasite of the Japanese beetle (*Popillia japonica* Newm.). Science 70:16–17

Grenier E, Bonifassi E, Abad P et al (1996) Use of species specific satellite DNAs as diagnostic probes in the identification of Steinernematidae and Heterorhabditidae entomopathogenic nematodes. Parasitology 113:483–489

Grewal PS, Martin WR, Miller RW et al (1997) Suppression of plant-parasitic nematode populations in turfgrass by application of entomopathogenic nematodes. Biocontrol Sci Tech 7:393–399

Grewal PS, Lewis EE, Venkatachari S (1999) Allelopathy: a possible mechanism of suppression of plant-parasitic nematodes by entomopathogenic nematodes. Nematol 1:735–743

Griffin CT, Boemare NE, Lewis EE (2005) Biology and behaviour. In: Grewal PS, Ehlers R-U, Shapiro-Ilan D (eds) Nematodes as biocontrol agents. CAB International, Wallingford, pp 47–64

Harington JS (1953) Observations on the biology, the parasites and the taxonomic position of the black maize beetle – *Heteronychus sanctae-helenae* Blanch. S Afr J Sci 50:11–14

Hatting J, Stock PS, Hazir S (2009) Diversity and distribution of entomopathogenic nematodes (Steinernematidae, Heterorhabditidae) in South Africa. J Invert Pathol 102:120–128

Hazir S, Stock SP, Keskin N (2003) Entomopathogenic nematodes (Steinernematidae and Heterorhabditidae) for biological control of soil pests. Turk J Biol 27:181–202

Hominick WM (2002) Biogeography. In: Gaugler R (ed) Entomopathogenic nematology. CAB International, Wallingford, pp 115–143

Hominick WM, Reid AP, Bohan DA et al (1996) Entomopathogenic nematodes: biodiversity, geographical distribution and the convention on biological diversity. Biocontrol Sci Tech 6:317–331

Hominick WM, Briscoe BR, Del Pino FG et al (1997) Biosystematics of entomopathogenic nematodes: current status, protocols and definitions. J Helminthol 71:271–298

Ishibashi N, Choi D-R (1991) Biological control of soil pests by mixed application of entomopathogenic and fungivorous nematodes. J Nematol 23:175–181

Ishibashi N, Kondo E (1986) *Steinernema feltiae* (DD-136) and *S. glaseri*: persistence in soil and bark and compost and their influence on native nematodes. J Nematol 18:310–316

Jagdale GB, Somasekhar N, Grewal PS et al (2002) Suppression of plant-parasitic nematodes by application of live and dead infective juveniles of an entomopathogenic nematode, *Steinernema carpocapsae*, on boxwood (*Buxus* spp.). Biol Control 24:42–49

Kaya HK, Gaugler R (1993) Entomopathogenic nematodes. Ann Rev Entomol 38:181–206

Koppenhöfer AM (2002) Nematodes. In: Lacey LA, Kaya LA (eds) Field manual of techniques in invertebrate pathology. Kluwer Academic Publishers, Dortrecht, pp 249–264

Koppenhöfer HS (2007) Bacterial symbionts of *Steinernema* and *Heterorhabditis*. In: Nguyen KB, Hunt DJ (eds) Entomopathogenic nematodes: systematics, phylogeny and bacterial symbionts. Nematology monographs and perspectives 5 (Series editors Hunt DJ, Perry RN). Brill, Leiden, p 735–808

Le Vieux PD, Malan AP (2013a) An overview of the vine mealybug (*Planococcus ficus*) in South African Vineyards and the use of entomopathogenic nematodes as potential biocontrol agents. S Afr J Enol Vitic 34:34–45

Le Vieux PD, Malan AP (2013b) The potential use of entomopathogenic nematodes to control *Planococcus ficus* (Signoret) (Hemiptera: Pseudococcidae). S Afr J Enol Vitic 34:296–306

Le Vieux PD, Malan AP (2015) Prospects for using entomopathogenic nematodes to control the vine mealybug, *Planococcus ficus*, in South African Vineyards. S Afr J Enol Vitic 36:59–70

Lewis EE, Grewal PS (2005) Interactions with plant-parasitic nematodes. In: Grewal PS, Ehlers R-U, Shapiro-Ilan DI (eds) Nematodes as biocontrol agents. CAB International, Wallingford, pp 349–361

Malan AP, Hatting J (2015) Entomopathogenic nematode exploitation: case studies in laboratory and field applications from South Africa. In: Campos-Herrera R (ed) Nematode pathogenesis of insects and other pests. Sustainability in plant and crop protection: ecology and applied technologies for sustainable plant and crop protection. Springer International Publishing, Cham, pp 475–506

Malan AP, Manrakhan A (2009) Susceptibility of the Mediterranean fruit fly (*Ceratitis capitata*) and the Natal fruit fly (*Ceratitis rosa*) to entomopathogenic nematodes. J Invert Pathol 100:47–49

Malan AP, Nguyen KB, Addison MF (2006) Entomopathogenic nematodes (Steinernematidae and Heterorhabditidae) from the southwestern parts of South Africa. Afr Plant Prot 12:65–69

Malan AP, Nguyen KB, De Waal JY et al (2008) *Heterorhabditis safricana* n. sp. (Rhabditida: Heterorhabditidae), a new entomopathogenic nematode from South Africa. Nematol 10:381–396

Malan AP, Knoetze R, Moore SD (2011) Isolation and identification of entomopathogenic nematodes (Heterorhabditidae and Steinernematidae) in citrus orchards in South Africa. J Invert Pathol 108:115–125

Malan AP, Knoetze R, Tiedt LR (2014) *Heterorhabditis noenieputensis* n. sp. (Rhabditida: Heterorhabditidae), a new entomopathogenic nematode from South Africa. J Helminthol 88:138–151

Malan AP, Knoetze R, Tiedt LR (2015) A new entomopathogenic nematode from South Africa. *Steinernema jeffreyense* n sp J Helminthol 90:262–278

Malan AP, Knoetze R, Tiedt LR (2016) A new entomopathogenic nematode from South Africa. *Steinernema nguyeni* n sp Nematology. doi:10.1163/15685411-00002978

Mekete T, Gaugler R, Nguyen KB et al (2005) Biogeography of entomopathogenic nematodes in Ethiopia. Nematropica 35:31–35

Molotsane R, Ngoma L, Gray V (2007) A biogeographic survey of entomopathogenic nematodes of Gauteng and the North West Province of South Africa. S Afr J Plant Soil 24:256

Mwaniki SW, Nderitu JH, Olubayo F et al (2008) Factors influencing the occurrence of entomopathogenic nematodes in the Central Rift Valley Region of Kenya. Afr J Ecol 46:79–84

Nguyen KB (2007) Methodology, morphology and identification. In: Nguyen KB, Hunt DJ (eds) Entomopathogenic nematodes: systematics, phylogeny and bacterial symbionts. Nematology Monographs and Perspectives 5 (Series editors Hunt DJ, Perry RN). Brill, Leiden, p 59–119

Nguyen KB, Smart GC Jr (1994) *Neosteinernema longicurvicauda* n. gen. n. sp. (Rhabditida: Steinernematidae) a new parasite of the termite *Reticulitermes flaviceps* (Koller). J Nematol 26:162–174

Nguyen KB, Tesfamariam M, Gozel U et al (2004) *Steinernema yirgalemense* n. sp. (Rhabditida : Steinernematidae) from Ethiopia. Nematology 6:839–856

Nguyen KB, Malan AP, Gozel U (2006) *Steinernema khoisanae* n. sp. (Rhabditida: Steinernematidae), a new entomopathogenic nematode from South Africa. Nematol 8:157–175

Nthenga I, Knoetze R, Berry S et al (2014) *Steinernema sacchari* n. sp. (Rhabditida: Steinernematidae), a new entomopathogenic nematode from South Africa. Nematol 16:475–494

Odendaal D, Addison MF, Malan AP (2015a) Control of codling moth (*Cydia pomonella*) with special emphasis on using entomopathogenic nematodes. Afr Entomol 23:259–274

Odendaal D, Addison MF, Malan AP (2015b) Evaluation of above-ground application of entomopathogenic nematodes for the control of diapausing codling moth (*Cydia pomonella* L.) under natural conditions. J Helminth. doi:10.1017/S0022149X15000887

Odendaal D, Addison MF, Malan AP (2016) Evaluation of above-ground application of entomopathogenic nematodes for the control of diapausing codling moth (*Cydia pomonella* L.) under natural conditions. African Entomol 24:61–74

Onstad DW, Fuxa JR, Humber RA et al (2006) An abridged glossary of terms used in invertebrate pathology, 3rd edn. USDA, Forest Service, Pacific Northwest Forest and Range Experimental Station. http://www.sipweb.org/resources/glossary.html

Perez EE, Lewis EE (2002) Effects of entomopathogenic nematode application to *Meloidogyne incognita* penetration and egg production in laboratory and greenhouse experiments. J Nematol 34:171–174

Perez EE, Lewis EE (2004) Suppression of *Meloidogyne incognita* and *Meloidogyne hapla* with entomopathogenic nematodes on greenhouse peanuts and tomatoes. Biol Control 30:336–341

Poinar GO (1976) Description and biology of a new insect parasitic rhabditoid, *Heterorhabditis bacteriophora* n. gen. n. sp. (Rhabditida; Heterorhabditidae n. fam.). Nematologica 21:463–470

Poinar GO (1990) Taxonomy and biology of Steinernematidae and Heterorhabditidae. In: Gaugler R, Kaya HK (eds) Entomopathogenic nematodes in biological control. CRC Press, Boca Raton, pp 23–61

Rizvi SA, Hennessey P, Knott D (1996) Legislation on the introduction of exotic nematodes in the US. Biocontrol Sci Tech 6:477–480

Somasekhar N, Grewal PS, De Nardo EAB et al (2002) Non-target effects of entomopathogenic nematodes on the soil nematode community. J Appl Ecol 39:735–744

Spaull VW (1988) A preliminary evaluation of entomogenous nematodes to control the African sugarcane stalk borer *Eldana saccharina* (Lepidoptera: Pyralidae). Pro S Afr Sugar Technol Assoc 62:120–123

Spaull VW (1990) Field tests to control the pyralid, *Eldana saccharina*, with an entomogenous nematode, *Heterorhabditis* sp. Pro S Afr Sugar Technol Assoc 64:103–106

Spaull VW (1991) *Heterorhabditis* and *Steinernema* species (Nematoda: Rhabditida) for the control of a sugar cane stalk borer in South Africa. Phytophylactica 23:213–215

Spaull VW (1992) On the use of a methylcellulose polymer to increase the effectiveness of a *Heterorhabditis* species against the sugarcane stalk borer, *Eldana saccharina*. Fund Appl Nematol 15:457–461

Steiner G (1923) *Aplectana kraussei* n. sp. eine in der Blattwespe *Lyda* sp. parasitierende Nematodenform, nebst Bomerkungen uber das Seitenorgaan der parasitischen Nematoden. Zentralblatt für Bakteriologie Parasitenkunde Infektionskrankbeiten und Hygiene Abteilung I Originale 59:14–18

Steiner G (1929) *Neoaplectana glaseri* n.g., n. sp. (Oxyuridae), a new nemic parasite of the Japanese beetle (*Popillia japonica* Newm.). J Wash Acad Sci 19:436–440

Stokwe NF (2009) Entomopathogenic nematodes: characterization of a new species, long term storage and control of obscure mealybug *Pseudococcus viburni* (Hemiptera: Pseudococcidae) using laboratory bioassays. MSc Agric thesis, Department of Conservation Ecology and Entomology/Stellenbosch University, Stellenbosch

Stokwe NF, Malan AP (2016) Susceptibility of the obscure mealybug, *Pseudococcus* viburni (Signoret) (Pseudococcidae), to South African isolates of entomopathogenic nematodes. Int J Pest Manage 62:119–128

Stokwe NF, Malan AP, Khuong BN et al (2011) *Steinernema citrae* sp. n. (Rhabditida: Steinernematidae), a new entomopathogenic nematode from South Africa. Nematol 13:569–587

Swart A, Heyns J (1997) Nematodes from *Pinus radiata* in South Africa (Nematoda: Cephalobidae, Rabditinae, Aphelenchoidinae and Ektaphelenchidae). J Afr Zool 111:381–395

Tarasco E, Triggiani O, Sai K et al (2009) Survey on entomopathogenic nematodes in Algerian soils and their activity at different temperatures. Frustula Entomol 32:31–42

Tribe G (2006) A wasp counterattack to save pine trees. Village Life 8:30–40

Van Niekerk S, Malan AP (2012) Potential of South African entomopathogenic nematodes (Heterorhabditidae and Steinernematidae) for control of the citrus mealybug, *Planococcus citri* (Pseudococcidae). J Invert Pathol 111:166–176

Van Niekerk S, Malan AP (2013) Adjuvants to improve control of *Planococcus citri* (Hemiptera: Pseudococcidae) using entomopathogenic nematodes (Heterorhabditidae and Steinernematidae). J Helminthol 89:189–195

Van Niekerk S, Malan AP (2014a) Evaluating the efficacy of a polymer–surfactant formulation to improve control of *Planococcus citri* (Risso) (Hemiptera: Pseudococcidae) under simulated natural conditions. Afr Plant Prot 17:1–8

Van Niekerk S, Malan AP (2014b) Compatibility of *Heterorhabditis zealandica* and *Steinernema yirgalemense* with agrochemicals and biological control agents. Afr Entomol 22:49–56

Van Zyl C, Malan AP (2014a) The role of entomopathogenic nematodes as biological control agents of insect pests, with emphasis on the history of their mass culturing and *in vivo* production. Afr Entomol 22:235–249

Van Zyl C, Malan AP (2014b) Optimisation of inoculation techniques for in vivo mass culture of entomopathogenic nematodes through nematode and insect host manipulation. Afr Entomol 24:405–416

Van Zyl C, Malan AP (2015) Cost-effective culturing of *Galleria mellonella* and *Tenebrio molitor* and nematode production in various hosts. Afr Entomol 23:361–375

Waturu CN (1998) Entomopathogenic nematodes (Steinernematidae and Heterorhabditidae) from Kenya. PhD dissertation, University of Reading, Reading

Waturu CN, Reid AP (1997) *Steinernema karii* sp. n. (Nematoda: Steinernematidae), a new entomopathogenic nematode from Kenya. Int J Nematol 7:67–75

Weiser J (1955) *Neoaplectana carpocapsae* n. sp. (Anguillulata, Steinernematidae) novy cizopasnik housenek obalece jableneho, *Carpocapsae pomonella* L. Vestnik Ceskoslovenske Spolecnosti Zoologicke 19:44–52

Chapter 23
Nematodes Associated with Terrestrial Slugs

Jenna L. Ross and Antoinette P. Malan

23.1 Introduction

Nematodes associated with terrestrial slugs (Mollusca: Gastropoda) are understudied. However, their host specificity and diversity of specialisations indicate that their relationship is ancient and widespread (Morand et al. 2004). Current knowledge of the slug-nematode affiliation is based on surveys conducted in Europe, North America, Australasia and Africa (Ross et al. 2016a). According to this research, seven nematode families are associated with terrestrial slugs, namely, the Agfidae, Alloionematidae, Angiostomatidae, Cosmocercidae, Diplogastridae, Mermithidae and Rhabditidae (Pieterse et al. 2016). These nematodes have several different relationships with slugs including parasitic (specialist or generalist), phoretic and necromenic associations. To date, the only nematode that has been commercially developed as a biological molluscicide is *Phasmarhabditis hermaphrodita* (Schneider, 1859) Andrássy, 1983 that belongs to the family Rhabditidae.

J.L. Ross (✉)
Department of Conservation Ecology and Entomology, Faculty of AgriSciences, Stellenbosch University, Private Bag X1, Matieland 7602, South Africa

Institute of Biological and Environmental Sciences, University of Aberdeen, St Machar Drive, AB24 3UU, Aberdeen, UK
e-mail: jross@sun.ac.uk; jenna.ross@abdn.ac.uk

A.P. Malan
Department of Conservation Ecology and Entomology, Faculty of AgriSciences, Stellenbosch University, Private Bag X1, Matieland 7602, South Africa

H. Fourie et al. (eds.), *Nematology in South Africa: A View from the 21st Century*, DOI 10.1007/978-3-319-44210-5_23

23.2 Slug Pests

Terrestrial slugs, which have colonised all inhabited continents, are important economic pests of a number of different crop types, including arable, pasture, ornamental and vegetable crops (Glen and Moens 2002; Moens and Glen 2002; Port and Ester 2002; Wilson and Barker 2011). Slugs reduce the vigour of crops by damaging stems and growing points, killing seeds and seedlings and reducing leaf area (Port and Port 1986; South 1992). In extreme conditions, the damage to germinating seeds might be so severe that entire fields have to be resown, resulting in huge economic costs to the producer concerned (Willis et al. 2006). Reduction in value can also occur to harvested crops due to feeding damage, mucus trails, eggs, faeces and the presence of slugs (Iglesias et al. 2002).

Invasive European slugs were introduced to South Africa (SA) during the eighteenth and early nineteenth centuries (Smith 1992), having successfully established themselves, especially in the Western Cape as a result of favourable climatic conditions (Herbert and Kilburn 2004). One crop particularly targeted is canola (*Brassica napus*), which is a winter crop that is commercially produced in SA for use in cooking, baking, food processing, fuels, plastics, pet food, fertilisers and animal feed. The most susceptible phase of the crop is the seedling stage, which usually occurs 4 weeks after planting. However, slug numbers tend to increase steadily during the season, with peaks occurring in September and October (J. Ross, unpublished data). The three slug species that have been identified as being pestiferous to canola crops are the grey field slug (*Deroceras reticulatum*), the brown field slug (*Deroceras panormitanum*) and the keeled slug (*Milax gagates*) (Fig. 23.1a–c, Table 23.1). In addition to having the ability to feed both above and below ground, these slugs have a short life cycle, during which they can, nevertheless, lay hundreds of eggs.

A total of 11 exotic slug species, representing five different families, have been introduced into SA (Table 23.1) (Herbert and Kilburn 2004). These species occur in a range of habitats, including agricultural land, woodland, gardens and glasshouses. Moreover, they also have a knock-on effect on the agri-, horti- and viticultural industries in SA. One particular family, Arionidae, is known to be adept travellers, with *Arion hortensis* and *Arion intermedius* having been recorded in North America, Australia, New Zealand and SA. Care should be taken with arionids, especially with the invasive Iberian slug (*Arion vulgaris*), which originates from the Atlantic regions of south-west France but which has now invaded all parts of Europe as well as some areas of Russia (Sysoev and Schileyko 2009). *Arion vulgaris* thrives in gardens, recreational areas and agricultural land, resulting in major damage to ornamentals, vegetable plots and agricultural crops (Frank 1998; Hofsvang et al. 2008). Once established, populations of *A. vulgaris* are difficult to manage, especially in years with favourable slug weather. Therefore, in SA, ongoing vigilance is necessary to avoid its introduction into the country.

One possible explanation for the success of exotic terrestrial slugs in SA is that of the 'enemy release' hypothesis, which is regarded as a theory, stating that organisms are freed from the influence of their co-evolved natural enemies when

Fig. 23.1 (**a–c**) European pestiferous slugs found in canola fields in South Africa: the grey field slug (*Deroceras reticulatum*) (**a**), brown field slug (*Deroceras panormitanum*) (**b**) and keeled slug (*Milax gagates*) (**c**) (Jenna Ross, Stellenbosch University, Stellenbosch, South Africa and Geoff Tribe, Protein Research Foundation, Johannesburg, South Africa)

invading new areas. This provides them with a competitive advantage over the native species of the areas concerned (Torchin et al. 2001). Ross et al. (2010a) demonstrated that enemy release plays an important role in the invasion of North America by European slugs. Similar results for SA have been generated with parasite prevalence and species richness being relatively low in the invasive range (Ross et al. 2012).

23.3 Slug Control

Current methods for controlling slugs in SA rely on chemical molluscicide pellets containing a combination of 30 g kg^{-1} metaldehyde and 20 g kg^{-1} carbaryl at a recommended application rate of 6–12 kg ha^{-1}. Many producers opt for an application rate of 8 kg ha^{-1} (G Tribe, Protein Research Foundation, Johannesburg, 2015, personal communication). Metaldehyde and carbamate compounds cause a range of symptoms, including convulsions, paralysis and mucus secretion, with death usually occurring 1–2 days after toxin exposure (Booze and Oehme 1986). An

Table 23.1 Exotic terrestrial slugs of South Africa (Herbert and Kilburn 2004)

Family	Species	Origin	Habitat
Agriolimacidae	*Deroceras laeve*	Europe	Agricultural land, woodland, gardens
	Deroceras panormitanum	Europe	Agricultural land, woodland, gardens
	Deroceras reticulatum	Europe	Agricultural land, woodland, gardens
Arionidae	*Arion hortensis*	Europe	Agricultural land, gardens
	Arion intermedius	Europe	Agricultural land, woodland, gardens
Limacidae	*Limax flavus*	Europe, North Africa	Gardens, compost heaps
	Limax maximus	Europe	Woodlands, hedgerows, gardens, waste ground
	Lehmannia valentiana	Iberian Peninsula	Glasshouses, gardens
	Lehmannia nyctelia	North Africa	Glasshouses, gardens
Milacidae	*Milax gagates*	Europe	Agricultural land, gardens
Testacellidae	*Testacella maugei*	Europe	Agricultural land, gardens

alternative to molluscicide pellets is seed treatment, where seeds are coated with an appropriate toxicant (e.g. metaldehyde or carbamate compounds). Although seed treatments have been successfully used for controlling slugs in winter wheat (Ester et al. 1996), they do not perform as well as metaldehyde pellets, possibly due to the breakdown of the active toxicants involved (Simms et al. 2006).

Chemical molluscicides carry a number of non-target and environmental risks, with metaldehyde and carbamate compounds being toxic to a range of vertebrates (Fletcher et al. 1994) and isopods (Santos et al. 2010). In addition, slug pellets are often overused by farmers who are unsure of the severity of the slug problem, resulting in the amplification of such environmental problems as surface wash-off after heavy rainfall (O'Brien et al. 2008). As a result, the use of chemical molluscicides has come under scrutiny by the European Union, after dangerously high levels of the toxins were recorded in water systems across Europe.

An alternative to chemical molluscicides is cultural control, in which physical barriers, irritants, antifeedants, chemical repellents, ploughing, trapping, crop rotation, seedbed preparation and deep drilling are used to control slugs (Glen 2000; Schüder et al. 2003). However, many of these methods are not economical for use in large-scale agriculture.

In addition to chemical and cultural control, slugs can also be controlled using biological agents. To date, the most efficient commercial agent for the biocontrol of slugs in Europe has been the slug-parasitic nematode *Phasmarhabditis hermaphrodita* Schneider, 1859.

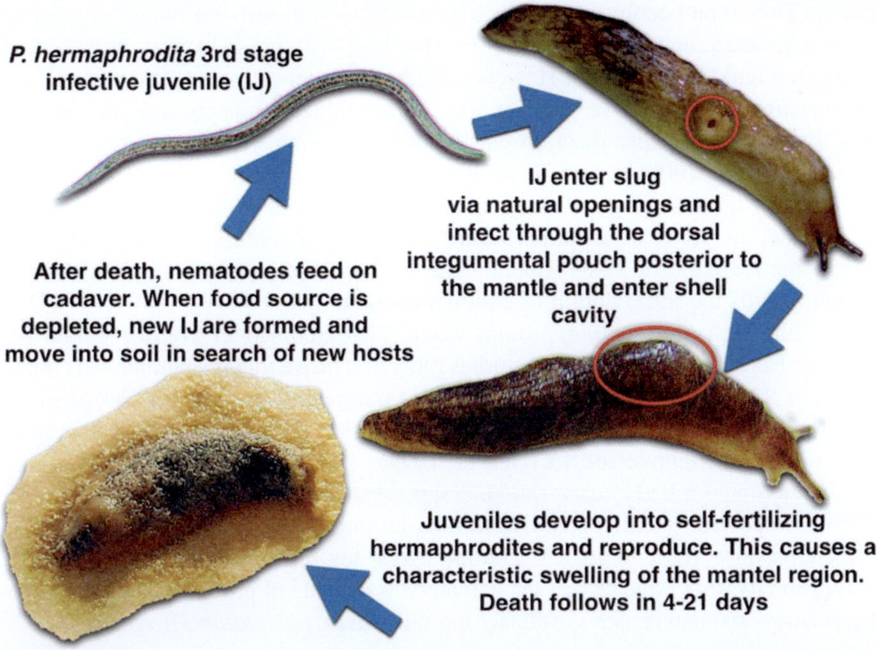

Fig. 23.2 The parasitic life cycle of *Phasmarhabditis hermaphrodita* (Photographs: Jenna Ross and A Pieterse, University of Stellenbosch, Stellenbosch, South Africa; Illustration: EC McGawley, Louisiana State University, Louisiana, United States of America)

23.4 Phasmarhabditis

The *Phasmarhabditis* Andrássy, 1976, genus currently contains six species (*P. hermaphrodita*, *Phasmarhabditis neopapillosa* (Mengert, 1952) Andrássy, 1983, *Phasmarhabditis papillosa* (Schneider, 1866) Andrássy, 1983, *Phasmarhabditis tawfiki* Azzam, 2003, *Phasmarhabditis huizhouensis* Huang, Ye, Ren and Zhao, 2015, and *Phasmarhabditis californica* Tandingan De Ley, Holovachov, Mc Donnell, Bert, Paine and De Ley, 2016) (Tandingan De Ley et al. 2016). However, the majority of research has focused on the biocontrol potential of *P. hermaphrodita*, which is pathogenic to a range of slug families including Agriolimacidae, Arionidae, Milacidae, Limacidae and Vagnulidae (Rae et al. 2007). In 1994, *P. hermaphrodita* was made commercially available by MicroBio Ltd (acquired by Becker Underwood and taken over by BASF in 2012) under the trade name Nemaslug®. This product is currently sold in 15 European countries, having a retail value of approximately 1 million Euro (G Martin, BASF, Littlehampton, United Kingdom, 2015, personal communication).

The life cycle of *P. hermaphrodita* (Fig. 23.2) is very similar to that of entomopathogenic nematodes (EPN) (see Fig. 22.1). On entering the slug *via* the dorsal integumental pouch, third-stage infective juveniles (IJ) move towards the shell cavity. They then become a feeding stage, develop into self-fertilising hermaphrodites, reproduce and eventually kill the slug host within 4–21 days (Wilson et al. 1993; Tan and Grewal 2001). However, the key difference from EPN is that *P. hermaphrodita* is a facultative parasite that can reproduce while feeding on compost, leaf litter, slug faeces, dead insects and earthworms (Tan and Grewal 2001; MacMillan et al. 2009; Nermut' et al. 2014) and it can exhibit a necromenic approach in relatively large slug species (e.g. *Arion ater*) (Rae et al. 2009).

To date, the use of *P. hermaphrodita* as a biological control agent is regulated in SA by the Agricultural Pests Amendment Act No. 18 of 1989 (Anonymous 1989), which forbids the introduction of exotic animals to the country (Ross et al. 2012). The application of *P. hermaphrodita* would be welcomed in SA to regulate key invasive European slug pests, but studies must first prove that the nematode is indigenous to SA and that it is non-pathogenic to indigenous mollusc species as many, e.g. Purcell's hunter slug, *Laevicaulis haroldi*, are included in the Database of Threatened Invertebrates of SA (Herbert 1997).

Various undescribed *Phasmarhabditis* spp., identified in the Western Cape Province of SA, are currently undergoing description. Of these, *Phasmarhabditis* sp. SA2 has shown promise as a biological control agent for terrestrial slugs (Ross 2010). Pathogenicity tests indicate that the nematode is capable of profuse growth on modified kidney agar and causes significant mortality to *D. reticulatum* (Ross 2010).

23.5 Other Nematodes Associated with Terrestrial Slugs

The seven nematode families associated with terrestrial slugs (see Sect. 23.1) represent four out of the five clades of Nematoda, indicating multiple origins of parasitism in slugs (Fig 23.3). Of these, Agfidae, Angiostomatidae and Rhabditidae form a tight monophyletic clade, indicating that the groups involved have evolved from a single slug-colonising ancestor, despite their present morphologically diverse features (Ross et al. 2010b)((Fig. 23.4).

The Agfidae, which are represented by three species, are characterised by their very thin neck region. They are obligate parasites that are usually found in their juvenile or adult stages in the salivary gland or genital tract of their mollusc hosts (Morand et al. 2004). The Angiostomatidae is represented by two genera, Angiostoma Dujardin, 1845 and Aulacnema Pham Van Luc, Spiridonov & Wilson, 2005. Molluscan angiostomatids are obligate parasites that have been isolated, as adults or fourth-stage juveniles, from the intestine, hepatopancreas, oesophagus, buccal mass, crop, mantle cavity, salivary gland and pallial cavity (Ross et al. 2016a, b). In addition to being molluscan angiostomatids, several species have been described from the intestine and bronchi of amphibian and reptile hosts (Falcón-Ordaz et al. 2008). The Alloionematidae family have two genera that are associated with molluscs, including *Alloionema* Schneider, 1859, and *Neoalloionema* Ivanova, Pham Van Luc & Spiridonov, 2016. *Alloionema* is represented by *Alloionema appendiculatum* Schneider, 1859, with third- and fourth-stage juveniles being associated with the

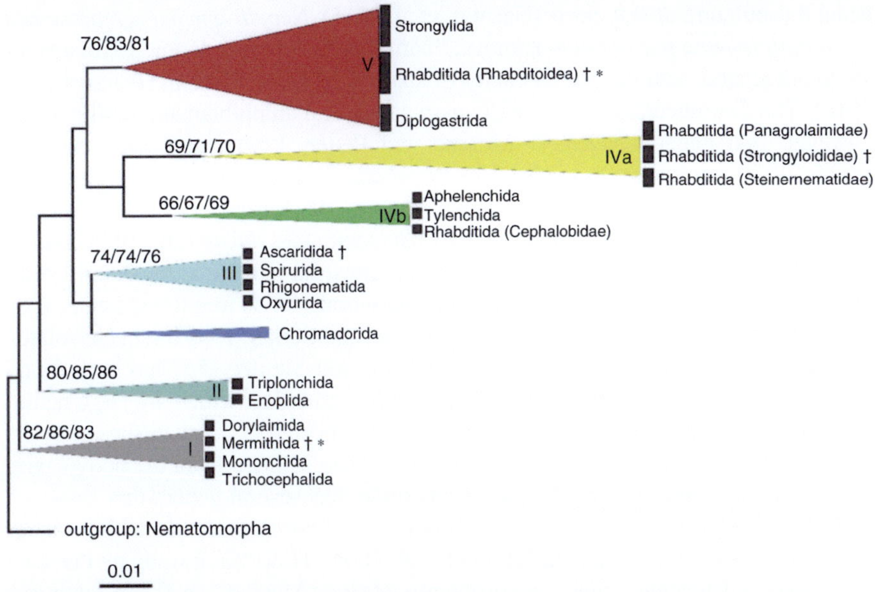

Fig. 23.3 Maximum-likelihood phylogenetic tree containing ML, distance and MP analyses of 18S rRNA gene sequences from nematode taxa sampled across the phylum (Adapted from Ross et al. 2010b); names marked with † represent nematodes that the authors considered to be truly parasitic to slugs as opposed to phoretic or necromenic associations, while those marked with * signifying nematodes isolated from slugs in South Africa. Significant bootstrap support values (>65 %) are shown near the nodes (ML/Distance/MP); Nematomorpha was defined as an out-group

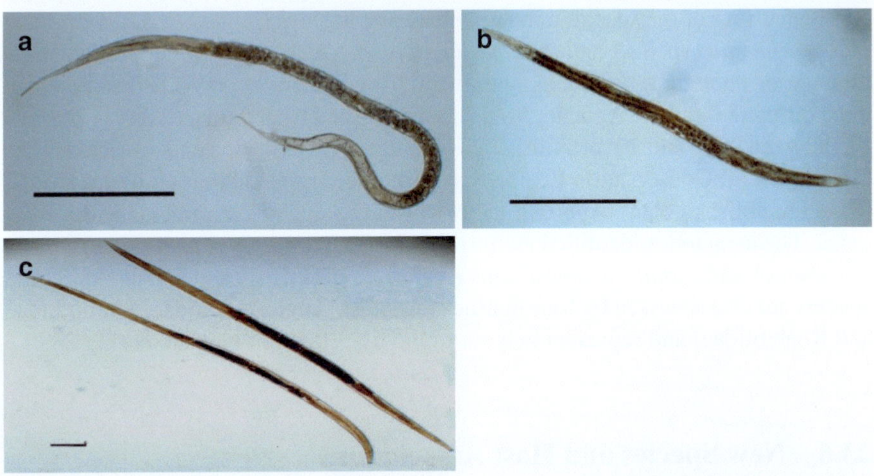

Fig. 23.4 (**a–c**) Female, or hermaphrodite, adult slug-parasitic nematodes, (**a**) *Agfa flexilis*, (**b**) *Phasmarhabditis hermaphrodita* and (**c**) *Angiostoma limacis* (scale = 1 mm) (Jenna Ross, University of Stellenbosch, Stellenbosch, South Africa)

pedal musculature of slug hosts (Nermut' et al. 2015). *Neoalloionema* is represented by *Neoalloionema tricaudatum* Ivanova, Pham Van Luc & Spiridonov, 2016, a nematode associated with the pallial cavity of snail hosts from Vietnam (Ivanova et al. 2016). The Cosmocercidae are usually associated with amphibian and reptile hosts; however, two genera, *Nemhelix* Morand and Petter, 1986 and *Cosmocercoides* Wilkie, 1930, are known from molluscs. *Nemhelix*, which is known from three species, associates with the reproductive organs of European snails, whereas *Cosmocercoides*, which is represented by *Cosmocercoides dukae* Hall, 1928, associates with the pallial cavity of North American terrestrial molluscs (Anderson 1960). Diplogastridae are usually associated with invertebrates or as free-living nematodes. However, *Hugotdiplogaster neozelandia* Morand and Barker, 1995 has been isolated in its adult stage from mollusc hosts, and *Diplogaster* Shultze 1857 has been found in its juvenile stage (Morand et al. 2004). The Mermithidae, which are large nematodes, ranging from 1 to 10 cm in size, have usually been found as parasites of invertebrates (Poinar 1983). Pre-adult juveniles have been isolated from the body cavity of mollusc hosts (Ross et al. 2010b). Rhabditidae has several genera that associate with slugs, including *Rhabditis*, *Caenorhabditis* and *Phasmarhabditis*, with the latter genus considered a true parasite (Morand et al. 2004). However, a study by Petersen et al. (2015) determined that *Caenorhabditis elegans* Maupas, 1900 can infect the intestine of molluscs and exit alive through the faeces.

In their survey of nematodes associated with terrestrial slugs in the Western Cape Province, Ross et al. (2012) found that nematodes were parasitising slugs at 14 (40 %) of the 35 sample sites investigated and that 6 % of all the slugs were infected with nematodes (Fig. 23.5, Table 23.2). During the survey, a total of seven nematode species were identified, including *Agfa flexilis* Dujardin, 1845, *Angiostoma margaretae* Ross, Malan and Ivanova, 2011, *Phasmarhabditis* sp. SA1, *Phasmarhabditis* sp. SA2, *C. elegans*, *Panagrolaimus* sp. and *Rhabditis* sp. During this survey the authors only considered four of the species to be truly parasitic to slugs (*A. flexilis*, *A. margaretae*, *Phasmarhabditis* sp. SA1 and *Phasmarhabditis* sp. SA2), as opposed to forming necromenic or phoretic associations with them. The nematodes concerned represented three different families, Agfidae, Angiostomatidae and Rhabditidae.

Between 2012 and 2016, a further 2,876 slugs were collected and dissected at the Department of Conservation Ecology and Entomology, Stellenbosch University (J. Ross, unpublished data). Approximately 8 % of the slugs were infected with nematodes. The nematodes identified included *A. flexilis*, *A. margaretae*, *Angiostoma* sp., *Phasmarhabditis* spp., *Caenorhabditis elegans*, *Rhabditis* sp. and *Mermis* sp. These species are characterised by four families (Agfidae, Angiostomatidae, Mermithidae and Rhabditidae) and represent two out of the five clades of Nematoda (Fig 23.3).

23.6 New Species and Host Associations

Two species of *Angiostoma* have been identified in the Western Cape Province of which one has been described as *Angiostoma margaretae* Ross, Ivanova and Malan, 2011, while the other is currently undergoing description. *Angiostoma*

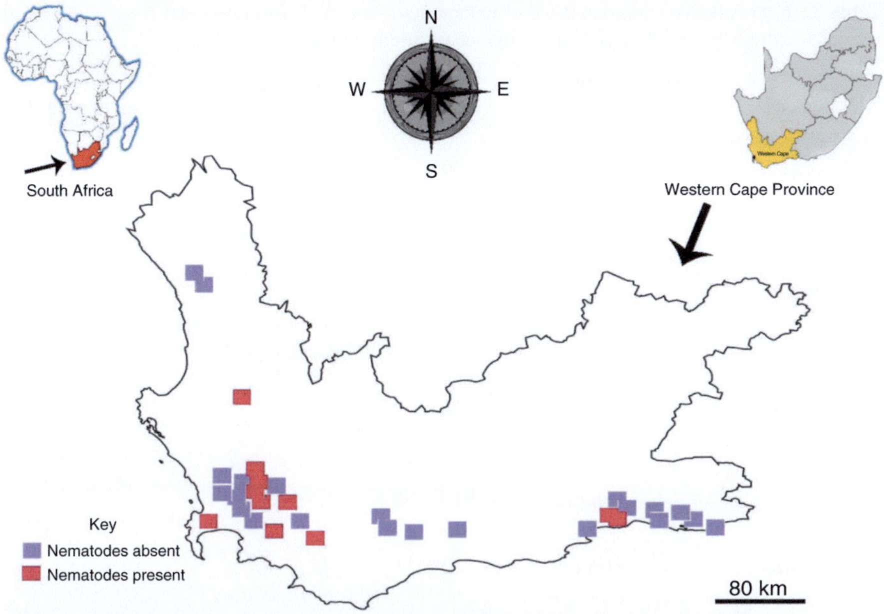

Fig. 23.5 Sites where slugs were sampled in the Western Cape Province to determine the presence of nematodes associated with slugs (Jenna Ross, University of Stellenbosch, Stellenbosch, South Africa; Redrawn by Ebrahim Shokoohi, North-West University, Potchefstroom, South Africa)

margaretae was originally isolated from the oesophagus of the keeled slug *M. gagates* that was collected near Caledon (Fig. 23.5). The nematode closely resembles *Angiostoma milacis* Ivanova and Wilson, 2009, which is another parasite of milacid slugs (Ross et al. 2011). More recently, *A. margaretae* has been found in association with the invasive slug species, *D. reticulatum*, and with the indigenous species, *Oopelta polypunctata* (J. Ross, unpublished data).

23.7 Future Research

Future research should focus on collecting indigenous slug-parasitic nematode isolates throughout SA, since the majority of work that has been undertaken thus far has focused on the Western Cape Province. The pathogenicity of any new isolates ought to be tested against different slug species, especially those that have been identified as pestiferous in canola. Research should in future focus on developing suitable mass-rearing protocols for indigenous isolates using monoxenic cultures, as they offer predictable results and reduce contamination problems (Ehlers and Shapiro-Ilan 2005). Effective monoxenic bacteria-nematode combinations should show high yields and reproducible infectivity of slug hosts (Wilson et al. 1995). A suitable formulation and different application techniques

Table 23.2 Prevalence and abundance of nematodes associated with terrestrial slugs collected at various localities in the Western Cape Province of South Africa

Host species	Native or introduced	Number collected	Parasite species recovered	Parasite prevalence (%)
Agriolimacidae				
Deroceras reticulatum	Introduced	69	*Phasmarhabditis* spp. (2)[a] *Rhabditis* sp.	4.3 4.3
Deroceras panormitanum	Introduced	26		–
Arionidae				
Ariopelta capensis	Native	1	*Rhabditis* sp.	100
Ariostralis nebulosa	Native	4	*Phasmarhabditis* sp. (1)[a]	75.0
Oopelta granulosa	Native	14		
Oopelta polypunctata	Native	4		
Chlamydephoridae				
Chlamydephorus gibbonsi	Native	1		–
Limacidae				
Limax flavus	Introduced	32	*Agfa flexilis*[a] *Caenorhabditis elegans* *Panagrolaimus* sp.	12.5 3.0 3.0
Lehmannia valentiana	Introduced	313	*Agfa flexilis*[a] *Caenorhabditis elegans* *Panagrolaimus* sp.	2.2 0.6 0.3
Milacidae				
Milax gagates	Introduced	44	*Angiostoma margaretae*[a] *Panagrolaimus* sp.	6.8 2.3
Testacellidae				
Testacella maugei	Introduced	4		–
Veronicellidae				
Laevicaulis alte	Native	9		–

Adapted with permission from Ross et al. (2012)
[a]New host association

will then need to be developed in order to allow the slug-parasitic nematodes to be applied effectively to the crop. In addition to local isolate research, studies must also look at whether *P. hermaphrodita* is indigenous to SA or non-pathogenic to protected molluscs, in order to import the commercial Nemaslug® product (Herbert 1997).

23.8 Conclusions

Nematodes associated with slugs have to date been relatively understudied, with the majority of work that has been undertaken in this area having focused on *P. hermaphrodita*. The latter has been successfully developed as a biological molluscicide in Europe. However, the introduction of this biological control agent is regulated in SA by the Agricultural Pests Amendment Act No. 18 of 1989 which prohibits the introduction of exotic organisms (Ross et al. 2012). Therefore, evidence is still required to prove that *P. hermaphrodita* is indigenous to SA and that it is non-toxic to endangered molluscs. Alternatively, research could focus on the more risk-adverse approach of developing indigenous nematode isolates as a biological control agent for slugs in SA.

References

Anderson RC (1960) On the development and transmission of *Cosmocercoides dukae* of terrestrial molluscs in Ontario. Can J Zool 64:1375–1384

Anonymous (1989) Agricultural Pests Amendment Act, No. 18 of 1989. Government Printers, Pretoria

Booze F, Oehme FW (1986) An investigation of metaldehyde and acetaldehyde toxicities in dogs. Fundam Appl Toxicol 6:440–446

Ehlers R-U, Shapiro-Ilan DI (2005) Mass production. In: Grewal PS, Ehlers R-U, Shapiro-Ilan DI (eds) Nematodes as biocontrol agents. CAB International, Wallingford, pp 65–78

Ester A, Darwinkel A, Floot HWG et al (1996) Control of the field slugs (*Deroceras reticulatum*) in winter wheat using seeds treated with pesticides. In: Henderson IF (ed) Slug and snail pests in agriculture, BCPC symposium proceedings. British Crop Protection Council, Farnham, pp 165–172

Falcón-Ordaz J, Mendoza-Garfias B, Windfield-Perez JC et al (2008) *Angiostoma lamotheargumedoi* n. sp. (Nematoda: Angiostomatidae) from the intestine of *Pseudoeurycea mixteca* (Caudata: Plethodontidae) in central Mexico. Rev Mex Biodivers 79:107–112

Fletcher MR, Hunter K, Barnet EA (1994) Pesticide poisoning of animals 1993. MAFF Publications, London

Frank T (1998) Slug damage and numbers of slug pests, *Arion lusitanicus* and *Deroceras reticulatum*, in oilseed rape grown beside sown wildflower strips. Agric Ecosyst Environ 67:67–78

Glen DM (2000) The effects of cultural measures on cereal pests and their role in integrated pest management. Integr Pest Manag Rev 5:25–40

Glen DM, Moens R (2002) Agriolimacidae, arionidae and milacidae as pests in West European cereals. In: Barker GM (ed) Molluscs as crop pests. CAB International, Wallingford, pp 271–300

Herbert DG (1997) The terrestrial slugs of KwaZulu-Natal: diversity, biogeography and conservation (Mollusca: Pulmonata). Ann Natal Mus 38:197–239

Herbert DG, Kilburn D (2004) Field guide to the land snails and slugs of Eastern South Africa. Natal Museum, Pietermaritzburg

Hofsvang T, Haukeland S, Andersen A (2008) Snegler som skadedyr på planter. Bioforsk Tema 24:1–12

Iglesias J, Castillejo J, Ester A et al (2002) Susceptibility of the eggs of the field slug *Deroceras reticulatum* to contact with pesticides and substances of biological origin on artificial soil. Ann Appl Biol 140:53–59

Ivanova ES, Van Luc P, Spiridonov SE (2016) *Neoalloionema tricaudatum* gen. n., sp. n. (Nematoda: Alloionematidae) associated with a cyclophorid snail in Cuc Phuong Natural Park, Vietnam. Nematol 18:109–120
MacMillan K, Haukeland S, Rae RG et al (2009) Dispersal patterns and behaviour of the nematode *Phasmarhabditis hermaphrodita* in mineral soils and organic media. Soil Biol Biochem 41:1483–1490
Moens R, Glen DM (2002) Agriolimacidae, arionidae and milacidae as pests in West European oilseed rape. In: Barker GM (ed) Molluscs as crop pests. CAB International, Wallingford, pp 301–314
Morand S, Wilson MJ, Glen DM (2004) Nematodes (Nematoda) parasitic in terrestrial gastropods. In: Barker GM (ed) Natural enemies of terrestrial molluscs. CAB International, Wallingford, pp 525–557
Nermut' J, Půža V, Mráček Z (2014) The effect of different growing substrates on the development and quality of *Phasmarhabditis hermaphrodita* (Nematoda: Rhabditidae). Biocontrol Sci Technol 24:1026–1038
Nermut' J, Půža V, Mráček Z (2015) Re-description of the slug-parasitic nematode *Alloionema appendiculatum* Schneider, 1859 (Rhabditida: Alloionematidae). Nematol 17:897–910
O'Brien M, Spillane C, Meade C et al (2008) An insight into the impact of arable farming on Irish biodiversity: a scarcity of studies hinders a rigorous assessment. Biol Environ 108:97–108
Petersen C, Hermann RJ, Barg MC et al (2015) Travelling at a slug's pace: possible invertebrate vectors of *Caenorhabditis* nematodes. BMC Ecol 15:19
Pieterse A, Malan AP, Ross JL (2016) Nematodes that associate with terrestrial molluscs as definitive hosts, including *Phasmarhabditis hermaphrodita* (Rhabditida: Rhabditidae) and its development as a biological molluscicide. J Helminthol. doi:10.1017/S0022149X16000572
Poinar GO (1983) The natural history of nematodes. Prentice Hall, Princeton, New Jersey
Port G, Ester A (2002) Gastropods as pests in vegetables and ornamental crops in Western Europe. In: Barker GM (ed) Molluscs as crop pests. CAB International, Wallingford, pp 337–352
Port CM, Port G (1986) The biology and behaviour of slugs in relation to crop damage and control. Agric Zool Rev 1:255–299
Rae RG, Verdun C, Grewal PS et al (2007) Biological control of terrestrial molluscs using *Phasmarhabditis hermaphrodita* – progress and prospects. Pest Manag Sci 63:1153–1164
Rae RG, Robertson JF, Wilson MJ (2009) Chemoattraction and host preference of the gastropod parasitic nematode *Phasmarhabditis hermaphrodita*. J Parasitol 95:517–526
Ross JL (2010) Diversity and mass production of slug-parasitic nematodes. PhD thesis, University of Aberdeen, Aberdeen
Ross JL, Ivanova ES, Severns PM et al (2010a) The role of parasite release in invasion of the USA by European slugs. Biol Invasions 12:603–610
Ross JL, Ivanova ES, Spiridonov SE et al (2010b) Molecular phylogeny of slug-parasitic nematodes inferred from 18S rRNA gene sequences. Mol Phylogent Evol 55:738–743
Ross JL, Ivanova ES, Malan AP (2011) *Angiostoma margaretae* n. sp. (Nematoda: Angiostomatidae) a parasite of the milacid slug *Milax gagates* Draparnaud collected near Caledon, South Africa. Syst Parasitol 79:71–76
Ross JL, Ivanova ES, Sirgel WF et al (2012) Diversity and distribution of nematodes associated with terrestrial slugs in the Western Cape Province of South Africa. J Helminthol 86:215–221
Ross JL, Ivanova ES, Hatteland BA et al (2016a) Survey of nematodes associated with terrestrial slugs in Norway. J Helminthol 90:582–587. doi:10.1017/S0022149X15000784
Ross JL, Haukeland S, Hatteland BA et al (2016b) *Angiostoma norvegicum* n. sp. (Nematoda: Angiostomatidae) a parasite of arionid slugs in Norway. Syst Parasitol. doi: 10.1007/s11230-016-9674-4
Santos MJG, Ferreira NGC, Soares AMVM et al (2010) Toxic effects of molluscicidal baits to the terrestrial isopod *Porcellionides pruinosus* (Brandt, 1833). J Soils Sedim 10:1335–1343
Schüder I, Port G, Bennison J (2003) Barriers, repellents and antifeedants for slug and snail control. Crop Prot 22:1033–1038

Simms LC, Glen D, Wilson MJ (2006) Mini-plot field experiments using seed dressings to control slug damage to oilseed rape. Crop Prot 25:890–892

Smith VR (1992) Terrestrial slug recorded from sub-Antarctic Marion Island. J Mollusc Stud 58:80–81

South A (1992) Terrestrial slugs: biology, ecology and control. Chapman and Hall, London

Sysoev A, Schileyko A (2009) Land snails and slugs of Russia and adjacent countries. Pensoft, Series Faunistica No. 87, Sofia-Moscow

Tan L, Grewal PS (2001) Infection behaviour of the rhabditid nematode *Phasmarhabditis hermaphrodita* to the grey garden slug *Deroceras reticulatum*. J Parasitol 87:1349–1354

Tandingan De Ley I, Holovachov O, McDonnell RJ, Bert W, Paine TD, De Ley P (2016) Description of *Phasmarhabditis californica* n. sp. and first report of *P. papillosa* (Nematoda: Rhabditidae) from invasive slugs in the USA. Nematol 18:175–193. doi:10.1163/15685411-00002952

Torchin ME, Lafferty KD, Kuris AM (2001) Release from parasites as natural enemies: increased performance of a globally introduced marine crab. Biol Invasions 3:333–345

Willis JC, Bohan DA, Choi YH et al (2006) Use of an individual-based model to forecast the effect of climate change on the dynamic, abundance and geographical range of the pest slug *Deroceras reticulatum* in the UK. Glob Chang Biol 12:1643–1657

Wilson MJ, Barker GM (2011) Slugs as pasture pests. Proc N Z Grassland Assoc 72:241–246

Wilson MJ, Glen DM, George SK (1993) The rhabditid nematode *Phasmarhabditis hermaphrodita* as a potential biological control agent for slugs. Biocontrol Sci Tech 3:503–511

Wilson MJ, Glen DM, Pearce JD et al (1995) Monoxenic culture of the slug parasite *Phasmarhabditis hermaphrodita* (Nematoda: Rhabditidae) with different bacteria in liquid and solid phase. Fundam Appl Nematol 18:159–166

Chapter 24
Marine and Estuarine Nematodes in South Africa

Mathys C. Vosloo and Martin G.J. Hendricks

24.1 Introduction

Free-living nematodes are present in terrestrial (see Chap. 20) and aquatic environments. Such nematodes occur in all marine and estuarine environments (Fig. 24.1) but are typically most common in sediments where they may comprise up to 92% of living carbon, a contribution over ten times more important than that of bacteria (Sikora et al. 1977). The meanings of a few of the typical terms used in aquatic, meiofaunal studies are given in Box 24.1.

Nematodes occupy the psammal or interstitial environment and are generally the most abundant component of the meiofauna (Dye 1977; 1978a, b; 1979; Dye et al. 1978; Dye and Furstenberg 1978; Moens and Vincx 1997). Meiofauna are of intermediate size between microbenthic and larger macrobenthic taxa. Some authors have classified meiobenthic organisms as larger than 63 μm but smaller than 2 mm (Fleeger et al. 1988), whilst others regard them as those animals passing through a 0.5-mm mesh sieve but are retained by a sieve of 45 μm (Kennedy and Jacoby 1999). General consensus, however, is that meiofauna pass through a 1-mm mesh sieve and are retained on a 42- or 63-μm mesh sieve (Gibbons 1991; Higgins and Thiel 1988; Mare 1942).

M.C. Vosloo (✉)
Zitholele Consulting, PO Box 6002, Halfway House, Midrand 1685, South Africa
e-mail: mathysv@zitholele.co.za

M.G.J. Hendrikcks
Department of Biodiversity and Conservation Biology, University of the Western Cape, Private Bag X17, Bellville 7535, South Africa

H. Fourie et al. (eds.), *Nematology in South Africa: A View from the 21st Century*, DOI 10.1007/978-3-319-44210-5_24

Fig. 24.1 Gamtoos River Estuary near Port Elizabeth where research on estuarine nematodes has been conducted (Kirk West, Port Elizabeth, South Africa)

Box. 24.1 Glossary of Typical Terms Used in Aquatic, Meiofaunal Research

Meiofauna: Organisms that pass through a 1-mm mesh sieve and are retained on a 42- or 63-μm mesh sieve

Psammal: Sandy soil environment or habitat

Interstitial: Spaces in-between grains of sand along beaches, river banks or estuaries

Microbenthic: Relating to invertebrate organisms living within or on the surface of aquatic sediments and that are less than 63 μm in size

Macrobenthic: Relating to invertebrate organisms living within or on the surface of aquatic sediments and that are larger than 1 mm in size

Depauperate: Lacking in numbers or variety of species in an ecosystem

Abyssal: Relating to the depth or bed of the ocean

A concise summary about the origin of marine meiofaunal research is given in Box 24.2. Marine meiofauna comprises many different taxa, including Nematoda, Sarcomastigophora, Ciliophora, Cnidaria, Turbellaria, Gastrotricha, Rotifera, Polychaeta, Oligochaeta, Copepoda and Ostracoda. Higgins and Thiel (1988) have usefully condensed aspects of the study of meiofauna into a standard reference work, which includes comprehensive biological and descriptive accounts of all permanent (those that remain meiofaunal in size their whole life) and temporary (those that are meiofaunal in size for a portion of their life) members of the meiofaunal community. This book also deals with field and laboratory methods of sample collection and analysis and has been a valuable resource for both novice and experienced meiobenthologists.

Box. 24.2 Pioneer Work in Terms of Meiofaunal Nematode Research
The earliest pioneering work on the study of meiofaunal communities originates from the Northern Hemisphere, primarily North America, Britain and Western Europe and dates back to the beginning of the nineteenth century. A number of recent marine nematode reviews include those by Gray (1994; 2002), Gage (1996), Levin et al. (2001), Lambshead and Boucher (2003) and Vanreusel et al. (2010). Heip et al. (1982) estimated that there are 450 genera of free-living marine nematodes, whilst Warwick et al. (1998) suggested that about 4,000 species have been described worldwide. Most free-living marine nematodes belong to the class Adenophorea, with only a few categorised under the Secernentea. The main diagnostic differences between the two classes are the presence of caudal glands, bristles and conspicuous amphids in the Adenophorea (Higgins and Thiel 1988).

Meiofaunal community structure may be influenced by a myriad of different physical and biological factors, which may act independently, or more commonly, in synergy. Such factors include sediment characteristics and organic matter loadings (Higgins and Thiel 1988), salinity (Moens and Vincx 2000), oxygen concentrations (Dye 1983a), behaviour (Moens et al. 2000), predation (Moens and Vincx 1997; Moens et al. 2000) and competition (Ólafsson 2003). Particle size is considered to be one of the most influential factors structuring communities (Higgins and Thiel 1988). Coarse sediments have large interstitial spaces; they are generally associated with well-oxygenated porewaters and are well flushed. Such sediments allow large wriggling nematode species, such as taxa belonging to the family Epsilonematidae (Hendricks and Gibbons 2010), to persist. By contrast, silty or muddy sediments have small interstitial spaces; they drain poorly and are often not well oxygenated. Such sediments are dominated by small-sized nematode species (Elliot et al. 1980). Dye and Furstenberg (1978) noted that meiofaunal densities correlated with sediment particle size in the Swartkops Estuary, Port Elizabeth. In estuarine environments, salinity has been observed to have a negative impact on the survival of nematode juveniles at both very low and very high levels (Dye and Furstenberg 1978). Low temperatures affect development time of some species to the point that reproduction can be halted (Moens and Vincx 2000). Nozias et al. (2005) found that temperature and estuary state (open or closed) were important controlling factors at the Mdloti Estuary mouth, on the east coast of South Africa (SA). On sandy beaches, oxygen concentration is one of the major controlling factors governing the distribution of meiofauna, although desiccation higher up on the intertidal regions also seems to play an important role (McLachlan 1977a; Dye and Furstenberg 1978). Dye (1983a) found that the redox potential of estuarine sediment strongly influenced distribution and abundance of meiofauna in the Mngazana Estuary, also on the east coast of SA. Nematode communities were depauperate below the discontinuity layer, which is generally a thin grey layer where the redox potential changes rapidly over a small distance. Anaerobic conditions prevail below the discontinuity layer. Hooge (1999) observed that sediment saturation on sandy beaches was strongly correlated to meiofaunal abundance at the mid- and low-tide levels. Meiofaunal abundance and distribution on exposed sandy beaches are also influenced by wave action and availability of food (McLachlan 1977a).

Food availability and prey selectivity play an important role in structuring nematode communities in the estuarine environment. Moens et al. (2000) observed that some predacious nematode species have prey density-dependent predation rates making them prey-limited in some microhabitats. These authors also showed that predation rates were strongly affected by temperature and light intensity and that some predators exhibited clear preferences for different prey species. Ólafsson (2003) concluded that most groups of macrofauna increase the diversity of the meiofaunal community through predation, physical disturbance, competition for food and biogenic structures.

24.2 Nematode Diversity

Large-scale surveys of marine meiobenthos and of benthic nematodes are limited (Gage 1996) due to the vastness of both the oceans and the cost related to such research. Nonetheless, a number of studies from different oceans and seas have been conducted during the last 25 years. In general, deep-sea marine taxa, including nematodes, exhibit high species richness in the Northern Hemisphere with a gradient decreasing from the tropics to the Arctic. This gradient can be ascribed to a decrease in the abundance of organic matter sinking to the ocean floor, with the highest rate occurring at the equator. This pattern may be altered by ecology, bathymetry or sediment structure (Lambshead et al. 2000, 2001; Brown et al. 2001; Rex et al. 2001). For instance, Lambshead et al. (2000) pointed out that the North Atlantic consists of a number of separate basins, each with its unique diversity. Highest diversity in the Southern Hemisphere is recorded in the temperate regions (Rex et al. 1993; Willig et al. 2003), decreasing to the north and south for South American Pacific coast molluscs (Rivadeneira et al. 2002). Levin et al. (2001) concluded that deep-sea diversity is high on both local and regional scales and that variation in diversity was caused by sediment structure, food availability, oxygen content, hydrological conditions and disturbance events.

Boucher (1990) and Boucher and Lambshead (1995) compared the ecological diversity of marine nematodes from temperate, tropical and deep-sea regions, and they found a non-linear relationship between depth and species diversities. Some of the deep-sea stations in the abyssal and bathyal depths were the most diverse. The non-linear relationship may result from the interaction between productivity, disturbance and depth. Grassle and Morse-Porteus (1987) used this relationship to propose the non-equilibrium spatiotemporal mosaic theory, whereby local species richness is enhanced because of the patchwork of phytochemical resources, of variable age, that are available on the ocean bed. The findings of Boucher and Lambshead (1995) and Brown et al. (2001) all tend to support this theory.

Gage (1996) postulated that meiofaunal communities reach peak sizes and attain the greatest biomass and/or abundance in shallow-water environments. Local diversity is affected by disturbances such as chemical pollution and dredging and landscape alterations resulting in changes in diversity patterns often accompanied by changes in abundance and biomass. Warwick (1986, 1993) demonstrated that abundance and biomass curves for disturbed environments revealed a dominance of the abundance over biomass curves, due to the presence of colonising organisms. This tends to agree with the findings of Gray (1994), Lambshead et al. (1994) and Vincx et al. (1994), who reported that a diversity gradient exists from shallow coastal waters to deep water. Shallow-water sites are often very small in extent, and diversity may be modest in comparison to off-shore sites. However, shallow-water environments at a regional scale will tend to yield a greater diversity of nematodes because of the complex array of small habitats, each with its unique diversity.

The nematodes and meiofauna of the North Sea are reasonably well known and a number of papers have been published in recent years. As early as 1975, Anker and Elmgren (1975) studied the benthic ecology of the macro- and meiofauna of the Askö-Landsort area, and they made a valuable contribution in the determination of abundance-biomass relationship in meiofauna communities. Later studies investigated a wide range of ecological and morphological questions. Some of these studies include those by Vincx (1990), Steyaert et al. (1999, 2003), Tita et al. (1999), Boyd et al. (2000), Vanaverbeke et al. (2002), Rzeznik-Orignac et al. (2003), Danovaro et al. (2004) and Braeckman et al. (2011). A few examples of global nematode studies since 1982 include:

(i) Deep-sea studies (Thistle and Sherman 1985; Lambshead et al. 2000; Lambshead and Boucher 2003; Gambi et al. 2003; Vanhove et al. 2004)
(ii) Shallow-water studies (Gheskiere et al. 2004, 2005; Armenteros et al. 2009; Sandulli et al. 2010)
(iii) Studies on the effect of effluents and heavy metals in marine environments or harbours (Dalto et al. 2006)
(iv) Ecological disasters, such as oil spills or low oxygen events (Boucher 1980; Gourbault 1987; Cook et al. 1999; Neira et al. 2001)
(v) Effects caused by aquaculture activities, such as fish farming (Mirto et al. 2002) or mussel farms (Danovaro et al. 2004)

24.3 Taxonomy and Classification of Free-Living Marine and Estuarine Nematodes

Regrettably, the systematics of free-living nematodes has always been turbulent, with classification systems changing frequently (De Ley and Blaxter 2002). Several authors have attempted to classify nematodes into distinct taxonomic groups since the turn of the twentieth century (see De Ley and Blaxter (2002), for a short history of nematode systematics). Lorenzen (1981, 1994), for example, classified free-living marine nematodes into the four orders Enoplida, Trefusiida, Chromadorida and Monhysterida, based on morphology. The Trefusiida has since been grouped into a different taxonomic level (De Ley and Blaxter 2002). Based on molecular, morphological and ontological features, De Ley and Blaxter (2002) reordered the classification system of Lorenzen (1981) and others into seven orders of free-living nematode species (Plectida, Araeolaimida, Monhysterida, Desmodorida, Chromadorida, Enoplida and Triplonchida). Unfortunately, as De Ley and Blaxter (2002) themselves eloquently stated: 'The latest (classification) system is never the last one'.

The Linnean Society of London, in association with the Estuarine and Coastal Sciences Association, published in three volumes an influential, illustrated identification guide to the most common British, free-living marine nematodes. These include Part I Enoplids (Platt and Warwick 1983, currently out of print), Part II Chromadorids (Platt and Warwick 1988) and Part III Monhysterids (Warwick et al. 1998). These identification guides employ pictorial keys for identification to genus level, together with identification notes on the most common British species. These books represent the most up-to-date and comprehensive guides available and are widely used by those studying meiobenthic nematodes. That notwithstanding, nematologists in developing countries, who are working with very limited resources, have often erroneously identified species found in their countries as common British species whilst using these guides.

Recognising the problem, Ghent University (Ghent, Belgium) has developed a series of more accessible tools (see Box 24.3) to provide such nematologists with the necessary skills and modern resources, which should allow them to correctly identify nematode species.

Box. 24.3 Recent Valuable Resources Available for the Identification of Marine Nematodes

NeMys is a generic online species information system (http://nemys.ugent.be) containing several databases on a variety of meiobenthic taxa, with the nematode database being only one of them (Deprez et al. 2005). *NeMys* offers an online identification guide based on the guides of Platt, Warwick and colleagues. It also offers several identification keys to families, genera or species contributed by experienced nematologists. New data and information can be published online, and past literature and taxonomic papers on all species can be downloaded free of charge.

The Plymouth Marine Laboratory has developed another internet resource to aid in the identification of marine nematodes (Austen et al. 2006). It offers a downloadable nematode identification key that is based on the pictorial keys of Platt and Warwick. This key offers the user identification of nematodes to genus level only using four identifiable, morphological characteristics such as the tail, mouth (buccal cavity), cuticle and amphid. Although a useful tool to aid identification of nematodes to genus level, it is limited by the small number of identifiable characteristics used by the key, especially when trying to identify species that are morphologically very similar, such as the members of the family Xyalidae, which often only differ in the position and number of cephalic setae.

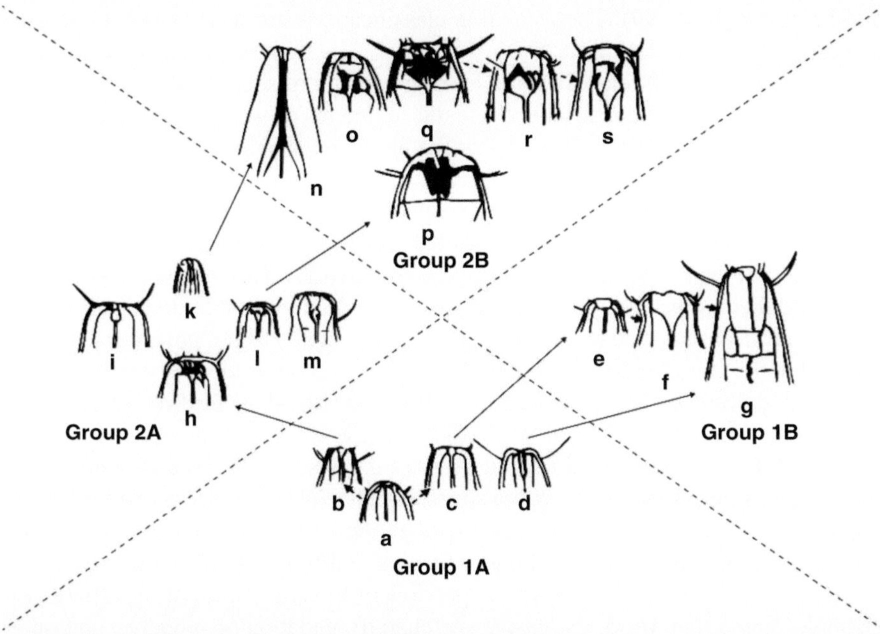

Fig. 24.2 The four main morphological feeding groups for free-living marine nematodes, *a* Oxystomatina, *b* Anticoma, *c* Terschellingia, *d* Parachromagasteriella, *e* Sabatiera, *f* Paramonhystera, *g* Bathylaimus, *h* Paracanthonchus, *i* Linhomoeus, *k* Onchium, *l* Chromadorea, *m* Microlaimus, *n* Siphonolaimus, *o* Halichoanolaimus, *p* Enoplus, *q* Oxyonchus, *r* Oncholaimus, *s* Eurystomatina (Adapted from Wieser 1953)

24.4 The Feeding Ecology of Estuarine and Marine Nematodes

One of the earliest works, detailing the feeding ecology of free-living marine nematodes is that of Wieser (1953). This classification is entirely based on the morphology of the buccal cavity and divides the nematodes into four groups, viz. Groups 1A, 1B, 2A and 2B (Fig. 24.2).

In short, Group 1A represents nematodes with small buccal cavities containing no teeth. Nematodes in this group were perceived to be selective deposit feeders, sucking soft food into the intestine. Group 1B includes nematodes with cup-shaped, conical or cylindrical buccal cavities with no teeth. Feeding is thought to be nonselective deposit feeding, aided by the lips and anterior part of the buccal cavity. Nematodes classified in Group 2A have buccal cavities armed with small teeth. Food is likely scraped off surfaces or cells are pierced and the cell fluid sucked out. A large buccal cavity, armed with large teeth or stylet, defines nematodes in Group 2B. These nematodes are the predacious species, feeding by ingesting the prey whole or by piercing or ripping the prey over the large teeth whilst sucking out the cell or body contents. Variations to the feeding classification scheme of Wieser can be found in Romeyn and Bouwman (1983), Jensen (1987), Moens and Vincx (1997) and Moens et al. (1999). These feeding classifications are widely used by marine and estuarine nematologists.

Yeates et al. (1993) reassessed the feeding ecology of soil nematodes and classified the nematodes of terrestrial and aquatic sediments as well as those parasitic on plants, according to their food source or feeding habit as inferred from the ecological setting of a particular nematode family, genus or species. Eight feeding groups were identified (see Sect. 20.2.1), viz. (i) plant feeders, (ii) fungal/hyphal feeders, (iii) bacterial feeders, (iv) substrate ingestion feeders, (v) carnivores, (vi) unicellular eucaryote feeders, (vii) parasites and (viii) omnivores. This restructuring of soil nematode feeding ecology from the literature resulted in a more universal feeding classification system used by nematologists. Yeates et al. (1993), however, acknowledged that the classification of nematodes according to their feeding habit 'may not be sharply delimited'. Since then, other works have refined the feeding classification (Moens and Vincx 1997; Moens et al. 2004). These works highlighted the fact that the current feeding classification of nematodes is revised as the results of new studies come to light. Moens et al. (2004) stressed that nematode feeding habit is largely dependent on the ecological setting of a species, genus or family of nematodes. They further suggested that food switching or feeding at different trophic levels may be a more common phenomenon that was previously thought. Elliot et al. (1980), for example, found that *Mesodiplogaster* switched to predation on amoebae and other protozoa when soil pores became too small for access to the bacteria, its primary food source. For this reason, Moens et al. (2004) found terminology used in the earlier feeding classifications such as 'nonselective deposit feeders' (Wieser 1953) and the feeding type 'omnivore' (Yeates et al. 1993) restrictive. Substrate ingestion, in the widest sense, is applicable to all feeding habits where some degree of sediment is swallowed together with prey. It is most likely not the sediment particles themselves that are of nutritional value, but the prey items ingested with the sediment or detritus. Moens et al. (2004) suggested that the use of the term 'omnivores' is limiting since many nematode groups at some point in their life cycle feed on more than one trophic level. Moens and Vincx (1997) further highlighted that many of the nematode species assembled in Wieser's Group 2B had diverse feeding habits. Many of the species are mainly predatory, whilst the remainder were considered omnivorous or scavengers, feeding on dead or decaying matter. It was also suggested that no compelling reasons exist why terrestrial, marine and estuarine nematodes should not be grouped together as was done by Yeates et al. (1993). In fact, the classification scheme of Yeates et al. (1993) has been regarded as an improvement as it allows for the assignment of nematodes to more than one trophic level, thereby embracing their flexible feeding strategies. Further, classification of feeding guilds based on morphological characteristics gives information on a nematode's ability to handle food rather than on any feeding preference or habit (Moens and Vincx 1997).

Other less well-understood factors that may influence our understanding of nematode feeding ecology include the levels of food specialisation, vulnerability to predators, orientation of food items, intra-guild resource partitioning, habitat utilisation and burrowing capacity, tolerance ranges of nematodes, response to changing environment, energy needs and budgets, quality and quantity of food source and functional contribution (Moens and Vincx 1997; Moens et al. 2004).

Understanding of nematode feeding ecology is constantly evolving as is demonstrated in Box 24.4.

Box. 24.4 The Complex, Feeding Ecology of Marine Nematodes: An Evolving Phenomenon

Evidence suggests that species such as *Adoncholaimus thalassophygas* 'garden' their own organic food sources (Riemann and Schrage 1978; Jensen 1986, 1987), whilst *Pseudochromadora* is also suspected of 'gardening' their own food source, as these species tend to dominate marine sediment devoid of food. One of the least documented feeding habits of nematodes is their possible trans-epidermal uptake of dissolved organic carbon (DOC) in detritus-rich estuarine habitats (Jensen 1986, 1987). Uptake of DOC was demonstrated by two oncholaimid and one comesomatid nematode species (Chia and Warwick 1969; Lopez et al. 1979; Riemann et al. 1990).

An understanding of the feeding ecology of marine and estuarine nematodes is important if scientists are to put the animals in the context of their environment (Moens et al. 2004). Marked changes in the structure of nematode feeding guilds are often linked to changes in their psammal habitat (Moens and Vincx 1997), and much work has been conducted in this area. Nematode feeding habit, and the accuracy of their subsequently derived ecological function in a particular ecosystem, therefore ultimately depends on the extent of the taxonomic, morphological, ecological species-specific information that is available. Most of the recent studies on nematode feeding ecology have come to a similar conclusion: Although a good indication of a particular nematode species, genus or family's feeding type can be inferred from the classification system of Wieser (1953) or Yeates et al. (1993), an accurate description of a particular nematode's feeding habit can often only be obtained through species or genus specific taxonomic and ecological studies.

24.5 Studies on Marine and Estuarine Nematodes from Southern Africa

24.5.1 Southern African Contributions to the Descriptive Taxonomy of Nematodes

The first recorded investigation of nematodes along the coast of southern Africa was a description of three species, *Euchromadora africana* Von Linstow, 1908; *Oncholaimus spiralis* Von Linstow, 1908; and *Enoplus bisetosus* Von Linstow, 1908, from Lüderitzbucht, Namibia. This study was followed by that of Steiner (1918) describing four new species, i.e. *Desmodora michaelseni* Steiner, 1918; *Euchromadora dubia* Steiner, 1918; *Euchromodora longicaudata* Steiner, 1918; and *Euchromadora luderitzi* Steiner, 1918, from the same area. IInglis (1961, 1963, 1966) described 24 new species, including five new genera, from off the coast of SA

Table 24.1 Taxonomic contributions primarily by European taxonomists (prior to 1980) and South African taxonomists in collaboration with European taxonomists (1980 to present), with authorities for each species and locations where type specimens were found

Prior to 1980	
Bathylaimus deconincki Inglis 1966 (Durban, South Africa)	*Mesacanthoides magna* Coles 1977 (Coast of southern Africa)
Deontostoma hopei Coles 1977 (Coast of southern Africa)	*Metoncholaimus murphyi* Inglis 1966 (Durban, South Africa)
Deontostoma parantarcticum Vitiello 1975 (Lamberts Bay, South Africa)	*Nudora omercooperi* Inglis 1966 (Durban, South Africa)
Desmodora cuddlesae Inglis 1963 (Coast of South Africa)	*Oncholaimus problematicus* Coles 1977 (Coast of southern Africa)
Desmodora michaelseni Steiner 1918 (Luderitz Bay, Namibia)	*Oncholaimus spiralis* Von Linstow 1908 (Luderitz Bay, Namibia)
Dorylaimopsis janetae Inglis 1963 (Coast of South Africa)	*Pheronus ogdeni* Inglis 1966 (Durban, South Africa)
Dorylaimopsis nini Inglis 1961 (Greater Kleinemonde Estuary, South Africa)	*Platycoma sudafricana* Inglis 1966 (Durban, South Africa)
Enoploides mandibularis Coles 1977 (Coast of southern Africa)	*Platycomopsis dayi* Coles 1977 (Coast of southern Africa)
Enoplus bisetosus Von Linstow 1908 (Luderitz Bay, Namibia)	*Plectolaimus juliani* Inglis 1966 (Durban, South Africa)
Epacanthion oliffi Inglis 1966 (Durban)	*Polygastrophora omercooperi* Inglis 1961 (Greater Kleinemonde Estuary, South Africa)
Euchromadora africana Von Linstow 1908 (Luderitz Bay, Namibia)	*Rhabdodemania dura* Inglis 1966 (Durban, South Africa)
Euchromadora dubia Steiner 1918 (Luderitz Bay, Namibia)	*Sigmophoranema brevispiculatum* Inglis 1963 (Coast of South Africa)
Euchromadora longicaudata Steiner 1918 (Luderitz Bay, Namibia)	*Sphaerolaimus anterides* Inglis 1961 (Greater Kleinemonde Estuary, South Africa)
Euchromadora luderitzi Steiner 1918 (Luderitz Bay, Namibia)	*Symplocostomella trichostoma* Coles 1977 (Coast of southern Africa)
Filoncholaimus capensis Coles 1977 Coast of Southern Africa)	*Thoracostoma ancorarium* Vitiello 1975 (Lamberts Bay, South Africa)
Hypodontolaimus angelae Inglis 1961 (Greater Kleinemonde Estuary, South Africa)	*Trileptium longisetosum* Inglis 1966 (Durban, South Africa)
Linhomoeus timmi Inglis 1963 (Coast of South Africa)	*Trissonchulus janetae* Inglis 1961 (Greater Kleinemonde Estuary, South Africa)
Marilynia dayi Inglis 1963 (Coast of South Africa)	*Wiesoncholaimus mawsonae* Inglis 1966 (Durban, South Africa)
Marilynia wieseri Inglis 1963 (Coast of South Africa)	*Xenodesmodora nini* Inglis 1963 (Coast of South Africa)
Mesacanthion fricum Inglis 1966 (Durban, South Africa)	*Xyzzors fitzgeraldae* Inglis 1963 (Coast of South Africa)

Table 24.1 (continued)

1980 to present	
Africanema interstitialis Vincx and Furstenberg 1988a (Algoa Bay, Beach, South Africa)	*Gonionchus africanus* Vincx and Furstenberg 1988b (Algoa Bay, Beach, South Africa)
Algoanema aestuariense Heyns and Furstenberg 1987 (Swartkops River, Estuary, South Africa)	*Leptepsilonema saldanha* Hendricks and Gibbons 2010 (Saldanha Bay)
Axonolaimus deconincki Vincx and Furstenberg 1989 (Sundays River, Estuary)	*Microlaimus africanensis* Furstenberg and Vincx 1992 (Algoa Bay, Harbour, South Africa)
Ceramonema africana Furstenberg and Vincx 1993 (Algoa Bay, Estuary, South Africa)	*Molgolaimus typicus* Furstenberg and Vincx 1992 (Algoa Bay, Rocky Shore, South Africa)
Ceramonema algoensis Furstenberg and Vincx 1993 (Algoa Bay, Estuary, South Africa)	*Namibnema papillata* Vincx and Furstenberg 1989 (Namibia, Beach)
Ceramonema kromensis Furstenberg and Vincx 1993 (St Francis Bay, Beach, South Africa)	*Perepsilonema benguelae* Hendricks and Gibbons 2010 (Saldanha Bay, South Africa)
Chromadoropsis granulosus Furstenberg and Vincx 1988b (Algoa Bay, Beach, South Africa)	*Procamacolaimus africanus* Furstenberg and Vincx 1988a (Algoa Bay, Beach, South Africa)
Chromadoropsis namibiensis Furstenberg and Vincx 1988b (Namibia, Beach)	*Pterygonema papenkuili* Furstenberg and Vincx 1993 (Algoa Bay, Estuary, South Africa)
Cobbionema capense Furstenberg and Heyns 1987 (Swartkops River, Estuary, South Africa)	*Viscosia erasmi* Furstenberg and Vincx 1989 (Sundays River, Estuary, South Africa)
Dasynemoides tenuis Furstenberg and Vincx 1993 (Algoa Bay, Estuary, South Africa)	*Xyala aestuariensis* Vincx and Furstenberg 1988b (Algoa Bay, Estuary, South Africa)
Eontolaimus capensis Furstenberg and Vincx 1988a (Algoa Bay, Beach, South Africa)	*Xyala psammonalis* Vincx and Furstenberg 1988b (Algoa Bay, Beach, South Africa)

and included nematodes from Durban (KwaZulu-Natal Province) harbour and the Great Kleinemonde Estuary (Eastern Cape Province) (see Table 24.1). These descriptions were mostly based on specimens that were collected during the University of Cape Town Ecological Surveys and by the National Institute for Water Research along the east and west coasts of SA. Vitiello (1975) described two new species of the family Leptosomatidae in Lamberts Bay (Western Cape Province), north of Saldanha Bay, along the west coast. Coles (1977) described seven new species of nematodes (order: Enoplida) and identified a further 25, (belonging to 19 genera and 8 families) from collections along the SA coast.

Free-living marine and estuarine nematology experienced a revival in the early 1980s. During the 1980s and early 1990s, J.P. Furstenberg, a local nematologist by training, together with his co-authors, described species from several families including Ceramonematidae, Desmodoridae, Microlaimidae, Oncholaimidae, Spirinidae, Trefusiidae, Selachinematidae, Chromadoridae and Xyalidae (Table 24.1). A total of 62 species new to science have been formally described from around Southern Africa (Table 24.1). Including these new species, 338 species have been recorded in SA, with 30 species being endemic to the region. The SA species represent 8.45 % of the global species (Gibbons et al. 1999).

The involvement of British and European experts (such as Furstenberg's co-author, M. Vincx) in the study of regional nematodes has been substantial. This was not only true in terms of SA but also in other African countries such as Kenya. Vincx and Muthumbi have published several works describing new species along the African east coast (Muthumbi and Vincx 1996, 1997, 1998, 1999; Muthumbi et al. 1995, 1997; Verschelde and Vincx 1992, 1993, 1995). Works describing new nematode species along the east and west coasts of Africa are of relevance to nematologists working along the coastline of southern Africa, as many of the more cosmopolitan species identified are likely to occur along the South African coastline.

Furstenberg and his co-workers' contribution to marine and estuarine nematode taxonomy in SA has been invaluable. In an early paper by Furstenberg and Dye (1982), it was noted that studies on free-living marine nematodes had largely been neglected in Africa. More than a decade later, Gibbons et al. (1999) still lamented the lack of trained taxonomists, especially those that work with meiofauna, and this compromises our ability to move the science of nematology forward.

24.5.2 Sandy Shore Meiofauna

Research on meiofauna and, in particular, nematodes by local scientists went through a very promising development period during the late 1970s and early 1980s with the emergence of a series of peer-reviewed contributions (McLachlan et al. 1977a, b; McLachlan and Furstenberg 1977; McLachlan 1977a, b, c; Dye et al. 1978; Dye and Furstenberg 1978; Dye 1977, 1978a, b, 1979, 1983a, b). These studies were all conducted by researchers based in and around Algoa Bay, Port Elizabeth (Eastern Cape Province) and investigated ecological aspects of the meiofauna, including distribution, composition and abundance in estuaries and sandy beaches of the east and south coasts of SA. They also included observations on the ecophysiology of meiofauna and the response and influence of physical conditions on meiofaunal ecology (see Furstenberg and Dye 1982). Only one of the papers (McLachlan and Furstenberg 1977) investigated and reported on nematode genera found in the sediment samples of Kings Beach and Sundays River beach in Algoa Bay. These samples contained 29 species from 23 genera and represented the only accounts with particular focus on the nematode diversity conducted during the 1980s.

McLachlan et al. (1977b) undertook one of the first studies on the vertical and horizontal distributions of sublittoral meiofauna in Algoa Bay, SA. They reported on the correlation between nematode abundance and dominance along unpolluted and sewerage exposed beaches. Abundance correlated strongly with particle size and level of desiccation on all the beaches, whilst results suggested higher nematode (and meiofauna) abundances at more polluted beaches. Nematodes dominated the meiofauna communities investigated in the study by number but did not make up more than 56 % of the meiofauna community on any of the beaches (McLachlan et al. 1977b). Unfortunately, none of the specimens was described, and there is no information available on the nematode diversity within the assemblage. Other studies included those of polluted and unpolluted sandy beaches along the west, south and south-western coastlines of SA (Fricke et al. 1981; Hennig et al. 1982, 1983; Fricke and Flemming 1983; Warwick 1984), work on vertical profiles of meiofauna in sandy subtidal environments (Malan and McLachlan 1985) and studies on a large artificial reef (Fricke et al. 1986). Furstenberg and Dye (1982) found that nematode abundance increased from the east to the west coasts of SA. They reported that nematodes accounted for 20 % of meiofauna assemblages between St Lucia and Blythedale on the KwaZulu-Natal north coast, whilst approximately 60 % of meiofauna consisted of nematodes in the Port Elizabeth area. A recent study on the ecosystem health of Nelson Mandela Bay sandy beaches in the Port Elizabeth area using, amongst others, meiofauna community composition as an indicator of change (Yani 2009) revealed that the nematode dominance in the meiofauna community structure of 13 beaches ranged between 47 and 84 %. Along the west coast, nematodes dominated the meiofaunal samples; for example, a baseline survey of meiofauna at Robben Island (Cape Town area in the Western Cape Province) showed almost complete dominance (>86 %) of nematodes over other meiofaunal taxa (International Oceanographic Institute, Prochazka 2003).

24.5.3 Rocky Shore Meiofauna

Both on the local and international fronts, recognition that meiofauna could be found on hard as well as soft substrata came late. Much of the pioneering work on rocky shore meiofauna in SA was undertaken by MJ Gibbons (Gibbons and Griffiths 1986, 1988; Gibbons 1988a, b, c) who published a brief overview of knowledge in 1991 (Gibbons 1991). Rocky shore meiofauna occurs in a variety of habitats including bare rock and rock crevices as well as on sessile macrofauna and in association with algae (see literature cited by Gibbons 1991). Factors influencing meiofaunal distribution, abundance and diversity in the rocky intertidal region include desiccation, wave exposure, algal architecture, accumulated sand, position on the shore gradient and the availability of refugia. Gibbons' work did not go into taxonomic detail. However, a study on the taxonomy of two new nematode species living on macroalgae has been conducted by Furstenberg and Vincx (1992). They described two new *Microlaimus* spp. from a sand sediment environment from a sand bank within the Port Elizabeth harbour and from brown algae collected from the subtidal zones at Flat Rocks, a flat rock feature located 5 km south of the Port Elizabeth harbour (Furstenberg and Vincx 1992).

Studies with the main focus on the nematode communities of rocky shores have been few and far between, and to the best of our knowledge, the only work is that of Garner (1995). She investigated the nematodes on subtidal macroalgae at a sheltered, moderately exposed and fully exposed rocky shore. Nematode densities at the three sites did not differ significantly, but densities were found to be significantly higher in coralline turf than on the algae, *Amphiroa ephedraea* and *Plocamium corallorhiza*. The nematode communities were maintained through continuous reproduction. Epigrowth feeders (Wieser's type 2A) and selective deposit feeders (Wieser's type 1A) dominated the communities. The different algal species and seasonal climatic variation were found to be the greatest factors influencing distribution and abundance within the nematode communities (Garner 1995).

24.5.4 Estuarine Meiofauna

One hundred and eighty-four of the 250 estuaries in SA are temporary open/closed estuaries (TOCEs) (Day 1981), and they function very differently to those that are permanently open. Permanently open estuaries maintain a connection to the sea at all times, whilst TOCEs are closed off from the sea periodically during the dry season. Most of the research conducted on meiofauna and nematodes in SA estuaries has been conducted in permanently open estuaries. Dye and co-workers published a series of papers on the ecophysiology of meiofauna of the permanently open Swartkops Estuary in the Port Elizabeth area (Dye 1978b; Dye et al. 1978; Dye and Furstenberg 1978). These studies concluded that the sand prawn, *Callianassa kraussi* Stebbing, 1900, appeared to oxygenate the deeper sediment layers around their burrows, thereby creating a suitable habitat for meiofaunal communities. Meiofaunal communities further showed strong correlation with particle size, oxygen and desiccation in the estuary. Oxygen and temperature also seemed to govern spring and autumn peaks in community abundance. It was finally concluded that the meiofauna was qualitatively important in the sand and mudflats within the Swartkops Estuary as they were responsible for more than 75 % of the secondary production, which represent production by faunal groups at the second trophic level.

Dye (1983a, b, c) also investigated the composition, distribution and seasonal fluctuations of meiofauna in the permanently open Mngazana River Estuary. This estuary falls within the transition zone between the tropical and sub-tropical ocean biogeographical regions and is characterised by the presence of large mangrove stands. Mean abundance was positively correlated with redox potential but showed poor relationships with pH and temperature. Maximum meiofauna densities were observed between the low and high tide levels, and, numerically, the nematodes dominated (80 %) the meiofauna community. Wynberg and Branch (1994) investigated the long-term effects of the disturbance associated with bait collecting on the biota of an intertidal sandflat in the Langebaan Lagoon, on the west coast of SA. Sediment compaction, associated with bait digging, seemingly caused the protracted recovery (18 months) of burrowing prawns, *Upogebia africana* and

Callianassa kraussi. In addition, meiofaunal numbers declined immediately after disturbance of the burrowing prawn species but rapidly recovered to control levels. Nematodes dominated the meiofaunal communities at this site.

More recently, Nozias et al. (2005) investigated the seasonal dynamics of meiofauna in the temporary open/closed sub-tropical Mdloti Estuary near Umdloti in the KwaZulu-Natal Province. Their study found, inter alia, that nematodes, mites and harpacticoid copepods dominated the meiofaunal community. Even though the average consumption of microphytobenthos for the whole estuary was low (11 %), meiofauna showed the potential to be a major consumer of microphytobenthos in the estuary, with potential consumption rates reaching as high as 254 % of the standing stock in some areas (Nozias et al. 2005). The meiofaunal community could thus be an important driver shaping the structure of the benthic food web. Further, meiofauna communities in TOCEs seem to be governed by temperature and the estuary mouth condition, i.e. whether open or closed, as meiofauna abundance peaked after prolonged mouth closure but decreased considerably after the estuary mouth breached.

24.6 Nematodes as Biomonitors

Despite the obvious potential of meiofauna and nematode communities as monitors of ecosystem health and pollution (Bongers 1990; Roberts 1996; Gyedu-Ababio et al. 1999; Gyedu-Ababio and Baird 2006; Blair 2007; Yani 2009), very little work of this sort has been conducted in SA. This is the case for both aquatic and terrestrial (see Chap. 20) ecosystems. Researchers have rather opted for other, perhaps easier, indicators of change such as macrofauna (Resh et al. 1995). South Africa is not alone in this regard as avoiding meiofauna as indicators is considered a worldwide dilemma (Kennedy and Jacoby 1999).

The first investigation into the ecological effects of pollution on benthic communities in SA was conducted by Emmerson et al. (1983). Since then studies of the effect of pollution on the meiofauna and nematode communities in the Swartkops Estuary have been conducted by Binning (1999), Gyedu-Ababio et al. (1999), Gyedu-Ababio and Baird (2006) and Gyedu-Ababio (2011). Gyedu-Ababio et al. (1999) reported that the richness of nematode genera showed significant negative correlations with the concentrations of iron (Fe) and zinc (Zn). Generally the density of nematodes decreased along the salinity gradient. These results contradict previous observations made in a sub-tropical Eastern Cape TOCE where nematode communities showed higher densities towards the head of the estuary (Vosloo, personal observation). In the Swartkops Estuary, the greatest diversity occurred towards the mouth. Fifty nematode genera belonging to 19 families were identified in the sediment of the estuary (Gyedu-Ababio et al. 1999). The dominant genera included *Monhystera*, *Theristus*, *Viscosia*, *Adoncholaimus* and *Metalinhomoeus*. Gyedu-Ababio et al. (1999) also found genera such as *Onyx*, *Calomicrolaimus* and *Belbolla* to be seasonal in abundance and localised. Nematode communities at polluted and unpolluted sites differed significantly, and *Theristus* spp. and *Monhystera* spp. were found to be good indicators of community change at polluted sites in the Swartkops Estuary. A combination of the Maturity Index (Bongers 1990), Shannon-Wiener Diversity Index (H') and the c-p (%) triangle (see Sect. 20.2), derived from the nematode data, were found to be good tools in pollution monitoring, especially organic pollution (Gyedu-Ababio et al. 1999).

The potential use of meiofauna and nematodes as indicators of polluted environments was confirmed by microcosm experiments conducted by Gyedu-Ababio and Baird (2006). In laboratory experiments, the genera *Axonolaimus*, *Theristus* and *Paramonohystera* were found to be tolerant to metal pollution. These authors also observed that nematode communities reacted differently to metal and organic pollutions in the polluted Swartkops Estuary. Roberts (1996) investigated the effect of aquaculture and industrial effluents on the meiofaunal communities between the Port Elizabeth harbour wall and the polluted Papenkuils River mouth further north. She found that nematodes dominated the meiofauna and that species of the genus *Oxystomina* were most abundant. Overall, Roberts (1996) noted that the organic content of the sediment was the greatest determinant of the meiofaunal community. This contradicts findings in the Swartkops Estuary, where sediment particle size and salinity were the most significant variables influencing meiofaunal distribution (Binning 1999), although copper, manganese and titanium were also important. Organic pollution due to mussel raft aquaculture did not seem to have any significant influence on the meiofauna, except in isolated cases (Joubert 2011).

Bollmohr et al. (2011) found that the presence of pesticides, most notably endosulfan, p,p-DDE and nitrate concentrations, showed a strong correlation with meiofaunal communities in the Lourens Estuary (False Bay, Cape Town). In the Rooiels Estuary (between Rooiels and Gordon's Bay, Western Cape Province), larger grain size and higher salinity at the bottom of the estuary seemed to influence variance in meiofaunal abundance negatively.

24.7 Recent and Ongoing Research

Three major marine nematode projects have been completed in SA. One of these was conducted by the Nelson Mandela Metropolitan University (NMMU), in Port Elizabeth, and represents the pilot phase of a project to assess the impact of open water fish farming on the meiofaunal and nematode communities under the cages. Results from the study indicated the presence of 14 nematode genera, with *Bathylaimus* spp. and *Linhomaeus* spp. displaying significant changes in abundance over time. The observed trend was in part suspected to be in response to the presence of the stocked fish cages at the sample sites. A minor shift in dominance from *Bathylaimus* spp. to *Terschellingia* spp. was further observed over time. The nematode/copepod ratio (nematode abundance relative to copepod abundance) increased with distance from the stocked cages and peaked during the May 2008 period after which a steady decrease was noted (Joubert 2011).

The second study has been completed (Hendricks 2013) at the University of the Western Cape (UWC). It describes the nematode communities in Saldanha Bay, with particular reference to the influence of eutrophication and pollution on community structure. Saldanha Bay (33 °S, 18 °E) is situated on the west coast of SA and lies within the nutrient-rich Benguela upwelling system. It is a vital and busy port, and the harbour system has been modified a number of times over the years. The addition of jetties and causeways has had a profound impact on water circulation patterns (Monteiro et al. 1990). The system is influenced by a number of other anthropogenic activities such as commercial shellfish (*Mytilus galloprovincialis*) culture, fish-processing factories and general urban development. There is also a significant anthropogenic organic load on the system (Monteiro et al. 1990, 1997; Jackson and McGibbon 1991). The annual yield of mussels from the Bay was approximately 27 metric tonnes (MT) wet weight raft^{-1} in 1999 (Stenton-Dozey et al. 1999) and increased to 700 MT (wet weight) in 2008 (Clark et al. 2011). Each raft generates huge quantities of debris in the form of fallen carcasses, faeces and pseudofaeces, i.e. mucus-covered particles unsuitable for digestion, which may result in increased levels of localised eutrophication. Studies conducted by Jackson and McGibbon (1991) and Stenton-Dozey et al. (1999) have revealed that these mariculture operations impact negatively on the macrobenthic fauna of the Saldanha Bay system. The change in macrofauna community structure was mainly caused by a significant increase in organic input from faeces and mussel debris.

Hendricks (2013) collected 12 (six in summer and six in winter) cores from each of six sampling stations along a transect line radiating outwards a distance of 3,000 m from beneath the mussel farm. A total of 157,932 individual nematodes were counted in all samples, and a total of 4,488 specimens have been identified following standard protocols. Thirty-five families, 117 genera and 136 nominal species have been identified. The results of Hendricks' study suggest that species richness was lowest in the stations immediately below the mussel raft (44 species) and that it generally increased with increasing distance along the transect (100 species were noted in communities furthest from the mussel raft). Sediments below the raft comprised mainly silt and organic matter and were characterised by very high concentrations of heavy metals. The nematode fauna in these sediments was dominated by species of *Sabatieria* (>74 %) and *Parodontophora* (7 %). Sediments furthest from the mussel raft ranged between 63 and 2,000 μm. They contained low concentrations of mud (Flemming 1977; Stenton-Dozey et al. 1999) and had significantly lower concentrations of heavy metals. *Microlaimus* was the dominant genus, followed by *Paralinhomoeus*.

A third study on the nematode community of the temporary open/closed river is Mngazi estuary, on the SA east coast, was being undertaken by MC Vosloo, also at the NMMU. Preliminary results indicate that the nematode communities are governed by the different sediment types at three sampling sites (mouth, middle reaches and head of the estuary). Dominant genera included *Pseudochromadora*, *Microlaimus* and *Trypiloides*. The genus *Terschellingia* was found in increasing abundance from the mid to upper reaches of the estuary, possibly indicating biological or agricultural pollution entering the upper reaches of the estuary. Nematode diversity was high at the mouth but with fewer individuals when compared to the upper reaches, indicating contrasting diversity and abundance gradients within the estuary.

24.8 The Future: Recommendations

The research projects at UWC and NMMU enhanced our understanding of regional nematode diversity and the role that shallow-water nematodes play in ecological processes, but they are just three projects. Despite the fact that excellent research on marine and estuarine nematodes has been conducted in SA in the past, there is so much more to do. Unfortunately, we must still consider SA marine and estuarine nematology to be in its infancy. Sample analysis and species identification are time consuming and we lack the expertise to identify most species. However, alternative approaches to traditional taxonomy exist internationally that could be usefully employed locally. Ambitious molecular bar-coding projects have been initiated (Bhadury et al. 2006, 2008), whilst Bik et al. (2010) used a sample of SA sandy shore nematodes in a study that investigated the relationship between deep-sea and shallow-water nematodes as well as the cosmopolitan nature of taxa.

Nematodes are excellent indicators of ecosystem health and pollution, they play key roles in biogeochemical processes and recycling, they are simple and fascinating models for phylogenetic studies, and they can be used to test a plethora of fundamental ecological questions. It is time that SA marine nematology joined the twenty-first century.

Acknowledgements The authors would like to graciously thank Professor Mark Gibbons at the University of the Western Cape (UWC) for reviewing this manuscript and providing valuable guidance on its structure and content.

References

Anker S, Elmgren R (1975) A survey of the benthic macro-and meiofauna of the Askö-Landsort area. Merentutkimuslaitoksen Julkaisu Havsforskningsinstitutes. Skrift 239:257–264

Armenteros M, Ruiz-Abierno A, Fernández-Garcá R et al (2009) Biodiversity patterns of free-living marine nematodes in a tropical bay: Cienfuegos, Caribbean Sea. Estuar Coast Shelf Sci 85:179–189

Austen M, Warwick R, Somerfield P et al (2006) The Darwin nematode project. Plymouth Marine Laboratory. http://web.pml.ac.uk/nematode. Accessed 15 Sept 2015

Bhadury P, Austen MC, Bilton DT et al (2006) Development and evaluation of a DNA-barcoding approach for the rapid identification of nematodes. Mar Ecol Prog Ser 320:1–9

Bhadury P, Austen MC, Bilton DT et al (2008) Evaluation of combined morphological and molecular techniques for marine nematode (*Terschellingia* spp.) identification. Mar Biol 154:509–518

Bik HM, Thomas WK, Lunt DH et al (2010) Low endemism, continued deep-shallow interchanges, and evidence for cosmopolitan distributions in free-living marine nematodes (order Enoplida). BMC Evol Biol 10:389. doi:10.1186/1471-2148-10-389

Binning KA (1999) An investigation into the relationship between water quality, sediment, and meiofaunal communities of the Swartkops River ecosystem. MSc Dissertation, University of Port Elizabeth, Port Elizabeth

Blair AG (2007) The impact of a sugar by-products effluent on the beach meiofauna at Sezela Beach, KwaZulu-Natal. MSc Dissertation, University of KwaZulu-Natal, Durban

Bollmohr S, van den Brink PJ, Wade PW et al (2011) Environmental variables, pesticide pollution and meiofaunal community structure in two contrasting temporarily open/closed False Bay estuaries. Water SA 37:391–400

Bongers T (1990) The maturity index: an ecological measure of environmental disturbance based on nematode species composition. Oecologia 83:14–19

Boucher G (1980) Impact of *Amoco Cadiz* oil spill on intertidal and sublittoral meiofauna. Mar Pollut Bull 11:95–101

Boucher G (1990) Pattern of nematode species diversity in temperate and tropical subtidal sediments. Mar Ecol 11:133–146

Boucher G, Lambshead PJD (1995) Ecological biodiversity of marine nematodes in samples from temperate, tropical and deep-sea regions. Conserv Biol 9:1594–1604

Boyd SE, Rees HL, Richardson CA (2000) Nematodes as sensitive indicators of change at dredged material disposal sites. Estuar Coast Shelf Sci 51:805–819

Braeckman U, Van Colen C, Soetaert K et al (2011) Contrasting macrobenthic activities differentially affect nematode density and diversity in a shallow subtidal environment. Mar Ecol Prog Ser 422:179–191. doi:10.3354/meps08910

Brown CJ, Lambshead PJD, Smith CR et al (2001) Phytodetritus and the abundance and biomass of abyssal nematodes in the central, equatorial Pacific. Deep-Sea Res I-48:555–565

Chia FS, Warwick RM (1969) Assimilation of labelled glucose from seawater by marine nematodes. Nature 224:720–721

Clark BM, Tunley K, Angel A et al (2011) State of the Bay 2010: Saldanha Bay and Langebaan Lagoon. Saldanha Bay Water Quality Trust, Saldanha Bay

Coles JW (1977) Free-living marine nematodes from southern Africa. Bull Brit Mus (Nat Hist) 31:1–49

Cook AA, Lambshead PJD, Hawkins LE et al (1999) Nematode abundance at the oxygen minimum zone in the Arabian Sea. Deep-Sea Res 47:75–85

Dalto AG, Grémare A, Dinet A et al (2006) Muddy-bottom meiofauna responses to metal concentrations and organic enrichment in New Caledonia South-West Lagoon. Estuar Coast Shelf Sci 67:629–644

Danovaro R, Gambi C, Luna GM et al (2004) Sustainable impact of mussel farming in the Adriatic Sea (Mediterranean Sea): evidence from biochemical, microbial and meiofaunal indicators. Mar Pollut Bull 49:325–333

Day JH (1981) The nature, origin and classification of estuaries. In: Day JH (ed) Estuarine ecology with particular reference to southern Africa. A. A. Balkema, Cape Town, pp 1–6

De Ley P, Blaxter ML (2002) Systematic position and phylogeny. In: Lee DL (ed) The biology of nematodes. Taylor and Francis, London, pp 1–30

Deprez T, Vanden Berghe E, Vincx M (2005) NeMys. World Wide Web electronic publication. www.nemys.ugent.be. (version 4/2009)

Dye AH (1977) An ecophysiological study of the benthic meiofauna of the Swartkops Estuary. MSc thesis, University of Port Elizabeth, Port Elizabeth

Dye AH (1978a) Diurnal vertical migration of meiofauna in an estuarine sand flat. Zool Afr 13:201–205

Dye AH (1978b) Seasonal fluctuations in the vertical distribution of meiofauna in estuarine sediments. Zool Afr 13:207–212

Dye AH (1979) Aspects of the ecology of meiofauna in the Mngazana Estuary, Transkei. S Afr J Zool 14:67–74

Dye AH (1983a) Oxygen consumption by sediments in a southern African mangrove swamp. Estuar Coast Shelf Sci 17:473–478

Dye AH (1983b) Composition and seasonal fluctuations of meiofauna in a southern African mangrove estuary. Mar Biol 73:165–170

Dye AH (1983c) Vertical and horizontal distribution of meiofauna in mangrove sediments in Transkei, southern Africa. Estuar Coast Shelf Sci 16:591–598

Dye AH, Furstenberg JP (1978) An ecophysiological study of the meiofauna of the Swartkops Estuary. II. The meiofauna: composition, distribution, seasonal fluctuations and biomass. Zool Afr 13:19–32

Dye AH, Erasmus T, Furstenberg JP (1978) An ecophysiological study of the meiofauna of the Swartkops Estuary. III. Partition of benthic oxygen consumption and the relative importance of the meiofauna. Zool Afr 13:187–199

Elliot ET, Anderson RV, Coleman DC et al (1980) Habitable pore space and microbial trophic interactions. Oikos 35:327–335

Emmerson WD, McLachlan A, Watling AR et al (1983) Some ecological effects of two sewage outfalls in Algoa Bay. Water SA 9:23–30

Fleeger JW, Thistle D, Thiel H (1988) Sampling equipment. In: Higgins RP, Thiel H (eds) Introduction to the study of Meiofauna. Smithsonian Institute, Washington DC

Flemming BW (1977) Distribution of recent sediments in Saldanha Bay and Langebaan Lagoon. Trans R Soc S Afr 42:317–340

Fricke AH, Flemming BW (1983) Selective microhabitat colonization by interstitial meiofauna as a function of grain size. In: Mclachlan A, Erasmus T (eds) Sandy beaches as ecosystems. Junk, The Hague, pp 421–431

Fricke AH, Hennig HF-KO, Orren MJ (1981) Relationship between oil pollution and psammolittoral meiofauna density of two South African beaches. Mar Environ Res 5:59–77

Fricke AH, Koop K, Cliff G (1986) Modification of sediment texture and enhancement of interstitial meiofauna by an artificial reef. Trans R Soc S Afr 46:27–34

Furstenberg JP, Dye AH (1982) Marine nematodes. In: Keetch DP, Heyns J (ed) Nematology in Southern Africa. Science bulletin. Department of Agriculture and Fisheries, Pretoria, pp 145–148

Furstenberg JP, Heyns J (1987) Cobbionema capense n.sp. from Swartkops estuary, Port Elizabeth (Nematoda: Selachinematidae). Canad J Zool 65:80–82

Furstenberg JP, Vincx M (1988a) Procamacolaimus tubifer Gerlach, 1953, Procamacolaimus africanus sp.nov. and Eontolaimus capensis gen.nov., sp.nov, (Nematoda, Leptolaimidae) from South Africa. S Afr Tydsk Dierk 23:208–214

Furstenberg JP, Vincx M (1988b) Three new Chromadoropsis species (Nematoda, Desmodoridae) from Southern Africa and the North Sea. S Afr J Zool 23:215–223

Furstenberg JP, Vincx M (1989) Two oncholaimid species from a South African estuary (Nematoda, Oncholaimidae). Hydrobiol 184:43–50

Furstenberg JP, Vincx M (1992) Two new species of the family Microlaimidae (Nematoda: Order Chromadorida) from South Africa. Cah Biol Mar 33:245–251

Furstenberg JP, Vincx M (1993) Five new species of the family Ceramonematidae (Nematoda: order Chromadorida) from South Africa. Nematologica 39:139–152

Gage JD (1996) Why are there so many species in deep-sea sediments? J Exp Mar Biol Ecol 200:257–286

Gambi C, Vanreusel A, Danovaro R (2003) Biodiversity of nematode assemblages from the deep-sea sediments of the Atacama Slope and Trench (South Pacific Ocean). Deep Sea Res I 50: 103–117

Garner CJ (1995) The ecological study of nematode assemblages associated with subtidal macroalgae of three rocky shore sites in the Eastern Cape. MSc Dissertation, University of Port Elizabeth, Port Elizabeth

Gheskiere T, Hoste E, Vanaverbeke J et al (2004) Horizontal zonation patterns and feeding structures of marine nematode assemblages on a macrotidal, ultra-dissipative sandy beach (De Panne, Belgium). J Sea Res 52:211–226

Gheskiere T, Vincx M, Urban-Maling B et al (2005) Nematodes from wave-dominated sandy beaches: diversity, zonation patterns and testing of the isocommunities concept. Estuar Coast Shelf Sci 62:365–375

Gibbons MJ (1988a) Studies on the meiofauna of rocky shores. PhD thesis, University of Cape Town

Gibbons MJ (1988b) The impact of sediment accumulations, relative habitat complexity and elevation on rocky shore meiofauna. J Exp Mar Biol Ecol 122:225–241

Gibbons MJ (1988c) The impact of wave exposure on the meiofauna of *Gelidium pristoides* (Turner) Kuetzing (Gelidiales: Rhodophyta) Estuar Coast Shelf Sci 27:581–593

Gibbons MJ (1991) Rocky shore meiofauna: a brief overview. Trans R Soc S Afr 47:595–603

Gibbons MJ, Griffiths CL (1986) A comparison of macrofaunal and meiofaunal distribution and standing stock across a rocky shore, with an estimate of their productivities. Mar Biol 93:181–188

Gibbons MJ, Griffiths CL (1988) An improved quantitative method for estimating intertidal meiofaunal standing stocks on an exposed rocky shore. S Afr J Mar Sci 6:55–58

Gibbons MJ, Abiahy BB, Angel M et al (1999) The taxonomic richness of South Africa's marine fauna: a crisis at hand. S Afr J Sci 95:8–12

Gourbault NE (1987) Long-term monitoring of marine nematode assemblages in the Morlaix estuary (France) following the *Amoco Cadiz* oil spill. Estuar Coast Shelf Sci 24:657–670

Grassle JF, Morse-Porteus LS (1987) Macrofaunal colonization of disturbed deep-sea environments and the structure of deep-sea benthic communities. Deep-Sea Res 34:1911–1950

Gray JS (1994) Is deep-sea biodiversity really that high? Species diversity of the Norwegian continental shelf. Mar Ecol Prog Ser 112:205–209

Gray JS (2002) Species richness of marine soft sediments. Mar Ecol Prog Ser 244:285–297

Gyedu-Ababio TK (2011) Pollution status of two river estuaries in the Eastern Cape, South Africa, based on benthic meiofauna analyses. J Water Resour Prot 3:473–486

Gyedu-Ababio TK, Baird D (2006) Response of meiofauna and nematode communities to increased levels of contaminants in a laboratory microcosm experiment. Ecotoxicol Environ Saf 63:443–450

Gyedu-Ababio TK, Furstenberg JP, Baird D et al (1999) Nematodes as indicators of pollution: a case study from the Swartkops River system, South Africa. Hydrobiologia 397:155–169

Heip C, Vincx M, Smol N et al (1982) The systematics and ecology of free-living marine nematodes. Helminthol Abst (Series E) 51:1–31

Hendricks MGJ (2013) Studies on the ecology and taxonomy of nematodes of Saldanha Bay, South Africa. PhD thesis, University of the Western Cape, Cape Town

Hendricks MGJ, Gibbons MJ (2010) Two new nematode species from Saldanha Bay, South Africa: *Perepsilonema benguelae* sp. nov. and *Leptepsilonema saldanhae* sp. nov. (Desmodorida, Epsilonematidae). Zootaxa 2504:20–30

Hennig HF-KO, Fricke AH, Greenwood PJ et al (1982) Relationships between meiofaunal population densities and physico-chemical properties of unpolluted sandy beaches. Environ Monit Assess 1:337–344

Hennig HF-KO, Eagle GA, Fielder L et al (1983) Ratio and population density of psammolittoral meiofauna as a perturbation indicator of sandy beaches in South Africa. Environ Monit Assess 3:45–60

Heyns J, Furstenberg JP (1987) Algoanema aestuariense n.gen. n.sp. from the Swartkops estuary, Port Elizabeth (Nematoda: Chromadoridae). S Afr J Zool 22:107–109

Higgins RP, Thiel H (1988) Introduction to the study of meiofauna. Smithsonian Institute Press, Washington DC

Hooge MD (1999) Abundance and horizontal distribution of meiofauna on a northern California beach. Pac Sci 53:305–315

Inglis WG (1961) Free-living marine nematodes from South Africa. Bull Br Mus (Nat Hist) 7:291–319

Inglis WG (1963) New marine nematodes from the coast of South Africa. Bull Br Mus (Nat Hist) 8:207–287

Inglis WG (1966) Marine nematodes from Durban, South Africa. Bull Br Mus (Nat Hist) 14:81–106

Jackson LF, McGibbon S (1991) Human activities and factors affecting the distribution of macrobenthic fauna in Saldanha Bay. S Afr J Aquat Sci 17:89–102

Jensen P (1986) Nematode fauna in the sulphide-rich brine seep and adjacent bottoms of the East Flower Garden, NW Gulf of Mexico, IV. Ecological aspects. Mar Biol 92:489–503
Jensen P (1987) Feeding ecology of free-living aquatic nematodes. Mar Ecol Prog Ser 35:187–196
Joubert CS (2011) Environmental monitoring of finfish cage farm in Algoa Bay (South Africa): a pilot study. MSc Dissertation, University of Port Elizabeth, Port Elizabeth
Kennedy AD, Jacoby CA (1999) Biological indicators of marine environmental health: Meiofauna – a neglected benthic component? Environ Monit Assess 54:47–68
Lambshead PJD, Boucher G (2003) Marine nematode deep-sea biodiversity- hyperdiverse or hype? J Biogeogr 30:475–485
Lambshead PJD, Elce BJ, Thistle J et al (1994) A comparison of the diversity of deep-sea marine nematodes from three stations in the Rockall Trough, Northeast Atlantic, and one station in the San Diego Trough, Northeast Pacific. Biodivers Lett 2:95–107
Lambshead PJD, Tietjen J, Ferrero T et al (2000) Latitudinal gradients in the deep-sea with special reference to North Atlantic nematodes. Mar Ecol Prog Ser 194:159–167
Lambshead PJD, Tietjen J, Glover A et al (2001) Impact of large-scale natural physical disturbance on the diversity of deep-sea North Atlantic nematodes. Mar Ecol Prog Ser 214:121–126
Levin LA, Etter RJ, Rex MA et al (2001) Environmental influences on regional deep-sea species diversity. Annu Rev Ecol Syst 32:51–93
Lopez G, Riemann F, Schrage M (1979) Feeding ecology of the brackish-water oncholaimid nematode *Adoncholaimus thalassophygas*. Mar Biol 54:311–318
Lorenzen S (1981) Entwurfeines phylogenetischen systems der freilebende nematoden. Veroeffentlichungen des Instituts fuer Meeresfforschung in Bremerhaven Suppl 7:1–472
Lorenzen S (1994) The phylogenetic systematics of freeliving nematodes. The Ray Society, London
Malan DE, McLachlan A (1985) Vertical gradients of meiofauna and bacteria in subtidal sandy sediments from two high-energy surf zones in Algoa Bay, South Africa. S Afr J Mar Sci 3:43–53
Mare MF (1942) A study of a marine benthic community with special reference to the micro-organisms. J Mar Biol Assess 25:517–554
McLachlan A (1977a) Studies on the psammolittoral meiofauna of Algoa Bay. II. The distribution, composition and biomass of the meiofauna and macrofauna. Zool Afr 12:33–60
McLachlan A (1977b) Composition, distribution and abundance of macrofauna and meiofauna of four sandy beaches. Zool Afr 12:279–306
McLachlan A (1977c) The effects of ore dust pollution on the physical and chemical features and on the meiofauna and microfauna, of a sandy beach. Zool Afr 12:73–88
McLachlan A, Furstenberg JP (1977) Studies on the psammolittora of Algoa Bay. III. A quantitative analysis of the nematode and crustacean communities. Zool Afr 12:61–71
McLachlan A, Erasmus T, Furstenberg JP (1977a) Migration of sandy beach meiofauna. Zool Afr 12:257–277
McLachlan A, Winter PED, Botha L (1977b) Vertical and horizontal distribution of sub-littoral meiofauna in Algoa Bay, South Africa. Mar Biol 40:355–364
Mirto S, La Rosa T, Gambi C et al (2002) Nematode community response to fish-farm impact in the western Mediterranean. Environ Pollut 116:203–214
Moens T, Vincx M (1997) Observations on the feeding ecology of estuarine nematodes. J Mar Biol Assoc UK 77:211–227
Moens T, Vincx M (2000) Temperature and salinity constraints on the life cycle of two brackish-water nematode species. J Exp Mar Biol Ecol 243:115–135
Moens T, van Gansbeke D, Vincx M (1999) Linking estuarine nematodes to their suspected food. A case study from the Westerschelde Estuary (south-west Netherlands). J Mar Biol Assoc UK 79:1017–1027
Moens T, Herman P, Verbeeck L et al (2000) Predation rates and prey selectivity in two predacious nematode species. Mar Ecol Prog Ser 205:185–193

Moens T, Yeates GW, De Ley P (2004) Use of carbon and energy sources by nematodes. Nematol Monogr Perspect 2:529–545

Monteiro PMS, McGibbon S, Henry JL (1990) A decade of change in Saldanha Bay: natural or anthropogenic? S Afr J Sci 86:454–456

Monteiro PMS, Anderson RJ, Woodbourne S (1997) δ^{15} as a tool to demonstrate the concentration of fish-waste-derived nitrogen to an *Ulva* bloom in Saldanha Bay, South Africa. S Afr J Mar Sci 18:1–9

Muthumbi AW, Verschelde D, Vincx M (1995) New Desmodoridae (Nematoda: Desmodoroidea): three new species from *Ceriops* mangrove sediments (Kenya) and one related new species from the North Sea. Cah Biol Mar 36:181–195

Muthumbi AW, Vincx M (1996) Nematodes from the Indian Ocean: description of six new species of the genus *Molgolaimus* Ditlevsen, 1921 (Nematoda: Desmodoridae). Biol 66: 17–28

Muthumbi AW, Vincx M (1997) *Acantholaimus* (Chromadoridae: Nematoda) from the Indian Ocean: description of seven species. Hydrobiologia 346:59–76

Muthumbi AW, Vincx M (1998) Chromadoridae (Chromadorida: Nematoda) from the Indian Ocean: description of new and known species. Hydrobiologia 364:119–153

Muthumbi AW, Vincx M (1999) Microlaimidae (Microlaimoidea: Nematoda) from the Indian Ocean: description of nine new and known species. Hydrobiologia 397:39–58

Muthumbi AW, Soetaert K, Vincx M (1997) Deep-sea nematodes from the Indian Ocean: new and known species of the family Comesomatidae. Hydrobiologia 346:25–57

Neira C, Sellanes J, Levin LA et al (2001) Meiofaunal distributions on the Peru margin: relationship to oxygen and organic matter availability. Deep Sea Res I 48:2453–2472

Nozias C, Perissinotto R, Tita G (2005) Seasonal dynamics of meiofauna in a South African temporarily open/closed estuary (Mdloti Estuary, Indian Ocean). Estuar Coast Shelf Sci 62:325–338

Ólafsson E (2003) Do macrofauna structure meiofauna assemblages in marine soft-bottoms. A review of experimental studies. Vie Milieu 53:249–265

Platt HM, Warwick RM (1983) Free-living marine nematodes Part I British Enoplids. Cambridge University Press, Cambridge

Platt HM, Warwick RM (1988) Free-living marine nematodes Part II British Chromadorids. Brill, Leiden

Prochazka K (2003) Marine biological monitoring of Robben Island marine sewage outfall. Final Report, International Oceanographic Institute, Malta

Resh VH, Norris RH, Barbour MT (1995) Design and implementation of rapid assessment approaches for water resources monitoring using benthic macroinvertebrates. Austral Ecol 20(1):108–121

Rex MA, Stuart CT, Hessler RR et al (1993) Global-scale latitudinal patterns of species diversity in the deep-sea benthos. Nature 365:636–639

Rex MA, Stuart CT, Etter RJ (2001) Do deep-sea nematodes show a positive latitudinal gradient of species diversity? The potential role of depth. Mar Ecol Prog Ser 210:297–298

Riemann F, Schrage M (1978) The mucus-trap hypothesis on feeding of aquatic nematodes and implications for biodegradation and sediment texture. Oecologia 34:75–88

Riemann F, Ernst W, Ernst R (1990) Acetate uptake from ambient water by the free-living marine nematode *Adoncholaimus thalassophygas*. Mar Biol 104:453–457

Rivadeneira MM, Fernández M, Navarrete SA (2002) Latitudinal trends of species diversity in rocky intertidal herbivore assemblages: spatial scale and the relationship between local and regional species richness. Mar Ecol Prog Ser 245:123–131

Roberts M (1996) The effect of pollution on meiofauna in the subtidal area between the harbour and Papenkuils River, Algoa Bay. MSc Dissertation, University of Port Elizabeth, Port Elizabeth

Romeyn L, Bouwman LA (1983) Food selection and consumption by estuarine nematodes. Hydrobiol Bull 17:103–109

Rzeznik-Orignac J, Fichet D, Boucher G (2003) Spatio-temporal structure of the nematode assemblages of the Brouage mudflat (Marennes Oléron, France). Estuar Coast Shelf Sci 58:77–88

Sandulli R, De Leonardis N, Vanaverbeke J (2010) Meiobenthic communities in the shallow subtidal of three Italian protected areas. Ital J Zool iFirst 1:1–11

Sikora JP, Sikora WB, Erkenbrecher CW et al (1977) Significance of ATP, carbon and caloric content of meiobenthic nematodes in partitioning benthic biomass. Mar Biol 44:7–14

Steiner G (1918) Neue und wenig bekannte nematoden von der Westküste Afrikas. Zoolog Anzeiger 50:4–18

Stenton-Dozey JME, Jackson LF, Busby AJ (1999) Impact of mussel culture on macrobenthic community structure in Saldanha bay, South Africa. Mar Pollut Bull 39:1–12

Steyaert M, Garner N, van Gansbeke D et al (1999) Nematode communities from the North Sea: environmental control on species diversity and vertical distribution within the sediment. J Mar Biol Assoc UK 79:253–264

Steyaert M, Vanaverbeke J, Vanreusel A et al (2003) The importance of fine-scale, vertical profiles in characterizing nematode community structure. Estuar Coast Shelf Sci 58:353–366

Thistle J, Sherman KM (1985) The nematode fauna of a deep-sea site exposed to strong near bottom currents. Deep-Sea Res 32:1077–1088

Tita G, Vincx M, Desrosiers G (1999) Size spectra, body width and morphotypes of intertidal nematodes: an ecological interpretation. J Mar Biol Assoc UK 79:1007–1015

Vanaverbeke J, Gheskiere T, Steyaert M et al (2002) Nematode assemblages from subtidal sandbanks in the Southern Bight of the North Sea: effect of small sedimentological differences. J Sea Res 48:197–207

Vanhove S, Vermeeren H, Vanreusel A (2004) Meiofauna towards the South Sandwich Trench (750–6300 m), focus on nematodes. Deep Sea Res II 51:1665–1687

Vanreusel A, Fonseca G, Danovaro R et al (2010) The contribution of deep-sea macrohabitat heterogeneity to global nematode diversity. Mar Ecol 31:6–20. doi:10.1111/j.1439-0485.2009.00352.x

Verschelde D, Vincx M (1992) Freeliving marine nematodes from East African coasts, *Bathyepsilonema anulosum* sp.n. and *Leptepsilonema richardi* sp.n. Hydrobiologia 239:179–186

Verschelde D, Vincx M (1993) *Polkepsilonema mombasae* gen. et sp.n. and *Pternepsilonema servaesae* gen. et sp.n. (Nematoda, Epsilonematidae) from East African coasts. Hydrobiologia 257:129–142

Verschelde D, Vincx M (1995) *Psammonema* gen.n. and *Pseudochromadora* Daday, 1889 (Nematoda, Desmodoridae) from sandy sediments of Gazi, Kenya. Biol 65:11–39

Vincx M (1990) Diversity of the nematode communities in the Southern Bight of the North Sea. Neth J Sea Res 25:181–188

Vincx M, Furstenberg JP (1988a) Africanema interstitialis gen.nov., sp.nov., a species which indicated the relationship between the Trefusiidae (Halanonchinae) and the Tripyloididae (Nematoda). Stygologia 4:10–16

Vincx M, Furstenberg JP (1988b) Three new Xyalidae species (Nematoda) from South Africa, with a redefinition of the genus Xyala Cobb, 1920. Cah Biol Mar 29:497–512

Vincx M, Furstenberg JP (1989) Namibnema papillata gen. et sp.n. and Axonolaimus deconincki sp.n. (Nematoda, Axonolaimoidae) from marine and estuarine beaches of southern Africa. Zool Scripta 18:231–237

Vincx M, Bett BJ, Dinet A et al (1994) Meiobenthos of the deep northeast Atlantic. Adv Mar Biol 30:1–88

Vitiello P (1975) *Deontostoma parantarcticum* n.sp. et *Thoracostoma ancorarium* n.sp., nouvelles espèces de Leptosomatidae, (Nematoda) d'Afrique du Sud. Trans R Soc S Afr 41:339–349

Von Linstow O (1908) II. Helminthes. Nematoden und Acanthocephalen. Denkschriften der Medicinisch-Naturwissenschaftlichen Gesellschaft zu Jena 13:19–28

Warwick RM (1984) Species size distributions in marine benthic communities. Oecologia 61:32–41

Warwick RM (1986) A new method for detecting pollution effects on marine macrobenthic communities. Mar Biol 92:557–562

Warwick RM (1993) Environmental impact studies on marine communities: pragmatical considerations. Aust J Ecol 18:63–80

Warwick RM, Platt HM, Somerfield PJ (1998) Free-living marine nematodes Part III Monhysterids. Field studies Council, Shrewsbury

Wieser W (1953) Die beziehung zwishen mundhöhlengestalt, ernährungsweise und vorkommen bei freilebenden marinen nematoden. Arkiv Zool 4:439–484

Willig MR, Kaufman DM, Stevens RD (2003) Latitudinal gradients of biodiversity: pattern, process, scale, and synthesis. Ann Rev Ecol Syst 34:273–309

Wynberg RP, Branch GM (1994) Disturbance associated with bait-collecting for sandprawns (*Callianassa kraussi*) and mudprawns (*Upogebia africana*): long term effects on the biota of intertidal sandflats. J Mar Res 52:523–528

Yani VV (2009) Ecosystem health of the Nelson Mandela Bay sandy beaches. MSc Dissertation, University of Port Elizabeth, Port Elizabeth

Yeates GW, Bongers T, De Goede RGM et al (1993) Feeding habits in soil nematode families and genera – an outline for soil ecologists. J Nematol 25:315–331

Chapter 25
Nematode Parasites of Humans in South Africa

Christopher C. Appleton

25.1 Introduction

Parasitic nematodes infect a wide range of animal phyla, including invertebrates such as annelids, molluscs and arthropods. Together with parasitic flatworms (cestodes and trematodes) and acanthocephalans, they form a group collectively called the helminths. Parasitic nematodes are common in humans, particularly children, and constitute a substantial economic burden, especially to developing countries. A large number of nematodes parasitise stock animals, causing widespread economic losses to agriculture, and are thus of enormous veterinary importance in southern Africa and elsewhere. There are too many to discuss adequately in the present chapter, but interested readers are referred to Reinecke (1983) for information on these veterinary nematodes. This chapter describes the parasitic nematodes infecting humans in southern Africa, focusing on those that are important to public health. Some of the species infecting other mammals such as non-human primates and rats in the region are also discussed because they may cause zoonotic infections in people. The lesions caused by these parasites can be misdiagnosed, often as tumours. Morphological details of these nematode taxa are not included because they are discussed in detail in most textbooks on medical parasitology. Life cycles are described but are not illustrated with diagrams.

C.C. Appleton
School of Life Sciences, University of KwaZulu–Natal,
Westville Campus, Durban 4000, South Africa
e-mail: appletonc@ukzn.ac.za

H. Fourie et al. (eds.), *Nematology in South Africa: A View from the 21st Century*, DOI 10.1007/978-3-319-44210-5_25

Table 25.1 Estimated disability-adjusted life year (DALY) totals lost annually due to selected human-nematode infections compared with those lost to tuberculosis, malaria and motor car accidents (Chan 1997; World Health Organization 2002)

Cause of morbidity/mortality	DALYs lost ($\times 10^6$)
Trichuris trichiura (Linnaeus, 1771) Stiles, 1901	6.4
Ascaris lumbricoides (Linnaeus, 1758)	10.5
Hookworms (*Ancylostoma duodenale* (Dubini, 1843) and *Necator americanus* (Stiles, 1902))	22.1
Total geohelminths	39.0
Wuchereria bancrofti (Cobbold 1877) Seurat 1921	5.6
Onchocerca volvulus (Leukart 1893) Railliet and Henry 1910	1.0
Total nematode diseases	45.6
Tuberculosis	36.0
Malaria	35.7
Motor vehicle accidents	31.7

25.2 Nematode Parasites of Humans

In his classic review of human helminth infections, 'This Wormy World', Stoll (1947) estimated that globally 2010 million people were infected by parasitic nematodes, with 1456 million of these (72 %) by the four most important of the soil-transmitted species or geohelminths (see Box 25.1). These four represent the common roundworm, *Ascaris lumbricoides* Linnaeus, 1758; the whipworm, *Trichuris trichiura* (Linnaeus, 1771) Stiles, 1901; and the hookworms, *Necator americanus* (Stiles, 1902) and *Ancylostoma duodenale* (Dubini, 1843) Creplin, 1845. Fifty years later in 1997, Stoll's study was repeated by Chan who estimated that 3452 million people were infected with these same parasites. He pointed out that while the number of infections had increased, so had the world's human population, but the infected proportion of the population remained more or less constant at 24–30 % for *A. lumbricoides,* 16–17 % for *T. trichiura* and 21–24 % for *N. americanus* and *A. duodenale*. An updated assessment by the World Health Organization (WHO) in 2002 was that 3,800 million people were infected with geohelminths worldwide.

The international measurement of the burden of morbidity and mortality, the disability-adjusted life year (DALY), is based on weightings given to disabilities due to particular causes and the number of years lived with the disabilities. Table 25.1 shows that, using 1997 data, the annual global number of DALYs lost to nematode diseases exceeded those lost to three other major causes of morbidity/mortality, tuberculosis, malaria and motor vehicle accidents. Indeed, the DALYs lost to the geohelminths alone exceeded those lost to these three other causes. The very high morbidity caused by the geohelminths is a major reason for the recent inclusion of geohelminthiasis as one of the world's 'neglected tropical diseases' by the WHO. Hookworms cause greater morbidity than the other geohelminths because they feed on blood and so contribute significantly to anaemia.

Although the first South African textbook on human parasitology *Parasiete en Parasitiese Siektes van die Mens* by Mönnig was published in 1944, it was broadly based, with limited local coverage because, as Mönnig noted in his preface, medical parasitology was not well developed in the country at the time (Mönnig, 1944). This was preceded by 1 year by the first edition of Gelfand's (1943) *The Sick African,* a clinical guide to disease in Africa, especially Zimbabwe, Zambia and Malawi, and which included the common nematode infections.

Occasional parasite surveys were carried out in South Africa (SA) between the late 1940s and early 1960s, but these focused more on identifiable groups such as migrant workers, dysentery and mental patients rather than the population at large. These surveys were prompted, in part at least by concern over children with heavy *A. lumbricoides* infections being admitted to hospital with intestinal bolus obstructions, as well as hepatic and biliary complications, sometimes requiring surgery. It was, however, only in the 1980s and 1990s that parasitological surveys were done in a purposeful way as the association between heavy intestinal worm burdens in children and severe morbidity was more fully appreciated. Though many of these surveys were carried out in SA, others were done as early as the 1930s in Zimbabwe, Zambia and Namibia, with the result that the distribution pattern of these nematode infections across the subcontinent has started to become clear.

Six of the nine nematode species commonly infecting people in Africa are widespread, particularly across the developing world. These six, *A. lumbricoides*; *T. trichiura; N. americanus; A. duodenale*; *Strongyloides stercoralis* (Bavay, 1876) Stiles and Hassall, 1902 (a threadworm); and *Enterobius vermicularis* (Linnaeus, 1758) Leach, 1853 (a pinworm), all occur as adults in the human intestinal tract. Four of these species, *A. lumbricoides*, *N. americanus*, *A. duodenale* and *S. stercoralis*, undergo a visceral phase in their life cycles in which migratory larvae (= juveniles) pass through the host's heart and lungs before lodging in the intestine. The purpose of this larval migration is the subject of debate, but there are two main hypotheses, viz. (i) that it is simply a relic of earlier life cycles in which alternative routes were used and (ii) that it allows the parasite to avoid specific host immune defence mechanisms at mucosal surfaces (Mulcahy et al. 2005). Except for their eggs, the life cycles of *T. trichiura* and *E. vermicularis* occur entirely within the host's intestine. The remaining three species, viz. *Wuchereria bancrofti* (Cobbold, 1877) Seurat, 1921; *Mansonella perstans* (Manson, 1891) Yorke and Maplestone, 1926; and *Onchocerca volvulus* (Leukart, 1893) Railliet and Henry, 1910, are all filarial nematodes. They are transmitted from person to person by mosquitoes and other bloodsucking flies and reach adulthood in the host's lymphatics or subcutaneous tissue. They occur in people of all ages. Several other species of filarial worms affect people further north in Africa, but these are not discussed here though they are occasionally reported in migrants. Still others are zoonoses transmitted to people from animal reservoirs by arthropod vectors, notably flies. Many fail to develop in the human host and remain as larval infections.

Three of the most common species affecting humans, viz. *A. lumbricoides, T. trichiura* and *N. americanus*, are, together with the less common *S. stercoralis*, transmitted to people via eggs or larvae that develop in faecally contaminated soil. Ingestion of such soil is referred to as geophagy and is elaborated on in Box 25.1. Together these geohelminths (often called soil-transmitted nematodes) constitute a serious public health problem which, in SA, is particularly severe in urban informal settlements (slums) where hygiene is poor and the most heavily infected children may pass >100,000 eggs g^{-1} of faeces (Appleton et al. 2009). Even pre-school children become severely infected. They are exposed to infection as soon as they can crawl and have direct contact with soil. Prevalences of both *T. trichiura* and *A. lumbricoides* at coastal crèches in KwaZulu-Natal are high (40–90 %), while those from inland crèches at higher altitudes are only 0–10 % (Colleen Archer, Durban, 2012, personal communication). Following the criteria given by the WHO (2002), intensities of infection in these crèches were generally 'moderate' for *T. trichiura* and 'light' for *A. lumbricoides*. Infants as young as 4 months are sometimes found with heavy worm burdens and may even cough up adult *A. lumbricoides*. In such cases infection must have occurred at least 1–2 months earlier, therefore shortly after birth. The pinworm *E. vermicularis* is not transmitted via soil but relies instead on infecting new hosts through its sticky eggs adhering to hands, clothing, bedding, etc. and being ingested via the hand-to-mouth route. It infects people of all age groups.

The larvae of several of the nematodes that undergo a visceral migration in the host are able to enter a state of arrested development (hypobiosis) within their host. This is seen as a strategy for evading unfavourable environmental conditions or the host's immune defences. Migrating worms can go astray, and a remarkable case of an 18-month-old child from Durban (KwaZulu-Natal Province) who presented herself with a fourth-stage larva of *A. lumbricoides* protruding from the tear duct of her right eye was reported by Kaplan et al. (1956). Measuring 50 × 1 mm, the larva must have moved up the nasolachrymal duct from the oesophagus into the tear duct. Another form of migration is vertical transmission, i.e. transmission via either the transplacental or trans-mammary routes from the mother to the foetus or the mother to the suckling infant, respectively. Vertical transmission has been demonstrated in all geohelminths except *T. trichiura* and the filarial nematodes. In an exceptional case of transplacental transmission, viable *A. lumbricoides* eggs were found in the faeces of an infant only 40 days old in Brazil (Da Costa-Macedo and Rey 1990).

Box. 25.1 Geohelminths and Geophagy

Geohelminths are intestinal parasites (phylum Nematoda) that are transmitted primarily through contaminated soil. Such parasites have a direct life cycle that requires no intermediate hosts or vectors.

Geophagy or soil eating is a widespread habit in South Africa and elsewhere, particularly amongst children and pregnant women some of whom eat soil daily. Such is the demand that bags of soil can often be bought at taxi ranks. This phenomenon is a proven risk factor in the transmission of geohelminths whose eggs are resistant to desiccation, viz. *Ascaris lumbricoides* and *Trichuris trichiura*, particularly the former (Fig. 25.1). Studies in rural KwaZulu-Natal showed geophagy to be more often practised by girls than boys and that *A. lumbricoides* prevalences were higher and infections heavier in children who regularly ate soil than in those who did not eat soil (Saathoff et al. 2002). Interestingly eating soil from free-standing termite mounds was correlated with high *A. lumbricoides* prevalences and heavy infections, whereas children who ate soil from tree termite mounds had significantly lower and lighter infections (Saathoff et al. 2005a). The termites involved are likely to be fungus-growing species of the family Termitidae. It is tempting to wonder whether these differences in *A. lumbricoides* infection are related to the symbiotic fungi the termites feed on, the indigestible fractions of which are excreted and form part of the 'glue' or 'cement' they use to build the walls of their mounds and which children eat. Curiously no marked differences were found for *T. trichiura* though children who eat soil regularly were found to be at increased risk of re-infection by both species after treatment. Pebsworth et al. (2012) showed that baboons in the Western Cape also eat soil and that they ingest *Trichuris* eggs (probably *T. trichiura*) with soil eaten at clay-rich geophagy sites and also accidentally with contaminated soil while foraging for roots. Troops from several localities in this fynbos-dominated area have been found to harbour *T. trichiura* infection rates above 90 %.

Geohelminths are of major public health importance because, following the typical age-prevalence curve for these nematodes, children between 5 and 10 years old comprise the most frequently infected age group. In addition, while approximately 90 % carry light or moderate infections that are largely asymptomatic, the remaining 10 % carry heavy infections. As a result, they not only experience severe morbidity but also contribute the most to ongoing transmission. For this reason parasitological surveys in southern Africa have, as elsewhere, concentrated on this child age group.

Prevalence data on adults are scarce, but a survey in KwaZulu-Natal showed that geohelminth prevalences amongst people over the age of 18 years were

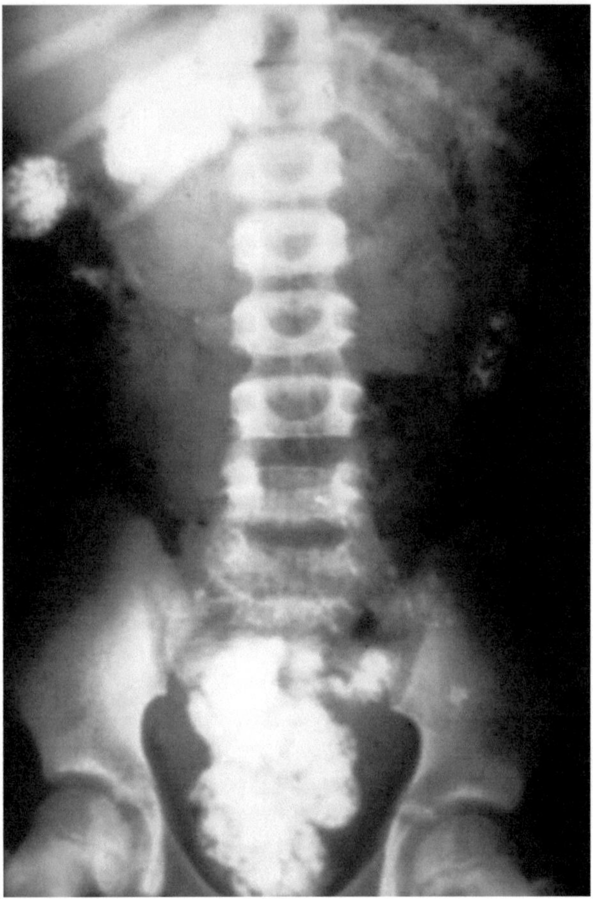

Fig. 25.1 An X-ray showing a child's intestine filled with soil, which is visible as an opaque white mass (Larry G. Hadley, Nelson Mandela School of Medicine, University of KwaZulu-Natal, Durban, South Africa)

consistently below 11 % (Kwitshana et al. 2008). As expected from the known distribution of geohelminths in children (see below), adults in the coastal districts are more frequently infected than those in inland areas such as in the Limpopo Province where Roodt et al. (1995) found prevalences in adults to be below 1 %. Hookworms are long-lived so that heavy infections in children may extend into adulthood.

The spectrum of nematodes infecting people can conveniently be thought of as comprising both common species for which humans are the principal final host and others that do so opportunistically. These latter are zoonotic infections, often larvae that do not develop further. Their natural hosts include non-human primates (monkeys, baboons and chimpanzees) and rats. Generally the common species have high prevalences in people, particularly children, with a small proportion having high intensities as well. The opportunistic parasites generally have low prevalences and low intensities with subclinical effects and are seldom important to public health.

Evidence for the transmission of nematodes from non-human primates to humans is growing as it is for transmission in the opposite direction. Both are driven by increasing contact between the two groups and have become a concern to primatologists working on primate conservation in Africa. Nematodes of primates in southern Africa that are transmissible to humans include *T. trichiura*; two spirurids *Physaloptera caucasica* von Linstow, 1902, and *Streptopharagus pigmentatus* von Linstow, 1897; the strongylids *Ternidens deminutus* Railliet and Henry, 1909, *Strongyloides fuelleborni* von Linstow, 1905 and *Oesophagostomum bifurcum* Creplin, 1849; and several species of *Trichostrongylus*. Of these, *T. trichiura* is the only one that is also common in people. The primate form of *T. trichiura* is morphologically indistinguishable from the human form, although the spicules of the males may differ slightly. The size ranges of their eggs are similar. New evidence (Ravasi et al. 2012) suggests that two different genotypes of *T. trichiura* occur in humans and baboons in Africa with both being found in either host. Whether these genotypes represent distinct species needs further research, but it does appear that transmission of the two genotypes has occurred between humans and non-human primates over millions of years and will presumably continue to do so.

However, the species that primatologists are most concerned about is *A. lumbricoides*. In southern Africa, increasingly frequent reports of *Ascaris* eggs (sometimes identified as *A. lumbricoides*) are being found in baboon faecal samples. Egg counts are, however, usually low. There is no evidence that *A. lumbricoides* can establish patent infections in non-human primates, but rather it is thought that the baboons are ingesting *Ascaris* eggs while foraging at rubbish dumps and that they pass through the gut without hatching. It appears that baboons, like urban rats in Durban (KwaZulu-Natal Province) which have also been found with *Ascaris* eggs in their faeces (Appleton and Archer 2006), play an unrecognised role in the epidemiology of ascariasis.

Box. 25.2 Human Nematode Parasites in Archaeoparasitology

Archaeoparasitology is a multidisciplinary field that refers to the study of parasites in archaeological contexts and includes studies on protozoan and metazoan parasites of humans in the past, as well as parasites that may have affected past human societies, including those of domesticated animals that may infect people as zoonoses.

Human nematode infections have a long history, and the eggs of several species, notably the geohelminths, have been recovered from coprolites found in archaeological deposits and from mummies. Probably the best known example is the finding of *Trichuris trichiura* eggs in the colon of Őtzi the Iceman, discovered frozen in the Austrian Alps in 1991 and dated to 3300 years before present (BP). Older finds have been made however. Eggs of *Ascaris lumbricoides, T. trichiura* and hookworm (unidentified) have been recovered from coprolites from Europe, Asia and the Americas, dating back to 30,000–24,000, 8,000–7,000 and 5,000 years BP, respectively (Gonçalves et al. 2003). Less frequently found are the eggs of *Enterobius vermicularis*, which go back to 10,000 years BP, and *Strongyloides* sp., apparently only recently acquired by humans, to just 3,000 years BP. The only study from South Africa reported eggs of *T. trichiura* and an ascarid, possibly a decorticated *A. lumbricoides* egg, in a coprolite from a late Stone-Age site in Gauteng and dated to between 10,000 and 7,000 years BP (Evans et al. 1996).

25.3 Systematic List of Nematodes Infecting People in Southern Africa

Life cycles, clinical features and distributions of common, opportunistic nematode parasites of humans in southern Africa are discussed below, using the classification system by Weischer and Brown (2002).

25.3.1 Class Adenophorea, Order Trichocephalida

25.3.1.1 *Trichuris trichiura* (Linnaeus, 1771) Stiles 1901 (Whipworm)

This species is called the whipworm because the anterior two thirds of both sexes are markedly thinner than the posterior third. Females are longer than males, 35–50 mm and 30–45 mm, respectively.

Life Cycle

The life cycle is direct. Adults occur in the caecum and large intestine of hosts, with their thin anterior ends anchored within the intestinal mucosa. Females lay eggs measuring 50–54 × 20–23 μm, with a plug at either end giving them a characteristic barrel shape. They are unembryonated when laid and are passed out of the host's intestine at a rate of up to 20,000 eggs female^{-1} day^{-1}. Depending on the ambient temperature, these eggs take 2–5 weeks to embryonate and develop to the rhabditiform larval stage. This is the infective stage. The survival time of embryonated eggs in the soil is not known, but they probably remain viable for several months.

Electron microscope studies of the egg of *T. trichiura* have shown that the egg's polar plugs differ in structure from the rest of the eggshell and are susceptible to action by digestive enzymes (Wharton and Jenkins 1978; Appleton and White 1989). When an embryonated egg is swallowed by a host, one or both of these plugs break open to release the enclosed larva, but the details of their early life are the subject of debate. There is evidence that *T. trichiura* larvae develop inside tunnels in the host's intestinal mucosa but that after the final moult, the adults keep their thin anterior ends embedded in these tunnels, while the thicker posterior part, including the gonopore and anus, protrudes into the lumen of the colon. The protection afforded to the larvae by these tunnels may explain the low cure rates commonly reported for *T. trichiura* after treatment with benzimidazole drugs (see Box 25.4, Sect. 25.4.1). Maturity in *T. trichiura* is reached after a prepatent period of approximately 2–3 months, and the estimated life span of adult worms is 1–3 years (Bundy and Cooper 1989).

Clinical Features

Light *T. trichiura* infections are usually asymptomatic, but severely infected individuals, i.e. with burdens >500 worms, are likely to develop disease (trichuriasis). This usually presents as '*Trichuris* dysentery syndrome' (Stephenson et al. 2000), a suite of non-specific symptoms including colitis, diarrhoea, dysentery, mucoid stools, stunting and occasionally finger clubbing and loss of tone of the anal sphincter muscle leading to rectal prolapse. The last of these is a consequence of heavy infections involving the wall of the rectum. Heavy infections may also be associated with appendicitis. An estimated 0.005 ml blood is lost adult^{-1} *T. trichiura* day^{-1}, contributing in some cases to anaemia.

The mucosal tunnels referred to above and the associated inflammation disturb the integrity of the mucosal architecture, especially in heavy infections, but this resolves after treatment with benzimidazole drugs (Hotez 2000). An interesting case from baboons in a reserve in the Western Cape Province suggests that ulcerous lesions in the intestinal mucosa caused by the invasive protozoan *Balantidium coli* (Malmsten, 1857) Stein, 1862, were exacerbated by a high concomitant *T. trichiura* burden. Secondary infection of these lesions by food-borne bacteria may have led to the severe mortalities reported in two affected troops by Barrett and Henzi (1998).

Distribution

Trichuris trichiura occurs widely across the subcontinent and has a similar altitudinal distribution pattern to *A. lumbricoides* (see Sect. 25.3.4.1). Both are common, though patchy in some areas, along the lowlands of the subtropical east and temperate south coasts of southern Africa, with prevalences of up to 15 % in pre-school children (<1–5 years) and >70 % in primary school children (6–15 years), with complications already noted. Prevalences are lower (<35 %) and complications correspondingly rare along the dry west coast, >± 1,100 m on the Drakensberg foothills, on the interior highveld plateau and into the semi-desert of north-eastern Namibia. Even in low-prevalence areas such as mountainous QwaQwa in the eastern Free State (5 %), cases of *Ascaris* bolus obstruction have been recorded, showing that intense transmission can occur, perhaps at the family or community level, in an otherwise inhospitable environment (Mosala and Appleton 2003).

The altitudinal difference between the areas of high and low prevalence for *A. lumbricoides* and *T. trichiura* in SA suggests that climatic factors influence the occurrence of the two parasites. Studies by Appleton and Gouws (1996) and Appleton et al. (1999) suggest that on the Drakensberg escarpment and coastal plain of KwaZulu-Natal, temperature and rainfall (or a combination of the two) play important roles in determining survival of *A lumbricoides* and *T. trichiura*, presumably by affecting the development of the larvae to the infective stage inside the eggs or the length of time these eggs can remain infective. There is evidence that temperature-related variables are more influential than rainfall-related variables at high altitudes but that the reverse may be the case at low altitudes. The available data also indicate that, although infections are often clustered and though they are often found together, the transmission requirements of *T. trichiura* and *A. lumbricoides* are different. Working on the Pongola River floodplain in north-eastern KwaZulu-Natal, Saathoff et al. (2005a) found a strong, positive correlation for *A. lumbricoides* with vegetation density and a negative one with exposure of the soil to sun. The clay content of the soil is not as influential in determining transmission as it is for hookworm and threadworm (see Sect. 25.3.1.3). Research is needed to identify more precisely the factors or combinations of factors involved in determining the distribution of *A. lumbricoides* and *T. trichiura*, why the former is dominant in some areas and the latter in others and the reasons behind the observed clustering of infections.

25.3.1.2 *Calodium hepaticum* (Bancroft, 1813) Moravec, 1982

More commonly known by its old name *Capillaria hepatica*, this is a 'tissue' parasite of mammals including dogs, cats and, most commonly, rodents. The adult stage occurs in the liver of its final host, but since its eggs cannot escape, they remain trapped in the liver parenchyma. These eggs will only be liberated when the infected host is eaten by a predator, its liver digested and the eggs passed out with the predator's faeces or when the infected host dies, its carcass decomposes and the eggs are ingested by a scavenger. In either case freed eggs develop outside the original host to the infective first larval stage but will only hatch when they have been eaten by a new host and reach its duodenum. Similar to *T. trichiura*, the eggs are barrel shaped, but the polar plugs are less prominent and the eggshell surface is characteristically pitted. This pitting is readily seen under a light microscope. The eggs measure 51–67 × 30–35 μm. The second-stage larva is infective when it hatches and breaks through the new host's intestinal wall and migrates via the portal circulation to the liver, where it takes a month or more to reach sexual maturity.

Approximately 3 % of brown rats (*Rattus norvegicus*) from Durban (KwaZulu-Natal Province) were found with livers infected with *Calodium hepaticum* Bancroft, 1893 (Colleen Archer, University of KwaZulu-Natal, Durban, 2012, personal communication), but the parasite is undoubtedly more widespread and involves the indigenous rodent community as well. Human *C. hepaticum* infections have occasionally been reported in SA (Cochrane et al. 1957; Kallichurum and Elsdon-Dew 1961; Silverman et al. 1973) with one in a 17-month-old child. Hira (1976) reported on infections in Zimbabwe and Zambia. Apart from a sometimes subacute hepatitis, such *C. hepaticum* infections are usually asymptomatic and are only diagnosed on liver biopsy or at autopsy. A high proportion of infections in young children have proved fatal (Cochrane and Skinstad 1960). Infected livers (Fig. 25.2) have characteristic white marks on the surface, indicating the presence of worms and cirrhosis associated with accumulations of trapped eggs. Infections are probably more com-

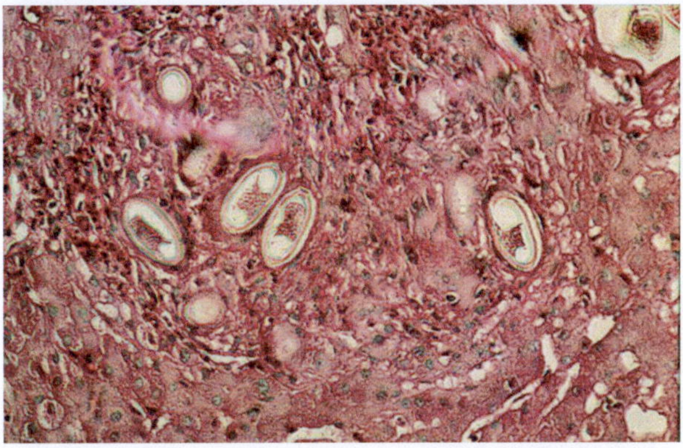

Fig. 25.2 Section of a human liver biopsy showing typical barrel-shaped *Calodium hepaticum* eggs (measurements of 51–67 × 30–35 μm) lodged in the parenchyma tissue (Christopher Appleton, School of Life Sciences, University of KwaZulu-Natal, Durban, South Africa)

mon in people than the evidence suggests since rodents are eaten in some parts of southern Africa.

The presence of *C. hepaticum* eggs in the faeces of a small proportion (1 %) of *R. norvegicus* from Durban (Colleen Archer, University of KwaZulu-Natal, Durban, 2012, personal communication) together with occasional reports of eggs in the faeces of people and baboons is important. As noted above, *C. hepaticum* eggs are not voided in faeces so that in these cases they must have been liberated from the livers of infected hosts that had been eaten. It thus seems that agents as varied as people, baboons and rats all contribute to the spread of *C. hepaticum* eggs within the environment.

25.3.1.3 *Trichinella spiralis* (Owen, 1835) (Trichina Worm)

The trichina worm is an example of a nematode parasite that uses the same animal as both a final and intermediate host. *Trichinella spiralis* sensu stricto, the nominate species of the *T. spiralis* complex, lives in its host's intestine and is viviparous, giving birth to larvae that encyst inside fibrous sheaths in striated muscle of the same host. These are termed 'encapsulated' cysts. Omnivores such as pigs and rats are the usual hosts from which people contract the infection by eating raw or undercooked pork or rat meat containing infective larvae. Other species of the *T. spiralis* complex occur in wild animal hosts in southern Africa but do not appear to infect people. Trichinosis due to *T. spiralis* is common in many parts of the world, particularly the northern hemisphere, but is rare in Africa and has not been reported from people in southern Africa. A common trichinellid in southern Africa is *T. zimbabwensis* Pozio, Foggin, Marucci, La Rosa, Sacchi, Corona, Rossi and Mukaratirwa, 2002, which belongs to a group of *Trichinella* spp. whose larvae do not become encapsulated when they encyst. It is common in crocodiles and poses a danger to people eating raw or undercooked crocodile meat. Mukaratirwa et al. showed in 2008 that baboons and monkeys were susceptible to infection by *T. zimbabwensis* and that the resulting pathology of their intestinal mucosa, diaphragm, tongue and muscles, particularly the psoas muscle, was severe, comparable with that produced by *T. spiralis* in humans. Clinical symptoms were also similar.

25.3.2 Class Secernentea, Order Rhabditida

25.3.2.1 *Strongyloides stercoralis* Bavay, 1876 (Threadworm)

Prevalences of *S. stercoralis* based on faecal analysis are generally low compared with other soil-transmitted nematodes (<25 %). This might, however, be shown to be an underestimate if coproculture were to be done routinely on unpreserved samples during faecal surveys (see Sect. 25.3.1.3).

Life Cycle

Strongyloides stercoralis is the usual threadworm affecting people. Its life cycle is complex and exhibits an alternation of generations. It is able to switch between a homogonic phase in which mitotically parthenogenetic females are parasitic but whose eggs produce larvae that develop in the soil, either to the infective filariform stage, which then becomes parasitic and infects a new host, or into free-living male and female larvae that become free-living adults (the heterogonic phase). The free-living males mate with free-living females but actual reproduction is by meiotic parthenogenesis. The resulting rhabditiform larvae either moult to become infective filariforms and resume the parasitic cycle or they continue the free-living cycle. When these infective larvae find a host, they penetrate its skin, usually the feet where they enter the capillaries of the peripheral circulation. They are then carried by the blood to the heart and lungs where they break out into the alveoli, ascend the trachea and are swallowed. In the intestine they moult again to become parthenogenetic females. These are parasitic and enter the duodenal mucosa where they lay eggs from which rhabditiform larvae hatch, usually while still inside the host's intestine so that it is usually the emerging rhabditiforms that pass out with the faeces. Some eggs may also pass out but larvae hatch within a few hours of being voided. Larvae hatching inside the intestine may reinfect the host either through the intestinal mucosa before being voided or through the perianal skin immediately after being voided (autoinfection). This can lead to heavy worm loads. Males do not become parasitic.

The cues triggering the switch between homogonic and heterogonic phases are unclear but probably relate to environmental conditions. When these are favourable, the rhabditiform larvae may follow the indirect heterogonic cycle and become free-living adult males and females. If the conditions are unfavourable, the larvae may enter the homogonic phase by moulting to become infective filariforms, which need to find a host and become parasitic females. The availability of suitable hosts must also play a role here. Under certain conditions the rhabditiforms moult into 'dwarf' infective filariforms while still in the host's intestine and reinvade the mucosa, a process known as hyper-infection. These various larval and adult forms are all small, measuring less than 2.2×0.5 mm.

The larvae of a second species of *Strongyloides*, *S. fuelleborni* normally a parasite of baboons and monkeys (see Sect. 25.3.3.1; *N. americanus*), hatch in the intestine and are usually passed in the faeces to develop in the soil outside. *Strongyloides fuelleborni* was found to be common with a prevalence of 79 % in San communities in the north-western Namibia (Evans et al. 1991).

Clinical Features

Penetrating *S. stercoralis* larvae may induce a transient dermatitis. When large numbers of larvae pass through the lungs during their visceral migration, the host may experience an eosinophilic pneumonia, an acute response to infection known as Löeffler's syndrome. Colonisation of the gut by the parthenogenetic females generally produces non-specific symptoms including diarrhoea, abdominal pain, weight loss and eosinophilia. As these worms typically burrow into the intestinal mucosa, they cause lesions, which may be invaded secondarily by bacteria, especially when worm loads are high. Up to 100,000 adult females have been recovered from a single patient after purgative treatment (Lee and Terry 1989).

Distribution

The free-living phases of the *S. stercoralis* life cycle occur in sandy soils, and their survival and distribution are, therefore, subject to the same environmental constraints as the hookworms and are similar (see Sect. 25.3.3.1; *N. americanus*). *Strongyloides fuelleborni* is distributed from Zimbabwe westwards to Zambia and northern Namibia and southwards into the KwaZulu-Natal Province of SA.

25.3.3 Class Secernentea, Order Strongylida

25.3.3.1 *Necator americanus* (Stiles, 1902) and *Ancylostoma duodenale* (Dubini, 1843) Creplin, 1845 (Hookworms)

The common hookworm of sub-Saharan Africa is *N. americanus*. Its specific name is a reminder that, although it was first described in the scientific literature from specimens collected in the southern United States of America (USA), it was in fact translocated there from Africa by slaves during the seventeenth and eighteenth centuries. The other common human hookworm *A. duodenale* is native to Europe, Asia and Africa north of the Sahara but has been reported at low prevalences from several parts of southern Africa, viz. Zimbabwe and Mozambique, where it occurs sympatrically with *N. americanus*. Its presence there may be due to introductions via foreign military personnel during the continent's turbulent, post-colonial history. Adults of both species are similar in size (males 7–9 × 0.3 mm and females 9–11 × 0.4–0.6 mm) but can be identified by *N. americanus* having two pairs of curved, chitinous cutting plates in its buccal cavity, whereas *A. duodenale* has two pairs of pointed 'teeth'. The morphology of the copulatory bursa of the male also differs between the two species.

Life Cycle

Adult male and female *N. americanus* live attached to the mucosal lining of the upper part of the small intestine of their host by means of their buccal cutting plates. Eggs laid by the female contain a four- to eight-celled embryo called a morula. These eggs are passed out in large numbers (up to 5,000 female^{-1} day^{-1}) via the host's faeces, and when they fall on moist, shady soil, the morulae develop to the first rhabditiform stage after only 48 h (at 25 °C). These larvae (±0.3 mm long) moult to the second rhabditiform stage and, by the fifth to eighth day after being voided from the host, moult again to the filariform stage. Neither of the rhabditiform larval stages is infective, but they are microphagous, feeding on bacteria and organic debris in the soil.

The third-stage filariform larva retains the cuticle from the second moult as a protective sheath that, although it prevents the larva from feeding, allows it to survive for a week or two in the soil while it waits for a suitable host. This is the infective stage, and it needs to locate and actively penetrate a human host before it can develop further. This delay in completion of the second moult is an example of hypobiosis or 'arrested development' referred to earlier. It enables the infective larvae to take advantage of suitable opportunities for transfer to a new host. Extended hypobiosis over the dry season may result in the accumulation of infective larvae at the start of the rains and explains the seasonality reported in hookworm transmission in West Africa. No evidence has been found for seasonal hookworm transmission in southern Africa, but further studies may yet show it to be part of the parasite's epidemiology here too. Filariform larvae have been recovered from soil at depths to a remarkable 10 cm, and this behaviour may be associated with hypobiosis.

When it is ready to infect a new host, the infective third-stage larva becomes negatively geotropic and positively thigmotropic so that it moves to the top few millimetres of the soil where it can respond to vibrations and locate the moist, unbroken skin of its host. It achieves this via a series of behavioural changes in response to cues of increasing specificity, from broad thermal gradients to the presence of skin and serum proteins of particular molecular weights. These cues differ between the different species of hookworm (Granzer and Haas 1991; Haas et al. 2005). Penetration of the host is either via hair follicles or through soft skin like the sole of the foot or between the toes. The larvae remain in the vicinity of the penetration site for approximately 40 h before starting on their visceral migration – entering the peripheral circulation, which carries them via the heart to the lungs. They stay in the lungs for more or less a week during which time they moult into fourth-stage larvae. These then break out of the alveolar capillaries, ascend the bronchi and trachea and then descend the alimentary tract into the intestine. Once there they attach to the mucosal lining and undergo their final moult into adult hookworms. This developmental cycle takes 4–6 weeks. Adult hookworms are long-lived, up to 5 years or even longer, and this longevity may account for the high prevalences reported in adult farm workers in Zimbabwe and Namibia (Chandiwana et al. 1989; Evans et al. 1990). As elsewhere in southern Africa, the intensities of these hookworm infections in adult hosts were low.

Clinical Features

Although a papular, itchy dermatitis ('ground itch') may result from penetration by infective filariform larvae in the soil, there are few symptoms or immune responses during the visceral migration unless large numbers of larvae are involved. In such cases, as with *S. stercoralis*, an acute-onset pulmonary pneumonia (Löeffler's syndrome) may occur.

As noted above, the filariform larvae and adult *N. americanus* are equipped with a set of chitinous cutting plates in the buccal cavity, while *A. duodenale* has several sharper 'teeth'. Both are in fact anterior extensions of the cavity's chitinous lining and serve to attach the worm to the mucosal epithelium of the host's intestine and to bite into it. Once attached they feed on blood with each adult *N. americanus*, ingesting approximately 0.04 ml of blood day^{-1}, and each *A. duodenale* approximately five times as much (Albonico et al. 1998). More host blood is lost, however, through bleeding since the worms move freely from one attachment site to another, leaving open wounds that continue to bleed after actual feeding has stopped. This bleeding is exacerbated by an anticoagulant introduced into the wound with the worms' salivary secretions as they bite.

Non-specific symptoms such as lassitude and abdominal pains frequently accompany hookworm infections, but the main damage is undoubtedly haemorrhage from the intestinal wall, in many cases leading to iron deficiency anaemia. People infected with hookworm are, however, often coinfected with other blood-feeding parasites such as *Plasmodium* (malaria) and *Schistosoma* (bilharzia) and also suffer from malnutrition, so that it is difficult to apportion damage. Nevertheless, the prevalence of anaemia has been shown to increase with both prevalence and intensity of hookworm infection. Van der Werf and De Vlas (2001) recognised two types of anaemia based on haemoglobin levels in the blood. Baseline-level anaemia with haemoglobin levels between <8–14 hb dl^{-1} is usual at lower hookworm prevalences of ± 60 % and intensities of ± 500 eggs g^{-1} faeces. Severe anaemia (≤8 g hb dl^{-1}) is usually found in cases where prevalences approach 80–100 % and intensities exceed 500 eggs g^{-1} faeces. Interestingly, anaemia (<11 g hb dl^{-1}) was common on the Makhathini Flats in north-eastern KwaZulu-Natal, but a study by Mayet et al. (1985) failed to show any statistical association with the *N. americanus* prevalence of 65 % and suggested a dietary cause instead. Malaria and bilharzia are also endemic to the area.

Distribution

The eggs of *N. americanus* are thin walled and susceptible to desiccation. They and the larvae of *S. stercoralis* both require sandy soils with a clay content of less than 15 % for optimal development to the infective stage. In other words they favour well-drained soils that allow aeration but retain enough moisture to prevent the larvae from drying out. Temperature is also important. In SA this restricts transmission of *N. americanus* (and *S. stercoralis*) to parts of the Mpumalanga lowveld and the KwaZulu-Natal coastal plain below an altitude of 150 m (Evans et al. 1987; Mabaso et al. 2003; Saathoff et al. 2005a). In fact prevalences of *N. americanus* decrease over this plain, with increasing south latitude from >90 % in the north-eastern corner of KwaZulu-Natal to approximately 40 % near the border with Eastern Cape. Prevalences of *S. stercoralis* are typically lower and range between <1 and 18 % but do not show any latitudinal gradient (Appleton et al. 1999). Neither *N. americanus* nor *S. stercoralis* occurs south of latitude 31 °S, surprisingly not even in the extensive sandy coastal areas of the Eastern and Western Cape provinces. This is probably because rain, which is necessary for the survival of larvae in the soil, falls in the cold winter. Saathoff et al. (2005b) found *N. americanus* infections on the Pongola River floodplain, north-eastern KwaZulu-Natal, to be clustered, and this is probably true elsewhere as well (see Sect. 25.3.4.1; *A. lumbricoides*). This may be a consequence of fine-scale differences in the soil or immediate environment.

Most of the above-mentioned studies showing high hookworm infection rates were done in low-lying rural areas in the lowlands of KwaZulu-Natal and contrast with studies done in slums (informal settlements) in peri-urban areas in cities on the same coastal plain. Enosse et al. (1995) and Appleton et al. (2009) reported hookworm prevalences of 6.4 % and 4.7 % in Maputo, Mozambique and Durban (SA), respectively. This was unexpected since human population densities are likely to be higher and levels of hygiene lower in peri-urban than in rural areas. Soil type may be an important factor here. The influence of urbanisation on nematode transmission needs to be investigated in greater detail.

A hookworm problem in the lowveld of south-eastern Zimbabwe, which has been well documented by Goldsmid et al. (1974), Chandiwana et al. (1989) and Midzi et al. (2008), is especially interesting because it involves both *N. americanus* and *A. duodenale* as well as the 'false hookworm' *Ternidens deminutus* (Railliet and Henry, 1909). All three parasites are transmitted together in the sandy soils of Burma Valley in the south of the Zimbabwean lowveld. Goldsmid et al. (1974) measured the following prevalences for these three nematodes in people in this region: *N. americanus* 82 %, *A. duodenale* 45 % and *T. deminutus* 13 %. Egg output was generally low as it is for *N. americanus* on the Makhathini Flats of KwaZulu-Natal. Unlike other geohelminths where intensities peak in adolescent years before declining, hookworm intensities (species cannot be differentiated in faecal analyses because eggs are identical) continued to rise to the age of 60 years and above. *Ternidens deminutus* is primarily a parasite of baboons and is also known from north-eastern SA.

Mention was made earlier of the adaptive reproductive strategies used by some nematodes to improve their chances of successful reproduction. Hookworm infections were found in two members of a small baboon troop living in the Namib Desert (Namibia), a stressful environment for baboons and nematodes alike (Appleton and Brain 1995). These infected baboons were in fact mother and daughter, which is circumstantial evidence that the hookworm (not identified) can be transmitted vertically, a behaviour that has seldom been documented for any hookworm species. Similarly person-to-person (vertical) transmission has been reported for *S. fuelleborni* in Zambia (Hira 1978). The high *S. fuelleborni* prevalence reported by Evans et al. (1991) in San communities in the dry north-east of Namibia may be an example of this. Taking into account the paucity of primate hosts in this area and its aridity, vertical transmission is probably responsible for maintaining this *S. fuelleborni* population in an inhospitable environment.

25.3.3.2 Other Hookworm Species

The infective larvae of two species of dog hookworm, *Ancylostoma braziliensis* de Faria, 1910, and *Ancylostoma caninum* (Ercolani, 1859) Hall, 1913, penetrate human skin but do not undergo any visceral migration and so fail to develop. Instead they remain in the subcutaneous tissue causing a sometimes painful condition known as creeping eruption, sandworm or 'cutaneous larva migrans'. Like other species of hookworm, they are limited to sandy areas such as the coastal plain of KwaZulu-Natal and the lowveld of Mpumalanga in SA, the lowlands of south-eastern Zimbabwe and the Caprivi Strip in Namibia. An analysis by Whiting (1976) of 50 cases from KwaZulu-Natal beaches showed that most infections (60 %) involved the feet, 30 % the buttocks and the rest the thighs and trunk. Usually only a single larval track was visible but as many as 30 were recorded. Although the larva is only 0.5 mm long and difficult to see, individual tracks can advance by as much as 50 mm daily. Albendazole is the drug of choice for 'cutaneous larva migrans'.

25.3.3.3 *Angiostrongylus*

The rat lungworm *Angiostrongylus cantonensis* (Chen, 1935) Dougherty, 1946, is a public health problem in Asia and parts of Latin America. This is because human infections by *A. cantonensis* usually involve the brain where the parasite's larvae cause an eosinophilic meningitis. There are very few records from Africa. It was however recently found to occur in 14 % of urban rats in Durban (SA) by Archer et al. (2011) and may, therefore, be more widespread across the region. The parasite uses a snail or slug as its intermediate host so that human infections rely on eating raw or undercooked infected snails or slugs. Larvae may also be present in the mucus trails left by infected snails on vegetables. A recent outbreak amongst tourists in the West Indies was traced to a badly washed Caesar salad (Slom et al. 2002).

25.3.3.4 *Trichostrongylus*

Several species of *Trichostrongylus* are common, pathogenic parasites of domestic stock in southern Africa, but they show little host specificity and infect both people and non-human primates opportunistically. The adults are generally small (5–10 mm long) and live in the small intestine of their host. Human infections are uncommon but prevalences in non-human primates are often high. Unidentified trichostrongylid eggs have been reported at low prevalence rates (<5 %) in faecal analyses of children in KwaZulu-Natal and probably occur elsewhere as well. Judging by the small numbers of eggs seen in these analyses, burdens are light and probably asymptomatic. The parasites are normally acquired when the host ingests infective larvae on vegetation grazed by stock animals.

25.3.3.5 *Oesophagostomum*

Isolated *Oesophagostomum* infections have been diagnosed in people across sub-Saharan Africa as far south as Zimbabwe. Morphological and molecular techniques have identified the parasite as *Oesophagostomum bifurcum* Creplin, 1849, one of the species commonly affecting domestic stock (Reinecke 1983) and also non-human primates. A localised *O. bifurcum* epidemic occurred in people between the 1960s and 1980s in the Sahel region of West Africa, producing a recognisable syndrome of granulomatous nodules containing larvae on the abdominal wall and inside the abdominal cavity, especially the wall of the colon. This was accompanied by diarrhoea and other non-specific, usually subclinical symptoms (Krepel 1994; Storey 2001). Surgery was required when these nodules became inflamed and blocked the colon's lumen. Although the origin of the human infections was not proven, evidence suggests that while people may have contracted the parasite from monkeys at the start of the outbreak, transmission was maintained subsequently by person-to-person contact. The epidemic has since died out. Eggs of *Oesophagostomum* spp. are frequently found in scat analyses from non-human primates but may be difficult to distinguish from the eggs of other strongyle parasites (see Box 25.3).

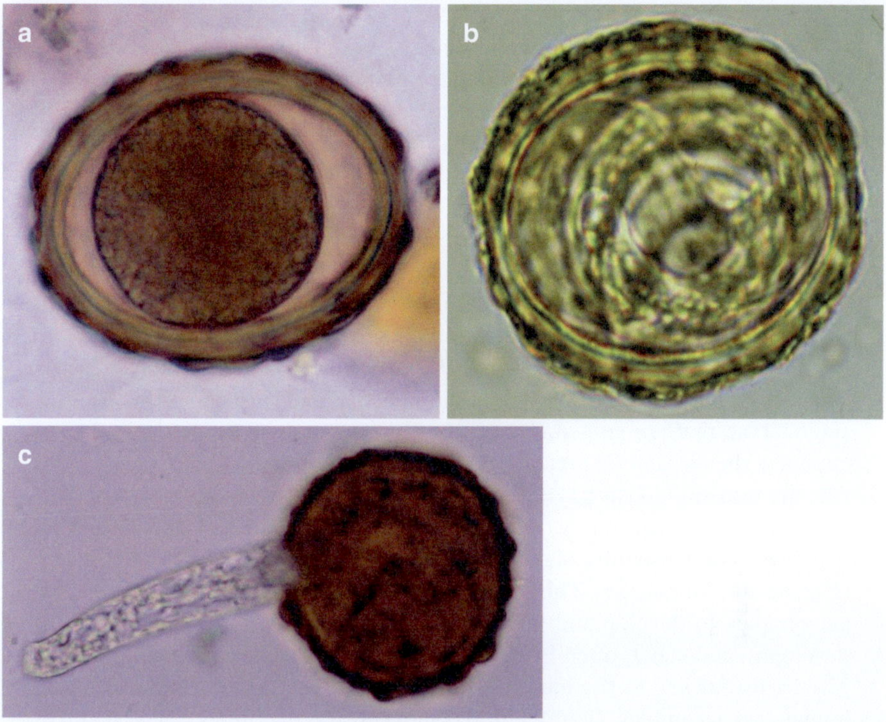

Fig. 25.3 (**a**–**c**) Eggs of *Ascaris lumbricoides* (measurements in the order of 50–70 × 40–50 µm), in unembryonated state (**a**), with a first-stage larva (L1) visible within the egg (**b**) and hatching of the infective second-stage larva (L2) (**c**) (Thabang Mosala, School of Life Sciences, University of KwaZulu-Natal, Durban, South Africa)

25.3.4 Class Secernentea, Order Ascaridida

25.3.4.1 *Ascaris lumbricoides* (Linnaeus, 1758) (Common Roundworm)

This is the largest and most common nematode parasite of people, such that a quarter of the world's population is estimated to be infected (Chan, 1997). Female worms measure up to 350 × 5 mm and males up to 300 × 3 mm.

Life Cycle

The female worm has a high fecundity rate, laying up to 200,000 eggs into its host's intestine each day. These eggs (Fig. 25.3a–c), which measure 50–70 × 40–50 µm, pass out with the faeces and unlike hookworm eggs have a thick wall that serves to protect them from adverse environmental conditions. The eggshell has four layers of which the innermost is composed largely of a lipid called ascaroside produced by the embryo itself.

Box. 25.3 Identification of Strongyle Eggs and Third-Stage Larvae

Shape and size are useful aids to identifying the eggs of strongyle nematodes in faecal analyses. They are all thin walled but their size ranges are typically broad and often overlap one another. Thus, the eggs of *Oesophagostomum*, which are similar in shape, are difficult to distinguish from those of *Necator americanus* and *Ancylostoma duodenale*. The eggs of *Strongyloides fuelleborni* are consistently smaller, whereas those of *Ternidens deminutus* and *Trichostrongylus* are conspicuously larger. Identification must therefore rely on careful size measurements, and here the diagrams of Goldsmid (1968), Goldsmid and Rogers (1978) and Appleton et al. (1991) are useful guides.

The eggs of *Strongyloides stercoralis* are seldom found in human faeces because larvae usually hatch within the intestine and it is the infective third-stage larvae that are voided. These filariform larvae characteristically lack a cuticular sheath, and this readily distinguishes them from the larvae of other faecally transmitted strongyle nematodes which do have a sheath.

Strongyle filariform larvae can be more reliably identified by microscopy than eggs, and these third-stage larvae can easily be obtained by coproculture (Harada and Mori 1955). This technique allows eggs in unpreserved fresh faecal samples to develop and larvae to hatch and grow to the filariform third developmental stage, often in large numbers, in test tubes. Goldsmid (1967) gives a useful key to the identification of the larvae of nematodes found in human faecal samples. Coproculture is inexpensive, and tubes can be set up in the field and transported to laboratories before larval development is complete, particularly if they can be kept cool. The technique should be used more often in human helminth surveys in South Africa than it has been. It may reveal that more species of parasitic nematodes infect people than are currently recognised or that some uncommon species are actually common.

This ascaroside layer is unique to the family Ascarididae and is responsible for the extraordinary ability of *A. lumbricoides* eggs to survive and remain infective in the soil for years, perhaps up to 10 in exceptional cases. The outermost layer is sticky due to a mucopolysaccharide derived from the female worm's uterus and explains the ease with which the eggs adhere to all sorts of household surfaces, such as door handles, banknotes and sand grains. Eggs are unembryonated when laid but contain a four- to eight-cell morula. Once passed out of the host's body, this morula develops over the next 3 weeks or so, usually on the ground, to the infective second larval stage. The first moult takes place inside the egg. This development is dependent on ambient temperature. Infective eggs are typically brown in colour due not to bile in the host's intestine as is commonly supposed but to a protective tanned protein in the eggshell.

Three 'types' of *A. lumbricoides* egg are seen in faecal analyses, viz. the normal mammillated fertilised egg referred to above; the decorticated, fertilised egg that lacks the distinctive rough outer layer; and, less commonly, the unfertilised egg, which is typically more elongate (60–100×40–60 μm) and lacks an embryo or larva. Further development of an embryonated egg will only take place after it has been ingested by a person and thus takes place at the constant temperature of the host's body rather than at the variable conditions outside. Transmission usually occurs via eggs carried to the mouth on hands, contaminated soil or unwashed vegetables, in water or even via wind-borne dust. Second-stage larvae hatch from eggs when they reach the duodenum (Fig. 25.3c).

These larvae are 1–2 mm long and start their visceral migration by burrowing through the intestine wall to enter the portal circulation or the lymphatics, which allows them to reach the liver, heart and then the lungs. Larvae remain in the lungs for about 9 days, during which time they moult twice and double their size to become filariform. The fourth-stage larvae migrate up the host's trachea to be swallowed down the oesophagus into the intestine again. Here they undergo their final moult to give rise to adult worms, which are typically white to pinkish and lodge mostly in the jejunum. Smaller numbers occur in the ileum, duodenum and stomach, especially when worm burdens are high. The period of development from ingestion as a second-stage larva to sexual maturity takes 8–12 weeks. Adult *A. lumbricoides* may live for 16–20 months but most die after about a year.

Unlike other, smaller intestinal nematodes, the filariform larva and adult of *A. lumbricoides* do not attach themselves to the mucosa but instead are surprisingly mobile. The need for mobility to forage, resist peristalsis and find mates is facilitated by the worms' longitudinal musculature. This generates undulating dorsoventral body waves, which allow alternating periods of quiescence and movement with or against the flow of the host intestine's contents.

Clinical Features

As for *T. trichiura*, light *A. lumbricoides* worm loads generally go unnoticed. Heavier infections, which Chan (1997) estimated at >15 worms depending on patient age, can result in colic, abdominal pain, nausea and loss of weight, while very heavy infections lead to complications such as bolus formation and acute appendicitis. Boluses typically comprise up to 200 worms but burdens of up to about 5000 worms are known. Such very high burdens account for more than one case of intestinal bolus obstruction each month in hospitals along the Eastern Cape and KwaZulu-Natal coasts, falling to half that number at higher altitudes such as the KwaZulu-Natal midlands (1,200–1,700 m) and less on the highveld plateau. This type of morbidity often presents as a surgical emergency and carries with it a case fatality rate of 5 % (Van der Werf and De Vlas 2001).

Hepatobiliary complications due to worms moving up the bile duct are less frequent, as are bowel penetration (worms entering the peritoneal cavity and laying eggs) and volvulus (twisting of the intestine due to the weight of worms). The latter may result in gangrene due to the twist point cutting off the blood supply and may also require surgery. The fatality rate is 11–15 % (Madiba and Hadley 1996; Van der Werf and De Vlas 2001) (Fig. 25.4a–c).

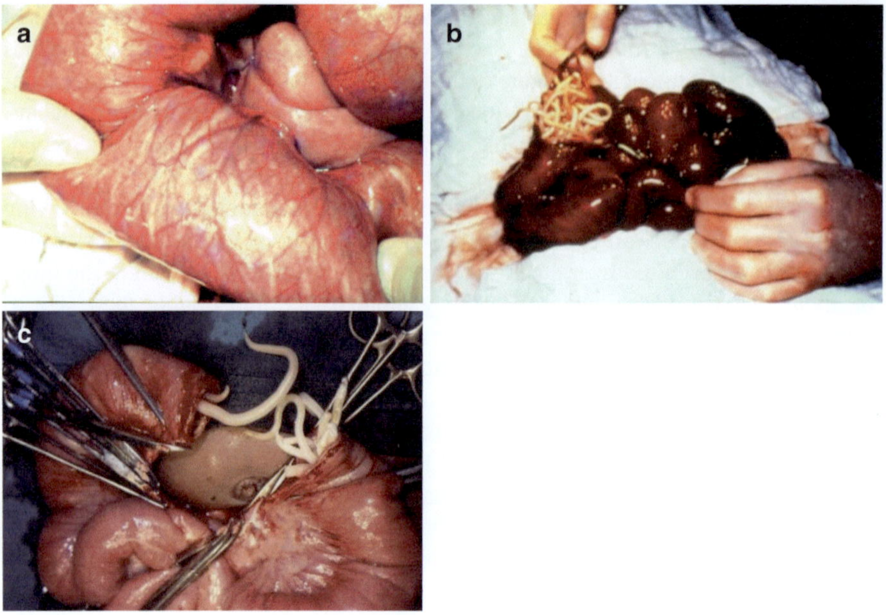

Fig. 25.4 (**a–c**) Complications due to heavy *Ascaris lumbricoides* infection, with close-ups of a bolus obstructing the intestine of a small child and individual worms clearly visible as packed tightly against the gut wall (volvulus **a**); the twisted, gangrenous section of the intestine that is packed with adult *A. lumbricoides* has been removed (**b**); and the two healthy sections of the intestine joined during a difficult surgical procedure (**c**) (Larry Hadley, Nelson Mandela School of Medicine, University of KwaZulu-Natal, Durban, South Africa)

The visceral migration by *A. lumbricoides* larvae seldom produces symptoms unless, as with *Strongyloides* and hookworm, large numbers pass through the lungs where they may cause allergen-based, asthmatic coughing, often called *Ascaris* pneumonia or more broadly Löeffler's syndrome. This may also trigger the formation of inflammatory granulomas around the larval worms, resulting in periportal fibroses. These may then block or pressurise nearby capillaries or other vessels, causing pulmonary hypertension.

Distribution

See Sect. 25.3.1.1 under *T. trichiura*.

25.3.4.2 *Toxocara canis* (Werner, 1782) Johnston, 1916

This is the common intestinal roundworm of dogs. *T. canis* is widespread in SA, with prevalences between 6 and 64 % in domestic dogs in cities across the country (Verster 1980; Fourie et al. 2002). The available data suggest that it has a similar distribution pattern to *A. lumbricoides*. Puppies and lactating bitches are most commonly infected. Unembryonated when laid, the eggs of *T. canis* develop over the subsequent 1–2 weeks to the infective second-stage and, like those of *A. lumbricoides*, can remain viable for years. When ingested accidentally by humans, second-stage larvae hatch in the intestine, break through the intestine wall and use the circulatory system to invade the lungs. However, they do not develop further, and instead of returning to the gut as they would in their natural canine hosts, they invade other organs, usually the liver but also the heart, lungs and brain as 'visceral larva migrans' (VLM). In some cases they lodge in the eyes as 'ocular larva migrans' (OLM) or ocular toxocariasis. *Toxocara canis* does not, therefore, complete its life cycle in humans since it does not develop beyond the rhabditiform second stage. Technically people who become infected with VLM or OLM are acting as paratenic hosts.

There are no statistics on the prevalence of human toxocariasis in SA, but both VLM and OLM undoubtedly occur (Zini and Wadee 1989). Serological (ELISA) surveys suggest that these infections are more common than generally realised, particularly the former. Although VLM may occur in the tissues of any organ, it is usually asymptomatic except for a persistent, elevated eosinophilia. Depending on the organ(s) involved, this may be accompanied by abdominal pain, hepatomegaly, pneumonia, allergic lung infections and diarrhoea. Typically the migrating larvae, which measure 400×20 μm, elicit a much larger area of inflammation, often encapsulating themselves in granuloma-like lesions. Lesions caused by OLM are also eosinophilic granulomas but usually occur in the vitreous humour of the eye and may be more frequent in adults than generally realised. Eye involvement may lead to blindness or loss of the infected eye. Clinically OLM is difficult to distinguish from retinoblastoma, the most common kind of retinal tumour in children. Accurate diagnosis of toxocariasis often depends on locating and identifying excised larvae by means of characteristic paired lateral alae close to the tail. These diagnostic structures can sometimes be seen in cross sections of juveniles in biopsies.

Children aged 1–5 years are the group most frequently infected with *T. canis*. The usual source of infection is thought to be playground sandpits contaminated with dog faeces where viable eggs can be ingested with sand. Beaches might also be important sites of infection. *Toxocara* eggs have been found in up to 55 % of soil samples from suburban gardens, public parks, pre-schools and roadside rest stops in different parts of SA (Markus 1982, Hayward et al. 2006). Direct contact with infected dogs is less likely to result in infection because any eggs on the dog will probably be freshly laid and unembryonated and therefore not infective. Eggs require several weeks outside the host to become infective but, once developed, can remain viable for several years.

Another source of infection may be the concentrated sludge from domestic sewage and waste from outdoor toilets such as the urine diversion toilet being used in several cities in SA. This waste is used as fertiliser and has been found to contain viable eggs of both *T. canis* and *A. lumbricoides*.

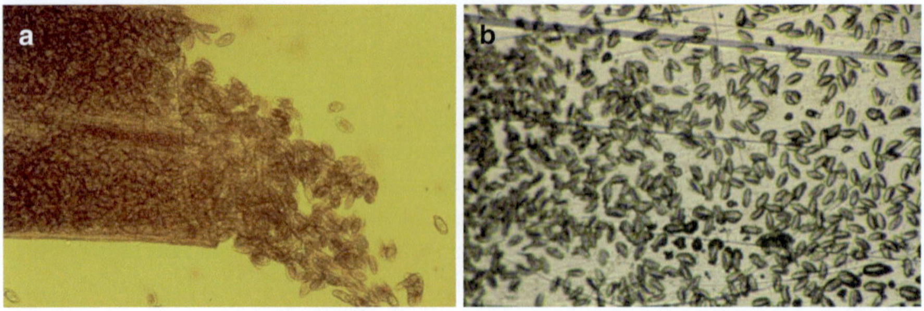

Fig. 25.5 (**a**, **b**) *Enterobius vermicularis* with an adult female worm cut open to show the enormous number of embryonated eggs (measuring approximately 50–60 × 25–32 μm) in her uterus (**a**) and an illustration of the Scotch tape test during which eggs stuck to a piece of Scotch tape applied to the perianal skin of a child were placed on a microscope slide for examination (**b**) (for best results, this should be done in the early morning) (Christopher Appleton, School of Life Sciences, University of KwaZulu-Natal, Durban, South Africa)

25.3.4.3 *Enterobius vermicularis* (Linnaeus, 1758) Leach, 1853 (Pinworm)

The human pinworm is a common parasite of all age groups but is most often found in children. Males measure 5.0 × 0.2 mm and are smaller than females, which measure 12 × 0.4 mm.

Life Cycle

Enterobius vermicularis has a direct life cycle in which all stages occur in the large intestine and the adults mature in the caecum of a host. The uterus of the gravid female contains enormous numbers of embryonated eggs (Fig. 25.5a), and she migrates to the rectum and anus at night to lay her eggs, 10,000–15,000 in all, on the host's perianal skin before she dies (Fig. 25.5b).

The eggs of *E. vermicularis* measure 50–60 × 25–32 μm and have a typically asymmetrical shape. They are also sticky and adhere to any surface with which they come into contact, viz. fingers, clothing, bedding, towels and toilet seats. When the female pinworm moves over the perianal skin of a host to lay her eggs, she elicits a slight irritation (probably exacerbated by fluids from her body if she dies), causing the host to scratch the area and picking up the sticky eggs on fingers or under fingernails. In this way the eggs are easily transferred to the mouth and swallowed, and because they are embryonated, they can hatch within 4 h of being laid. This may be thought of as an example of a parasite manipulating its host's behaviour to increase its chances of successful transmission.

Female *E. vermicularis* are sensitive to the slight drop in body temperature that occurs when an infected person sleeps and causes them to follow their nocturnal pattern of moving out of the anus to lay eggs. This explains why children attending crèches and pre-schools where they sleep in the afternoons are so readily infected. After the children wake up, their bedding is shaken out, and the eggs, which are very light, become airborne, can be inhaled and infect anyone nearby. Egg laying by pinworms may sometimes occur during the day.

Infective larvae hatch in the upper part of the small intestine, and the larvae then migrate towards the caecum, developing as they do so. The life cycle is normally completed in 4–6 weeks. Reinfection via the host's anus by larvae hatching on the perianal skin may occur.

Clinical Features

Pinworm infections are generally asymptomatic, but the perianal itching referred to above may lead to restlessness and fitful sleep, tiredness during the day, scratching and secondary infection. Complications do occur, usually when the worms penetrate the peritoneal cavity and move to ectopic (abnormal) sites, but this is rare. In such cases they become encapsulated in granulomatous lesions. Infections may also be associated with appendicitis.

Distribution

Enterobius vermicularis infection is common in schools, in other crowded institutions and in family groups where eggs are easily distributed amongst people. Little is known about the environmental requirements of its eggs, the only stage of its life cycle exposed to the outside environment. These eggs are, however, thin walled and susceptible to desiccation so that infection rates are likely to be influenced by climate. In moist conditions the eggs can remain infective in dust or on clothing for several months. In SA, *E. vermicularis* seems to be widespread, but faecal surveys have shown prevalences to generally be low (<10 %) because the eggs are seldom incorporated into faeces. It may in fact be the only nematode parasitising people in Lesotho since Mosala and Appleton (2003) showed it to be by far the most common species in the mountains of adjacent QwaQwa (Free State Province). At an altitude of 2,000 m in the latter mountainous areas, stool analysis showed a prevalence of 3 % compared to 45 % recorded as a result of a Scotch tape survey by Mosala and Appleton (2003). The only other Scotch tape surveys in SA (Leary et al. 1973) found prevalences between 26 and 38 % in the Western Cape Province. Vermund and Wilson (2000), however, maintain that surveys such as these, which use a single Scotch tape procedure, identify only about 50 % of infected patients. If true, *E. vermicularis* prevalences in southern Africa are higher, between 50 and 90 % in at least some situations. Research is needed on the influence that the often air-conditioned, domestic environment has on the distribution of *E. vermicularis* across the region.

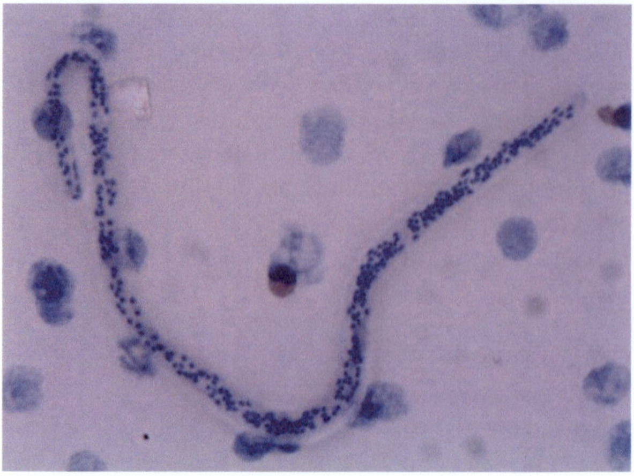

Fig. 25.6 Stained human blood smear showing a microfilaria of *Wuchereria bancrofti*, which are usually between 244 and 296 μm long (Colleen Archer, University of KwaZulu-Natal, Durban, South Africa)

25.3.5 Class Secernentea, Order Spirurida

The typical spirurid life cycle involves larval development to the infective stage in an arthropod intermediate host. The larva is then passed on to the final vertebrate predator host for the parasite to develop to maturity in various ways. This order includes the important filarial nematodes that are typically viviparous, releasing first-stage larvae called microfilariae directly into the host's blood, from where they are taken up by bloodsucking insect vectors. These microfilariae are generally between 200 and 300 μm in length (Fig. 25.6). Three species of filarial nematodes, *W. bancrofti*, *M. perstans* and *O. volvulus* affect humans in southern Africa but only in the northern parts.

25.3.5.1 *Wuchereria bancrofti* (Cobbold, 1877) Seurat, 1921 (Cause of Lymphatic Filariasis)

This is the best known of the filarial nematodes that infect people because advanced infections produce the grotesquely swollen legs and scrotum known as elephantiasis. Adult males measure 40×0.5 mm and females 80–100×0.2–0.3 mm. The microfilariae of *W. bancrofti* (Fig. 25.6) are ingested by a variety of mosquito species (family Culicidae) belonging to several genera. In tropical Africa these include members of the *Anopheles gambiae* complex and the *Anopheles funestus* group (the same mosquitoes that transmit malaria), as well as the *Culex pipiens* complex. Sasa (1976) differentiated between 'rural' and 'urban' transmission cycles based on the breeding requirements of the different mosquito vectors. Members of the *A. gambiae* and *A. funestus* complexes, which breed in rain pools and small unpolluted streams, are the common vectors in rural areas, while members of the *C. pipiens* complex, which favour more polluted habitats, fill this role in urban areas.

Clinical Features

Adult *W. bancrofti* live in the nodes and canals of the host's lymphatic system. Typical symptoms in the acute stage of infection are intermittent 'filarial fevers' and lymphangitis due to inflammatory reactions to the presence of worms, usually in the hands, legs and scrotum. In the later chronic stage when lymph vessels become blocked by worms, the nodes swell, producing the characteristic, disfiguring symptoms of elephantiasis, hydrocoele and chyluria, depending on the site of the blockage.

Distribution

Lymphatic filariasis is present in northern Angola, Zambia, the northern half of Zimbabwe including the Zambezi River basin, the Shire basin in southern Malawi as well as the western shore of Lake Malawi and central and northern Mozambique. It does not occur south of approximately latitude 17 °S (Simonsen et al. 2008).

25.3.5.2 *Mansonella perstans* (Manson, 1891) Yorke and Maplestone, 1926 (Cause of Benign Filariasis)

Previously called *Dipetalonema perstans*, this is an enigmatic parasite because not only have adults seldom been found but infections, which are common in some areas, seem to be largely asymptomatic. Males measure approximately 45 mm × 60 µm and females 70–80 mm × 60 µm. Infective larvae are transmitted to people by the biting midge *Culicoides austeni* (family Ceratopogonidae).

Clinical Features

Non-specific symptoms, including swellings of the arms and face, itching, joint pains and tiredness, have been reported in some cases.

Distribution

Infections by *M. perstans* are known from Angola, Zambia and Zimbabwe, but as for *W. bancrofti*, not south of latitude 17 °S. Orihel (1973) suggested that several brain infections ascribed to *M. perstans* microfilariae in people from the Zambezi Valley, Zimbabwe, were probably due to *Meningonema peruzzi* Orihel and Esslinger, 1973, a filarial parasite infecting the central nervous system of African monkeys.

25.3.5.3 *Onchocerca volvulus* (Leuckart, 1893) Railliet and Henry, 1910

This filarial nematode is notorious for causing onchocerciasis or 'river blindness' which is one of the world's major blinding infections. Female *O. volvulus* are much longer (335–500 × 0.27–0.4 mm) than males (18–32 × 0.13–0.21 mm). Blackflies of the genus *Simulium* (family Simuliidae) serve as the vectors of *O. volvulus*. Blackfly larvae breed and develop in fast-flowing water – hence the name 'river blindness'.

Clinical Features

Adults live in subcutaneous tissue throughout the body of their host and in chronic infections, often inside fibrous tumours and in large numbers. Microfilariae produced by female worms disperse through the subcutaneous and lymphatic tissues of the host's body and may migrate to the eyes. Here they may invade just about every structure of the eye, causing mechanical damage and affecting vision, viz. the optic nerve, cornea and ciliary body. These complications are often accompanied by calcification and inflammatory damage due to immunological responses to dead microfilariae. Skin lesions are itchy and often confined to the trunk. In chronic infections these skin lesions may show abnormal pigmentation and hypertrophic thickening, leading to a loss of elasticity. Inguinal lymph glands may become enlarged.

Distribution

Onchocerciasis is widespread in West and Central Africa, extending southwards into the coastal river basins of north-western Angola with a few outlying records from south-west Zambia. In some West African villages, up to 50 % of people suffer from onchocerciasis. Iconic photographs of children leading blind adults by holding the ends of a stick are well known.

25.3.5.4 *Dirofilaria*

The heartworms of dogs, wild canids and cats, *Dirofilaria* spp., are transmitted by mosquitoes. Two species, *Dirofilaria repens* Railliet and Henry, 1911, and *Dirofilaria immitis* Leidy, 1856, are also known to infect people in Africa and are usually present as inflammatory cutaneous or, less frequently, pulmonary lesions. Several cases involving both species have been reported from southern Africa (Moodley et al. 2015; Colleen Archer, unpublished data, 1990; M van der Linden, unpublished data, 2013). Although not commonly seen in people in SA, human dirofilariasis is considered by some as an emerging zoonosis.

25.3.5.5 *Physaloptera caucasica* von Linstow, 1902, and *Streptopharagus pigmentatus* (von Linstow, 1897) Railliet and Henry, 1918

These two spirurids are common parasites of monkeys and baboons in southern Africa and are routinely found infecting people. Goldsmid (1974) and Hira (1976) reported *P. caucasica* infections in people in Zimbabwe and Zambia but did not describe the conditions in which the infections were acquired. Evans et al. (1990) reported *P. caucasica* in a San community in the arid north-east of Namibia, an unusual environment for nematode transmission (see Sect. 25.3.2.1 for *Strongyloides fuelleborni*). During the drought of 1982–1983, numerous children living along the Pongola River floodplain in KwaZulu-Natal, in the Mpumalanga lowveld and in the former Transkei (situated in Eastern Cape Province) were found infected with *P. caucasica* and/or *S. pigmentatus*. Spirurid nematodes typically use orthopteran insects such as locusts and grasshoppers as intermediate hosts in their life cycles. These infections must therefore have been acquired by the children eating infected insects, perhaps locusts, to supplement their diet.

Most diagnoses of human infection by *Physaloptera* and *Streptopharagus* are based on eggs found in faecal analyses. Goldsmid (1974) and Hira (1976, 1978) noted, however, that these eggs were often present for a brief period only and they were probably contained in food and simply voided in the faeces. Repeat stool analyses should be done to confirm an infection. Nevertheless, Goldsmid (1974) showed that patent infections do occur in people by recovering adult *P. caucasica* up to 50 mm long from patients in Zimbabwe on several occasions after treatment with thiabendazole.

25.3.5.6 *Gnathostoma spinigerum* Owen, 1836

Species of the genus *Gnathostoma* reach maturity in the stomach of fish-eating carnivores, but their larvae use freshwater copepods and fish as intermediate hosts during their life cycle. Human infections (gnathostomiasis) are usually due to *G. spinigerum* and typically occur in Asia and parts of Latin America. Recently several tourists to northern Botswana and southern Zambia contracted infections by *G. spinigerum* larvae after eating raw bream (Hale et al. 2003; Herman et al. 2009). This led to the proposal that gnathostomiasis should be considered an emergent imported pathogen in southern Africa. Eating bream sushi should probably be avoided in these areas. How this parasite reached southern Africa is not clear, but it is known that between 1990 and 2000, the Nile tilapia was introduced to the Kafue River for aquaculture purposes. It is likely that these fish, which must have originated in North Africa, came from established cultures in Asia where they became infected with *G. spinigerum*. It is possible that they could then have made their way from the Kafue to the Zambezi and so to the Okavango Delta, but this would have taken several years at least.

Initial symptoms following the ingestion of infective filariform larvae are non-specific and include nausea, vomiting and diarrhoea. Subsequently the larvae migrate from the intestine and are seen as 'larva migrans' in subcutaneous tissue, often on the trunk and upper legs after passing through the liver and lungs. Local inflammation, bleeding and sometimes necrosis are also seen. The larvae, which are recognisable by the four rows of spines on a cephalic bulb at the anterior end, vary considerably in size, from 10–20 cm. *Gnathostoma* does not develop to maturity in humans.

25.4 Effects of Geohelminths on Growth and Cognition Development in Children

Studies into the effects of geohelminth infections on children have confirmed that growth is compromised, but none has reached firm conclusions on the mechanisms involved. Children typically become infected with geohelminths early in life, and infection rates have usually started to rise by the age of 3–4 years, a time when they are growing and developing rapidly. Indeed, different parasite species may be acquired concurrently, successively and continuously during these 'formative years', leading to multiple parasite infections, a phenomenon known as polyparasitism (Kvalsvig 2002). In SA children have been reported with as many as eight different intestinal parasites, including the common geohelminths. Polyparasitism may be even higher in areas that are endemic for blood-borne parasites such as malaria, trypanosomiasis and filariasis. Although these parasites impact collectively on nutrition via a range of factors, the geohelminths are the largest, and their digestion and absorption of host nutrients must contribute significantly to the loss of proteins and micronutrients by the host and hence retard growth as well. Geohelminths clearly play a major role in impairing their host's physical growth, but the size of their contribution relative to other intestinal parasites is difficult to evaluate.

To illustrate the effect of *T. trichiura* and *A. lumbricoides* on growth in children in SA, Fincham (2001) compared the height-for-age Z-scores for infected children under 12 years from a rural part of the Western Cape Province with a reference population. He found that infected children had markedly lower scores than the controls, such that 17 % were classified as stunted (Z-scores below −2) compared to 2.5 % in the reference population. Parasitism is unlikely to be the only cause of such stunting. Using differences between weights and heights of children before and after treatment with anthelmintics in a variety of countries, Van der Werf and De Vlas (2001) calculated that the average annual losses in weight and height due to geohelminth infections were 1.12 kg and 0.48 cm, respectively.

There is also convincing evidence that soil-transmitted nematodes, particularly when present at high worm burdens, impair or delay the development of cognitive function in children and contribute to iron deficiency anaemia (Kvalsvig 2002, 2003). This relationship is, however, complex and confounded by factors such as concomitant parasitic infections, malnutrition, poor sanitation and level of education.

25.4.1 Control of Nematode Parasites

25.4.1.1 Geohelminthiasis Control

The enormous number of children infected with geohelminths means that community-based control programmes will be more cost-effective than ad hoc treatment of individuals at clinics or health centres. The aim of such programmes is no longer to eliminate the disease but rather to reduce the intensity of infections to a level where they no longer constitute a public health problem and then to keep intensities and hence morbidity low (Box 25.5). Guidelines developed by the WHO (1987, 2002) recommend a three-phase approach, viz. planning followed by attack and finally maintenance.

The planning phase brings together all stakeholders to design and plan a programme to suit the situation and the available resources. Operational work begins with the attack phase, and the main tool here is chemotherapy, representing the use of anthelmintic drugs quickly reducing the intensity of transmission. School-based programmes are cost-effective because working through primary schools allows large numbers of the most susceptible child age group to be treated at the same place and time. Longer-term interventions such as improvements in sanitation, water supplies and health education should be started during this phase. Once the intensity of transmission and to a lesser extent prevalence has been reduced to a manageable level, the slower interventions introduced during the attack phase are developed further during an open-ended maintenance phase. The objective here is to keep intensities low and integrate the programme into primary health-care services. Programmes of this kind are inevitably long-term operations and should be expected to run for decades before permanent control is achieved.

Intestinal nematode infections, single or mixed, can be successfully treated with single-dose anthelmintic drugs, notably the benzimidazoles and in particular mebendazole or albendazole, with expected cure and egg reduction rates both between 80 and 100 % (except for *T. trichiura* as discussed below and highlighted in Box 25.4). Reinfection by geohelminths following chemotherapy is, however, rapid (within 3–4 months) (Appleton et al. 2009), so that repeat treatments are necessary. Although albendazole is preferred by WHO for geohelminth control purposes, mebendazole is easier to obtain and administer in SA because it is a schedule-one drug and can be bought over the counter in pharmacies. It is, however, produced in polymorphs A, B and C of which only polymorph C is active against nematodes (Liebenberg et al. 1998). It is important, therefore, that if mebendazole is the drug of choice for a control programme, a formulation incorporating polymorph C is used. Albendazole is a schedule-four drug and can only be obtained on prescription and administered by medically qualified personnel. It is, however, widely and successfully used and administered by school teachers and community appointees in other African countries and should be descheduled in SA to schedule one or zero (Fincham et al. 2005). Both drugs are well tolerated, even by pre-schoolers and pregnant women.

A concern for any geohelminth control programme is the low cure rate (15–40 %) routinely reported in southern Africa and elsewhere for *T. trichiura,* although egg reduction rates generally remain high (>90 %). This has given rise to suggestions that *T. trichiura* is tolerant or resistant to benzimidazole drugs. An equally plausible explanation is the fact that, as noted earlier, larval development occurs inside tunnels in the colonic mucosa. Larvae developing inside tunnels inside the mucosa will be shielded from anthelmintics taken orally. Only when they have matured and the adult worms' thick posterior regions protrude into the lumen of the intestine from the tunnels will they be exposed to the drugs (see Box 25.4). In such a scenario, treatment may kill the partially exposed adults and perhaps advanced larvae as well, but they will quickly be replaced by newly maturing larvae already developing in the mucosa. These will probably have started to lay eggs by the time post-treatment surveys are done so that monitoring of drug efficacy by means of faecal egg counts is unlikely to record anything more than a transient drop in egg output. In an attempt to overcome the poor drug efficacy against *T. trichiura*, a study in the Western Cape Province by Adams et al. (2004) showed that repeated treatments with albendazole at 400 mg on consecutive days improved the cure rate by up to 67 %. But this was still not as high as the cure rate expected from a single dose against the other geohelminths (80–100 %). Anyway, from a public health point of view, such repeat treatments will be both expensive and logistically impractical.

Resistance to anthelmintic drugs is a real concern. Fortunately there have been very few reports of resistance to drugs used against medically important nematodes. Nevertheless, the severe resistance problems experienced with drugs used against veterinary nematodes, particularly in the southern hemisphere, make it essential continually to monitor cure rates. Falling cure rates might signal the development of resistance and allow measures to be taken to prevent it progressing unnoticed. One strategy to delay the onset of resistance that can perhaps be borrowed from veterinary nematologists is the use of combinations of unrelated drugs.

No treatment has been recommended for the minor intestinal nematodes such as *Ternidens* and *Trichostrongylus*, but benzimidazole drugs will probably be effective. Similarly, benzimidazoles, especially albendazole and thiabendazole, can be used successfully to treat cutaneous larva migrans due to *A. braziliensis*, *A. caninum*, *T. canis* and *G. spinigerum*.

Box. 25.4 Low *Trichuris trichiura* Cure Rates After Treatment

The failure of chemotherapy to produce the expected high *T. trichiura* cure rates in people is a matter of concern to control programme managers. This is probably explained by the unusual larval biology of whipworms, but since the larval development of *T. trichiura* has not been studied in any detail, it is pertinent to review the literature on other species of the genus, namely, *T. muris*, *T. vulpis* and *T. suis* (Beer 1973; Lee and Wright 1978; Kirkova and Dinev 2005).

Although these accounts differ in some details, there is agreement on the general pattern of development, and it seems reasonable to assume that this pattern is true for *T. trichiura* as well. Although *Trichuris* larvae hatch in the host's intestine, there are conflicting views on exactly where this happens. It, however, seems that most hatch in the caecum and colon. Regardless of the actual site, the first-stage larvae penetrate the crypts of Lieberkühn where they begin what Beer (1973) calls their histiotrophic phase by penetrating the gland (goblet) cells of the epithelium lining both the crypts and the villi themselves. After entering via the gland cell's opening, they spend their first few days coiled inside the gland's lumen. However, as they moult and grow, they occupy more of the cell's volume before disrupting them as well as adjacent enterocytes to form tunnel-like syncytia which may extend deep into the lamina propria. Estimates of the larval life of *T. trichiura* vary from 1–2 months though considerably longer for non-human species. As the fourth-stage larvae become sexually mature, their posterior ends protrude through the mucosal epithelium into the lumen of the caecum or colon, while the anterior ends remain embedded within the lamina propria. The adult worms remain in this position for the rest of their lives and are able to feed and lay their eggs directly into the intestine for passage to the outside in the faeces. Surprisingly even heavy *Trichuris* infections do not by themselves severely damage the mucosa. Cases of severe pathology are thought due to coinfection by invasive protozoans such as *Balantidium coli* or *Entamoeba histolytica*. Blood loss from damaged mucosae may contribute to anaemia.

Box. 25.5 World Health Assembly Resolution WHA54-19

The importance attached by the international community to the morbidity caused by geohelminth infections was demonstrated by Resolution WHA54-19 *Schistosomiasis and Soil-Transmitted Helminth Infections* adopted by the 54th World Health Assembly in Geneva in May 2001 (World Health Assembly 2001). The key point of this resolution from the point of view of geohelminth control is clause 2.2 which urges member states to ensure access to essential drugs against soil-transmitted helminth infections. Its goal is a minimum target of regular chemotherapy administration to at least 75 % and up to 100 % of all school-aged children at risk of morbidity by 2010. South Africa was a signatory to WHA54-19. Resolutions of this nature are not binding, but they do place a moral obligation on signatories to carry out their recommendations.

25.4.1.2 Filariasis Control

Following active surveillance, door to door in some cases and mapping to determine the extent and severity of infection in a community, targeted chemotherapy is the main tool used against the filariases. Vector control is seen as playing a supportive role. The current drugs of choice for treating these diseases are albendazole, diethylcarbamazine (DEC) and ivermectin, with expected cure rates between 70 and 90 %.

The strategy recommended by WHO for controlling lymphatic filariasis is two pronged and is having beneficial effects where the problem is endemic. Transmission control is achieved by annual, mass, community-wide treatment with a combination of albendazole and either DEC or ivermectin. Morbidity control by reducing suffering is achieved by improving awareness, hygiene and capacity building at the community level, coupled with surgery in cases of severe urogenital infection (World Health Organization 2000).

Cairncross et al. (1988) described the transmission of *W. bancrofti* as 'a remarkably inefficient process' because of high mortalities amongst infected mosquitoes and amongst the developing larvae in both the mosquito and human hosts. He estimated that between 2700 and 1 million bites by infective mosquitoes (generally <1 % of the mosquito population) were required to produce a single human infection. Transmission of *W. bancrofti*, therefore, requires a very high man-biting rate so that simply reducing the vector mosquito population rather than trying to eliminate it would probably be sufficient. For this reason the use of bed nets impregnated with a residual pyrethroid insecticide to repel mosquitoes is currently seen as a supporting option in control programmes rather than a major one.

Onchocerciasis control programmes have relied largely on the use of the organophosphate insecticide, abate against the larvae of the blackfly *Simulium* sp., which typically breed in fast-flowing water. Spraying from aircraft allows large areas to be covered.

Acknowledgements I am grateful to Prof Larry G. Hadley (Nelson Mandela School of Medicine, University of KwaZulu-Natal, UKZN), Mrs Colleen Archer and Dr Thabang Mosala (School of Life Sciences, UKZN) for the use of their photographs. Mrs Archer prepared Fig. 25.4 and kindly read the manuscript.

References

Adams VJ, Lombard CJ, Dhansay MA et al (2004) Efficacy of albendazole against the whipworm *Trichuris trichiura* – a randomised, controlled trial. S Afr Med J 94:972–976

Albonico M, Stoltzfus RJ, Savioli L et al (1998) Epidemiological evidence for a differential effect of hookworm species *Ancylostoma duodenale* or *Necator americanus* on iron status in children. Int J Epidemiol 27:530–537

Appleton CC, Archer CE (2006) Intestinal parasites of rats from Durban and implications for human health. In: Abstracts of an International Workshop on Rodent-Borne Diseases and the Ratzooman Research Project: 15–16, Malelane, South Africa, 3–6 May

Appleton CC, Brain C (1995) Gastro-intestinal parasites of *Papio ursinus* living in the central Namib Desert, Namibia. Afr J Ecol 33:257–265

Appleton CC, Gouws E (1996) The distribution of common intestinal nematodes along an altitudinal transect in KwaZulu-Natal, South Africa. Ann Trop Med Parasitol 90: 181–188

Appleton CC, White BJ (1989) The structure of the shell and polar plugs of the egg of the whipworm *Trichuris trichiura* (Nematoda: Trichuridae) from the Samango monkey (*Cercopithecus albogularis*). Onderstepoort J Vet Res 56:219–221

Appleton CC, Henzi SP, Whitehead SI (1991) Gastro-intestinal helminth parasites of the chacma baboon, *Papio cynocephalus ursinus*, from the coastal lowlands of Zululand, South Africa. Afr J Ecol 29:149–156

Appleton CC, Maurihungirire M, Gouws E (1999) The distribution of helminth infections along the coastal plain of KwaZulu-Natal Province, South Africa. Ann Trop Med Parasitol 93:859–868

Appleton CC, Mosala TI, Levin J et al (2009) Geohelminth infection and reinfection after chemotherapy among slum-dwelling children in Durban, South Africa. Ann Trop Med Parasitol 103:249–261

Archer CE, Appleton CC, Mukaratirwa S et al (2011) The rat lung-worm *Angiostrongylus cantonensis*: a first report in South Africa. S Afr Med J 101:174–175

Barrett L, Henzi SP (1998) Epidemic deaths in a chacma baboon population. S Afr J Sci 94:441

Beer RJS (1973) Studies on the biology of the life-cycle of *Trichuris suis* Schrank, 1788. Parasitol 67:253–262

Bundy DAP, Cooper ES (1989) *Trichuris* and trichuriasis in humans. Adv Parasitol 28:107–173

Cairncross S, Rajavel AR, Vanamail P et al (1988) Engineering, mosquitoes and filariasis: a case report. J Trop Med Hyg 91:101–106

Chan MS (1997) The global burden of intestinal nematode infections – fifty years on. Parasitol Today 13:438–443

Chandiwana SK, Bradley M, Chombo F (1989) Hookworm and roundworm infections in farm-worker communities in the large-scale agricultural sector in Zimbabwe. J Trop Med Hyg 92:338–344

Cochrane JC, Sagorin L, Wilcocks MG (1957) *Capillaria hepatica* infection in man. S Afr Med J 31:751–755

Cochrane JC, Skinstad EE (1960) *Capillaria hepatica* in man – follow-up of a case. S Afr Med J 34:21–22

Da Costa-Macedo LM, Rey L (1990) *Ascaris lumbricoides* in neonate: evidence of congenital transmission of intestinal nematodes. Rev Inst Med Trop Sao Paulo 32:351–354

Enosse SM, Gama R, Schwalbach J (1995) *Ancylostoma duodenale* and other intestinal and vesical parasites in the Infulene valley and Mahotas, Maputo. Rev Med Moamb 6:40–43

Evans AC, Du Preez L, Maziya SP et al (1987) Observations of the helminth infections in black pupils of the Eastern Transvaal lowveld of South Africa. S Afr J Epidemiol Infect 2:7–14

Evans AC, Markus MB, Steyn E (1990) A survey of the intestinal nematodes of Bushmen in Namibia. Am J Trop Med Hyg 42:243–247

Evans AC, Markus MB, Joubert JJ et al (1991) Bushmen children infected with *Strongyloides fuelleborni*. S Afr Med J 80:410–411

Evans AC, Markus MB, Mason RJ et al (1996) Late stone-age coprolite reveals evidence of prehistoric parasitism. S Afr Med J 86:274–275

Fincham JE (2001) Intestinal parasites, growth of children, sanitation and water quality at primary schools in the Boland/Overberg Health Region. Epidemiological Comments 4:5–16

Fincham JE, Markus MB, Ngobeni JT et al (2005) Synchronized and regular deworming of children and women in South Africa: policy and practice. S Afr J Sci 101:13–17
Fourie JJ, Williams E, Kok DJ et al (2002) Nematode infections of dogs from peri-urban resource-limited communities in the Eastern Cape. J S Afr Vet Assoc 74:97
Gelfand M (1943) The sick African, 1st edn. Juta and Co., Johannesburg
Goldsmid JM (1967) *Ternidens deminutus* Railliet and Henry (Nematoda), a diagnostic problem in Rhodesia. Cent Afr J Med 13:54–58
Goldsmid JM (1968) The differentiation of *Ternidens deminutus* and hookworm ova in human infections. Trans R Soc Trop Med Hyg 62:109–116
Goldsmid JM (1974) The intestinal helminthozoonoses of primates in Rhodesia. Ann Soc Belge Med Trop 54:87–101
Goldsmid JM, Rogers S (1978) A parasitological study on the chacma baboon (*Papio ursinus*) from the northern Transvaal. J S Afr Vet Assoc 49:101–111
Goldsmid JM, Morrison JF, Saunders CR (1974) The hookworm problem in the Rhodesian lowveld: its importance and treatment. Cent Afr J Med 20:97–100
Gonçalves MLC, Araújo A, Ferreira LF (2003) Human intestinal parasites in the past: new findings and a review. Mem Inst Oswaldo Cruz 98(Suppl):103–118
Granzer M, Haas W (1991) Host finding and host recognition of infective *Ancylostoma caninum* larvae. Int J Parasitol 21:429–440
Haas W, Haberl B, Idris SI et al (2005) Infective larvae of the human hookworms *Necator americanus* and *Ancylostoma duodenale* differ in their orientation behaviour when crawling on surfaces. Parasitol Res 95:25–29
Hale DC, Blumberg L, Frean J (2003) Case report: gnathostomiasis in two travellers to Zambia. Am J Trop Med Hyg 68:707–709
Harada Y, Mori O (1955) A new method for culturing hookworm. Yonago Acta Med 1:177–179
Hayward C, Knight R, Rossum N et al (2006) Nematode contamination in sandpits of registered pre-school facilities in Bloemfontein. S Afr J Epidemiol Infect 21:173–177
Herman JS, Wall EC, Van Tullekan C et al (2009) Gnathostomiasis acquired by British tourists in Botswana. Emerg Infect Dis 15:594–597
Hira PR (1976) Observations on helminthozoonoses in Zambia. East Afr Med J 53:278–286
Hira PR (1978) Some helminthozoonotic infections in Zambia. Afr J Med Sci 7:1–7
Hotez PJ (2000) Pediatric geohelminth infections: trichuriasis, ascariasis and hookworm. Semin Pediatr Infect Dis 11:236–244
Kallichurum S, Elsdon-Dew R (1961) *Capillaria* in man. A case report. S Afr Med J 35: 860–861
Kaplan CS, Freedman L, Elsdon-Dew R (1956) A worm in the eye. S Afr Med J 30:791–792
Kirkova Z, Dinev I (2005) Morphological changes in the intestine of dogs, experimentally infected with *Trichuris vulpis*. Bulg J Vet Med 8:239–243
Krepel HP (1994) *Oesophagostomum bifircum* infection in man: a study on the taxonomy, diagnosis, epidemiology and drug treatment in northern Togo and Ghana. PhD Thesis, University of Leiden, Leiden
Kvalsvig JD (2002) Intestinal nematodes and cognitive development. In: Holland C, Kennedy M (eds) World class parasites, Vol. 2. The Geohelminths, *Ascaris*, *Trichuris* and Hookworm. Kluwer Academic Publishers, Amsterdam, pp 63–73
Kvalsvig JD (2003) Parasites, nutrition, child development and public policy. In: Crompton DWT, Montresor A, Nesheim MC et al (eds) Controlling disease due to infections. World Health Organization, Geneva, pp 55–65
Kwitshana ZL, Tsoka JM, Mabaso MLH (2008) Intestinal parasitic infections in adult patients in KwaZulu-Natal. S Afr Med J 98:709–711
Leary PM, Elsworth M, Maytham D et al (1973) Difetarsone in the treatment of *Trichuris trichiura*. S Afr Med J 47:1389
Lee MG, Terry SI (1989) Arteriomesenteric duodenal occlusions associated with strongyloidiasis. J Trop Med Hyg 92:41–45
Lee TDG, Wright KA (1978) The morphology of the attachment of and probable feeding site of the nematode *Trichuris muris* (Schrank, 1788) Hall, 1916. Can J Zool 56:1889–1905

Liebenberg W, Dekker TG, Lötter AP et al (1998) Identification of the mebendazole polymorphic form present in raw materials and tablets available in South Africa. Drug Dev Ind Pharm 24:485–488

Mabaso MLH, Appleton CC, Hughes JC et al (2003) The effect of soil type and climate on hookworm (*Necator americanus*) distribution in KwaZulu-Natal, South Africa. Trop Med Int Health 8:722–727

Madiba TE, Hadley GP (1996) Surgical management of worm volvulus. S Afr J Surg 34:33–36

Markus MB (1982) Man's best friend a threat. Sci Prog 15:3–4

Mayet FGH, Schutte CHJ, Reinach SG (1985) Anaemia among the inhabitants of a rural area in northern Natal. S Afr Med J 67:458–462

Midzi N, Sangweme D, Zinyowera S et al (2008) The burden of polyparasitism among primary schoolchildren in rural and farming areas in Zimbabwe. Trans R Soc Trop Med Hyg 102:1039–1045

Mönnig HO (1944) Parasiete en Parasitiese Siektes van die Mens. Voortrekkerpers Bpk, Pretoria

Moodley K, Govind CN, Peer AKC (2015) First detection of dirofilariasis in South Africa. Infect Dis Rep 7:5726

Mosala TI, Appleton CC (2003) True prevalence of pinworm (*Enterobius vermicularis*) infection among children in Qwa-Qwa, South Africa. S Afr J Sci 99:465–466

Mukaratirwa S, Dzoma BM, Matenga E et al (2008) Experimental infection of baboons (*Papio* spp.) and vervet monkeys (*Cercopithecus aethiops*) with *Trichinella zimbabwensis* and successful treatment with ivermectin. Onderstepoort J Vet Res 75:181–188

Mulcahy G, O'Neill S, Fanning J et al (2005) Tissue migration by parasitic helminths – an immunoevasive strategy? Trends Parasitol 21:273–277

Orihel TC (1973) Cerebral filariasis in Rhodesia – a zoonotic infection? Am J Trop Med Hyg 22:596–599

Pebsworth PA, Archer CE, Appleton CC et al (2012) Parasite transmission risk from geophagic and foraging behavior in Chacma baboons. Am J Primatol 74:940–947. doi:10.1002/ajp.22046

Ravasi DF, O'Riain MJ, Davids F (2012) Phylogenetic evidence that two distinct *Trichuris* genotypes infect both humans and non-human primates. PLoS One 7(8):e44187. doi:10.1371/journal.pone.0044187

Reinecke RK (1983) Veterinary helminthology. Butterworths, Pretoria

Roodt L, Felix M, Wadee AA (1995) Total IgE levels in a rural Pedi population in the north eastern Transvaal. S Afr J Epidemiol Infect 10:79–82

Saathoff E, Olsen A, Kvalsvig JD et al (2002) Geophagy and its association with geohelminth infection in rural schoolchildren from northern KwaZulu-Natal, South Africa. Trans R Soc Trop Med Hyg 96:485–490

Saathoff E, Olsen A, Kvalsvig JD et al (2005a) Ecological covariates of *Ascaris lumbricoides* infection in schoolchildren from a small area within rural KwaZulu-Natal, South Africa. Trop Med Int Health 10:412–422

Saathoff E, Olsen A, Sharp B et al (2005b) Ecological covariates of hookworm infection and reinfection in rural KwaZulu-Natal, South Africa: a geographic information system-based study. Am J Trop Med Hyg 72:384–391

Sasa M (1976) Human filariasis – a global survey of epidemiology and control. University Park Press, Tokyo

Silverman NH, Katz JS, Levin SE (1973) *Capillaria hepatica* infestation in a child. S Afr Med J 47:219–221

Simonsen PE, Malecela MN, Michael E et al (2008) Lymphatic filariasis, research and control in eastern and southern Africa. DBL-Centre for Health Research and Development. Copenhagen

Slom TJ, Cortese MM, Gerber SI et al (2002) An outbreak of eosinophilic meningitis caused by *Angiostrongylus cantonensis* in travellers returning from the Caribbean. N Engl J Med 346:668–675

Stephenson LS, Holland CV, Cooper ES (2000) The public health significance of *Trichuris trichiura*. Parasitol 121:S73–S95

Stoll N (1947) This wormy world. J Parasitol 33:1–18

Storey P (2001) Human Oesophagostomiasis: clinical presentations and subclinical colonic pathology. PhD Thesis, University of Leiden, Leiden
Van der Werf MJ, De Vlas S (2001) Morbidity and infection with schistosomes or soil-transmitted helminths. Final report, Department of Public Health, Erasmus University, Rotterdam
Vermund SH, Wilson CM (2000) Pinworm (*Enterobius vermicularis*). Semin Pediatr Infect Dis 11:252–256
Verster A (1980) The prevalence and control of gastro-intestinal helminths of domestic dogs in the Republic of South Africa. J S Afr Biol Soc 21:25–31
Weischer B, Brown DJF (2002) An introduction to nematodes: general nematology. Pensoft Publishers, Sofia
Wharton DA, Jenkins T (1978) Structure and chemistry of the egg-shell of a nematode (*Trichuris suis*). Tissue Cell 10:427–440
Whiting DA (1976) The successful treatment of creeping eruption with topical thiabendazole. S Afr Med J 50:253–255
World Health Assembly (2001) Agenda Item 13.3, Resolution WHA 54.19, Annex 1: Schistosomiasis and soil-transmitted helminth infections, pp 50–51
World Health Organization (1987) Prevention and control of intestinal parasitic infections. World Health Organ Tech Rep Ser 749:1–86
World Health Organization (2000) Lymphatic filariasis. WHO Fact Sheet No. 102 (revised), pp 1–3
World Health Organization (2002) Prevention and control of schistosomiasis and soil-transmitted helminthiasis. World Health Organ Tech Rep Ser 912:1–57
Zini D, Wadee AA (1989) An ELISA for the diagnosis of human *Toxocara canis* infections. South Afr J Epidemiol Infect 4:57–59

Chapter 26
Achievements and Challenges

Hendrika Fourie, Robin K. Jones, Vaughan W. Spaull, Mieke S. Daneel, and Dirk De Waele

The book, *Nematology in South Africa: A View from the 21st Century*, was born out of the need to update an earlier book, published 34 years ago. This new book brings into a single source, the current knowledge of the science of Nematology in South Africa. It testifies to the achievements of the many authors of the chapters, both on topics included in the earlier book and on new subjects that were only ideas in 1982.

It can be confidently stated that researchers have identified the principal nematode pest problems within South Africa, that the basic environmental interactions in a multitude of habitats are better known and that both as a 'pure and applied' science, the South African knowledge base has reached a level for which South African nematologists can be proud. But for every achievement met, a new challenge arises. So this concluding chapter seeks to chronicle the significant achievements of the last three decades and set these achievements alongside the major challenges to be faced in the coming years.

H. Fourie (✉)
North-West University, Unit for Environmental Sciences and Management,
Private Bag X6001, Potchefstroom 2520, South Africa
e-mail: driekie.fourie@nwu.ac.za

R.K. Jones
Robin Jones Nematological Services, 24 Victor Street, Murrayfield, South Africa

V.W. Spaull
170 Edmonds Road, Glenwood, Durban 4001, South Africa

M.S. Daneel
Agricultural Research Council–Institute for Tropical and Subtropical Crops, Private Bag X11208, Mbombela 1200, South Africa

D. De Waele
North-West University, Unit for Environmental Sciences and Management,
Private Bag X6001, Potchefstroom 2520, South Africa

Laboratory of Tropical Crop Improvement, Department of Biosystems, University of Leuven, Willem de Croylaan, 42, 3001 Heverlee, Belgium

H. Fourie et al. (eds.), *Nematology in South Africa: A View from the 21st Century*, DOI 10.1007/978-3-319-44210-5_26

One fundamental past achievement, on which all recent and future achievements have depended, has been to create opportunities to train and employ future nematologists. This past success remains the central challenge of the future so that a stream of competent researchers can be groomed to take on the mantel left by their predecessors.

On a broader point, the editors believe by highlighting these issues they will contribute to the future growth and health of the discipline.

26.1 Achievements

26.1.1 Research: Principal Achievements

- The major nematode crop problems identified in the 20th century are now well researched from taxonomic, pest-crop pathology and pest management perspectives. This does not exclude the possibility of new nematode-crop problems arising in the future
- South African nematologists made significant contributions with regard to agricultural practices including legislative control, the utilisation of clean seed, accredited seedling nurseries, precision agriculture, conservation agriculture, the use of entomopathogenic nematodes (EPN) and companies to manufacture and promote these and other biocontrol agents
- Research into the nematode problems experienced by smallholding farmers was identified as a major responsibility for nematologists, and significant data have been published and results disseminated to the smallholding farming community.
- The research community has established and maintains a close partnership with the crop protection industry in quickly bringing to the fore new technologies to limit nematode pest problems
- The research community has established channels to disseminate new research results to the various crop commodity organisations and their growers, e.g. citrus, deciduous and subtropical fruit crops, potato, sugar and summer row crops.
- The Nematological Society of Southern Africa (NSSA) thrives and is now a globally recognised and active society

26.1.2 Milestones

- The identification of the most important plant-parasitic nematodes on the key crops, including the role of *Meloidogyne* spp. on maize
- The identification of new crop pests, e.g. new seed- and leaf-gall nematodes, *Aphelenchoides arachidis* and *Ditylenchus africanus* in groundnut kernels, new *Globodera* spp., *Meloidogyne enterolobii* in guava and vegetables and many others

- The advances in citrus nematode control, namely:
  - In nurseries where certified nematode-free seedlings are now available to establish orchards free of the citrus nematode
  - The use of multiple nematicide applications to provide long-term nematode control
  - The identification of accelerated microbial degradation (AMD) as a problem in citrus orchards and the development of recommendations to limit AMD
- The research into and development of viable broad range control options on all the major nematode-induced crop loss situations
- The establishment of EPN as a key research discipline and the initiation of commercialisation of EPN-based biocontrol solutions for several pests of diverse crops, notably codling moth and false codling moth
- The research and dissemination to smallholding farmers of suitable crop rotations and possible bionematicide options for nematode-induced problems
- The dedication of outstanding taxonomists active in morphological and molecular identification and the utilisation of improved technologies
- The creation of the South African Plant-Parasitic Nematode Survey (SAPPNS) in 1987
- The creation and ongoing maintenance of the National Collection of Nematodes (NCN) with over 200,000 specimens listed
- The extension of knowledge to producers and the commercial industry. Currently there are nine diagnostic laboratories staffed in SA
- Collaborative international networking, e.g. quarantine diagnostic protocols for *D. africanus*, *Ditylenchus dipsaci*, *Aphelenchoides ritzemabosi*, *Aphelenchoides fragariae* and *Aphelenchoides besseyi*
- The growth of the NSSA from a small society focused on southern African problems to a society with local and international membership of over 150 individuals.
- The hosting of the 6th International Congress of Nematology by the NSSA in March 2014 in Cape Town, a congress attended by 450 delegates (of which 100 were students) from 38 countries
- Following a successful helminth control programme piloted in KwaZulu-Natal from 1998 to 2000, a national deworming programme was launched in February 2016

26.1.3 Training: Principal Achievements

- The expansion of the number of institutions that train or postgraduate students and the growth of the net inflow of new researchers in Nematology
- The regular availability of short courses to train new nematologists
- Collaboration with international institutions in training by means of internationally funded projects, e.g. EUMAINE, a project involving a consortium of European Universities, the University of Stellenbosch and North-West University

26.1.4 Training: Milestones

- The formal training of students for honours, masters and PhD degrees in Nematology at the Universities of the Free State, Pretoria, KwaZulu-Natal, Limpopo, North-West and Stellenbosch
- The running of Nematology short courses at universities notably North-West University (accredited by the South African Qualifications Authority (SAQA)), KwaZulu-Natal University and University of Stellenbosch
- PhD degrees awarded to South African nematologists from the University of Claude Bernard, Lyon and the University of Montpellier in France and the University of Leuven in Belgium

26.2 Challenges

The world faces increasing demands for the production of more food, an increase in food choices and improved quality. These demands are set against a reduction in available resources (soil, land and water), a changing environment (global warming, agronomic practices) and socio-economic conditions. Nematologists need to participate in finding solutions to these issues.

26.2.1 Research: Challenges

- Reduction in crop losses. Currently nematicides can, and in a number of situations do, limit crop losses by as much as 20 % or more. However, because of poor financial returns (in rain-fed crops in particular), producers rarely apply nematode control practices. Thus, a major challenge is the development of management practices that address the gap between costs of inputs and the returns producers need to justify their use. This will lead to a far greater use of nematode management practices to enhance yields
- The promotion of centres of expertise that monitor local and global opportunities for the screening of genotypes of crops to assess their host status to target nematode pests. The use of non- or poor-host crop genotypes is one of the most viable strategies to reduce population densities of nematode pests. Alongside this effort, crop rotations must be developed that limit losses even in adverse abiotic or biotic conditions, with or without the use of chemical control inputs
- To manage crop production sequences and systems with the objective of reducing pathogens and enhancing soil function. This will be facilitated through the interaction between nematologists and soil scientists, agronomists, plant

breeders, entomologists and pathologists, etc. The promotion of multidisciplinary and multi-institutional research programmes into nematode interactions with environmental conditions, edaphic factors, plant hosts and microorganisms to improve root health is thus crucial

- The promotion of research programmes to advance environmentally-friendly nematode management options, soil ecosystem management, the role of nematodes as indicators of biodiversity and, in particular, the return of degraded soils to full health, season by season
- The impact of genetically modified crops on nematode pest populations
- Assessments of the occurrence and spread of nematode pests through water bodies, e.g. irrigation water and boreholes
- The impact of climate change on nematode-induced crop losses and on human infection by parasitic nematodes
- Expansion of research on EPN and on other nematode pathogens of invertebrate pests
- The implementation of new plant quarantine systems and technologies
- Effective staff succession planning to feed in skilled taxonomists for morphological and molecular identification and to maintain the NCN

26.2.2 Training: Challenges

- Raising the level of awareness about nematodes and their associated problems, notably to both smallholding and commercial farmers, extension staff and crop commodity organisations, university and research institute administrators and research funding organisations
- Maintaining a constant inflow of new researchers into the science of Nematology
- Maintaining a constant output of quality publications in international peer-reviewed journals is imperative when facing the above challenges

26.3 Policy: Challenges

- Coordination of research and training
- Increasing and encouraging international collaboration in terms of research and training and strengthening the NSSA beyond the borders of southern Africa
- Increasing the accessibility of published information

Index

A

Ancylostoma duodenale, 522, 523, 533–537, 540
Anguina, 10, 33, 54, 56, 80, 103, 410–413
Aphelenchoide, 50, 79, 146, 156, 430
Aquatic, 19, 34, 35, 65, 148, 419–446, 495, 496, 502, 509
Ascaris lumbricoides, 522–525, 527, 529, 530, 536, 539–543, 550
Avocado, 168, 373–375

B

Bacteria, 19, 20, 25, 57, 107, 141, 147, 202, 206, 236, 267, 270, 273, 274, 277, 290, 292, 341, 346, 352, 353, 355, 390, 410, 413, 420, 422, 423, 426, 429, 459, 461, 465–466, 495, 502, 529, 533, 534
Banana, 16, 20–22, 24, 27, 29, 65, 80, 122, 141, 359–370, 375, 385, 389
Barley, 184, 196, 280
Beetroot, 92, 156, 167, 170, 252, 254, 255
Beneficial nematodes, 269, 276, 277
Biofumigation, 19, 20, 161, 173–175, 223, 339
Biological control, 130, 147, 210, 219, 273–274, 290, 297, 337, 340–341, 350, 369, 397, 459, 460, 463, 472, 473, 486, 491
Biomonitors, 509–510
Body, 34–42, 48, 50, 51, 60, 61, 96, 98–100, 102, 135, 145, 241, 333, 421, 423, 428, 430, 488, 501, 540, 541, 544, 548

C

Cabbage, 61, 170–172, 174, 223, 251–252, 254, 276, 424
Canola, 11, 174, 339, 482, 483, 489
Carrot, 20, 58, 61, 92, 106, 170–172, 252–255
Cavendish banana, 359
Cavernicolous nematodes, 430–432
Certification schemes, 120, 124, 235, 240, 245, 334
Challenges, 7, 108, 130, 136, 138, 147, 196, 341, 401, 405, 432, 474, 559–563
Chemical control, 25, 129, 130, 136, 142, 147, 188–190, 211, 217–218, 234, 235, 239, 245, 246, 250, 252, 271–273, 289–290, 296–297, 302, 319–321, 340, 370, 401, 403–405, 459, 471, 562
Citrus, 7, 8, 14, 16, 18–20, 24, 27, 58, 64, 69, 93, 97, 122, 129, 135, 141, 159, 168, 311–322, 332, 364, 366, 367, 463, 465, 469–470, 472, 560, 561
Citrus Improvement Scheme (CIS), 314, 315, 318, 319, 322
Community management, 274, 277–279
Control strategies, 131, 139, 147, 237, 245, 250, 291, 349, 401, 413, 414
Cotton, 7, 10, 11, 14–16, 22, 27, 58, 110, 133, 135, 237, 238, 285, 291–300, 302, 303, 377, 452
Criconemoides xenoplax, 10, 327, 330, 338, 339, 352, 353, 355, 374, 379
Crops, 1, 15, 61, 76, 123, 129, 152, 156, 184, 201–226, 231–256, 273, 285, 339, 345, 377, 388, 397, 409, 561

H. Fourie et al. (eds.), *Nematology in South Africa: A View from the 21st Century*, DOI 10.1007/978-3-319-44210-5

Crop production, 129, 130, 144, 154, 167, 183–184, 202, 226, 301, 376, 405, 417, 562
Crop rotation, 129–131, 139, 147, 154, 167, 170–173, 191–192, 205, 210, 219, 224, 225, 235, 237, 243, 250, 290–291, 298–299, 303, 339, 384, 401, 402, 484, 561, 562
Cucumis, 20, 21, 159, 161, 164, 165
Cultivars (cvs), 20, 22, 23, 105, 171–173, 175, 184, 190, 191, 202, 205, 207–210, 218, 220, 223, 225, 226, 232, 237–240, 243, 247, 249, 250, 261, 265, 272–275, 277, 278, 286, 288, 290, 296–299, 301, 303, 304, 312, 315, 359, 364, 377–379, 395, 396, 425
Cultural control, 218–219, 245–247, 250, 274–279, 369, 387, 401–402, 484
Cuticle, 36–39, 41, 48, 51, 60, 98–100, 142, 461, 462, 500, 534

D
Dagger nematodes, 11, 15, 124, 295, 327, 331, 338, 347, 348, 350, 352, 353, 355
Damage, 15, 26, 58, 77, 79, 80, 87, 98, 108, 120, 124, 126, 131, 146, 148, 153, 154, 174, 186–190, 192, 196, 197, 204–206, 211–213, 215–220, 222, 224–226, 232–240, 243–246, 248, 252–254, 264, 268–271, 273–275, 277, 280, 286–289, 292–294, 296, 298, 299, 317–319, 321, 327, 329, 330, 332–334, 336–338, 346–348, 351, 353–356, 360, 362–364, 367, 370, 374–376, 378, 380–390, 397–400, 429, 472, 482, 535, 548, 553
Database, 16, 184, 202, 211, 232, 264, 286, 312, 327, 346, 360, 367, 374, 396, 410, 414, 451, 452, 454, 486, 500
Deciduous fruit, 14, 15, 19, 28, 129, 133, 135, 345–356, 462, 463, 469
Department of Agriculture, Forestry and Fisheries (DAFF), 14, 15, 22, 23, 26, 119, 143, 144, 155, 194–196, 211, 219, 222, 235, 237, 247, 251–254, 360, 452
Ditylenchus africanus, 10, 18, 21, 23, 66, 79, 80, 91, 102–106, 194, 211–219, 416, 422, 560, 561
Dry bean, 170–172, 201, 220–222, 451

E
Entomopathogenic nematodes (EPN), 15, 19, 27–28, 34, 88, 103, 106–108, 459–475, 485, 486, 560, 561, 563
Environmentally-friendly nematode management, 563
Estuary, 433, 496, 497, 504, 505, 508–512
Exotic slugs, 482
Extraction, 27, 76, 80–93, 103, 110, 125, 185, 261, 265, 330, 428, 454

F
Feeding sites, 7, 12
Field application, 167
Flax, 16, 285, 303–304
Free-living nematodes, 1, 39, 43, 52, 175, 466, 488, 495, 499, 500

G
Geohelminths, 522–527, 537, 550–553
Geohelminthiasis, 522–523, 551–552
Globodera, 10, 12, 15, 16, 55, 60, 61, 100, 102, 103, 122, 124, 232, 233, 239, 240, 242, 560
Grain sorghum, 11, 183, 184, 193–194, 220, 451
Grapevine, 10, 11, 14, 15, 18, 19, 58, 59, 62, 65, 68, 75, 76, 122, 124, 146, 325–341, 367, 452, 464, 469, 472
Grasses, 2, 4, 10, 22, 56–59, 63, 65, 67–69, 91, 111, 190, 219, 237, 263, 276, 298, 367, 390, 397, 402, 409–417, 464, 466
Green pea, 253–254
Groundnut, 6, 9–11, 19–21, 23, 57, 66, 76, 79, 80, 91, 92, 102, 104, 105, 133, 135, 172, 192, 201, 211–219, 222–224, 235, 416, 424, 451, 452, 560
Guava, 10, 11, 24, 61, 68, 133, 168, 287, 373, 376–379, 390, 464, 560

H
Helicotylenchus dihystera, 24, 25, 156, 195, 196, 265, 267–269, 277, 278, 301, 303, 350, 362, 374, 376, 385, 397, 416
Hemicriconemoides, 24, 64, 156, 195, 375
Hemp, 16, 20, 58, 276, 285, 291, 299–302, 402

Heterodera, 12, 33, 35, 60, 100, 102, 221, 265, 433
Heterorhabditis, 103, 104, 424, 460–469, 471–474

I
Insect pests, 15, 273, 355, 460, 468, 473, 474
Interaction, 20, 21, 25, 129, 131, 136, 147, 176, 190, 191, 193, 197, 236, 238, 268–269, 271, 292, 293, 299, 317, 333, 346, 356, 375, 377–379, 381, 383, 384, 432, 466–467, 498, 559, 562, 563

K
Kenaf, 11, 16, 285, 302–303

L
Laboratory assays, 339, 471–473
Larva migrans, 537, 543, 550, 552
Legislation, 119, 120, 126, 366, 368
Legumes, 10, 170, 232, 424
Lesion nematodes, 11, 16, 19, 23, 80, 106, 185–188, 190, 192–196, 203, 204, 210, 217, 218, 220, 232, 238–239, 245–247, 288, 291, 294, 299, 302, 327, 329–330, 339, 347, 348, 350, 354, 355, 366–367, 396–402, 405, 417, 425
Litchi, 10, 11, 16, 24, 133, 373, 375, 381–383, 390

M
Maize, 6, 10, 11, 18, 21, 23, 57–59, 61, 62, 66, 68, 75, 76, 105, 133, 135, 146, 156, 167, 169–172, 183–197, 205, 220, 224, 225, 237–239, 288, 298, 346, 369, 415, 417, 424, 451, 452, 460, 469, 560
Management strategies, 23, 131, 147, 151–176, 188–193, 205–210, 217–220, 222–226, 237–239, 243–245, 250, 289–291, 296, 299, 303, 304, 317–321, 337–341, 368–370, 374–375, 377–381, 383–390, 401–405, 417
Marine environment, 499
Meiofauna, 495–499, 506–510
Meloidogyne, 11, 13, 14, 18, 20, 21, 24, 33–35, 77, 89, 90, 103, 105, 139, 146, 152, 155–157, 159, 160, 171, 184–187, 189–191, 194, 196, 202, 207, 210, 211, 216, 217, 219, 221, 223–225, 232–237, 244, 245, 250, 251, 253–256, 265, 267, 270, 273, 274, 279, 286, 290, 292, 298, 302, 304, 327–329, 338, 352, 353, 360, 361, 369, 376, 377, 379–382, 385, 387–389, 397, 402, 414, 415, 422, 431, 433, 455, 560
M. javanica, 20, 22, 25, 26, 156, 157, 167, 169–173, 185, 187, 189, 191, 192, 194–196, 202, 205, 207, 210, 216, 220, 222–226, 233–235, 237, 246, 247, 249, 250, 252, 255, 264, 265, 268, 269, 272, 274–276, 280, 286, 290, 301, 303, 304, 327, 328, 338, 339, 353, 354, 360, 362, 374, 376, 379, 384–386, 396–398, 401, 414–416, 455
Millet, 57, 66, 184, 190, 195–196, 237, 276, 277
Molecular identification, 23, 25, 561, 563
Mounting, 93–95, 97–100, 110
Musa, 359

N
National collection, 16, 34, 366, 452, 561
Necator americanus, 522–524, 533–537, 540
Nematicides, 21, 110, 129–148, 153, 158, 188, 205, 210, 233, 245, 252, 269, 287, 317, 329, 348, 369, 374, 386, 398, 413, 561
Non-human primates, 521, 526, 527, 538

O
Onion, 62, 91–92, 146, 251, 254, 424

P
Papaya, 24, 133, 373, 375, 380–381
Pecan, 10, 11, 122, 133, 373, 379–380
Permaculture, 23, 167, 168, 250
Phasmarhabditis, 481, 484–488, 490
Photorhabdus, 461, 465, 466
Phytonematicides, 20, 21, 154, 158–167
Phytosanitary, 119–122, 124–126
Pineapple, 10, 11, 22, 68, 69, 133, 139, 146, 147, 395–405, 452
Potato, 9, 13–14, 16, 18, 22, 26, 58, 60, 61, 64, 69, 77, 79, 80, 92, 121, 123–125, 135–137, 139, 146, 156, 173, 188, 211–212, 231–256, 285, 287, 346, 377, 414, 424, 425, 560

Pratylenchus, 11, 13, 16, 18, 21, 35, 54, 58, 80, 103, 106, 156, 172, 184–187, 189, 190, 193–194, 196, 202–204, 211, 217, 219–221, 224, 232, 233, 237–239, 244–246, 251, 253–254, 265–267, 273, 288, 291, 292, 294, 299, 301, 302, 327–330, 347, 349, 350, 352, 353, 355, 360, 361, 366, 367, 374, 381, 382, 387, 388, 396–401, 416, 425, 431, 455
P brachyurus, 16, 22, 106, 185, 189, 193–196, 217, 238, 288, 294, 301, 366, 388, 396–399, 401, 402

Q

Quarantine, 15, 61, 119–126, 130, 148, 239, 240, 242–244, 364, 410, 413, 467, 469, 561, 563

R

Radopholus similis, 11, 20, 27, 80, 106, 122, 136, 313, 360, 361, 364–369, 385, 389
Ratoon crop, 264, 268–270, 272, 273, 275, 277, 396, 398, 404
Resistance, 21–23, 105, 129–131, 147, 148, 154, 170–172, 190–191, 202, 205–210, 218, 224, 225, 235, 237, 239, 243, 249, 250, 255, 273, 275, 277, 280, 286, 290, 297–299, 301, 303, 304, 315, 319, 329, 337–339, 350, 369, 377–379, 384, 415, 424, 471, 525, 552
Rice, 58, 65, 66, 184, 195, 280
Rootstocks, 21, 75, 124, 164, 312, 317, 319, 320, 322, 329, 332, 336–339, 341, 348, 350–352, 355, 356, 384
Ryegrass toxicity, 19, 410, 411

S

Sampling, 74–77, 88, 123, 175, 236, 299, 318, 430, 452, 454, 511
Seed-and leaf-gall nematode (SLGN), 22, 80, 410–414, 416, 560
Slugs, 1, 19, 481–491, 538
Smallholding farmers, 16, 19, 22, 152–158, 161, 167–170, 173, 176, 191, 243, 248, 285, 311, 414–416, 560, 561
Soil-transmitted nematodes, 524, 532, 550
Sorghum, 14, 20, 23, 58, 66, 167, 183, 184, 193–194, 220, 237–238, 276, 451
Soybean, 6, 9, 11, 21–23, 33, 58, 76, 100, 105, 147, 156, 167, 171, 172, 190–192, 202–210, 239, 298, 346, 377, 425, 451, 454
Species distribution, 186
Steinernema, 103, 104, 459–467, 469
Stubby-root nematodes, 5, 11, 288, 295, 327, 331, 348, 350, 351, 354–356
Stylet, 35, 42–47, 61, 96, 334, 335, 423, 501
Subanguina, 10, 52, 54, 80, 103, 410–412
Sugarcane, 7, 25, 27, 58, 59, 63, 66, 67, 129, 138, 142, 172, 173, 261–280, 297, 369, 390, 425, 460, 462, 468–469
Sugar production, 262
Survey, 16, 21, 23, 135, 152, 156, 185, 193, 194, 202, 203, 213, 219, 232, 234, 240, 264, 265, 267, 273, 286, 292, 301, 312, 327, 330, 336, 346–348, 350–353, 360–362, 364, 367, 368, 374, 375, 381, 382, 387, 396, 410, 413, 414, 419, 424, 428, 429, 433, 451–455, 461–463, 465, 469, 470, 481, 488, 498, 507, 523, 525, 532, 540, 543, 545, 552, 561
Sweet potato, 69, 156, 172–173, 252–253, 377, 424
Symptoms, 66, 77, 79, 80, 93, 123, 153, 186–187, 193, 203–204, 213–217, 220, 222–225, 235–236, 238, 240, 243, 248, 251, 253, 254, 269–271, 286, 294, 296, 304, 312, 316, 328–331, 333, 346, 353–355, 360, 362, 363, 367, 374, 377–385, 387–390, 398–399, 410, 483, 529–533, 538, 542, 547, 550

T

Terrestrial habitats, 429, 434
Tobacco, 7, 10, 11, 15–16, 26, 58, 134, 135, 138, 139, 147, 190, 234, 237, 285–304, 336, 452
Tolerance, 130, 154, 162, 206, 207, 224, 235, 236, 244, 275, 299, 301, 376, 378, 379, 428, 432, 502
Tomato, 10, 11, 20, 21, 42, 43, 57, 61, 68, 69, 78, 105, 125, 134, 156, 157, 160,

161, 163, 164, 166–170, 172, 173, 191, 231, 237, 240, 243, 247–251, 256, 280, 287, 336, 348, 377, 414, 425
Trichuris trichiura, 522–525, 527–530, 541, 542, 550–553
Trophic groups, 10, 42, 422–427, 430
Tylenchulus semipenetrans, 11, 14, 20, 21, 90, 93, 122, 136, 159, 160, 312–314, 317, 319–321, 332

V
Vegetable crops, 11, 18, 22, 157, 167, 170, 231–256, 346, 415, 482
Virus vector, 333–337

W
Weeds, 22, 138, 148, 152, 167, 192, 212, 219, 237, 240, 280, 302, 318, 328, 334, 340, 351, 362, 367, 399, 403, 409–417, 424, 425
Wheat, 6, 10–12, 23, 33, 134, 183, 184, 190, 194–195, 276, 280, 298, 299, 410, 412, 451, 452, 470, 484

X
Xenorhabdus, 461, 465, 466
Xiphinema, 11, 15, 17, 18, 21, 25, 50, 56, 67, 68, 83, 103, 124, 147, 156, 184, 190, 194, 195, 203, 211, 220, 221, 223, 224, 255, 264–267, 270, 295, 301, 313, 327, 328, 330–332, 334–337, 347–350, 352–355, 368, 374, 379, 381, 382, 389, 416, 431, 455

Z
Zoonoses, 524, 527

MIX
Papier aus verantwortungsvollen Quellen
Paper from responsible sources
FSC® C105338

If you have any concerns about our products,
you can contact us on
ProductSafety@springernature.com

In case Publisher is established outside the EU,
the EU authorized representative is:
Springer Nature Customer Service Center GmbH
Europaplatz 3, 69115 Heidelberg, Germany

Printed by Libri Plureos GmbH
in Hamburg, Germany